AF753136

Renewable Energy

Renewable Energy

Edited by
T J Hammons

Renewable Energy
Edited by T J Hammons

CBS, Edition 2016

Published by InTech

Janeza Trdine 9, 51000 Rijeka, Croatia

All chapters are distributed under the Creative Commons Attribution 3.0 license, which allows users to download, copy and build upon published articles even for commercial purposes, as long as the author and publisher are properly credited, which ensures maximum dissemination and a wider impact of our publications. After this work has been published by InTech, authors have the right to republish it, in whole or part, in any publication of which they are the author, and to make other personal use of the work. Any republication, referencing or personal use of the work must explicitly identify the original source.

Copyright © 2009 InTech

As for readers, this license allows users to download, copy and build upon published chapters even for commercial purposes, as long as the author and publisher are properly credited, which ensures maximum dissemination and a wider impact of our publications.

Notice

Statements and opinions expressed in the chapters are these of the individual contributors and not necessarily those of the editors or publisher. No responsibility is accepted for the accuracy of information contained in the published chapters. The publisher assumes no responsibility for any damage or injury to persons or property arising out of the use of any materials, instructions, methods or ideas contained in the book.

Additional hard copies can be obtained from orders@intechopen.com

Renewable Energy,
Edited by T J Hammons
p. cm.
ISBN 978-953-7619-52-7

Preface

Our goal in preparing this book was to discuss and publish new discoveries and improvements, innovative ideas and concepts, as well as novel and further applications and business models which are related to the field of Renewable Energy.

Renewable Energy is energy generated from natural resources—such as sunlight, wind, rain, tides and geothermal heat—which are naturally replenished. In 2008, about 18% of global final energy consumption came from renewables, with 13% coming from traditional biomass, such as wood burning. Hydroelectricity was the next largest renewable source, providing 3% (15% of global electricity generation), followed by solar hot water/heating, which contributed 1.3%. Modern technologies, such as geothermal energy, wind power, solar power, and ocean energy together provided some 0.8% of final energy consumption.

Alternative energy includes all sources and technologies that minimize environmental impacts relative to conventional hydrocarbon resources and economic issues related to fossil fuel resources. Fuel cells and natural gas might be alternatives to coal or nuclear power. Throughout the book, the fundamentals of the technologies related to integration of such alternative and renewable energy sources are reviewed and described with authority, skill, and from critical engineering aspects for the end user of energy. Climate change concerns coupled with oil prices with its uncertainty and increasing government support is driving increasing renewable energy legislation, incentives and commercialization. Investment capital flowing into renewable energy climbed from $80 billion (US) in 2005 to $100 billion in 2006.

The book provides the forum for dissemination and exchange of up-to-date scientific information on theoretical, generic and applied areas of knowledge.

The topics deal with new devices and circuits for energy systems, photovoltaic and solar thermal, wind energy systems, tidal and wave energy, fuel cell systems, bio energy and geo-energy, sustainable energy resources and systems, energy storage systems, energy market management and economics, off-grid isolated energy systems, energy in transportation systems, energy resources for portable electronics, intelligent energy power transmission, distribution and inter-connectors, energy efficient utilization, environmental issues, energy harvesting, nanotechnology in energy, policy issues on renewable energy, building design, power electronics in energy conversion, new materials for energy resources, and RF and magnetic field energy devices.

Open Access is a new direction in academic publishing where all chapters are available for full free access online. The book is published as open access fully searchable by anyone anywhere.

We believe that immediate, worldwide, barrier-free, open access to the full text of research articles is in the best interests of the scientific community.

Free online availability substantially increases an article's impact. The mean number of citations to offline articles has been shown to be 2.5~3.0 % smaller. The book is also available in printed edition (hardcopy} and has been distributed to major university and learned societies libraries, etc. free worldwide.

The timely initiatives taken by the authors in this book to cover closely renewable energy is applauded. I am confident that this work will contribute to our better understanding of how to integrate renewable energy sources into our electricity grids for commercial, domestic and industrial applications. I strongly recommend this work to a wide audience, including environments, engineering educators, students, industrialists, consultants, and those concerned in reducing greenhouse emissions that is affecting our planet.

December 2009

T J Hammons
University of Glasgow, UK

Contents

1

A Model for Greener Power Generation for North-east Sri Lanka based on Stand-alone Renewable Energy Systems

Reggie Davidrajuh
University of Stavanger
Norway

1. Introduction

In Northern and Eastern Sri Lanka (NE-SL), electricity is not available in most of the places due to the civil war. There is practically no electricity production and distribution system available or planned. Since Sri Lanka, as a whole, is suffering from power-shortage, it will be not possible for her to supply electricity to NE-SL, by the time the civil war comes to an end. An alternative generation and distribution system should, therefore, be designed to supply electricity to NE-SL.

The need for electric power may be analyzed by a fairly intensive and deep study, which invariably requires life cycle analysis of the society in terms of energy use and conservation, industrial and household development, population and industry distribution, export and import of electricity, etc. In contemporary NE-SL, however, the very basics in a societal infrastructure such as industry, highways and education system, are either non-existent or in primitive form. And due to the on-going war in this region, important data for electricity power sector decision-making (such as income and purchasing power of the population, planned transportation and highway infrastructure, environmentally sensitive areas, meteorological data, etc.) are either not available or inadequate.

The objective of this chapter is to demonstrate that to achieve a sustainable electricity generation for NE-SL, in addition to utilizing the abundant intermittent resources NE-SL has, namely solar and wind, biomass production must be given utmost priority. The scope of this chapter is limited to design of a mixture of power plants in the cogeneration system, not involving the other aspects of electricity sector such as transmission and distribution. Limitation of this work: due to the ongoing civil war in the region, there is no recent data available about energy production and consumption in this region; the data used in this chapter is from 1994 to 1996.

2. Needs and Requirements for Sustainable Power Supply

The long-term life cycle scenarios for energy to the planet Earth points to massive utilization of solar power, or its derivatives as hydro-, wind- or biomass power (Weinberg, 1990). Power from tidal waves, geothermal and ocean thermal power is insignificant as far as NE-SL is concerned. Each type of energy has its own economy, life cycle and complications. Direct solar power and wind power is complicated by its diluted form and its variation with time and location. However, some of these variations are predictable and may be used in optimal predictive control of power cogeneration systems (Asbjornsen, 1984). Example of such cogeneration system is a combination of solar, wind and biomass power. The power generating systems interconnected by the power grid is an existing technology. But incorporating intermittent power supplies (like solar, wind) to the grid requires careful planning. Nuclear power or hydropower systems acting as the base-load supplier of the utilities, try to even out any fluctuations or failures in the intermittent power supplies, if the intermittent power supplies are connected to the grid. With absence of fossil fuels, hydropower or nuclear power, a sustainable cogeneration system for NE-SL has to use biomass as buffer in incorporating intermittent power supplies. The other option is to include energy storage in the cogeneration system. The energy storage will store energy when the generated electricity from intermittent supply is in excess and it will contribute electricity to the grid when demand exceeds supply. However, studies show that energy storage systems are generally expensive (Friberg, 1993).

2.1 Basic Needs Analysis

The total system needs may be formulated as a combined need for electric power by the society and a need for global environmental protection, formulated as follows:

- There is a need for electric power, fairly distributed, to the society population
- There is a need to replace fossil fuel as energy source for industrial and domestic electric power generation on economic and environmental perspective.

2.2 Basic Requirements Definition

The system needs lead to requirements, which are consequences partly of the needs, partly of the customer and user situation, and partly of conceptual solutions. The definition of system requirements is an iterative process, which expands in detail, as the system baseline concept is designed. At the present level of details, and the present system analysis, a set of requirements may be defined fairly simply:

- The cost of electric power generation and distribution in NE-SL shall be within the price range for customers and users in the rest Sri Lanka
- Compared to the rest of Sri Lanka, the power generation system designed for NE-SL shall have the same or better, reliability, availability and fair distribution of electric power. Cogeneration system should also meet the customers' power requirement at any time
- The power generation system shall have none or minimal harmful pollution effects, on the air, soil or water environment, or on the social environment.

There are some other requirements too, due to the varying nature of the intermittent energy sources, and due to the economic requirement on conversion technology:

- In case of intermittent primary energy source is used, energy back-up shall be provided by the system
- Existing technology for energy conversion shall be utilized to a degree, which is the most feasible economically
- The economy of scale shall be utilized optimally in terms of size and capacity of units, modularization and standardization, distribution and centralization

3. System Integration

Some thoughts are given below on integrating generators into the electric grid, by considering the technology's social and environmental effects.

3.1 System Response to Customers' Power Requirements

Changes in customers' power requirements will automatically be taken care of by the demand management system through control of the turbines in the power generation systems. The optimal strategy will be to let the solar power generation run at its maximum capacity at any time. This capacity will change during the day and go to zero when the sun sets. It is also optimal to let the wind power run at its maximum capacity, which will also change with the intensity of the wind. The remaining power requirement will be taken care of by the biomass and fossil fuel power generation systems. As long as the system is designed to tackle all possible situations, as described, the customer will see the power generation system as totally reliable. The key issue in the whole system is to match capacities of the total cogeneration system to the customers' power requirements and to the availability of the energy sources, solar radiation, and wind speed.

Because of the intermittent nature of solar and wind power, when incorporating these generators into the grid, the total plant capacity must always exceed the maximum expected demand by a large margin (penalty for intermittence), in-order to increase system reliability. In NE-SL, where wind and solar power will be contributing most of the time, thermal plants that have higher operating fuel costs but cheaper to build become more attractive (such as natural gas plant), because the reduced operating time will make fuel costs less important. There is a limitation on the extent to which the intermittent sources like solar and wind contribute to the total power generation. As the percentage of power generation of wind and solar increases, there is steady decline in value, because adding intermittent sources reduce the reliability. Contribution from wind energy ranging from 0% to 50% of overall installed capacity is feasible before operational losses become prohibitive (Grubb and Meyer, 1994).

3.2 Smaller the Better

For a developing country like NE-SL, which does not have any large-scale hydrologic resource for electricity generation, the selection of optimum power plant mix should be based on small, affordable power generators. Reliability of the power generation and distribution system will be increased if large number of small generators, scattered throughout the nation, is used rather than few large plants. By using a large number of smaller plants, the potential danger of over building or commissioning less cost effective large plants could be avoided. Small plants can be added quickly as they are needed and even disassembled and moved if loads decline; the economics of the larger plants depend heavily on long-term forecasts (Johannson et al, 1993; Kelly and Weinberg, 1994).

In case of biomass-generated electricity, power plants are modest in scale (less than 100MW) to assist collection and transportation of dispersed biomass supplies (Williams and Larson, 1994). If the small generators are scattered throughout the region, a number of other benefits emerges: heavy pollution of a region (congestion) is avoided; employment (less rural exodus) and community participation (awareness) is also increased in distributed power generation. Ideally, each region should become self-sufficient in electricity generation; if the demand exceeds its supply, the utility in that region imports electricity from the neighboring utilities, and similarly, the excess capacity is also sold to the neighboring utilities (Johannson et al, 1993; Kelly and Weinberg, 1994).

3.3 Participatory Development

For a sustainable cogeneration system to succeed, the participation of the community becomes essential. Take for example, production of electricity from residues form paddy: today rice husk is normally sold at the rice mills, for insignificant price for trivial use. Now, the rice mill is expected to sell the husk to the utility. The mill management will certainly hesitate to sell if high expenses are incurred in doing so (transportation). However the question is, even if the extra expenses incurred are negligible, whether the management will be willing to sell. The answer to this question depends on the participation of the management in the community's affairs. This 'bottom up' approach where grass roots organize, assist and take responsibilities, and solve (save energy) the energy problem will certainly promote the efficient supply of electricity to the whole nation.

3.4 Household Energy Situation

The primary energy consumed in Sri Lanka consists mainly of (Vijayanathan and Lamasena, 1997):

- Bio-mass fuel (67%),
- Petroleum (22%), and
- Hydro-electricity (11%).

With the absence fossil fuel resources, biomass is used for both cooking and for small industries, and hence biomass is of vital importance to the economy. Biomass is used by 94% of households use for cooking (Vijayanathan and Lamasena, 1997).

Table 1 given below depicts household energy resources for the whole Sri Lanka. Though data for NE-SL is not available, one can safely assume that both fuelwood and kerosene are the crucial energy resources for cooking. Figure 1 show parts of the world where basic remedies like dung and wood are still used for domestic cooking.

Type of Energy	%
Fuelwood and other biomass	87.5
Electricity	6.6
Kerosene	4.8
LAG(LPG)	1.1

Table 1. Biomass resources used for cooking by type

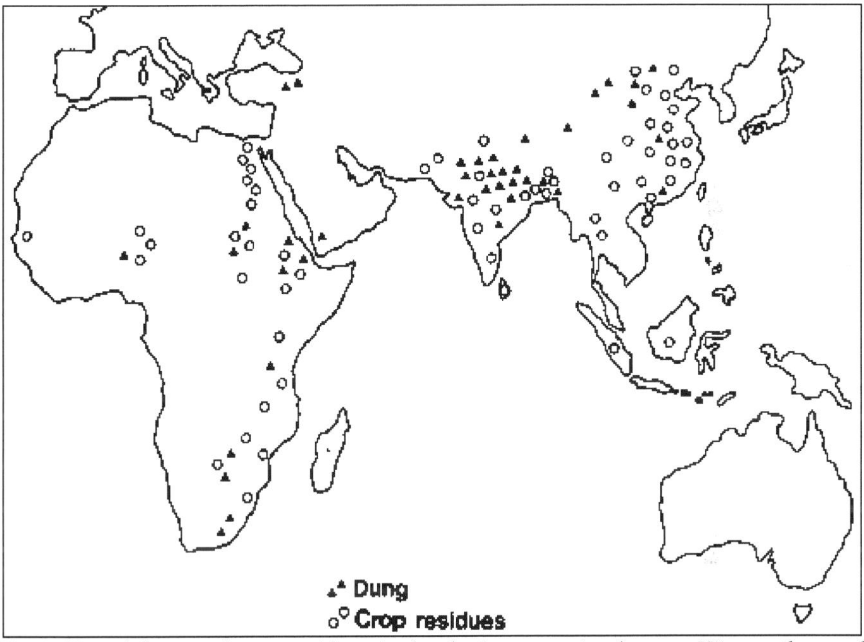

Fig. 1. Use of dung and crop residues in developing countries (source: Vijayanathan and Lamasena, 1997)

4. Estimating Electricity Demands of NE-SL by 2010

Since no energy-macroeconomics data for NE-SL is available, the basic assumption made here is that, the energy needs of her population will be the same as the needs of the rest of Sri Lanka. Sri Lanka, with a population of 18 million, today has a per capita electricity consumption of 168 kWh (CIA, 1996). The electricity needs of the NE-SL with a population of 3 million, is therefore 504 GWh per year.

In recent years, total firewood consumption in Sri Lanka has been around 5 million tons per year. Of this value, industries consumed approx. 1.05 million tons (Hall et al, 1994). It is reported that the only half of the firewood supply is obtained in a sustainable way (such as forest regeneration, rubber re-plantation cycle etc) (Munasinghe and Meier, 1993). Therefore decline in the natural forest area in Sri Lanka is inevitable. The natural forest area has fallen to 24% in 1980 from 44% in 1955. Surely NE-SL wants to keep its very limited natural forests therefore the amount of energy obtained in Sri Lanka through firewood supply has to be compensated for through electricity supply. Since firewood is generally the most economic biomass fuel for industrial heating (if used by efficient combustion), only the portion of firewood (ca. 4 million tons) consumed for domestic purposes has to be substituted by electricity. For NE-SL this is equivalent to 0.7 million tons. Therefore additional electricity needed to compensate for use of firewood is 355 GWh.

In 1986, Sri Lanka consumed 154.2 thousand tons (kt) of Kerosene, which is solely used for domestic lighting and cooking and 129.0 kt of Fuel oil for industrial use and electricity generation. A portion of fuel oil used for electricity generation is assumed to be 50%. This portion of fuel oil and the amount of kerosene used (totaling 218.7 kt) could be substituted by electricity. Assuming again consumption is proportional to population, this will be equivalent to 130 GWh. The total electricity needed today is therefore approximately 989 GWh.

By 2010, the energy needs of the population are expected to be three folded (assuming 8% rate of increase in the standard of living (ca. 6.9%) and population growth (ca. 1.2%)) (CIA, 1996). The total electricity needs per year will then be 2967 GWh. With these data, general performance requirements for the cogeneration system could be set up as shown on Table 2.

Functional Requirement	The cogeneration system should be able to supply 2967 GWh of electricity per year.
Operational Requirement	The cogeneration system should be able to support quickly varying demands; should be able to deliver peak power of 564 MW.
Physical Requirement	The system should be smaller, distributed throughout the nation. Building materials and construction, operation and scrapping conform to international and local standards.

Table 2. General requirements of the cogeneration system for the year 2010

5. Energy Resources in NE-SL

By 2010, relevant power generation technologies to NE-SL will be solar, wind, biomass, coal and natural gas.

5.1 Biomass

Energy from biomass is divided into two general categories 1) agricultural crops and residue, wood and wood waste, animal waste, aquatic plants and 2) plantation biomass:

Crop residues:

Sugarcane cultivation is a well-established industry in NE-SL. In 1995, NE-SL produced 1.2 million tons of sugarcane (FAOSTAT, 1995). Each ton of sugarcane produce about 0.43 ton of bagasse and leaves. BIG/ISTIG power plant produces 677 kWh of electricity per ton bagasse (Williams and Larson, 1994). If production is assumed to grow at 3.3% (corresponding to Sri Lankas current growth in agricultural production) and all the residues from the sugar industry are used for electricity production, they could produce 568 GWh of electricity by 2010. This is about 19% of the anticipated electricity needs of NE-SL. Residues from the pulp industry, sawdust and bark from de-barking and chipping of logs, could also be used as biomass energy source to the power plant; It is therefore possible for a collateral generation of electricity - pulp - sugar production, especially suited for Eastern Sri Lanka.

In NE-SL, agriculture is and will be the main source of income. In 1995, Sri Lanka produced 2.2 million tons of rice (FAOSTAT, 1995). Before the civil war broke out, NE-SL used to produce about 33% of the total rice production in Sri Lanka (TEEDOR, 1992). Each ton of rice produced is accompanied by 2.25 tons of residues in terms of paddy straws and rice husks. With the same rate of growth in production (3.3%), and assuming NE-SL will be able to produce 33% of total production, the residues will have an energy equivalent of 900 GWh of electricity which is about 30% of the electricity needed by year 2010 (paddy rice is a C_3 plant, which has about 50% energy contents of a C_4 plant such as sugarcane). In NE-SL, Groundnut, Maize, Soybean, Cowpea are also cultivated. Substantial amounts of energy can be generated from residues of these crops.

In summary, the large amounts of crop residues that are normally not fully utilized could be used to generate more than 49% of electricity needs.

Plantation biomass:
One of the tropical legumes, *Leucaena leucocephala* ("Ipil-Ipil") has already been successfully planted in Sri Lanka. In NE-SL, experimental *Casuarina* plantation gave promising results (TEEDOR, 1994; TEEDOR, 1995). Plantation biomass energy is a solid option for electricity production in NE-SL. For plantation biomass to be successful plantation wood should be cheaper than the wood purchased on the open market and its cost of energy should be lower than that of conventional energy systems. Also, energy crops should not have long-term environmental, ecological damages. So far no serious study has been done on the economics of harnessing energy crops in Sri Lanka or NE-SL.

5.2 Solar Energy
The utilization of solar energy (Photovoltaic or solar thermal) is a major option to satisfy the energy needs of NE-SL. NE-SL, being a tropical country receives annual average solar insulation of 200 W/m^2. The solar power system is renewable, it answers all the serious global as well as local environmental issues such as, zero emission of Green House Gases (GHG) and no solid waste.

Already a number of micro projects are being operated in many parts of NE-SL, funded by Non-governmental organizations (NGOs) like TECH-German (Tech, 2009a). Figures 2 shows a solar micro power plant in operation to help education of the youth in most rural areas of NE-SL; In order to improve the education of the rural students, NGOs like TECH-German provide solar panels to schools, libraries, community centers where the students can gather at night and study.

Fig. 2. Solar panel installed in NE-SL

5.3 Wind Energy

Wind power generation offers many benefits too such as, environmentally friendly (no gaseous emissions), rapid modular addition of capacity, wide range of capacities (from few kW to hundreds of MW), easy to integrate with existing power grid, coexistence with other use of land, no fuel needed and economically competitive (NRC, 1991). NE-SL confines to a region where the wind energy is attractive for power generation with average wind speeds over 5 m/s, giving an average generation of over 33% of rated power (Twidell and Weir, 1986).

Figures 3 and 4 show a wind power plant in operation at the integrated model farm in Vaddakatchchi, Kilinochchi. Figure 3 shows the wind turbine and figure 4 shows the battery storage system. The 6KW wind mill powers a Dairy Processing unit and a computer center (TECH, 2009b).

Fig. 3. Wind Turbine in Vaddakatchchi, Kilinochchi

Fig. 4. Battery Storage System

6. Estimating the Costs of Energy Production

There are a huge number of parameters involved in the equations for calculating costs of energy production. In this section, only a very limited set of parameters is considered. However, the parameters taken into consideration are the most important ones, thus the model developed in this section and in the following section 7, can be considered as representative.

6.1 Estimated Energy Costs in the Year 2010

Fixed annual cost (FAC) of the total installed cost:

$$FAC = ICC \times CRF \tag{1}$$

Where *ICC* is the total installed capital cost and *CRF* is the capital recovery factor.

CRF is equal to:

$$\frac{r}{1-(1+r)^{-n}} \tag{2}$$

Where *r* represents the interest rate and *n* is the lifetime in years.

The cost of electricity (COE) is then,

$$COE = \frac{(FAC + OMC)}{P_r \times CF \times 8766} \text{ \$/kWh} \tag{3}$$

Where P_r is the rated capacity, *OMC* is the Operational and Maintenance (O&M) costs per year, and *CF* is the average capacity factor and *8,766* is the number of hours per year (*365.25* × *24*).

Table 3 presents values for the parameters involved in the calculations for costs of energy production. All values are given in 1994-US$. The capital recovery factor (0.0888) is for 8% interest rate and a 30-year plant life. All taxes and insurance costs are neglected. Table 4 summarizes the electricity generation costs for different technologies:

	Biomass (Heaps and Hill, 1996)	Solar (De Laquil, 1994)	Wind (Cavallo et al, 1994)	Coal (Johannson, 1994)
System rating (MW)	111	200	500	109
Installed capital cost ($ /kW)	974	3200	850	1122
Variable O&M costs ($/kWh)	0.0009	0.0065	0.008	0.0010
Fixed O&M costs ($ /kW-year)	34.2	-	-	48.0
Fuel costs ($ /GJ)	2.43	-	-	2.00
CF	0.75	0.59	0.33	0.75
Efficiency	0.43	-	-	0.42

Table 3. Parameter values for calculating costs of energy production

	Wind	Biomass	Coal	Solar
Capital (fixed)	2.60	1.32	1.52	5.50
Fuel	-	1.18	1.71	-
Operation & Maintenance	0.80	0.61	0.83	0.65
Total	3.40	3.11	4.06	6.15

Table 4. Electricity generation costs for different technologies (US cents / kWh).

6.2 Environmental Costs Associated with Electricity Generation

It is important to note that the costs of energy given in Table 4 represent only a part of the costs the society has to pay for electricity generation. The costs of electricity generation can be divided into two, namely the direct or internal costs that are borne by the generating company (capital costs and recurrent or operating costs) and the external costs ('externalities') such as the costs associated with environmental damage due to electricity production (Hill et al, 1995). Environmental externalities are becoming the most important external costs in electricity generation. There are various economic methods available to internalize (incorporate into the total cost of energy) environmental externalities. In this chapter, internalization values based on US and German studies are used as presented in (Seshadri, 1978).

Estimation for air emissions from a coal-fired power plant is given below, in Table 5 (Munasinghe and Meier, 1993). Also given are the values to monetize emissions (Fritsche, 1994).

	Emission (kg per kWh)	*Value \$/kg*	*Environmental cost \$/kWh*
SO_2	0.00640	4.060	0.0260
CO_2	0.32000	0.013	0.0042
NO_X	0.00370	1.640	0.0061
Particulate	0.00050	2.380	0.0012
HC	0.00004	0.680	0.0000
Solid ashes	NA	NA	NA
Limestone	NA	NA	NA
Total			0.0375

Table 5. Environmental costs due to coal-fired power generation.

7. Design of Optimal Power Cogeneration

Computer modeling and simulations are done to find the optimal mix of power plants in the cogeneration system; most of these computer models are based on linear programming formulation, and the question is to find out "*the mix of different types of plants that will satisfy the power demand with minimum costs*".

The objective function: Minimize S = "The cost of power generation by different plants" subject to constraints:

1. The total generation should meet the demand
2. Each plant's generation cannot exceed its physical limitation.

In mathematical notation (simplified):
Minimize:

$$\{S\} = \sum x_k \cdot c_k \;\; [\$ / \text{year}] \tag{4}$$

Subject to:

- Constraint 1:

$$\sum x_k \geq \eta \tag{5}$$

- Constraint 2:

$$x_k \leq X_k \cdot CF_k \tag{6}$$

Where $x, c, \eta, X,$ and CF are the generation of a specific plant per year, the Coefficient of Energy (COE) of the specific plant type, the peak power of that plant, and the average capacity factor (CF), respectively.

7.1 Basic Assumptions

Two basic assumptions are given below:

1. In order to prevent over-relying on the intermittent solar and wind energy systems, penetration of intermittent power supplies is limited to 33%, if wind power alone is used. If wind and solar power are used (with the assumption that, owing to subsidy solar power costs are now equal to wind power's) then 40% penetration of these two combined, is chosen.
2. Biomass: Contribution from biomass is taken to be 1468 GWh /year or 170 MW, which is 49% of the total electricity needed.

7.2 Typical Solutions for Cogeneration System

Depending on whether solar power is subsided or not, two solutions are proposed:

- Case - A
  In case-A, the environmental damages are neglected. Also, it is assumed that there is no form of subsidy to consider solar power:
  In this case, a mix of wind farm(s) of 186 MW and biomass power plants totaling to 170 MW could act as the base-load power supplier. Fossil fuel-fired power plants (coal and natural gas) providing 303 MW of power act as the intermediate and peak suppliers. Solar power is too expensive to consider under any category. Total rated power of the plants put together amounts to 659 MW. So there is a reasonable margin (17%) to take care of the intermittent nature of wind energy and also to increase the reliability of the cogeneration system.

- Case - B
  In case-B, the environmental costs are included in the production costs. Also, it is assumed that there are enough subsidies available to consider solar power:
  In this case, the cogeneration mix becomes: intermittent solar and wind could contribute 226 MW together (40% penetration) to the base load with the addition of 170 MW biomass power plants. Fossil fuel power should provide 281 MW. The ratio of Solar/Wind depends on the subsidy available for solar power generation and meteorological-statistical data. Total rated power of the plants put together amounts to 677 MW. Here too, a reasonable margin (20%) is available to improve the reliability.

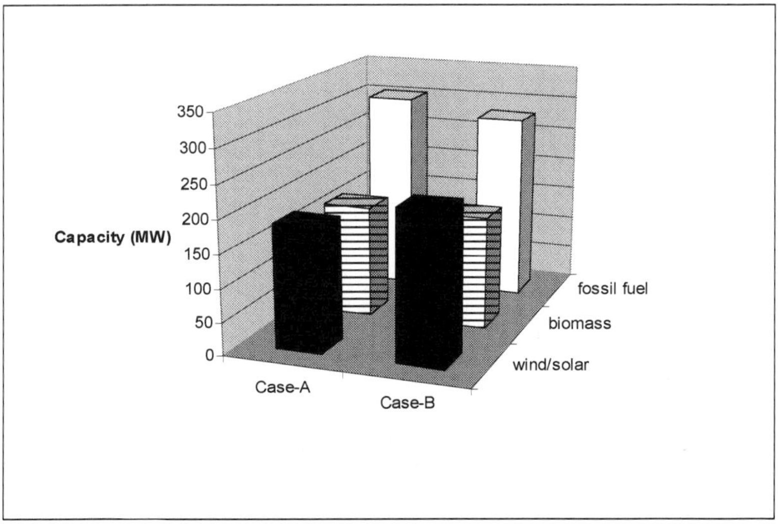

Fig. 5. Suggestion for an optimal mix of power plants.

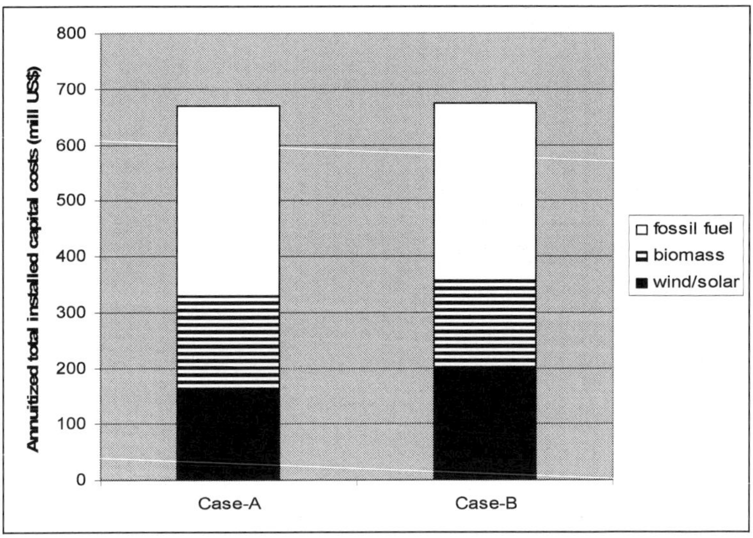

Fig. 6. Cost of the cogeneration system.

Figures 5 and 6 shows optimal mix of plants in terms of capacity and costs respectively. In Figure 6, coal power replaces fossil fuel power (coal and natural gas) for simplicity. Also, the reduced capital cost of solar power (in Case - B) is assumed to be equal to that of wind power.

8. Conclusion

In the typical cogeneration system for NE-SL, the contribution of fossil fuel-fired power is too high: when considering peak rate, it is 53% in case-A, and 50% in case-B (Figure 5). The growing demand for fossil fuel resulting in the rising price of oil, together with the serious environmental damages due to fossil fuel burning indicates that NE-SL should investigate the use of more sustainable resources for her electricity generation. Though intermittent energy resources solar and wind is abundant, their percentage of contribution cannot go beyond certain limit. Therefore the only way out is to engage large-scale plantation biomass energy.

In case-B, fossil fuel contributes to 281 MW of electricity generation. This is equivalent to 6.1 million tons of wood. Assuming that one hectare of biomass plantation yields 25 tons of wood, a total area of 109 thousand hectare (5.4% of total area of NE-SL) of biomass plantation is needed for complete sustainable electricity generation.

Normally, unproductive or degraded land (unsuitable for field crops) is allocated for plantation thereby avoiding land for food versus fuel conflict. Plantation biomass species like Ipil-Ipil and Casuarina has rapid growth, even on degraded land. In Tamil Nadu, South India, biomass plantation yields an optimal value of 55 dry tons of wood per hectare per year (Seshadri, 1978). Also, Ipil-Ipil and Casuarina requires minimal management and care. Ipil-Ipil has the ability to fix nitrogen from atmosphere and thus has the 'self-fertilizing' ability too (a by product of Ipil-Ipil, high-protein seeds and leaves could be used as animal feed and green fertilizer).

9. References

Asbornsen, O. (1984) Feed forward predictive and adaptive control by the dynamic matrix. Proceedings of the ACC 1984, pp. 1864-1869

Cavallo, I, Hock, S., and Smith, D. (1994) Wind Energy: Technology and Economics. In *Renewable Energy - Sources for Fuels and Electricity*, ed. Thomas B. Johansson et. al, Island Press, Washington

CIA (1996) *World Factbook*. CIA, US

De Laquil, P. Kearney, D., Geyer, M., and Diver, R. (1994) Solar-thermal Electric Technology. In *Renewable Energy - Sources for Fuels and Electricity*, edit. Thomas B. Johansson et. al, Island Press, Washington

FAOSTAT (1995) Database Food and Agricultural Organisation of the United Nations.
FAO, UN

Friberg, R. (1993) A photvoltaic solar-hydrogen power plant for rural electrification in India. Part 1. *Int. J. Hydrogen Energy, Vol.18, No.10, pp.853-882* (1993)

Fritsche, U. (1994) Modelling Externalities: Cost-Effectiveness of Reducing Environmental Impacts. In *Renewable Energy - Sources for Fuels and Electricity* , edit. Thomas B. Johansson et. al, Island Press, Washington

Grubb, M. and Meyer, N. (1994) Wind Energy: Resources systems, and regional strategies. In *Renewable Energy - Sources for Fuels and Electricity*, edit. Thomas B. Johansson et. al, Island Press, Washington

Hall, D. Rosillo-Calle, F., Williams, R. and Woods, J. (1994) Biomass for energy: supply prospects. In *Renewable Energy - Sources for Fuels and Electricity*, edit. T. B. Johansson et. al, Island Press, Washington

Heaps, C. and Hill, D. (1996) Applying Fuel Chain Analysis in Venezuela and Sri Lanka. In *SEI-B Energy & Environment News, Nr.1, Feb 1996*

Hill, R., O'Keefe, P. and Snape, C. (1995) *The Future of Energy Use,* Earthscan Publications Limited, London

Johannson, T., Kelly, H., Reddy, A., and Williams, R. (1993) Renewable fuels and electricity for a growing world economy. In *Renewable Energy - Sources for Fuels and Electricity,* edit. Thomas B. Johansson et. al, Island Press, Washington

Kelly, H. and Weinberg, C. (1994) Utility Strategies for using Renewables. In *Renewable Energy - Sources for Fuels and Electricity* , edit. Thomas B. Johansson et. al, Island Press, Washington

Munasinghe, M. and Meier, P. (1993) *Energy policy analysis and modeling.* Cambridge University Press, Cambridge, UK

NRC (1991) *Assessment of Research Needs for Wind Turbine Rotor Materials Technology.* Committee on Assessment of Research Needs for Wind Turbine Rotor Materials Technology, National Research Council. National Academy Press, Washington, D.C.

Seshadri, C. (1978) *Energy Plantation - a case study of the Coromandel Littoral.* Shri A. M. M. Murugappa Chettiar Research Centre, Madras, India

TECH (2009a): Solar Power Micro Power Plant, http://www.techonnet.org/projects/Project-EP-01.htm

TECH (2009b): Wind Power Micro Power Plant, http://www.techonnet.org/projects/Project-EP-02.htm

TEEDOR (1995) *Casuarina Plantation - Project description.* (In Tamil language). TEEDOR, Thavady, Sri Lanka

TEEDOR (1994) *Aatharam. Mar-Apr 94.* (In Tamil language). TEEDOR, Thavady, Sri Lanka

TEEDOR (1992) *Food needs and goals to achieve* (In Tamil language). TEEDOR, Thavady, Sri Lanka

Twidell, J. and Weir, A. (1986) *Renewable Energy Sources.* E. & F. Spon Ltd., New York

Vijayanathan, L., Lamasena, L. (1997): Boiling Point 39: Using biomass residues for energy. Autumn 1997

Weinberg, C. (1990) Energy from the Sun. In: *Energy for the Planet Earth.* Readings from Scientific American, W. H. Freeman and Co., NY

Williams, R. and Larson, E. (1994) Advanced gasification based biomass power generation. In *Renewable Energy - Sources for Fuels and Electricity,* edit. Thomas B. Johansson et. al, Island Press, Washington

2

Automatic Sun-Tracker System for Photo-Voltaic Plants

Joao M. G. Figueiredo
CEM-IDMEC, Universidade Évora, Mechatronics Group
R. Romão Ramalho, 59; 7000-671 Évora, Portugal
Phone/Fax number: +00351 266 745 300, e-mail: jfig@uevora.pt

1. Introduction

According to market economy, the increasing worldwide demand for energy, forces a continuous rise on the price of fossil combustibles. In fact, it is expected in the near future, that the demand for energy will grow faster than the finding out of new available fossil resources (Khan et al., 2007).

This market behaviour brings a positive challenge to the scientific community as more funds are allocated for the research and development of new alternatives to the usual main energetic sources (fossil combustibles). In this context we have seen, in the last decades, to a concentrated focus on renewable energy research. Among these renewable energetic sources, the international scientific community has devoted intense efforts to wind, solar photovoltaic and biomass. Some investigations and hardware developments on wave energy have been led by Great Britain and Portugal (Wave-Energy-Centre). In this paper an intelligent sun-tracking system for efficiency maximization referring photovoltaic energy production is developed.

Nowadays photovoltaic energy has a low efficiency ratio concerning the complete distribution chain from production to consumption (ca. 12%). In optimized environments (materials, electric inverters, tracking systems, etc) an input of 1000W of solar incident energy can bring ca. 190W in electricity (efficiency of 19%). This low performance ratio implies big Earth surface consumption when it is intended to install industrial photovoltaic units with significant production impact (50MW - 100MW). Today it is being built in south Portugal a photovoltaic plant with 64MW production capacity which occupies an huge area of ca. 400 ha (4 Km^2).

The more relevant side effect of the low efficiency of photovoltaic systems is its poor competition related to traditional combustibles in both economical and financial aspects.

Owing to changes in the solar radiation energy and in the cell operating temperature, the output power of a solar array is not constant at all times. Consequently, a maximum solar power tracking controller is always needed in any scheme with solar cell arrays to ensure maximum utilization. Therefore, works to solve the problems on maximum power point (MPP) tracking have always been a hot topic for photovoltaic array utilization systems. A logical MPP tracking search algorithm using normalized current, voltage and power at the

work points, that corresponds to the maximum power point values for different operating conditions was early tested (Atlas, 1992), (Atlas, 1996). A on-line controller to track the MPPs under changing illumination was described in (Hua & Lin, 2003).
An optimization approach using fuzzy was given in (Benlarbi et al., 2004) for PV water pumping systems. Other MPP tracking controllers can be found in (Hua & Lin, 2004) and (Chen et al., 2004).
This paper focuses on the optimization of the electric energy production by photovoltaic cells through the development of an intelligent sun-tracking system. The developed tracking system is innovative in relation to the usual sun tracking systems available in the market.
The usual available solutions for tracking systems rely on the knowledge of the geographical position of the solar panel on the earth surface. With this knowledge it is possible to know the relative position of the sun, on a time basis, according to the well known solar tables (Solardat). Modern solutions incorporate a GPS system to calculate the position of the solar panel on the Earth surface. The orientations to be followed by the photovoltaic panel, on a regular time-base, are then pre-programmed, on an open loop approach.
There are significant efforts on the optimization of sun tracking systems as it is documented by several registered international patents. These solutions are based either on the above described principle either on the quantification of the received solar energy, either on the maximization of the solar incident radiation through the use of light concentration lens or mirrors (Biee & Chace, 2009), (Rubio et al., 2007). The solution developed in this paper is innovative related to the above referred approaches as this system is autonomous regarding the information needed to process the optimal orientation and it is intelligent in a way that it monitors, on a real-time base, the photovoltaic energy production and it avoids systematic failures coming from changes on the assumed blind values (position, initial infrastructure orientation, cleanness of the photovoltaic cells, etc.).

2. System Description

2.1 Overall System Presentation

The overall system is presented in fig. 1. The complete strategy is composed by 5 sub-systems: 1) Electro-Mechanical Structure; 2) Control Unit; 3) Supervisory System; 4) Wind-meter; 5) Photovoltaic Park.
The developed tracking system searches the optimal orientation of a surface, related to the sun incident radiation. The global performance of the system is described below. The planar surface is composed by a photovoltaic cell which is motorized by 2-orthogonal axis. These two controlled DOF (Degrees Of Freedom) are managed by a PLC (Programmable Logic Controller) according to a search programme that compares the electric power produced by the photovoltaic cell in each correspondent orientation. The maximal power value is stored and the correspondent orientations on both motorized axis are stored. This new optimal orientation of the tracking system is then communicated to the industrial photovoltaic park in order to transfer the new optimal orientation to all PV-production panels.

2.2 Electro-Mechanical Structure

The operational subset of the tracking system, named Electro-Mechanical System, is presented in figs 2 and 3.

This structure has two DOF, motorized by stepper motors with incorporated encoders, in order to track exactly the prescribed path.
The mechanical system was designed using standard industrial Aluminium profiles in order to obtain a simple and economic structure.
The mechanical structure is mainly composed by Bosch-Profiles and Aluminium plates.
The two motorized axis are composed by Step-motors assembled to Aluminium shafts.
Figure 2 illustrates the several main components of the mechatronic system:

Part n. 6 = Step-Motor to control axis 1;
Part n. 7 = Step-Motor to control axis 2;
Part n. 8 = Photovoltaic cell (150mmx150mm).

Figure 3 details the two designed degrees of freedom (DOF).

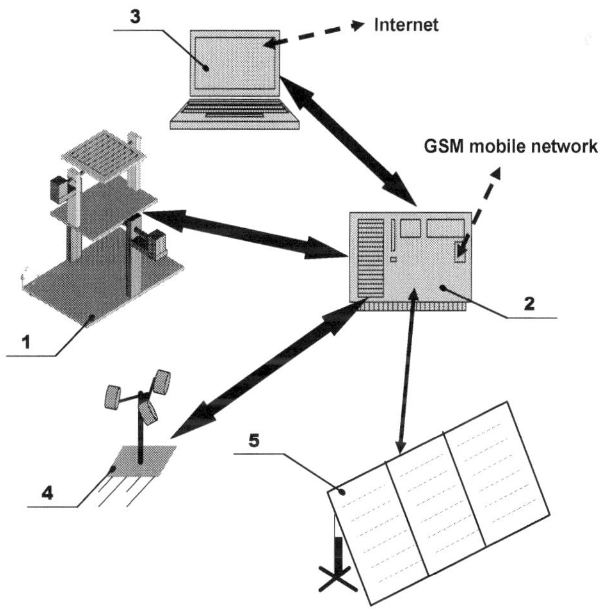

Fig. 1. Overall System Presentation

2.3 Control Unit

The control unit is composed by a PLC system (Programmable Logic Controller). This control system has the complete operational management of the tracking system.
The main tasks performed by the system are:

- Control of the two step motors;
- Processing the data from both encoders;
- Processing the voltage signal coming from the PV-Cell;

- Processing the data from the external proximity sensors that informs the system about the hard-home position reference.

This PLC controls directly the tracking system and commands all other PV-Panels, from the solar Park, through a Profibus-DP network (Siemens, 2001a).
Figure 4 shows an example of a solar park with motorized PV-Panels. Figure 5 illustrates the Profibus network implemented in this study.

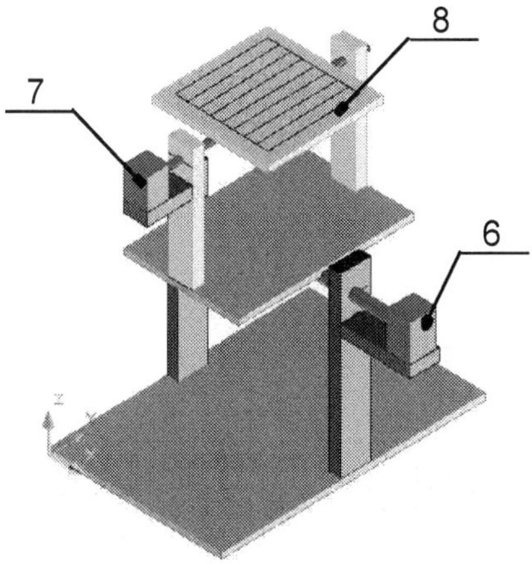

Fig. 2. Electro-Mechanical System: Main Components

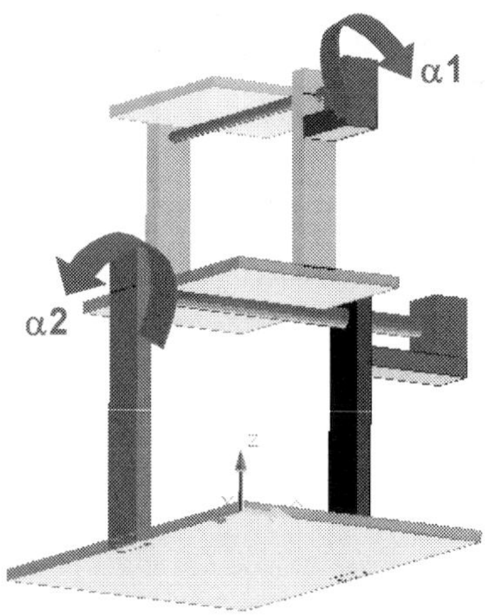

Fig. 3. Electro-Mechanical System: Axis 1 and 2

2.4 Supervisory System

A SCADA system (Supervisory Control And Data Acquisition) is implemented to monitor and supervise the tracking system.
A Supervisory Control and Data Acquisition (SCADA) System is used as an application development tool that enables system integrators to create sophisticated supervisory and control applications for a variety of technological domains, mainly in the industry field. The main feature of a SCADA system is its ability to communicate with control equipment in the field, through the PLC network. As the equipment is monitored and data is recorded, a SCADA application responds according to system logic requirements or operator requests.
In the developed supervisory system the SCADA application manages the overall system dynamics.
The Communication flux between the supervisory system and the control unit is illustrated in fig. 5.
The SCADA PC is simultaneously a SCADA server and an internet server, as the implemented SCADA application is web enabled.

Fig. 4. Solar Park - motorized PV panels

3. Experimental Prototype

3.1 Physical Description

The prototype built followed the design presented in figure 2. This system incorporates a PV-cell 150mmx150mm, Pmax=1,12W, (Polycrystalline Silicon wafer) and the whole structure is made of aluminium alloy. In fig. 6 the global developed prototype is shown. The

control unit was developed using an industrial Siemens S7-300 PLC (Programmable Logic Controller).
The selected PLC system is a modular device that is constituted by the following modules:

Slot1 = Power supply PS 307-2A

Slot2 = Processor CPU 315-2DP

Slot4 = Communication module CP 342 -5

Slot5 = Digital card DI8/DO8xDC24V/0,5A

Slot6 = Analog card AI8 x12bit

Slot7 = Analog card AO4 x12bit

Slot8 = FM card - Counter Module (FM350)

Slot9 = FM card - Counter Module (FM350)

Slot10 = FM card - Stepper Motor (FM353)

Slot11 = FM card - Stepper Motor (FM353)

Additionally, the PLC-tracker has a modem for GSM communication that provides the system capacity to communicate through the mobile phone network.

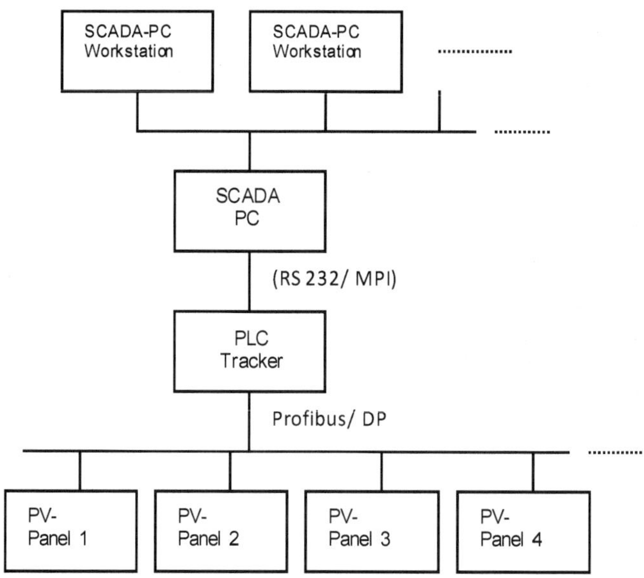

Fig. 5. Communication Strategy (SCADA - PLC Tracker)

The driving unit is composed by two motorized axis, with the following characteristics:

i) Axis 1
- Step motor: Nanotec ST4018L0804, 50Ncm;
- Opt. Encoder: HP HEDL-5540 A14, 500 Pulses
- Coupling unit: Oldham D5
- Proximity sensor: Omrom EA2 M8 PNP

ii) Axis 2
- Step motor: Nanotec ST5918L1008, 170Ncm;
- Gear box: Nanotec PLE40-1S-4
- Opt. Encoder: HP HEDL-5540 A14, 500 Pulses
- Coupling unit: Oldham D25
- Proximity sensor: Omron EA2 M8 PNP

Figure 6 shows a global view of the built Prototype.

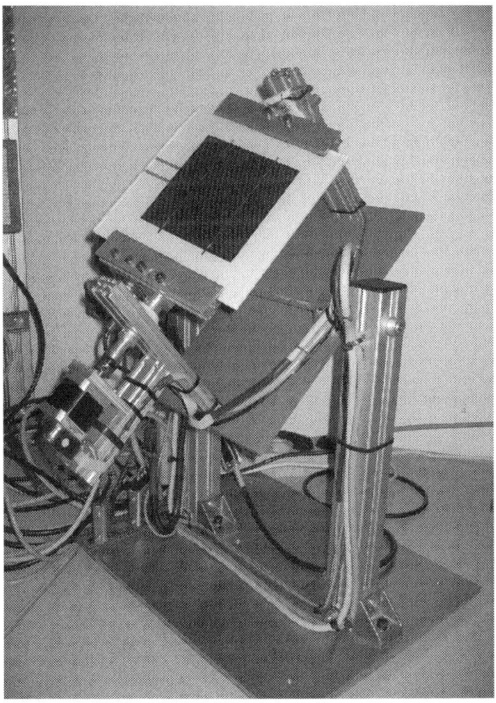

Fig. 6. Built Prototype: Global view

3.2 Implemented Control Algorithm

The software used for the PLC programming was the Siemens Simatic Step 7 (Siemens, 2000). The designed control algorithm was implemented using the Ladder Logic - LAD (Siemens, 2001b). The developed control algorithm is illustrated in fig. 7.

Box0: After reset is activated, the system stores the PV-power generated in the actual position, P_{actual}, in the variable P_{in}. The system searchs its reference-null position. It moves until it finds the hardhome position (both external proximity sensors on). In this position the system assumes the absolute orientation angles for both axis equal zero ($\alpha1 = \alpha2 = 0$). The maximal Power, Pmax is set to zero. Both counters, C1, C2, are loaded;

Box1: After start is activated, the system iniciates the search for the maximal power generated in axis 1, with an angle increment $\alpha10$. The system stores the power generated in variable P1.

Box2: If P1 < Pmax, the system goes to Box 4, and follows for a new position;

Box3: If P1 > Pmax, this position is stored in the variables: $\alpha1max$, $\alpha2max$. The max. Power value, Pmax is actualized with the new Power value P1;

Box4: Counter for axis 1 is updated;

Box5: After all orientations for axis 1 are evaluated, regarding a fixed orientation for axis 2, axis 2 is positioned in a new position, with an angle increment $\alpha20$, and axis 1 returns to its initial position $\alpha1=0$. The system re-initiates the search for the optimal orientation of axis 1, regarding the new position of axis 2. The information flux returns to box 1.

Box6: After all orientations for axis 1 are evaluated, regarding all different positions of axis 2, the system compares the maximal power found (Pmax) with the initial Power generated, before the search process had begun (Pin). If the new Power value is greater than a pre-defined gain, this new correspondent orientation ($\alpha1max$, $\alpha2max$) is sent to all park panels. If the power gain is not enough, the new found position is not to follow by the other PV-panels.

Box7: After a pre-defined time interval (K) the tracker system initiates a new complete search process in both axis. The information flux returns to box 0.

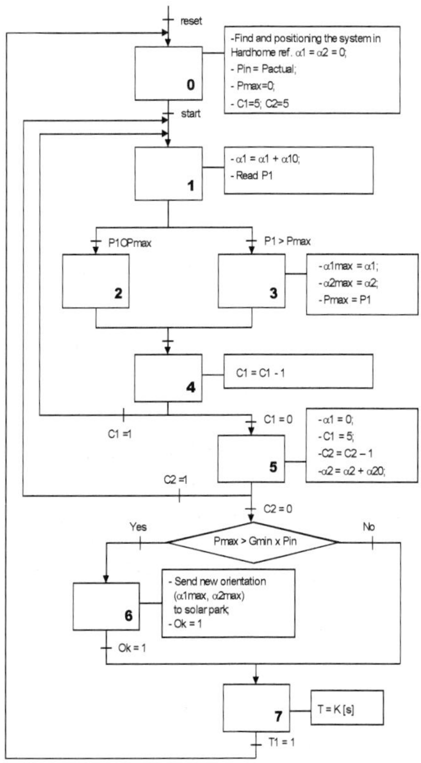

Fig. 7. Control Algorithm for the Tracking System

A short description of the tasks performed by the tracker controller, regarding the above referred algorithm, is described together with fig. 7.

3.3 SCADA Supervisory System

The Scada system was developed over the platform Siemens WinCC (Siemens, 2005).
The SCADA system used to implement this monitoring and control strategy permits the selective access to the application, depending on the user's responsibility degree. In this paper we developed three user levels: Operators, Supervisors and Administrators.
Several SCADA menus were built. The main characteristic of a SCADA Menu is to be simple, explicit and quick on transmitting the information to the operator or to the System administrator.
One of the developed Graphical User Interfaces (GUI) is shown in fig 8.
As this SCADA platform is web enabled, all the GUI displayed data is also on-line accessible through the internet.
In fig. 9 it is shown the developed main menu for the sun-tracker system. The on-line available information, referring actual data from the tracker unit is: actual position for both axis, actual PV-power generated, max. daily PV-power generated, actual efficiency ratio.

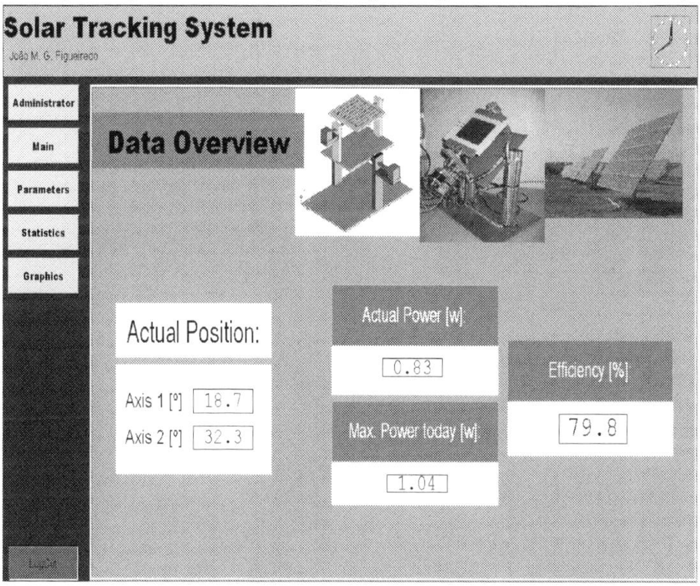

Fig. 8. Sun Tracker System: SCADA main Menu

4. Conclusion

This paper focus the optimization of the electric energy production by photovoltaic cells through the development of an intelligent sun-tracking system. The developed tracking system is innovative in relation to the usual sun tracking systems available in the market. In fact, the developed solution has many advantages in relation to similar existing devices, as this system is autonomous regarding the information needed to process the optimal orientation and is intelligent in a way that it performs on-line monitoring of the photovoltaic energy production.

The increase in power generation, in relation to other PV-systems, without tracking devices, is of similar magnitude (ca. 25%) as for other usual tracking solutions. However, this system has a relative advantage, as it measures exactly the controlled variable: the actual PV-power generation.

5. References

Atlas I, Sharaf A. (1992). "A Fuzzy Logic Power Tracking Controller for a Photovoltaic Energy Conversion Scheme"; Electr. Power Syst. Res. J., 1992; 25 (3); pp. 227-238

Atlas I, Sharaf A. (1996); "A Novel on-line MPP search algorithm for PV arrays"; IEEE Trans. Energy Convers., 1996; 11 (4); pp. 748-754

Benlarbi K., Mokrani L., Nait-Said M. (2004), "A Fuzzy Global Efficiency Optimization of a Photovoltaic Water Pumping System"; Sol Energy 2004; 77; pp. 203-216.

Biee, S.; Chace JR, A. (2009). "Solar Tracking Reflector System for Structure Lighting". European Patent Office, IPC: H01L31/052, EC: E04D13/03; US2009084431 (A1); 2009-04-02

Chen Y., Liu Y., Wu F. (2004) "Multi-Input Converter with Power Factor Correction, Maximum Power Point Tracking, and Ripple-free Input Currents"; IEEE Trans. Power Electron. 2004, 19 (3), pp. 631-639

Hua, C., Lin, J. (2003); "An on-line MPPT algorithm for rapidly changing illuminations of Solar arrays"; Renew Energy 2003; 28; pp. 1129-1142.

Hua C., Lin J. (2004) "A modified tracking algorithm for maximum power tracking of solar array"; Energy Conversion and Management 2004, Vol. 45, pp. 911-925.

Khan, N., Mariun, Z., Saleem, N., Abas, N. (2007). "Fossil Fuels, New Energy Sources and the Great Energy Crisis". Renewable and Sustainable Energy Rev (2007), doi:10.1016/j.rser.2007.11 .011

http://solardat.uoregon.edu/ SolarPositionCalculatorhtml

http://www.wave-energy-centre.org

Rubio, F., Ortega, M., Gordillo, F., López-Martínez, M. (2007). "Application of new control strategy for sun tracking"; Energy Conversion and Management, Vol. 48, Issue 7, July 2007, Pages 2174-2184.

Siemens (2000). *System Software for S7-300 and S7-400 – Reference Manual*, SIEMENS 08/2000; A5E00069892-02

Siemens (2001a). *Simatic Net – NCM S7 for Profibus/ FMS*. SIEMENS 12/2001.

Siemens (2001b). *Simatic S7-300 – Ladder Logic (LAD) for S7-300*, SIEMENS, 2001.

Siemens (2005). *Simatic WinCC V6.0 SP2*, SIEMENS, 2005.

3

Development of Space-Based Solar Power

Lyle M. Jenkins
Jenkins Enterprises
Houston, Texas
Jenkins1166@sbcglobal.net

1. Overview

The potential of space-based solar power (SBSP) as a resource to alleviate world energy needs has not been effective in obtaining the political support for a comprehensive program of evaluation and demonstration. An alternative approach is to emphasize the environmental benefits. Not fully understanding the stability of the Earth System and the specific feedback mechanisms controlling our climate, scientists are unable to effectively predict the course of change to the global environment. Geologic records show a potential rate of change that leaves civilization vulnerable to severe economic effects in a period of significant population growth.

Solar energy as an alternative to fossil fuel reduces stress on the Earth's environmental system. Cost of solar power, particularly from space, is not competitive with current prices of fossil fuels. Collecting the energy in space provides significant advantages in continuity of supply over terrestrial solar, but there is large initial cost prior to getting a return on the investment. The "Fresh Look at Space Solar Power" study shows that concepts needing less investment in an operational system may be feasible. Resources are needed to develop technology and to demonstrate practicality.

2. SBSP Concept Definition

The SBSP concept is to collect energy from the Sun in Earth orbit. The electrical energy is converted to microwave frequency for transmission to the surface of the Earth. There it is converted back in to electricity for use. Possible usages are base-load power, fuel conversion or direct delivery to consumers in isolated locations. The available potential of solar energy is greater than energy in petroleum reserves. The primary issue is defining the path to development of SBSP capability.

No scientific or technological breakthroughs are needed to develop SBSP. Certain technology may require demonstration such as microwave power transmission through the atmosphere. The primary questions relate to beam focus and efficiency. The challenge will to limit the cost of the required hardware. One of the requirements will be a minimum weight for the particular subsystems that make up the orbital system. This is directly related to the cost of launching the total mass into orbit.

Cheap, reliable access to space is a key issue in making SBSP economically viable. The mass to be deployed will mandate a reusable launch system. Trades of the number of stages will be needed to optimize the efficiency. One evaluation may be the air launch concept being developed by Burt Rutan. It enables the launch of the upper stages above much of the sensible atmosphere. This reduces aerodynamic loads, but may be limited by a reasonable takeoff weight.

Current assembly concepts have assumed construction in low earth orbit. After completion, the solar power satellite would be transferred to a higher orbit. Propulsion to accomplish this is a critical issue. One concept that has the specific impulse to make transfer practical is Variable Specific Impulse Magnetoplasma Rocket (VASIMR). NASA spinoff firm, the Ad Astra Rocket Company, has announced a key milestone in ground testing of its prototype plasma drive technology,

The VASIMR "helicon first stage" - which generates the plasma for acceleration by the rest of the drive has achieved its full rated power of 30 kilowatts using Argon propellant, according to the company. This paves the way for further trials in which the ion-cyclotron second stage will get to boost the helicon plasma stream to the target power of 200 kW.

The idea of the plasma drives is to use electric power to blast reaction mass (in this case Argon) from its rocket nozzles at a much higher speed than regular chemical rockets can achieve. This means that the carrying spacecraft gets a lot more acceleration from a given amount of fuel. A potential demonstration for VASIMR is maintaining the orbit of the international space station (ISS) without the need to burn large amounts of chemical rocket fuel. This serves as a demonstration of the transfer large structures between orbits.

Since the solar power satellite was studied in the late 1970's there have been many advancements in subsystem technology. These advances have included (a) improvements in photo-voltaic efficiency from about 10% (1970s) to more than 40% (2007); (b) increases in robotics capabilities from simple teleoperated manipulators in a few degrees of freedom (1970s) to fully autonomous robotics with insect-class intelligence and 30-100 degrees of freedom (2007); (c) increases in the efficiency of solid-state devices from around 20% (1970s) to as much as 70%-90% (2007); (d) improvements in materials for structures from simple aluminum (1970s) to advanced composites including nanotechnology composites (2007); (e) the application to large space structures; (f) high temperature super-conductors and many other technologies may be integrated into the design.

Microwave beams are constant and conversion efficiencies high. They can be beamed at densities substantially lower than that of sunlight. This delivers more energy per area than terrestrial solar energy. The peak density of the beam can be significantly less than noon sunlight, and at the edge of the rectenna equivalent to the leakage of a microwave oven. This low energy density and choice of wavelength also means that biological effects are likely to be low. The safety of wild life wandering into the beam is not expected to be an issue.

The physics of electromagnetic energy beaming is uncompromising. The size of the antenna makes microwave beaming unsuitable as a “secret” weapon. The distance from the geostationary belt is so great that beams diverge beyond the coherence and power concentration needed for a weapon. The beam is likely to be designed to require a pilot beam transmitted from the rectenna site. Absent the pilot signal, the system can be programmed to go into an incoherent mode. Concerns may also be addressed through an inspection regime. The likelihood of the beam wandering over a city is extremely low. Even if it occurred, it would not be a hazard.

Wireless energy transfer by laser beam represents a different set of requirements. To achieve comparable efficiency, the beam must be more intense. The clouds in the atmosphere will reduce the transfer. The intense beam may produce a hazardous level to be avoided by aircraft and satellites. Still for the application to military power supply, it may be a manageable method.

At present, the United States has very limited capabilities to build large structures, very large aperture antennas or very high power systems in orbit. The capability to control and maneuver these systems in space must be developed and demonstrated. Presently, the ability to translate large mass between Earth orbits will be required for deployment SBSP. Eventually, the capability for in-space manufacturing and construction or in-situ space resource utilization may be developed, but at this point it is a challenge that should not be incorporated into the program. One critical item to be demonstrated is capability for beamed power and application to propulsion of large space systems.

The Thunderstorm Solar Power Satellite (TSPS) is a concept for interacting with thunderstorms to prevent formation of tornadoes. TSPS benefits are saving lives and reducing property. These benefits are not as sensitive to the system economics as the commercial solar power satellite and justify government investment in space solar power. The TSPS can develop and demonstrate the technology and operations critical to understanding the cost of space solar power. Consequently, there is no direct competition with fossil fuel based power supplies until SSP technology and operations have been demonstrated. Before weather modification can be safely attempted, the fine structure of thunderstorms must be simulated and related to tornadogenesis

3. Environmental Benefits

Advocates of space solar power have been presenting the concepts as a means to help meet world energy needs. This argument has not been effective in garnering support for even basic research and technology development. Fossil fuel alternatives have been too cheap and near term effect on the "economy" inhibits action by policy makers. Concern for the environment is greater than the policy makers realize.

The key to getting support for space solar power may be the growing awareness of the threat of rapid global environmental change. Scientists are extending their traditional role of theory and observation to emphasize the risks of global change. The risks provide the context for action by policy makers to move toward sustainable systems. The transition to power from space is responsive to the environmental concerns and the need to stabilize the Global environment and consequently the Earth's economic and social stability.

The "overview effect" from space has played a major role in developing a public sense of the fragile nature of the global environment. Stress on the Earth's environmental system is increasing due to the buildup of carbon dioxide and other greenhouse gases. Models predicting the response to this buildup have not performed well in projecting the effect on the Earth's climate because of the complexity of the system and the feed- backs within the system. Even the direction of climate change has not been predictable. An enhanced greenhouse effect has not been detected by temperature measurements. There may be interactions that are not well defined by the computer models, but that are reducing the stability of the Earth system. Because of the potential influence on the stability of the ocean currents that transport heat from the tropics to the higher latitudes, there is even a risk of returning the Earth to a glacial period rather than the global warming that is the present paradigm. Analysis of glacial ice cores indicates that such a shift can take place in less than a few decades.

The likely effects of rapid climate change are increases in storm intensities, flooding, droughts, regional cropping shifts and sea level rise. These effects will have severe social and economic consequences.

The rate of change and it's direction leave civilization vulnerable to severe economic change in a period of significant population growth. Sustainable development has become the mantra for dealing with the potential global crises that are facing civilization. Clean, renewable energy is a resource that meets the criteria of sustainability. Collecting solar energy is prime candidate. Collecting the energy in space provides significant advantages in continuity of supply, although it's development represents many challenges. A primary challenge is the issue of large initial cost prior to generating a return on that investment. The NASA Fresh Look at Space Solar Power study shows that concepts needing less initial investment are feasible. Even so, early SSP systems are not likely to be price competitive unless fossil fuel pricing incorporates the long range economic impact.

The risks identified through the rigor of the U. S. Global Change Research Program (USGCRP) must provide the motivation for action toward sustainable systems. The USGCRP is an integrated program documenting the Earth system, understanding Earth system processes and developing computer models to predict the course of changes induced by humans or as the result of natural variations. The program is beginning to analyze the environmental, socioeconomic and health consequences of global change. The obvious next step is to assess means for mitigation of the effects of global change.

The prosperity of future generations is dependent on a stable global environment. To ensure environmental stability, a continued effort to understand the effect of human activities must be a priority. Just understanding may not be sufficient because of the complex relationships of greenhouse gases, wind circulation, ocean currents and atmospheric water vapor. It is undisputed that carbon dioxide in the atmosphere has increased by over twenty percent since the beginning of the industrial age. Fossil fuels are certainly a major contributor to that increase. By replacing fossil fuel use, SSP could reduce the buildup of CO_2 in the atmosphere and the consequent climate changes from an enhanced greenhouse effect. There are economic returns from a space-based power source that will lead to commercial management and operation of the system.

There will continue to be an element of the political community that is committed to the short-term view because of the immediate economic impact. This reality is a factor that will have to be dealt with through facts and risk assessment for the long term view. The anticipated benefit to the Earth's environment is the overarching objective that may provide support for technology development and demonstration toward space solar power for use on Earth.

4. Summary

Space-Based Solar Power is a huge project. It might be considered comparable in scale to the national railroads, highway system, or electrification project rather than the Manhattan or Apollo endeavors. However, unlike such purely national projects, this project also has components that are analogous to the development of the high-volume international civil aviation system. Such a large endeavor includes significant international and environmental implications. As such it would require a corresponding amount of political will to realize its benefits.

Most of America's spending in space does not provide any direct monetary revenue. SBSP will create new markets and produce new products. Great powers have historically succeeded by finding or inventing products and services not just to sell to themselves, but to sell to others. Today, investments in space are measured in billions of dollars. The energy market is trillions of dollars and will generate substantial new wealth for our nation and our world. Investments to develop SBSP have significant economic spin-offs. They open up or enable other new industries such as space industrial processes, space tourism, enhanced telecommunications, and use of off-world resources.

After the fundamental technological risks have been defined, shifting SBSP from a research and development project to a financial and production program is needed. Several major challenges will need to be overcome to make SBSP a reality, including the creation of low-cost space access and a supporting infrastructure system on Earth and in space. The opportunity to export energy as the first marketable commodity from space will motivate commercial sector solutions to these challenges. The delivered commodity can be used for base-load terrestrial electrical power, wide-area broadcast power, carbon-neutral synthetic fuels production, military tactical support or as an in-space satellite energy utility.

Solving these space access and operations challenges for SBSP will in turn also open space for space tourism, manufacturing, lunar or asteroid resource utilization, and eventually expansion of human presence and permanent settlement within our solar system.

Space-based geoengineering concepts for environmental countermeasures are a potential supplement to earth-based actions. By defining options and benefits, SBSP may alert decision-makers to the potential of space operations as more than a tool to monitor the course of global change. Within the envelope of environmental protection is the preventing tornadoes concept. It promises early benefits by saving lives and reducing property damage. The principal payoff is projected to be the demonstration of space solar power technology and operations. This can lead to investment by the commercial energy organizations when their technical and operational risk is reduced. Once the potential for clean renewable energy from space is demonstrated, the way will be opened for further exploration and development of space.

5. References

Stephen L. Klineberg, 'Trends in Stakeholder Opinions: Findings from the Texas Environmental Survey (1990-1996)', DeLange-Woodlands Conference, 1997.

'Our Changing Planet, The FY 1997 U. S. Global Change Research Program', Supplement to the President's Fiscal Year 1997 Budget.

National Security Space Office, "Space-Based Solar Power as an Opportunity for Strategic Security, Phase 0 Architecture Feasibility Study", Report to Director, 10 October 2007.

Dr. Bernard J. Eastlund, Lyle M. Jenkins, "Taming Tornadoes: Storm Abatement from Space", IEEE Aerospace Conference, ID 205, March 7, 2001.

"Policy Implications of Greenhouse Warming: Mitigation, Adaptation and the Science Base", National Academy Press, Washington, D. C., 1992.

B.J. Eastlund, "Systems Considerations of Weather Modification Experiments Using High Power Electromagnetic Radiation", Workshop on Space Exploration and Resources Exploitation - Explospace, European Space Agency, October 20-22, 1998.

Glaser, P. E.. Feasibility Study of a Satellite Solar Power Station, *NASA Contact. Rep. CR-2357, NTIS N74-17784. Nat. Tech. Inform. Serv., Springfield, Va., USA*,1974.

M. Xue and K. K. Droegemeier. "The Advanced Regional Prediction System (ARPS)- A Multi-Scale Nonhydrostatic Atmospheric Simulation and Prediction Tool", Model Dynamics, *Monthly Weather Review.*

Eastlund, Bernard J., and Jenkins, Lyle M., "Space-based Concepts for Taming Tornadoes", *51st International Astronautical Congress, Rio de Janeiro, Brazil* October 2, 2000

Ming Sue, "Tornadogenesis within a Simulated Supercell Storm", 22nd Severe Local Storms Conference, 6 October 2004.

S. Businger, S. Chiswell, M. Bevis, J. Duan, R. Anthes, C. Rocken, R. Ware, M. Exner, T. van Hove, and S. Solheim - The Promise of GPS Atmospheric Monitoring, 1996

Mankins, John, "A Fresh Look at the Concept of Space Solar Power", *SPS '97, Energy and Space for Humanity*, August 26, 1997

Dr. Bernard J. Eastlund, Lyle M. Jenkins, "Taming Tornadoes: Storm Abatement from Space", IEEE Aerospace Conference, ID 205, March 7, 2001.

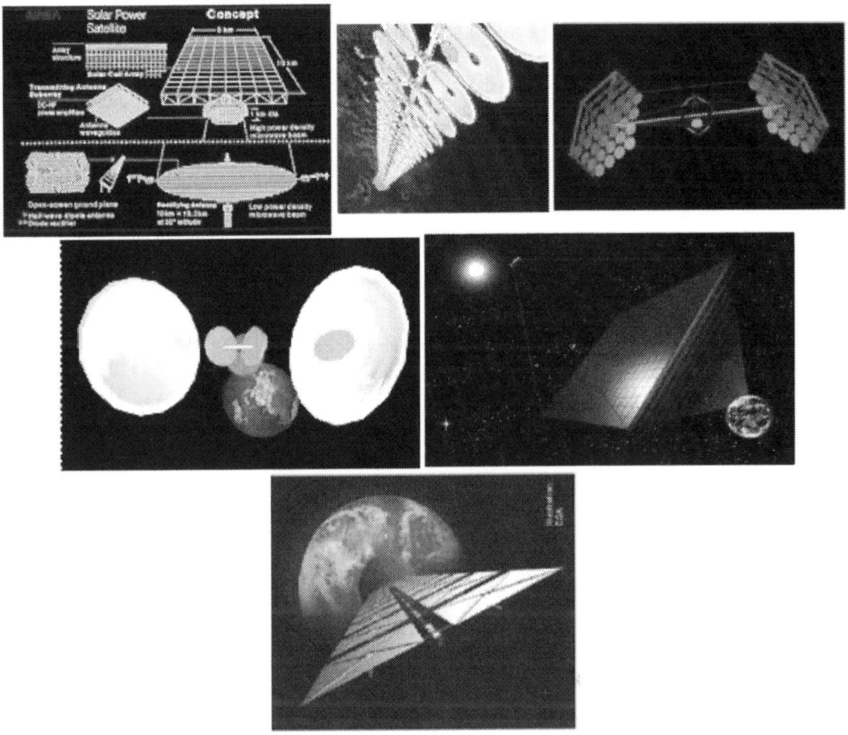

Fig. 1. Top left-NASA/DOE reference 5GW; top center-NASA Sun Tower 200MW; top right-Integrated Symmetrical Concentrator; center left-JAXA Free Flyer Model; center right-USEF Tethered SPS; bottom- ESA Sail Tower 400MW

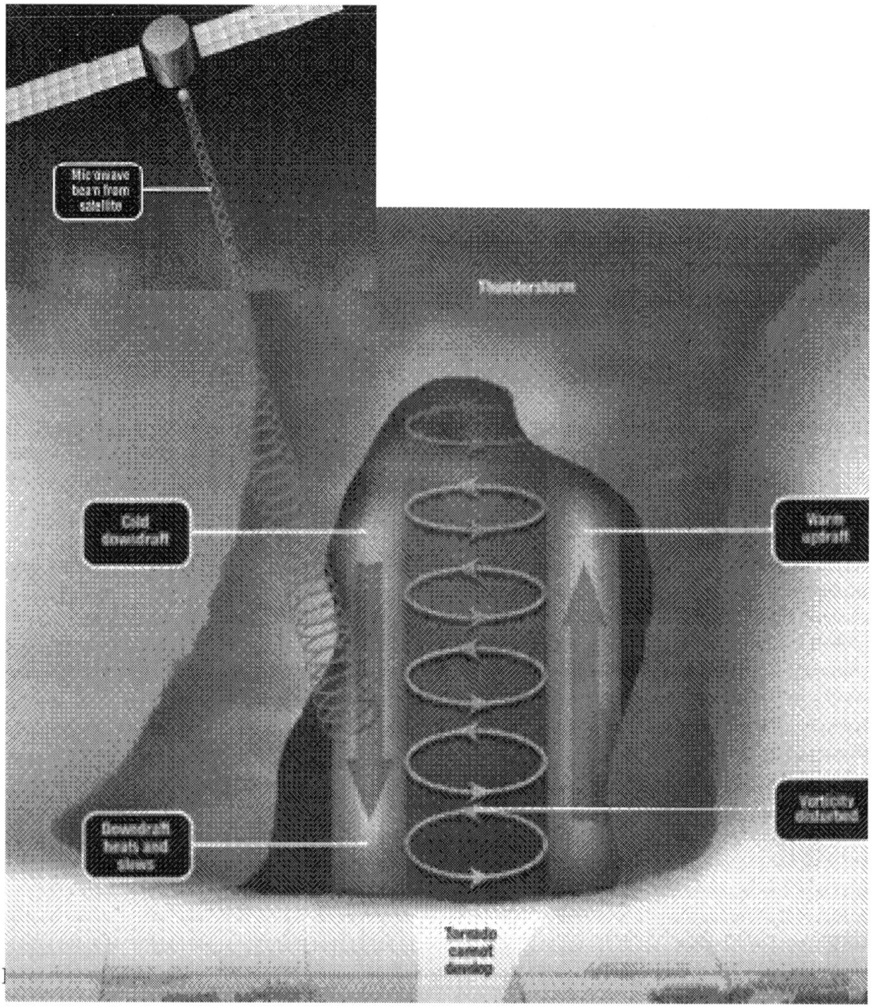

Fig. 2. Thunderstorm Solar Power Satellite Concept for preventing tornadoes.

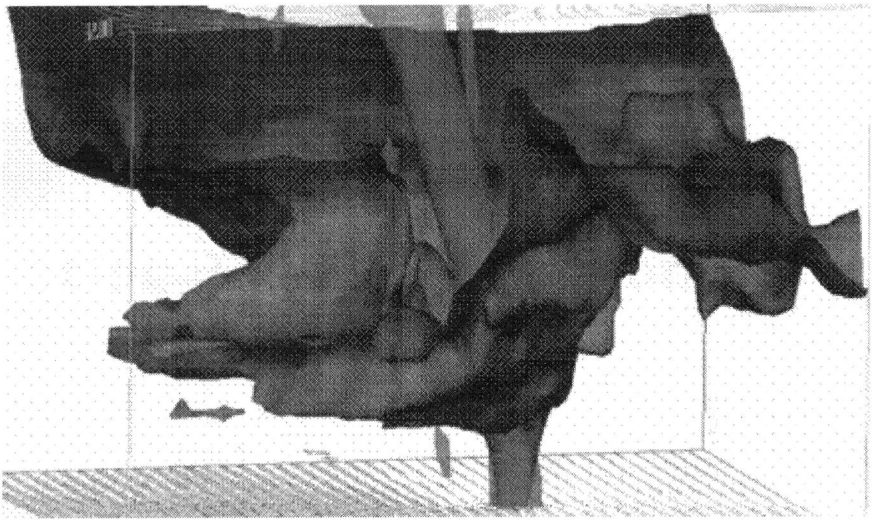

Fig. 3. Computer simulation of tornadogenesis.

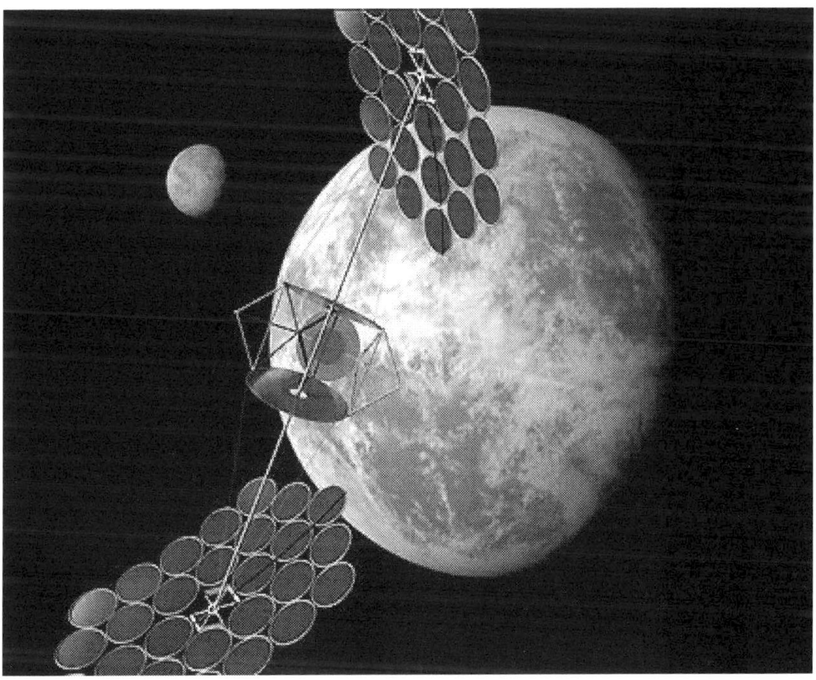

Fig. 4. SBSP concept.

4

Increasing the energy yield of generation from new and renewable energy resources

Samuel C. E. Jupe, Andrea Michiorri and Philip C. Taylor
Durham University
UK

1. Introduction

The impetus of governments, on an international scale, to move towards low-carbon economy targets has brought about the proliferation of electricity (and heat) generation from new and renewable energy (RE) resources. This, coupled with increasing consumer energy demands, has caused distribution network operators (DNOs) to seek methods of increasing the utilisation of their existing power system assets. The increased utilisation of assets must be realised cautiously such that the security of supply to customers is not reduced, particularly when the age of distribution network assets is taken into account. A developer that is seeking to connect generation of significant capacity may be offered a firm connection by the DNO on the condition that an investment is made (by the developer) in the necessary network reinforcements. However, the developer may not be able to justify the expense of the required reinforcement and may negotiate a non-firm or 'constrained' connection agreement, whereby the generation installation is tripped off or constrained back under certain network operating conditions. Furthermore, difficulties may be encountered when attempting to gain permission to build network infrastructure, in order to accommodate new generation installations, due to planning problems and environmental objections (Fox-Penner, 2001). One potential solution or means of deferring these problems is the adoption of real-time thermal rating systems which have the potential, in certain circumstances, to be both less invasive and more cost effective when compared to network reinforcement options. Non-firm generation connections are expected to occur more frequently as network power flow congestion occurs. Therefore the deployment of a power output control system, informed by real-time thermal ratings, may be required to increase the energy yield of generation from new and RE resources.

The stages in the development of an output control system for generation installations are illustrated. Section 2 provides a comprehensive literature review in order to provide the context for the research presented. Section 3 describes the assessment of the location of power flow congestion within the power system (due to the proliferation of generation from new and RE resources) so as to facilitate the targeted development of thermal models for thermally vulnerable components. This is achieved through the calculation of thermal vulnerability factors that relate power flow sensitivity factors (derived from governing alternating current (AC) power flow equations) to component steady-state thermal limits. Section 4 describes models for the steady-state assessment of power system component real-time thermal ratings. Industrial standard lumped parameter models are described for overhead lines, electric cables and power transformers. In a consistent manner, these models allow the influence of environmental conditions (such as wind speed) on component real-time thermal ratings to be assessed. Section 5 describes thermal state estimation techniques that allow the rating of components, which are not directly monitored within the power system, to be assessed. Thermal state estimations facilitate the precise and reliable assessment of environmental conditions whereby limited meteorological monitoring allows the thermal state of components within a wide area to be assessed. This may then be validated through the carefully selected monitoring of component operating temperatures. In Section 6, the power flow sensitivity factors are brought together with component real-time thermal ratings and candidate strategies are presented for the power output control of single or multiple generation installations. In Section 7, a case study is used to illustrate the developmental stages described above. In Section 8 the strengths and weaknesses of the proposed output control system for generation installations are discussed.

The research described in this chapter forms part of a UK government part-funded project (Neumann *et al*, 2008) which aims to develop and deploy an online power output controller for wind generation installations through the exploitation of component real-time thermal properties. This is based on the concept that high power flows resulting from wind generation at high wind speeds could be accommodated since the same wind speed has a positive effect on component cooling mechanisms. In this project the control system compares component real-time thermal ratings with network power flows and produces set points that are fed back to the generation scheme operator for implementation, as shown in Figure 1.

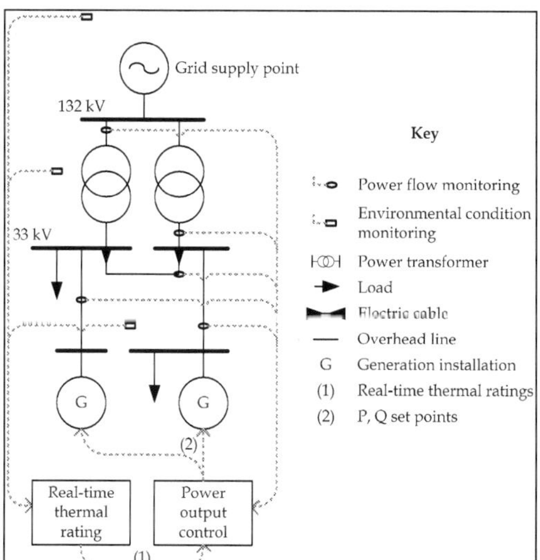

Fig. 1. Generation power output controller informed by real-time thermal ratings

2. Background

Installed capacity assessments for generation from new and RE resources are the current research focus of numerous institutions in order to determine the impact of voltage regulations, operational economics, fault levels, losses and thermal limits as constraining parameters. Voltage limitations and the wind farm installed capacity relative to the system fault level in 33kV networks are considered (Dinic *et al.*, 2006) and it is concluded that capacitive compensation can allow capacity maximisation within operational limits. The economics of generation connections to exploit multiple new and RE resources are considered (Currie *et al.*, 2006) with a methodology that facilitates greater generation installed capacities. In order to manage power flows within prescribed voltage and thermal limits, operating margins are utilised with an active power output control technique termed 'trim then trip'. An optimal power flow (OPF) technique is developed (Vovos *et al.*, 2005) along with an iterative procedure to calculate the possible installed capacity of generation at nodes based on fault level limitations. The impact of increased generation installed capacities on electrical losses within the IEEE 34-node test network is examined (Mendez Quezada *et al.*, 2006) and it is concluded that losses follow a U-shaped trajectory when plotted as a function of the generation penetration. An OPF formulation is presented (Harrison & Wallace, 2005) to determine the maximum generation installed capacity based on thermal limits and statutory voltage regulation. The 'reverse load-ability' methodology coupled with OPF software models generators as loads with a fixed power factor and creates an analysis tool that could allow additional constraints (such as fault-level limitations) to be incorporated into the formulation if necessary.

Significant research has been carried out at the transmission level for real-time thermal rating applications. Research tends to focus on overhead lines which, due to their exposure to the environment, exhibit the greatest rating variability. A description of the cost and suitability of different uprating techniques for overhead lines is described (Stephen, 2004) taking into account different operating conditions. This work shows how real-time thermal ratings can be a more appropriate solution than network reinforcement when connecting new customers to the network who are able to curtail their generation output or reduce their power demand requirement at short notice. Similarly, experience regarding thermal uprating in the UK is reported (Hoffmann & Clark, 2004) where it was suggested that real-time thermal ratings could give overhead lines an average uprating of 5% for 50% of the year. An example of a real-time thermal rating application for transmission overhead lines of Red Eléctrica de España is described (Soto *et al.*, 1998) where a limited number of weather stations are used to gather real-time data. The data is then processed using a meteorological model based on the Wind Atlas Analysis and Application Program (WAsP), taking into account the effect of obstacles and ground roughness, and the thermal rating is calculated. A similar system was developed in the USA by EPRI (Douglass & Edris, 1996) which considered overhead lines, power transformers, electric cables and substation equipment. Preliminary results of field tests (Douglass *et al.*, 1997) show that up to 12 hours of low wind speeds ($<0.76\ ms^{-1}$) were observed during the field tests which therefore suggests that overhead line real-time thermal ratings may be lower than seasonal ratings for extended periods of time. Furthermore, a strong correlation was found to exist between independent air temperature measurements distributed along the lengths of the overhead lines. At the distribution level, a real-time thermal rating project carried out by the Dutch companies

NUON and KEMA (Nuijten & Geschiere, 2005) demonstrates the operating temperature monitoring of overhead lines, electric cables and power transformers.

The advantages of a real-time thermal rating system for the connection of generation from new and RE resources are reported in various sources, each of which considers only single power system components. It is demonstrated (Helmer, 2000) that the rating of transformers positioned at the base of wind turbines may presently be oversized by up to 20%. Moreover, the power flowing in an overhead line close to a wind farm is compared to its real-time thermal rating using WAsP (Belben & Ziesler, 2002). In this research it was highlighted that high power flows resulting from wind generation at high wind speeds could be accommodated since the same wind speed has a positive effect on the line cooling. This observation makes the adoption of real-time thermal rating systems relevant in applications where strong correlations exist between the cooling effect of environmental conditions and electrical power flow transfers. Moreover, the influence of component thermal model input errors on the accuracy of real-time thermal rating systems is studied (Piccolo *et al.*, 2004; Ippolito *et al.*, 2004; Villacci & Vaccaro, 2007). The application of different state estimation techniques, such as affine arithmetic, interval arithmetic and Montecarlo simulations was studied for overhead lines, electric cables and power transformers. Errors of up to ±20% for an operating point of 75°C, ±29% for an operating point of 60°C and ±15% for an operating point of 65°C were found when estimating the operating temperature of overhead lines, electric cables and power transformers respectively. This highlights the necessity to have reliable and accurate environmental condition monitoring. The thermal models, used to estimate real-time thermal ratings for different types of power system components, are fundamental to this research as the accuracy of the models influence significantly the accuracy of real-time thermal ratings obtained. Particular attention was given to industrial standards because of their wide application and validation both in industry and academia. For overhead lines, the models (House & Tuttle, 1959; Morgan, 1982) have been developed into industrial standards by the IEC (IEC, 1995), CIGRE (WG 22.12, 1992) and IEEE (IEEE, 1993). Static seasonal ratings for different standard conductors and for calculated risks are provided by the Electricity Network Association (ENA, 1986). Thermal model calculation methods for electric cable ratings are described (Neher & McGrath, 1957) and developed into an industrial standard by the IEC (IEC, 1994). The same models are used by the IEEE (IEEE, 1994) and the ENA (ENA, 2004) to produce tables of calculated ratings for particular operating conditions. Power transformer thermal behaviour is described (Susa *et al.*, 2005) with further models described in the industrial standards by the IEC (IEC, 2008), the IEEE (ANSI/IEEE, 1981) and the ENA (ENA, 1971).

The work detailed in this chapter moves beyond the offline assessment of generation installed capacities to outline the development stages in the online power output control of generation installations. The thermal vulnerability factor assessments presented in this chapter complement network characterisation methods (Berende *et al.*, 2005) by first identifying the type (overhead line, electric cable, power transformer) and geographical location of thermally vulnerable components. The assessments may be used to give a holistic network view of the impact of multiple generation installations in concurrent operation on accumulated power flows and hence vulnerable component locations. This facilitates the targeted development of component thermal models. Moreover, (Michiorri *et al.*, 2009) describes the influence of environmental conditions on multiple power system component types simultaneously. This is of particular relevance in situations where the increased power flow resulting from the alleviation of the thermal constraint on one power system component may cause an entirely different component to constrain power flows. Whilst OPF is acknowledged as a powerful tool for the offline planning of electrical networks, there is an emerging requirement to manage non-firm generation connections in an online manner. This requires the deployment of a system which has the capability of utilising real-time information about the thermal status of the network and, in reaching a control decision, guarantees that the secure operation of the distribution network is maintained. The rapid processing time, reduced memory requirements and robustness associated with embedding predetermined power flow sensitivity factors in a power output control system for generation installations make it attractive for substation and online applications. This is strengthened further by the ability of the power output control system to readily integrate component real-time thermal ratings in the management of network power flows for increased new and renewable energy yields. Moreover, since this research project aims to develop and deploy an economically viable real-time thermal rating system, it is important that algorithms are developed with fast computational speeds using limited environmental condition monitoring. Thus an inverse distance interpolation technique is used for modelling environmental conditions across a wide geographical area, which offers faster computational speeds than applications such as WAsP. Beyond the research described above, this chapter also suggests potential annual energy yields that may be gained through the deployment of an output control system for generation installations.

3. Power flow sensitivity factors

Once the inverse Jacobian has been evaluated in the full AC power flow solution, perturbations about a given set of system conditions may be calculated using Eq.1 (Wood & Wollenberg, 1996). This gives the changes expected in bus voltage angles and voltage magnitudes due to injections of real or reactive power.

$$\begin{bmatrix} \Delta\theta_i \\ \Delta|V_i| \\ \Delta\theta_k \\ \Delta|V_k| \\ \vdots \end{bmatrix} = [J]^{-1} \begin{bmatrix} \Delta P_i \\ \Delta Q_i \\ \Delta P_k \\ \Delta Q_k \\ \vdots \end{bmatrix} \tag{1}$$

Where θ_i and θ_k (rad) represent voltage angles at nodes i and k respectively, $|V_i|$ and $|V_k|$ (kV) represent nodal voltages, J is the Jacobian matrix, P_i and P_k (MW) represent real power injections at nodes i and k respectively and Q_i and Q_k (MVAr) represent reactive power injections at nodes i and k respectively.

The work presented in this paper is specifically concerned with calculating the effect of a perturbation of ΔP_m – that is an injection of power at unity power factor (real power) into node m. Since the generation shifts, the reference (slack) bus compensates for the increase in power. The $\Delta\theta$ and $\Delta|V|$ values in Eq.2 are thus equal to the derivative of the bus angles and voltage magnitudes with respect to a change in power at bus m.

$$\begin{bmatrix} \Delta\theta \\ \Delta|V| \end{bmatrix} = [J]^{-1} \begin{bmatrix} \Delta P_m \\ \Delta Q_m \\ \vdots \\ \Delta P_{ref} \\ \Delta Q_{ref} \end{bmatrix} \tag{2}$$

Thus the sensitivity factors for a real power injection at node m are given in Eq.3-6:

$$f(\theta): \frac{dP_{i,k}}{dG_{P,m}} = \left(\frac{\partial P}{\partial \theta}\right)_{i,k} \left(\frac{d\theta_k}{dG_{P,m}} - \frac{d\theta_i}{dG_{P,m}}\right) \tag{3}$$

$$f(|V|): \frac{dP_{i,k}}{dG_{P,m}} = \left(\frac{\partial P}{\partial |V|}\right)_{i,k} \left(\frac{d|V_k|}{dG_{P,m}} - \frac{d|V_i|}{dG_{P,m}}\right) \tag{4}$$

$$f(\theta): \frac{dQ_{i,k}}{dG_{P,m}} = \left(\frac{\partial Q}{\partial \theta}\right)_{i,k} \left(\frac{d\theta_k}{dG_{P,m}} - \frac{d\theta_i}{dG_{P,m}}\right) \tag{5}$$

$$f(|V|): \frac{dQ_{i,k}}{dG_{P,m}} = \left(\frac{\partial Q}{\partial |V|}\right)_{i,k} \left(\frac{d|V_k|}{dG_{P,m}} - \frac{d|V_i|}{dG_{P,m}}\right) \tag{6}$$

Where f(θ) and f(V) represent functions of voltage angles and voltage magnitudes respectively, $(\partial P/\partial\theta)_{i,k}$, $(\partial P/\partial V)_{i,k}$, $(\partial Q/\partial\theta)_{i,k}$ and $(\partial Q/\partial V)_{i,k}$, and represent elements within the Jacobian matrix and $d\theta_k/dG_{P,m}$, $d\theta_i/dG_{P,m}$, $dV_k/dG_{P,m}$ and $dV_i/dG_{P,m}$ represent elements corresponding to the relevant $\Delta\theta$ and $\Delta|V|$ values evaluated in Eq.2. This gives an overall power flow sensitivity factor ($S_{i,k,m}$) in the component from node i to node k, due to an injection of real power, at node m, as in Eq.7.

$$\begin{aligned} SSF^{c}_{i,k,m} = & \left[\left(\left(\frac{\partial P}{\partial\theta}\right)_{i,k}\left(\frac{d\theta_k}{dG_{P,m}} - \frac{d\theta_i}{dG_{P,m}}\right)\right) + \left(\left(\frac{\partial P}{\partial|E|/|E|}\right)_{i,k}\left(\frac{d|E_k|/|E_k|}{dG_{P,m}} - \frac{d|E_i|/|E_i|}{dG_{P,m}}\right)\right)\right] \\ & + j\left[\left(\left(\frac{\partial Q}{\partial\theta}\right)_{i,k}\left(\frac{d\theta_k}{dG_{P,m}} - \frac{d\theta_i}{dG_{P,m}}\right)\right) + \left(\left(\frac{\partial Q}{\partial|E|/|E|}\right)_{i,k}\left(\frac{d|E_k|/|E_k|}{dG_{P,m}} - \frac{d|E_i|/|E_i|}{dG_{P,m}}\right)\right)\right] \end{aligned} \tag{7}$$

Simplified versions of the power flow sensitivity factor theory (focusing on the P-θ sensitivity) are used at the transmission level for real power flow sensitivity analyses. The generation shift factor (GSF) technique (Wood & Wollenberg, 1996) is acceptable for use in DC representations of AC systems where the network behaviour is approximated by neglecting MVAr flow and assuming voltage to be constant. However, in distribution networks those assumptions do not always hold since, in some cases, the electrical reactance (R_E) and electrical resistance (X) of components is approximately equal (i.e. $X/R_E \approx 1$). Thus reactive power flow may contribute to a significant portion of the resultant power flowing in components. In these situations it is important that both real and reactive power flows are considered when assessing the locations of thermally vulnerable components and developing techniques for the online power output control of generation from new and RE resources.

3.1 Thermal vulnerability factors

Eq.7 may be combined with the relevant component thermal rating and the resulting thermal vulnerability factor, as given in Eq.8, is standardised by conversion to a per unit term on the base MVA

$$TVF_{i,k,m} = \left(\frac{SSF_{i,k,m}}{S_{lim}} \right) \text{ on } S_{base} \tag{8}$$

where $TVF_{i,k,m}$ represents the thermal vulnerability factor of the component from node i to node k due to a real power injection at node m, $SSF_{i,k,m}$ represents the power flow sensitivity factor in the component from node i to node k, due to a real power injection at node m, S_{lim} (MVA) represents the thermal limit of the component and S_{base} is a predefined MVA base. This gives a consistent measure of component thermal vulnerabilities, relative to one another and accounts for different nodal real power injections, for a particular network operating condition. It can also be seen in Eq.9 that the sensitivity factor relative to the component rating is equivalent to the change in utilisation of a particular component from node i to node k, due to an injection of real power at node m

$$\left(\frac{SSF_{i,k,m}}{S_{lim}} \right) = \left(\frac{\Delta S_{i,k}}{\Delta G_{P,m} \times S_{lim}} \right) \equiv \left(\frac{\Delta U_{i,k}}{\Delta G_{P,m}} \right) \tag{9}$$

where $SSF_{i,k,m}$ represents the power flow sensitivity factor of the component, from node i to node k, due to a real power injection at node m, S_{lim} (MVA) represents the thermal limit of component, $\Delta S_{i,k}$ (MVA) represents the change in apparent power flow in the component from node i to node k, $\Delta G_{P,m}$ (MW) represents the change in real power injection at node m and $\Delta U_{i,k}$ represents the change in capacity utilisation of the component from node i to node k.

Power flow sensitivity factors indicate the extent to which power flow changes within components due to nodal power injections. However, a large change in power flow, indicated by high sensitivity, does not necessarily mean a component is thermally vulnerable unless its rating is taken into account. A large power flow change in a component with a large thermal rating could be less critical than a small power flow change in a component with a small rating. By calculating the apparent power sensitivity relative to rating for each component, the thermally vulnerable components are identified and can be ranked for single nodal power injections or accumulated for multiple injections.

3.2 Factor assessments

An empirical procedure (Jupe & Taylor, 2009b) to assess power flow sensitivity factors and generate lists of thermally vulnerable components for different network topologies has been developed as follows: Initially a 'base case' AC load flow is run in the power system simulation package to establish real, reactive and apparent power flows for each component. The procedure iterates by injecting 1pu of real power at each node of interest and recording the new component power flows. The initial flow, final flow and thermal rating of each component are used to relate component power flow sensitivity factors to nodal injections and ratings. The resulting power flow sensitivity factors and thermal vulnerability factors are efficiently stored in matrix form and, with the thermal vulnerability factors represented graphically, a visual identification of the most thermally vulnerable components is given.

4. Thermal modelling approach

In order to assess, in a consistent manner, component real-time thermal ratings due to the influence of environmental conditions, thermal models were developed based on IEC standards for overhead lines (IEC, 1995), electric cables (IEC, 1994) and power transformers (IEC, 2008). Where necessary, refinements were made to the models (WG22.12, 1992; ENA, 2004). Steady-state models have been used in preference to dynamic models since this would provide a maximum allowable rating for long term power system operation.

4.1 Overhead lines

Overhead line ratings are constrained by a necessity to maintain statutory clearances between the conductor and other objects. The temperature rise causes conductor elongation which, in turn, causes an increase in sag. The line sag, ψ (m), depends on the tension, H (N), the weight, mg (N) applied to the conductor inclusive of the dynamic force of the wind and the length of the span, L (m). The sag can be calculated as a catenary or its parabolic approximation, as given in Eq.10.

$$\psi = \frac{H}{mg}\left[\cosh\left(\frac{mgL}{2H}\right) - 1\right] \approx \frac{mgL^2}{8H} \tag{10}$$

To calculate the tension, it is necessary to consider the thermal-tensional equilibrium of the conductor, as shown in Eq.11, where E represents the Young's modulus of the conductor (Pa), A represents the cross-sectional area of the conductor (m^2) and β represents the conductor's thermal expansion coefficient (K^{-1}).

$$EA\beta\left(T_{c,2} - T_{c,1}\right) + \left(\frac{m_1^2 g^2 L^2 EA}{24H_1^2}\right) - H_1 = \left(\frac{m_2^2 g^2 L^2 EA}{24H_2^2}\right) - H_2 \tag{11}$$

For calculating the conductor operating temperature at a given current, or the maximum current for a given operating temperature, it is necessary to solve the energy balance between the heat dissipated in the conductor by the current, and the thermal exchange on its surface, as given in Eq.12

$$q_c + q_r = q_s + I^2 R_E \tag{12}$$

where q_c represents convective heat exchange (Wm^{-1}), q_r represents radiative heat exchange (Wm^{-1}), q_s represents solar radiation (Wm^{-1}) and I^2R_E represents the heat dissipated in the conductor due to the Joule effect (Wm^{-1}). The proposed formulae (IEC, 1995) were used for the calculation of the contribution of solar radiation, radiative heat exchange and convective heat exchange as given in Eq.13-15 respectively,

$$q_s = \alpha \cdot D \cdot Sr \tag{13}$$

$$q_r = \varepsilon \cdot \sigma_{SB} \cdot \left(T_c^4 - T_a^4\right) \cdot \pi \cdot D \tag{14}$$

$$q_c = \pi \cdot Nu \cdot \lambda \cdot \left(T_c - T_a\right) \tag{15}$$

where α represents the absorption coefficient, D represents the external diameter of the conductor (m) and Sr represents solar radiation (Wm^{-2}), ε represents the emission coefficient, σ_{SB} represents the Stefan-Boltzmann constant ($Wm^{-2}K^{-4}$), and T_c and T_a represent the respective conductor and ambient temperatures (K), Nu represents the Nusselt number and λ represents the air thermal conductivity ($Wm^{-1}K^{-1}$).

The influences of wind direction and natural convection on convective heat exchange are not considered in the IEC standard model (IEC, 1995). However, in this research these effects were considered to be important, particularly as a wind direction perpendicular to the conductor would maximise the turbulence around the conductor and hence the heat exchange on its surface whereas a wind direction parallel to the conductor would reduce the heat exchange with respect to perpendicular wind direction. Therefore the modifications (WG22.12, 1992) given in Eq.16 and Eq.19 were used. It is possible to calculate the Nusselt number, Nu, from the Reynolds number, Re, as shown in Eq.17. The Reynolds number can be calculated using Eq.18

$$K_{dir} = K_{dir,1} + K_{dir,2} \cdot (\sin Wd)^{K_{dir,3}} \tag{16}$$

$$Nu = K_{dir} \cdot \left(0.65 \cdot Re^{0.2} + 0.23 \cdot Re^{0.61}\right) \tag{17}$$

$$Re = 1.644 \cdot 10^9 \cdot Ws \cdot D \cdot \left(\frac{T_c + T_a}{2}\right)^{-1.78} \tag{18}$$

where K_{dir} represents the wind direction influence coefficient, $K_{dir,1-3}$ represent wind direction coefficient constants, Wd represents the wind-conductor angle (rad) and Ws represents the wind speed (ms^{-1}).

For null wind speeds, the Nusselt number must be calculated as in Eq.19 where Gr is the Grashof number, calculated as in Eq.20, and Pr is the Prandtl number

$$Nu = K_{nat,1} \cdot (Gr \cdot Pr)^{K_{nat,2}} \tag{19}$$

$$Gr = \frac{D^3 \cdot (T_c - T_a) \cdot g}{\frac{(T_c + T_a)}{2} \cdot \upsilon^2} \tag{20}$$

where $K_{nat,1\text{-}2}$ represent natural convection coefficients and υ represents kinematic viscosity (m^2s^{-1}). It should be noted that for wind speeds between 0-0.5ms^{-1} the larger of the Nusselt numbers resulting from Eq.17 and Eq.19 should be used.

4.2 Electric cables

The current carrying capacity (ampacity) of electric cables is limited by the maximum operating temperature of the insulation. Sustained high currents may generate temperatures in exceedance of the maximum operating temperature, causing irreversible damage to the cable. In extreme cases this may result in complete insulation deterioration and cable destruction. The conductor temperature in steady-state conditions was modelled (Neher & McGrath, 1957; IEC, 1994; IEEE, 1994) to account for the heat balance between the power dissipated in the conductor by the Joule effect, and the heat dissipated in the environment through the thermal resistance, R_T (mKW^{-1}), of the insulation and the soil, due to the temperature difference, ΔT (K), as shown in Eq.21. The electrical current rating, I (A), may then be calculated, as shown in Eq.22.

$$I^2 R_E = \Delta T / R_T \tag{21}$$

$$I = \sqrt{\frac{\Delta T}{R_E \cdot R_T}} \tag{22}$$

Refinements incorporating dielectric losses, q_d (Wm^{-1}), the number of conductors in the cable, n, eddy currents and circulating currents in metallic sheaths, ($\lambda_{1,2}$), electrical resistance variation with conductor temperature, $R_E(T_c)$ (Ω), skin and proximity effects and the thermal resistance of each insulating layer, $R_{T,1\text{-}4}$, of lead to the more complex expression for calculating the cable ampacity as given in Eq.23.

$$I = \sqrt{\frac{\Delta T - q_d[\frac{1}{2} R_{T,1} + n(R_{T,2} + R_{T,3} + R_{T,4})]}{R_E(T_c)[R_{T,1} + n(1+\lambda_1)R_{T,2} + n(1+\lambda_1+\lambda_2)(R_{T,3} + R_{T,4})]}} \tag{23}$$

Thermal resistances, $R_{T,1\text{-}3}$, for cylindrical layers are calculated using Eq.24 and soil thermal resistance, $R_{T,4}$, is modelled using Eq.25

$$R_{T-1,2,3} = \frac{\rho_{s-T}}{2\pi} \ln\left(1 + 2\frac{D-d}{d}\right) \tag{24}$$

$$R_{T-4} = \frac{\rho_{s-T}}{2\pi} \ln\left(\frac{2z_b}{D} + \sqrt{\left(\frac{2z_b}{D}\right)^2 + 1}\right) \tag{25}$$

where $\rho_{s\text{-}T}$ represents the soil thermal resistivity ($Wm^{-1}K^{-1}$), D and d represent the respective external and internal diameters (m) for the calculated layer and z_b represents the burial depth of the cable (m).

Other calculation methods (IEC, 1994) have to be utilised when operating conditions differ from those stated above (for example when the cable is in a duct or in open air). The model described above requires detailed knowledge of the electric cable installation. However, this information may not always be available and therefore it is difficult to make practical use of the model. In these circumstances an alternative model given in Eq.26 may be used (ENA, 2004). The rated current of electric cables, I_0 (A), is given in tables depending on the nominal voltage level, V (kV), the standardised cable cross-sectional area, A (m^2) and laying conditions (trefoil, flat formation; in air, in ducts or directly buried). The dependence of the cable ampacity on the actual soil temperature, T_s (K) away from the rated soil temperature $T_{s\ rated}$ (K) as well as the actual soil thermal resistivity away from the rated soil thermal resistivity is made linear through the coefficients ξ_T and ξ_ρ respectively.

$$I = I_0(A, V, \text{laying}) \cdot \left[\xi_T \cdot \left(T_s - T_{s\ rated}\right)\right] \cdot \left[\xi_\rho \cdot \left(\rho_{s,T} - \rho_{s,T\ rated}\right)\right] \tag{26}$$

Since this research concerns the influence of environmental conditions on component ratings, the effect of the nominal voltage level which influences the dielectric loss, q_d in Eq.23 is not considered. The effect of the heating given by adjacent components is also neglected as it is assumed that each cable has already been de-rated to take this effect into account.

4.3 Power transformers

The thermal model (IEC, 2008) can be used to calculate the winding hot spot temperature for power transformers. This is the most important parameter since hotspot temperature exceedance can damage the transformer in two ways. Firstly, a temperature exceedance of 120°C-140°C can induce the formation of bubbles in the coolant oil, which in turn is liable to cause an insulation breakdown due to the local reduction of dielectric insulation strength. Secondly, high temperatures increase the ageing rate of the winding insulation. For this reason the maximum operating temperature should not exceed the rated value. The thermal model consists of a heat balance between the power dissipated in the winding and iron core, and the heat transferred to the environment via the refrigerating circuit. Considering the ambient temperature, T_a (K), thermal resistance between the windings and the oil, $R_{T,W}$ (mKW^{-1}), the thermal resistance between the heat exchanger and the air, $R_{T,HE}$ (mKW^{-1}) and the power dissipated into the core, $I^2R_{E,\ windings}$ (W), it is possible to calculate the hot spot temperature, T_{HS} (K) as in Eq.27.

$$T_{HS} = T_a + I^2 R_{E,windings} \cdot (R_{T,W} + R_{T,HE}) \quad (27)$$

Eq.27 is discussed (Susa *et al.*, 2005) and leads to the IEC standard model for rating oil-filled power transformers as shown in Eq.28

$$T_{HS} = T_a + (T_{TO} - T_a)\left(\frac{1 + rK^2}{1 + r}\right)^x + (T_{HS} - T_{TO})K^y \quad (28)$$

where T_{TO} represents the top oil temperature (K), r represents the ratio between winding and core losses, K represents the loading ratio of the transformer, x represents the transformer oil exponent and y represents the transformer winding exponent.

The maximum rating can be obtained by iteration, once the hot spot temperature has been set, and tabulated values for the parameters are given (IEC, 2008) for transformers with different types cooling system. Correction factors (IEC, 2008) can be used to model other operating conditions such as transformers operating within enclosures. Transformer cooling systems are classified with an acronym summarising (i) the coolant fluid: oil (O) or air (A), (ii) the convection around the core: natural (N), forced (F) or direct (D), (iii) the external refrigerating fluid: air (A) or water (W) and (iv) the external convection method: natural (N) or forced (F). Typically distribution transformers have ONAN or ONAF cooling systems.

5. Thermal state estimation

This section describes the approach adopted to estimate, correct and interpolate environmental conditions to represent more accurately the actual environmental operating conditions for areas of the distribution network. This may then be used together with the models described in Section 4 to provide an estimation of power system component real-time thermal ratings.

5.1 Environmental condition estimations

The inverse distance interpolation technique (Shepard, 1968) allows environmental conditions to be determined over a wide geographical area using a reduced set of inputs. This is attractive for situations where a large amount of installed measurements may be financially unattractive to the DNO or generator developer. The technique is also computationally efficient and allows the input locations to be readily adapted. The wind speed correction process is described below, as is the soil parameter correction process. Wind direction, air temperature and solar radiation values are included within interpolations but do not require the application of a correction factor. At each point in the geographical area, κ, the value of the parameter, Z, representing the environmental condition can be estimated as a weighted average of the parameter values known at ι points. The weighting factor is a function of the distance between the points as shown in Eq.29.

$$Z_{\kappa} = \frac{\sum_{\iota} \frac{1}{d_{\iota,\kappa}^{2}} Z_{\iota}}{\sum_{\iota} \frac{1}{d_{\iota,\kappa}^{2}}} \tag{29}$$

Ground roughness influences wind speed profiles and may lead to differences between the wind speed recorded by anemometers and the actual wind speed passing across an overhead line, particularly if the anemometer and overhead line are installed at different heights. This may be corrected using the wind profile power law given in Eq.30. The wind speed at two different heights is linked with the ground roughness through the exponent Kshear. Values of Kshear for different ground types are given in (IEC, 1991).

$$Ws = Ws_a \cdot \left(\frac{z_{ref}}{z_a}\right)^{Kshear_a} \cdot \left(\frac{z_c}{z_{ref}}\right)^{Kshear_c} \tag{30}$$

Using Eq.30, the anemometer wind speed, Ws_a (ms^{-1}) at the weather station height, z_a (m) is extrapolated to a reference height, z_{ref} (m), to remove ground roughness dependence represented by the parameter $Kshear_a$. The values from different anemometer locations may then be interpolated, using Eq.29 to provide a wind speed estimate at the reference height for a particular geographical location. The ground roughness at this location is then taken into account through the coefficient $Kshear_c$ along with the conductor height, z_c (m) in Eq.30 to estimate the wind speed, Ws (ms^{-1}) across the overhead line.

Electric cable ratings are dependent on soil temperature and soil thermal resistivity, as well as cable construction, burial layout and burial depth (which is typically 0.8–1m). UK MetOffice datasets contain information regarding soil temperatures at a depth of 0.3m. However, no information is currently available from this source regarding soil thermal resistivity. Depth-dependent soil temperature distributions may be calculated using the Fourier law (Nairen *et al.*, 2004) as shown in Eq.31

$$\frac{dT_s}{dt} = \frac{d}{dz}\left(\delta_{s-T}(\theta)\frac{dT_s}{dz}\right) \tag{31}$$

where T_s represents the soil temperature (K), t represents time (s), z represents soil depth (m) and $\delta_{s\text{-}T}(\theta)$ represents the soil thermal diffusivity (m^2s^{-1}) as a function of gravimetric water content. Boundary conditions are set up for the lower soil layer (for example a constant temperature of 10°C at a depth of 2m) and soil temperature readings are used for the upper layer. Soil thermal resistivity $\rho_{s\text{-}T}$ (mKW^{-1}) may be calculated from Eq.32 using the soil thermal diffusivity, $\delta_{s\text{-}T}$ (m^2s^{-1}), the dry soil density $\rho_{s\text{-}density}$ (kgm^{-3}), and the soil thermal capacitance, $C_{s\text{-}T}$ ($Jkg^{-1}K^{-1}$).

$$\rho_{s-T} = \left(\delta_{s-T} \cdot \rho_{s-density} \cdot C_{s-T}\right)^{-1} \tag{32}$$

Soil thermal diffusivity and soil thermal capacity are influenced by soil composition, N (%), and soil gravimetric water content, θ and can be calculated using Eq.33-34 (Nidal, 2003).

$$\delta_{s-T}(\theta) = -14.8 + 0.209 \cdot N + 4.79 \cdot \theta \tag{33}$$

$$C_{s-T} = -0.224 - 0.00561 \cdot N + 0.753 \cdot \rho_{s-density} + 5.81 \cdot \theta \tag{34}$$

Ground water content may be determined using the closed form of Richard's equation (Celia *et al.,* 1990) as described in Eq.35 after the calculation of the unsaturated hydraulic diffusivity, $\delta_{s\text{-}\theta}(\theta)$ (m^2s^{-1}) and the unsaturated hydraulic conductivity, $k_{s\text{-}\theta}(\theta)$ (ms^{-1}) (Van Genuchten, 1980).

$$\frac{d\theta}{dt} = \frac{d}{dz}\left[\delta_{s-\theta}(\theta)\frac{d\theta}{dz} + k_{s-\theta}(\theta)\right] \tag{35}$$

In order to solve Eq.35, boundary and initial conditions must be specified. A constant water content equal to the saturation value can be set at a depth of 2.5 metres, corresponding to the UK water table. Furthermore, the ground-level water content can be linked to MetOffice rainfall values, l_r (mm) using the model described in Eq.36, where $Krain_1$ represents the normalized soil water loss (day^{-1}) and $Krain_2$ represents the normalized net rainfall coefficient ($day^{-1}mm^{-1}$) (Rodriguez-Iturbe, 1996).

$$\frac{d\theta}{dt} = -Krain_1 \cdot \theta(t) + Krain_2 \cdot l_r(t) \tag{36}$$

5.2 Component rating estimations

In this section a description of the algorithm responsible for the state estimation is given. The primary aim of the thermal state estimation algorithm is to allow the rating of components, which are not directly monitored within the power system, to be assessed. Thermal state estimations facilitate the precise and reliable assessment of environmental conditions whereby a minimal amount of meteorological monitoring installations facilitate the assessment of component thermal ratings within a wide area. This may then be validated through the carefully selected monitoring of component operating temperatures. The algorithm provides a reliable estimation of power system component thermal ratings described by an appropriate cumulative probability function. A state estimation technique based on the Montecarlo method is used, giving a more complete description of the possible states of the system. The minimum, maximum, average and standard deviation of component rating forecasts may be calculated according to the possible forecasted weather conditions. As necessary for overhead lines and electric cables, each component is divided into sections to take into account different thermal operating conditions such as overhead line orientations and changes in electric cable installation conditions. The section resulting in the lowest rating values is then used to provide a rating for the entire component. Furthermore, the deployment of a real-time thermal rating system underpinned by thermal state estimation techniques has the potential to reduce the necessity of auxiliary communications infrastructure whilst simultaneously increasing the reliability of the system if measurement or communication failures occur.

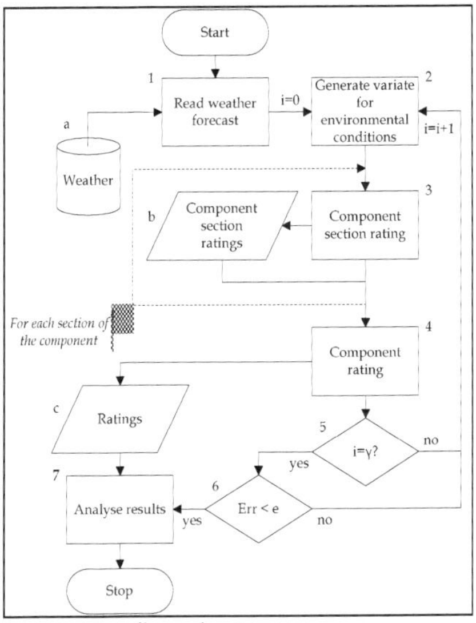

Fig. 2. Component rating estimation flow chart

The thermal state estimation algorithm is illustrated in Figure 2, where it is possible to see the following steps:

1. Forecasted weather data is read from an external source (in this case the database "a"), where real-time readings from different meteorological station locations within the distribution network are stored.
2. A set of values for weather parameters is calculated in the following way: From the data read in "1" the parameters of a cumulative probability function are calculated. In this case the Beta probability function is used. A random value for the probability is selected and from the cumulative probability function the corresponding parameter value is found. This is repeated for each weather parameter.
3. For each section of the component the values of local environmental condition are calculated with the methodology described in Section 5.1, then the rating is calculated using the models described in Section 4. The result is stored temporarily in "b".
4. The component rating is calculated by selecting the minimum rating of each section. The results are temporarily stored in "c".
5. The steps from 2 to 4 are repeated for a fixed number of times γ.
6. The precision of the result is compared with a predefined value. If the result is not acceptable, a new value for γ is calculated and the steps from 2 to 5 are repeated
7. Component ratings stored in "c" are analysed in order to calculate the minimum, maximum, average value and standard deviation for each component of the network.

6. Generator output control strategies

This section describes strategies for the power output control of single and multiple generation installations.

6.1 Single generation installations

A number of different solutions are presented (ENA, 2003) that may be developed to manage, in an active manner, the power flows associated with the connection of a single generation installation. The most basic systems involve the disconnection of generation in the event that network power flow congestion occurs when utilising component static thermal ratings. This solution may be developed further by actively switching between seasonal ratings based on fixed meteorological assumptions and adjusting the number of disconnected generation units accordingly. More sophisticated solutions are developed from the principle of generator power output control, utilising technologies such as the pitch control of wind turbine blades to capture a desired amount of wind energy. In this approach the powers flowing in the critical feeders of the network are monitored, taking load demand into account, and the power exported from the generator is controlled to ensure the capability of the network is not exceeded. This may be developed as a solution utilising component static ratings, with demand-following control of the generator output, or as a solution utilising component real-time thermal ratings, with demand-following control of the generator output. Although generators may, at times, be requested to operate in voltage control mode, it is in the interest of the operators to maximise annual exported active energy (MWh) as that directly relates to the revenue of the generation installation owner. Thus it is assumed that the control system outlined in this chapter is utilised for generators exporting real power at unity power factor. An assessment of the real power output adjustment required by a generator may be calculated in Eq. 37-38, relating to the component power flow sensitivity factor and the real-time thermal rating of the component

$$\Delta G_{P,m} = \frac{\Delta P_{i,k}}{\left(dP_{i,k} / dG_{P,m}\right)} \tag{37}$$

$$\Delta P_{i,k} = \sqrt{\left(\left(U_{Tar} \cdot S_{lim}\right)^2 - \left({}''Q_{i,k}\right)^2\right)} - \sqrt{\left(\left({}'S_{i,k}\right)^2 - \left({}'Q_{i,k}\right)^2\right)} \tag{38}$$

where $\Delta G_{P,m}$ (MW) is the required change in real power output of the generator at node m, $\Delta P_{i,k}$ (MW) is the required change in real power flow through the component, from node i to node k in order to bring the resultant power flow back within thermal limits, $dP_{i,k}/dG_{P,m}$ is the power flow sensitivity factor that relates the change in nodal real power injection at m with the change in real power flow seen in the component from node i to node k, $'S_{i,k}$ (MVA) is the apparent power flow in the component before control actions are implemented, $'Q_{i,k}$ (MVAr) is the reactive power flow in the component from node i to node k before the control actions are implemented, U_{Tar} is the target utilisation limit of the component after the control actions have been implemented, S_{lim} (MVA) is the static or real-time thermal rating of the component and $''Q_{i,k}$ (MVAr) is the reactive power flow in the component from node i to node k after the control actions have been implemented. Rather than neglecting MVAr flow, it may be assumed constant for a particular operating condition. Thus $dP_{i,k}/dG_{P,m} >> dQ_{i,k}/dG_{P,m}$ and a simplification can be made to Eq.38 that $''Q_{i,k} \approx {}'Q_{i,k}$.

6.2 Multiple generation installations

Present last-in first-off control strategies include the complete disconnection of generation installations or the power output reduction of generators in discrete intervals (Roberts, 2004). A step beyond this could be to implement a last-in first-off 'sensitivity-based' control strategy whereby generator power outputs are adjusted (in contractual order) based on Eq.37-38. Moreover, these equations may be developed further to define the power output adjustment of multiple generation installations (assuming contracts are in place to allow generator operation outside of a last-in first-off constraint priority) in order to maximise the annual energy yield of the aggregated generation installations. Two candidate strategies are outlined below:

1. Equal percentage reduction of generator power output

The equal percentage reduction signal, Φ, broadcast to multiple generators in order to proportionally reduce their power output (taking power flow sensitivity factors into account) may be assessed as in Eq.39-40.

$$\Delta G_{P,m} = G_{P,m} \times \Phi \tag{39}$$

$$\Phi = \frac{\Delta P_{ik}}{\sum_{m=0}^{m=\infty} \left(G_{P,m} \times \left(dP_{ik} / dG_{P,m} \right) \right)} \tag{40}$$

This strategy may be seen as the most 'fair' option since each generator is constrained as an equal proportion of their present power output.

2. Most appropriate technical strategy

In order to minimise the overall generator constraint and thus maximise the annual energy yield of the aggregated generators, an assessment of the required generator power output adjustment is given in Eq.41. In this case, rather than adjusting generators in contractual order according to the historical order of connection agreements, the generator constraint order is prioritised according to the magnitude of the power flow sensitivity factors. Thus the generator with the greatest technical ability to solve the power flow congestion is adjusted first.

$$\left\{ \Delta G_{P,m} \right\}_{min} = \frac{\Delta P_{ik}}{\{ dP_{i,k} / dG_{P,m} \}_{max}} \tag{41}$$

7. Case study: Power output control of multiple generators informed by component thermal ratings

The following case study illustrates various aspects of the development of a power output control system for generation installations. An assessment of power flow sensitivity factors is made which allows thermal vulnerability factors to be calculated. This informs the development of a real-time thermal rating algorithm for an electric cable section within the case study network. An offline simulation is conducted which allows the energy yield of candidate strategies for the power output control of multiple generators (informed by static and real-time thermal ratings) to be assessed.

7.1 Case study network description

The case study network shown in Figure 3 was derived from a section of the United Kingdom generic distribution system (UKGDS) 'EHV3' network (Foote *et al.*, 2006). A hydro generator and two combined heat and power (CHP) generators were connected to the network at 33kV nodes. A summary of the generator properties is given in Table 1.

Generator node, m	Generator type	Installed capacity (MW)	Last-in first-off constraint order	$dP_{353,357}/dG_{P,m}$
352	Hydro	18	2	0.475
353	CHP	40	3	0.477
354	CHP	30	1	0.327

Table 1. Generator properties

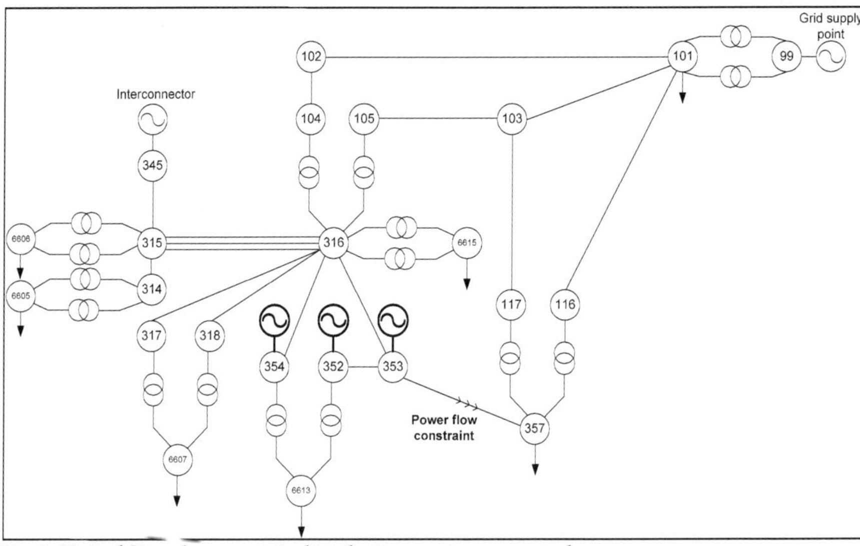

Fig. 3. United Kingdom generic distribution system case study

7.2 Thermal vulnerability factor assessment

The most thermally vulnerable components within the case study network were identified using a thermal vulnerability factor assessment as given in Figure 4 (and detailed in Sections 3.1 and 3.2). From this assessment it may be seen that a generator real power injection at node 352 increases the thermal vulnerability of the components between nodes 352 and 353, and between nodes 353 and 357 but reduces the thermal vulnerability of the component between nodes 316 and 353. Moreover, a generator real power injection at node 353 increases the thermal vulnerability of the component between nodes 353 and 357 but reduces the thermal vulnerability of the component between nodes 316 and 353. Furthermore, a generator real power injection at node 354 increases the thermal vulnerability of the components between nodes 316 and 353, 354 and 316, and 353 and 357. However, this injection has little impact on the thermal vulnerability of the component between nodes 352 and 353. In this particular case, multiple generator injections cause the thermal vulnerability factors to accumulate for the electric cable between nodes 353 and 357, making it the most thermally vulnerable component. Having identified this component as being a potential thermal pinch-point and causing power flow congestion within the network, the targeted development of a real-time thermal rating system is informed.

7.3 Component real-time thermal rating assessment

A summary of the characteristics of the thermally vulnerable electric cable are given in Table 2. The 'Valley' UK MetOffice dataset for the calendar year 2006 was used with the thermal state estimation technique described by Eq.31-36 (Section 5.1) to estimate environmental operating conditions local to the electric cable. This information was used to populate the electric cable steady-state thermal model described by Eq.26 (Section 4.2). The resulting real-time thermal rating for the electric cable, together with the static rating, is given in Figure 5. It was found that the average uprating for the electric cable, based on minimum daily ratings was 6.0% across the year.

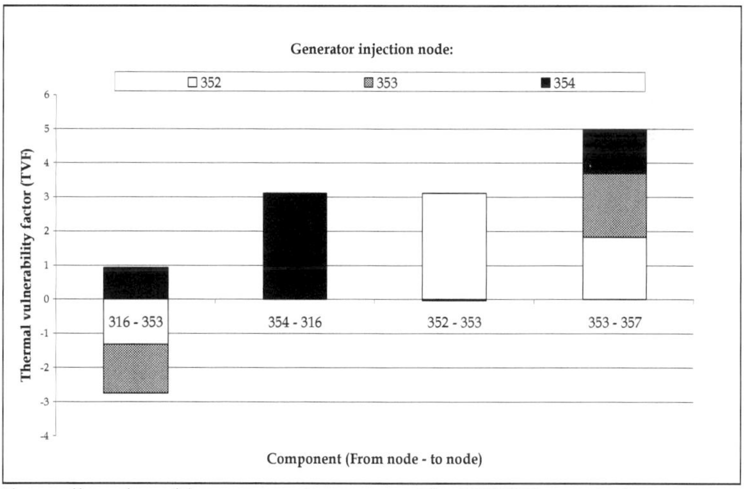

Fig. 4. Thermally vulnerable component assessment for UKGDS case study

Electric cable characteristic	Value / category
Resistance, R_E (pu)	0.01489
Reactance, X (pu)	0.01296
Cross-sectional area, A (mm^2)	150
Standard thermal rating, S_{lim} (MVA)	30
Nominal line-to-line voltage, V (kV)	33
Standardised phase current rating, I_0 (A)	525
Installation conditions	Flat formation Buried in a duct

Table 2. Electric cable characteristics

7.4 Electro-thermal simulation approach

This section describes the offline electro-thermal simulation that was used to quantify the individual and aggregated generator annual energy yields (as seen in Section 7.6) for the candidate multiple generator control strategies. UKGDS annual ½ hour loading and generation profiles were utilised for the simulation of network power flows and busbar voltages through a load flow software tool. The open-loop power output control system compared network power flows to component static and real-time thermal limits for each ½ hour interval within the simulated year. When power flow congestion was detected (signified in this case by an electric cable utilisation, $U_{353,357} > 1$) the controller implemented the relevant generator power output control algorithm as described below. A validating load flow simulation was conducted as part of the control solution to ensure that the updated generator power outputs solved the power flow congestion and did not breach busbar voltage limits (set to ±6% of nominal).

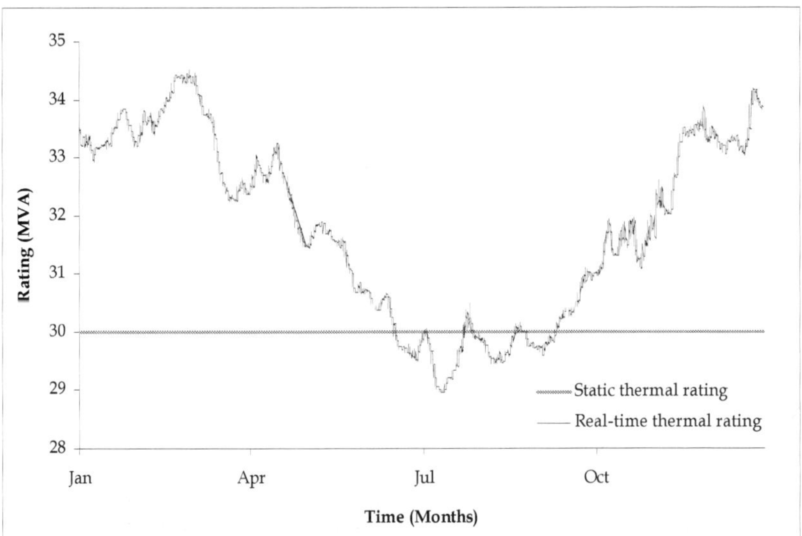

Fig. 5. Electric cable static and real-time thermal rating

The individual and aggregated generator power outputs were integrated across the simulated year in order to calculate the potential annual energy yields resulting from the candidate multiple generator control strategies, informed by component static and real-time thermal limits. The results are summarised in Section 7.6.

1. Last-in first-off discrete interval generator output reduction

In order to solve the power flow congestion, the power output of the generators was reduced in discrete intervals of 33% according to the contractual priority order given in Table 1. Thus the power output of the generator at node 354 was reduced by 33% then 66% followed by generator disconnection and the 33% power output reduction of the generator at node 352.

2. Last-in first off sensitivity-based generator output reduction

Using Eq.38 with U_{Tar} = 0.99, the required real power reduction in the electric cable, $\Delta P_{353,357}$ in order to bring the resultant power flow back within the desired thermal limit was calculated. The required reduction in the power output of the first generator (contractually) to be constrained, $\Delta G_{P,354}$, was calculated using Eq.42 for $G_{P,354} - \Delta G_{P,354}>0$ and the output from $G_{P,353}$ and $G_{P,352}$ was unconstrained. On occasions when the required constraint of the generator at node 354 would result in $G_{P,354} - \Delta G_{P,354}<0$, the generator was disconnected and the remaining power flow congestion was solved by the next generator (at node 352) to be contractually constrained.

$$\Delta G_{P,354} = \frac{\Delta P_{353,357}}{0.327} \tag{42}$$

3. Equal percentage reduction of generator power outputs

Using Eq.38 with U_{Tar} = 0.99, the required real power reduction in the electric cable, $\Delta P_{353,357}$ in order to bring the resultant power flow back within the desired thermal limit was calculated. The equal percentage reduction signal, Φ, broadcast to all generators in order to solve the power flow congestion was calculated and implemented using Eq.43-46.

$$\Phi = \frac{\Delta P_{353,357}}{0.475G_{P,352} + 0.477G_{P,353} + 0.327G_{P,354}} \tag{43}$$

$$\Delta G_{P,352} = G_{P,352} \times \Phi \tag{44}$$

$$\Delta G_{P,353} = G_{P,353} \times \Phi \tag{45}$$

$$\Delta G_{P,354} = G_{P,354} \times \Phi \tag{46}$$

4. Most appropriate technical strategy

Using Eq.38 with $U_{Tar} = 0.99$, the required real power reduction in the electric cable, $\Delta P_{353,357}$ in order to bring the resultant power flow back within the desired thermal limit was calculated. Since, in this case, $G_{p,353}$ has the greatest technical ability to solve power flow congestion (as assessed by the comparative magnitude of power flow sensitivity factors in Table 1), Eq.47 was implemented to minimise the overall generation constraint and thus maximise the annual energy yield of the aggregated generators. On occasions when the required constraint of the generator at node 353 would result in $G_{P,353} - \Delta G_{P,353} < 0$, the generator was disconnected and the remaining power flow congestion was solved by the generator with the next-greatest power flow sensitivity factor (in this case at node 352).

$$\Delta G_{P,353} = \frac{\Delta P_{353,357}}{0.477} \tag{47}$$

7.5 Findings

The marginal annual energy yields resulting from the applied candidate control strategies informed by the electric cable real-time thermal ratings are summarised in Table 3. A similar table is given (Jupe & Taylor, 2009a) for the marginal energy yields resulting from static component ratings. The last-in first-off 'discrete interval' strategy with static component ratings was used as a datum with annual energy yields of 106.38GWh, 298.26GWh, 120.84 GWh and 525.48GWh for $G_{P,352}$, $G_{P,353}$, $G_{P,354}$ and $G_{P,aggregated}$ respectively.

Generator control strategy	Generation marginal annual energy yield (%) with electric cable real-time thermal rating			
	$G_{P,352}$	$G_{P,353}$	$G_{P,354}$	$G_{P,aggregated}$
Last-in first-off (Discrete-interval)	1.7	0.0	25.8	6.3
Last-in first-off (Sensitivity-based)	1.7	0.0	41.9	10.0
Equal % reduction of power output	-5.5	-6.2	73.6	12.3
Most appropriate technical strategy	1.7	-12.0	85.1	13.1

Table 3. Generation marginal annual energy yields

8. Discussion

Considering Table 3, for the case study network it can be seen that the adoption of both real-time thermal rating systems and sensitivity factor-based generator power output control strategies have the potential to unlock gains in the aggregated annual energy yield contribution of multiple generation installations. Moving from a static rating system to a real-time rating system has the potential to unlock an extra 6.3% aggregated annual energy yield for the last-in first-off 'discrete interval' power output reduction strategy. If a last-in first-off contractual priority is retained but a sensitivity-based power output control strategy is adopted this has the potential to unlock an extra 3.7% marginal aggregated annual energy yield. Moreover, a further 3.1% annual energy yield gain may be achieved by utilising the 'most appropriate technical' strategy that minimises overall generator constraint. For the 'equal percentage reduction of generator power output' strategy, it can be seen that the relative annual energy yields of $G_{P,352}$ and $G_{P,353}$ are reduced in order to achieve an aggregated generator annual energy yield gain. This phenomenon is even more pronounced in the 'most appropriate technical' strategy where annual energy yield gains of 1.7% and 85.1% for $G_{P,352}$ and $G_{P,354}$ respectively are facilitated by the 12.0% reduction in the annual energy yield of $G_{P,353}$.

The aim of this research was to outline the developmental stages required in the output control of generation installations for improved new and renewable energy yield penetrations. The purpose of the thermal vulnerability factor assessment was to identify thermally vulnerable components within distribution networks. It has been shown that the thermal vulnerability factor assessment is appropriate for use in a strategic way to assess longer term and wide-spread generation growth scenarios. Therefore the thermal vulnerability factor assessment procedure would be valuable for DNOs looking to develop long-term generation accommodation strategies for areas of their network. Furthermore, the thermal vulnerability factor assessment identifies those components that would most benefit from being thermally monitored to unlock latent power flow capacity through a real-time thermal rating system. The derived power flow sensitivity factors are network configuration-specific and assume that the network configuration will not be frequently changing. It is feasible, however, to develop an online control system that makes use of alternative sets of the above mentioned predetermined power flow sensitivity factors based on network switch status information. In order to deploy the control strategies presented there clearly needs to be a contractual mechanism in place that gives an incentive to separately owned generators to have their power output reduced in order to maximise the overall annual energy yield. One such mechanism could be to set up cross-payments between generators whereby the generators that are constrained to maximise the aggregated energy yield are rewarded by payments from the other generators.

9. Conclusion

It is anticipated that distribution networks will continue to see a proliferation of generation from new and RE resources in coming years. In some cases this will result in power flow congestion with thermally vulnerable components restricting the installed capacity and energy yield penetration of generators. Therefore a system that can be developed for the management of power flows within distribution networks, through the power output control of generators, could be of great benefit. The development stages of such a system were outlined in Sections 3 - 6. These included: (i) the identification of thermally vulnerable components through a thermal vulnerability factor assessment that related power flow sensitivity factors to component thermal limits, (ii) the targeted development of thermal models to allow latent capacity in the network assets to be unlocked through a real-time thermal rating system, (iii) the development of thermal state estimation techniques that would allow component steady-state thermal models to be populated with real-time data from meteorological monitoring equipment installed in the distribution network (iv) the use of the real-time thermal ratings together with power flow sensitivity factors to control the power output of multiple generators. The development stages were illustrated with a case study and the energy yields resulting from the offline application of strategies were quantified. It was demonstrated for the particular case study scenario that the adoption of both a real-time thermal rating system and power flow sensitivity factor-based control strategies have the potential to unlock gains in the aggregated annual energy yield of multiple generation installations. Work is continuing in this area to realise the potential of generator output control systems for increased new and renewable energy yield penetrations.

10. References

Abu-Hamdeh, N.H. (2003). Thermal properties of soil as affected by density and water content. *Biosystems engineering,* Vol. 86, No.1 (September 2003) 97-102

ANSI/IEEE. (1981). *C57.92 Guide for loading mineral oil-immersed power transformers up to and including 100 MVA with 55°C or 65°C average winding rise,* IEEE

Belben, P.D. & Ziesler, C.D. (2002). Aeolian uprating: how wind farms can solve their own transmission problems, *Proceedings of 1st World Wind Energy Conference and Exhibition,* Berlin, Germany, July 2002

Berende, M.J.C., Slootweg, J.G. & Clemens, G.J.M.B. (2005). Incorporating Weather Statistics in Determining Overhead Line Ampacity, *Proceedings of International Conference on Future Power Systems,* 1-8, ISSN 90-7805-02-4, Amsterdam, Netherlands, November 2005, FPS, Amsterdam

Celia, M.; Boulotas, E. & Zarba, R. (1990). A general mass-conservative numerical solution for the unsaturated flow problem. *Water resources research,* Vol. 26, No.7, 1483-1496

Currie, R. A. F.; Ault, G. W. & McDonald, J. R. (2006). Methodology for determination of economic connection capacity for renewable generator connections to distribution networks optimised by active power flow management. *IEE Generation, Transmission, Distribution,* Vol. 153, No. 4, (July 2006), 456-462, ISSN 1350-2360

Dinic, N.; Flynn, D.; Xu, L. & Kennedy, A. (2006). Increasing Wind Farm Capacity. *IEE Generation, Transmission, Distribution,* Vol. 153, No. 4, (July 2006), 493-498, 1350-2360

Douglass, D. A. & Edris A. A. (1996). Real-time monitoring and dynamic thermal rating of power transmission circuits. *IEEE Transactions on Power Delivery,* Vol. 11, No. 3, (July 1996), 1407-1418, ISSN 0885-8977

Douglass D. A.; Edris A. A. & Pritchard G. A. (1997). Field application of a dynamic thermal circuit rating method. *IEEE Transactions on Power Delivery,* Vol. 12, No. 2, (April 1997), 823-831, ISSN 0885-8977

Electricity Networks Association. (1971). *Engineering Recommendation P15: Transformers loading guide,* ENA, London

Electricity Networks Association. (1986). *Engineering Recommendation P27: Current rating guide for high voltage overhead lines operating in the UK distribution system,* ENA, London

Electricity Networks Association. (2003). *Engineering Technical Report 124: Guidelines for actively managing power flows associated with the connection of a single distributed generation plant,* ENA, London

Electricity Networks Association. (2004). *Engineering Recommendation P17: Current rating for distribution cables,* ENA, London

Foote, C.; Djapic, P.; Ault, G.; Mutale, J.; Burt G. & Strbac, G. (2006). *United Kingdom Generic Distribution System (UKGDS) - Summary of EHV Networks,* DTI Centre for Distributed Generation and Sustainable Electrical Energy, UK

Fox-Penner, P.S. (2001). Easing Gridlock on the Grid - Electricity Planning and Siting Compacts. *The Electricity Journal,* Vol. 14, No.9 (November 2001), 11-30

Harrison, G.P. & Wallace, A.R. (2005). Optimal Power Flow Evaluation of Distribution Network Capacity for the Connection of Distributed Generation. *IEE Generation, Transmission, Distribution,* Vol. 152, No. 1, (January 2005), 115-122, ISSN 1350-2360

Helmer, M. (2000). Optimized size of wind power plant transformer and parallel operation, *Proceedings of Wind Power for the 21st century*, Kassel, Germany, September 2000

Hoffmann, S.P. & Clark, A.M. (2004). The approach to thermal uprating of transmission lines in the UK. *Electra, (*B2-317), CIGRE, Paris

House, H.E. & Tuttle, P. D. (1959). Current Carrying Capacity of ACSR. *IEEE Transactions on Power Apparatus Systems,* Vol. 78, No. 3, (April 1958), 1169-1177, ISSN 0018-9510

IEC. (1991). *IEC 60826: Loading and strength of overhead transmission lines,* IEC, Switzerland

IEC. (1994). *IEC 60287: Electric cables – calculation of the current rating,* IEC, Switzerland

IEC. (1995). *IEC TR 1597: Overhead electrical conductors – calculation methods for stranded bare conductors,* IEC, Switzerland

IEC. (2008). *IEC 60076-7: Power transformers - Part 7: Loading guide for oil-immersed power transformers,* IEC, Switzerland

IEEE. (1993). *IEEE 738: Standard for calculating the current-temperature relationship of bare overhead conductors,* IEEE

IEEE. (1994). *IEEE 835, IEEE Standard Power Cable Ampacity Tables,* IEEE

Ippolito, L., Vaccaro, A., Villacci, D. (2004). The use of affine arithmetic for thermal state estimation of substation distribution transformers. *International Journal for Computation and Mathematics in Electrical and Electronic Engineering,* Vol. 23, No. 1, 237–249, ISSN 0332-1649

Jupe, S.C.E. & Taylor, P.C. (2009a). Strategies for the control of multiple distributed generation schemes, *Proceedings of 20th International Conference on Electricity Distribution*, Prague, Czech Republic, June 2009, (*In press*)

Jupe, S.C.E. & Taylor, P.C. (2009b). Distributed generation output control for network power flow management. *IET Renewable Power Generation,* (*In press*)

Mendez Quezada, V.H.; Abbad, J.R. & San Romain, G.T. (2006). Assessment of Energy Distribution Losses for Increasing Penetration of Distributed Generation. *IEEE Transactions on Power Systems,* Vol. 21, No. 2, (May 2006), 533-540, ISSN 0885-8950

Michiorri, A.; Taylor, P.C.; Jupe, S.C.E. & Berry, C (2009). An investigation into the effect of environmental conditions on power system ratings. *IMechE: Journal of Power and Energy,* (*In press*)

Morgan, V.T. (1982). The thermal rating of overhead-line conductors. *Electric Power Systems Research,* Vol. 5, No. 2, (June 1982), 119-139, ISSN 0378-7796

Nairen, D.; Qinyun, L. & Zhaohong, F. (2004). Heat transfer in ground heat exchangers with groundwater advection. *International Journal of Thermal Sciences,* Vol. 43, (December 2004) 1203-1211

Neher, J.H. & McGrath, M.H. (1957). The calculation of the temperature rise and load capability of cable systems. *AIEE Transactions,* Vol. 76, No. 3, (April 1957), 752-772, ISSN 0018-9510

Neumann, A.; Taylor, P.; Jupe, S.; Michiorri, A.; Goode, A.; Curry, D. & Roberts, D. (2008). Dynamic thermal rating and active control for improved distribution network utilisation, *Proceedings of PowerGrid Europe,* Milan, Italy, June 2008, PennWell, Milan

Nuijten, J.M.A. & Geschiere, A. (2005). Future network planning and grid control, *Proceedings of International Conference on Future Power Systems,* pp. 7-13, ISSN 90-78205-02-4, Amsterdam, Netherlands, November 2005, FPS, Amsterdam

Piccolo, A.; Vaccaro, A. & Villacci, D. (2004). Thermal rating assessment of overhead lines by Affine Arithmetic. *Electric Power Systems Research,* Vol. 71, No. 3, (November 2004), 275-283, ISSN 0378-7796

Roberts, D.A. (2004). *Network Management Systems for Active Distribution Networks – A Feasibility Study,* Scottish Power Plc, Glasgow

Rodriguez-Iturbe, I.; Isham, I.; Cox D.R. *et al.* (2006). Space-time modelling of soil moisture: stochastic rainfall forcing with heterogeneous vegetation. *Water resources research,* Vol. 42, (March 2006)

Shepard D. (1968). A two-dimensional interpolation function for irregularly-spaced data, *Proceedings of 23rd ACM national conference,* 517-524, August 1968, ACM, New York

Soto, F., Latorre, J. & Wagensberg, M. (1998). Increasing the capacity of overhead lines in the 400 kV Spanish transmission network: real time thermal ratings. *Electra,* (22-211), CIGRE, Paris

Stephen, R. (2004). Description and evaluation of options relating to uprating of overhead transmission lines. *Electra,* (B2-201), CIGRE, Paris

Susa, D.; Lehtonen, M. & Nordman, H. (2005). Dynamic thermal modelling of power transformers. *IEEE Transactions on Power Delivery,* Vol. 20, No. 1, (January 2005), 197-204, ISSN 0885-8977

Van Genuchten M. (1980). A closed for equation for predicting the hydraulic conductivity of unsaturated soils. *Soil science society of America Journal,* Vol. 44, (May 1980), 892-898

Villacci, D. & Vaccaro, A. (2007). Transient tolerance analysis of power cables thermal dynamic by interval mathematic. *Electric Power Systems Research,* Vol. 77, No. 3-4, (March 2007), 308–314, ISSN 0378-7796

Vovos, P. N.; Harrison, G.P.; Wallace, A.R. & Bialek, J. W. (2005). Optimal Power Flow as a Tool for Fault Level-Constrained Network Capacity Analysis. *IEEE Transactions on Power Systems,* Vol. 20, No. 2, (May 2005), 734-741, ISSN 0885-8950

WG 22.12. (1992). The Thermal Behaviour of Overhead Conductors - Section 1. *Electra,* Vol. 144, No. 3, CIGRE, Paris

Wood, A.J. & Wollenberg, B.F. (1996). *Power Generation, Operation and Control,* Wiley, ISBN 0-471-58699-4, New York, USA

5

Embedded Energy Storage Systems in the Power Grid for Renewable Energy Sources Integration

Sérgio Faias*, Jorge Sousa* and Rui Castro**

** ISEL –Instituto Superior de Engenharia de Lisboa*

*** Cie3 / IST – Technical University of Lisbon*

Portugal

1. Introduction

The increasing integration of renewable energy sources in the power system allows a reduction of the fossil fuels consumption and on their consequent environmental impacts, such as the climate change and air pollution. Moreover, a high share of renewable power allows other significant advantages for the countries promoting these technologies, such as the improvement of the security of supply and the decrease of the energy dependency from other countries.

Nevertheless, as the most renewable technologies ultimately derive energy from unpredictable natural sources or sources that vary in their availability over different timescales, the integration of larger amounts of renewable energy raises some issues related to the power systems operation (IEA, 2005).

In this regard, the power demand and generation balancing becomes one of most important issues that result from the increasing integration of renewable sources. This issue will emerge especially in periods with high renewable generation and low demand and implies an action from the power system operator in order to maintain the system stability. Apparently, there are three different solutions to face the power demand and generation balancing; the first one is the curtailment of the renewable generation, resulting in a waste of this power source, the second one is to export power to an interconnected grid (Hammons, 2006), not disregarding the possibility of that grid being also submitted to the same power balance issue, and finally the third solution, that avoids the disadvantages of the preceding ones, is to store the excess of renewable power for use when it is needed, keeping the stability of the power grid (Barton & Infield, 2004 and Faias et al., 2007).

The concept of energy storage systems embedded in the power grid, which will be presented in the current chapter, allows decoupling partially the demand from the supply, facilitating the penetration of renewable sources and improving the flexibility and efficiency of the power grid. This concept will be applied to the Portuguese power system case, considering the predictions for the load and for the generation installed capacity, including renewable, for the year 2011.

The methodology used along this chapter begins with the description of the power unbalance between generation and demand issue and the solutions for its mitigation. After that, the available energy storage technologies are presented as well as the energy storage system design process. Following, a brief discussion on the different perspectives for the location of embedded energy storage system in the power grid is presented. Finally, the results for Portuguese power system case study are presented, analysed and discussed.

2. Balancing Power Demand and Generation

Traditionally the power system operation is based on the principle that at each moment in time the power demanded by the load is generated, at that moment, by the set of power plants in the system. For that reason, if either the demand or the generation experience a sudden and sharp increase or decrease, power unbalances would occur resulting in power system instability. As, in general, the power demand is not controlled by the power system operator, its action is focused on the power generation dispatching.

Power generation dispatch is also known as the unit commitment, and it consists of defining which power units should be operating at a specific moment and the power generated by each unit (dispatch) in order to satisfy the power demand.

Power demand varies along time and according to different drivers, such as seasonality, weather condition, geographical location and the society economic activity. Extreme weather temperatures usually induce a higher power demand. On the other hand, there is a reduction on the power demand during the night and during the weekends, as a result of lower industrial activity and people's specific daily routine. Figure 1 presents different time cycles variations imposed by each one of the causes referred above.

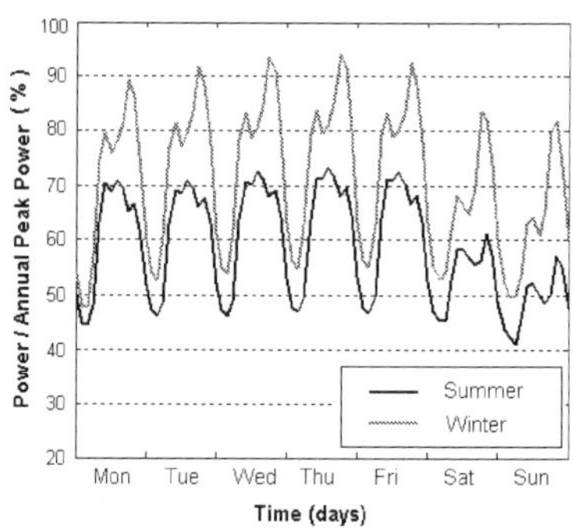

Fig. 1. Power demand variation along different time cycles (REN, 2008)

In the past, from the power system operator perspective, the uncertainly was mainly on the demand-side. However, the increasing integration of non-dispatchable renewable power sources results in an uncertainly that is also endogenous to the generation-side.

Renewable energy, which is dependent on natural resources like wind, rain and sun, present a much variable availability and can not be easily committed and dispatched by the power system operator. This variability of renewable power is apparent in Figure 2, where an example of the aggregated wind power generation along the period of one week is presented.

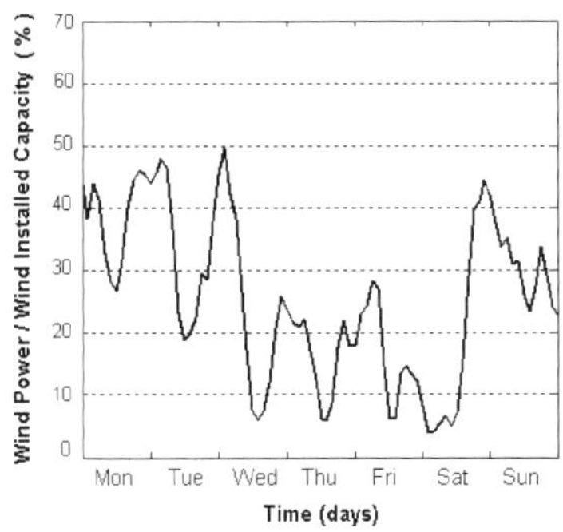

Fig. 2. Aggregated wind power generation along the period of one week (REN, 2008)

Therefore, the system operator usually considers the renewable generation as a negative power demand, satisfying the remaining demand through the unit commitment of the conventional (thermal and large hydro) power plants (Ortega-Vazquez & Kirschen, 2009).

2.1 Power Unbalance

Besides the power system operation constraints induced by the renewable power sources availability, there is an additional issue that should be taken into account. This issue is the power unbalance between power generation and demand, resulting from the coincidence of the higher availability of renewable sources (wind power, for instance) with the periods of lower demand. This fact is confirmed in Figure 3, where the equivalent average day of an annual aggregated wind power generation and the average day of an annual aggregated power demand are compared.

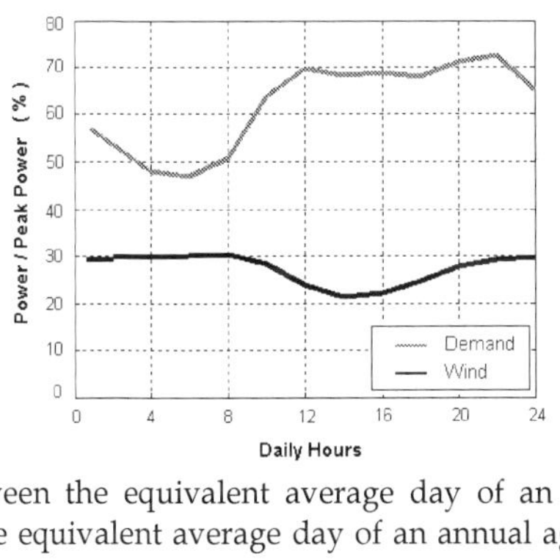

Fig. 3. Comparison between the equivalent average day of an annual aggregated wind power generation and the equivalent average day of an annual aggregated power demand (REN, 2008)

The coincidence of the wind peak power with the demand off-peak power may origin some moments where the non-dispatchable generation is greater than the demand. The resulting power unbalance can be characterized by:

$$P_{Unb}(t) = P_{Gen}(t) - P_{Dem}(t) \quad (1)$$

where $P_{Unb}(t)$ is the power unbalance, $P_{Gen}(t)$ is the power generated and $P_{Dem}(t)$ is the power demand, at time t.

Positive power unbalances correspond to an excess of generation, while negative power unbalances correspond to a generation shortage.

Positive power unbalances can occur in power systems with high penetration of non-dispatchable renewable power sources, like wind power. High hydro power availability and the minimal spinning reserve of the thermal power units also contribute to the occurrence of positive power unbalances.

Negative power unbalances only occur in moments when the available power capacity is not enough to cover all the power demands needs.

2.2 Solutions for Power Unbalance Mitigation

In practice no power unbalances should occur in a power system, otherwise, it would not be possible to the power system operator to keep the system in perfect operation and maintaining the standard power quality levels.

In moments when the power system tends to be unbalanced, namely driven by an increasing renewable power generation, the power system operator must act in order to mitigate that power unbalance and its consequences.

Following, it is presented a sort of solutions that can be adopted, individually or complementarily, to mitigate the power unbalance issue, such as curtailment of renewable generation, interconnections with other power systems and energy storage

(a) Curtailment of renewable generation

The curtailment of renewable generation is one of the possible solutions to avoid unbalances between power generation and power demand. Such solution consists of an order emitted by the power system operator to the renewable power producers to cut partially or totally their generation.

Renewable generation curtailment usually implies the waste of an environmental friendly natural resource and an increase on the fossil fuels consumption. As so, this power unbalance mitigation solution should only be considered in case of extreme contingencies.

(b) Interconnections with other power systems

Strong interconnections between different power systems are today an important advantage in terms of energy management and compensation of local power unbalances (Hammons, 2006).

The recent increasing integration of renewable power sources has been one of the drivers to reinforce the interconnection capacity of the power grids, enabling each country or region to best exploit the endogenous renewable resources creating together a diversified generation mix.

In spite of the interconnection between different power systems being one of the best solutions to mitigate the power unbalance, there are some constraints to its application, namely the one related to the potential of existing simultaneous power unbalances in the interconnected regions.

(c) Energy storage
Energy storage offers additional benefits in utility settings because it can decouple demand from supply, thereby mitigating the unbalances on the power system and allowing increased asset utilization, facilitating the penetration of renewable sources, and improving the flexibility, reliability, and efficiency of the electrical network (Schoenung, 1996).
The option for energy storage solution involves an investment on an energy storage system, but, on the other hand, it avoids the disadvantage of the renewable power curtailment and the constraints related to exporting power through the grid interconnections.

3. Energy Storage Technologies

Today, there are several high performance storage technologies available or at an advanced state of development, demanded by a new range of the energy storage applications. The singularities of each storage technology, dependent on their operation fundamentals, turn it unique and difficult to compare them.
A typical method used for the storage technology comparison, based on the power and energy capacities of commercialized devices, is presented as example in Figure 4. This power and energy range comparison of the technologies allows the identification of the devices that are best suited for a specific application.

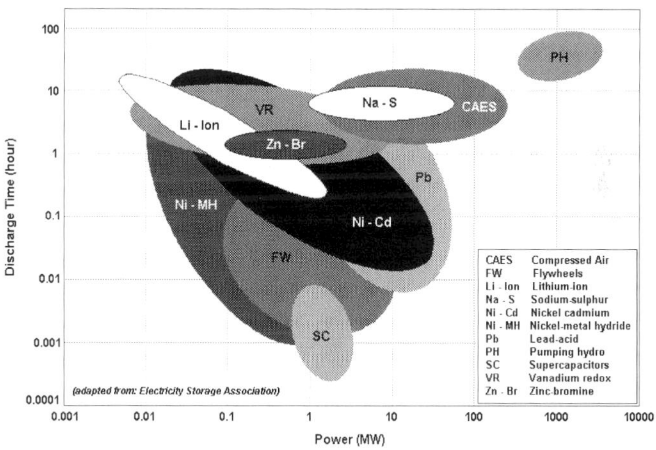

Fig. 4. Power and energy capacity comparison for different energy storage technologies (ESA, 2009)

In order to best distinguish the energy storage technologies applications and considering their placement on the ordinates axe of the figure above, storage technologies are here classified into two different categories, based on their discharge time. These categories are: short-term discharge energy storage devices and long-term discharge energy storage devices.
Short-term discharge energy storage devices present a very fast response to the power system needs. However, they just can supply their rated power for short periods, which

vary from milliseconds to few minutes. The short-term discharge energy storage devices are usually applied to improve power quality, to cover load during start-up and synchronization of backup generators and to compensate transient response of renewable power sources (Tande, 2003).
Long-term discharge energy storage devices are able to supply power from some seconds to many hours. Their response to the power system needs is usually slower than the short-term discharge energy storage devices, and much dependent on the technology. Long-term discharge energy storage devices are usually applied on the energy management, renewable energy sources integration and power grid congestion management (Price et al., 1999).

3.1 Short-term discharge energy storage devices

Short-term discharge energy storage devices should be used to aid power systems during the transient period after a system disturbance, such as line switching, load changes and fault clearance. Their application prevents collapse of power systems due to loss of synchronism or voltage instability, improving its reliability and quality.
Short-term discharge energy storage devices use is getting common in power systems with important renewable energy penetration (like wind, for instance) and weak interconnections or in islands, avoiding temporary faults and contributing to the provision of important system services such as momentary reserves and short-circuit capacity (Hamsic et al., 2007).
The main short-term discharge energy storage devices and their operation are presented below.

(a) Flywheels
Flywheels store kinetic energy in a rotating mass. Such equipments have typically been used as short-term energy storage devices for propulsion applications such as powering train engines and road vehicles, and in centrifuges. In these applications, the flywheel smoothes the power load during deceleration by dynamic braking action and then provides a boost during acceleration (Lazarewicz and Rojas, 2004). Figure 5 presents the operating diagram of a flywheel energy storage system.

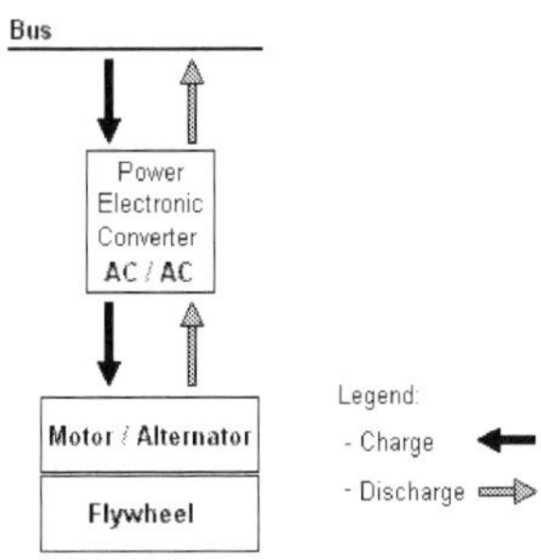

Fig. 5. Flywheel energy storage device operation diagram

(b) Supercapacitors
Supercapacitors are the latest innovative devices in the field of electrical energy storage. In comparison with a battery or a traditional capacitor, the supercapacitor allows a much powerful power and energy density (Zhai et al., 2006).

Supercapacitors are electrochemical double layer capacitors that store energy as electric charge between two plates, metal or conductive, separated by a dielectric, when a voltage differential is applied across the plates (Rufer et al., 2004). As like battery systems, capacitors work in direct current. This fact imposes the use of electronic power systems, as presented in Figure 6.

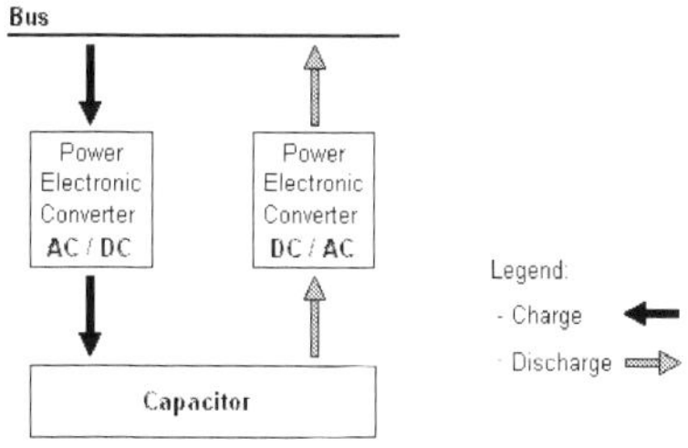

Fig. 6. Supercapacitor energy storage device operation diagram

(b) Magnetic Superconducting
Superconducting magnetic energy storage devices store energy in the form of a magnetic field, through a direct current flowing in a superconducting coil. The alternate current from a power bus is converted to direct current and injected in the coil. When necessary, the stored energy can be released, through a direct current that is converted to alternate current and injected in the power bus (Hsu and Lee, 1992). The interface between the power bus and the superconducting coil uses power electronic converters (Nomura et al., 2006).
The Superconducting Magnetic Energy Storage (SMES) device operation diagram is presented in Figure 7.

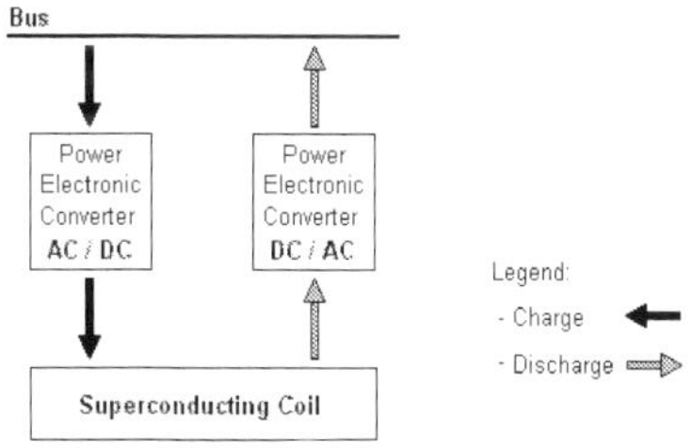

Fig. 7. SMES device operation diagram

The conductor for carrying the direct current operates at cryogenic temperatures where it behaves as a superconductor and thus has virtually no resistive losses as it produces the magnetic field. Consequently, the energy can be stored in a persistent mode, until required. The most important advantage of SMES device is that the time delay during charge and discharge is quite short. Power is available almost instantaneously and very high power output can be provided for a brief period of time (Mito et al., 2004).

3.2 Long-term discharge energy storage devices

The so-called long-term discharge energy storage devices have the capability to supply or absorb electrical energy during hours.
Sort of different long-term discharge energy storage technologies are already available today and their use is expected to rise in the next years because of the increasing integration of non-dispatchable renewable energy generation in the power systems (IEA, 2005).
A brief description of the main long-term discharge energy storage technologies is presented below.

(a) Pumping Hydro
In pumping hydro energy storage, a body of water at a relatively high elevation represents a potential or stored energy. When generation is needed, the water in the upper reservoir is lead through a pipe downhill into a hydroelectric generator and stored in the lower reservoir. To recharge the storage system, the water is pumped back up to the upper reservoir and the power plant acts like a load as far as the power system is concerned.
Pumping hydro energy storage system is constituted by two water reservoirs, an electric machine (motor/generator) and a reversible hydro pump-turbine unit. The system can be started-up in few minutes and its autonomy depends on the volume of stored water.
There are three possible configurations for the pumping hydro systems. The first one, the pure pumping hydro, corresponds to a power plant that is specifically set-up for storage, where the only turbinated/pumped water is the one stored in the upper and lower reservoirs. The second configuration corresponds to a reservoir hydro power plant, integrated in a river course, equipped with a lower reservoir and a reversible pump-turbine unit. The third configuration corresponds to a cascade of hydro power plants, where some reservoirs act simultaneously like upper and lower reservoir for the different power plants. In second and third configurations, the most common, the power plant operation is more complex because of the coordination of the different power plants and the reservoir inflows resultant from the river.
The operation of a pumping hydro system is presented in Figure 8.

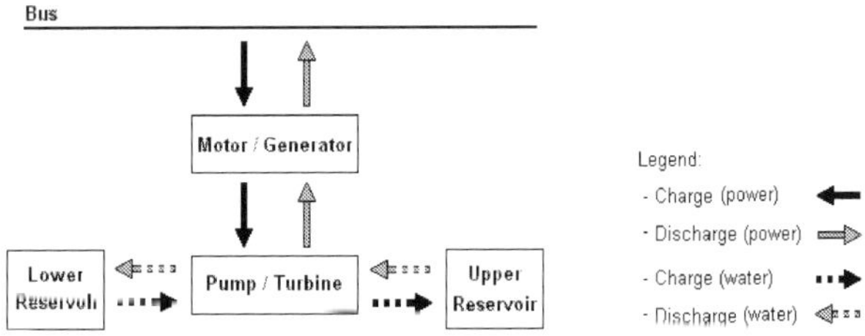

Fig. 8. Pumping hydro system operation diagram

Pumping hydro energy storage system operation is constrained by the weather conditions, reducing its storage capacity in periods extremely wet or dry.
The main restrictions to pumping hydro energy storage implementation are related with geographical constraints.

(b) Batteries

Batteries store energy in electrochemical form creating electrically charged ions. When the battery charges, a direct current is converted in chemical potential energy, when discharges, the chemical energy is converted back into a flow of electrons in direct current form (Hunt, 1998). The connection of the system to the grid, as presented in Figure 9, implies the use of power electronic converters in order to rectify the alternate current during the battery charge periods and to invert the direct current during the battery discharge periods.

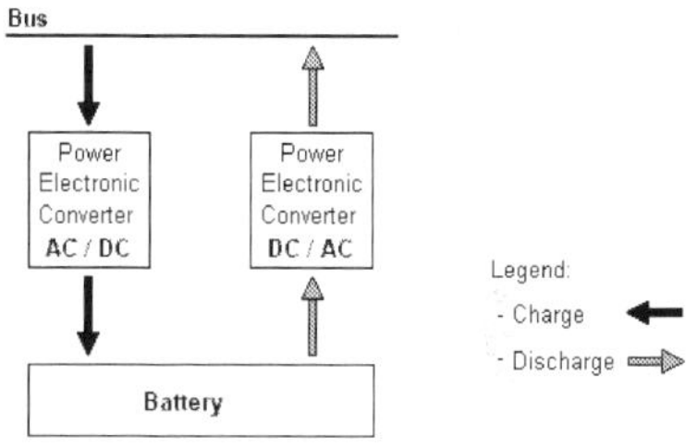

Fig. 9. Battery device operation diagram

Batteries are the most popular energy storage devices. However, the term battery comprises a sort of several technologies applying different operation principals and materials. As so, the distinction between two important battery concepts, electrochemical and redox-flow, is hereby emphasized.

Electrochemical

Electrochemical batteries use electrodes both as part of the electron transfer process and store the products or reactants via electrode solid-state reactions (Price et al., 1999).

There are a number of battery technologies under consideration for energy storage, the main being:

- Lead acid
- Nickel cadmium
- Nickel metal-hydride
- Sodium sulphur
- Lithium ion

Redox-Flow

Redox-flow batteries are storage devices that convert electrical energy into chemical potential energy by charging two liquid electrolyte solutions and subsequently releasing the stored energy during discharge (Ponce de León et al., 2006).

The name redox-flow battery is based on the redox reaction between the two electrolytes in the system. These reactions include all chemical processes in which atoms have their oxidation number changed. In a redox flow cell the two electrolytes are separated by a semi-permeable membrane. This membrane allows ion flow, but prevents mixing of the liquids. Electrical contact is made through inert conductors in the liquids. As the ions flow across the membrane, an electrical current is induced in the conductors (EPRI, 2007).

Over the past few years three types of redox-flow batteries had been developed up to the stage of demonstration and commercialization. These types are vanadium redox batteries (VRB), the polysulphide bromide batteries (PSB) and the zinc bromine (ZnBr).

(c) Compressed air

Compressed air energy storage is a device based on as gas turbine where the compression and the combustion processes are divided. During charging, the compressor is coupled to the electrical machine, working as a motor, compressing the air. After the compression, the air is stored into a sealed underground cavern. Discharging the device consists in generating power through the coupling of the gas turbine with the electrical machine, working as generator, and supplying the stored compressed air to the combustion process (Lerch, 2007). A compressed air energy storage system operation diagram is presented in Figure 9.

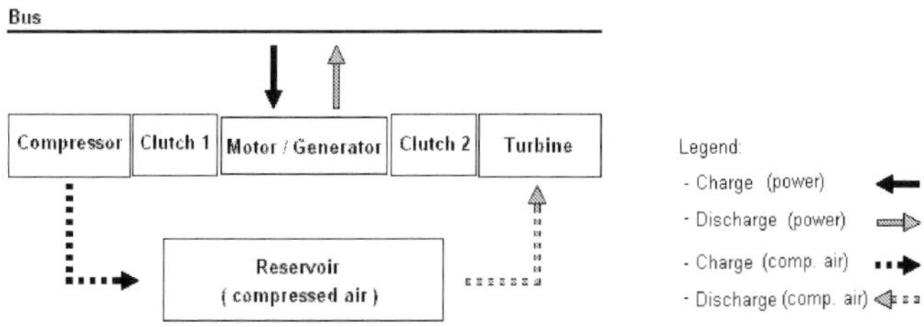

Fig. 9. Compressed air energy storage system operation diagram

Three air reservoir types are generally considered: naturally occurring aquifers (such as those used for natural gas storage), solution-mined salt caverns, and mechanically formed reservoirs in rock formations. Main compressed air energy storage system implementation constraints are related with reservoirs achievement (Schoenung, 1996)

(d) Hydrogen fuel cell

A fuel cell is an energy conversion device that is closely related to a battery. Both are electrochemical devices for the conversion of chemical to electrical energy. In a battery the chemical energy is stored internally, whereas in a fuel cell the chemical energy (fuel and oxidant) is supplied externally and can be continuously replenished (Hoogers, 2003).
The overall reaction in a fuel cell is the spontaneous reaction of hydrogen and oxygen to produce electricity and water. During the operation of a fuel cell, hydrogen is ionized into protons and electrons at the anode, the hydrogen ions are transported through the electrolyte to the cathode by an external circuit (load). At the cathode, oxygen combines with the hydrogen ions and electrons to produce water.
The hydrogen fuel cell system can be reversible, allowing electric power consumption for the production of hydrogen and that hydrogen can be stored for later use in the fuel cell (Agbossou, 2004).
The operation diagram of a hydrogen fuel cell energy storage system is presented in Figure 11.
Hydrogen volatility and its atoms reduced dimension put the hydrogen storage reservoir as the critical element in this device. Last research place Metallic Hydrates as one of most efficient (Ogden, 1999).

In the last years, hydrogen fuel cell systems become one of the most referred storage technologies to set up renewable energy integration issue. Price and charge/discharge efficiency about the 30% are its main constraints.

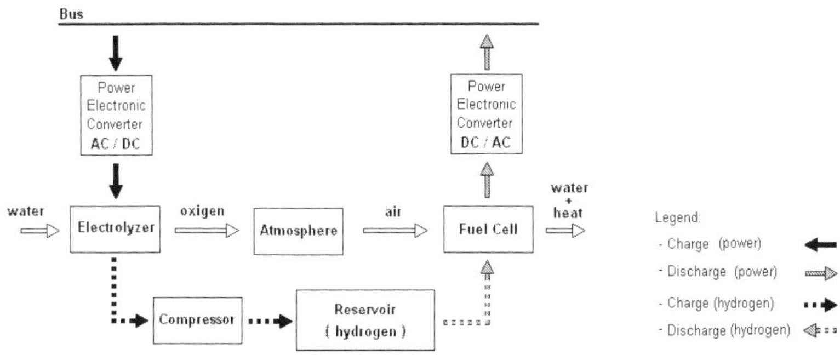

Fig. 11. Hydrogen fuel cell energy storage system operation diagram

4. Energy Storage System Design

The energy storage system design process consists off the determination of the storage power and energy capacity and the technologies that allow a better integration of the renewable sources of energy and the minimization of the thermal units fuel consumption and greenhouse effect gaseous emissions.

The energy storage system design process is divided in two different phases. The first phase consists of the implementation of an unit commitment, including the energy storage system, in order to enable the technical evaluation of several power and energy storage capacity combinations and optimize their operation. Besides the operation optimization and the feasibility evaluation of each power and energy storage capacity combination, in the first phase of the design process are also determined the needs for renewable energy curtailment and the total thermal power units operation cost.

The second phase of the energy storage system design process is based on an economical evaluation, where the costs and benefits associated to each technically feasible power and energy storage capacity combination are considered and the best techno-economical energy storage solution is determined.

4.1 Optimization of the energy storage system operation

The energy storage system operation, along a time-series, is determined throughout an optimization process that manages the system in order to best integrate the renewable energy generation, allowing, consequently, a minimization of the thermal power plants costs. The thermal power plants costs are computed by adding up the fuel costs with the emission costs due to CO_2.

The optimization process is based on a power plant commitment problem, where the load, the renewable generation and the interconnections with other power grids are considered as input data, forecasted in a previous process. The energy storage system is integrated and operated as an additional power generation unit. The specificity of that unit is related to its ability to absorb power, especially in moments when the power system has no capacity to accommodate all the renewable generation. A negative power of that power generation unit corresponds to charging the storage system and a positive power corresponds to its discharge.

The power and energy limits of the energy storage system are imposed and considered as an input of the optimization problem. Therefore, it may be possible that, for some moments of the time series, the need for power storage overloads the storage capacity, and renewable power generation must be curtailed. In order to quantify the global renewable power and energy curtailed, a power unit that represents the excess of renewable generation is considered.

The main objective of the energy storage system implementation is to support the integration of the renewable generation, being available to that integration as long as possible. For that reason, a penalty on the cost function has been introduced for the use of the storage system power charging, avoiding the temptation of the optimization process to exploit that system to attain the cheapest thermal power plants operation, when their minimal technical limits do not allow it. The penalty on the cost function does not affect the energy storage system negative power, enabling the system to be discharged anytime it is necessary or advantageous to avoid thermal power generation costs.

The curtailment power has also a penalty in the cost function, greater than the energy storage system charging one, enabling its use only when it is strictly necessary.

The objective function that is intended to be minimized considers the penalties presented above and the variable costs of the thermal power units, corresponding to their fuel costs and externalities resulting from the greenhouse effect gaseous emissions. Next, one presents the objective function:

$$\min z: \sum_{t=1}^{T}\left[\delta_1 \times C^{ESS} \times P^{ESS}(t) + C^{Ctm} \times P^{Ctm}(t) + \sum_{j=1}^{J} C_V^j \times \tau \times P^j(t)\right] \tag{2}$$

where T is the total time series period, τ is the time interval between t and $t+1$, J is the number of thermal power units, $P^j(t)$ is the power generated by the thermal power unit j at moment t, C_V^j is the variable cost of generated energy of the thermal power unit j, $P^{Ctm}(t)$ is the renewable power curtailment at moment t, C^{Ctm} is the renewable power curtailment penalty (<0), $P^{ESS}(t)$ is the power of the energy storage system at moment t, C^{ESS} is the power charging penalty of the storage system (<0), finally δ_1 is used to distinguish positive and negative energy storage system power:

$$\delta_1 = \begin{cases} 0, & P^{ESS}(t) > 0 \\ 1, & P^{ESS}(t) \leq 0 \end{cases}$$

The balance between the power demand and the power generation is maintained by the following equation:

$$P^{ESS}(t) + P^{Ctm}(t) + \sum_{j=1}^{J} P^j(t) = (1 + l_{AV}) \times L(t) - Q(t) \tag{3}$$

where $L(t)$ is the load power at moment t, $Q(t)$ is the sum of the renewable and the interconnections power and l_{AV} is the average value of the losses in the power network.
The commitment of the power units must consider the spinning reserve as presented in the next restriction:

$$P_{Max}^{ESS} + \sum_{j=1}^{J} P_{Max}^{j} \times v^{j}(t) \geq (1 + l_{AV}) \times L(t) - Q(t) \tag{4}$$

In (4) P_{Max}^{j} is the maximum power generated by the thermal power unit j, $v^{j}(t)$ is a binary variable that indicates, at moment t, if the thermal power unit j is running or if it is inactive. The term P_{Max}^{ESS} corresponds to the maximum discharging power of the energy storage system.
Considering the increasing integration of the wind power generation on the power systems and taking into account the volatility of that renewable energy source, there are some restrictions, introduced by the power systems operators, to the share of the wind power in the load power satisfaction. Therefore, in the moments when the wind power generated overcomes the level considered safe by the power system operator, the excess of wind power must be stored and the thermal power generation must be increased in order to fit the power demand. The following restriction reflects that concern:

$$\sum_{j=1}^{J} P^{j}(t) \geq \delta_2 \times \left[(1 + w_{Max}) \times (1 + l_{AV}) \times L(t) - Q(t) + W(t)\right] \tag{5}$$

where w_{Max} is the maximum share of wind power admitted by the power system operator, $W(t)$ is wind power generated at moment t, and δ_2 distinguish the moments when the maximum share of wind power is overloaded:

$$\delta_2 = \begin{cases} 0, & W(t) \leq w_{Max} \times (1 + l_{AV}) \times L(t) \\ 1, & W(t) > w_{Max} \times (1 + l_{AV}) \times L(t) \end{cases}$$

In power system generation scheduling, the reflection of actual operating processes needs the use of ramp-rate constraints to simulate the thermal unit generation changes (Wang & Shahidepour, 1993). The thermal power units ramp-rates are considered and imposed by restrictions (6) and (7), corresponding to the ramp-up power rate and to the ramp-down power rate, respectively. No ramp-rates had been considered for the ESS.

$$P^{j}(t) - P^{j}(t-1) \leq P_{up}^{j} \tag{6}$$

$$P^{j}(t-1) - P^{j}(t) \leq P_{down}^{j} \tag{7}$$

where P_{up}^{j} and P_{down}^{j} correspond to the ramp-up power rate and to the ramp-down power rate of the thermal power unit j.
The limits to maximum and minimal power generation of each power unit are restricted by the conditions below:

$$P^{j}(t) \leq P_{Max}^{j} \times v^{j}(t) \tag{8}$$

$$P^j(t) \geq P^j_{Min} \times v^j(t) \tag{9}$$

In (9) P^j_{Min} corresponds to the technical minimum power that can be generated by the thermal power unit j.
The energy storage system operation is limited by:

$$P^{ESS}(t) \leq P^{ESS}_{Max} \tag{10}$$

$$P^{ESS}(t) \geq P^{ESS}_{Min} \tag{11}$$

The P^{ESS}_{Max}, as already referred, is the maximum discharging power of the storage system, and the P^{ESS}_{Min}, corresponds to the its maximum charging power (negative), or, alternatively, to the energy storage system minimum power.
The renewable curtailment power intervention just makes sense when the energy storage system has no ability to absorb more renewable power generation. Therefore, as presented in (13), its value is never positive.

$$P^{Ctm}(t) \leq 0 \tag{13}$$

As referred above, the storage system power and energy capacity are imposed and act like inputs of the problem. In the restrictions (10) and (11), the energy storage system power limits had been presented, and, next, one presents restrictions related to its stored energy limits.
The stored energy at each moment of the time-series results from the integral of the energy storage system power in the preceding moments, and can be determined by the following equation:

$$E^{ESS}(t) = E^{ESS}(t-1) - \tau \times P^{ESS}(t) \times \left[\delta_1 + (1-\delta_1) \times \frac{1}{\eta_{ESS}} \right] \tag{14}$$

In (14) $E^{ESS}(t)$ is the stored energy in the storage system at moment t, and η_{ESS} corresponds to the energy storage system charge/discharge power efficiency.
The stored energy in the storage system is limited by restrictions (15) and (16).

$$E^{ESS}(t) \leq E^{ESS}_{Max} \tag{15}$$

$$E^{ESS}(t) \geq E^{ESS}_{Min} \tag{16}$$

where E^{ESS}_{Max} and E^{ESS}_{Min} correspond to the maximum and minimum stored energy, respectively.
Imposing that the initial state-of-charge of the energy storage system must be the same as at the end of period T, it comes:

$$\sum_{t=1}^{T} P^{ESS}(t) \times \left[\delta_1 + (1-\delta_1) \times \frac{1}{\eta_{ESS}} \right] = 0 \tag{17}$$

Once considered all the restrictions of the problem, the minimization of the objective function, because of the on-off solution for the thermal power units, can be provided by a mixed integer programming (Gollmer et al., 2000).

4.2 Cost of the energy storage system

The costs associated with the acquisition of an energy storage system are much diversified and dependent on the technology adopted.

The total energy storage system acquisition cost is composed by two different contributions; the contribution associated with peak power capacity of the storage device and the contribution associated to the amount of energy that can be stored. In Figure 12 a comparison of the acquisition costs for the commercialized energy storage technologies is presented.

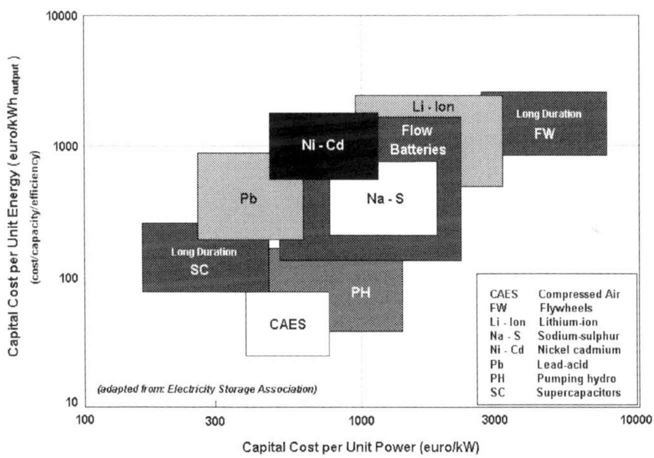

Fig. 12. Capital cost for the acquisition of different energy storage technologies (ESA, 2009)

The long-term discharge technologies applicable at large storage systems, like pumping hydro, compressed air or flow batteries, present a low dependency between the energy storage capacity and the peak power. In those cases, the capital cost for the energy storage device acquisition can be expressed as the sum of the power capital cost with the energy capital cost (Barton & Infield, 2004 and Chacra et al., 2005):

$$C^{ESS} = a_d \times P_{\max}^{ESS} + \beta_d \times E_{\max}^{ESS} \tag{18}$$

where a_d is the capital cost per unit power of the technology d and β_d is the capital cost per unit energy of the technology d.

Variable costs associated to the energy storage system operation and maintenance are not considered in the present work.
Considering that the investment on the energy storage system acquisition can be amortized along its lifetime, it comes (Riggs et al., 1998):

$$C^{ESS} = \sum_{n=1}^{N} \frac{A^{ESS}}{(1+i)^n} \tag{19}$$

where C^{ESS} is the initial investment on the energy storage system acquisition, n is one of the years from the total lifetime N, A^{ESS} is the constant annual amortization of the initial investment and i is the interest rate.
Developing the series (19), comes:

$$C^{ESS} = A^{ESS}\left[(1+i)^{-1} + (1+i)^{-2} + \ldots + (1+i)^{-N+1} + (1+i)^{-N}\right] \tag{20}$$

Manipulating (20), the constant annuity A^{ESS} can be obtained by:

$$A^{ESS} = C^{ESS} \frac{i(1+i)^N}{(1+i)^N - 1} \tag{21}$$

Annuity A^{ESS} should be computed for each energy storage solution achieved and used for its economical evaluation.

5. Location of the Energy Storage Systems in the Power Grid

The implementation of the energy storage systems on the power grid allows a better management of the power flows due to the partial decoupling of the power demand and generation.
Through the energy storage use it is possible to absorb the excess of renewable generation that occurs during the off-peak periods and inject that stored energy into the grid during peak hours, avoiding the operation of the most expensive and pollutant power thermal units.
The carefully chosen location of the energy storage systems in the power grid can avoid non-desired power flows or congestions, improving the flexibility and the efficiency of the power grid.
Following, three different perspectives about the power grid energy storage systems location are presented and discussed.

5.1 Distributed energy storage at generation

The concept of distributed energy storage at generation considers that the energy storage systems are located near the most important non-dispatchable renewable power producers, such as the wind farms. Distributed energy storage at generation is the most intuitive perspective, because it mitigates the power unbalance issue at the original source.
In fact, as far as energy management is concerned, the distributed energy storage at generation perspective is as suitable as any other approach. However, from this energy

storage location, a considerable contribution in terms of global power grid congestions reduction is not expected. In this case, the energy storage systems are located near the renewable power generation and considering that the positive unbalance is prominent during off-peak hours, the stored energy is discharged into the system, leading to an increase of the power flows when the grid is eventually already congested.

One of the advantages of the distributed energy storage at generation is related to the possibility of downsizing the equipment used for the connection of the renewable producers to the power grid. This is due to the contribution of the energy storage system for the modulation of the renewable generation shape, enabling a less variable power output and avoiding the power peaks.

The implementation of distributed energy storage at generation depends on the availability of modular long-term discharge energy storage devices adapted to the specific characteristics of each renewable power plant (Faias et al., 2008).

5.2 Distributed energy storage at demand

Distributed energy storage at demand is a concept that considers the energy storage devices located near the demand, eventually in the interface between the transmission and distribution networks.

Distributed energy storage at demand promotes power flows at off-peak periods and relieves the system from power flows at peak periods. As the energy storage devices are disposed close to the regions where the energy is going to be consumed, the power grid congestions can be avoided, because the power flows between the renewable energy producers and the energy storage system is going to occur mainly during the off-peak periods. During peak hours, when the energy storage system is discharged, the power loads are partially supplied by that power and the needs for high power flows on the transmission network will be reduced.

Like the distributed energy storage at generation concept, the distributed energy storage at demand concept implies the use of modular medium-size long-term discharge energy storage devices.

5.3 Centralized energy storage

The concept of centralized energy storage is based on the idea of installing some (few) large energy storage devices on the power system in order to better manage the power unbalances induced by the increasing integration of the renewable power sources.

It is expected that the energy storage, being centralized in a small number of devices, does not contribute for an improvement on the power flows and congestions for all power grid topologies, because the power flows resulting from the excess of renewable generation will converge to the places where the centralized energy storage devices are located. In addition, during the discharge periods, the power flows on the energy storage devices neighbourhood are also going to rise, increasing the congestions and resulting in a potential negative contribution for the overall power system efficiency.

The main centralized energy storage advantages come from the energy storage technologies that can be applied as far as this concept is concerned. Large-size long-term discharge energy storage devices like pumping-hydro and compressed air are best suited for this kind of centralized application. These technologies present the lower capital costs per power and energy units. Despite the geographical constraints associated to each one of these storage devices, as the number of units to be implemented is small, it is not expected that it would be much difficult to find some location presenting adequate physical conditions for their installation.

6. Case Study: Portuguese Power System at 2011

The case study hereby presented corresponds to the Portuguese power system, which has experienced an increasing integration of renewable power sources during the last years.

The currently installed capacity in Portugal is about 15 GW and comprises coal (12%), natural gas (15%), fuel oil (11%) and diesel fuel utilities (1%), referred as thermal technologies; reservoir (16%) and run-of-river (14%), referred as hydro with more than 10 MW; cogeneration (6%), wind (18%), small hydro that account for less than 10 MW (2%), biomass (2%), municipal solid waste (0.6%) and photovoltaic (0.4%), referred as renewable (REN, 2008).

2008 was a year in which hydro generation presented a value 44% lower than the usual. That is why the renewable generation was only about 27% of the consumed energy. The main contributors for the renewable generation were the hydro power (12%), the wind power (11%) and the biomass (4%). For the year 2010, Portuguese authorities estimate that the renewable generation will correspond to 45% of the consumed energy.

Wind power is the renewable source of energy that has experienced the highest increase in the Portuguese power system. At the end of the year 2008, the wind power installed capacity was about 2800 MW, 11 times higher than at the 2003. According to the national power system operator predictions, by year 2010, wind power will correspond to 25% of the installed capacity.

As already referred in this chapter, such significant share of a variable and non-dispatchable renewable power source could induce some problems related to the power system unbalance. In the sequence, one presents an evaluation of the power unbalance issue in the Portuguese power system, by year 2011, and discusses possible solutions for that problem.

6.1 Power unbalance forecast

The power unbalance forecast intends to evaluate the impact of the increasing integration of renewable power sources, namely wind power, in the power unbalance between generation and demand on the Portuguese power system by year 2011.

The forecast procedure is based on the Portuguese power system historic data of 2007 by power plant and by aggregated demand buses at transmission level.

Forecasts for the demand, renewable generation and interconnections with other power grids use the historic data adjusted to the predicted installed capacity and energy consumption by the year 2011.

Thermal power generation predictions have been computed using the above mentioned (section 4.2) thermal unit commitment methodology, based on the power demand needs and considering the fuel and pollutant emissions costs. Thermal power spinning reserve was determined also through the historic data analysis. An average power grid losses value have been also considered and added to the power demand.

Therefore, considering all conditions referred above, the power unbalance had been forecasted and the obtained results are presented in Figure 13.

Forecast results show that the power generation will overcome the power demand during 25% of the year, corresponding to an energy equivalent to 9.5% of the annual wind power generation. It is also possible to verify that, in selected moments, the power unbalance reaches almost 50% of the annual average load.

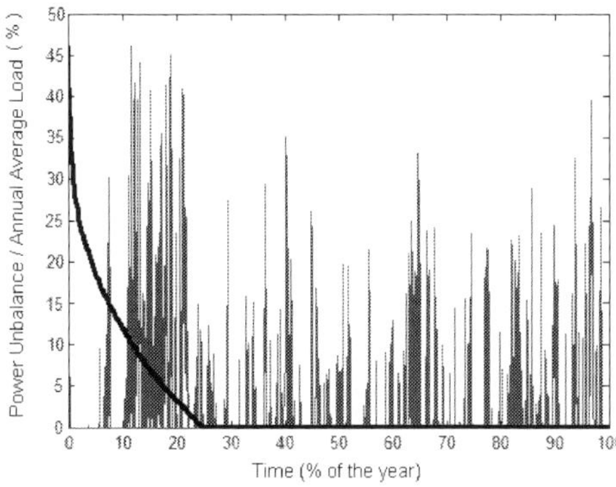

Fig. 13. Time-series and duration curve of the power unbalance forecast (year 2011)

In order to better understand the hourly distribution of the power unbalance along the day, the equivalent average day of the forecasted annual power unbalance time-series was determined. The result is presented in Figure 14 where, as expected, the emphasis goes to the power unbalance prominence during the off-peak hours. This fact results from the time coincidence of the lowest power demand with the highest wind power availability.

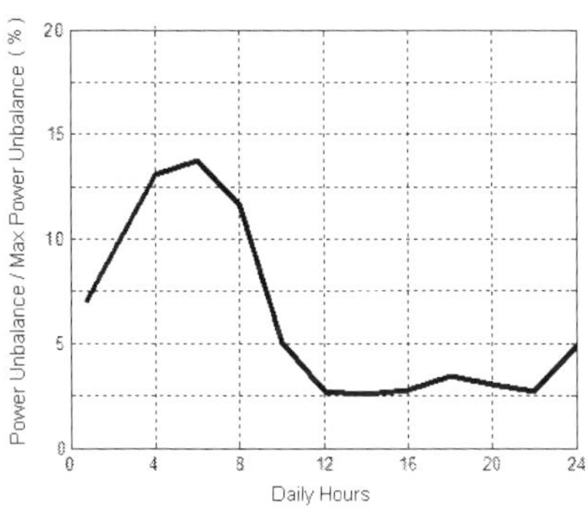

Fig. 14. Equivalent average day of the forecasted annual power unbalance

Regarding the excess of power generation estimated by the forecast and the need to mitigate that power unbalance, two different, but complementary, solutions can be adopted. The first one corresponds to the implementation of an energy storage system able to absorb the excess of generation, and the second one, the curtailment of the wind power generation.
Taking into account that the renewable power generation should not be wasted because of its several benefits, the wind power curtailment must be assumed as the last option.

The use of the interconnections with the Spanish power system is not considered as a solution to mitigate the power unbalance. Spanish power system also presents an important share of renewable energy, namely wind power, and probably the power unbalance issue affects similarly both power systems. Moreover, historic interconnection data shows that during the off-peak periods, when the forecasted power unbalance tends to be more intense, the fluxes of energy are typically from Spain to Portugal.

6.2 Energy storage system design

The energy storage system design follows the procedure previously described in section 4. Different power and energy combinations for the energy storage system are introduced in the unit commitment problem and their feasibility is evaluated. The economical benefit that result from the cost reduction of the fuel consumption and greenhouse gases emissions due to the lower renewable power curtailment are also computed during that procedure. Results are presented in Figure 15.

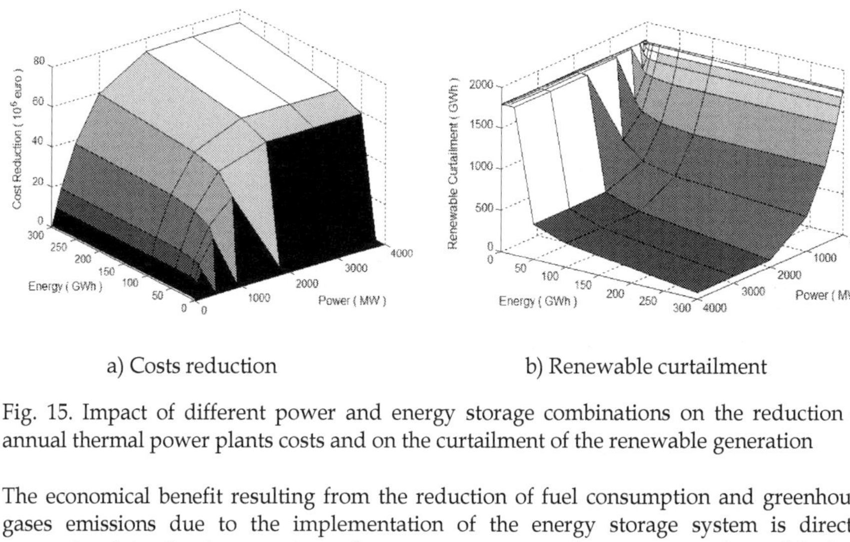

a) Costs reduction b) Renewable curtailment

Fig. 15. Impact of different power and energy storage combinations on the reduction of annual thermal power plants costs and on the curtailment of the renewable generation

The economical benefit resulting from the reduction of fuel consumption and greenhouse gases emissions due to the implementation of the energy storage system is directly proportional to the increment on the storage power and energy capacity, while it is inversely proportional to the need of renewable generation curtailment. The energy storage system able to completely match the power unbalance would need a global capacity of 3270 MW and 395 GWh.

Together with the thermal power plants economical benefits induced by the energy storage system, also the capital costs associated with the energy storage system acquisition must be considered. Therefore, the capital costs of the different technologies are deduced to the benefits that result from their implementation.

The net economic benefit was computed for the long-term discharge energy storage technologies like compressed air, pumping hydro and redox-flow batteries, assuming the interval between the maximum and minimum capital cost of the energy storage system acquisition (Figure 12) and an interest rate of 5%. The lifetime assumed for each technology was respectively 30 years, 40 years and 15 years (Schoenung et al., 1996) (EPRI, 2007).

Hydrogen fuel cell energy storage technology was not considered in this procedure due to its low charge/discharge efficiency.
From the three technologies considered, it was concluded that only compressed air energy storage presents some positive net economic benefits, when assuming the minimal value of the acquisition cost interval.
The best economic result corresponds to an annual benefit of 1.2 million euros for a 100 MW and 1500 MWh storage system. Despite the contribution of this storage system to the power unbalance mitigation, its small capacity forces a renewable generation curtailment equivalent to 8.6% of the annual wind power generation.
The solution that presents the second best economic result corresponds to a 50 MW and 500 MWh energy storage system and allows an annual benefit of 690000 euros. The renewable generation curtailment corresponding to this system is equivalent to 9% of the annual wind power generation.
An energy storage system of 200 MW and 5000 MWh capacity presents the third best economic result, allowing a benefit of 570000 euros and a renewable generation curtailment equivalent to 7.7% of the annual wind power generation. This system, from all the energy storage solutions with positive economic results, is the one that better absorb the power unbalance.
The 200 MW and 5000 MWh energy storage system operation along the period of one typical week, and its contribution for balancing the power generation and demand is presented in Figure 16.

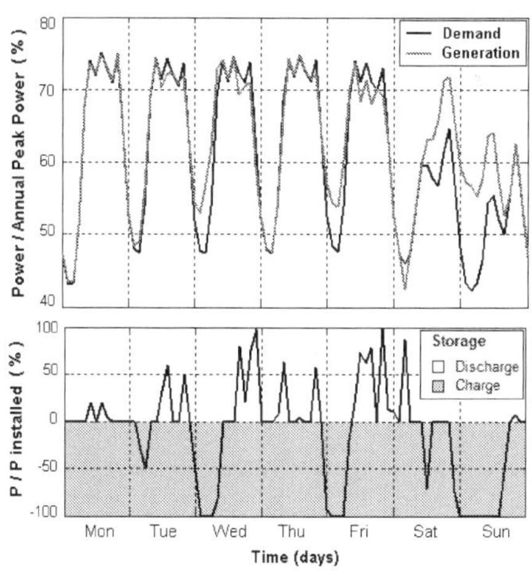

Fig. 16. Energy storage system (200 MW and 5000 MWh) operation and its contribution for balancing the power generation and demand along the period of one week

Through the analysis of the figure, it is possible to identify, during the off-peak hours, the moments when the energy storage is charging, in order to absorb the excess of power generation. Also the energy storage system discharging moments are easily identified, during the peak hours, when the energy storage system is contributing to avoid additional thermal power generation.

The results obtained from the energy storage system design procedure, applied to the Portuguese case study, lead to the conclusion that the economically viable solutions only enable the integration of a small percentage of the power unbalance induced by the renewable generation. However, it should be emphasized that assuming higher fuel and greenhouse gases emissions costs, as well as lower storage systems capital costs, the economic viability of larger integration of the power unbalance would be improved.

6.3 Power system with integrated pumping hydro

Some power systems with large amount of hydro installed capacity, such as the Portuguese system, comprise reversible hydro power plants with energy storage capacity that are usually used for the energy management improvement.

In the present scenario it is assumed that all of the pumping hydro installed capacity is available for helping the integration of the renewable power generation (in opposition to the previous scenario where the pumping hydro system was operated according to the historic data).

The estimated Portuguese pumping hydro capacity corresponds to 1051 MW and 25.2 GWh, and as far as the power unbalance forecast evaluation is concerned, it had been introduced in unit commitment problem as an additional energy storage system. As the Portuguese pumping hydro power plants are integrated in river flows, besides the usual constraints considered for all energy storage systems, an additional constraint was introduced to the pumping hydro operation, in order to consider the forced power discharges that result from the predicted up-side reservoir water inflows.

Therefore, the power unbalance forecast procedure had been applied assuming that the pumping hydro system is completely available to cooperate in the integration of the renewable generation. As expected, that availability results in a lower power unbalance, reducing the renewable curtailment need for a value equivalent to 4% of the annual wind power generation.

In order to evaluate the viability of an additional energy storage capacity for reducing the need for renewable power curtailment, the energy storage design process was again applied. Results show that, in this scenario, there is only one economically viable power and energy storage capacity combination. The energy storage system corresponds to a single compressed air energy storage unit with 10 MW, and 100 MWh. The annual net economic benefit of the energy storage system implementation is 780000 euros, and it is only viable assuming the minimum capital costs for the compressed air energy storage technology. The benefit introduced by this storage system, in terms of the integration of renewable power generation, is marginal. The renewable power generation curtailment, before the implementation of the energy storage system, was equivalent to 4% of the annual wind power generation, and after that it was 3.9%.

One of the objectives of the design process economical evaluation is to identify a particular energy storage system capacity and technology, however, the specificities of that technology should be carefully considered in order to validate its technical feasibility.

In Figure 17, where contribution of the two different energy storage systems for the power demand and generation balancing is presented, it is possible to follow the desired response for the compressed air energy storage system, referred as the other storage. Remembering that the compressed energy storage system operation is based on a gas turbine, difficultly it will respond to such demanding operation conditions.

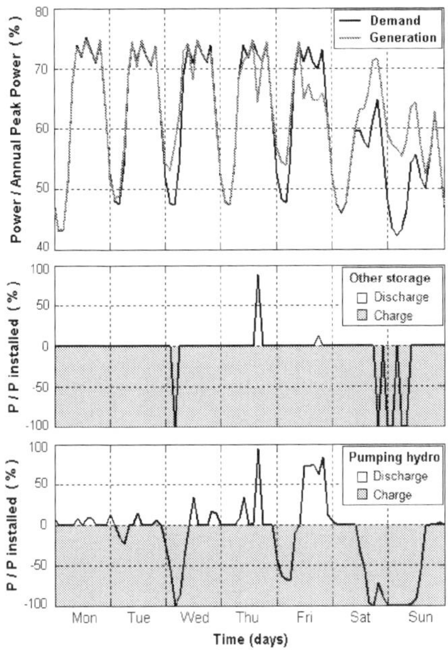

Fig. 17. Pumping hydro and the designed energy storage system operation and their contribution for balancing the power generation and demand along the period of one week

7. Conclusion

In the last years the world has assisted to an impressing development of the renewable energies, driven by the need for reducing negative environmental impacts that resulted from an intense exploitation of fossil fuels. Besides the environmental concern, as the renewable energy sources are endogenous to many countries, also the security of supply has been one of the key factors responsible for that development of the renewable energy.

The electricity generation sector, in particular, is the one that more intensely has experienced the replacement of fossil fuel based technologies by technologies that make use of renewable energy sources. As a result of that change, new challenges and issues have emerged on the operation of the power systems.

As most of renewable technologies ultimately derive energy from unpredictable natural sources or sources that vary in their availability over different timescales, balancing power generation and demand is one of the issues that is raised with an increasing integration of renewable energy in the power system. Balancing power generation and demand is an issue that occurs especially during the off-peak periods, when typically the power demand is lower and the non-dispatchable renewable power can become higher than the demand.

There are three different approaches to mitigate the power unbalance resulting from an excess of renewable power generation; curtailment of the renewable generation, interconnections with other grids and the energy storage. However, in order to avoid the waste of renewable generation and considering that different interconnected power systems can be simultaneously with similar difficulties for balancing their systems, the use of an energy storage system seems to be the best suited solution.

At the moment, there are several energy storage technologies available for helping the integration of renewable energy sources in the power system. From those technologies one distinguishes among the pumping hydro energy storage, the compressed air energy storage and redox-flow batteries. Different technologies correspond to different power and energy capacity ranges and to different capital costs. Pumping hydro and compressed air energy storage systems present the larger power and energy capacity ranges and the lower investment costs. However, there are some restrictions to their implementation due to the need for specific geographical conditions. The modularity of the redox-flow batteries put great expectative in its use, nevertheless their investment costs make them few competitive at present.

An energy storage system design methodology had been developed along the present chapter. The main goal of this design methodology was to choose an energy storage system able to integrate as much renewable generation as possible in the power system, taking into account the trade off between economic benefits and its capital costs.

The first phase of the energy storage system design procedure consists of the evaluation of the economic benefit of the lower thermal power plants generation that results from the operation of several power and energy capacity storage combinations. During the second phase, the capital costs of the several power and energy capacity of the storage systems are computed and the best solutions are identified.

In the case study presented for the Portuguese system it is showed that the implementation of the energy storage equipment in power systems with high penetration of renewable power sources, improves the power demand and generation balance and reduces the need for renewable generation curtailment.

Some power systems with large amount of hydro installed capacity, such as the Portuguese system, comprise reversible hydro power plants with energy storage capacity, usually used for the energy management improvement. Therefore, two opposite scenarios were considered: the first scenario is related to the pumping hydro system operation accordingly to the historic data; in the second scenario it was assumed that all of the pumping hydro installed capacity was available for helping the integration of the renewable power generation. As far the energy storage systems implementation is concerned, those scenarios allowed the identification of their maximum and minimum capacity requirements.

The availability of the installed pumping hydro systems, to cooperate in the integration of the renewable generation appears as a very positive solution for the power unbalance mitigation, especially when considering the costs of completely new energy storage systems implementation.

The energy storage system design process needs then to consider the physical constraints of power grid. Therefore, in order to validate the results from that process, namely the power flows imposed by the unit commitment problem outputs, a power system simulator was used (PSS/E - Siemens) for the simulation of the Portuguese transmission network model.

All the solutions were validated using the centralized energy storage concept. The solution that does not consider the pumping hydro availability had been validated also with the distributed energy storage at demand and the distributed energy storage at generation concepts. Regarding to the power grid losses, in this specific case study, the three different energy storage locations presented very similar results.

8. References

Agbossou, K.; Kolhe, M.; Hamelin, J. & Bose, T. K. (2004). Performance of a Stand-Alone Renewable Energy System Based on Energy Storage as Hydrogen. *IEEE Transactions on Energy Conversion,* Vol. 19, No. 3, (Sept. 2004), page numbers (633-640), ISSN 0885-8969

Barton, J. P. & Infield, D. G. (2004). Energy Storage and Its Use with Intermittent Renewable Energy. *IEEE Transactions on Energy Conversion,* Vol. 19, No. 2, (June 2004), page numbers (441-448), ISSN 0885-8969

Chacra, F.; Bastard, P.; Fleury, G. & Clavreul, R. (2005). Impact of Energy Storage Costs on Economical Performance in a Distribution Substation. *IEEE Transactions on Power Systems,* Vol. 20, No. 2, (May 2005), page numbers (684-691), ISSN 0885-8950

EPRI (2007). *Vanadium Redox Flow Batteries: An In-Depth Analysis.* EPRI, Palo Alto, CA: 2007. 1014836

ESA (2009). Technology Comparisons, Electricity Storage Association (Online), Available: http://www.electricitystorage.org

Faias, S.; Sousa, J.; Castro, R. (2007). Contribution of energy storage systems for power generation and demand balancing with increasing integration of renewable sources: application to the Portuguese power system, *Proceedings of the 12th European Conference on Power Electronics and Applications, EPE 2007,* pp. 1-10, Vol. 1, ISBN 978-92-75815-10-8, Aalborg, September 2007

Faias, S.; Santos, P.; Matos, F.; Sousa, J.; Castro, R. (2008). Evaluation of energy storage devices for renewable energies integration: Application to a Portuguese wind farm, *Proceedings of the 5th International Conference on the European Electricity Market, EEM08,* pp. 1-7, Vol. 1, ISBN 978-1-4244-1743-8, Lisbon, May 2008

Gollmer, R.; Nowak, M.; Römisch, W. & Schultz, R. (2000). Unit commitment in power generation - a basic model and some extensions. *Annals of Operations Research,* Vol. 96, No. 1-4, (November 2000) page numbers (167-189), ISSN 0254-5330

Hammons, T. J. (2006). Integrating Renewable Energy Sources into European Grids, *Proceedings of the 41st International Universities Power Engineering Conference,* pp. 142-151, Vol. 1, ISBN 978-186135-342-9, Newcastle upon Tyne, September 2006

Hamsic, N.; Schmelter, A.; Mohd, A.; Ortjohann, E.; Schultze, E.; Tuckey, A. & Zimmermann, J. (2007). Increasing Renewable Energy Penetration in Isolated Grids Using a Flywheel Energy Storage System, *Proceedings of Powereng 2007 - International Conference on Power Engineering, Energy and Electrical Drives,* pp. 195-200, ISBN 978-1-4244-0895-5, Setúbal - Portugal, April 2007

Hoogers, G. (2003). *Fuel Cell Technology Handbook,* CRC Press, ISBN 0-84-930877-1

Hsu, C. & Lee, W. (1992). Superconducting Magnetic Energy Storage for Power System Applications. *IEEE Transactions on Industrial Applications,* Vol. 29, No. 5, (Sept.-Oct. 1993), page numbers (990-996), ISSN 0093-9994

Hunt, G. L. (1998). The Great Battery Search. *Spectrum IEEE,* Vol. 35, No. 11, (Nov. 1998), page numbers (21-28), ISSN 0018-9235

IEA (2005). *Variability of Wind Power and other Renewables: Management Options and Strategies.* International Energy Agency Publications, Paris

Lazarewicz, M. & Rojas, A. (2004). Grid Frequency Regulation by Recycling Electrical Energy in Flywheels, *Proceedings of IEEE-Power Engineering Society General Meeting,* Vol.2, pp. 2038- 2042, ISBN 0-7803-8465-2, Denver, CO, June 2004

Lerch, E. (2007). Storage of Fluctuating Wind Energy. *Proceedings of the 12th European Conference on Power Electronics and Applications, EPE 2007*, pp. 1-8, ISBN 978-92-75815-10-8, Aalborg, Denmark, September 2007

Mito, T.; Kawagoe, A.; Chikaraishi, H.; Okumura, K.; Abe, R.; Baba, T.; Yamauchi, K.; Yokota, M.; Henmi, T.; Seo, K.; Hayashi, K.; Iwakuma, M. & Sumiyoshi, F. (2004). Development of UPS-SMES as a Protection From Momentary Voltage Drop. *IEEE Transactions on Applied Superconductivity*, Vol. 14, No. 2, (June 2004) page numbers (721-726), ISSN 1051-8223

Nomura, S.; Tsutsui, H.; Tsuji-Iio, S. & Shimada, R. (2006). Flexible Power Interconnection With SMES. *IEEE Transactions on Applied Superconductivity*, Vol. 16, No. 2, (June 2006) page numbers (616-619), ISSN 1051-8223

Ogden, J. M. (1999). Prospects for Building a Hydrogen Infrastructure. *Annual Review of Energy and the Environment*, Vol. 24, No. 1, page numbers (227-279), ISSN 1056-3466

Ortega-Vazquez, M.; Kirschen, D. (2009). Estimating the Spinning Reserve Requirments in Systems with Significant Wind Power Generation Penetration. *IEEE Transactions on Power Systems*, Vol. 24, No. 1, (Feruary 2009), page numbers (114-124), ISSN 0885-8950

Ponce de León, C.; Ferrer, A.; García, J. & Walsh, F. (2006). Redox Flow Cells for Energy Conversion. *Journal of Power Sources*, Vol. 160, No. 1, (Sep. 2006), page numbers (716-732), ISSN 0378-7753

Price, A.; Bartley, S.; Male, S. & Cooley, G. (1999). A Novel Approach to Utility Scale Energy Storage. *Power Engineering Journal*, Vol. 13, No. 3, (June 1999) page numbers (122-129), ISSN 0950-3366

REN (2008). Electricity - Real Time Information: Technical Data, REN (Online), Available: http://www.centrodeinformacao.ren.pt/EN

Riggs, J.; Bedworth, D. & Randhawa, S. (1998). *Engineering Economics – Fourth Edition*, McGraw-Hill International Publications, ISBN 0-07-115526-0

Rufer, A.; Hotellier, D. & Barrade, P. (2004). A Supercapacitor-Based Energy Storage Substation for Voltage Compensation in Weak Transportation Networks. *IEEE Transactions on Power Delivery*, Vol. 19, No. 2, (April 2004) page numbers (629-636), ISSN 0885-8977

Schoenung, S.; Eyer, J.; Iannucci, J. & Horgan, S. (1996). Energy Storage for a Competitive Power Market. *Annual Review of Energy and the Environment*, Vol. 21, No. 1, (November 1996) page numbers (347-370), ISSN 1056-3466

Tande, J. O. G. (2003). Grid Integration of Wind Farms. *Wind Energy*, Vol. 6, No. 3, (July/September 2003) page numbers (281-295), ISSN 1095-4244

Wang, C. & Shahidehpour, S. M. (1993). Effects of Ramp-Rate Limits on Unit Commitment and Economic Dispatch. *IEEE Transactions on Power Systems*, Vol. 8, No. 3, (August 1993) page numbers (1341-1350), ISSN 0885-8950

Zhai, N.; Yao, Y.; Zhang, D. & Xu, D. (2006). Design and Optimization for a Supercapacitor Application System, *Proceedings of International Conference on Power System Technology*, pp. 1-4, ISBN 1-4244-0110-0, Chongqing, October 2006

6

Single-Phase Grid Connected Converters for Photovoltaic Plants

Emilio Lorenzani[1], Giovanni Franceschini[1], Alberto Bellini[2] and Carla Tassoni[1]
[1]*University of Parma, Italy*
[2]*University of Modena and Reggio Emilia, Italy*

1. Introduction

Among renewable energy sources photovoltaic systems are one of the most up to date solutions. However their diffusion is limited by relatively high cost in comparison with traditional energy sources. The downward tendency in the price of the PV (Photovoltaic) modules, together with their increasing efficiency, put solid-state inverters under the spot lights as the enabling technology for integrating PV systems into the grid.

This chapter will analyze the single-phase grid connected converter usually used in PV domestic rooftop applications.

There are two mandatory tasks in grid connected PV systems: the maximization of the energy extracted from the PV panels (1) and the use of a high efficiency topology for the power converter able to inject only active current into the grid (2), i.e. a pure sinusoidal current in phase with the grid voltage.

The first task will not be discussed in depth in this chapter and will be mentioned in section 2. The second task is accomplished by using a PLL (Phase Looked Loop) for the grid synchronisation (Chung, 2000), (Arruda et al., 2001), (Silva et al., 2004) and by designing a suitable current controller.

In three-phase converters current controllers in a rotating reference frame (d-q frame) are used that provide zero steady-state error and superior disturbance rejection. In single-phase converters achieving zero steady state error at grid frequency is not a simple task. As known, the poor performances of the integral action at frequencies different from zero leads to steady state error and to poor disturbance rejection, making the PI controller not suitable to track a sinusoidal current set point. Interesting solutions were presented in (Yuan et al., 2002), (Teodorescu et al., 2004). In these papers the authors develop the P+resonant (PR) controller for reference tracking in the stationary frame.

The use of control system in synchronous reference frame (d-q control), also for single-phase converters, was proposed for stand alone (Ryan & Lorenz, 1997), (Miranda et al., 2005) and for grid connected operations (Cacciato et al., 2008), (Bellini et al., 2008). In section 4 the d-q control is presented and the experimental results will show its nice performances.

Section 3 presents the topologies of PV systems, past and present technology, and the possible evolutions oriented towards the increase of the energy extracted by each PV module.

A very common topology for single-phase converters is the full-bridge with a line frequency transformer on the AC-side for galvanic isolation purposes. In this configuration the current injected into the grid is the transformer output current. Adopting this solution the DC current component at the transformer input is not directly controlled, and core saturation may appear producing current distortion. A simple solution to avoid core saturation issues is to control the current at transformer input instead of the injected grid current (Ciobotaru et al., 2005). However, this solution leads to lower power factor because of the presence of the transformer magnetizing current, especially at light load conditions. On the other hand, the direct control of the grid current requires a solution to saturation problems without introducing input transformer current measurement. In section 5 an original solution based only on injected current measurement is presented (Bellini et al. 2008). Experimental results show the effectiveness of the proposed method.

In new converter designs the tendency is to abandon line frequency transformers because of size, weight and price in favour of high frequency transformers. The presence of high frequency transformer requires several power stages and, as a consequence, increasing efficiency and reducing cost may be a hard task. In low power PV applications it is possible to remove the transformer at all in order to reduce losses, costs and size. The latter configurations are usually referred to as transformerless power converters. Without transformer coupling galvanic isolation between the grid and the DC source is lost and the big parasitic capacitance of the PV panel (Calais & Agelidis, 1998), (Calais et al, 2001) may introduce ground leakage currents.

Ground leakage currents increase conducted and radiated electromagnetic emissions, harmonics injected in the grid, losses and the converter has to be disconnected from the grid if the ground leakage currents exceed prefixed limits.

Section 6 reports the root causes of the ground leakage currents and some solutions. The best way to reduce them is the use of a suitable converter topology. Inside this section an original full bridge topology with unipolar PWM is presented that ensures a strong reduction of the leakage ground currents. Simulation results are reported that show the effectiveness of the proposed topology.

2. Maximum power exploitation of PV cells

The power generation from PV modules suffers from two main shortcomings: the efficiency of electric power generation is very low, especially under low radiation states (1), and the amount of electric power generated by solar arrays is always changing with weather conditions (2). Figure 1 shows the typical variations of the current voltage characteristic of a PV module respectively for different values of irradiance and working temperature. Figure 2 shows the equivalent circuit of a PV cell. A maximum power point tracking (MPPT) algorithm must operate in real-time in PV generation systems. The ideal MPPT algorithm sets the operating point of the PV panel, usually forcing the panel voltage at the Maximum Power Point (MPP). Figure 3 shows an example of the power-voltage characteristic of a photovoltaic field, composed by PV modules connected in series and in parallel for different values of irradiance and partial shading conditions.

Most of the MPPT algorithms are based on perturb and observe and on incremental conductance methods. Many improvements to these methods have been presented in order to improve the dynamic performances and to reduce undesired oscillations around the

MPP, (Femia et al., 2005), (Fangrui et al, 2008). In high power PV systems the ripple correlation control method can be used that is based on measuring and processing the current and/or voltage ripple created by the switching stage of the power converter (Esram et al., 2006).

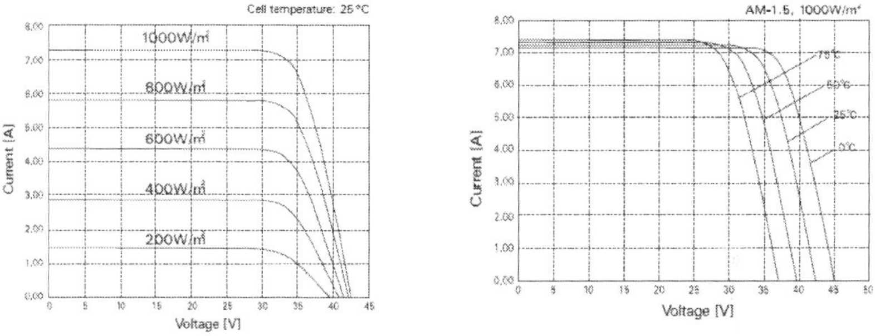

(a) (b)

Fig. 1. Electric characteristics of PV panels (Sanyo Photovoltaic HIP-230HDE1) for different values of irradiance and working temperature.

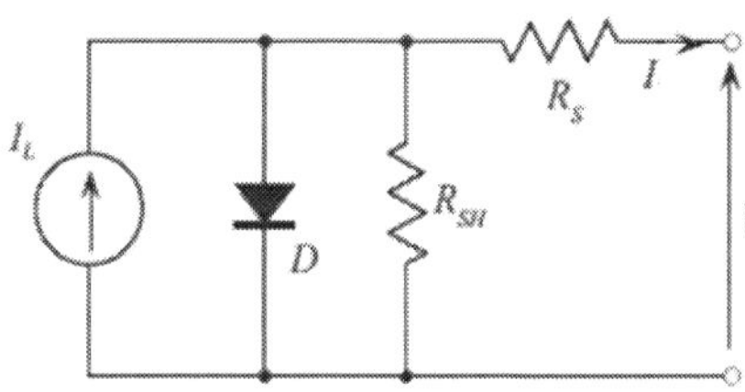

Fig. 2. Equivalent circuit of a PV cell.

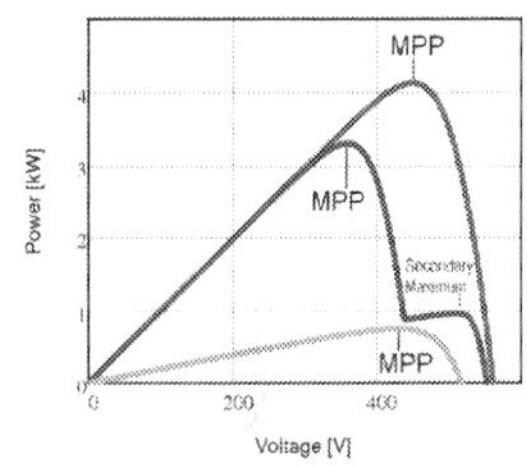

Fig. 3. Example of power-voltage characteristic of a photovoltaic field for different weather conditions.

In order to harvest the maximum energy the ripple at the terminals of the PV modules, must be minimized around the MPP. In single-phase converters since the instantaneous power does not equal the average power, there is an intrinsic fluctuation of the PV module output power at twice the frequency of the grid.

The reduction of the output ripple of the PV modules can be obtained only with a power decoupling device that is normally realized by means of large DC capacitors which decrease the lifetime and reliability of the whole system. The DC capacitors are either placed in parallel to the PV modules or to the dc link after a DC-DC converter used for voltage amplification.

In a three-phase system the instantaneous output power is constant and no DC large capacitors are needed leading to smaller cost and higher reliability with respect to single-phase inverters. The use of three-phase system also in low power applications could be an interesting solution for newer designs.

3. Topologies for grid connected photovoltaic systems

The basic elements of a PV system are the modules that are usually series-connected. A series of PV modules is usually called a PV string. If the voltage of the PV string is always higher than the peak voltage of the grid the PV converter does not require a step-up stage. In this case higher efficiency can be obtained because a single stage full-bridge converter can be used. Otherwise, a DC-DC converter or a transformer must be added for voltage amplification reducing efficiency. A PV system is the combination of PV fields and the related power converters.

The peak current that can be delivered by one string is determined by the PV module characteristics, figure 1. To achieve a higher power level several strings can be connected in parallel as shown in figure 4a. In this way, a single converter can be used reducing the cost and the losses of the static energy conversion. In this topology, usually referred to as Central Converter, the lack of individual MPPT for each string does not permit to harvest the maximum electric power from PV modules, especially when shading or different orientation of modules occurs. This major shortcoming results in avoiding this simple topology in newer photovoltaic system designs.

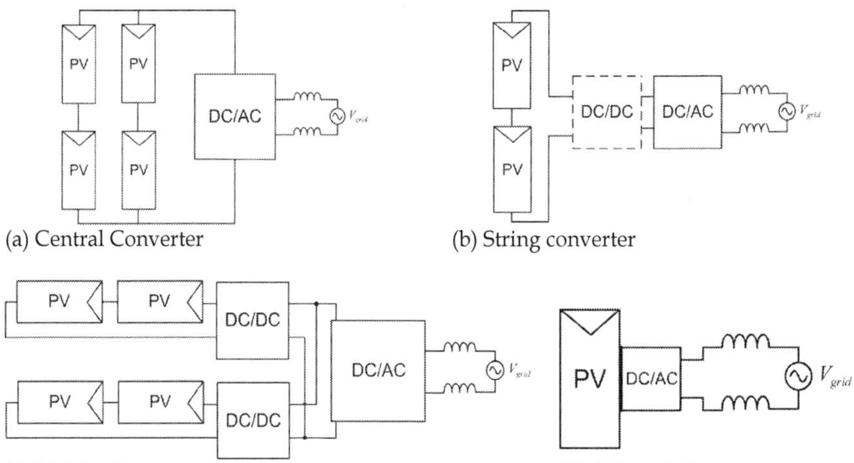

(a) Central Converter (b) String converter

(c) Multi-string converter (d) AC module

Fig. 4. Configurations of the power converter for a grid connected PV System, central converter (a), string converter (b), multi-string converter (c) and AC module (d).

Other options are possible as sketched in Figure 4. The string converter topology is shown in figure 4(b). This configuration does not employ the parallel connections of the strings. Each string has its own MPPT and is completely independent from each other. Therefore it is easy to build PV systems with different orientations, shading conditions and number of PV modules for each string.

A disadvantage of string-converters in comparison to central converters is the higher price per kW. String converters are often build only as single-phase converters due to the low power level. A very common classic topology is the full-bridge with a line frequency transformer on the AC-side for galvanic isolation and for voltage step-up. This architecture is analyzed in section 5.

The multi-string converter (figure 4(c)) manages two or three strings, and provides independent MPPT by different DC-DC converters. In this case a two-stage configuration is mandatory. Thanks to these additional DC-DC stages, used also in some string converter, it is possible to obtain a very wide input voltage range which gives to the user big freedom in designing of the photovoltaic field.

In these topologies the modules of one string have to be well matched and should be installed in the same orientation to achieve a high energy harvest. The photovoltaic energy harvesting can be maximized by using an individual MPPT for each PV module. The low power level permits to integrate the converter into the housing of the PV module, that is called AC module (figure 4(d)). The AC module is connected directly to the grid and no DC wirings are needed between PV modules.

Despite their simple use and installation the low power level of AC modules leads to higher cost per watt. The major issue of this solution is the lifetime of the actual converters that is smaller than the lifetime of a PV module (20 years and more). When it will be comparable this solution will become interesting.

4. Current Controllers for Single-Phase Grid Connected Converters

4.1 d-q current controller

The DC-AC stage of a PV system must inject active grid current only, i.e. a pure sinusoidal current in phase with the grid voltage. To satisfy this condition the steady state error between the desired grid current and the actual one must be tightly zero at grid frequency. To this aim many control architectures have been presented in literature to overcome the drawbacks of the simple PI controller. As known, the poor performances of the integral action at frequency different from zero lead to steady state error and to poor disturbance rejection making the PI controller not suitable to track a sinusoidal set point in a stationary reference frame. The performances of PI controllers can be improved adding a feed-forward compensation to its output. The feed-forward value is computed in order to obtain at the output of the power converter the same voltage imposed by the grid in case of a zero contribution of the PI controller. The PI + feed-forward controller is used in some experimental results shown in section 6.

An interesting alternative is the P+resonant (PR) controller that introduces an infinite gain at a selected resonant frequency providing a theoretical zero steady-state error at that frequency. Actually the gain at resonant frequency is limited by stability problems.

A further option is the use of a control system in a reference frame synchronous with the grid frequency (d-q reference frame). The d-q control allows an infinite control loop gain at grid frequency and a superior disturbance rejection. Hence, it can increase efficiency cancelling reactive current delivered to the grid.

Mimicking the technique used in case of the single-phase transport delay PLL (Arruda et al, 2001), (Silva et al., 2004) to obtain the d-q current components in synchronous reference a current $i_\beta(t) = I\sin(\omega t + \varphi_i)$, orthogonal to the grid current $i_\alpha(t) = I\cos(\omega t + \varphi_i)$, is introduced. Appling the Park transformation I_d and I_q can be easily computed.

Let $v_\alpha(t) = V\cos(\omega t + \varphi_v)$ and $i_\alpha(t) = I\cos(\omega t + \varphi_i)$ be the grid voltage and the grid current, where ω is the grid pulsation, φ_v and φ_i are respectively the voltage and the current displacement, it holds:

$$\begin{bmatrix} I_d \\ I_q \end{bmatrix} = \begin{bmatrix} \cos\vartheta(t) & \sin\vartheta(t) \\ -\sin\vartheta(t) & \cos\vartheta(t) \end{bmatrix} \begin{bmatrix} i_\alpha \\ i_\beta \end{bmatrix} \tag{1}$$

where the angle $\vartheta(t) = \omega t + \varphi_v$ is obtained once the PLL is locked to the grid voltage angle. Matrix computations lead to the following results:

$$\begin{aligned} I_d &= I\cos(\varphi_v - \varphi_i) = I\cos\varphi \\ I_q &= -I\sin(\varphi_v - \varphi_i) = -I\sin\varphi \end{aligned} \tag{2}$$

Hence, I_d and $-I_q$ are respectively the amplitude of the active grid current, in phase with the grid voltage, and the amplitude of the reactive grid current, in quadrature with the grid voltage; $\cos\varphi$ is the power factor.

Being the system a single-phase one, only one manipulated variable can be managed. This variable, named in Figure 5 V_β, depends, generally speaking, on park transformation reference choice, and drives the single-phase DC/AC converter in order to obtain the desired grid current. V_β is a output of the Park inverse transformation, whose inputs are: 1) the V_d voltage, i.e. the output of the active current (I_d) control loop, 2) the V_q voltage, i.e. the output of reactive current ($-I_q$) control loop. The set-point of the reactive current control loop is set to zero because in ideal conditions only active current must be supplied.

A simpler choice would require only one current loop, fixing V_q=0. However, if the I_q control loop is opened, the presence of reactive voltage drops and system non ideal behavior will introduce a reactive current component and then a wrong current angle, as it will be shown in the following by simulation.

The right angle is achieved only if the reactive current loop is closed with a zero set-point. In fact, in ideal conditions, only the I_d current component must appear and the actual I_q current represents a measure of the system error.

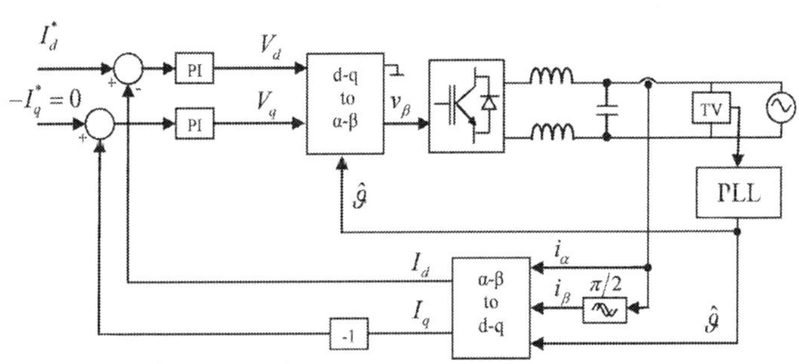

Fig. 5. Block diagram of d-q control structure.

4.2 Simulation Results

Numerical simulations were used to assess the performances of the proposed d-q control structure. The MATLAB simulink environment was used to model the whole system, where the power converter is modeled by PLECS toolbox. A fixed step time of 5 ms was used and all simulations refer to a step set-point of Id with an amplitude of 10A.

At first some simulations were made to verify the system performances in case of I_q current loop opened and V_q set to zero. In this case, the presence of the I_d control loop will fix the active grid current at the right value. However, because of reactive voltage drops and system non ideal behavior a reactive current component will appear and the grid current amplitude will increase beyond the desired level. Specifically, the synchronous reference frame control forces a zero steady state error in d-axis but this result is reached increasing the grid current amplitude instead of adjusting its angle.

Figure 6 shows the grid voltage and current under these conditions: the amplitude of the grid current is higher than requested because a remarkable amount of reactive current is injected into the grid. In these charts the grid current amplitude was amplified (x10) in order to identify easily the current displacement. The resulting power factor and then system performances are very poor.

Then simulations were made relying on the structure of figure 5. Thanks to the action of the I_q axis control loop the reactive current is forced to zero by integral action. Figure 7 shows clearly that only active current is now present. A unitary power factor is obtained and the amplitude of the grid current reaches its set-point (10A).

Figure 8 highlights the dynamic response of I_d and I_q: in a few periods d-q current control ensures zero steady state error as expected.

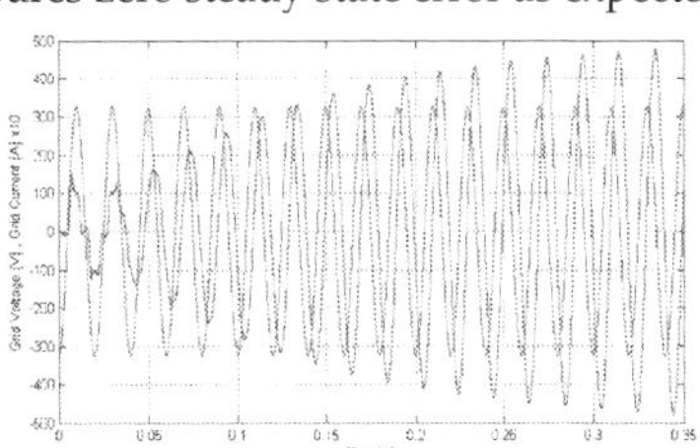

Fig. 6 Simulation results with simple control structure (V_q=0). Grid voltage (violet trace) and amplified (x10) grid current (red one).

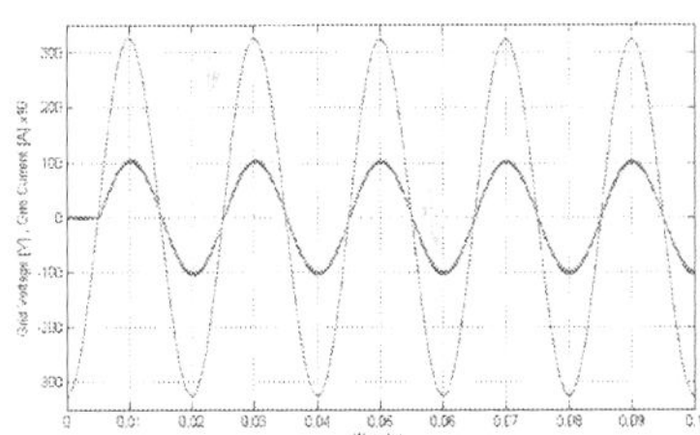

Fig. 7 Simulation results with d-q control. Grid voltage (violet trace) and amplified (x10) grid current (red one).

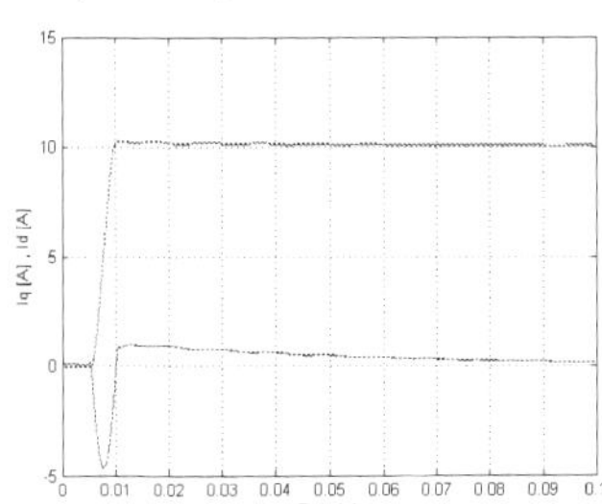

Fig. 8 Simulation results with d-q control. Dynamic response of I_d (green trace) and I_q (blue one).

4.3 Experimental Results

Several experiments were made in order to evaluate the performances of the d-q control. The designed prototype is based on a 56F8323 DSP that implements the control algorithm, the digital filtering, the grid synchronization, the reference frame transformation and the PWM modulation. Specifically a unipolar PWM modulation is adopted with a switching frequency of 10 kHz. The power stage is implemented on a 600 V, 25 A Intelligent Power Module (IPM). Using unipolar modulation a simple LC filter can be used with L = 2 mH, and C = 1.5 μF. The detailed design and optimization of the output filter is not investigated in this paper. A 400V low distortion DC voltage source was adopted.

Figure 9 and Figure 10 show the performances of the d-q control architecture. Figure 9 refers to a step set-point of I_d from 0 to 8A while the set point of I_q is fixed to zero. Under these conditions, the set-point is reached in about 10 ms, i.e. less than a period. The dynamic behavior of the currents is in a nice agreement with simulation results of Figure 8 and, as theoretically predicted, the d-q control keeps the current tightly at its set-point.

Figure 10 shows the steady state behaviour of the d-q control for a set-point of I_d = 7A. The injected grid current shows only a low distortion content in correspondence of voltage zero crossing.

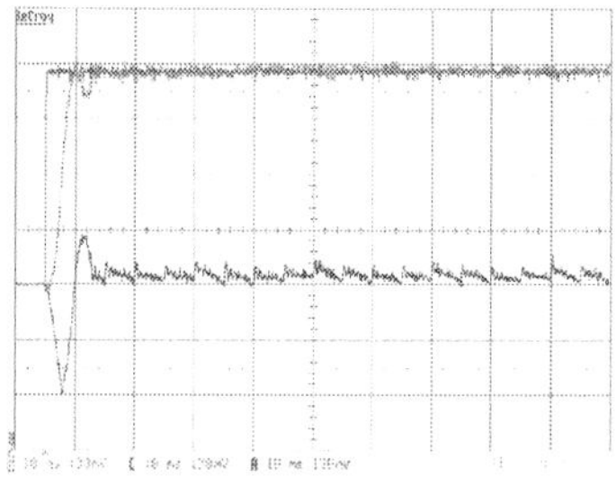

Fig. 9. Experimental Results. Step response: I_d (green solid line), I_q (blue solid line) and I_d^* set-point (red solid line).

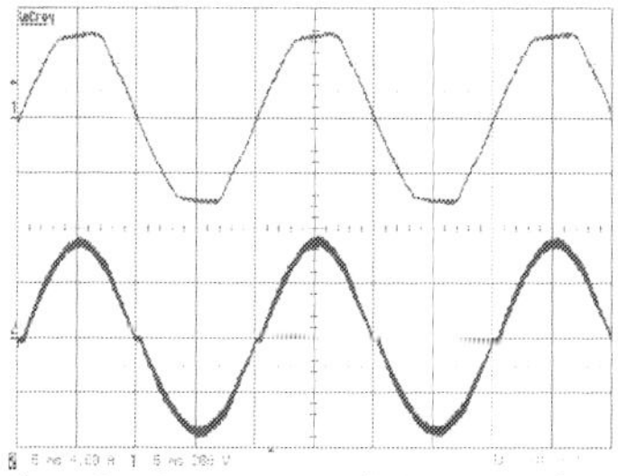

Fig. 10. Experimental Results. Stationary conditions: grid voltage (red solid line) and injected grid current (blue solid line).

5. Core saturation compensation strategies for PV power converters with line frequency transformers

5.1 Control of injected grid current in line frequency transformer PV converter

A very common topology for single-phase converters is the full-bridge voltage source inverter (VSI), driven by unipolar PWM, with a line frequency transformer on the AC-side for galvanic isolation (Figure 11). The line frequency transformer is a safe solution in order to match the electrical safety standards and to block out the DC current component. Moreover, it can be used to adjust the voltage level without requiring a DC/DC power converter. However, the transformer features increased size, weight, and price of the converter. The use of a toroidal transformer can partly reduce these drawbacks.
In order to achieve higher efficiency the inverter strategy control must inject only active grid current, i.e. a pure sinusoidal current in phase with the grid voltage. To this aim injected grid current must be controlled directly providing zero steady state error at grid frequency.
For a line frequency transformer PV converter, the control of the injected grid current stands for the control of the transformer output current. In this way, the DC current component at transformer input is not directly controlled and core saturation and thus current distortion may appear. A simple solution for avoiding core magnetic saturation, if current measurement is tuned properly, is to control the current at transformer input instead of the grid current. This solution leads to poorer performances at light loads conditions because of the transformer magnetizing current.
On the other hand, the direct control of the injected grid current without the measurement of current at transformer input is affected by saturation problems. In (Bellini et al. 2008) the authors presented a solution based only on injected grid current measurement.

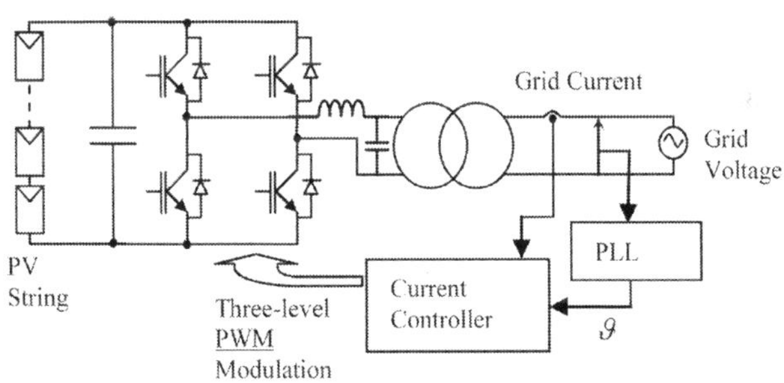

Fig. 11. Block diagram of a line frequency transformer PV converter.

If the control loop is closed at the transformer output a DC voltage and current bias may appear at the transformer input when a non ideal DC-AC converter feeds the transformer. In fact, the transformer blocks out the DC current component, hence the output current control is not able to manage the DC component at the transformer input. This bias can lead to magnetic core saturation and then to harmonic distortion of input and output currents. Figure 12 shows the effects of the DC voltage bias: the magnetic flux, that is, roughly, the integral of the voltage, will saturate asymmetrically. Depending on the sign of the bias, saturation will affect positive or negative semi-periods only. In correspondence of flux saturation the e.m.f. will decrease and the input current will increase accordingly, see Figure 12.

5.2 Saturation Compensation Strategy

In order to avoid core saturation and then current distortion a suitable algorithm was developed to compensate the DC bias effects. At first, the algorithm must identify the semi-period affected by saturation (i.e. the "sign" of saturation), then it must estimate its entity. Neglecting voltage drop across stray inductance the maximum and the minimum values of magnetic flux are synchronous with the zero crossings of the supply voltage waveform. Hence, it is expected that the effect of saturation will appear around zero crossings of supply voltage. Comparing the output current with its set-point value in correspondence of the zero crossings of supply voltage, two saturation indexes referred to as positive SI_P and negative SI_N values can be computed:

$$SI_P = \int_{\frac{T}{2}-\Delta T_{ZC}}^{\frac{T}{2}+\Delta T_{ZC}} \left| i^*_{grid} - i_{grid} \right| dt \qquad (3)$$

$$SI_N = -\int_{T-\Delta T_{ZC}}^{T+\Delta T_{ZC}} \left| i^*_{grid} - i_{grid} \right| dt \qquad (4)$$

where i_{grid} and i^*_{grid} are the actual grid current and the desired grid current waveform. From a theoretical point of view the time interval ΔT_{ZC} must be less then or, at least, equal to T/4. A time integration value of T/10 is sufficient to obtain a robust estimation of saturation core conditions.

The combination of SI_P and SI_N is a nice saturation estimator that can lead to an effective DC bias compensation as shown in the following. If a positive DC current component $I^+_{\mu 0}$ is superimposed to the current at transformer input, saturation affects the positive values of the flux, as reported in Figure 13. Hence the index SI_P will be positive while $SI_N = 0$. On the contrary, if a negative DC current component $I^-_{\mu 0}$ is present, saturation will affect the negative flux values. Hence $SI_P = 0$ and $SI_N > 0$. In case of absence of DC current component ($I_{\mu 0} = 0$), saturation is not present and $SI_P = SI_N = 0$. Computing and summing the two saturation indexes every period a rough estimation ($I_{\mu 0\_est}$) of DC current component can be made: $I_{\mu 0\_est} = SI_P + SI_N$.

Thanks to the estimation of $I_{\mu 0}$ the effect of the DC current component can be compensated with the simple feedback control detailed in Figure 14, where V^* is the output of the current regulator. The control loop forces a zero DC current component $I_{\mu 0}$ thus preventing magnetic saturation. This compensation strategy does not require DC current measurements, thus avoiding intrinsically transducer offset issues.

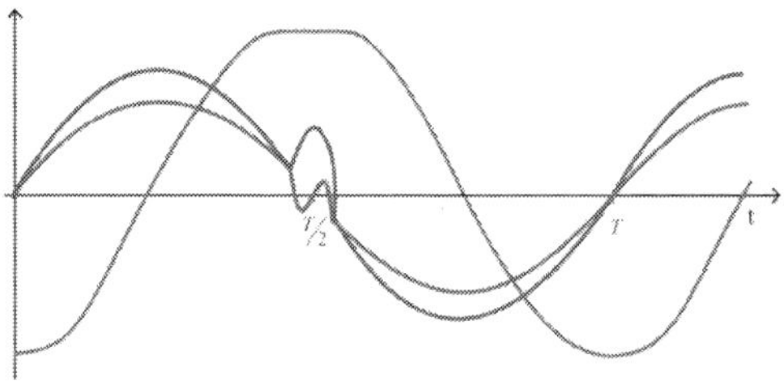

Fig. 12. Effect of a positive DC bias on the current at transformer input. Core flux (green solid line), transformer input (blue solid line) and output (red solid line) currents.

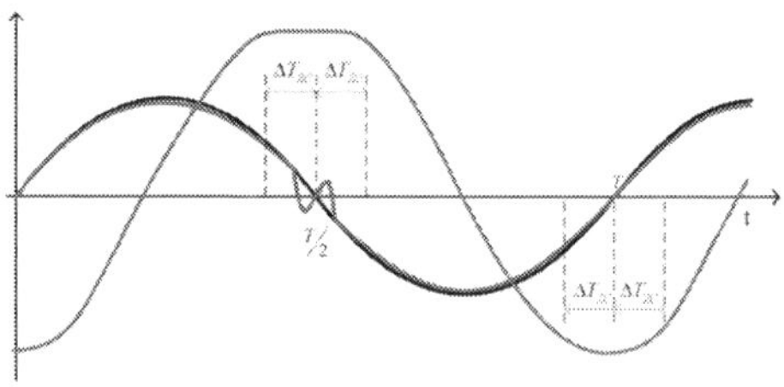

Fig. 13. Computation of saturation indexes. Core flux (green trace), current set point (black trace) and transformer output current (red one).

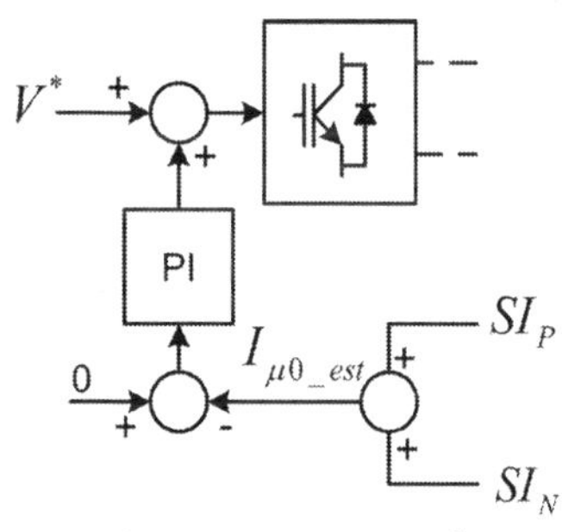

Fig. 14. Block diagram of the saturation compensation scheme.

5.3 Experimental Results

Several experiments were made in order to evaluate the performances of the saturation compensation strategy. Experiments were made with the prototype described in section 4.1. The saturation compensation strategy is applied to two different current controllers: PI+feed forward and d-q synchronous reference frame controllers. The amplitude of the desired grid current, i.e. the set-point of the current controller, is 8A in both cases.

Figure 15a shows the grid voltage, with an intrinsic distortion not correlated to the PV system, the grid current (transformer output current) and the transformer input current in case of PI regulator and feed-forward without saturation compensation. As it can be noted, current distortion appears in correspondence of the voltage zero crossing at the end of a positive voltage semi-period, hence saturation is due to a positive DC bias at the transformer input. Figure 15b shows the effects of the saturation compensation strategy.

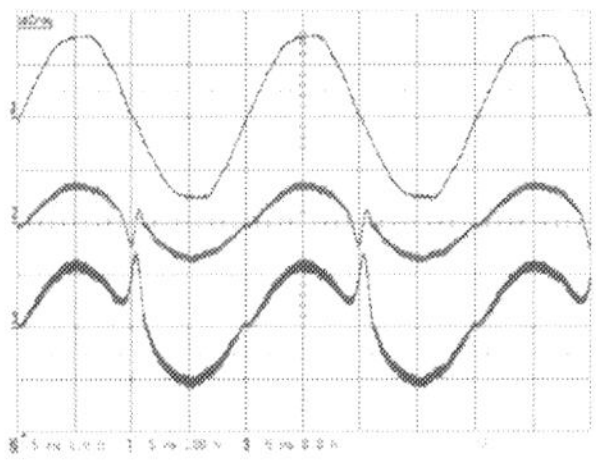

(a) PI regulator + feed forward control without saturation compensation

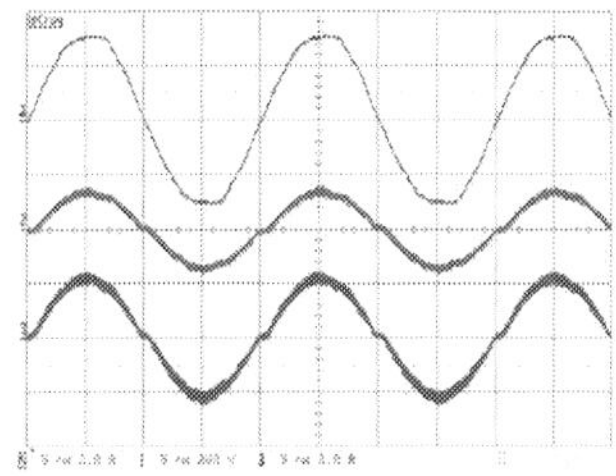

(b) PI regulator + feed forward control with saturation compensation

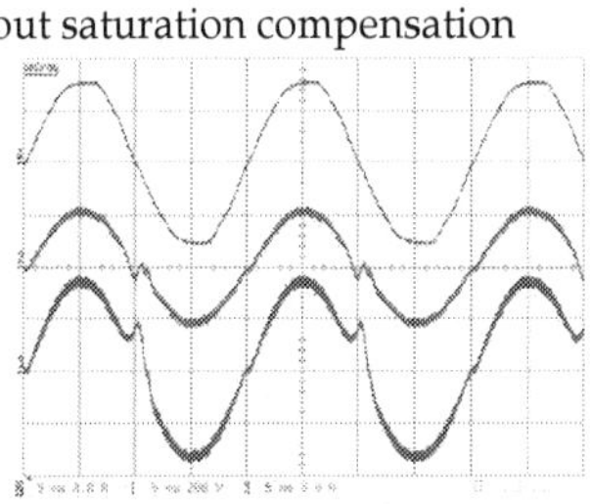

(c) d-q control without saturation compensation

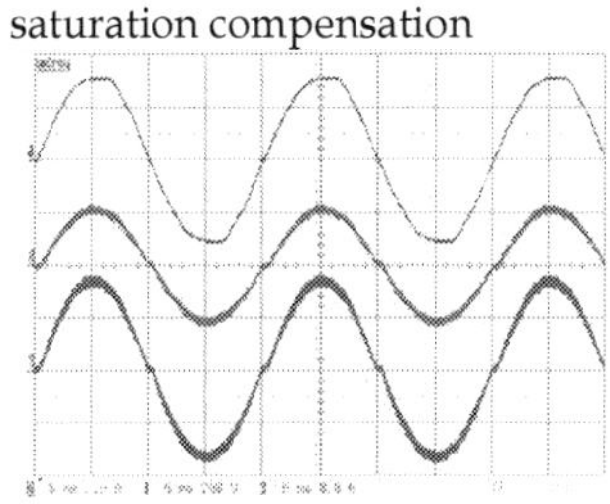

(d) d-q control with saturation compensation

Fig. 15. Experimental Results. Time waveforms of grid voltage (violet solid line), grid current (red solid line) and transformer input current (blue solid line).

The same experimental results are made using the d-q control without (Figure 15c) and with (Figure 15d) saturation compensation. As expected, in this case the steady-state error is zero and the better disturbance rejection of the d-q control damps harmonic distortion of the current. The proposed compensation strategy results very effective for both control schemes.

6. Transformerless grid-connected converters

In new converter designs the trend is to abandon line frequency transformers because of size, weight and price in favour of high frequency transformers. The presence of high frequency transformer requires several power stages and, as a consequence, increasing efficiency and reducing cost is a hard task.

In low power PV applications it is possible to remove transformer at all (transformerless configurations). The absence of the line frequency transformers does not ensure the elimination of DC grid current. The DC component is mainly caused by an offset in the grid current measurement that is usually realized through Hall effect sensors. As known, these sensors are strongly affected by DC offset drift. A current sensor based on fluxgate technology could be adopted to reduce the DC offset drift.

The absence of the line frequency transformer also causes two major drawbacks: safety issue in case of isolation faults, and wrong operation because of ground leakage currents.

The first issue occurs since the PV generator is not isolated from the grid, as a consequence isolation faults at the PV generator or at DC-wiring can get a safety hazard in case of indirect electric shock. Hence, transformerless inverters are equipped with Residual Current Devices (RCD), which monitor the ground leakage currents during operation. When a dangerous leakage current is recognized, the inverter is immediately disconnected from the AC-grid.
A typical configuration for a transformerless power converter, based on H-brige configuration, is reported in Figure 16.

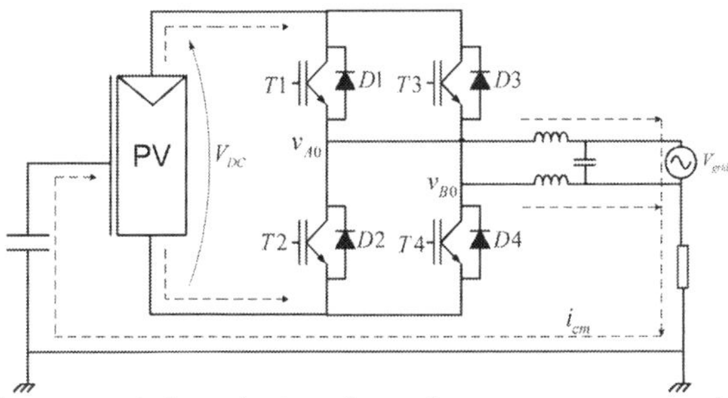

Fig. 16. Schematic representation of a transformerless power converter, based on H bridge, for PV systems. Leakage ground currents are shown as dashed lines.

The second issue is created by the large parasitic capacitance of the PV panel (10-100nF/kWp) that results in high ground leakage currents in case of variations of common mode voltage at the output of the power converter. Impulsive common mode voltages will result in large ground leakage currents because the power converter is coupled to a resonant circuit with a small damping coefficient. The resonant circuit is composed by the earth capacitance, the converter, the AC filter and the grid. A blocking filter is not easy to tune, since the earth capacitance changes with environmental conditions. The damping of the resonant circuit is small since it is mandatory to reduce dissipative terms and thus to achieve a high efficiency.
Therefore the best remedial strategy for ground leakage current is the reduction of the excitation, i.e. of variation of common mode voltage. To this aim it could be necessary to adopt topologies free from common-mode voltage variations (Lopez et al., 2007), (Gonzales et al., 2007). Converter topologies intrinsically safe from leakage currents are the Half-Bridge (HB) and the Neutral Point Clamped (NPC) ones (Gonzales et al., 2008), (Lopez et al., 2006). However, these configurations require twice the input voltage to obtain the same voltage level that is one of the major drawbacks for a PV system. On the other hand, a full-bridge converter driven by bipolar PWM (Gonzales et al 2007) exploits the full voltage rails but reduces common mode voltage with lower efficiency than HB of NPC configurations. From a power quality point of view all these solutions present current ripple at switching frequency.
For power converters with transformers the full bridge converter with unipolar PWM is widespread, since it represents a nice trade-off between efficiency, complexity and price. The biggest advantages of unipolar PWM are that the current ripple appears at twice the switching frequency and that the voltage across the AC output filter is unipolar, thus

reducing core losses. In summary, with unipolar PWM the maximum current ripple is four time smaller than the case of bipolar PWM. However, in terms of ground leakage currents the unipolar PWM generates common mode voltage at switching frequency with a peak to peak value equal to the DC voltage Bus. As a consequence, a big common-mode filter is required in photovoltaic transformerless applications relying on full bridge configurations with unipolar PWM.

Here an original architecture is proposed where a full bridge converter with a unipolar PWM is used with two additional blocks that ensure no common-mode voltage variations also in case of actual (asymmetrical) commutations.

The instantaneous common-mode voltage in the full-bridge converter of the Figure 16 can be computed from the two mid-points of both legs by:

$$v_{cm} = \frac{v_{A0} + v_{BO}}{2} \tag{5}$$

The most widespread single-phase topology is the full bridge one. This topology allows both bipolar and unipolar PWM. In the following sections the behavior of full bridge topologies with bipolar and unipolar PWM will be presented showing performances in terms of ground leakage currents and the original compensation circuitry.

6.1 Full bridge topologies with bipolar PWM for PV transformerless converters

Ideally, the full bridge converter with bipolar PWM does not generate a variation of common-mode output voltage. Hence, it is adopted in some commercial transformerless converters. With reference to the schematic of Figure 16 two configurations are sequentially operated in the switching cycle:

1) T1 and T4 ON (T2, T3 OFF): $v_{A0} = V_{DC}, v_{B0} = 0, v_{cm} = \frac{V_{DC}}{2}$

2) T2 and T3 ON (T1,T4 OFF): $v_{A0} = 0, v_{B0} = V_{DC}, v_{cm} = \frac{V_{DC}}{2}$

In case of ideally synchronous commutation of the switches the common-mode voltage is kept constant at half the DC bus level, and thus no ground leakage currents would appear. However, in actual converters, a small common-mode high-frequency filter is necessary to avoid ground leakage currents due to switching mismatch and asymmetries.

Recently modified version of the full bridge configuration with bipolar PWM were proposed in order to combine higher efficiency with reduced ground leakage currents in real operations.

Figure 17 presents a modified full-bridge topology with two additional blocks used alternatively: inserting only the DC decoupling block a former topology, known as H5, is obtained while, inserting the AC decoupling block a latter one, known as HERIC, is obtained (Kerekes et al., 2007), (Gonzalez et al., 2006). The use of DC or AC decoupling allows, during conduction of freewheeling diodes, the disconnection of the grid from photovoltaic panel.

If no conduction of freewheeling diodes occurs the power switch of DC Decoupling in H5 topology is on while the power switches of the AC decoupling in HERIC topology are off. The H bridge is driven according to 1) and 2) and common mode voltage will be equal to $V_{DC}/2$. On the other side during conduction of freewheeling diodes H5 topology modifies the PWM involving only the high side of the bridge. DC decoupling is OFF hence, in case of symmetrical commutations $v_{A0} = v_{B0} = V_{DC}/2$. In this way common mode voltage does not vary avoiding ground leakage currents increase.
The HERIC topology obtains the same result turning on one of the two added power switches of the AC decoupling block in function of the sign of the injected grid current and driving off the transistors of the H bridge. So doing the freewheeling will involve only the AC Decoupling.
In both topologies the output voltage is zero during freewheeling obtaining a three level output voltage as in case of unipolar modulation. However the output current ripple is still at switching frequency.
Again asymmetrical PWM commutations can determine little leakage ground currents.

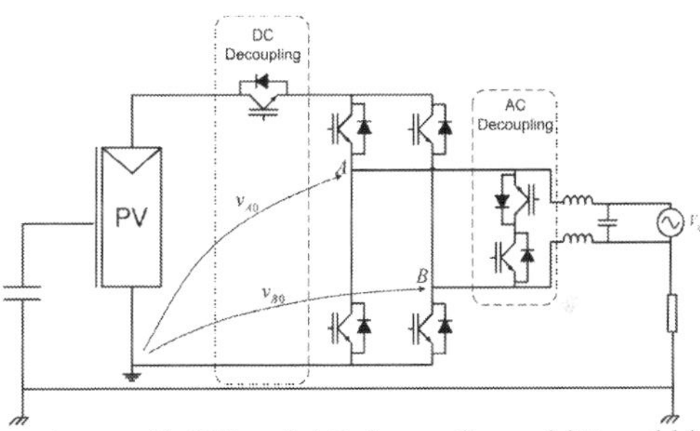

Fig. 17. Full Bridge topology with DC and AC decoupling additional blocks.

6.2 Full bridge topology with unipolar PWM for PV transformerless converters

The full bridge topology with unipolar PWM ensures a reduction of the current ripple, specifically the maximum current ripple is four times smaller then the bipolar PWM maximum current ripple.
Referring to Figure 16 the common-mode voltage resulting during a switching cycle in case of unipolar PWM is computed in the following. Reference is made to positive output voltage and current (first quadrant operations); output voltage is indicated as $v_{AB} = v_{A0} - v_{B0}$.
Four configurations are sequentially operated in the switching cycle:

1) T1, T4 On (T2, T3 OFF): $v_{AB} = V_{DC}$, $v_{cm} = \frac{V_{DC}}{2}$.

2) T2, T4, D2 ON: $v_{AB} = 0$. Low side current freewheeling through D2, T4, $v_{cm} = 0$.

3) T1,T4 On (T2, T3 OFF): $v_{AB} = V_{DC}$, $v_{cm} = \frac{V_{DC}}{2}$.

4) T1, T3, D3 ON: $v_{AB} = 0$. High side current freewheeling through T1, D3, $v_{cm} = V_{DC}$.

The common mode voltage v_{cm} varies from 0 to V_{DC} and thus a large ground leakage current will flow.
Figure 18 shows the original configuration that is proposed to reduce ground leakage currents. Two additional blocks are added: the former one, named DC decoupling, fixes the common mode voltage to $V_{DC}/2$, while the latter, named No-Ideality compensation, helps to keep a constant common mode voltage in case of non ideal (asymmetrical) commutations.
The DC decoupling disconnects the H-bridge from DC source during freewheeling. In particular during high side freewheeling T5 is switched off while during low side freewheeling T6 is switched off. In case of ideal commutations this block is sufficient to guarantee no variations of the common mode voltage v_{cm}. In actual converter, to fix the v_{cm} at $V_{DC}/2$ during asymmetrical commutations, two additional low power switches should be added. In order to avoid the complexity introduced by two controlled switches an attempt was made replacing them with two diodes. The idea relies on the following explanation. Once the decoupling switch T5 opens, the voltage of the high side of the H bridge, is floating. Then this voltage, due to parasitic capacitances, would decrease, however the inserted diode will clamp it to $V_{DC}/2$. Simulation results confirm this behavior.
Since the two diodes impose a maximum voltage equal to $V_{DC}/2$ across T5 and T6, latest generation Mosfets can be chosen as controlled switches, in order to reduce the power losses.
In figure 18 x and y represent the PWM signals used for driving the legs of the full bridge.
A simple combination of PWM signals can be used for driving the switches of DC decoupling block.
With reference to the schematic of Figure 18 four configuration are sequentially operated in the switching cycle:

1) T1, T4, T5,T6 ON (T2 and T3 OFF): $v_{AB} = V_{DC}$, $v_{cm} = \frac{V_{DC}}{2}$.

2) T2, T4, T5, D2 ON (T1, T3, T6 OFF): $v_{AB} = 0$. Low side current freewheeling through T4, D4, $v_{cm} = v_{A0} = v_{B0} = \frac{V_{DC}}{2}$.

3) T1, T4, T5, T6 ON (T2 and T3 OFF): $v_{AB} = V_{DC}$, $v_{cm} = \frac{V_{DC}}{2}$.

4) T1, T3, T6, D3 ON (T2, T4, T5 OFF): $v_{AB} = 0$. High side current freewheeling through T1,D3, $v_{cm} = v_{A0} = v_{D0} = \frac{V_{DC}}{2}$.

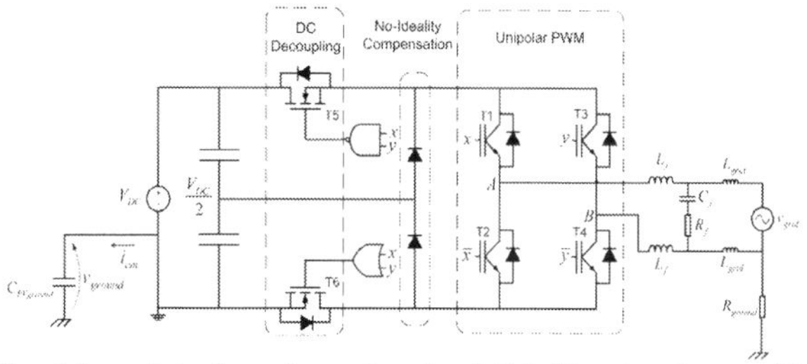

Fig. 18. Schematic of the original configuration for full bridge topology with unipolar PWM for transformerless converters.

In an actual converter the PWM modulation strategy of the DC decoupling block has to be modified because of the presence of dead times. A simple adaptation of the driving circuit can be made that ensures correct operations in actual conditions.

6.3 Simulation Results

The proposed topology (Figure 18) was validated thanks to the PLECS toolbox, which allows fast simulation of power electronic circuits under MATLAB/Simulink. The simulation parameters are: $V_{DC} = 450V$, $V_{grid} = 230V_{rms}$, $R_{ground} = 2\Omega$, $C_{PVground} = 14nF$, $L_f = 2mH$, $C_f = 2.2\mu F$, $R_f = 0.5\Omega$, $L_{grid} = 40\mu H$, the switching frequency is $f_{sw} = 10kHz$. A simple current controller is used to inject a sinusoidal current of $13A_{rms}$ into the grid with unitary power factor equivalent to 3kW of injected active power.

Extensive simulations were performed with the attempt to assess the performances of the proposed converter topology with special reference to common mode voltage and ground leakage currents. Results confirm that the proposed topology allows to keep the common voltage constant at $V_{DC}/2$. In this way, the resulting ground leakage currents are negligible.

Figures 19, 20, 21 show respectively the common mode voltage v_{cm}, the ground voltage v_{ground} and the ground leakage currents i_{cm} without DC decoupling and No-Ideality compensation additional blocks. These simulations were made to quantify the common mode voltage and the ground leakage currents arising in a transformerless full bridge topology with unipolar PWM. Figures 22, 23 show the simulation results obtained when the two additional blocks are added. Figure 22 shows that the common mode voltage v_{cm} is fixed to $V_{DC}/2$. Therefore the ground voltage v_{ground} does not contain high frequency components and only the fundamental grid frequency is present (Figure 22). The resulting leakage ground currents will be very small, as shown in Figure 23.

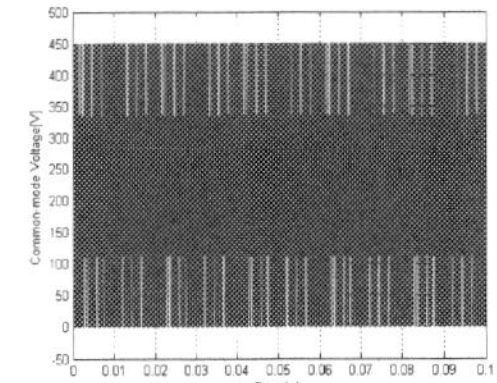

Fig. 19. Simulation results. Common mode voltage in a transformerless unipolar PWM full bridge converter.

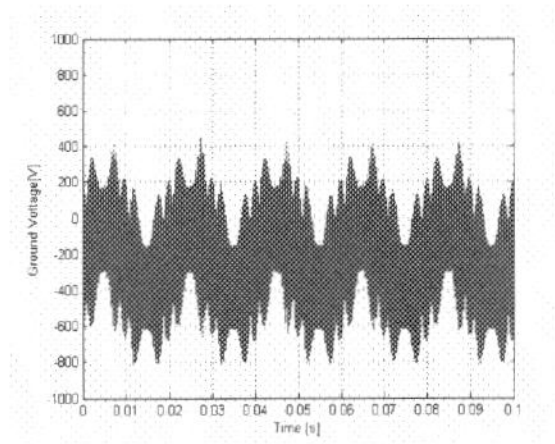

Fig. 20. Simulation results. Ground voltage in a transformerless unipolar PWM full bridge converter.

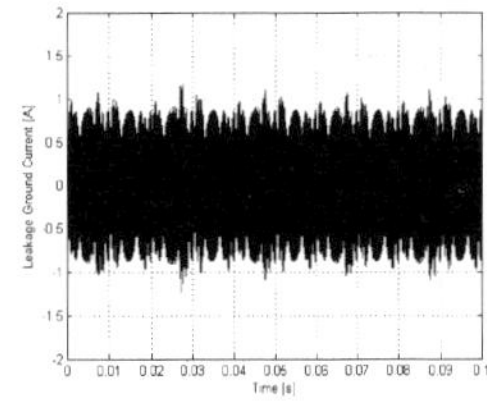

Fig. 21. Simulation results. Ground leakage current in a transformerless unipolar PWM full bridge converter.

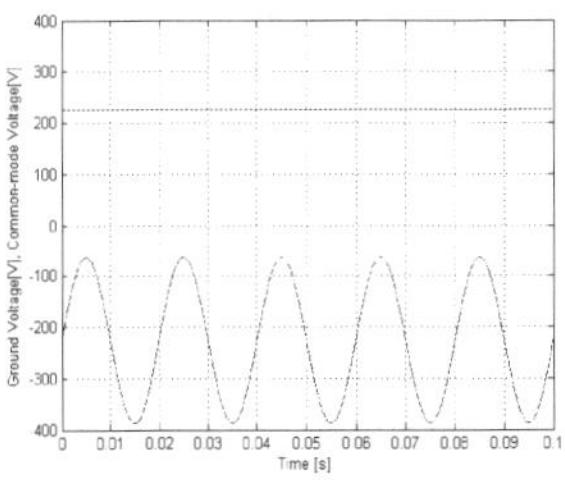

Fig. 22. Simulation results. Common mode voltage (red trace) and ground voltage (blue trace) with the proposed topology.

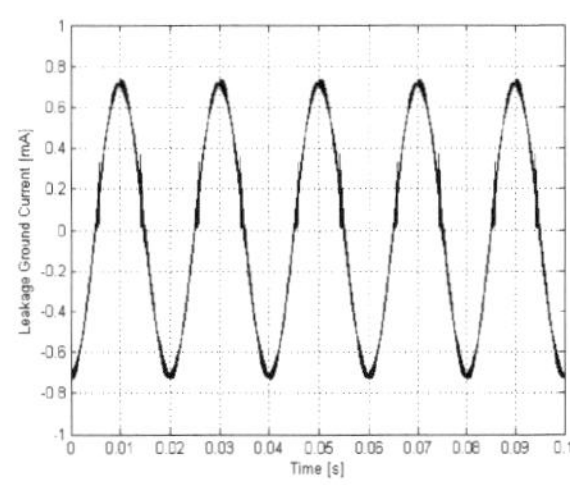

Fig. 23. Simulation results. Ground leakage currents [mA] with the proposed topology.

7. Conclusion

This chapter deals with single-phase grid connected converters used in photovoltaic plants. Some proposed improvements to the widespread full bridge topology are investigated in order to improve energy conversion.

PV power converters, based on full bridge topology, are distinguished by the presence or not of a line frequency transformer.

In both cases the control of the injected grid current is one of the most important issue. The d-q current controller was deeply analyzed and some experimental results are reported to show its superior power quality in comparison with traditional controllers.

In order to increase the performances, notably at light loads conditions, the full bridge topology was modified moving the feedback from the transformer input to the transformer output. In this way, the DC component at the transformer input is not directly controlled, thus an original core compensation strategy is proposed to avoid saturation. Experimental results are reported showing the effectiveness of the proposed compensation.

In low power applications the trend is to remove the transformer (transformerless configurations) in order to reduce losses, costs and size.

In transformerless configurations the major issues are the accurate control of DC component of injected grid current, the safety issues in case of isolation faults, and the reduction of ground leakage currents. Here the most common configurations are reviewed and an original solution is presented that achieves a nice trade-off between efficiency, power quality and reduction of ground leakage currents.

8. References

Chung, S.K. (2000). Phase-Locked Loop for Grid-Connected Three-Phase power conversion systems, *IEE Proc. - Electr. Power Appl.*, Vol. 147, No. 3, pp. 213-219.

Arruda, L. N.; Cardoso Filho, B. J.; Silva, S. M. & Diniz, S.A.C. (2001). Wide bandwidth single and three-phase PLL structures for grid-tied PV systems, *in Proc. of EPE'01*, pp. 1660-1663.

Silva, S.M.; Filho, B.C.; Campana, R.P. & Boaventura, W.C (2004). Performance evaluation of PLL algorithms for single- phase grid-connected systems, *in Proc. of IAS'04*, vol. 4, pp. 2259 - 2263. C. J. Kaufman, Rocky Mountain Research Lab.

Yuan, X.; Merk, W.; Stemmler, H. & Allmeling, J. (2002). Stationary-frame generalized integrators for current control of active power filters with zero steady-state error for current harmonics of concern under unbalanced and distorted operating conditions. *IEEE Transactions on Industry Applications*, vol. 38, no. 2, pp. 523–532, Mar./Apr. 2002.

Teodorescu, R.; Blaabjerg, F.; Borup, U. & Liserre M. (2004). A new control structure for grid-connected LCL PV inverters with zero steady-state error and selective harmonic compensation, *in Proc. of Applied Power Electronics Conference and Exposition, 2004.* APEC '04. Nineteenth Annual IEEE, vol. 1, pp. 580–586.

Ryan, M. J. & Lorenz R. D. (1997). A synchronous-frame controller for a single-phase sine waveinverter. *in Proc. Of Applied Power Electronics Conference and Exposition.* APEC '97 Conference Proceedings 1997., Twelfth Annual, vol. 2, Atlanta, GA, USA, Feb. 1997, pp. 813–819.

Miranda, U. A.; Rolim, L. G. B. & Aredes M. (2005). A DQ synchronous reference frame current control for single-phase converters, *in Proc. Of Power Electronics Specialists Conference.* PESC '05. IEEE 36th, 2005, pp. 1377–1381.

Cacciato, M.; Consoli, A.; Aiello, N.; Attanasio, R.; Gennaro, F. & Macina G. (2008). A digitally controlled double stage soft-switching converter for grid-connected photovoltaic applications, in Proc. *Of Applied Power Electronics Conference and Exposition.* APEC 2008. Twenty-Third Annual IEEE, Feb. 2008, pp. 141–147.

Bellini, A.; Franceschini, G.; Lorenzani, E. & Tassoni, C. (2006). Synchronous reference frame grid current control for single-phase photovoltaic converters, *in Proc. of IAS 2008*, pp 1 - 7, Edmonton (Canada), October 5-9 2008

Ciobotaru, M.; Teodorescu, R. & Blaabjerg, F. (2005). Control of single-stage single-phase PV inverter, *in Proc. of Power Electronics and Applications*, 2005 European Conference on, Sept. 2005.

Calais, M. & Agelidis, V. G. (1998). Multilevel converters for single-phase grid connected photovoltaic systems—An overview, *in Proc. IEEE Int. Symp. Ind. Electron.*, vol. 1, pp. 224–229.

Calais, M.; Myrzik, J. M. A. & Agelidis, V. G. (2001). Inverters for single phase grid connected photovoltaic systems—Overview and prospects, *in Proc. 17th Eur. Photovoltaic Solar Energy Conf.*, Munich, Germany, Oct. 22–26, pp. 437–440.

Kuo, Y. C.; Liang, T. J. & Chen, J. F. (2001). Novel maximum-power-pointtracking controller for photovoltaic energy conversion system, *IEEE Trans. Ind. Electron.*, vol. 48, no. 3, pp. 594–601, Jun. 2001.

Femia, N.; Petrone, G.; Spagnuolo, G. & Vitelli, M. (2005). Optimization of perturb and observe maximum power point tracking method, *Power Electronics, IEEE Transactions on* , vol.20, no.4, pp. 963-973, July 2005

Fangrui Liu; Shanxu Duan; Fei Liu; Bangyin Liu & Yong Kang,(2008). A Variable Step Size INC MPPT Method for PV Systems, *IEEE Transactions on Industrial Electronics,* vol. 55, no. 7, pp. 2622-2628, July 2008

Esram, T.; Kimball, J.W.; Krein, P.T.; Chapman, P.L. & Midya, P. (2006). Dynamic Maximum Power Point Tracking of Photovoltaic Arrays Using Ripple Correlation Control, *IEEE Transactions on Power Electronics,* vol. 21, no.5, pp. 1282-1291, Sept. 2006

Lopez, Oscar; Teodorescu, Remus; Freijedo, Francisco & Doval-Gandoy, Jesus (2007). Leakage current evaluation of a singlephase transformerless PV inverter connected to the grid, *Applied Power Electronics Conference, APEC 2007* - Twenty Second Annual IEEE ,pp. 907 - 912, Feb. 25 2007-March 1 2007

Gonzalez, R.; Lopez, J.; Sanchis, P. & Marroyo, L. (2007). Transformerless Inverter for Single-Phase Photovoltaic Systems, *IEEE Transactions on Power Electronics* Volume 22, Issue 2, March 2007, pp. 693 - 697.

Gonzalez, R.; Gubia, E.; Lopez, J. & Marroyo, L. (2008). Transformerless Single-Phase Multilevel-Based Photovoltaic Inverter, *IEEE Transactions on Industrial Electronics* Volume 55, Issue 7, July 2008, pp. 2694 - 2702.

Lopez, Oscar; Teodorescu, Remus & Doval-Gandoy, Jesus.(2006). Multilevel transformerless topologies for single-phase grid-connected converters, *IEEE IECON 2006* - 32nd Annual Conference on Nov. 2006, pp. 5191 - 5196.

Kerekes, T.; Teodorescu & R.; Borup, U. (2007). Transformerless Photovoltaic Inverters Connected to the Grid, *Applied Power Electronics Conference, APEC 2007* - Twenty Second Annual IEEE Feb. 25 2007-March 1 2007, pp.1733 - 1737.

Gonzalez, R.; Lopez, J.; Sanchis, P.; Gubia, E.; Ursua, A. & Marroyo, L. (2006). High-Efficiency Transformerless Single-phase Photovoltaic Inverter, *Power Electronics and Motion Control Conference, 2006. EPE-PEMC 2006,* 12th International Aug. 2006, pp. 1895 - 1900.

Grid Integration of Renewable Energy Systems

Athula Rajapakse[1], Dharshana Muthumuni[2] and Nuwan Perera[1]
University of Manitoba[1]
Manitoba HVDC Research Centre[2]
Manitoba, Canada

1. Introduction

Integration of renewable energy into the utility grid can be at either the transmission level or the distribution level, depending on the scale of generation. Large renewable energy generation such as wind farms are directly interconnected to the transmission system. Small scale distributed generation is generally interconnected to the medium or low voltage distribution systems. Both types of interconnections present different challenges that must be carefully analysed before systems are designed. This chapter outlines and discusses the most common issues encountered during grid integration of different types of renewable energy systems. The application of detailed simulation methods to identify problems and to find suitable solutions is also highlighted. Few case studies and simulation examples are presented to reinforce the concepts presented in the chapter.

2. Integration of small scale generation into distribution grids

Electricity generation using renewable energy resources is often taking place in small scale due to disperse nature of the resources. Good examples are small hydro, solar photovoltaics, biogas, biomass and small wind turbine based electricity generation systems. The size of these generators typically varies from a few hundreds of kilowatts to several megawatts. These small scale electricity generators are generally connected to the grid at the primary or secondary distribution level and are considered distributed generation (DG) or distributed resources (DR). Distributed resources include both renewable and non-renewable small scale generation as well as energy storage.

2.1 Different types of grid interfaces

Often, the small scale renewable generators are not directly connected to the grid. The generation technology or the operational characteristics requires the use of some interface between the generator and the utility distribution grid. For example, solar photovoltaics (PV) panels generate dc electricity and therefore, a power electronics based dc-to-ac converter is required between the grid and the generator. Some technologies such as induction generator based small hydro or wind can be directly connected to the ac grid. However, concerns such as starting transients, energy conversion efficiency and power quality issues make connecting them through a power electronics interface a better choice. Table-1 summarizes some of the common types of generation and their preferred interfacing technologies.

Type	Interfacing technology
Fuel cells	Power electronic converter
Wind	Induction generator/power electronic converter
Photovoltaics	Power electronic converter
Small hydro	Synchronous or induction generator, power electronic converter

Table 1. Interfacing technologies

2.2 Issues related to grid integration of small scale generation

Most electric distribution systems are designed, operated and protected on the premise of there being a single voltage source on each distribution feeder. Interconnection of small scale renewable generation to the distribution grid violates this fundamental assumption. Therefore, certain special requirements need to be satisfied when interconnecting distributed generation to the grid in order to ensure safe and reliable operation. Fault clearing, reclosing and inadvertent islanded operation are major protection related concerns [IEEE, 2004]. In many cases, the grid-DG interface is based on power electronics inverters or asynchronous generators. The utilities are concerned about their impact on power quality. The impacts include harmonics, voltage dips, overvoltages, and voltage flicker [IEEE, 2003]. Several studies [Aldferfer et al, 2000; Dugan & McDermott, 2002] have identified that the technical requirements imposed by utilities to address these concerns as a major technical barrier for grid integration of DG. The standards such as IEEE P1547 [IEEE, 2003] provide guidelines for developing the general technical requirements for interconnection of small scale generation.

2.2.1 Protection issues

Protection is one of the most problematic issues related to the interconnection of distributed generation. Radial systems are the most common form of distribution network configuration. They are usually protected using time graded overcurrent protection schemes. Interconnection of DGs may alter the coordination of the existing protection schemes. If not carefully addressed, this can lead to malfunction of protection equipment. The main protection issues related to the interconnection of DG are briefly discussed below.

(a) Change of Short Circuit Levels

Short circuit level is one of the main parameters used in the selection of circuit breakers, reclosers, fuses and current transformers (CTs), and the coordination between overcurrent relays. Short circuit level is characterized by the equivalent system impedance at the fault point and indicates the expected level of the fault current. Time variation of the fault current is influenced by the characteristics of the rotating machinery in the vicinity. Most distribution systems are initially designed as passive networks without DG. However, with the connection of DG, equivalent network impedance can decrease, resulting in an increase in the fault level. Thus, in case of a fault, there can be unexpectedly high fault currents

which may exceed the interrupting capacity of existing circuit breakers. High fault currents can also lead to CT saturation. Also, the changed fault levels can disrupt the coordination between overcurrent relays leading to unsatisfactory operation of protection systems.
To illustrate the concerns, consider the distribution system in Figure 1. When the DG is not present, the fault currents seen by the feeder breaker BK and the recloser RC are approximately equal to the fault current I_f ($I_s \approx I_r \approx I_f$). However, when the DG is connected, $I_r = I_s+I_{dg}$, and $I_r > I_s$. This is a condition normally not observed in passive radial networks. This doesn't cause a problem if the recloser interrupting capacity is sufficient to handle the increased fault current. However, it is likely that coordination between the recloser and any downstream fuses (for example between RC and FS_1) is lost. Because both the recloser and fuses operate faster at higher fault currents, the required margins between the recloser fast curve and the fuse minimum melt curve could be reduced to an extent that the coordination is lost.

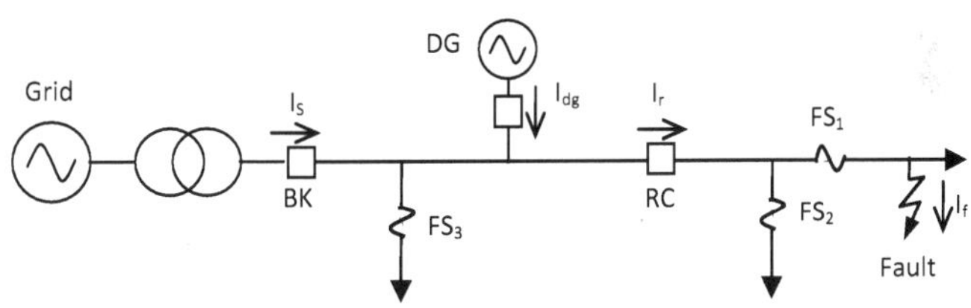

Fig. 1. Change of fault currents due to DG interconnection

(b) Reverse Power Flow
In radial distribution systems, the power flows are unidirectional. Protection schemes are initially designed based on these power flows. When a DG is connected, the power flows may be reversed [Lopes, 2002]. This could also alter the coordination of protection relays.

(c) Lack of Sustained Fault Current
In order for the protection relays to reliably detect and discriminate fault currents from the normal load currents, the faults must cause a significant and sustained increase in the currents measured by the relays. If the fault current contribution from a DG is limited, it becomes difficult for the overcurrent based protection relays to effectively detect faults. Renewable energy based generation often employs induction generators, small synchronous generators or power electronic converters. Induction generators cannot supply sustained fault currents to three-phase faults and make only a limited fault current contribution to asymmetrical faults. Small synchronous generators are usually not able to supply sustained fault currents that are significantly greater than the rated current. Power semiconductor devices cannot withstand significant overcurrent for sustained periods and therefore power electronic converters are designed to internally limit the output current. The lack of sustained fault current compromises the ability of relays to detect faults. [Barker & de Mello, 2000].

(d) Islanding
Islanding is the situation where a part of the utility network is disconnected from the main grid and operates as an independent system supplied with one or more generators. Islanding results in abnormal variations of frequency and voltage in the 'island'. Opening of an auto recloser during a fault may lead to the formation of two independent systems that

operate at two different frequencies [Jenkins et al, 2000]. Reclosing of the auto recloser while the two systems are out of phase could bring disastrous results. [Borbely & Kreider, 2000]. Further, islanding operation may create an ungrounded system depending on the transformer connection. An unidentified island would be hazardous for the repair crews. Due to the above reasons, islanding is recognized as an unsafe situation and immediate disconnection of the DGs from the grid is recommended upon formation of an island [IEEE, 2003].

2.2.2 Voltage control

Interconnection of DG results in changes in power flows and the voltage profile of the feeder, and generally results in overvoltages under low load or high (DG) production conditions. In weak networks, the DG capacity is generally determined by the voltage limits. Furthermore, the connection status of a DG is not controlled by the utility. Disconnection of a DG during the high load can cause undervoltages, while re-connection of a DG under low load conditions may cause overvoltages. This may lead to poor power quality situation and may result in operation of under/over voltage relays.

Selection of appropriate tap settings for the distribution transformers becomes difficult with the increased penetration of DG. This is specially difficult when the DGs are not equally distributed among the feeders supplied by the same transformer [Uchida et al, 2006]. Such a situation is illustrated in Figure 2 where there are two feeders supplied by the same transformer, but DGs are concentrated on only one of them. Generally, when the DGs are connected, the net current flow through the transformer is reduced because the DGs provide the power to the nearby loads. Consequently, the transformer tap needs to be changed to the light load setting. The resulting decline of the sending voltage can cause a voltage violation at the far end of the feeder without DGs as shown in Figure 2. Leaving the transformer tap at the heavy load setting risks overvoltages on the feeder with DGs. Switched capacitors and static VAr compensators can be used to control the feeder voltages, but these solutions are often too costly.

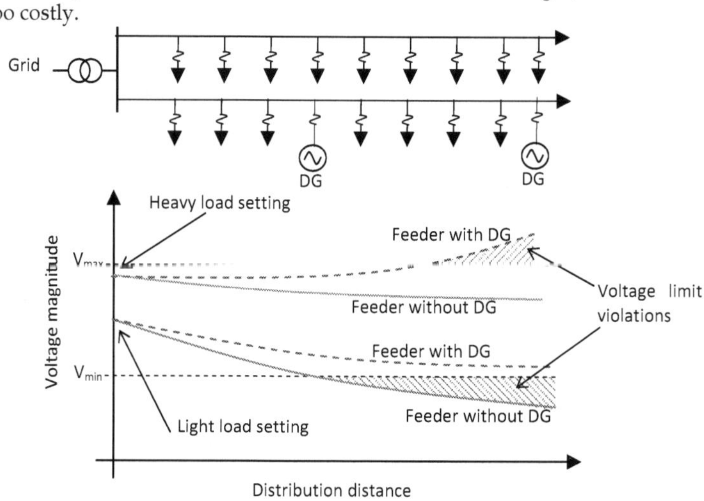

Fig. 2. Possible DG interconnection configurations

Another issue that affects the operation of DGs at distribution level is the unbalanced voltage profile. As shown in Figure 3, both the loads as well as DGs can be either three-phase or single-phase. Interconnection of single phase sources will increase the system unbalance. On the other hand, inherently unbalanced distribution systems can pose problems for the three-phase DGs connected to it: the resulting unbalance currents in the DG can cause overheating and frequent shutdowns.

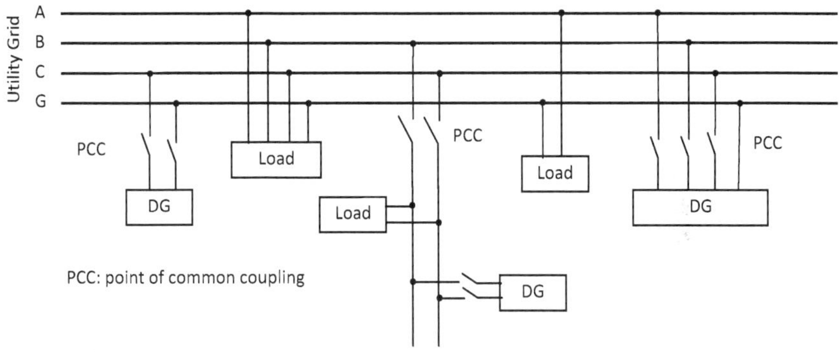

Fig. 3. Possible DG interconnection configurations

2.2.3 Harmonics and flicker

The quality of power supplied to the customers is one of the major concerns of service providers. The majority of the DGs are interfaced using power electronic converters. Power electronic devices inject harmonics into the system resulting in a poor power quality for the consumers. This can be overcome by using harmonic filters. Harmonic filters may be active, passive or hybrid and must be selected appropriately according to the requirement. In addition the starting of directly connected induction generators may cause voltage flicker. The soft-staring mechanisms are sometimes useful in reducing these voltage fluctuations. Wind generators can also introduce periodic fluctuations in the voltages due to tower shadow effect [Fadaeinedjad et al, 2009].

2.2.4 Interconnection standards and examples of grid codes

In order to address protection and safety issues, DG interconnection standards and guidelines have been introduced. The most widely applied standard is the IEEE Std. 1547-2003, "IEEE Standards for Interconnecting Distributed Recourses (DR) with Electric Power Systems 2003" [IEEE, 2003]. Most utilities have adopted the IEEE Std. 1547-2003 in formulating guidelines/ rules for interconnecting DG to their networks. A brief description of the salient features of IEEE Std. 1547-2003 is given below.
IEEE Std. 1547-2003 recommends disconnection of distributed energy resources when the voltage and/or frequency at the point of interconnection deviate from their base values due to faults and other disturbances. The allowable deviations are defined in Tables 2 and 3. Also, for an unintentional island in which the DG energizes a portion of the network, the

DG interconnection system must detect the island and disconnect the DG within two seconds of the formation of an island [IEEE, 2003].

Voltage range (% of base voltage [a])	Clearing time(s) [b]
V < 50 %	0.16
50 % ≤ V < 88 %	2.00
110 % ≤ V < 120 %	1.00
V ≥ 120 %	0.16

[a]Base voltages are the nominal system voltages stated in ANSI C84.1-1995.
[b]DR ≤ 30 kW, maximum clearing times; DR > 30kW, default clearing times.

Table 2. Interconnection System Response to Abnormal Voltages [IEEE, 2003]

DR size	Frequency range(Hz)	Clearing time(s) [a]
≤ 30kW	> 60.5	0.16
	< 59.3	0.16
> 30kW	> 60.5	0.16
	< {59.8-57.0} (adjustable set point)	0.16 to 300 (adjustable)
	< 57.0	0.16

[a]DR ≤ 30 kW, maximum clearing times; DR > 30kW, default clearing times.

Table 3. Interconnection System Response to Abnormal Frequencies [IEEE, 2003]

2.3 Islanding and anti-islanding protection methods

As discussed earlier, formation of unintentional power islands is unsafe an undesirable. Therefore, formation of islands must be swiftly detected and the generators energizing the island must be disconnected. This is commonly known as the anti-islanding protection or the loss-of-main protection. There are several methods that provide anti-islanding protection. They can be broadly categorized into three groups: active methods, passive methods and telecommunication based methods.

2.3.1 Passive methods

The passive islanding detection methods make decisions based on measured electrical quantities such as voltage and frequency. Voltage relay is one of the simplest passive methods used to detect islands. The reactive power imbalance between production and consumption, which occurs after the loss of mains, leads to a change in the voltage level. Thus, the voltage magnitude measured at the DG can be used to detect the island. Trip signals are generated if the measured voltage shows abnormal variation over a predefined period of time. Normally, voltage relays respond to both under-voltage and over-voltage situations. However, voltage collapse is a slow process and detection of islands using voltage relays could be very slow.

Voltage Surge relay (also known as Voltage Vector Shift relay or Voltage Phase Jump relay) is one of the methods used for fast detection of islands. The loss of grid connection results in a sudden change in load and thus, a sudden change in cycle length of the terminal voltage waveform. The cycle duration becomes either shorter or longer, depending on if there is an excess or a deficit of active power in the islanded system. The Vector Shift relay measures the length of each cycle of the voltage waveform. The cycle duration of the waveform is compared with the duration of the previous cycle. If the difference in the cycle duration exceeds the relay setting, a trip signal is issued. Usually these relays are provided with low voltage blocking facility to block the trip signal if the terminal voltage drops below a threshold level to avoid false operations [Freitas et al, 2005].

Frequency relay is another example of passive detection methods. Under-frequency can occur if the grid connection is lost at a situation where the local load exceeds the production of the generator, whereas over-frequency situations can arise if there is a surplus production at the time of grid disconnection. Thus, the frequency relays can take decisions based on the frequency of the voltage at the DG. The over-frequency or under-frequency elements are used to trip the generator from the system. If the real power in the island is almost balanced, the change of frequency of the islanded section will be low, making the relay ineffective. Rate of change of frequency (ROCOF), which is the time derivative of the frequency, is also frequently used to detect power islands. ROCOF relays are popular as they response much faster than the frequency relays.

2.3.2 Active methods

In the active detection methods, disturbances are injected into the network and islands are detected based on system responses to the injected disturbance [Freitas et al, 2005]. Reactive error export, fault level monitoring, system impedance monitoring and frequency-drift are some of the active island detection methods.

The reactive error export method controls the DG excitation current to generate a known value of reactive current. This cannot be supported when the DG is not connected to the grid. The change in the reactive power measured during the formation of power island is used to detect the situation [Redfern et al, 1993; Kane & Fox, 1997]. The fault level monitoring method uses a point-on-wave switched thyristor in series with a shunt inductor. The thyristor is triggered at a point close to the voltage zero crossing and the current through the shunt inductor is measured. It is possible to estimate the system impedance and the fault level using the measured current pulse. The fact that the fault level drops when the grid connection is lost, is used to detect the islanding. Usually the estimation of fault level is repeated at every half cycle [Redfern et al, 1993, Kane & Fox, 1997]. System impedance monitoring as its name implies, is a method that detects loss of mains (LOM) by actively monitoring the system impedance. A high frequency (HF) source is connected via a coupling capacitor at the point of interconnection (PCC). The capacitor is in series with the equivalent network impedance. With the presence of connection to the utility grid, the equivalent impedance at the PCC is low; therefore the HF-ripple at coupling point is negligible. However, after islanding, the equivalent impedance at the PCC increases dramatically and the divided HF-signal is clearly detectable [Redfern et al, 1993; Kane & Fox, 1997].

The frequency drift LOM detection method is specifically used in inverter interfaced DGs. The output current of the converter is controlled to a frequency which is slightly different to the nominal frequency of the system. Under normal conditions, the terminal frequency is dictated by the powerful bulk supply. If the mains supply is lost, frequency will drift until a certain shutdown level is exceeded [Ropp et al, 1999; Hung et al, 2003].

2.3.3 Telecommunication-based Methods

The telecommunication-based methods use communicated circuit breaker status signals to alert and trip DGs when islands are formed. Their performance is independent of the type of distributed energy resources involved. Telecommunication based systems can be implemented in conjunction with SCADA systems. In this approach, the states of the circuit breakers in the grid are continuously tracked. The circuit breaker state information can be used to determine whether a part of the system has become an island, using predetermined logic. Transfer tripping schemes can be considered as a decentralized version of the SCADA based system described above. In transfer trip schemes, a logic circuit uses information of circuit breaker states to determine if a part of the grid has been islanded.

Comparison of rate of change of frequency (COROCOF) at the substation and the DG location is another telecommunication-based method. The rate of change of frequency is measured at the substation and if it exceeds a certain limit, a block signal is sent to the DG end. If the rate of change of frequency at DG is greater than a set point, and if there is no block signal received from the substation end, the DG will be tripped.

Out of the above described methods under/over voltage, under/over frequency, rate of change of frequency (ROCOF) and voltage vector surge (VVS) relays are the most widely used methods for anti-islanding protection. One of the main disadvantages of these methods is the possibility of nuisance tripping of DGs during other system disturbances such as load switching, faults, etc.

2.3.4 Intentional islanding and Microgrids

Although islanded operation of DG has recognized as hazardous situation, a generator or cluster of generators operating as an island (microgrid) may be able to supply a part of the local load when grid supply is not available. This type of operation is commonly known as intentional islanding. Numerous technical issues have to be addressed to make intentional islanded operation a reality [Freitas et al, 2005]. The Power balance needs to be maintained between production and consumption. An effective protection coordination must be maintained under both grid connected and islanded modes. In order to reach such a high level of local autonomy, it requires solving advanced control and protection functions.

3. Interconnection studies

When interconnecting a renewable energy based generator, the grid interface needs to be designed to meet applicable regulatory requirements set by the local utility. In order to verify the requirements pertaining to protection, control and power quality, detailed simulation studies need to be performed. Time domain simulation using Electromagnetic Transient (EMT) programs is a powerful method that can be used for studies involving controller tuning, protection setting, power quality investigations and system validations. EMT programs typically come with a built-in library that include detailed rotating machine models (synchronous, induction, permanent magnet, etc.), transformers models (including iron core saturation), frequency dependent transmission line and cable models, measurement transformer (CT, VT and CCVT) models from which realistic secondary waveforms can be derived for protection system validation. In addition to electrical components, facilities exist for simulating complex control systems. These computer programs also offer the ability of modeling of power electronic systems in detail. Thus, they can be used for studying all type of DG interfaces:

- directly connected synchronous/induction generators,
- converter connected DC sources (fuel cell, PV),
- doubly fed induction generators (DFIG), and
- converter connected variable frequency synchronous generators.

Few simulation studies that demonstrate different interconnections interfaces and issues are discussed in the following sections. The simulations were carried out using the industry standard PSCAD/EMTDC software program.

3.1 Soft-starting of fixed speed induction generator wind turbines

The first example is a fixed speed wind generator interconnected to a medium voltage (MV) distribution system. The distribution system shown in Figure 4 is adopted from a benchmark MV distribution system proposed by CIGRE for DG integration studies [Rudion, 2005]. The network is normally operated as a radial system, although two disconnect switches (S2 and S3) are available for network reconfiguration.

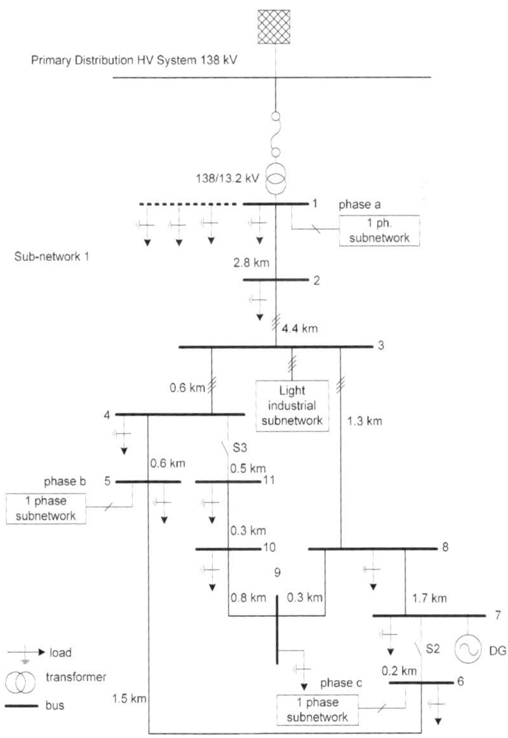

Fig. 4. Medium voltage distribution system

In detailed simulation studies, it is customary to use a reduced model of the network. In the simplified distribution network shown in Figure 5, a section of the network (upstream of bus 8) is replaced by its Thevenin's equivalent circuit. Parts of the network that are of interest for the study are represented in detail. In this example, it is assumed that bus 11 is supplying a critical load sensitive to power quality problems. Therefore, the feeder section from bus 8 to bus 11 is represented in detail. The direct coupled wind generator is represented using a detailed induction machine model. Although it is not critical for this study, a detailed wind turbine model with pitch control is also used.

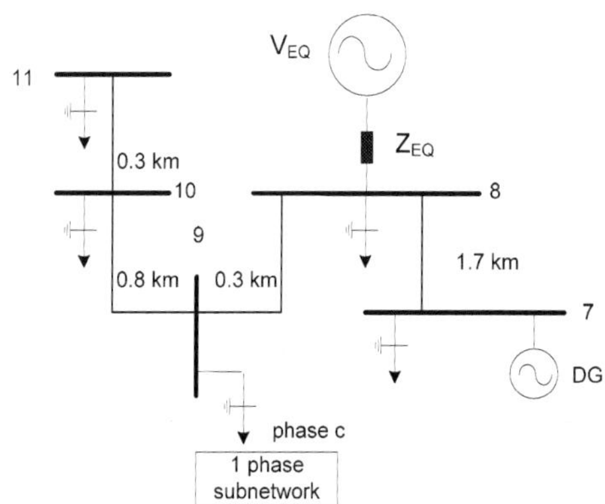

Fig. 5. Simplified distribution system

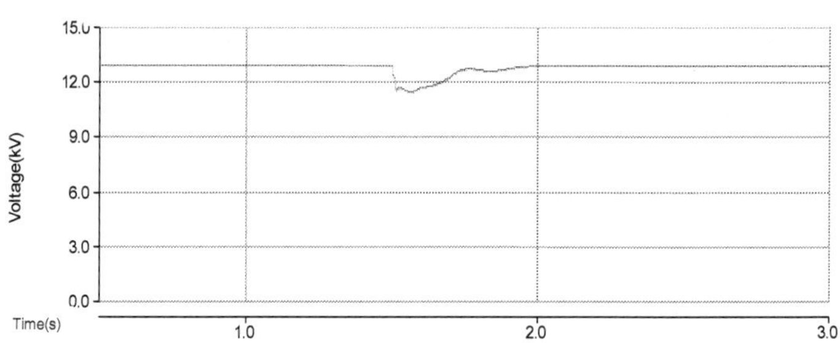

Fig. 6. Voltage at bus 11 during the wind generator start-up.

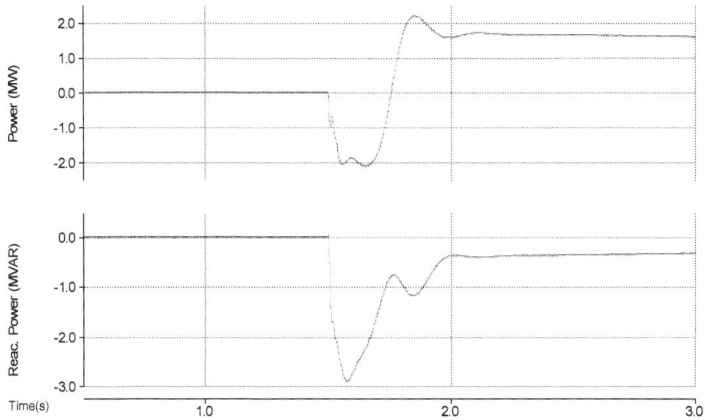

Fig. 7. Wind generator real and reactive power output during the start-up.

The usual starting procedure is to let the wind turbine accelerate the generator to a specific predetermined speed (close to the synchronous speed) and then connect the generator terminal to the grid supply. The machined speeds up, (in motoring mode), drawing both real and reactive power from the grid. This starting procedure has caused a large voltage dip at bus 11 as can be seen in Figure 6. This is a result of the machine initially darwing a significat amount of starting current from the system. The real and reactive power exchange between the induction generator and the grid during this period is shown in Figure 7. One low cost solution to this problem is the use of a thyrister based soft starter [Peters et al, 2006]. The variation of bus 11 voltage with the soft starter is shown in Figure 8. As evident from the simulaion results, the soft starter can significantly reduce the voltage dip. The real and reactive power exchanges with the soft starter is shown in Figure 9.

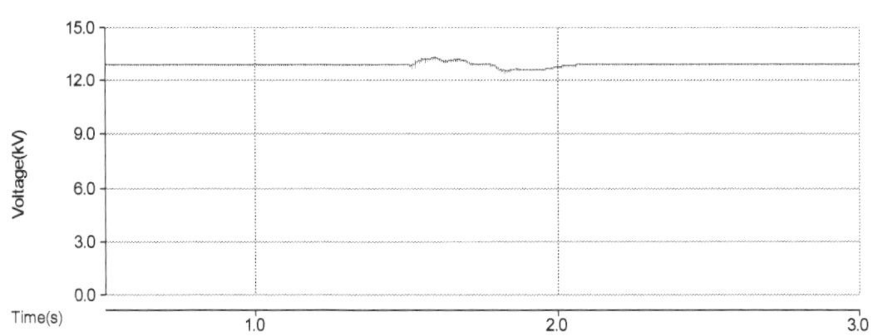

Fig. 8. Voltage at bus 11 during soft starting of the wind generator.

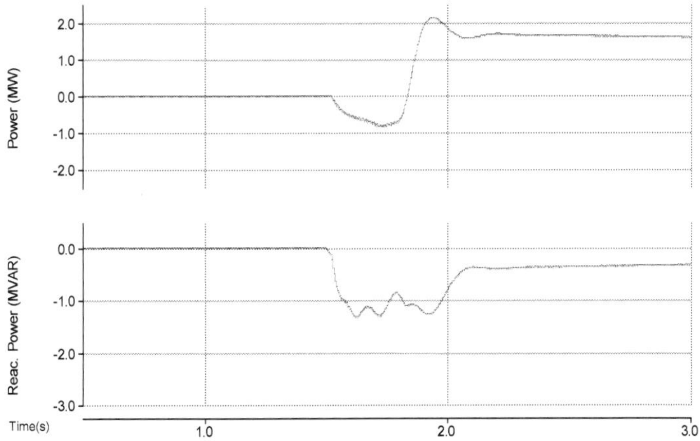

Fig. 9. Wind generator real and reactive power output with soft starting

3.2 Fault current contribution of different generators

The nature of fault current contribution of different DG interface technology can be accurately analysed through detalied electromagnetic transient simulations. The output current produced by a synchronous generator connected to bus 7, during a three-phase short circuit at bus 8 is shown in Figure 10. The output currents of (i) a direct coupled induction generator, (ii) a direct coupled induction generator with external rotor resistance, and (iii) a dobly fed induction generator (DFIG) for the same fault are shown in Figs. 11, 12, and 13 respectively. The lack of sustained fault current in the latter three cases is clearly vissible. It should be noted that with power electronic intrefacing (e.g. DFIG, PV), the shape of the fault current is influenced by the converter portection and control stratergy.

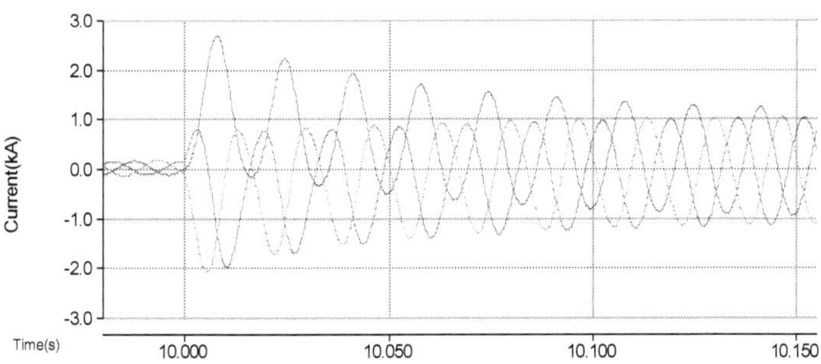

Fig. 10. Fault current contribution from a syncronous generator

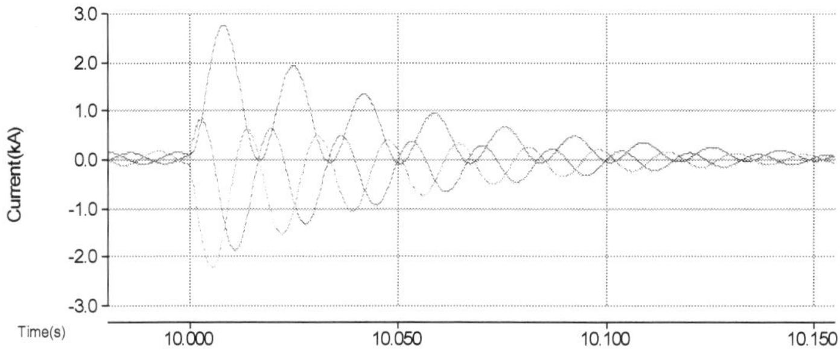

Fig. 11. Fault current contribution from an induction generator

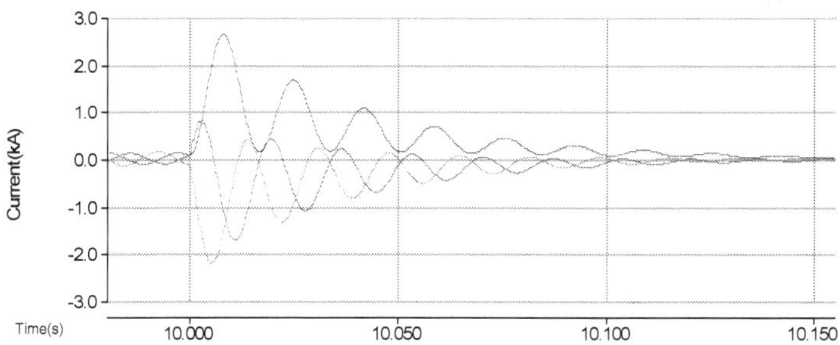

Fig. 12. Fault current contribution from an induction generator with external rotor resistance. Note the faster decay than in the normal induction generator case in Figure 11.

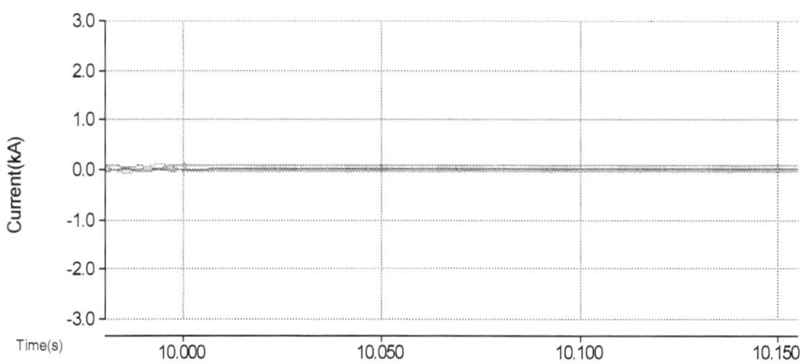

Fig. 13. Fault current contribution from a doubly fed induction generator (crow bar operation is not simulated).

3.3 PV grid integration

Interconnection of a 250 kW photovoltaic (PV) power system in place of the wind turbine is considered in this example. Detailed models of the PV array, power converter and its controller with maximum power point tracking (MPPT) are used in the simulation. The system behaviour under rapidly changing solar radiation intensity is examined. The schematic of the PV system in PSCAD/EMTDC is shown in Figure 14. Figure 15 shows the variation of the solar radiation intensity. The voltage at bus 14 under the simulated solar radiation variation is shown in Figure 16. The real and reactive power output of the PV inverter is shown in Figure 17. The inverter output current shown in Figure 18 is highly distorted, and its harmonic spectrum is shown in Figure 19. This distorted current causes voltage distortion. The harmonics spectrum of the voltage at bus 14 is shown in Figure 20. If the harmonic distortion is unacceptable, it is necessary to install a filter at the PCC of the PV power system.

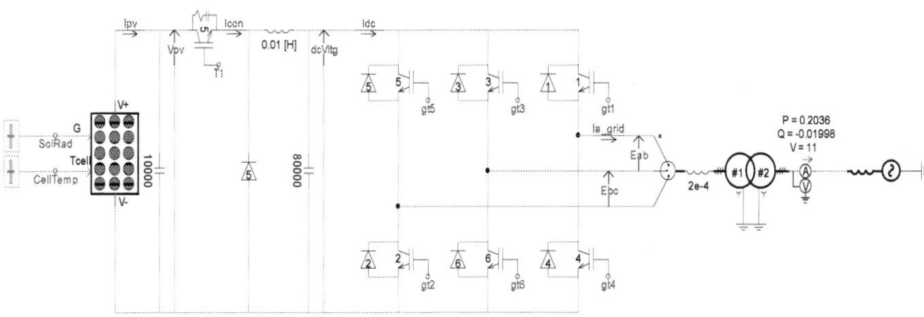

Fig. 14. Schematic of the grid connected PV system simulation (for the simplicity, the ac system is shown as an equivalent source).

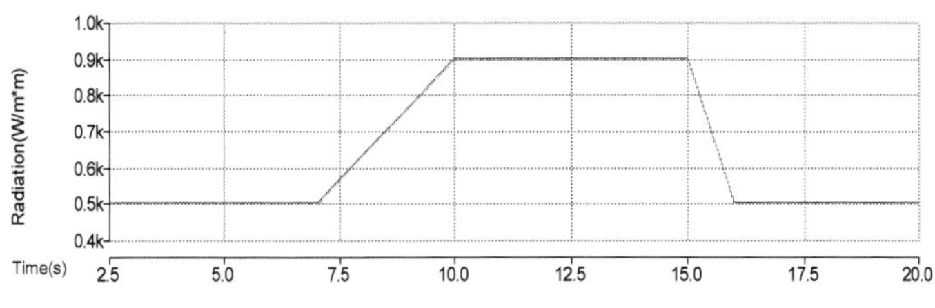

Fig. 15. Variation of the solar radiatin intensity

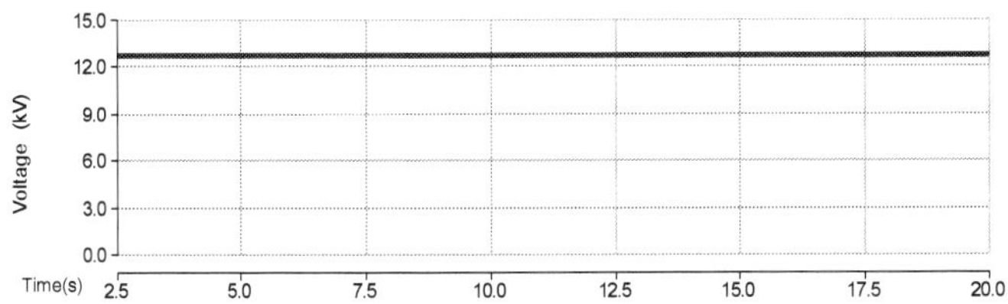

Fig. 16. Variation of the voltage at bus 7

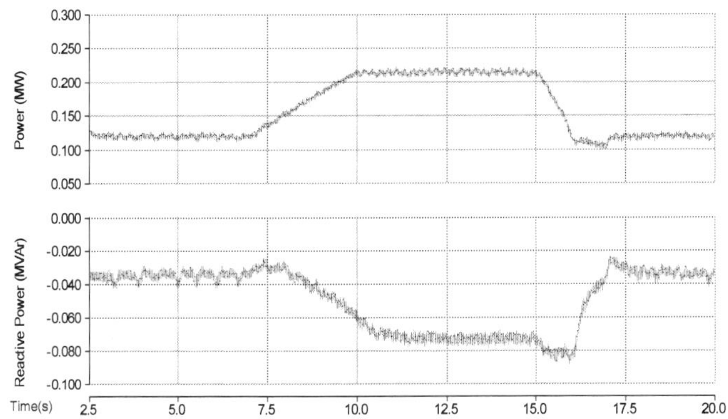

Fig. 17. Real and reactive power output of the inverter

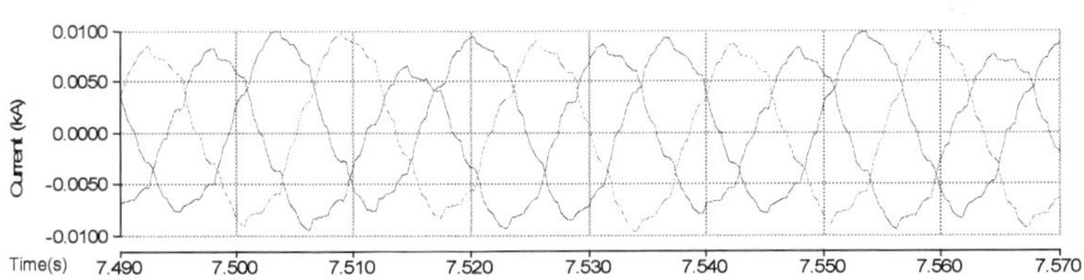

Fig. 18. Inverter output current

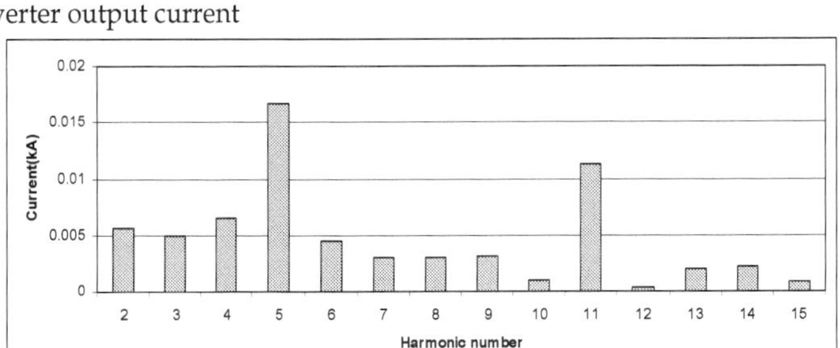

Fig. 19. Harmonic spectrum of the inverter current (Fundamental 0.28 kA)

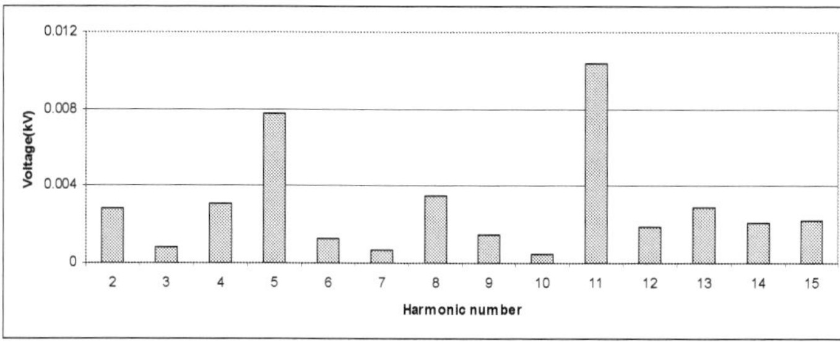

Fig. 20. Harmonic spectrum of the voltage at bus 7 (Fundamental 7.43kV)

4. Integration of large scale renewable energy generation

Large scale renewable energy based generation such as conventional hydro, steam turbines operated with biomass or geothermal energy use conventional technology based on synchronous generators. Utilities are well experienced with this established technology. However, interconnection of large scale wind farms poses entirely different set of challenges, mainly due to the intermittent nature of the wind and the desirability for variable speed operation of the generators.

4.1 Issues related to grid integration of large wind farms

Recent rapid growth of wind energy generation has resulted in the development of large wind farms with capacities in excess of 100 MW. Such large scale wind farms are generally interconnected to the transmission grid. In some countries like Germany and Spain, the installed wind generation capacity has reached several thousand megawatts and the installed wind capacity is continuously increasing. However, power output of the wind generators cannot be dispatched. Despite the advances in short-term (and long-term) wind forecasting technology, accurate forecasting of wind speed is difficult. Therefore, a significant spinning reserve and stand by capacity is required to cover the fluctuations in wind farm production. Although the distribution of wind farms over a wide geographic area provides some cushioning effect, the possibility of simultaneous low production from all wind farms cannot be entirely ruled out.

During a disturbance such as a short duration voltage dip caused by a short circuit in the high voltage transmission network, several thousand megawatts of wind generation can get disconnected depriving real power to the grid. Such a condition can result in serious stability problems within the power system. Thus, new connection requirements for large wind farms have been introduced by many system operators. These new connection regulations such as fault ride through capabilities compel wind turbines to have a supporting effect on grid operation.

A typical fault ride through capability requirement, expected from large, modern wind farms is shown in Figure 21. It specifies the time duration that the wind farm is required to stay connected, for different levels of voltage depressions. This is an attempt to ensure that the wind farm is capable and ready to supply power to the system immediately after clearing the fault to maintain system stability.

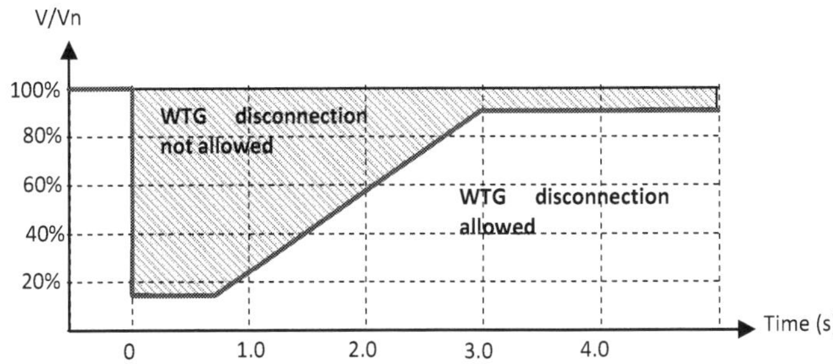

Fig. 21. Typical fault ride through requirements. The actual fault ride through curve can vary depending on the transmission operator.

The 'strict' fault ride through standards presents serious challenges to the wind turbine system designer.

- During the fault, the power transfer capability of the wind turbine generator (WTG) is diminished. This will result in mechanical over speeding of the machine and the turbine blades. This will place mechanical stress on blades and the other rotating parts.
- During the fault recovery period, depending on the generator technology employed, the WTG may draw increased amounts of reactive power from the system. This will result in a poor fault recovery response.
- The nature and the magnitude of the fault current contribution from the wind farm depend on the WTG technology. A good understanding of this is necessary to design equipment ratings and for protection and control settings.
- Turbine blade pitch control must be capable of regulating the input mechanical power. During faults, fast pitch control of blades will limit the mechanical over speeding of rotating elements.
- Application of appropriate FACTS based reactive power compensation devices can help improve the overall performance of the wind farm.

The requirement of reactive power for voltage support is one of the key issues related to wind power generation. Most wind generation systems use induction generators. Unlike synchronous machines, induction machines do not have reactive power control capability. Interconnection regulations require the wind farms to maintain the power factor within a narrow range, usually within ±0.95. If the wind generators are not capable of meeting this requirement, dynamic reactive support must be provided. Usually the reactive power support is provided using a combination of a capacitor bank and static var compensators (SVCs)/STATCOM. The sudden disconnection of the wind turbine generators while the capacitor bank is still connected can result in overvoltage at the generator terminals. This is a major protection problem related to the large wind farms.

The type of reactive power compensation method and devices used at a wind farm also plays an important role in the performance of the wind farm during the period immediately following the clearance of a system fault. Reactive power capability of a capacitor bank (and

SVC's) is directly related to the 'available' terminal voltage. Due to this reason, capacitor banks have a degraded reactive power capability during fault recovery when in fact, their reactive power capability is most needed. FACTS based voltage source converters (e.g STATCOM) does not have this inherent drawback and is generally the superior option, purely from a reactive power capability point of view. The selection of the reactive power compensation scheme and its control strategy must be carefully evaluated.
Synchronous condensers are another reactive power compensation option for large wind farms. They have a slower response compared to a STATCOM, but do provide inertia to the system. This may have advantages, especially when interconnecting to weak ac systems with dynamic stability concerns.
Other possible challenges that require detailed investigation may include:

- Turbine power electronic design and controller optimization
- The design of filters and protection systems
- Interactions between numerous turbines
- Problems of wind farms connected into series compensated systems
- Power quality issues including voltage flicker
- Starting and synchronizing of wind farms to the grid.
- Sub synchronous resonance issues due to interaction of the electric network and the complex shaft/gear system of the wind turbine.

Detailed studies, based on electromagnetic transient programs is essential to identify possible interconnection issues and to find optimum solutions.

4.2 Modeling and simulation requirements

4.2.1 Turbine characteristics

The power transferred by the turbine blades is given by (1)

$$p_{mech} = \frac{1}{2}\rho \cdot A_r \cdot C_p(\lambda, \beta) \cdot V_w^3 \tag{1}$$

$$\lambda = \frac{\Omega_r r_r}{V_w} \tag{2}$$

where C_p is the power coefficient, β is the pitch angle, λ is the tip speed ratio, V_w is the wind speed, Ω_r is the rotor speed (on the low-speed side of the gearbox), r_r is the rotor-plane radius, ρ is the air density and A_r is the area swept by the rotor. The power coefficient, C_p, represents the aerodynamic characteristics of the blade. Note that in Figure 22, C_p is plotted as a function of the 'tip speed ratio'. Tip speed ratio is a measure of the relative speed of wind compared to the rotational speed of the blades. The C_p characteristics can be altered through blade pitch angle (β) control, thereby controlling the amount of mechanical power available to the generator for power conversion. The exact characteristic of C_p depends on the particular design of the blade. The function $C_p(\lambda,\beta)$ can be approximated using analytical equations or determined by tests.

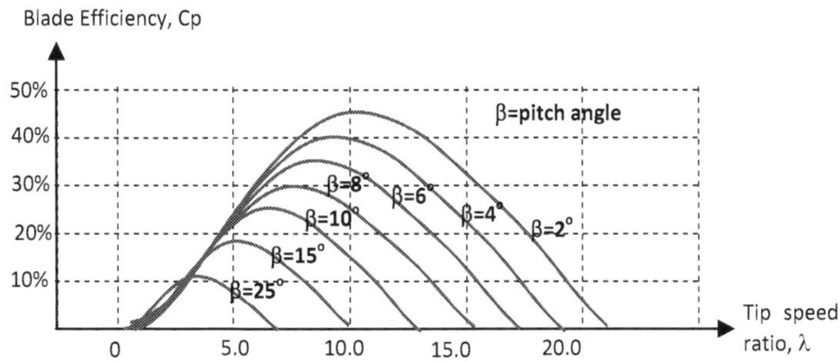

Fig. 22. Typical variation of C_p with wind speed

4.2.2 Turbine pitch control

The control system that orders the pitch angle (β) and the response of the actuators and other mechanical parts may need to be adequately represented in the simulation study. It should be noted that due to the large inertias involved (turbine blades), the overall response of pitch control is 'slow' and is in the order of a few seconds.

4.2.3 Wind generator technology

Most wind turbine generators at present are based on asynchronous generators. It is advantageous to operate these machines at variable speeds during low wind speed conditions. To produce constant voltage and frequency from these asynchronous generators operating at variable speed, they are often interfaced with the grid using various setups of power electronics and controls. The power electronic interface also allows for other desirable features. The controls are sophisticated and complex.

For example they possess the capability of controlling the voltage, real and reactive power output of the machine within some design range, while meeting other constraints such as limiting the positive power ramp rate of the machine. Controller design and settings must ensure low-voltage ride-through compliance. In the future, it is expected that wind turbines would be able to provide more functionalities such as advanced governor functions and controlled ramp down during high wind speed events.

There are several wind generator technologies with the more common WTG schemes being,

- Induction machines directly connected to the grid with external reactive power compensation
- Wound rotor induction machines with an external variable resistor in the rotor circuit to control slip (speed)
- Doubly-fed induction machines, which employ an ac-dc-ac converter between the stator and rotor circuits
- Conventional or permanent magnet synchronous machines connected to the grid through a fully-rated ac-dc-ac converter.

Detailed, high band width models of machines that does not neglect the stator flux transients are required for detailed analysis of such events as

- starting and synchronization wind generators,

- accurate fault ride through analysis
- DFIG control and protection (crow bar etc.)
- VSC based back to back ac-dc-ac converter design including dc line/cable resonance issues.

4.2.4 Power system model

Electromagnetic transient simulations allow for accurate representation of power system equipment. This may include nonlinear core models of transformers and reactors, induction motor and other dynamic loads, overhead lines and cables considering inter-conductor mutual effects, accurate representation of power electronic switching devices, FACTS devices and also control systems.

In general, grid integration requires understanding the needs of the power system, knowledge of machine capabilities, and application knowledge of the power converter and control capability to put the entire picture together. Thus a typical study needs modelling of the grid, wind resource, wind turbine, generator, power electronics, controllers and protection.

4.3 Simulation example

In this example, the application of an SVC to enhance voltage control and system stability is demonstrated. The single line diagram of the high voltage transmission system used for the study is shown in Figure23 [Jiang et al, 2005].

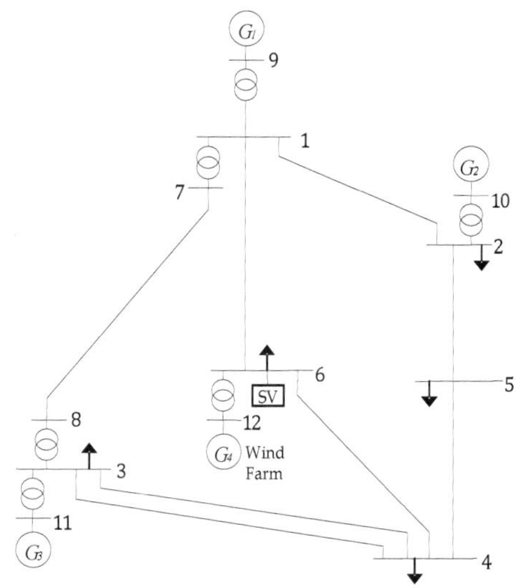

Fig. 23. High voltage transmission system model.

A wind farm is connected to Bus 12. For this study, all generators in the wind farm are represented by a single equivalent generator. In order to provide voltage support, an SVC is connected at bus 6. A fault is applied on the 230 kV transmission line connecting Bus 1 and

Bus 6. The fault is cleared by opening the line breakers after 9 cycles. The response of the wind generators with and without the SVC in operation is shown in Figs. 24 and 25. It is clear that the wind farm is unstable for this fault if not for the voltage support from the SVC.

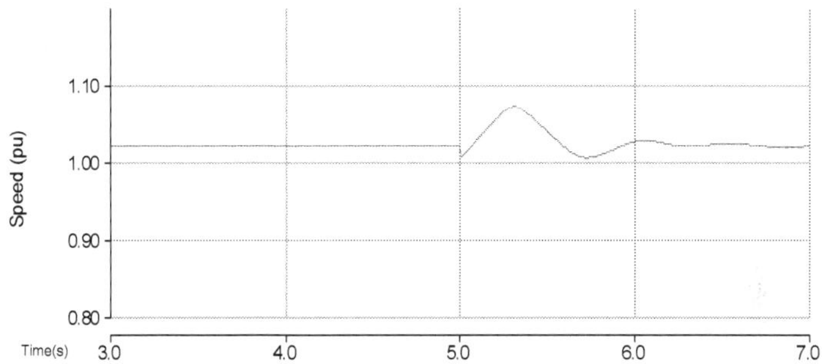

Fig. 24. With the SVC in operation, the generator speed reaches a stable steady state upon the removal of the fault

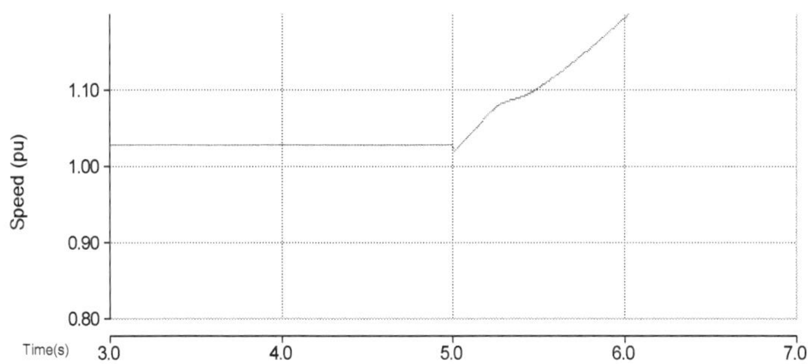

Fig. 25. Without the SVC in operation, the wind generator is unstable.

5. Conclusion

Various issues and challenges that need to be addressed in grid integration of renewable energy systems are discussed. The major concerns in interconnection of the small scale renewable energy generation at the distribution level are related to protection, voltage control and power quality. The reserve requirements, reactive power requirements and the grid support during the disturbances are among the major issues to be considered in grid interconnection of the large wind farms. The importance of the simulation tools in grid integration studies is demonstrated through several case studies.

6. References

Alderfer, R. B.; Eldridge, M. M. and Starrs, T. J. (2000), Making Connections: Case Studies of Interconnection Barriers and their Impact on Distributed Power Projects, Report No. NREL/SR-200-28053, *National Renewable Energy Laboratory*, Golden, Colorado, USA, May 2000.

Barker, P. & de Mello, R.W. (2000), Determining the Impact of Distributed Generation on Power Systems: Part 1- Radial Power Systems, *Proc. IEEE Power Eng. Soc. Summer Meeting*, pp.1645-1658, July 2000.

Borbely, A.M. & Kreider, J.F. (2000), Distributed Generation, *The Power Paradigm for the New Millennium*, LLC: CRC Press, 2000.

Dugan, B. & McDermott, T. (2002) Distributed Generation: Operating Conflicts for Distributed Generation Interconnected with Utility Distribution System, *IEEE Industry Applications Magazine*, March/April 2002, pp: 19-25.

Fadaeinedjad, R.; Moschopoulos, G. & Moallem, M. (2009), The Impact of Tower Shadow, Yaw Error, and Wind Shears on Power Quality in a Wind–Diesel System, *IEEE Transactions on Energy Conversion*, Vol. 24, No. 1, March 2009, pp: 102-111.

Freitas, W.; Huang, Z. & Xu W.(2005), A Practical Method for Assessing the Effectiveness of Vector Surge Relays for Distributed Generation Applications, *IEEE Transactions On Power Delivery*, Vol. 20, No. 1, January 2005, pp: 57-63.

Freitas, W.; Wilsun Xu; Affonso, C.M. & Zhenyu, Huang (2005), Comparative analysis between ROCOF and vector surge relays for distributed generation applications, *IEEE Transactions on Power Delivery*, April 2005. pp: 1315 - 1324.

Hung, Guo-Kiang.; Chang, Chih-Chang & Chen, Chern-Lin (2003), Automatic Phase-Shift Method for Islanding Detection of Grid-Connected Photovoltaic Inverters, *IEEE Transactions On Energy Conversion*, Vol. 18, No. 1, March 2003, pp: 169-173.

IEEE (2003), IEEE Standard for Interconnecting Distributed Resources with Electric Power Systems, Std. 1547-2003.

IEEE (2004), Impact of Distributed Resources on Distribution Relay Protection, Special report of the IEEE, *Power System Relay Committee Working Group D3*, August 2004.

Jenkins, N.; Allan, R.; Crossley, P.; Kirschen, D. & Strbac, G. (2000), Embedded Generation, *IEE Power and Energy Series 13*, 2000.

Jiang, S.; Annakkage, U.D. & Gole, A. M. (2005) , A platform for validation of FACTS models, *IEEE Trans. Power Delivery*, Vol. 20, No.4, October 2005.

Kane, P.O. & Fox, B. (1997), Loss of Mains Detection for Embedded Generation by System Impedance Monitoring Developments in Power System Protection, *IEE Conference Publication No. 434*, 25-27th March 1997, pp: 95-98.

Lopes, J.A.P. (2002), Integration of dispersed generation on distribution networks impact studies, *IEEE Power Engineering Society Winter Meeting*, Vol. 1, pages 323328, 2002.

Mozina, C.J. (2001), Interconnection protection of IPP generators at commercial/industrial facilities, *IEEE Transactions on Industry Applications*, pp 681-688, 2001.

Peters, R.R.; Muthumuni, D.; Bartel, T.; Salehfar, H. & Mann, M. (2006), Dynamic model development of a fixed speed stall control wind turbine at start-up, *IEEE Power Engineering Society General Meeting*, pp: June 2006, pp: 1-7.

Redfern, M. A.; Usta, O. & Fielding, G. (1993), Protection Against Loss Of Utility Grid Supply For A Dispersed Storage And Generation Unit, *IEEE Transactions on Power Delivery*, Vol. 8, No. 3, July 1993, pp: 948-954.

Ropp, M. E. ; Begovic, M. & Rohatgi, A. (1999), Analysis and Performance Assessment of the Active Frequency Drift Method of Islanding Prevention, *IEEE Transactions on Energy Conversion*, Vol. 14, No. 3, September 1999, pp: 810–816.

Rudion, K.; Orths, A.; Styczynski, Z. & Strunz K.,(2005), Design of benchmark of medium voltage distribution network for investigation of DG integration, *Proc. Conference on the Innovations*, Commercial Applications and Regulatory Frameworks of Distributed Generation. November 30, December 1-2, 2005, Boston, USA.

Uchida, A.; Watanabe,S. & Iwamoto, S. (2006), A Voltage Control Strategy for Distribution Networks with Dispersed Generations, *Proceeding of IEEE – PES General Meeting*, pp. 1-6, Tampa, Florida, USA, June 2007.

Hardware in the loop simulation of renewable distributed generation systems

Marco Mauri
Department of Mechanics, Politecnico di Milano
Italy

1. Introduction

The strong growth of Distributed Generation (DG) systems installed on Low Voltage (LV) electrical power networks (with a typical power range of 1 to 400 kW) has emphasized the need to perform preliminary tests before they are put into service to check their efficiency and validate Power Quality (PQ) functions. Typically, these tests are made on DG system prototypes and entail high experimental cost and effort. In addition, it is often very hard to perform significant tests in a short time because DG systems use renewable energy sources, like the wind or sun that are not always available. A hybrid approach, using Hardware in the Loop (HIL) simulation, has turned out to be an effective tool for the evaluation of these systems and various methods have been proposed in the literature, where Hardware in the Loop (HIL) simulation refers to a system in which parts of a pure simulation have been replaced with actual physical components.

Hardware in the loop simulations (HILS) are not new, but have been used extensively for controller assessment for a long time. The aerospace industry has been using this technique ever since software became a safety critical aspect of flight control systems. More recently, however, a combination of several factors has led to a sharp increase in its use, principally the reduced cost and greater availability of HILS products and the intense pressure to reduce development time and costs.

HIL simulation is now used more and more to develop new components and actuators in many fields. Vehicle component evaluation, drive control assessment, power electronics and electric grids, servo control and robotics, railway traction systems for trains and subways and educational applications can all be cited. More recently, HIL simulations have also been used to test distributed generation systems, e.g., to validate control algorithms or Maximum Power Point Tracking (MPPT) strategies. Generally, these HIL techniques are limited to control applications, i.e., there is only signal coupling between the hardware and the virtual system. This approach, even if it is very useful during the control design stage, presents some deficiencies when it is used to develop electrical drives for renewable low power energy generators.

In these systems, it is very important to evaluate the interactions between the electrical drives and the mechanical components in order to avoid dangerous working conditions and, moreover, to estimate precisely the system's energy efficiency. In particular, this parameter is an important variable in a renewable energy plant because these systems are quite expensive and a small increase in global efficiency can drastically reduce the payback time. For this reason, different kinds of HIL simulations can be used to meet these requirements.

2. HIL simulation concepts for Distributed Generation systems

DG systems can present different topologies, but a wide number of them require that a static power converter be efficiently connected to an LV network or that active power with a stable voltage and frequency is provided. Indeed, the power converter interface can perform different functions:

- Compensation of reactive power (with or without storage elements).
- Compensation of negative sequence components and components with different frequencies for the current.
- Compensation of voltage sags, micro interruptions and interruptions (with storage elements).

DG sources can rely on both renewable and non-renewable sources, such as wind turbines, photovoltaic generators, small hydro, fuel cells and gas powered combined heat and power (CHP) stations. Figure 1 shows a simplified schematic of a complete DG system.

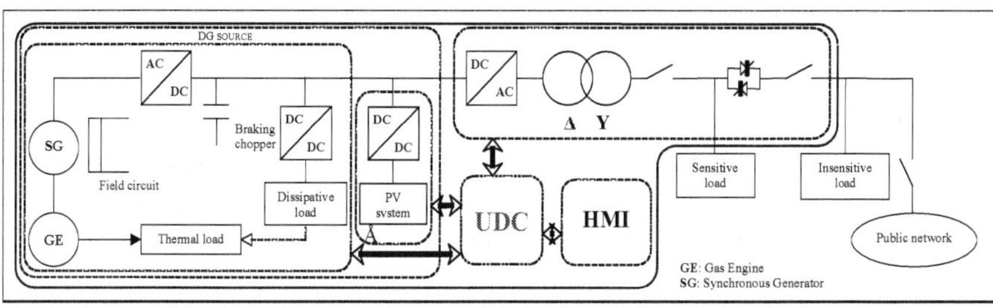

Fig. 1. The simplified architecture of a generic Distributed Generation (DG) system.

The Electrical Drive constitutes the main subsystem of a renewable DG system and the major development effort is often concentrated on this part. An electrical drive can be decomposed into several subsystems: the control software, the power electronics set, the electrical machine and the mechanical load to move (Fig. 2). Typically, a dedicated control board implements the control software and supplies the switching commands to the power electronics converter. Measurements of all the power parts are inputs for the controller board.

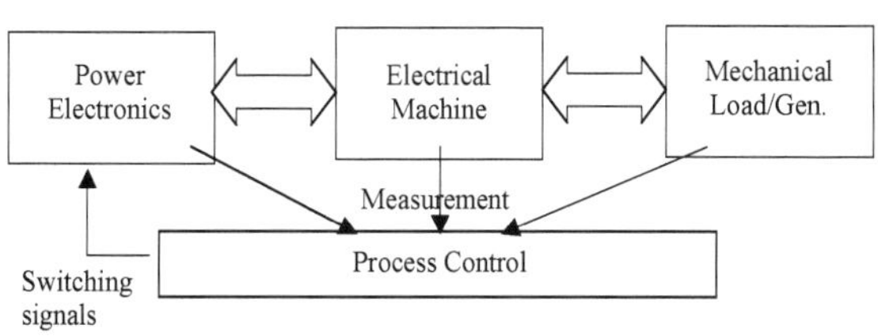

Fig. 2. The Electrical Drive structure.

In a traditional computer simulation, all of these parts are simulated in the same simulation software environment and, in order to reduce the computation time, simple models and other simplifications are considered (e.g., the sampling period of the control is often neglected). This approach is not always accurate enough to allow a real-time implementation of the control without other intermediate steps. HIL simulation could be a very useful intermediary step: one of the simulated parts can be replaced by its hardware component in order to take into account the real constraints in the simulation loop (Fig. 3).

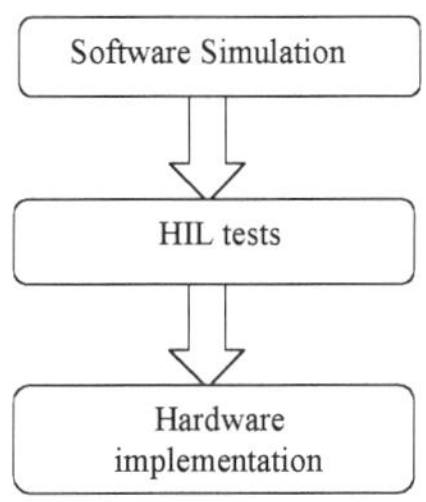

Fig. 3. Electrical Drive design procedure.

Different kinds of HIL simulations can be considered:

- Signal level HIL simulation, where only the control board is tested and the other parts are simulated in real-time. This kind of HIL simulation has very often been employed in aerospace and automotive applications for the assessment of controller boards (Fig. 4).
- Power level HIL simulation, where the control board and the power electronics converter are evaluated and the electrical machine and the mechanical load are simulated (Fig. 5).
- Mechanical level HIL simulation. In this case, the whole drive (control, power electronics and electrical machine) is tested and the mechanical part is simulated (Fig. 6).

Each of these simulations can be used to test a DG electrical drive, although obviously highlighting different aspects and problems.

2.1 Signal Level HIL simulation

In a Signal Level Hardware in the Loop Simulation, only the controller board (which contains the process control) is tested (Fig. 4). The other parts (power electronics, machine and mechanical load) are simulated in real-time. The simulation system must manage the inputs and outputs of the controller board under test. A second controller board is thus used to simulate in real-time the power parts of the system. Specific signal conditioning is required to impose the same inputs and outputs as imposed by the power parts. This method can be called "signal level HIL simulation" because only signals are used at the interface between the system under test and the simulation environment. This kind of HIL has very often been employed in aerospace and automotive applications for the assessment of controller boards.

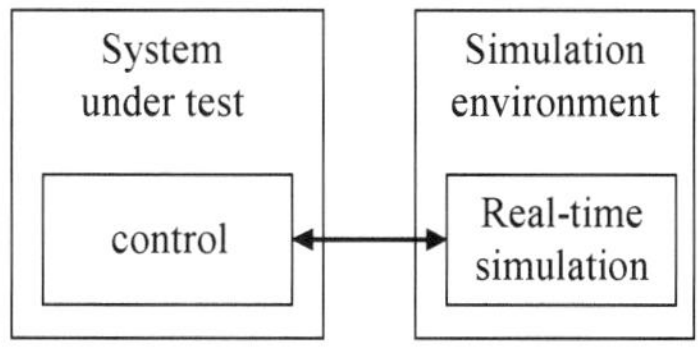

Fig. 4. Signal Level HIL simulation.

2.2 Power Level HIL simulation

In a Power Level HIL simulation, the actual controller board and the power electronics converter are evaluated. The other parts (electrical machine and mechanical load) are simulated. The simulation system must impose inputs and outputs for the power electronics and the controller board under test. The simulation environment is generally composed of a second power electronics set (electric load) and a second controller board (real-time simulation) (Fig. 5).

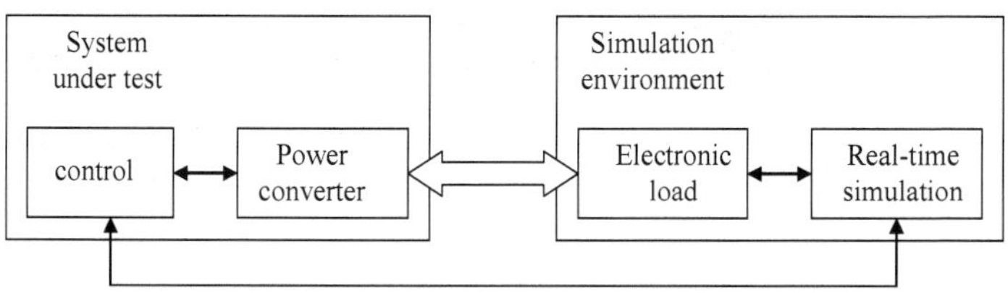

Fig. 5. Power Level HIL simulation.

2.3 Mechanical Level HIL simulation

In this case, the whole drive (control, power electronics and electric machine) is tested and the mechanical part is simulated. The simulation system must impose mechanical inputs and outputs on the electrical machine under test. Moreover, measurements on the mechanical part have to be sent to the controller board under test. Another electrical machine (load machine) is often used as a controlled mechanical load. It is supplied by a second power electronics set (load supply). A second controller board (real-time simulation) is required to control the load machine and to send fictitious mechanical "measurements" to the controller board under test (Fig. 6).

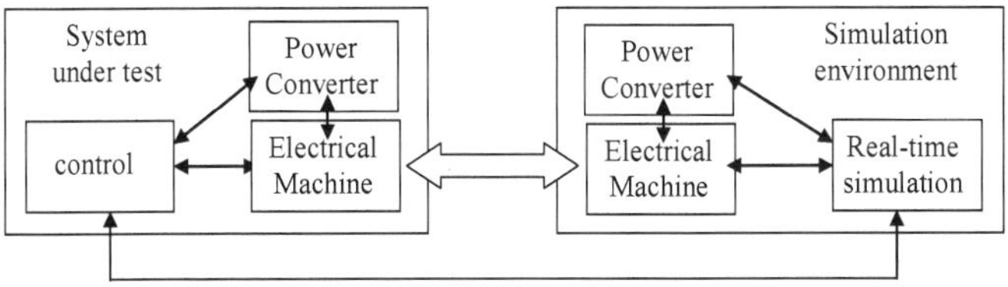

Fig. 6. Mechanical Level HIL simulation.

3. Real Time modeling concepts

The implementation of real time models of controlled devices is a well-known problem but becomes a more difficult challenge when the time constraints are very narrow. The simulation models should satisfy different requirements:

- precision and reliability to represent physical system behavior;
- a simple mathematical structure to allow for fast numerical integration and easy software implementation;
- a standard structure for the model equation in order to obtain a more flexible and easily configurable simulation platform.

An Object Oriented approach is typically followed to meet these characteristics during the implementation of emulation models, making it possible to improve the emulator flexibility. Each part of the complete system has been considered as a single object with inputs, outputs and internal states. In this way it is very simple to reconfigure the simulation system without changing the hardware and software structure.

3.1 Signal Level Modeling

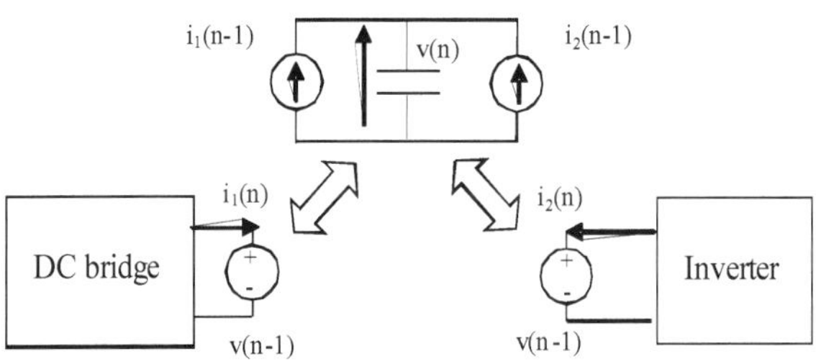

Fig. 7. Example of an Object-Oriented integration procedure for the DC Stage.

In signal level real-time modeling, the smallest bound is the time step necessary to implement the high dynamic behavior of the power electronics stage. This time step is very small (typically 10-100 μs), so it is necessary to make a great effort to develop suitable models.

However, a suitable choice for input/output object signals makes it possible to extend the Object Oriented approach during the simulation phase in order to improve the simulation speed because each object can be integrated separately and in a parallel way.
It is obvious that this approximation is acceptable only if input/output object signals change slowly. Signals that satisfy this requirement are the state variables of each subsystem. Figure 7 shows the proposed procedure for the connection between two power converters of a DG system (Fig. 1). The presence of a capacitor on the DC bus makes it possible to decouple the integration of the two power converter models.
Another big problem that is typical of this HIL simulation is the high sampling speed necessary in order to acquire control board switching commands. In order to decouple the acquisition time of the switching state and the integration, the average values of the switching functions are often used instead of the actual values. Different averaging methods are presented in the literature, according to the simulation requirements, especially in the emulation of system faults. An interesting averaging law is presented in (Bucca et al. 2006), where a two level inverter has been modeled using a dedicated acquisition board with two counters. In fact, the output voltage vector of the two-level inverter depends on the DC bus voltage value and on the gate signal sequence only (Fig. 8).
The counter increment is a gate signal function (Table 1) and the mean values of the d and q components of the output voltage space vector could be obtained using Equation (1).

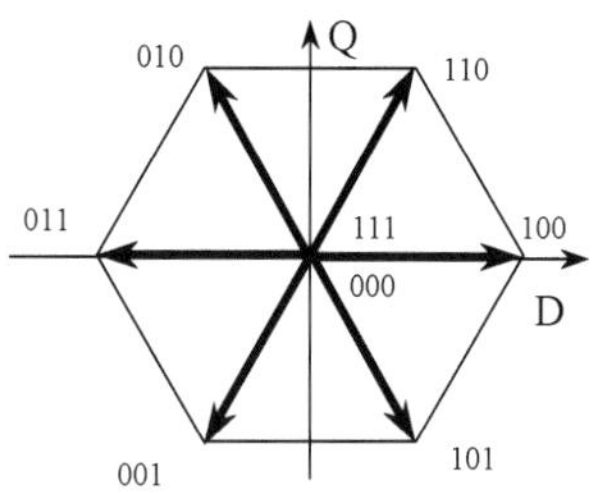

Fig. 8. Output voltage vectors for a two level inverter.

Gate signals	swd	swq increment
000	0	0
001	$-1/2$	$-\sqrt{3}/2$
010	$-1/2$	$\sqrt{3}/2$
011	–1	0
100	1	0
101	$1/2$	$-\sqrt{3}/2$
110	$1/2$	$\sqrt{3}/2$
111	0	0

Table 1. Counter increments.

$$v_{sd} = \sqrt{\frac{2}{3}} V_{DC} \frac{1}{N} \sum_{n=1}^{N} sw_d(n)$$
$$v_{sq} = \frac{\sqrt{2}}{2} V_{DC} \frac{1}{N} \sum_{n=1}^{N} sw_q(n) \qquad (1)$$

where:

- N is the number of gate signal samples during an integration step;
- $sw_d(n)$, $sw_q(n)$ are the n-th values of the counters according to Table 1;
- V_{DC} is the value of the DC bus voltage.

Because this voltage is typically used as an input of electrical machine models that are written on a dq axis, this averaging method reduces the number of calculations, with a remarkable improvement in the simulation speed.

3.2 Power Level Modeling

Power level HIL simulations have problems similar to Signal level HIL, because the time boundaries depend on the same electrical components.
A big problem in a Power level HIL simulation is that it is necessary to decouple the dynamics of an electronic load from the power electronic converter. Typically, this simulation has been used in DG HIL simulations to test the PV field controller in order to also evaluate the power converter efficiency.
A typical schematic of an HIL power level test bench is shown in Fig. 9.

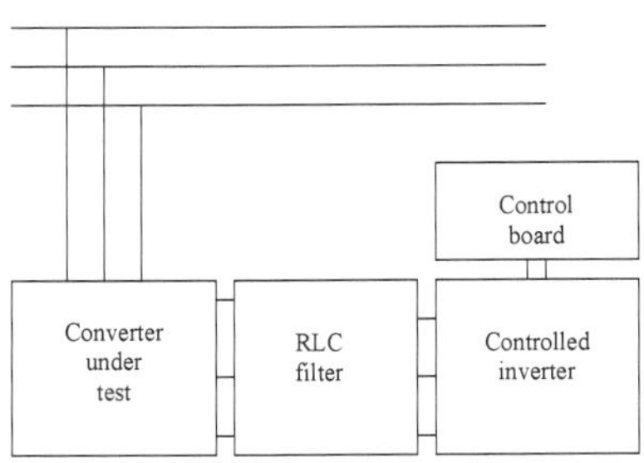

Fig. 9. Schematic of Power level HIL test bench.

A three-phase RLC filter connects the hardware under test with the controlled converter. The filter decouples the operation of the converter from the hardware under test and reduces the ripple introduced by the switching operations of the controlled inverter. In this configuration, the currents drained in each branch of the filter can be regulated by controlling the output voltages of the inverter in order to track the references computed by the real-time simulation.

A critical step in the realization of this test bench is the measurement of the voltages generated by the converter under test, because these voltages represent the inputs for the real-time simulation. A way to overcome this problem is to estimate these voltages by acquiring the switch signal in a way similar to the signal level HIL test bench, but this structure depends on the control electronics of the converter under test. In fact, the platform interface has to be specifically customized based on the characteristics of the signals that drive the switch gates. In addition, these signals may not be available in industrial converters.

The easiest solution if the gate signals cannot be used is to measure the average voltage of the converter output directly. The voltages are scaled and reduced to a signal level by means of suitable sensors (e.g., hall sensors) that provide galvanic isolation, and then they are averaged over a simulation period. The main advantage of this technique is that the voltage used in the real-time simulation is reconstructed from the power signal instead of the gate signal in order to include switching non-linearity and improve the HIL test bench standardization.

3.3 Mechanical Level Modeling

In a mechanical level HIL simulation, the presence of a real electrical drive removes the very small time step boundaries of signal level and power level HIL simulations and introduces new boundaries that depend on mechanical dynamics (2).

$$T_m = T_r + J\frac{d\omega}{dt}, \tag{2}$$

where T_m is the electromagnetic torque, T_r is the load torque and J is the equivalent inertia. Equation (2) shows that the electrical drive used to emulate the mechanical load should supply not only the load torque but also the inertial torque, which could present high values and dynamics. To satisfy this requirement, it is necessary to use a high performance drive, which is not very simple to implement and control.

To avoid this problem, a different approach can be used, as proposed in (). In particular, the DG power source presents a low dynamic speed variation due to a high damping factor and inertia, so it is possible to modify the mechanical emulator control of the electrical drive from torque to speed control. In this way, the mechanical model could be reduced to a quasi-steady-state model by considering the DG power source speed as a constant during the integration step. This is possible because the integration model step is very small compared with the mechanical time constants, due to the previous considerations.

Using this approach, the acceleration torque can be calculated by subtracting the load power (P_l) from the available mechanical source power (P_s) and dividing the result by the actual speed. The load power is measured by the electrical drive to take into account all the mechanical losses in the power train.

The new speed reference (ω_{ref}) for the electrical drive that emulates the DG power source is calculated by dividing the acceleration torque by the system equivalent inertia, as is indicated in (3) and (4) and represented in Fig. 10.

$$T_a = \frac{P_s - P_l}{\omega_{act}} \tag{3}$$

$$\frac{d(\Delta\omega_m)}{dt} = \frac{1}{J} T_{acc} - B\omega_{act} \tag{4}$$

$$\Delta\omega_m = \int_0^{\Delta t} \left(\frac{T_a}{J} + B\omega_{act} \right) dt \Rightarrow \omega_{ref} = \omega_{act} + \Delta\omega_m \tag{5}$$

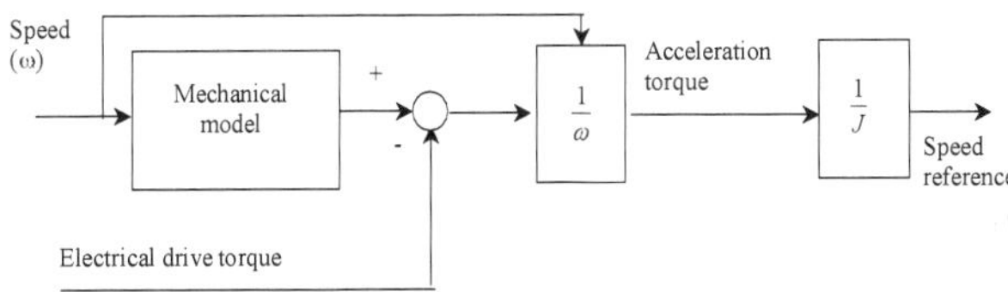

Fig. 10. Calculation of the new speed reference.

The proposed approach simplifies the modeling of high inertia loads, making it possible to test control strategies under dynamic conditions, while avoiding the measurement of load acceleration. Moreover, it is possible to use a less sophisticated electrical drive, control board and emulation board, thus reducing the test bench cost.
A different approach was proposed in (Kuperman and Rabinovici, 2005), where the simulation of load variation was introduced by adding a virtual load signal to the output of the electrical motor controller. Even though this approach is very interesting, it is very difficult to implement it, because it is necessary to modify the controller software.

4. Experimental results

In the following sections, some experimental results using different HIL configurations are presented.

Motor	
Rated power at 50 Hz	370 W
Rated torque	1.3 Nm
Rated speed	2820 rpm
Rated current	1.7 A
Starting current	4.56 A
Power factor	0.83
Starting Torque	3.9 Nm
Rotor inertia	3.5×10^{-4} kgm^2
Nr. of pole pairs	1
Statoric phase resistance	24.6 Ω
Rotoric phase resistance	16.1 Ω
Magnetizing inductance	1.46 H
Electrical Time constant	1.62×10^{-3} s

Converter	
Input voltage	single-phase 180-240 V
Output power	750W, 150% overl.1 min.
Output voltage	0-230 V
Power switches	IGBT

Table 2. Emulated Converter and Motor characteristics.

4.1 Signal Level HIL simulation

An example of a real-time emulator hardware structure for signal level HIL simulations is presented in Fig. 11. The HIL test bench emulates an asynchronous drive with the characteristics indicated in Table 2. The electrical drive used for the test was a Technosoft induction motor drive with a control board based on TMS320F240 DSP.

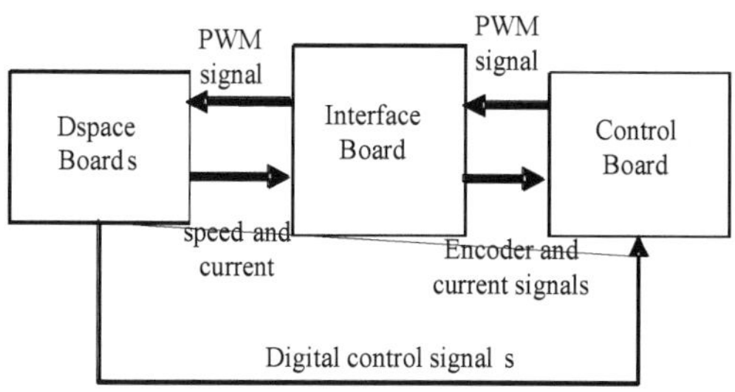

Fig. 11. Signal level HIL test bench architecture.

A dSpace system (ds1006 board based) was used to integrate the electrical drive models, acquire the PWM signal and generate speed and current signals. A dedicated interface board was used to adapt the analogue signal levels and generate the encoder signals.

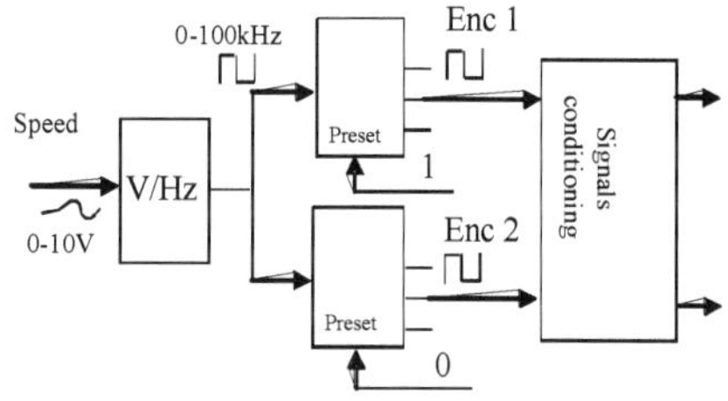

Fig. 12. Encoder emulation.

The encoder signal generation was done using a Voltage/Frequency converter and two digital counters (Fig. 12). The basic idea was based on the observation that if the counter

chips have been initialized to 0 and 1, respectively, the output 2nd bit signals are square signals with a π/2 phase shift.
Fig. 13 shows the phase current and speed signals acquired by the control board for different speed transients. The induction motor was controlled using close loop field oriented control.

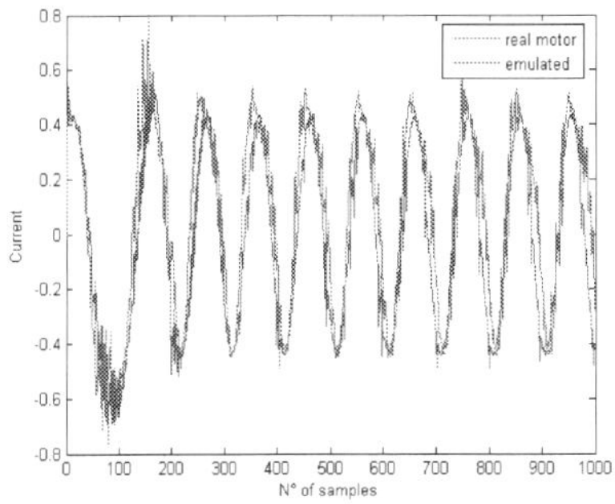

a) Phase current

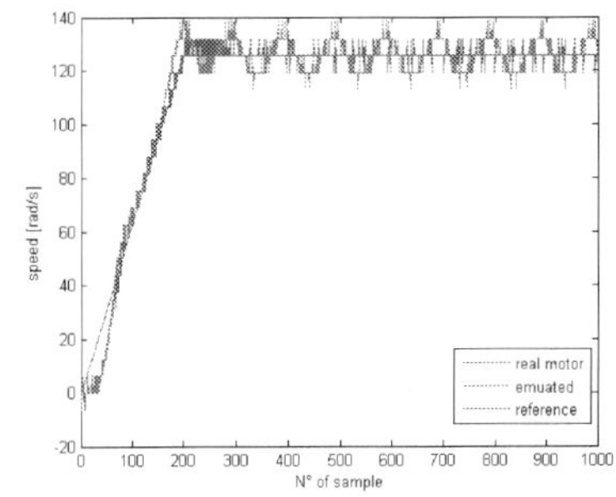

b) Speed

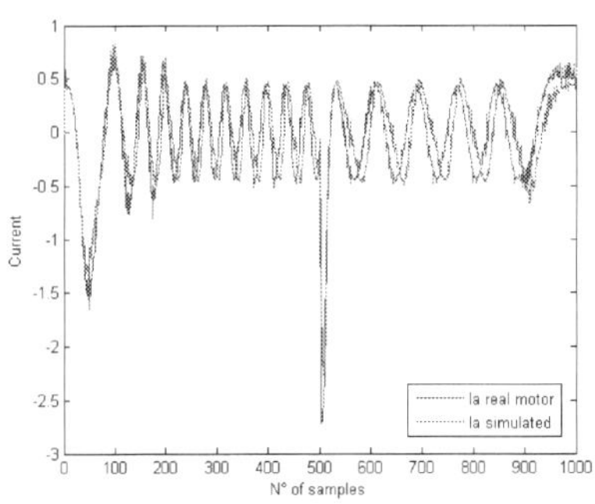

c) Phase current

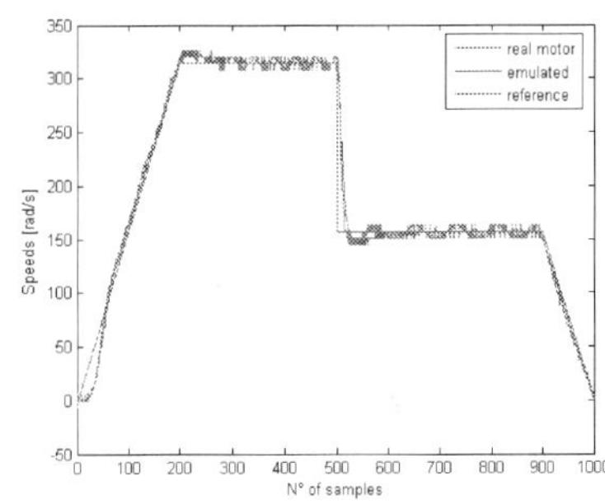

d) Speed

Fig. 13. Comparison between real motor acquisition and emulated system acquisition of phase current and speed for different speed transients.

It is interesting to note that the real transients and emulated transients are very similar. There was a little difference in the first speed acquisition (Fig. 13b), but it was practically limited to the less significant bit of the digital acquisition. This difference was principally caused by the absence of any additional conditioning stage between the encoder emulation board and the control board and was more significant at low speed (comparing Fig. 13b-d).

4.2 Power Level HIL simulation

A power level HIL simulation of a PV field was performed. A typical configuration for a photovoltaic (PV) generator is shown in Fig. 14.

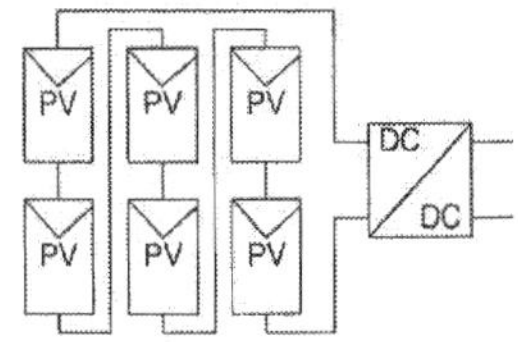

Fig. 14. Configuration of emulated PV field.

Normally a set of PV modules is series and parallel connected to form a PV multiple module or PV subfield, with suitable rated voltage and current. The PV subfield is then connected to a DC-DC converter. It is also possible to connect individual PV strings to their own DC-DC converters. The PV generator can work either in a grid connection mode or in stand alone operation without a storage system, thanks to a suitable control strategy. When a photovoltaic system works in the grid connection mode, a MPPT control technique is implemented in order to extract the maximum electrical energy from the PV panel.
Using the HIL test bench, a DC/DC converter for a PV field with a Perturb and Observe MPPT algorithm was tested. The behavior of a SHELL SQ8 PV panel was emulated. Fig. 15 shows the behavior of a SHELL SQ8 PV panel for different irradiance levels and temperatures.

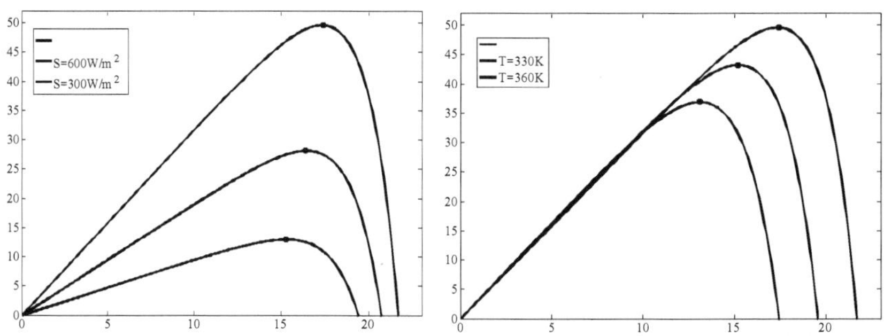

Fig. 15. SHELL Q8 PV V/I characteristics.

The implemented P&O algorithm operates by periodically perturbing (i.e., incrementing or decrementing) the array terminal voltage and comparing the PV output power with that of the previous perturbation cycle. If the PV array operating voltage changes and the power increases ($dP/dV_{PV}>0$), the control system moves the PV array operating point in that direction; otherwise the operating point is moved in the opposite direction. In the next perturbation cycle, the algorithm continues in the same way.
A common problem in P&O algorithms is that the array terminal voltage is perturbed every MPPT cycle; therefore when the MPP is reached, the output power oscillates around the maximum, resulting in power loss in the PV system. This is especially true in constant or slowly-varying atmospheric conditions, but also occurs under rapidly changing atmospheric conditions (in this case the perturbation direction can be wrong).

There are many different P&O methods available in the scientific literature. In this case, a modified P&O algorithm was adopted in which the perturbation amplitude of the PV operating point has a constantly variable magnitude. The magnitude of the perturbation is a function of the power variation necessary to reach the target.

a)

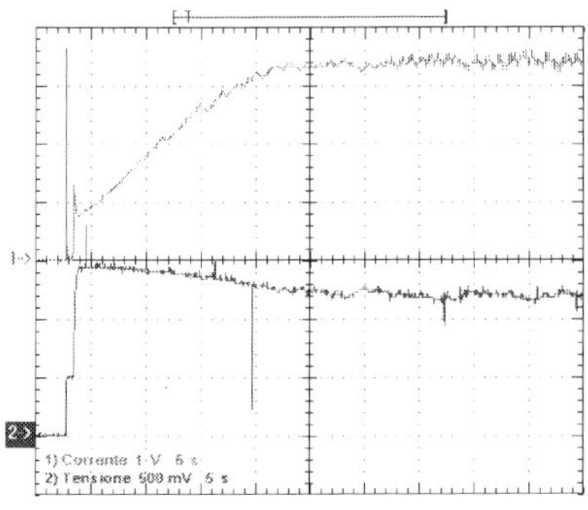

b)

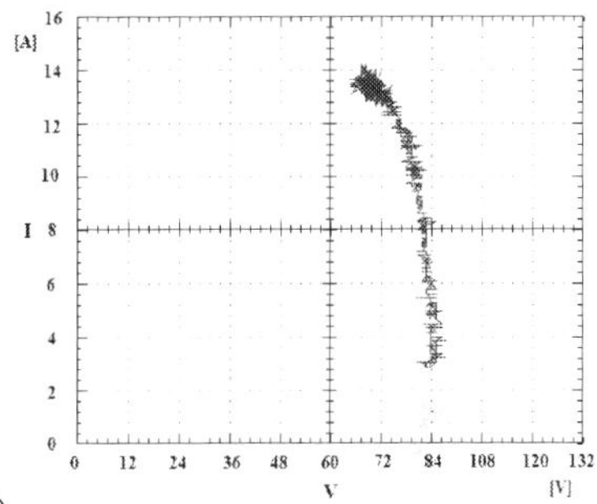

Fig. 16. a) 1 kW PV subfield I/V output characteristics during start up phase (I=20A => 5V; V=300V=>5V) b) I/V characteristics during start up phase of MPPT control.

a)

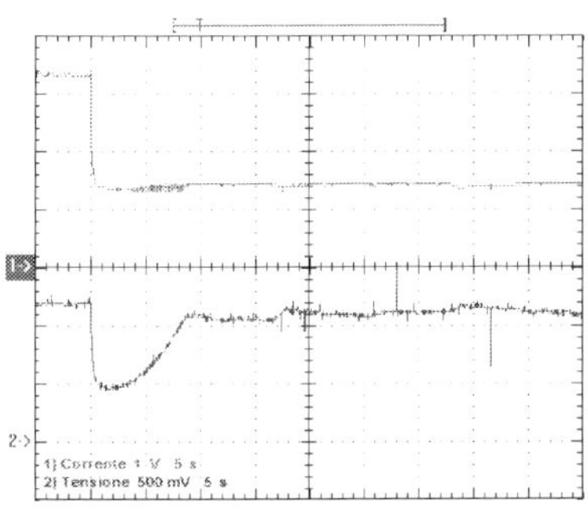

b)

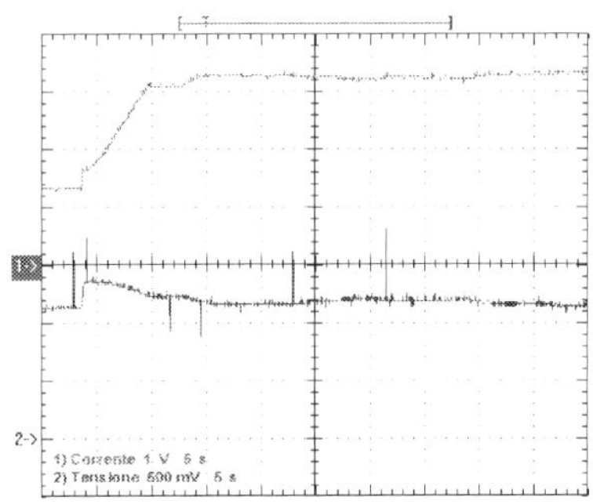

Fig. 17. 1 kW PV subfield I/V output characteristics when variable step P&O algorithm is used (I=20A => 5V; V=300V=>5V), a) during power discharging (from 1000W to 400 W) and b) during power changing (from 400 W to 1000 W).

Fig. 16 presents the V/I characteristics during the PV subfield start up simulation. The PV subfield current starts from a null value with the maximum voltage value and reaches the MPP in 30 seconds, with constant current ripple amplitude.

An instantaneous reduction in the PV subfield power from 1000 W to 400 W was simulated. The MPP changes from 74V-13A to 69V-5,5A. Fig. 17 represents the algorithm behavior during an instantaneous increase from 400 W to 1000 W. The MPP changes from 69V-5,5A to 74V-13A.

4.3 Mechanical Level HIL simulation

Two different DG systems were emulated using an HIL mechanical level test bench.

- small CHP Generating Set
- Small-scale hydropower generator

4.3.1 Small CHP generating set.

Small power CHP (Combined Heat and Power) generating sets, based on internal combustion engines, are normally grid-connected and have a power range of 10 kW to 100 kW. These small CHP generating sets can be equipped with both diesel engines and gas fuelled engines. The advantage of using combined heat and power generation is that the global efficiency can be increased to 85 to 90%, which make them a very favorable solution for Distributed Generation systems. The most convenient arrangement is to use an asynchronous generator, even if it shows some drawbacks. In particular, it requires reactive power for the excitation of its magnetic circuits and, in the case of failure of the mains, it does not work in a stand alone mode to provide electrical power to the on-site ac load.

Conventional CHP sets work at a fixed speed in order to generate electrical power at a fixed frequency, but they show low energy efficiency when they operate at partial load. An improvement can be obtained by using a variable speed gas engine to drive the electrical generator at a variable frequency. Operating the CHP set at a variable speed makes it possible to achieve the highest energy conversion efficiency at any partial load of more than 20%, but it is necessary to use an AC/AC frequency converter in order to connect the variable speed CHP set to the public grid.

The behavior of a variable speed CHP set can be further enhanced by introducing an additional DC/DC static converter, in order to increase the amount of electrical power that can be transformed into heat (dissipating electrical power on a resistor). In this way, a control strategy with heating priority can meet a temporary high demand for heat, reducing the production of electricity.

4.3.2 Small-scale hydropower generator

Hydroelectric generation has been well recognized as environmentally friendly and socially beneficial for many applications, but large-scale hydroelectric plant developments have become more and more difficult because of the shortage of undeveloped suitable sites and concerns about global environmental protection. However, there are many sites where it is possible to install a small-scale hydroelectric plant.

These systems normally adopt expensive Kaplan turbines that allow optimal control of the water flow. In order to reduce plant costs in small-scale plants, it is preferable to use fixed-blade propeller water turbines.

This choice can reduce the plant power efficiency because typically in traditional hydroelectric power plants, the speed of the generating unit remains constant to keep it synchronized with the grid, neglecting turbine discharge variation over time. Moreover, this situation becomes critical in stand-alone operation mode, where keeping the speed of the generation unit constant reduces the power plant efficiency drastically. In these cases, a variable speed operation mode and maximum power efficiency operation point tracking is welcome, but it presents numerous design problems regarding the definition of the hardware architecture and power management strategies. Such small-scale hydroelectric plants generally use a PM synchronous machine as an electrical generator.

4.3.3 The HIL test bench

Each of these applications presents similar problems. In particular, it is very important to evaluate the mechanical performance and, moreover, it is necessary to evaluate the global efficiency (converter and motor) with a good approximation. This it is very difficult to do using an HIL environment where the power converter is also emulated: the calculation effort necessary to simulate the electrical drive characteristics and losses does not allow simple and low cost real-time implementation.

Because of these considerations, mechanical emulation is the better choice. An example of mechanical test bench hardware architecture is shown in Fig. 18, which includes:

- A motor drive, including an inverter controlled squirrel cage induction motor. This motor was used to emulate the static and dynamic mechanical behavior of the DG power source. An incremental optical encoder was used to feed back the speed signal for control and emulation purposes.
- A DC motor was used to emulate the generator side of the DG system. The rated parameters of the induction motor and generator are listed in Table I.

Induction Motor		DC Motor	
Rated voltage	380 V	Rated voltage	240 V
Rated current	35.1 A	Rated current	60 A
Rated Power	18.5 kW	Rated power	14 kW
Power factor	0.88		
Rated speed	1465 rpm	Rated speed	1500 rpm
Wound connection	D		
N° of poles	4	N° of poles	4

Table 3. Induction Motor and DC Motor rated Parameters.

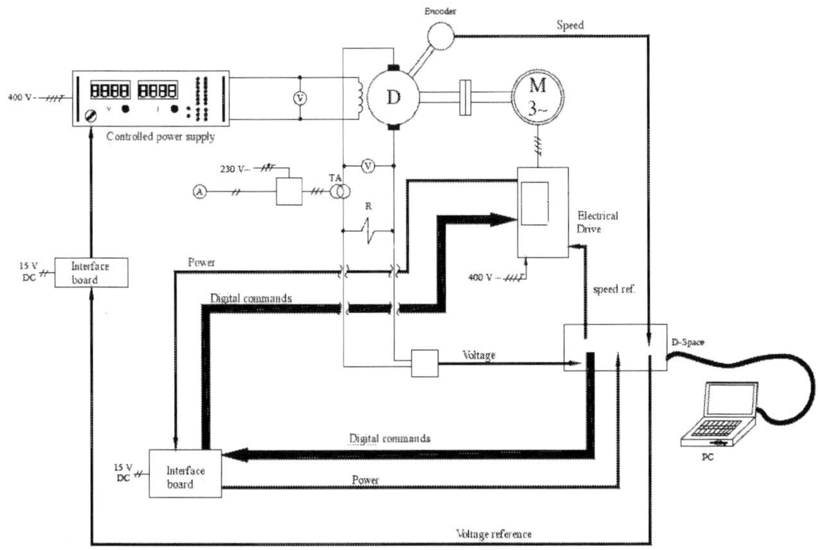

Fig. 18. Example of mechanical HIL test bench

- A desktop PC with a dSpace 1104 board that implements the DG power source model. A dedicated interface board was used to acquire measurement signals from the field and the motor drive and to transmit control reference signals from the dSpace board to the motor drive.
- A passive load to dissipate the generated power.

4.3.4 Experimental results

According to the previous discussion, when simulating the gas engine of a CHP system, it is necessary to evaluate the equivalent inertia and damping factor, along with the generated power vs. rotational speed curve. In a gas engine, this curve is proportional to the throttle valve opening.

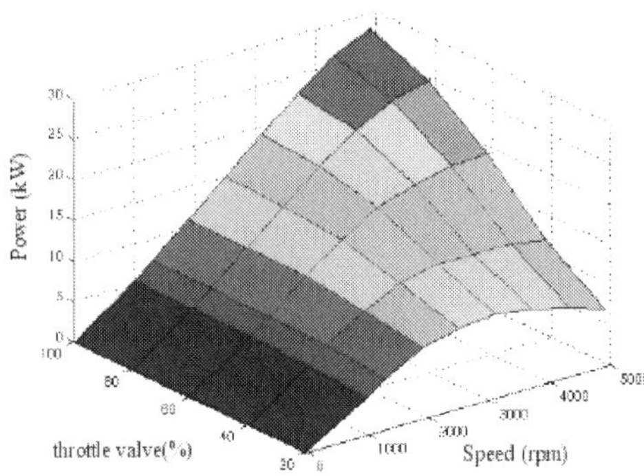

Fig. 19. Generated power vs. speed and throttle valve opening.

Fig. 19 represents the generated power vs. rotational speed for different throttle valve openings (in percentage) for a 30 kW gas engine. In order to validate different control strategies, it was also necessary to emulate the fuel consumption under different load conditions. Fig. 20 shows the fuel flow in kg/h vs. speed and the throttle valve openings for the same motor.

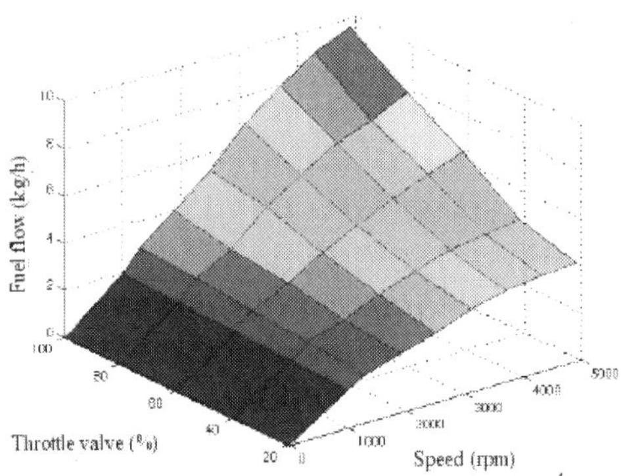

Fig. 20. Fuel flow vs. speed and throttle valve opening.

$$J = 4\ kg / m^2 \qquad (6)$$
$$B = 3\ J / (rad / s)$$

The estimated inertia and damping factor of the motor are indicated in (6).

The fixed blade turbine of a small-scale hydropower plant can be modeled in the same way as the gas engine. In this case, the turbine discharge (Q) and efficiency (η) vs. speed are obtained using the steady-state curves and these values are used in combination with the turbine head (H) to calculate the available mechanical power (6). Table 4 shows the parameters used.

$$P_{hyd} = HQg \cdot \eta \tag{7}$$

Rated power	3.2 kW
Rated discharge	0.1 m^3/s
Design head	5.2 m
Rated Rotational Speed	180 rpm

Table 4. Hydro turbine technical data.

$$\begin{aligned} J &= 10\ kg/m^2 \\ B &= 50\ J/(rad/s) \end{aligned} \tag{8}$$

The efficiency of the system depends on the opening of the distributing valve and the speed as indicated in Fig. 21, and the turbine speed is considered to have a constant value during the integration step (in this case it was set to 100 ms).

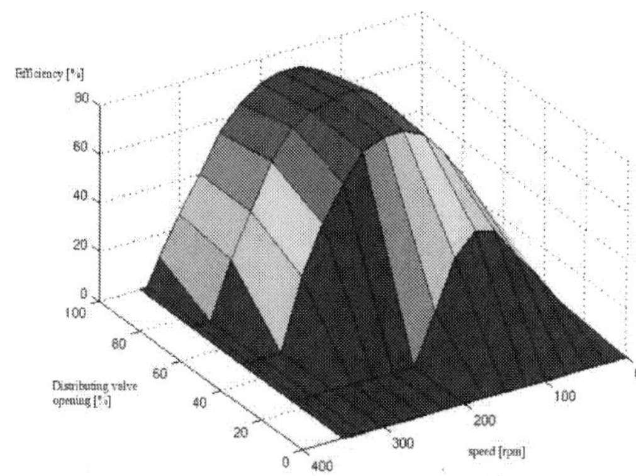

Fig. 21. Efficiency vs. turbine speed and distributing valve opening.

Fig. 22 shows the generated power reference vs. time and the real generated power vs. time using the hydropower turbine model, emulating a grid connection working mode. Fig. 13 compares the estimated generated power using the HIL hydropower model and the real power dissipated on the resistive load of the test bench. The power reference was calculated based on the load variations, while the generated power should also take into account the plant losses. These losses produced the stationary difference between the generated power and power reference.

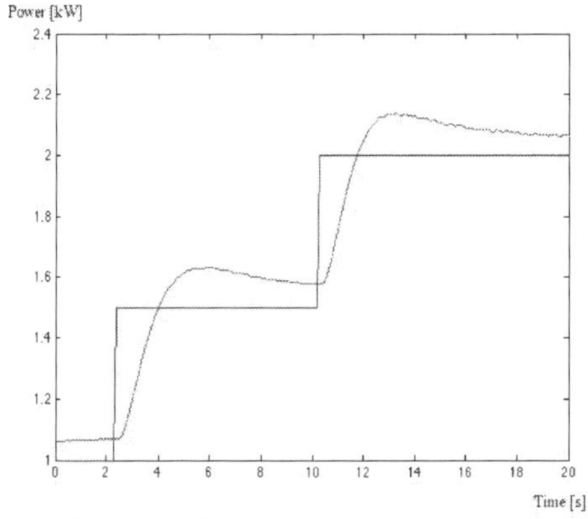

Fig. 22. Power reference and generated power vs. time.

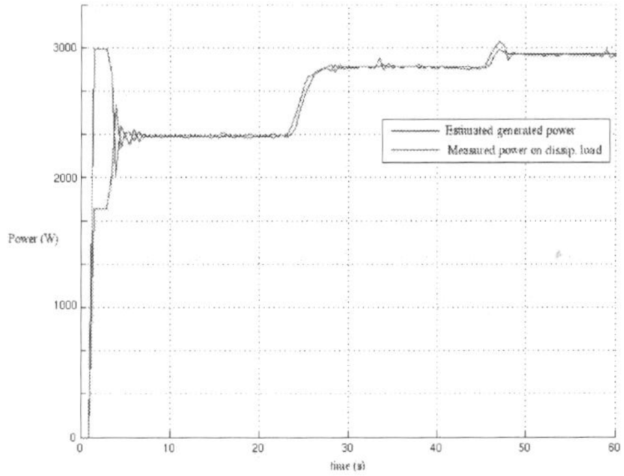

Fig. 23. Estimated generated power (from HIL test bench) and dissipated power on the resistive load.

5. Conclusion

Hardware in the loop simulations (HILS) are a powerful method to test DG system components, such as control system electronics, without connection to the real equipment under control.

The use of this technique does, however, raise challenges, such as complex interfacing requirements and the execution of an accurate equipment simulation that must operate in real-time.

The increasing adoption of HILS has been driven by a number of factors, including the ability to perform system testing before the complete system is available and to simulate failure conditions that would be dangerous or expensive to create in a real plant.

HILS is an important tool to reduce development time and ensure the safety and reliability of DG components. The availability of cost effective tools to perform HILS should lead to an increase in the use of this technique as a standard method to test DG components.

6. References

Bouscayrol (2008). Different types of Hardware-In-the-Loop simulation for electric drives. *Proceedings of IEEE International Symposium on Industrial Electronics, 2008. ISIE 2008.* pp. 2146 - 2151, ISBN 978-1-4244-1665-3, Cambridge, July 2008, IEEE

Bin Lu; Xin Wu; Figueroa, H. & Monti, A. (2007). A Low-Cost Real-Time Hardware-in-the-Loop Testing Approach of Power Electronics Controls. *Industrial Electronics, IEEE Transactions on*, vol. 54 n. 2, pp. 919 - 931, ISSN 0278-0046

Bucca G.; Mapelli F. & Mauri M. (2006). A Real-time environment for industrial electrical drive control board testing. *Proceedings of International Conference on Electrical Machines*, ICEM2006

Facchinetti A. & Mauri M. (2008). Hardware in the loop test-rig for pantograph active control evaluation. *Proceedings of IEEE International Symposium on Industrial Electronics, 2008. ISIE 2008*, pp. 2171 - 2176, ISBN 978-1-4244-1665-3, Cambridge, July 2008, IEEE

Jian Wu; Yong Cheng; Srivastava A.K.; Schulz N.N.; Ginn, H.L.. (2006). Hardware in the Loop Test for Power System Modeling and Simulation. *Proceedings of Power Systems Conference and Exposition, 2006. PSCE '06*, pp. 1892-1897, ISBN 1-4244-0177-1, Atlanta GA, November 2006, IEEE PES

Kuperman A. & Rabinovici R. (2005) Virtual torque and inertia loading of controlled electric drive. *Education, IEEE Transactions on*, vol. 48 n.1, pp. 47- 52, ISSN 0018-9359.

Maclay D. (1997). Simulation gets into the loop. *IEE Review* vol. 43 n. 3, pp. 109 - 112, ISSN 0953-5683

Mohan, N.; Robbins, W.P.; Undeland, T.M.; Nilssen, R. & Mo, O. (1994) Simulation of power electronic and motion control systems-an overview. *Proceedings of the IEEE*,vol. 82 n. 8, pp. 1287 - 1302, ISSN 0018-9219

Mauri M; Castelli Dezza F. & G. Marchegiani (2007). A novel small-scale variable speed hydropower emulator using an inverter-controlled induction motor. *Proceedings of European Conference on Power Electronics and Applications, 2007* pp. 1-7, ISBN: 978-92-75815-10-8, Aalborg, September 2007, IEEE

Mauri M.; Castelli Dezza F. & Marchegiani G. (2008). Hardware in the Loop (HIL) test bench for small-scale Distributed Generation systems. *Proceedings of IEEE International Symposium on Industrial Electronics, 2008. ISIE 2008*, pp. 2177 - 2182, ISBN 978-1-4244-1665-3, Cambridge, July 2008, IEEE

Shi Y. & Monti A. (2008). FPGA-based fast real time simulation of power systems. *Proceedings of Power and Energy Society General Meeting - Conversion and Delivery of Electrical Energy in the 21st Century*, pp. 1 - 5, ISBN 978-1-4244-1905-0, July 2008, IEEE

9

Harmonics Reduction Techniques in Renewable Energy Interfacing Converters

Ali M. Eltamaly, Ph.D
King Saud University
Saudi Arabia

1. Introduction

In the recent years there has been growing interest in renewable energy generation along with growing demands for development of a suitable utility interface system. Low harmonics in the line currents of the converters used in the utility interface of the renewable energy system is the highest challenge. Controlled and uncontrolled converters are used widely in the utility interface of renewable energy systems to reduce cost, harmonic contents in line currents and to increase the reliability. These converters suffer from high level of Total Harmonic Distortion (THD) in the utility line currents which creates a lot of problems in the power system. A detailed third harmonic current injection technique has been introduced in details to reduce the harmonic contents in line currents of controlled and uncontrolled converters that used in renewable energy applications. The approach is based on circulating third harmonic current from DC link current to line currents of controlled and uncontrolled converters to reduce its THD. A new injected current shape is computed from analysis to achieve sinusoidal utility line currents. A controllable single switch boost converter connected in shunt is employed to circulate the new injected current shape. A method to implement the proposed approach under varying load condition is shown. Analysis, design, limitations, simulation and experimental result are presented.

In modern power electronics converters, three-phase controlled converter is commonly used especially as a rectifier or as an inverter in interfacing renewable energy systems with electric utility (Naik et al., 1995; Mohan, 1992, and Eltamaly, 2008). Three-phase controlled converter has simple construction and control, low cost, and low acoustic noise compared with other solutions like a full-bridge IGBT line side inverters or with Vienna rectifiers. Line current of controlled converter and its FFT components are shown in Fig. 1. It is clear from this waveform that this converter generates high harmonics in the line currents which distort the voltage at the point of common coupling in the power system. A lot of efforts have been performed to reduce harmonic contents in the utility line currents of controlled converter (Pacheco et al., 1994 - Arrillaga et al., 2006). Passive filters have been used in many researches with different configurations (Pacheco et al., 1994 and Bhattacharya et al., 1993), but this technique suffers from bulky, heavy filter elements and sometimes causes resonance problems. Active filters have been used in many researches and it sounds an interesting option but this technique suffers from complexity and high cost (Ortega et al., 2005 and Lin & Ou, 2004). Hybrid solutions using active filters and passive filters in high power applications for improving passive filter performance (Bhattacharya et al., 1997). Increasing number of pulses (Lee et al., 1997 - Tinsley, 2003) reduces the harmonic contents in line current but this technique needs large in size, heavy in weight, high in cost, complex construction, and it will not be ready available from manufacturer (Lee, 1998). Early work in third harmonics injection techniques have been used in (Mohan, 1992 & Bhattacharya et al., 1993 and Arrillaga et al., 2006). Some other literatures use switches in the main pass of power flow which increase the switching losses and reduce the system reliability (Naik et al., 1995). Injection of third harmonic current to line currents can be achieved by using LC branches tuned around third harmonic frequency (Arrillaga et al., 2006). But, this scheme suffers from bulky construction, resonant problems, and the current in the injection branch is very sensitive to the deviation of L and C values. Most of three phase line current harmonics reduction techniques are summarized in (Mao et al., 19987 and Singh et al., 2004). Injection of third harmonic to line current using zigzag transformer has been shown in many researches and (Naik et al., 1995 and Pejovic, 2000). These researches and the ideas shown in this research show the superiority of this technique in reducing the harmonic contents of line currents and increasing the power factor of the controlled or uncontrolled converters.

2. Proposed Approach

Figure 2 shows the topology of the proposed approach to reduce harmonics generated by line commutated thyristor inverter used in interfacing renewable energy sources with electric utility. This approach contains a zigzag transformer that presents high impedance for fundamental frequency component and very low impedance for the injection current. A single-phase transformer is connected between the dc-link mid-point 'o' and the zigzag transformer neutral 'n'. The secondary of single-phase transformer is connected to a rectifier boost converter stage feeding the dc link. By operating single switch the injected current If can be regulated. The duty ratio of the boost converter is varying to control the injection current to be depending on the dc link voltage.

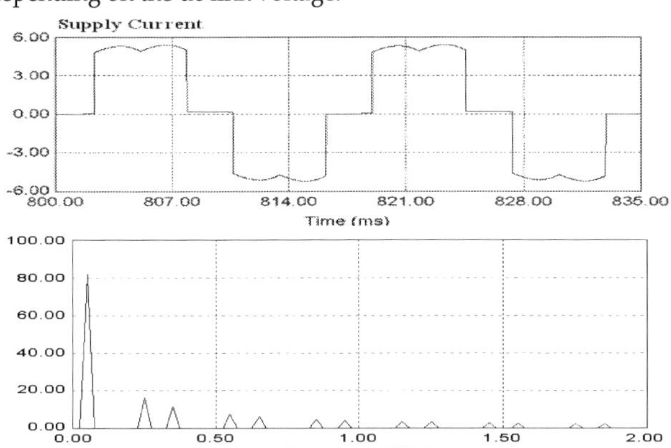

Fig. 1. Supply current and its FFT of controlled converter without harmonic reduction.

The optimal rms value and phase angle of injection current to get the minimum THD of supply current is the main issue in this chapter. Previous results say; the best rms value of 3rd harmonic current is equal to the average value of dc link current (Choi et al., 1996 and Lee, 1998. But it is not the only condition required, however, the angle of injection current plays a very important rule in the THD of the line current. So, it is important to determine the relation between the angle of injection current and the firing angle α.

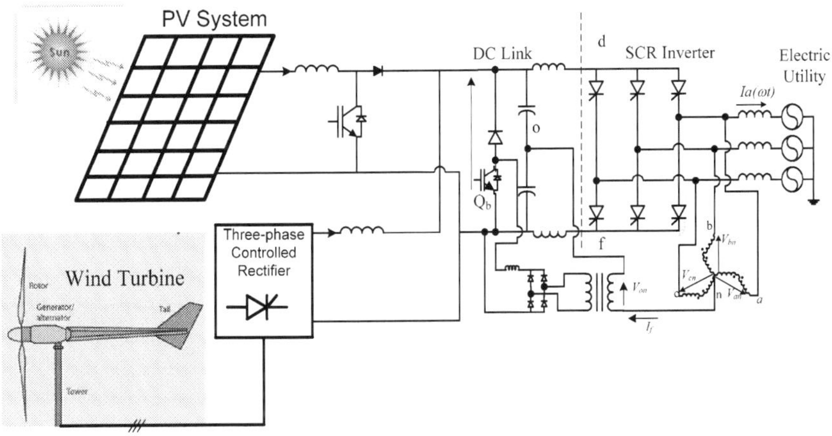

Fig. 2. The proposed approach

For firing angle α =20° (as an example), Fig. 3 (a) shows the utility line current with respect to the voltage of phase '*a*'. In Fig. 3 (b), the 3rd harmonic injection current with respect to phase *a* voltage. It is clear from this waveform that the injection current leads the phase a voltage by 120° which agree with the analysis and waveforms in the upcoming sections. In Fig. 3(c); the line current with using the 3rd harmonic current along with the phase voltage *a*. This figure reveals that the angle between I_f and I_a must be 180° with respect to 3rd harmonic frequency. In Fig. 3(d) the voltage between point '*d*' and '*n*'' V_{dn}, the voltage between point '*f*' and '*n*' V_{fn} and the voltage between '*o*' and '*n*' V_{on} along with phase '*a*' voltage. The third harmonic components of the voltages V_{dn}, V_{fn} and V_{on} have been used to inject third harmonic back to utility line current to reduce the harmonic contents in utility line currents. So a careful analysis for these voltages is required to get the optimum value of injection current and its angle. It is convenient to employ Fourier series in the analysis of the distorted waveforms. In general, a non-sinusoidal waveform $f(\omega t)$ can be expressed as follows:

$$f(\omega t) = a_0 + \sum_{n=1}^{\infty} \left(a_n \cos\left(n\omega t\right) + b_n \sin\left(n\omega t\right)\right)$$

where a_0, a_n, and b_n are Fourier coefficients.

By applying Fourier equations to the waveforms of V_{dn} only to get the third harmonic component as following:

$$a_3 = \frac{3}{\pi} \int_{\frac{\pi}{6}+\alpha}^{\frac{5\pi}{6}+\alpha} V_m \sin\omega t * \cos 3\omega t \; d\omega t = \frac{3\sqrt{3}\, V_m}{8\pi}\left[2\sin(2\alpha) - \sin(4\alpha)\right] \tag{1}$$

$$b_3 = \frac{3}{\pi} \int_{\frac{\pi}{6}+\alpha}^{\frac{5\pi}{6}+\alpha} V_m \sin\omega t * \sin 3\omega t \; d\omega t = \frac{3\sqrt{3}\, V_m}{8\pi}\left[\cos(4\alpha) - 2\cos(2\alpha)\right] \tag{2}$$

From (1) and (2), V_{dn3} and its angle can be obtained as in (3) and (4).

$$V_{dn3} = \frac{1}{\sqrt{2}} * \sqrt{a_3^2 + b_3^2} = \frac{3V_{LL}}{8\pi}\sqrt{1 + 8\sin^2\alpha} \tag{3}$$

where: V_m is the peak value of phase voltage,
V_{LL} is the rms value of line to line supply voltage, and,
V_{dn3} is the rms value of 3rd harmonic of the voltage between points *d* and *n*.
And angle of V_{dn3} is

$$\tan^{-1}\left(\frac{a_3}{b_3}\right) = \tan^{-1}\left(\frac{2\sin(2\alpha) - \sin(4\alpha)}{\cos(4\alpha) - 2\cos(2\alpha)}\right) \tag{4}$$

In the same way V_{fn3} can be obtained as following:

$$a_3 = \frac{3}{\pi} \int_{\frac{7\pi}{6}+\alpha}^{\frac{11\pi}{6}+\alpha} V_m \sin\omega t * \cos 3\omega t \; d\omega t = \frac{3\sqrt{3}\, V_m}{8\pi}\left[2\sin(2\alpha) - \sin(4\alpha)\right] \tag{5}$$

$$b_3 = \frac{3}{\pi} \int_{\frac{7\pi}{6}+\alpha}^{11\frac{5\pi}{6}+\alpha} V_m \sin\omega t * \sin 3\omega t \; d\omega t = \frac{3\sqrt{3}\, V_m}{8\pi}\left[\cos(4\alpha) - 2\cos(2\alpha)\right] \tag{6}$$

From (5) and (6), V_{fn3} and its angle as in (7) and (8).

$$V_{fn3} = \frac{1}{\sqrt{2}} * \sqrt{a_3^2 + b_3^2} = \frac{3V_{LL}}{8\pi}\sqrt{1 + 8\sin^2\alpha} \tag{7}$$

V_{fn3} is the rms value of 3rd harmonic of the voltage between points *f* and *n*.
And angle of V_{fn3} is;

$$\tan^{-1}\left(\frac{a_3}{b_3}\right) = \tan^{-1}\left(\frac{2\sin(2\alpha) - \sin(4\alpha)}{\cos(4\alpha) - 2\cos(2\alpha)}\right) \tag{8}$$

$$V_{on3} = \frac{V_{dn3} + V_{fn3}}{2} \tag{9}$$

Then;

$$V_{on3} = \frac{3V_{LL}}{8\pi}\sqrt{1+8\sin^2\alpha} \tag{10}$$

And angle of V_{on3}, is

$$\theta = \tan^{-1}\left(\frac{a_3}{b_3}\right) = \tan^{-1}\left(\frac{2\sin(2\alpha)-\sin(4\alpha)}{\cos(4\alpha)-2\cos(2\alpha)}\right) \tag{11}$$

The 3rd harmonic injection current is;

$$I_f\angle\varphi = \frac{V_{on3}\angle\theta}{Z\angle\psi} = \frac{V_{on3}}{Z}\angle(\theta-\psi) \tag{12}$$

where, V_{on3} is the rms value of 3rd harmonic of the voltage between points *o* and *n*, and, θ *is the angle of* V_{on3}, and, ψ is the angle between V_{on3} and I_f (the impedance angle of injection current path).

As shown in Fig. 3, the voltage V_a is taken as a reference, so the phase angle of fundamental and 3rd harmonic components of phase *a* current are $-\alpha$ and -3α respectively. As explained in (Eltamaly, 2003) and it is clear from Fig. 3, the optimum phase difference between phase *a* current and injection current is 180°. Then the difference between the angle of I_f (which is ϕ) and the angle of fundamental current which is ($-\alpha$ and -3α for fundamental and third harmonic frequencies respectively) is 180° as described in (13).

$$(\varphi_{opt}) - (-3\alpha) = 180 \tag{13}$$

So the optimum angle of the injection current, ϕ_{opt} can be obtained from the following equation;

$$(\varphi_{opt}) = 180 - 3\alpha \tag{14}$$

Also, the optimum angle of the impedance of injection current path, ψ_{opt} can be obtained from the following equation;

$$\psi_{opt} = \theta + 3\alpha - 180 \tag{15}$$

From (14) and (15) the phase difference between each vectors for various firing angles; α =20° and 40° (as an example for rectifier mode) and α =130° and 150° (as an example for inverter mode) is shown in Fig. 4 (a) and (b) respectively. From (10) the relation between V_{on3}/V_{LL} and the firing angle α is shown in Fig. 5. In the same way; from (14) and (15) the variation of injection current optimum angle, the angle of V_{on3}, θ and the angle between I_f, and V_{on3} (angle of impedance angle of injection path) along with the firing angle α is shown in Fig. 5.

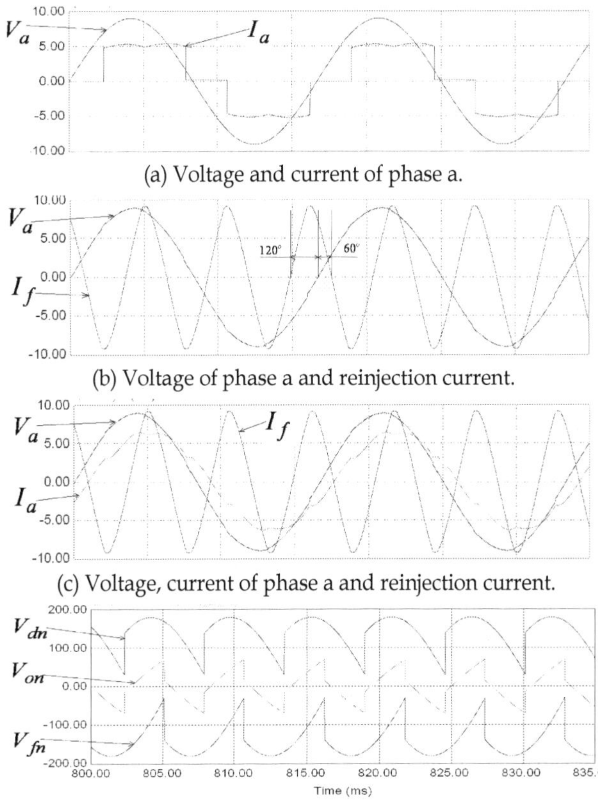

(a) Voltage and current of phase a.

(b) Voltage of phase a and reinjection current.

(c) Voltage, current of phase a and reinjection current.

(d) The voltages V_{dn}, V_{on}, and V_{fn}.

Fig. 3. Various voltages and currents of the proposed approach at α=20°.

3. Design Example

Experimental and simulation verification for this technique has been carried out. The prototype model is used in simulation and in the experiment. DC link is connected to three-phase controlled converter and electrical utility as shown in Fig. 2. For the controlled rectifier, the values of V_{on3} varies between $V_{on3}(\alpha=0)$=0.1194pu to $V_{on3}(\alpha=90°)$=0.358pu. At full load the dc current I_o= 0.6224 pu. The base voltage and current are 220V and 3.3A, respectively.

The simulation results in the next section reveals that the minimum THD occurs almost at I_f/I_o=1.15, so at full load; I_f=0.716pu. The rated third harmonic current passing through zigzag transformer is $I_f/3$=0.239pu. The value of I_f is controlled depending on the value of dc link current for minimum THD. The third harmonic injection current can be controlled by controlling the duty ratio of the boost converter. Current sensor is used to measure the actual dc link current and the 3rd harmonic injection current. The error signal between these two currents is used to control the duty ratio of the boost converter. 10 kHz switching frequency of the boost converter is used in simulation and in the practical prototype. The schematic of the controlled circuit of the boost converter is shown in Fig. 6.

Design of Single-Phase Transformer

The rated primary voltage of single phase transformer in 3rd harmonic injection pass is $V_{on3}(\alpha = 90) = 0.358\,pu$. Also the rated primary current of this transformer is $I_f = 0.716\,pu$. So, the rated kVA of this transformer is $0.716 * 0.358 = 0.256\,pu$.

The turns ratio of step up single-phase must handle the minimum value as well as the maximum value of V_{on3}. The minimum value of V_{on3} is at $\alpha = 0$ which is $V_{on3}(\alpha = 0) = 0.1194$ pu. Then, the corresponding value of V'_{on3} is:

$$V'_{on3}(\alpha = 0) = 0.1194 / n \text{ pu} \tag{16}$$

The output dc voltage of single-phase diode rectifier, V_{bi} is:

$$V_{bi} = 2 * \sqrt{2} * V'_{on} / \pi \tag{17}$$

The minimum value of V_{on3} is corresponding on the maximum value of the duty ratio of boost converter. Assume the maximum allowable value of the duty ratio is 0.8. Then the output voltage of the boost converter, V_{bo} is:

$$V_{bo} = V_{bi} / (1 - D) \tag{18}$$

Substitute the value of V_{bi} from (17) into (18) and D=0.8, the following equation can be obtained:

$$V_{bo}(D = 0.8) = \left(2 * \sqrt{2} * V'_{on} / \pi\right) / (1 - 0.8) = \left(10 * \sqrt{2} * V'_{on}\right) / \pi \tag{19}$$

The output voltage from the boost converter must equal the dc link voltage. The dc link voltage is kept constant at 1.15 pu by controlling the firing angle of the controlled converter and the modulation index of PWM inverter. So, by equating (19) with 2.2pu, the following equation can be obtained:

$10 * \sqrt{2} * V'_{on3} / \pi = 2.2\,pu$,

Then;

$$V'_{on3}(\alpha = 0) = 2.2 * \pi / \left(10\sqrt{2}\right) pu \tag{20}$$

By equating the values of V'_{on3} in (14) and (18), the following equation can be obtained:

$0.1194 / n = 2.2 * \pi / \left(10\sqrt{2}\right)$

Then the turns ratio of the single-phase transformer is:

$$n = 10\sqrt{2} * 0.1194 / (2.2 * \pi) = 0.24 \tag{21}$$

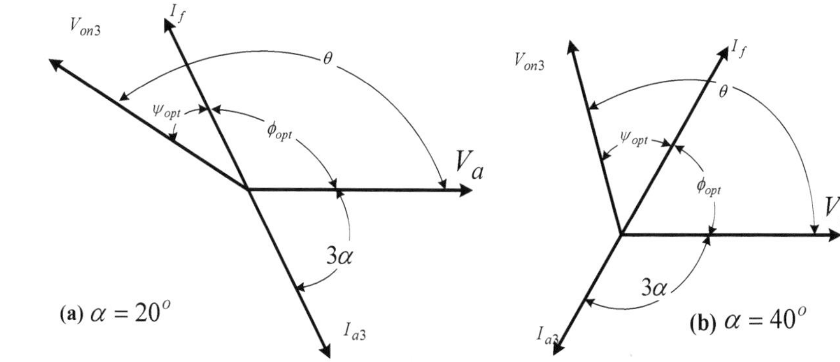

(a) Rectifier

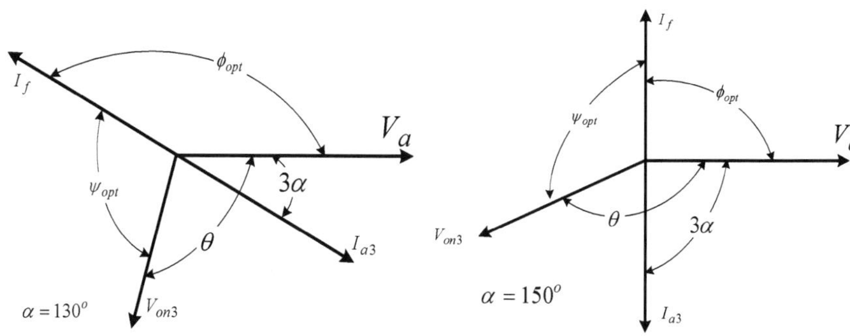

(b) Inverter

Fig. 4. Phase difference between each component in a rectifier and inverter

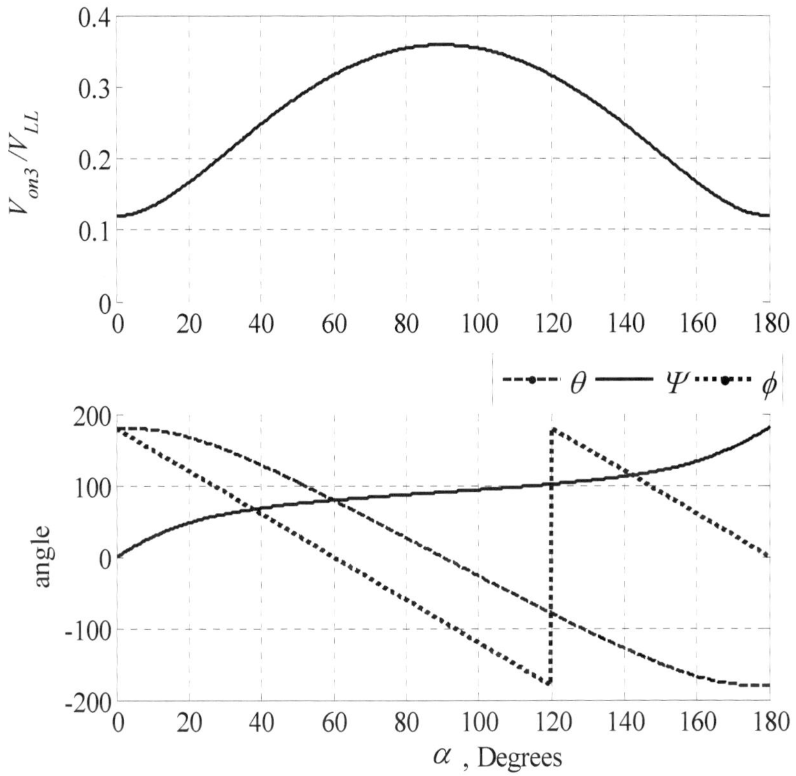

Fig. 5. The variation of V_{on3}, θ, ϕ_{opt}, ψ_{opt} along with α.

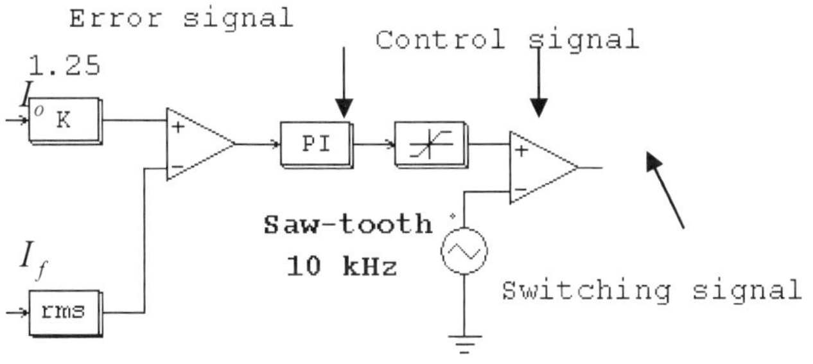

Fig. 6. The schematic of the control circuit of the boost converter.

4. Simulation and Experimental Results

The simulation of the proposed technique has been performed by using PSIM computer program (PSIM6.1). Same values of the components used in simulation program have been used in the experimental prototype to compare these results.

Fig. 7 shows the relation between THD and the value of I_f/I_o for different values of firing angle, α. This figure reveals that the optimal value of I_f/I_o is about 1.15 for the best THD. Also, This figure shows the importance of 3rd harmonic injection technique especially for high value of firing angle, α.

Fig. 8 shows the relation between THD and the angle of I_f with respect to V_{on}, ψ at I_f/I_o=1.15 for different values of firing angle, α. It is clear that the best THD occurs around ψ_{opt} which obtained from (13). The value of ψ_{opt} increase with increasing the firing angle α.
The simulation and experimental waveforms are shown for α=20° and 40° as an examples.

4.1 Simulation and experimental results at α=20°

Figure 9 shows the waveforms of voltage V_{df} and V_{ab}. Fig. 10 shows the supply current waveform and its FFT components with respect to phase a voltage without injection of 3rd harmonic current. It is clear from this figure that the supply current has very high THD mostly of 5th and 7th harmonics. Fig. 11 shows the voltage of V_{dn}, and V_{fn}. It is shown in the top of the experimental result waveform that the frequency of this voltage is 180° which is the 3rd harmonic voltage and the value of rms voltage is 39.9 that can be obtained from (10) at α=20°. Figure 12 shows the supply current waveform and its FFT components with respect to phase a voltage with optimum amplitude and angle of 3rd harmonic current injection. It is clear from this figure that the supply current becomes very near to sine-wave with 5% THD.

Figure 13 shows supply current waveform I_a and optimum value and angle of injection 3rd harmonic current with respect to phase a voltage. It is clear from these waveforms that the value of the angle between I_f and V_a is about 120° and the angle between I_a and I_f is 180° which agree with the vector diagram shown in Fig. 4 (a).

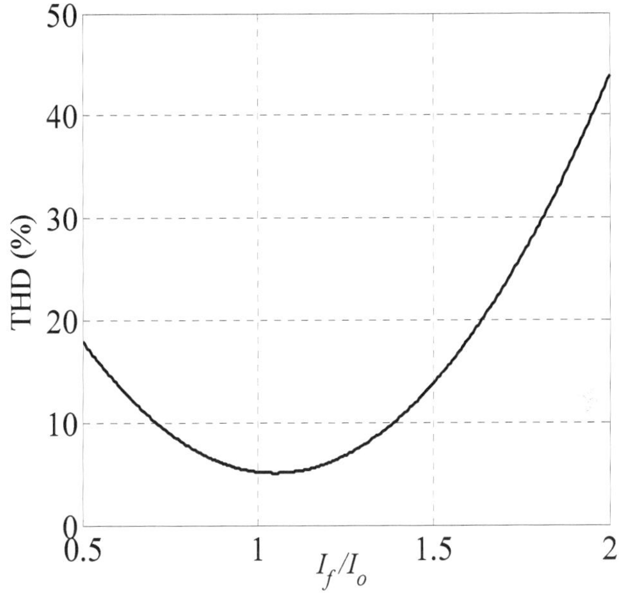

Fig. 7. The relation between THD and the value of I_f/I_o.

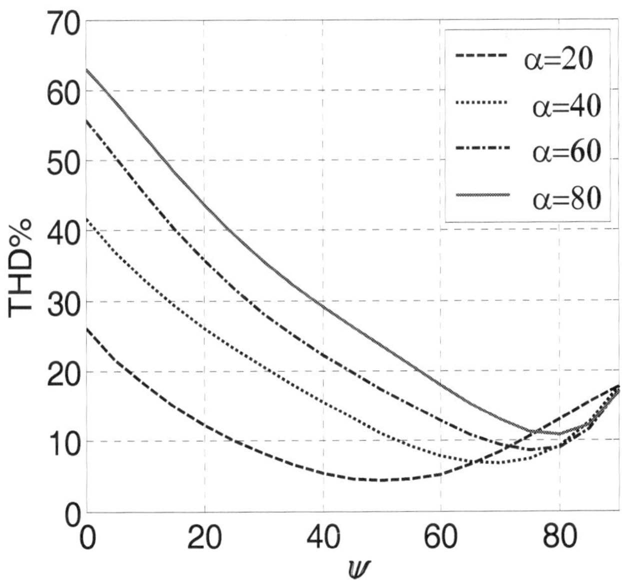

Fig. 8. The relation between THD and the angle of I_f with respect to V_{on} at I_f/I_o=1.15 for different values of firing angle, α.

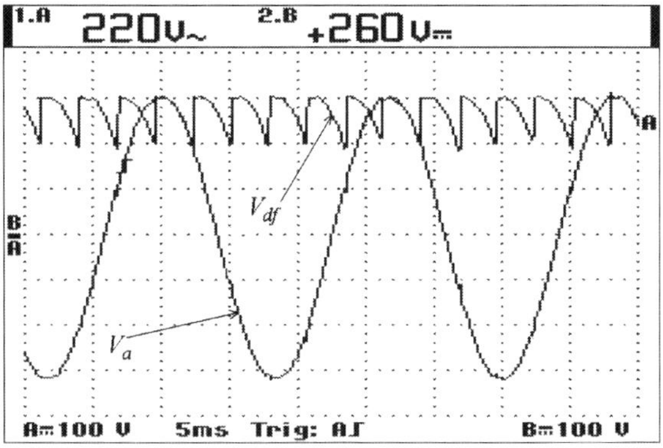

Fig. 9. The waveforms of voltage V_{df} and V_{ab}.

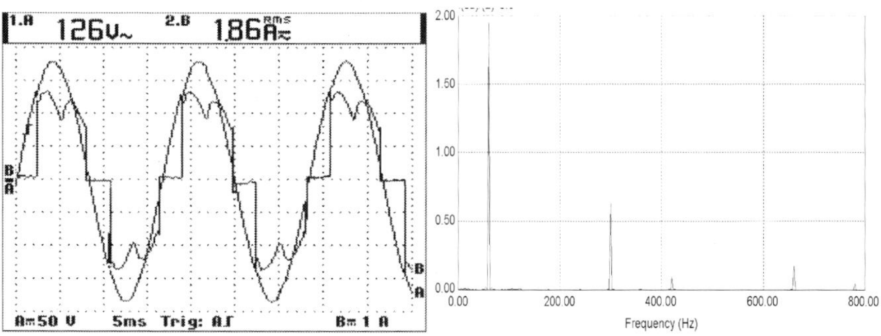

(a) Experimental result. (b) FFT components of I_a

Fig. 10. The supply current waveform and its FFT components with respect to phase *a* voltage without injection of 3rd harmonic current.

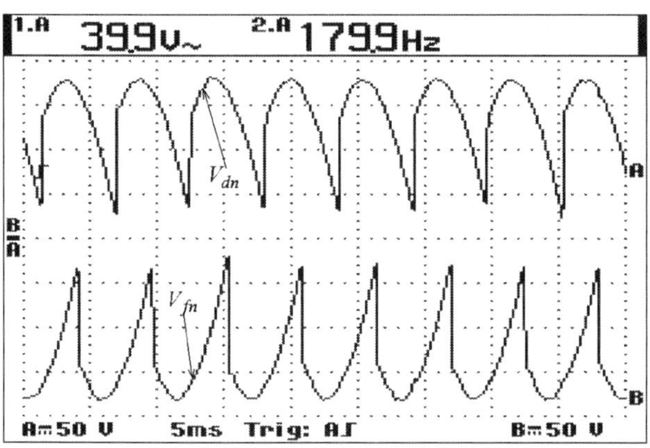

Fig. 11. The voltage of V_{dn}, and V_{fn}

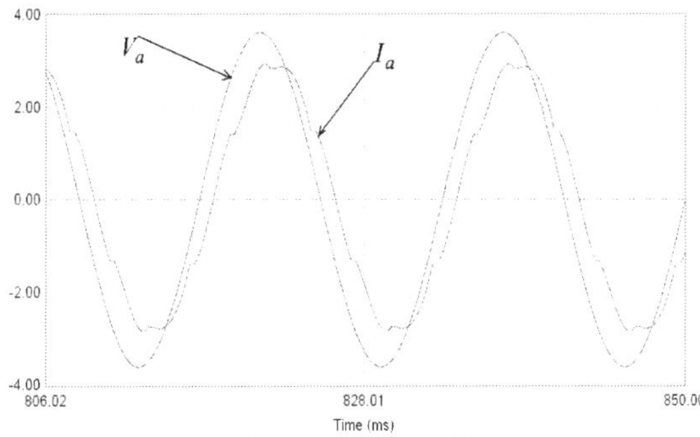

(a) Simulation result.

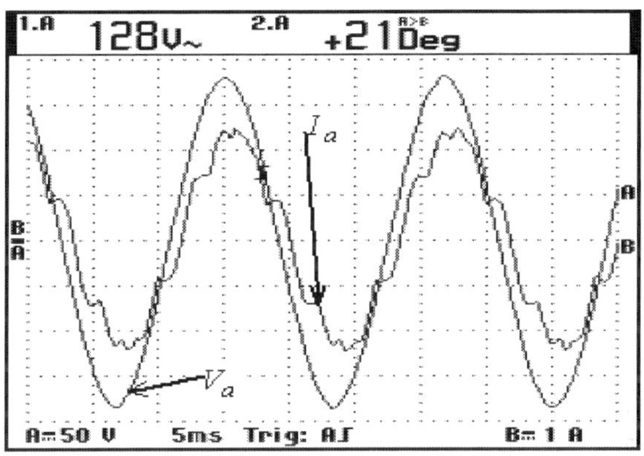

(b) Experimental result.

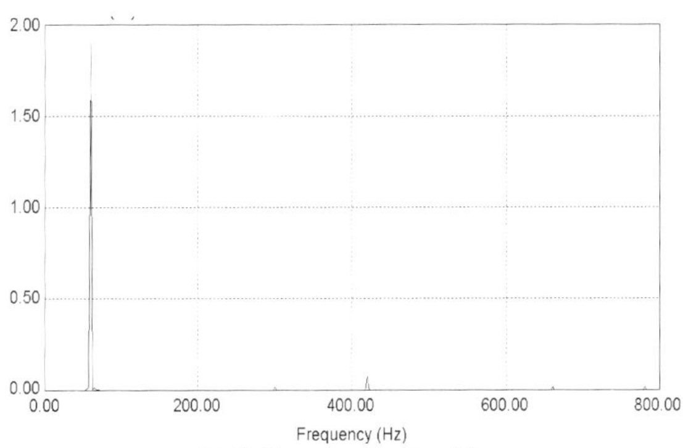

(c) FFT components of I_a.

Fig. 12. The supply current waveform and its FFT components with respect to phase *a* voltage at optimum 3rd harmonic current injection.

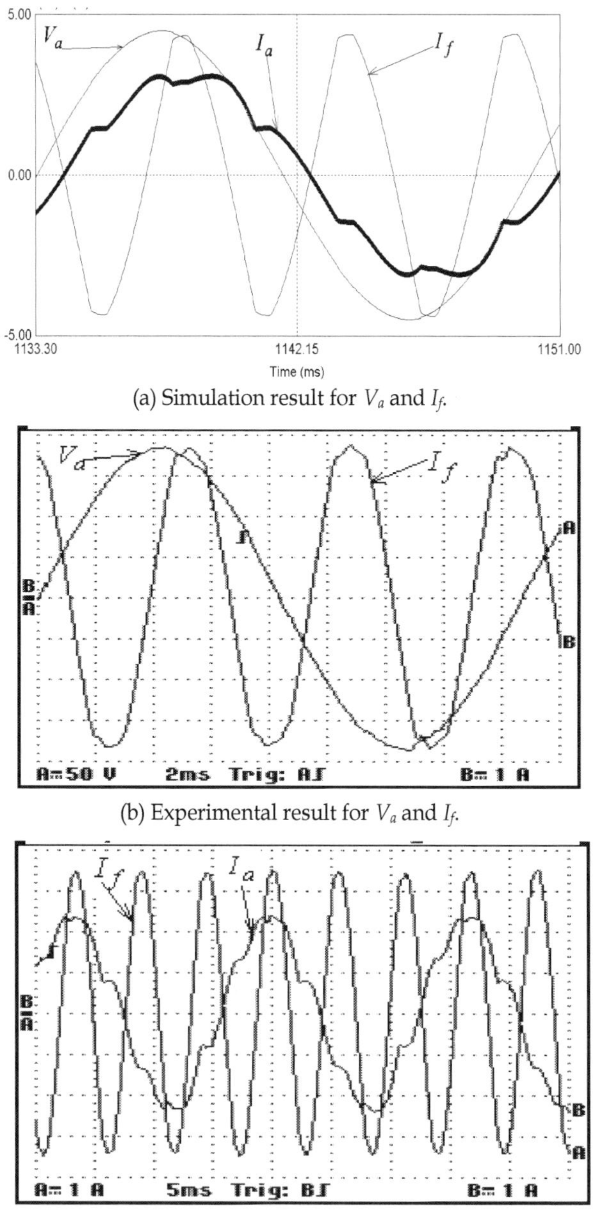

(a) Simulation result for V_a and I_f.

(b) Experimental result for V_a and I_f.

(c) Experimental result for I_a and I_f

Fig. 13. The supply current waveform I_a and optimum value and angle of injection 3rd harmonic current with respect to phase a voltage.

4.2 Simulation and experimental results at α=40o

Figure 14 shows the waveforms of voltage V_{df} and V_{ab}. Fig. 15 shows the supply current waveform and its FFT components along with phase *a* voltage without injection of 3rd harmonic current. It is clear from this figure that the supply current has very high THD mostly of 5th and 7th harmonics. It is clear that this THD is high than the one of α=40°.
Figure 16 shows the voltage of V_{dn}, and V_{fn} and their FFT components with respect to phase *a* voltage. It is clear from this figure that these voltages have triplex harmonics and the 3rd harmonic is the most dominant harmonic. It is also clear that both voltages V_{dn}, and V_{fn} have the same harmonics.

Figure 17 shows supply current waveform and its FFT components with optimum value and angle of injection 3rd harmonic current. It is clear from this figure that the supply current becomes very near to sine-wave with 6% THD.

Figure 18 shows the supply current waveform I_a and optimum value and angle of 3rd harmonic injection current with respect to phase *a* voltage. It is clear from these waveforms that the angle between I_f and V_a is about 60° which agree with the vector diagram in Fig. 4 (b).

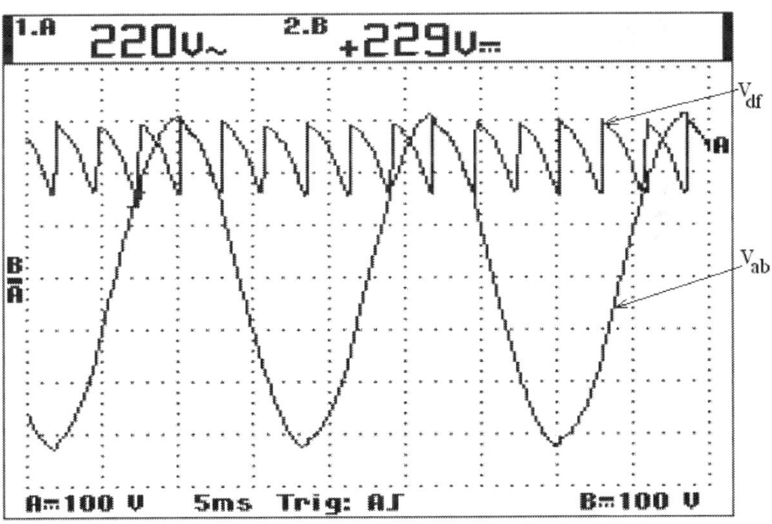

Fig. 14. The waveforms of voltage V_{df} and V_{ab}.

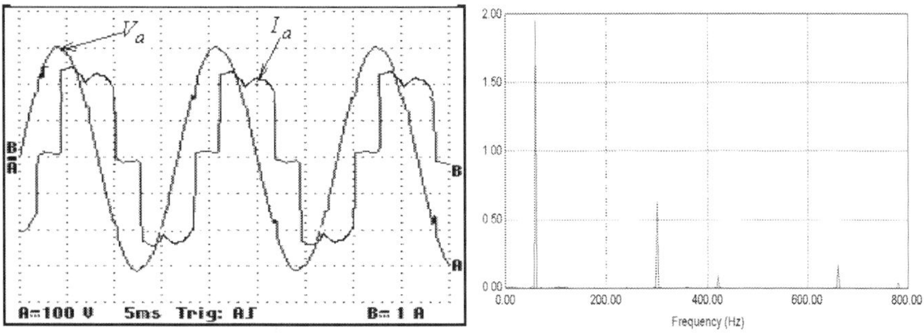

(a) Experimental result. (b) FFT components of I_a

Fig. 15. The supply current waveform and its FFT components along with phase a voltage without injection of 3rd harmonic current.

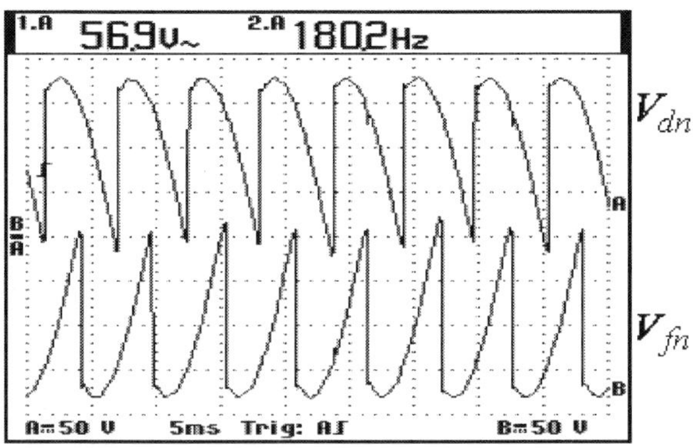

(a) Experimental result.

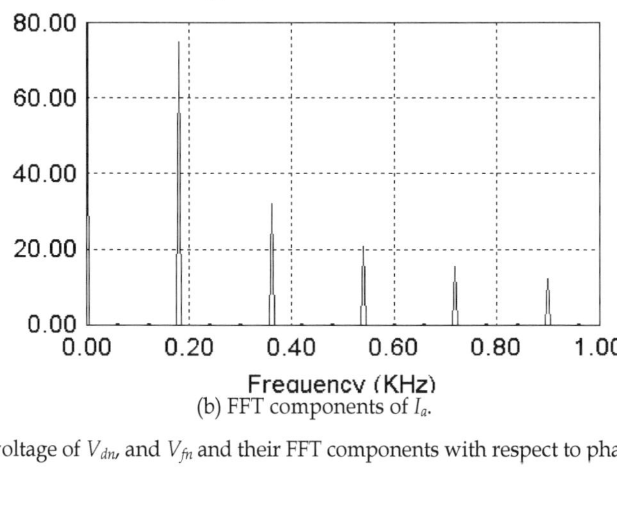

(b) FFT components of I_a.

Fig. 16. The voltage of V_{dn}, and V_{fn} and their FFT components with respect to phase a voltage.

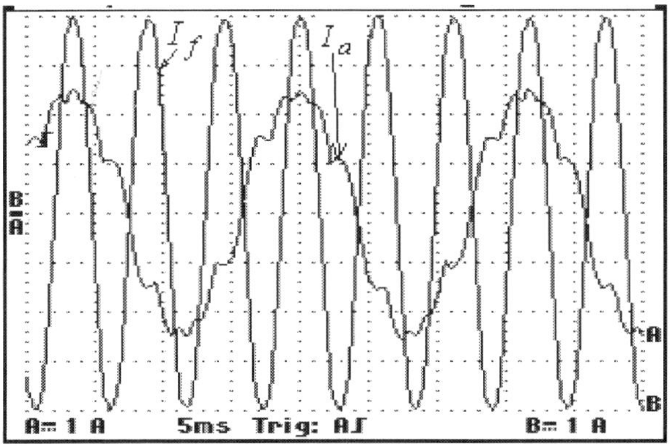

(a) Experimental result.

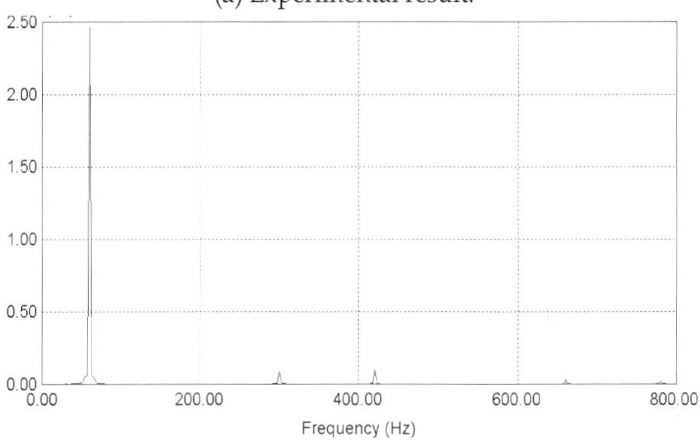

(b) FFT components of I_a.

Fig. 17. The supply current waveform and its FFT components with optimum value and angle of injection 3rd harmonic current.

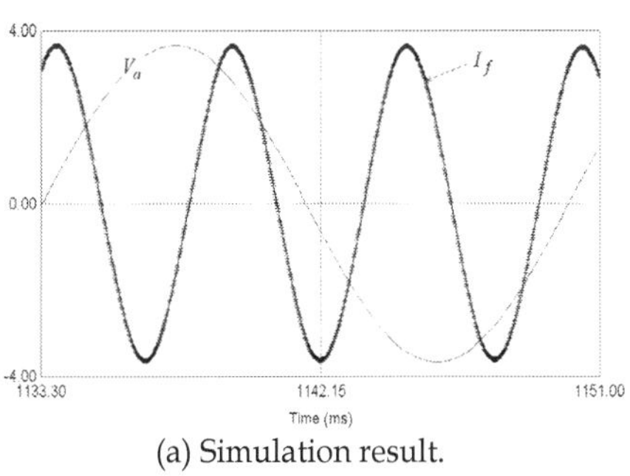

(a) Simulation result.

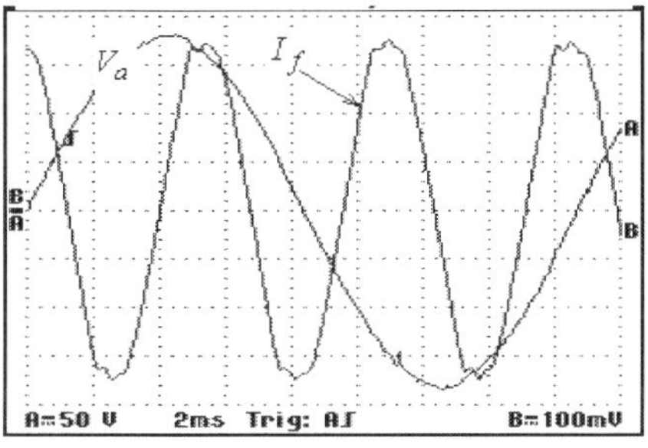

(b) Experimental result.

Fig. 18. The supply current waveform I_a and optimum value and angle of injection 3rd harmonic current with respect to phase a voltage.

4. Conclusions

Third harmonics injection technique plays a significant rule in reducing of the THD of the utility line current of three-phase controlled converters. The optimal amplitude and phase angle of injection current slightly changes with the firing angle. In this research, a detailed analysis for determining the relation between the optimal amplitude and phase angle of injection current along with the firing angle has been carried out. The THD of line current is highly affected by the amplitude of 3rd harmonic injection current as well as its phase angle with supply voltage. The optimum amplitude and angle of 3rd harmonic current have been determined by a detailed mathematical analysis of the system. The THD of the utility line current from simulation and the experimental results proves the mathematical results for this technique. The THD of the utility line current with optimal harmonic injection current is lower than limits of harmonics standards.

5. References

A. Ametani "Harmonic reduction in thyristor converters by harmonic current injection" *IEEE Trans. on Power Apparatus and Systems*, vol. PAS-95, no. 2 , March/April 1976, pp.441-450.

Ali M. Eltamaly "A Modified Harmonics Reduction Technique of Three-Phase Controlled Converter" IEEE Trans. Industrial Electronics, vol.55, No.3 March. 2008, pp. 1190-1198.

Ali M. Eltamaly" A new relation between firing angle of three-phase controlled converter and best angle of injection current " in *IEEE Conf. Proc. MEPCON'2003*, Minoufiya University, Shebin El-Kom, Egypt, Dec. 2003, pp. 793-798.

B. Lin, Y. Ou "Active power filter based on three-phase two-leg switch-clamped inverter" *Electric Power Systems Research*, vol. 72, June 2004, pp. 63–72.

B. S. Lee "New clean power reactor systems for utility interface of static converters", *Ph.D. Thesis Texas A&M University*, Aug. 1998.

B. S. Lee; P. N. Enjeti, I. J. Pitel "An optimized active interphase transformer for auto-connected 12-pulse rectifiers results in clean input power" in *IEEE Conf. Proc. APEC1997*, Atlanta, Georgia, USA, vol. 2, Feb. 1997, pp. 666-671.

Bhim Singh, Brij N. Singh, Ambrish Chandra, Kamal Al-Haddad, Ashish Pandey, Dwarka P. Kothari, "A review of three-phase improved power quality ac-dc converters" *IEEE Trans. on Industrial Electronics*, vol. 51, no. 3, June 2004.

C. T. Tinsley" Modeling of multi-pulse transformer rectifier units in power distribution systems" MS.c., Thesis Virginia Polytechnic Institute, USA, Aug. 2003.

Cassiano Rech, and José R. Pinheiro, "Line current harmonics reduction in multipulse connection of asymmetrically loaded rectifiers" *IEEE Trans. on Industrial Electronics*, vol. 52, no. 3, June 2005.

H. Mao, F. C. Y. Lee, D. Boroyevich, and S. Hiti, "Review of high-performance three-phase power-factor correction circuits" *IEEE Trans. on Industrial Electronics*, vol. 44, no.4, Aug. 1997, pp. 437-446.

H.A. Pacheco; G. Jimenez, J. Arau "Optimization method to design filters for harmonic current reduction in a three phase rectifier" in *IEEE Conf. proc. CIEP'94*, Puebla, Mexico, Aug. 1994, pp. 138-144.

J. Marafao, J. Pomilio, and G. Spiazz " Improved three-phase high-quality rectifier with line-commutated switches" *IEEE Trans. on Power Electronics*, vol.19, no.3, May 2004, pp. 640-648.

J. Ortega, M. Esteve, M. Payán, A. Gómez "Reference current computation methods for active power filters: accuracy assessment in the frequency domain" *IEEE Trans. on Power Electronics*, vol. 20, no. 2, March 2005, pp. 446-456.

Jos Arrillaga, Y. H. Liu, Lasantha B. Perera, and Neville R. Watson "A current reinjection scheme that adds self-commutation and pulse multiplication to the thyristor converter" *IEEE Trans. on Power Delivery*, vol.21, no. 3, July 2006.

M. Rastogi, N. Mohan, and C. P. Henze " Three-phase sinusoidal current rectifier with zero-current switching" *IEEE Trans. on Power Electronics*, vol. 10, no. 6, Nov. 1995, pp. 753-759.

N. Mohan "A novel Approach to minimize line current harmonics in interfacing renewable energy sources with 3-phase utility systems" in IEEE Conf. Proc. APEC'92, Boston Massachusetts, Feb. 1992, pp. 852 -858.

N. Mohan, M. Rastogi and R. Naik " Analysis of a new power electronics interface with approximately sinusoidal 3-phase utility currents and a regulated dc output" *IEEE Trans. on Power Delivery*, April 1993, pp. 540-546.

P. Pejovic, D. Shmilovitz " Low-harmonic thyristor rectifiers applying current injection" *IEEE TRANS. on Aerospace and Electronic Systems* vol. 39, no. 4, Oct. 2003, pp.1365-1374.

P. Pejovic, P. Bojzovic, and D. Shmilovitz "Low-harmonic, three-phase rectifier that applies current injection and a passive resistance emulator" *IEEE Power Electronics Letters*, vol. 3, no. 3, Sept. 2005, pp. 96-100.

P. Pejovic" Two three-phase high power factor rectifiers that apply the third harmonic current injection and passive resistance emulation", *IEEE Trans. on Power Electronics*, vol. 15, no. 6, Nov. 2000, pp. 1228-1240.

PSIM6.1, available: www.powersimtech.com.

R. Naik, N. Mohan, M. Rogers and A. Bulawka" A novel grid interface, optimized for utility-scale applications of photovoltaic, wind-electric, and fuel-cell systems" *IEEE Trns. on Power Delivery*, vol.10, no.4, Oct. 1995, pp.1920-1926.

S. Bhattacharya, D. M. Divan; B. B. Banerjee; "Control and reduction of terminal voltage harmonic distortion (THD) in hybrid series active and parallel passive filter system" in *IEEE Conf. Proc. PESC 93*, Seatle, USA, June, 1993, pp. 779-786.

S. Bhattacharya, P. Cheng, and D. M. Divan, " Hybrid solutions for improving passive filter performance in high power applications" *IEEE Trans. on Industry Applications*, vol. 33, no. 3, May/June 1997, pp.732-747.

S. Choi, P. N. Enjeti, H. Lee, and I. I. Pital " A new active interface reactor 12-pulse rectifiers provides clean power utility interface " *IEEE Trans. on Industry Applications*, vol. 32, No. 6, 1996, pp. 1304-1311.

S. Pal "Simulation of current mode control schemes for power factor correction circuits" MS.c *thesis, Faculty of Engineering and Applied Science, Memorial University of Newfoundland, St. John's Newfoundland Canada,* May 1998.

W. Mielczarski, W. B. Lawrance, R. Nowacki, and D. G. Holmes " Harmonic current reduction in three-phase bridge-rectifier circuits using controlled current injection" *IEEE Trans. on Industrial Electronics*, vol. 44, no. 5, Oct. 1997, pp. 604-611.

10

Hybrid Control of DC-DC Power Converters

Ilse Cervantes[*], Francisco J. Perez-Pinal[†] and Angelica Mendoza-Torres[*]
[*]*Applied Mathematics Division. Institute for Scientific and Technological Research of San Luis Potosi (IPICYT)*
[†]*National Polytechnic Institute, IPN*
Mexico

1. Introduction. Review of Renewable Energy

In recent years, there has been an increasing interest in applying renewable energy (RE) in **electricity generation and transportation.** In accordance to Enerdata, an independent energy consulting and information services company (Enerdata, 2008a), the global electricity needs in 2007 were closed to 19,900 TWh (almost 40 % of the overall energy consumption), where just 18 % of the total electricity production was coped by some source of RE. Additionally, it is estimated that the electricity demand will grow up to 60 % of the worldwide energy in 2040 (Enerdata, 2008b). This means that, in order to increase energy production almost 50 % in 30 years, some energy generation and distribution problems must be solved to obtain feasible and reliable applications. For example, the concept of distributed generation (DG) has to be revisited in view of the existence of multiple and highly disperse energy sources; as well as further investigation about High Voltage Direct Current (HVDC) energy transmission, Flexible AC Transmission Systems (FACTS) and Micro grids. Furthermore, power electronics can account for, as much as 30-40 % of a distributed generation facility cost (ECPE et al., 2007). This is due to the fact that most utility scale power electronics are custom designed, which generates an expensive installation. That is, there are also important concerns such as price reduction and system reliability that need further investigation.

On the other hand, *transportation* demands about 28 % of the worldwide energy and it is expected an increment up to 40% in 2040 (International Energy Outlook 2008). In order to reduce this tendency, during the past three decades, a rapid development of more efficient vehicles has been performed. Those efforts have been focused on two main areas: i) the development of high efficient and high-density power converters, and ii) the improvement of robust and flexible modern control techniques. However, in spite of some advances, there are still open questions regarding the achievement of three of the most important issues in RE applications: reliability, efficiency, and cost.

1.1 DC-DC Power converters in Renewable Energy

Power converters (PC) and in particular Direct Current to Direct Current (DC-DC) converters play an important role in RE since they serve as a link stage for energy conditioning (usually step-up); and they can be used in low and high power converters such as HVDC, FactS, Microgrids, Fuel cell generator, Fuel cell vehicles, Photo voltaic Panels (PVP) and hybrid generation systems*.

The connection of renewable energy sources to power grids is not possible without power electronics, *i.e.* DC-DC converters optimize the efficiency of solar panels, rectifiers and inverters which are needed in wind generators. Automotive applications such as electric and hybrid drive trains are only possible with efficient and intelligent power electronics. X-by-wire concepts operated by power converters may lead to power saving of more than 20% (ECPE et al., 2007). Different topologies of PC are used in RE, the final decision to use one or another is based on the range of the final application, the kind of RE source, and if the system requires mandatory isolation or not (this specification varies upon country regulations). The most used topologies of DC-DC power converters are,

1. Simple buck and boost converters.
2. Interleaved and cascaded converters.
3. Synchronous converters.
4. High Frequency (HF) transformers using an isolated push-pull boost converters.
5. Full-bridge isolated converters.
6. Single-Inductor push-pull converters (SIC).
7. Double-Inductor push-pull converters (SIC).
8. Full-bridge converters.
9. Self-commutated inverters using thyristors.

On the other hand, at the present time, the new directions in the design of DC-DC in RE deals with the following topics (ECPE *et al.*, 2007)

1. System cost reduction.
2. New interconnection technologies for ultra-high power density systems and high temperature electronics.
3. Advanced thermal management; high temperature magnetics, capacitors, sensors, control ICs.
4. In lighting smart and simple dimming concepts; smart control of street lighting; high efficient.
5. Light sources (LED/OLED) and their power electronic drivers.
6. Higher level of integration e.g. for more compact energy saving lamps.
7. New topologies for photo voltaic (PV) solar converters focused on more efficient PV solar cells.
8. Load management by power electronics in distributed energy generation networks.
9. Zero-defect design and improved system reliability including fault-tolerant systems.
10. Digital power conversion and smart power management.

---

*Here the term *"hybrid"* refers to the use of two different energy sources.

It is in the solution of the problems related to these new tendencies in power electronics and renewable energy, that control plays a crucial role. Robust and flexible controllers are capable to guarantee efficient energy transmission and processing, are able compensate some design limitations and to reduce operation and maintenance cost. Furthermore, they can be used also for increasing reliability and safety of operation.

In the past, traditional control schemes as direct duty ratio control (Mohan *et al.,* 1989), (Rashid, 1988), (Erickson & Maksimovic, 2001), voltage and current programming control (Capel *et al.*, 1978), (Redl & Sokal, 1985), (Middlebrook R., 1985), have been applied in the solution of some of the main operation concerns. However they are based on averaged models (Middlebrook & Cuk, 1976), (Erickson *et al.*, 1982).

Basically, an averaged model assumes high switching frequency, giving as a result continuous time models and controllers. Nevertheless, it has been shown that this procedure has some limitations in the exact estimation of the DC output voltage and harmonics components; in addition, averaged models fail to predict sub-harmonic and chaotic oscillations under some control techniques (Deane, 1990). In order to analyze this erratic behavior, it is used sometimes discrete-time maps, which relate state variables at every switching period. The main limitation of digital techniques is that obtained models are nonlinear with respect to the duty cycle (Kassakian *et al.,* 1991); this fact makes more complicated the control design task.

The highly need of flexibility and more robust controllers in areas such as motor controllers, renewable energy, and power electronics; has opened the niche to apply and develop a new family of more active control methods that are able to work in operation regions rather than in a single operation point. Control methodologies such as: fuzzy, linear quadratic, passivity based and pulse adjustment among others, are good examples of complex controllers usually designed to fulfill strict standards of performance. However, in spite of the advantages in terms of robustness and good performance, their applicability is limited by the complexity of their design and implementation.

That is, in spite of the last efforts of research in the field, there is still a truly need to develop new families of controllers, due to the highly demand environments of modern power electronics applications and the advances in digital processors. Therefore, hybrid-time controller has received a lot of attention due to its practical feasibility to achieve a high performance and flexibility as well as, natural digital implementation in signal processors.

There are several types of hybrid controllers that has been applied to power converters; but perhaps the most commonly found in the literature, in view of the switching nature of the converters, are the piece-wise continuous controllers. In this type of control, system commute by virtue of a discrete event, that is product of the satisfaction of a prescribed set of conditions. Due to this fact, the control is also known as switched (Liberzon, 2003), (Li *et. al*, 2005), (Sun & Ge, 2005) or boundary control (Ting-Ting & Chung, 2008), (Leung & Chung, 2004), (Quaicoe & Iqbal, 2008).

Most switched control algorithms are easy to implement with respect to other nonlinear methodologies and they have advantage that the switching criteria can be expressed by very intuitive (design) restrictions, *i.e.* maximum current switching criterion, maximum ripple switching criterion, etc. Moreover, some of these controllers have high flexibility and adaptability in a large set of operating conditions, including changes on power load and voltage source or under continuous and discontinuous operation mode. Furthermore, since their design is not based on an averaged behaviour of the converter they are suitable to compensate a wide range of nonlinear phenomena.

Following this trend in the development of flexible controllers, the main purpose of this chapter is to describe some of the most significant results regarding hybrid controllers used in DC-DC converters; to discuss their stability properties and their limitations, as well as their applicability and relevance.

The chapter begins (Section 2) with an introduction to hybrid systems (HS). Different definitions found in the literature are discussed and special attention is paid to switched and impulsive systems. Moreover, a detailed description of the switched characteristics is given. In Section 3, a description of the tools used to analyse stability of HS is performed, emphasis is made about the class of stability result obtained from each methodology. A general description of most relevant hybrid controllers proposed in the literature is given in Section 4; it is stated that the proposed methodologies can be classified in two different kinds and particularities of each one are discussed. General aspects as robustness and implementing advantages and drawbacks are discussed. Finally some concluding remarks are given (Section 5).

2. Hybrid Systems Description

The term hybrid denotes a mixed origin. In physics a hybrid electromagnetic wave is one having components of both electric and magnetic field vectors in the direction of propagation. A hybrid vehicle is one having two or more energy sources; for example electric and chemical (combustion).

Regarding dynamic systems, there is not a universal definition of hybrid systems, but most authors coincide that ***in such systems, it exists interaction and coexistence of two kinds of dynamics: continuous and discrete*** (Liberzon, 2003), (Li *et al.*, 2005) (Goebel *et al.*, 2009). In view of this idea, most real-time controlled systems can be called hybrid; for example, an electric motor or a generator controlled by a Digital Signal Processor (DSP). In this case, the continuous dynamics is given by the machine, which by the process of sampling and data processing is controlled at discrete times. Moreover, notice that the definition above can describe interaction between finite or infinite dimensional dynamic systems with discrete events or dynamics. It can be considered as hybrid, continuous systems manipulated by piece continuous controls as supervisory control, sliding mode control and gain scheduling control for example. On the other hand, controllers as current peak control (CPC) are also a good example of hybrid controllers.

Deriving a general model for hybrid systems (HS) may be quite complicated, since various types of interactions between discrete and continuous dynamics may exist. Models found in the literature vary mainly depending upon the type of HS studied and the way the author conceives such interaction. Among hybrid systems, **switched and impulsive systems** are probably the most studied and since the discrete interaction is well defined, general models can be defined.

At this point, it is worthy noticing that some authors (Sun & Ge, 2005) consider also as impulsive and switched, classes of systems that commute between discrete dynamics. That is, no continuous dynamics exists and the overall system is constituted only by different discrete behaviours. Such systems are not hybrid in the sense discussed here, even if they are truly switching or impulsive. In this way, the terms switched and impulsive cannot be taken as synonymous of hybrid.

In **Impulsive systems (IS)** the members of a family of continuous systems (**subsystems or modes**) interact between them by virtue of a discrete event or condition. Such discrete interaction changes the dynamics in such a way that the system trajectories are discontinuous. In Fig. 1 it is possible to observe the time evolution of a second order impulsive system (i.e. $x = [x_1,x_2]^T$) as well as the system trajectories in the phase plane. Observe that in a given time, the trajectories reset to a given new value. This is, the discrete interaction induces a jump on systems trajectories making the initial condition at the $k+1$ subsystem be different from the final condition of the k subsystem. In this way, the trajectories of an impulsive system are only piecewise continuous.

Examples of impulsive systems are those describing the population of insects and other living species in which, grow rate varies impulsively *(i.e.* birth rate may lead to instantaneously population differences). Some chemical reactors with fast continuous dynamics behave as IS if the reactive is fed impulsively. Some financial systems behave also impulsively. Some other examples can be found in (Yang, 2001 a,b), (Li *et al.*, 2005), (Sun & Ge, 2005).

Impulsive systems are almost inexistent in DC-DC power electronics, even if perturbations are usually modelled as impulsive. The reason resides in the fact that in most cases, there exist continuous dynamics in such applications that serve as a filter (slow dynamics), making continuous every trajectory subjected to impulsive phenomena. For example, the RC tank in DC-DC converters serves as a low pass filter to reduce voltage ripple originated by the (impulsive) transition of the switch. Models of impulsive systems can be found in (Yang, 2001a) and (Yang, 2001b).

2.1 Switching Systems

Since switching systems are of special interest in power electronics, this subsection will be devoted to gain insight in the most common description used to describe such systems. We begin by saying that switching systems are constituted mainly of two parts: i) The family of continuous subsystems or modes and, ii) the switching law. As in the case of hybrid impulsive systems; in switched systems, the members of a family of continuous systems (**subsystems or modes**) interact between them by virtue of a discrete event or condition.

Such discrete interaction changes the dynamics in such a way that the system trajectories **remain continuous** (see Fig 2). Good examples of switched systems are the piecewise linear systems as the following:

$$\dot{x} = A_m x + B_m u_m \quad m \in \{1,\ldots p\} \tag{1}$$

where $\boldsymbol{x} \in \Re^n$ is the state variable vector, $\boldsymbol{u_i}$ is the input vector (possible controlled) and matrices $\boldsymbol{A_m}$ and $\boldsymbol{B_m}$ have suitable dimensions. The family of continuous time systems (1) is subjected to a **discrete time event** $\boldsymbol{m}$ that defines the current election of the active subsystem $\boldsymbol{i}$, that is, m is an index that stands for the switch between stages. Such an index can be described by algebraic equations, difference equations or by a set of restrictions or conditions depending possible on systems states and/ or time. The expression that defines m at every instant is called **switching law**. In most cases, such an expression is a single valued function at every time (i.e. uniquely defined at every time).

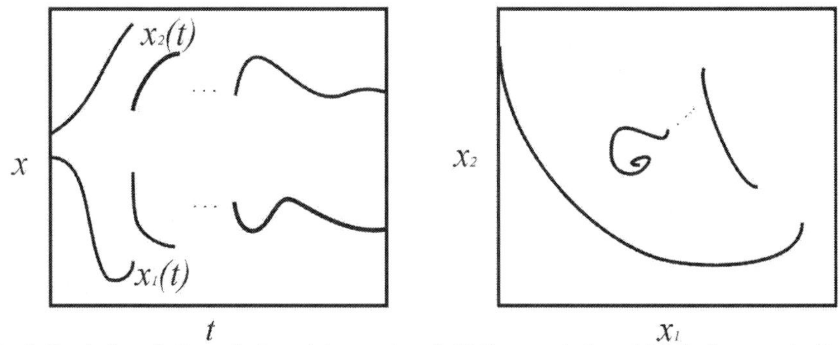

Fig. 1. Typical evolution of a impulsive system (left) time evolution, (right) phase portrait.

The switching law can be state dependent and/or time dependent (Liberzon, 2003). Usually, the switching law is given by a set of prescribed conditions leading to a division of the state space. As time evolves, the switching law generates a sequence of subsystems active for a certain time, that is

$$\sigma = \{(\theta_0, i_0), (\theta_1, i_1)\ldots(\theta_k, i_k)\} \tag{2}$$

where θ_k stands for the residence time in mode or subsystem i_k for $i_k = 1,2\ldots p$ and $k = 1,2,3\ldots$ Such a sequence is called **switching sequence** and it is used some times instead of the switching law to study the stability of the overall hybrid system (Xu, 2004), (Xu *et al.*, 2007).

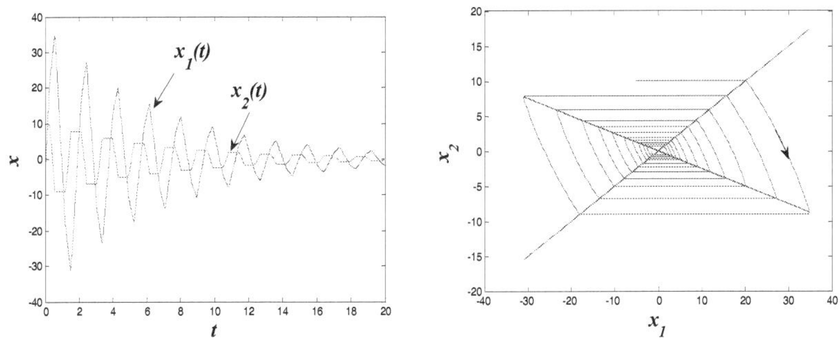

Fig. 2. Typical evolution of a switched system (left) time evolution, (right) phase portrait.

In general, switching systems can be described as:

$$\dot{x} = f_i(x,t,u) \quad i \in \{1,...p\} \tag{3}$$

where $\boldsymbol{x} \in \Re^n$ is the state variable vector and $\boldsymbol{u}$ is the input vector (possible controlled). On the other hand, the switching law is sometimes described as a piece-wise continuous function that in a general case depends on system states, time and the input vector as follows:

$$i(l+) = g(x(t), i(l), u(t), t) \tag{4}$$

where $i(l+) = i(l+1)$. Eq. (4) is not the only way of representing a switching law; a combination of a **switching criterion** and **mode assignment criterion** is also used. The switching criterion or **switching conditions** are usually formulated only upon a part of the state space (output) in the form of hyper-surfaces that are called **switching surfaces**. It is required that every region in the entire state space has assigned an active subsystem in order to have a well defined system (Liberzon, 2003). Once the system hits a pre-defined surface (Eq. (5)), the *mode assignment criterion* (Eq. (6)), decides based on the function ψ, which system is the next to be active in view of the current subsystem (Li *et al.*, 2005). In such a case, the switching law can be described as follows:

$$S = S(x(t), i(t), t) = 0 \tag{5}$$

$$i(l+) = \psi(i(l)) \tag{6}$$

Notice that in general, **a finite set of surfaces instead a single surface may be used to describe the switching criterion**. In this case, the state space is divided in a finite number of regions and the possible combinations of active subsystems can increase dramatically. It is worthy to notice that hybrid switched control can be seen as a generalization of sliding mode control. This last is a switching control that uses a single switching surface along with a mode assignment criterion that chooses only between two options (Barnejee & Vergese, 2001).

Notice that in Eqs. (3)-(4), there exists the possibility of using different *control* inputs in a switching system. That is, a control problem in a general switched system can be formulated as one of stabilization or tracking, using three kinds of control inputs: The switching criterion, the mode assignment criterion and the continuous control input u. To be precise, using hybrid control one can increase the number of degrees of freedom that can be used to accomplish a control objective.

There exist in the control literature, various authors leading with these control problems (Lin, 2008), (Lazar, *et al.*, 2006), (Tsinias, (2007), (Zhai & Michel, 2002), (Zhai & Michel, 2003), (Rodrigues & How, 2003). However much of this literature cannot be applied directly to power electronics due that the input u cannot be manipulated (consider converter description as in (3), (4)). In other words, the vector fields among which the converter is evolving between switches cannot be manipulated; hence, properties as stability and equilibrium points can not be modified. In view to this fact, a relevant problem in switching control applied to power electronics is how to choose the switching surfaces (possible moving surfaces) in order to stabilize the overall system. Observe that, given a mode assignment criterion, the manipulation of surfaces can be seen as equivalent to choose a set of switching times. Since in DC-DC Power electronics, the control actions are applied only at some instants (not necessarily synchronous), sometimes these systems are called ***partially open-loop systems***.

Another option of describing switched systems can be performed using **Hybrid Automata**. A hybrid automaton (HA) is a mathematical model, borrowed from computation theory, used to describe processes consisting in discrete state transition and continuous evolution. In this way, HA can be seen as a finite state machine that uses differential equations; in which the model is composed of a finite number of states, transitions between states and actions (differential equations) (Henzinger, 1995), (Henzinger, 1996), (Rajeev *et al.*, 1996).

A hybrid automaton H is composed by a finite set of variables ($X = \{x_1, x_2 ... x_n\}$), that describe the system during continuous evolution ($X = \{x_1, x_2 ... x_n\}$) and during discrete transitions ($X' = \{x'_1, x'_2 ... x'_n\}$). The system dynamics is represented by a multi-directed graph called a control graph (V,E). The vertices V are called **control modes**. The edges E are called **control switches or guards**. By virtue of a **jump or switch condition** it is assigned to each control switch $e \in E$ a given consequence which is detonated by an event.

In mathematics literature, the description of switched system by means of differential inclusions is common; however, in power electronics literature is rarely found. The interested reader is referred to (Filipov, 1998) for detailed description of this modelling tool.

2.2 Examples in DC_DC Power converters

In order to illustrate the modelling techniques mentioned above, the case of a conventional boost converter is used as a benchmark. In this way, the piece-wise continuous equations describing the converter dynamics are given by Eq (1) where $x = [i_L, v_c] \in \Re^2$ is the state vector constituted by the inductor current and the output voltage, $u \in \Re^2$ is the input vector that account for the translation due to the input voltage source. If the converter is working under continuous conduction mode (CCM) the number of modes or subsystems is $p = 2$ with

$$A_1 = \begin{bmatrix} 0 & 0 \\ 0 & -\frac{1}{RC} \end{bmatrix} \quad and \quad B_1 = \begin{bmatrix} \frac{E}{L} \\ 0 \end{bmatrix} \quad for\ m{=}1 \tag{7}$$

$$A_2 = \begin{bmatrix} 0 & -\frac{1}{L} \\ \frac{1}{C} & -\frac{1}{RC} \end{bmatrix} \quad and \quad B_2 = \begin{bmatrix} \frac{E}{L} \\ 0 \end{bmatrix} \quad for\ m{=}2$$

where E is the voltage input, L is the inductance, C the capacitance and R the load resistance. If the converter is working under discontinuous conduction mode (DCM) the number of modes or subsystems is $p = 3$ with $m = 1$ and $m = 2$ given as in Eq. (7) and

$$A_3 = \begin{bmatrix} 0 & 0 \\ 0 & -\frac{1}{RC} \end{bmatrix} \quad and \quad B_3 = \begin{bmatrix} 0 \\ 0 \end{bmatrix} \quad for\ m{=}3 \tag{8}$$

The description of the switching law will depend if the converter is open or closed loop and on the switching regime used. If a driven signal of frequency $f = 1/T$ is used, the switching law can be described as follows for a converter operating in CCM:

$$i(t+) = \begin{cases} 1 & if \quad 0 \le \mathrm{mod}(t,T) < \delta T \\ 2 & if \quad \delta T \le \mathrm{mod}(t,T) < T \end{cases} \tag{9}$$

where the function $mod(a,b)$ stands for the remainder of the division of a/b, and δ is the duty cycle. For the converter in DCM the switching law is given by

$$i(t+) = \begin{cases} 1 & if \quad 0 \le \mathrm{mod}(t,T) < \delta T \\ 2 & if \quad \delta T \le \mathrm{mod}(t,T) < T_2 \ \ and \ \ \mathrm{mod}(t,T) > \delta T \\ 3 & if \quad T_2 \le \mathrm{mod}(t,T) < T \ \ and \ \ \mathrm{mod}(t,T) > \delta T \end{cases} \tag{10}$$

where T-T_2 stands for the time length that the inductor has zero current.

On the other hand, the description of the converter in both CCM and DCM using hybrid automaton can be observed in the control graph shown in Fig. 3, which is a reproduction of Figure 3 in (Sreekumar & Agarwal, 2008), where the inductor currents rising phase is defined with σ_1†, the falling phase with σ_2 and zero phase with σ_3. The transition between stages (σ) is defined by the guards in Table 1. In Fig. 3, q are the discrete states, x are the continuous states, $I_{1,2,3}$ are invariant sets and G are the guards.

---

† $m = 1$ in Eq. (7)

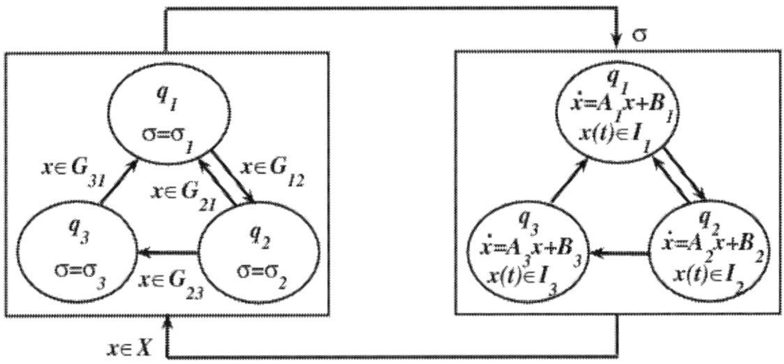

Fig. 3. Hybrid description of a conventional boost converter under CCM and DCM (Sreekumar & Agarwal, 2008).

where:

G_{12}	$0<mod(T,t)\leq\delta T$	G_{23}	$\delta T <mod(T,t) \leq T_2$ and $t>\delta T$
G_{21}	$\delta T <mod(T,t) \leq T$	G_{31}	$T_2<mod(T,t) \leq T$ and $t>\delta T$

Table 1. Guard description of Fig. 3.

Notice that under CCM the active guards are G_{12} and G_{21}. Moreover, under DCM the switching can only be performed unidirectionally; that is $G_{12} \rightarrow G_{23} \rightarrow G_{31} \rightarrow G_{12} \rightarrow \ldots$ (Senesky *et al.*, 2003).

2.3 Control Problems in DC-DC Power Converters. Hybrid Control Solution

This subsection is devoted to describe the control problems in DC-DC power converters that are best suitable to be addressed with hybrid control, as well as describe some advantages of this tool.

From conceiving the system as piecewise continuous one or by using a combination of classical averaged models with discrete transitions or discrete controllers is that hybrid control arises in power electronics. That is, the system under study has open and/or closed loop hybrid description and, as stated before, its mathematical description can be very vast due to the variety of the possible interactions.

In order to describe the control problems in DC-DC power converters, it is worthy noticing that the main performance concerns in these applications are:

a. **Voltage regulation.** Voltage regulation of a constant desired reference is the primary objective of control and a basic performance requirement that has to be fulfilled in every real life application.
b. **Current limitation**. For safety reasons and to endure converter components life, it is required a current limitation in the circuit.

c. **Maximum current and voltage ripple**. Between 10-20% of the reference current(s) signal(s) and about 1-2% of the desired voltage.

Furthermore, it is highly desirable that the control is designed to fulfil the following requirements:

- **Adaptability or Applicability to CCM or DCM regimes**. Flexibility to operate in both modes, mainly due to possible perturbations in the parameters and to face near- to- cero load resistance conditions.
- **Robustness against model uncertainty and disturbances**. For example, disturbances in load resistance, input voltage and voltage reference. That is, it is desirable that the converter operates in regions of the state space, rather than in a small neighborhood about the nominal point.
- **Output feedback**. It is desirable that the control uses only a subset of the state vector for feedback purposes (output feedback). This is of particular interest in high order converters, where measuring all signals becomes prohibitive due to the high costs.
- **Fixed Switched Frequency.** Among the main reasons to operate at fixed frequency conditions are: i) the possibility that harmonic content arise in varying frequency operation, which in turn may induce a poor steady-state performance and ii) It is easiest to implement fixed frequency controllers.

An advantage of hybrid control and in particular, of switched control is that some performance requirements can be formulated explicitly by defining suitable switching surfaces or control guards. For example, effective current limitation can be performed with a time varying current-dependent switching surface *i.e.* peak current control with stabilizing ramp; voltage or current ripple can be limited for a suitable hysteresis-like controller , which can be seen as a switching control with two voltage and/or current dependent switching surfaces (Cervantes *et al.*, 2009a), (Perez-Pinal & Cervantes, 2009a). Furthermore, due to the fact that the converters can be seen as switched systems, the switching criterion and the mode assignment criterion can be manipulated independently to obtain **multiple control objective**, for example, current limitation at the same time that voltage regulation (Austria-Gonzalez *et al.*, 2008) (Perez-Pinal & Cervantes, 2009b), or maximum voltage and current rippling as well as voltage regulation (Cervantes *et al.*, 2009b).

On the other hand, hybrid controllers represent an option to solve **optimal operation problems**. For example in (Almer *et al.*, 2007) a predictive control with load estimation is proposed, in which a performance index is minimized to obtain a prescribed performance (see also, Geyer *et al.*, 2008). In (Chen *et al.*, 2006) a control strategy is proposed that is able to compute optimal switching instances that delivered over one switching period allow to a series parallel resonant converter to achieve its required steady state in minimum time.

Hybrid controllers can also be designed to solve voltage regulation problems even if the converter changes from **CCM** to **DCM** regime and vice versa (Sreekumar &, Agarwal, 2006), (Sreekumar &, Agarwal, 2007), (Sreekumar &, Agarwal, 2008), (Senesky *et al.*, 2003). Furthermore, many of the hybrid control strategies proposed can be implemented using **fixed frequency switching** (Geyer *et al.*, 2008).

Moreover, an additional advantage of describing the converter dynamics using a piece-wise model allows a confident description of the system for the whole operating regime. This fact opens the possibility of designing control laws valid in every operation point.

3. Some Stability Issues

A main concern when dealing with converter control is the closed-loop stability of the system. Regarding stability of hybrid systems, it is possible to find various approximations in the literature (Aeyels & Peuteman, 1998), (Branicky, 1998), (Filipov, 1998), (Liberzon, 2003) among many others. However, a great percentage of those results cannot be used directly in power electronics due to the fact that the HS considered in such works, do not fit in a possible description of a converter. That is, it is considered one or more of the following assumptions,

- Vector fields of the continuous part of the HS can be modified.
- In switched systems,
  - Subsystems have a common equilibrium point
  - All subsystems have equilibrium points
  - Subsystems have a given stability property

The reasons of the applicability limitation are clear when one looks into the converter structure. Consider for example the boost converter in Eqs. (7)-(8) subsystems 1 and 3 do not have any equilibrium point. Theoretically, the current is able to grow unlimited, while subsystem 2 has an stable equilibrium point in $x = (E/R\ ,E)^T$. Therefore, common equilibrium point approaches cannot be applied. Furthermore, as discussed before, the vector fields among which, the converter is evolving (subsystems 1-3) cannot be manipulated, since an extra control input u in Eq. (3) is not available. That is, neither equilibrium points nor stability properties of the subsystems can be modified. In view of this fact and as stated before, admissible control inputs are the residence time in each subsystem and mode assignment criterion.

3.1 Asymptotic Stability vs. Practical Stability

It is possible to find various definitions of stability in the literature; each one describing different behaviour of the system. Boundedness, input-output stability, output-state stability, asymptotic stability, exponential stability, practical stability and absolute stability are examples of these concepts. Although the objective of this subsection is not giving an exhaust description of such notions (the interested reader is referred to (Sontag, 1998) and (Khalil, 2002), it is relevant to notice that the inexistence or existence of equilibrium points is crucial to determine the type of stability that can be achieved by the system.

For example, asymptotic stability, exponential stability and absolute stability can only be achieved by systems with equilibrium points. In fact, these concepts are related to the ability of system trajectories to approach and stay at the equilibrium point (EP). In this way, it is said that trajectories are attracted by the EP and the set of all initial conditions such that this happens is called **attractive domain** or **region of attraction**. Having a large region of attraction is of great interest to controller designers mainly because it guarantees the existence of a set about the equilibrium point such that system trajectories are confined. This fact gives to the system, natural robustness against some perturbations.

Although it would be desirable to obtain asymptotic (or exponential) stability in power converts, in most cases this cannot be achieved using a hybrid description. The reason resides in the fact that in most cases, a common equilibrium point can not be found, e.g. the hybrid system (3)-(4). At this point, it is interesting to notice that when an average converter model is used along linear or nonlinear continuous controllers, a common average equilibrium point can be obtained. From the continuous point of view, we may be able to achieve both asymptotic and exponential stability to this "**fictitious**" equilibrium point; however in the converter, attractivity to a neighbourhood of this averaged equilibrium can only be obtained (practical stability).

Due to this fact, many of the results in switched and hybrid systems deal with the concept of practical stability, rather than with asymptotic stability. Since these two concepts will appear frequently in this chapter, the definitions are stated below:

Definition 1. (Khalil, 2002). The equilibrium point $x = 0$ is of the system $\dot{x} = F(x,t)$ is

- stable if, for each $\varepsilon > 0$, there is $\delta = (\varepsilon, t_0) > 0$ such that $\|x(t_0)\| < \delta \Rightarrow \|x(t)\| < \varepsilon, \quad \forall t \geq t_0 \geq 0$
- asymptotically stable if it stable and there is a positive constant $c = c(t_0)$ such that $x(t) \to 0$ as $t \to \infty$ for all $\|x(t_0)\| < c$

Definition 2. (Xu *et al.*, 2007). Given $\varepsilon > 0$, system (3) with u=0,

- is ε-practically uniformly stable (ε-PUS) about the origin under (4), if there is a $\delta = \delta(\varepsilon)$ >0, independent of t_0 such that $x(t_0) \in B(0,\delta) \Rightarrow x(t) \in B(0,\varepsilon) \quad \forall t \geq t_0 \geq 0$
- is ε-practically uniformly asymptotically stable (ε-PUAS) about the origin under (4), if is ε-practically uniformly stable and there is a $T = T(c) > 0$ independent of t_0 such that $x(t_0) \in B(0,c) \cap D \Rightarrow x(t) \in B(0,\varepsilon), \quad \forall t \geq t_0 + T$

Notice that the stability definition 1 differs from the concept of ε-PUS in the fact that in Definition 1 the condition over ε is assumed for all $\varepsilon > 0$, while in Definition 2, is enough the existence of some value (usually design-fixed). Furthermore, notice that Definition 1 requires the existence of an equilibrium point and the stability property is attributed to it. In contrast, in Definition 2, is the system the one which is stable and the result basically establishes confinement of system trajectories about the origin and in ε-PUAS, also attractiveness. On the other hand, when dealing with switched control, the concept of practical stability is also related with the *controlled invariant* one (Senesky *et al.*, 2003).

Notice that Definition 1 concerns about continuous nonlinear, time varying systems. If the hybrid system can be described as in this definition, necessary and/or sufficient conditions can be found to deduce stability (Sontag, 1998), (Khalil, 2002) . On the other hand, conditions that lead us to conclude asymptotic stability of a general HS may be hard to find. However, for switched systems such a task is easier in some cases.‡

In the following subsection we will describe some contributions in the area of hybrid control applied to DC-DC power converter. The discussion about their stability properties will be explained in view of the concepts and ideas stated here.

4. Types of HC in Power Electronics

The types of HC found in the literature can be classified in two groups mainly; a first one uses either a switched description or a switching control action in the converter; in these papers, the control objective is to find a switching law such that voltage is tracked. A second group, which description of the converter may be continuous or piece wise continuous, in these strategies the control objective is aimed to solve optimally (time or using and index performance) the voltage regulation problem. We will refer to the first group as **Hybrid Switched Control** (HSC) and to the second as, **Hybrid Optimal Control** (HOBC).

4.1 Hybrid Switched Control

In this category we can find variable structure controllers such as: hysteresis, sliding and current peak mode controllers among the most common. **Sliding mode controllers** (SMC) are usually designed based on an averaged continuous model of the converter. In general, the converter is described by non-linear differential equations evolving in a *fictitious* averaged state space. Under this situation the control input is also continuous but restricted to $u \in [0,1]$. Once a continuous model is obtained, the next step is to define a suitable surface that resembles the desired steady state, usually such surface is a plane of the form:

$$S = \sum_{i=1}^{n} a_i \tilde{x}_i \tag{11}$$

where $\tilde{x} = x - x_{ref}$, and x_{ref} the desired state. However, the sliding surface can be also non-linear (Sira-Ramirez & Ríos-Bolivar, 1994), (Tokat *et al.*, 2002). Sliding mode control action can be divided in two operation modes: the **reaching mode** and the **sliding mode**. The reaching mode is related with the operation phase in which from an arbitrary initial condition the trajectories goes toward the surface S. The control input of SMC can be seen also as divided in two, u_{eq} that acts mostly on the reaching mode and u_d for the sliding. The input u_{eq} is usually continuous and parameter dependent, while u_d is a discontinuous input usually depending on the sign of the desired surface therefore, is parameter insensitive. There exist various papers in the literature dealing with the robust design of controllers for the reaching mode. For example in (Sira-Ramirez & Ríos-Bolivar, 1994) a passivity based controller is proposed to enhance the robustness properties of SMC. The controller make use of the passive relation between input voltage and inductor current of the converter, along with

‡There exist trivial cases regarding stability of switched systems that are not of our interest. For example, if there is a stable mode in the subsystems family, we can guarantee stability of the switched system just by staying indefinitely on it, among others cases.

Lyapunov based arguments to show that the desired tracking dynamics approaches the origin asymptotically (*i.e.* the equilibrium point is asymptotically stable). The use of non-linear and time-varying sliding surfaces has also been proposed to enhance the robustness of SMC (Sira-Ramirez & Ríos-Bolivar, 1994), (Sira-Ramirez et al., 1995), (Tokat *et al.*, 2002). On the other hand, sensorless SMC has also been proposed, as well as some techniques for fixed frequency operation, the interested reader is refer to (Tan *et al.*, 2008) for a comprehensive review of sliding mode control in power electronics.

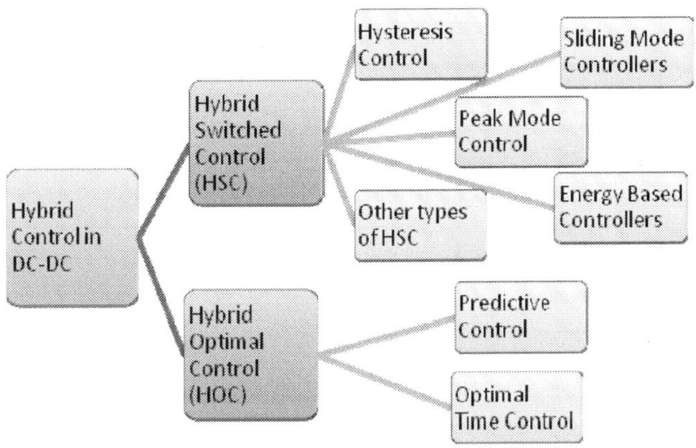

Fig. 4. Types of hybrid controllers used in DC-DC power converters.

Hysteresis control can be interpreted as a switched control with two output-dependent switching surfaces: turn-on and turn-off surfaces (Babaa *et al.*, 1968). In this control methodology, the converter is described as a piece wise linear system (as in Eq. (1)) and the control objective is to maintain voltage confined within a pre-specified region, given two ad-doc surfaces. For example, in buck converters such surfaces will be:

$$\begin{aligned} S_{off} &= V_{ref} - \Delta v = 0 \\ S_{on} &= V_{ref} + \Delta v = 0 \end{aligned} \tag{12}$$

For boost or fly-back converters (among others), current limitation has also be performed in order to avoid system instability. Some options to avoid such instability are the use of a current dependent switching surface or some kind of multi-loop control. On the other hand, since in this kind of controller the control input (switching surfaces and assign mode criterion) does not depend on system parameters, such control strategy results very robust (model free controller). However, a drawback of this controller is that it may lead to variable frequency operation. Since the first works about hysteresis control (Babaa *et al.*, 1968), various improvements have been performed. In (Leung & Chung, 2009), a time varying surface is proposed in order to reject large signal perturbation in two switching cycles. In (Schild *et al.*, 2009) a procedure that uses both clock-driven and event-driven switching criterion for controlling the stationary operation of a converter is proposed. The switching surfaces are planes that can be systematically computed using periodic control systems. Consequently, desired loop properties such as orbital stability of a limit cycle and a fast transient response are guaranteed at least in a local neighbourhood around a nominal set point. In (Tao *et al.*, 2006) variable hysteresis band control is proposed to achieve zero voltage switching for a multiple input converter.

On the other hand, in (Kimball, 2006) a sensorless hysteresis-type controller is proposed. Such strategy does not need current measurements and it can be implemented in converters operating under DCM conditions. In (Khaligh & Emadi, 2007) a hysteresis type controller that works under fixed frequency switching conditions is proposed. At this point, it is interesting to notice that although most hysteresis controllers have parameter insensitivity§ and their stability and performance have been evaluated experimentally (see for example (Hwu & Yau, 2004); there exist a little formal theoretical support that explain their robust capabilities. Formal results involving robust practical stability of buck and boost converters under hysteresis-like controllers have been presented in (Cervantes *et al.*, 2009a), (Cervantes *et al.*, 2009b), (Perez-Pinal & Cervantes, 2009b); however, there are a lot of questions that remain open, specially for time varying and non-linear hysteresis (switching) surfaces.

Peak Current Mode Control (PCMC) is a class of switched control that has both time dependent and state dependent surfaces. This control strategy was conceived to limit inductor current while guaranteeing a fixed frequency switching operation. For a buck converter the PCMC is the following

$$\textit{If } i > i_{max} \quad \textit{then switch goes off} \tag{13}$$

$$\textit{If } \quad 0 \le \mathrm{mod}(t,T) < \delta T \quad \textit{then switch goes on} \tag{14}$$

For PMCM to be well defined it is necessary that the switching condition (12) be of higher priority than the one in (11). It is well known that, for duty cycles greater than 50%, PCMC leads to converter instability and strange behaviour such as bifurcation, chaos (Barnejee & Verghese, 2001) and breathing** (Cervantes & Femat, 2008). However, a stabilizing ramp is used to avoid such undesirable behaviour. Such a ramp makes a switching surface be both state and time dependent (the one that limits current). In spite of the well-known capabilities of PMCM, as in the case of hysteresis control, formal studies regarding stability are rare; hence issues regarding local or global stability capabilities are not yet completely known, as well as robustness properties of the controller.

There exist other expressions of switching control in the literature, for example the so-called *"boundary"* †† control studied in (Leung & Cheun, 2004), (Song & Chung, 2008) and (Leung &

§ Due to this fact, hysteresis control can be applied directly in many topologies without need of adjustment. This control is also called by some authors universal (Khaligh & Emadi, 2007).

** If a low frequency variable voltage input is used.

††This term is rarely used since confusion arise with the so-called boundary control in distributed system (infinite order systems) in which the frontier or boundary condition is

Cheun, 2007). Such controllers use a piece-wise linear description of the converter and switching surfaces which can be either planes or second order surfaces. It is shown that second order surfaces display better transient behaviour than planes, however the implementation of such surfaces may be quite involved. The design of the surfaces is based on one hand, on the regulation requirements of the converter; and on the other, on natural trajectories of the system, that reduce the transient response of the systems (ideally in two switching actions). The switching surfaces depend on system parameters and although a sensitivity study is performed in (Leung & Cheun, 2007) with respect to system inductance and capacitance, the surfaces depend dramatically on load resistance; such fact may limit the advantages of the controller if such parameters are uncertain.

Computation of switching instants based on an energy balance of the converter is shown in (Gupta & Patra, 2005). In this controller, the modes in which the converter gains and looses energy are identified. The switching instants are chosen so the total energy change is cero. Although this strategy has a physical motivation, the proposed strategy has the disadvantage of leading to variable current ripple and variable frequency operation.

Another expression of switching control can be found in (Senesky *et al.*, 2003). In such a reference, switched systems are described as hybrid automata. Based on this description the stabilization problem is stated in terms of the existence (and design) of suitable guards that guarantee that converter states remain in a safe set. In this excellent work, two regulation strategies are designed, the minimum ripple control and the minimum switching control that guarantee that a set of interest is controlled invariant (*i.e.* the system is practically stable). The proposed strategies have the advantage of having rigorous mathematical support and the effect of sample and perturbation is analysed. Although such techniques are probably easy to implement, unfortunately the authors do not show any experimental evidence of controller performance. On the other hand, other controllers based on the hybrid automata can be found (Sreekumar & Agarwal, 2007) (Sreekumar & Agarwal, 2008). In such works, a controller that is able to operate under both CCM and DCM is introduced. The authors give experimental evidence of the controller performance but unfortunately, the stability analysis presented do not take into account dynamic behaviour.

4.2 Hybrid Optimal Control

As pointed out before, in HOC the description of the converter is continuous or piece wise continuous and the control strategy is aimed to solve an optimal problem. Within this category we find predictive control of converters e.g. (Geyer *et al.*, 2008), (Morari *et al.*, 2006). In this case, the control is usually operated in fixed frequency and the optimization problem can be solved: i) once every switching cycle, ii) every fixed number of switching cycles or iii) offline. The optimization problem is usually stated, based on a performance index function that penalizes voltage error tracking, control deviation from nominal, among other common criteria. Major concerns when dealing with optimization problems are the solution existence and its computation time. Due to these facts, usually some simplifications to the optimal problem are performed and off-line computation of the control law is preferred.

---

manipulated in order to control variables that are spatial dependent (You & Lee, 1989) (Morgul, 1992).

In (Geyer *et al.*, 2008) a predictive control of a step-down converter is proposed. The controller makes use of a piece-wise continuous model that is used to generate a discrete time approach of the system in order to simplify the solution of the constrained optimal control. The optimization problem is solved offline using dynamic programming and a Kalman filter is used to ensure voltage tracking and load estimation. The stability analysis is performed using piece-wise Lyapunov functions to show that a discrete equilibrium point is exponentially stable. Observe that in contrast to the discussed in Section 3, exponential stability in the discrete case can be guarantee since a discrete equilibrium point can be either i) a constant value or ii) a periodic orbit (which is the case in converters). A similar predictive control is also proposed in (Almer *et al.*, 2007) but in this case, a multi-loop robust controller that uses discrete integral actions to compensate uncertainty and track the reference voltage is used.

On the other hand, an optimum time controller has been proposed in (Chen *et al.*, 2006) for a series-parallel resonant converter. In this paper a piece-wise continuous model of the converter is used to analyze the state-state trajectories to compute optimal switching instants. The optimal control is able to compensate step voltage commands theoretically in one switching cycle. The controller makes use of voltage and current measurements as well as zero voltage crossing detection devices. Although the authors show the experimental performance of the controller, stability analysis is not complete and robustness of the proposed control has to be studied.

It can be observed that HOC can be much difficult to synthesize than switched controllers, such fact may affect adversely the complexity of actual implementations. Since these controllers are much complex, their application has to be well justified in terms of performance improvement or robust stability capabilities.

4.3 High Order Converter

Hybrid control and in particular, switched control has been gained more attention lately, among other things, due to its property of being easily implemented even in high order converters (interleaved or cascaded converters). These types of converters are mainly used in medium to high power applications in order to divide the converter power in cells, allowing managing lower currents and faster switching devices. In some topologies, there is a dramatic increasing in the number of switches. Incrementing the number of switches, increment also the number of degrees of freedom of the system, which in some cases may affect adversely the complexity of controller design. However, also open the possibility to solve more efficiently some operation problems. For example, in interleaved converters, the possibility of reducing substantially voltage rippling arises even at low switching frequencies. Moreover, perturbation rejection can be performed smoothly, since input actions and efforts can be divided among the branches (Batchvarov *et al.*, 2000).

In spite of these advantages, there exists in the literature very few works leading with voltage regulation and ripple limitation in higher order DC-DC power converters. In (Baja *et al.*, 2008) a stabilizing approach used to control capacitors voltage and current of a four-level three-cell DC-DC converter is proposed. The description of the converter is performed using bond graph formalism‡‡. Under the assumption that switching actions are performed in pairs, the notion of commutation cell is used to allow the converter description via bond graph formalism be topologic simple and with fixed causality in any mode. Three types of cells are detected: source, internal and load cells, for each one a static model is derived that relates inputs and outputs. The dynamic description of the converter takes also into account the dynamics of the Boolean control variables. The controller design is based on Lyapunov redesign technique and local atractivity of system trajectories is guarantee to a neighborhood of the origin.

On the other hand, a hysteresis like controller is proposed in (Cervantes *et al.*, 2009b) for the case of interleaved converters. The controlled derivation is based on an approximated switched description of the system. Closed-loop stability analysis of the converter is performed leading to a practical stability result. The proposed controller is able to regulate voltage, while warranting maximum current and voltage rippling, as well as current balancing. The performance of the controller is tested via both numerical simulations and experimental work; and it is shown that the controller can be readily implemented using integrated circuits.

In multi cell converters, the number of dynamic elements increases compared to traditional converters; therefore, it is convenient to use sensor-less control approaches, otherwise, the cost of the control implementation may increase dramatically. After a short review of the literature and in spite of the importance of developing such controllers, we could establish a truly need of investigation of this subject in the case of hybrid controllers. Furthermore, stability analysis of switching control in both parallel and cascaded topologies is also an open problem, as well as the development of a systematic procedure to design switching surfaces that lead to the solution of highly versatile control specifications.

5. Conclusions

In this work, a general description and classification of the most common hybrid controllers for DC-DC power converters is given. It is stated that in particular switched control has the advantage that some performance requirements can be formulated explicitly by defining suitable switching surfaces. Moreover, in some cases, the switching criterion and the mode assignment criterion can be manipulated independently to obtain multiple control input schemes. A main advantage of switched control is its inherent flexibility to operate in a wide operation region. Hybrid control also represents an option to solve **optimal operation problems,** for example, minimum switching control or predictive-type control. Furthermore, hybrid controllers have been designed to solve voltage regulation problems even if the converter changes from **CCM** to **DCM** regime and vice versa. Although some

‡‡ Bond graph modeling is a method for describing dynamic systems, which is based on the assumption that it is possible to define the characteristics of the subsystems and the connections between them without energy losses. Bond graph formalism may be perceived *some how close* to graph automata description, however there are substantial differences, see (Antic *et al.*, 1999) for more details.

HC methodologies are designed to operate at varying frequency (mostly transient response), there is an increasing tendency in the literature to investigate fixed frequency hybrid control schemes.

There are still a lot of open questions regarding stability and design of HC. We find a research opportunity in the formal stability study of some of the proposed control methodologies, in order to reveal advantages and limitations. We also find a research opportunity in the systematic design of switching laws based on time-varying nonlinear switching surfaces as well as on the design of sensor-less hybrid control of higher order converters.

Summarizing, hybrid controllers represent a versatile solution to guarantee energy quality in the renewable energy field, but the uncontroversial success of this control methodology will reside in the development of systematic approaches that guarantee both, simplicity in the implementation and robust stability properties.

6. References

Aeyels D. & Peuteman J. (1998). A new asymptotic stability criterion for nonlinear time-variant differential equations, *IEEE Trans. Automat. Contr.*, vol. 43, no. 7, pp. 968–971, July 1998, ISSN 0018-9286.

Almer S., Fujioka H. Jonsson U, Kao C.Y., Patino D., Riedinger R., Geyer T., Buccuti A., Papafotiou G., Morari M., Wernrud A & Rantzer A., (2007). Hybrid control techniques for switched-mode DC-DC converters. Part I. The step-down topology. *Proc. of American Control Conference*, pp. 5450-5457, ISBN 1-4244-0989-6, July11-13, New York.

Antic, D.; Vidojkovic, B.; Maldenovic, M. (1999). An introduction to bond graph modelling of dynamic systems, *Proc. of Telecommunications in Modern Satellite, Cable and Broadcasting Services Conference*, 1999. Vol. 2, No. 13-15 Oct. 1999 Page(s):661 - 664.

Austria-Gonzalez, L.F., Perez-Pinal, F.J. & Cervantes I.(2008). Hybrid Multi-objective Control of DC-DC. *Proc. of IEEE Vehicle Power and Propulsion Conference*. IEEE Harbin, China, República Popular. septiembre 2008.

Babaa, I.; Wilson, T. & Yuan Y. (1968). Analytic solutions of limit cycles in a feedback-regulated converter system with hysteresis; *IEEE Trans. Automatic Control*, Vol. 13, No. 5 Oct 1968 Page(s): 524 - 531, ISSN 0018-9286.

Baja, M., Cormerais H. & Buisson J. (2008). Modeling and Hybrid control of a four-level three cell DC-DC converter. *Proc of Industrial Electronics Conf.* 10-13 Nov. 2008, Page(s): 3278-3283. ISBN: 978-1-4244-1766-7

Barnejee, S. & Vergese G.C. (2001). *Nonlinear Phenomena in Power Electronics. Attractors, Bifurcations, Chaos, and Nonlinear Control*, Ed. IEEE Press, ISBN 0-7803-5883-8, New York.

Batchvarov, J.S.; Duarte, J.L. & Hendrix, M.A.M. (2000). Interleaved converters based on hysteresis current control; *Proc. of Power Electronics Specialists Conference, 2000. PESC 00. 2000* Vol. 2, 18-23 June 2000 pp: 655 - 661.

Branicky, M.S. (1998). Multiple Lyapunov functions and other analysis tools for switched and hybrid systems, *IEEE Trans. Automat. Contr.*, vol. 43, pp. 1679–1684, 1998, ISSN 0018-9286.

Capel A., Ferrante G., *et al.*, (1978) "Application of injected current mode control for the dynamic analysis of switching regulators with new concept of LC3 modulator," *IEEE PESC Record*, pp.135-147.

Cervantes I., Femat R. (2008). Intermittence in Linear Driven Switched Systems. *Int. J. Bif. Chaos*, vol. 18, pp. 495-508.

Cervantes, I. Mendoza-Torres, A. Garcia-Cuevas, R. & Perez-Pinal. J. (2009b) Switched Control of Interleaved Converters. *Proc. of IEEE Int. Vehicle Power and Propulsion Conference*. Dearborn, Michigan, September 7-11, 2009.

Cervantes, I., Mendoza-Torres, A. & Perez-Pinal. J. (2009a) High Performance Hybrid Control for Buck Converters. *Proc. 13th European Conference on Power Electronics and Applications - EPE 2009* 8-10 September 2009, Barcelona, Spain.

Chen H., Sng, E.K.K & Tseng K.J.T (2006). Optimum trajectory switching control for series-parallel resonant converter. *IEEE. Trans. Ind. Electronics*, vol. 53, No. 5, pp. 1555-1563, ISSN 0278-0046.

Deane J. H. B. and Hamill D. C. (1990) Instability, Subharmonics, and Chaos in Power Electronic Systems, *IEEE Transactions on Power Electronics*, 260-268, ISSN: 0885-8993.

ECPE, European Center for Power Electronics, EPE European Power Electronics and Drives Association (2007), "Position Paper on Energy Efficiency - the Role of Power Electronics," March. http://www.ecpe.org/download/power_electronic/Position_Paper_Energy_Efficiency.pdf, accessed February 1st 2009.

Enerdata, (2008), "The world energy demand in 2007," pp. 1-17

Enerdata, (2008b), "2020 Global Energy Forecasts", pp. 1-21.

Erickson R. & Maksimovic D. (2001), *"Principles of Power Electronics,"* Kluwer Academic Publishers.

Erickson R., Cuk S., and Middlebrook R. (1982), "Large signal modeling and analysis of switching regulators," Proc. of *IEEE Power Electronics Specialists Conference Records*, pp. 240-250

Filipov, A.F. (1998). *Differential Equations with Discontinuous Right-Hands Sid*es, Ed. Arscott, F.M., Kluwer Academic Publishers ISBN. 902772699X.

Geyer T., Papafotiou G. & Morari M. (2008). Hybrid model predictive control of the step-down DC-DC converter. *IEEE Trans. Control System Technology*, vol 16, No. 6, pp. 1112- 1124, ISSN. 1063-6536.

Goebel, R., Sanfelice R.G. & Teel, A.R. (2009). Hybrid dynamical systems. *IEEE Control Systems Magazine*, pp. 28-93, ISSN. 066-033x

Gupta, P. & Patra A.N. (2005) Super stable energy based switching control scheme for DC-DC buck converter circuits. *Proc of IEEE Int.Symp on Circuts and Systems*. pp. 3063-3066, ISBN 0-7803-8834-8

Henzinger, T. A. (1996). The Theory of Hybrid Automata, *Proc.11th Annual Symposium on Logic in Computer Science*, 278-292. IEEE Computer Society Press

Henzinger, T.A. (1995). *Hybrid Automata with Finite Bisimulations*. ICALP 95: Automata, Languages and Programming, Lecture Notes in Computer Science, 944, pp 225-238. Springer Verlag.

Hwu, K.I. & Yau Y.T. (2004). Perfomance improvement of hysteresis voltage-controlld forward converters using FPGA-based SR control. *Proc. of Int. Conf. Power System Technology*, Singapure 21-24 Nov, 2004. pp. 281-287, ISBN 07803-8610-8.

International Energy Outlook 2008, Energy Information Administration, US Government, Report#:DOE/EIA-0484(2008) Release Date: June.

Khaligh A. & Emadi A. (2007). Suitability of the pulse adjustment technique to control single DC/DC choppers feeding vehicular constant power loads in parallel with conventional loads. *Int. J. Electric and Hybrid Vehicles*. Vol. 1, No., 1, 2007.

Khalik H. S. (2002). Nonlinear Systems. Prentice Hall, ISBN 0-13-067389-7, New Jersey.

Kimball, J.W.; Krein, P.T. & Chen Y. (2006). Hysteresis and delta modulation control of converters using sensorless current mode; *IEEE Trans. Power Electronics*, Vol.21, Issue 4, July 2006 Page(s):1154 - 1158, ISSN: 0885-8993.

Lazar, M.; Heemels, W.P.M.H.; Weiland, S.; Bemporad (2006) Stabilizing Model Predictive Control of Hybrid Systems Lazar, *IEEE Trans. Automatic Control*, Vol. 51, No. 11, pp: 1813 - 1818, ISSN 0018-9286.

Leung, K.K.-S. & Chung, H.S-H. (2007). A comparative study of boundary control with first and second order switching surfaces for buck converters opertin in DCM. *IEEE Trans. Power Electr.* Vol. 22, No. 4, pp. 1196-1209 ISSN: 0885-8993.

Leung, K.K.S. & Chung, H.S.H. (2004) Derivation of a second-order switching surface in the boundary control of buck converters. *IEEE Power Electronics Letters*, Vol. 2, No. 2, Page numbers (63 - 67).

Leung, K.K-S & Chung H. S.-H. (2009) Dynamic hysteresis band control of the buck converter with fast transient response; *IEEE Trans. Circuits and Systems-II*, Vol. 52, No 7, July 2005 Page(s):398 - 402

Li, Z; Soh, Y. & Wen, C. (2005). *Switched and Impulsive Systems. Analysis, Design and Applications*, Springer-Verlag, ISBN 3-540-23952-9, Berlin.

Liberzon, D. (2003). *Switching in Systems and Control*, Birkhausen, ISBN 0-8176-4297-8, Boston.

Lin H. (2008). Hybrid Output Feedback Stabilization for LTI Systems With Single Output *IEEE Trans. Aut. Control*, Vol. 53, No 7, pp: 1736 - 1740, ISSN 0018-9286.

Middlebrook R. (1985), Topics in multi-loop regulators and current mode programming, *Proc. IEEE Power Electronics Specialists Conf. (PESC)*, pp. 716-732.

Middlebrook R. Cuk S. (1976), A general unified approach to modeling switching-converter power stages, *Proc. IEEE Power Electronics Specialists Conf. (PESC)*, pp. 18-34.

Mohan N. T. Undelead and P. Robbins William (1989), *Power Electronics: Converters, Applications and Design*, New York: John Wiley.

Morari, M., Papafotiou, G., Buisson, J. & De Schutter B. (2006). Final Report on Modelling Tools, Benchmarks and Control Methods. *Project report Hybrid Control: Taming Heterogeneity and Complexity of Networked Embedded Systems*, 13/03/2006, v. 1.1.

Morgul, O. (1992). Dynamic boundary control of a Euler-Bernoulli beam. *IEEE Trans. on Automatic Control*, Vol. 37, No.5, pp. 639-642, ISSN 0018-9286.

Perez-Pinal J. & Cervantes. I. (2009b). Multi-Objective Control for Cascaded Boost Converter with SingleActive Switch *Proc. IEEE International Electric Machines and Drives Conference: IEMDC* 09 Miami Florida

Perez-Pinal J. , Cervantes I., Leyva, J. & L.H. Diaz (2009a) Control of single active switch cascade boost converter for fuel cell applications *Proc. 13th European Conference on Power Electronics and Applications - EPE 2009* 8-10 September 2009, Barcelona, Spain.

Quaicoe, J. E. & Iqbal, M. T (2008). Advanced Boundary Control of Inverters Using the Natural Switching Surface: Normalized Geometrical Derivation Ordonez, *IEEE Trans. Power Electronics,* Vol. 23, No. 6, pp. 2915-2930, ISSN: 0885-8993.

Rajeev, A., Henzinger T.A. & Sontag, E.D. (1996). *Hybrid Systems III. Verification and Control.* Lecture Notes in Computer Science, 3, Springer Verlag. ISBN-13: 978-3540611554. New York.

Rashid M. (1988), *Power Electronics: Circuits, Devices and Applications,* Englewood Cliffs, NJ: Prentice Hall.

Redl, R. & Sokal, N. O. (1985) Current-Mode Control, Five Different Types, Used With the Three Basic Classes of Power Converters: Small-Signal AC and Large-Signal DC Characterization, Stability Requirements, and Implementation of Practical Circuits, *IEEE Power Electronics Specialists Conference - 1985 Record,* pp. 771-785.

Rodriges L. & How J.P. (2003). Observer-based control of piecewise-affine systems. *Int. J Control,* vol 76, No. 5, pp. 459-477.

Schild, A.; Lunze, J.; Krupar, J.; Schwarz, W. (2009) Design of Generalized Hysteresis Controllers for DC–DC Switching Power Converters.; *IEEE Trans. Power Electronics,* Vol. 24, No. 1, Jan. 2009, Page(s): 138 - 146, ISSN: 0885-8993.

Senesky, M., Eirea, G. & Koo, T.J. (2003). Hybrid Modeling and Control of Power Electronics, *Hybrid systems: computation and control, Lecture Notes in Computer Science* Eds. Maler, O. & Pnueli A. Springer, 2003, pp. 450-465. ISBN:3-540-00913-2.

Sira-Ramirez H. & Rios-Bolivar, M. (1994). Sliding Mode Control of dc-to-dc Power Converters via Extended Linearization, *IEEE Trans Circ Syst-I,* Vol. 41, No. 10 pp. 652-662.

Sira-Ramirez, H.; Ortega, R.; Perez-Moreno, R.; Garcia-Esteban, M (1995). A sliding mode controller-observer for DC-to-DC power converters: a passivity approach. *Proc. of IEEE Conf. Decision and Control,* Vol. 4 Date: 13-15 Dec 1995, Pages: 3379 - 3384.

Song T.-T. & Chung, H.S-H. (2008). Boundary control of boost converters using state-energy plane. *IEEE Trans. Power Electr.* Vol. 23, No. 2, pp. 551563 ISSN: 0885-8993.

Sontag E.D. (1998). *Mathematical Control Theory. Deterministic Finite Dimensional Systems.* Springer Verlag, ISBN 0-387-98489-5

Sreekumar C. & Agarwal V. (2007). Hybrid control approach for the output voltage regulation in buck type DC-DC converter, *IET Electr. Power Appl,* vol 1, No. 6, pp 897-906.

Sreekumar C. & Agarwal V. (2008). A hybrid control algorithm for voltage regulation in DC-DC Boost Converter, *IEEE Trans. Industrial Electronics,* vol 55, No. 6, pp 2530-2538. ISSN 0278-0046.

Sreekumar C. &, Agarwal V. (2006). A Circuit Theoretical Approach to Hybrid Mode Switching Control of a Pseudo CCM Boost Converter, *Proc IEEE International Conference on Industrial Technology,* pp. 791-795, ISBN 1-4244-0726-5 , Numbai, India, 15-17 Dec. 2006

Sun, Z. & Ge, S.S. (2005). *Switched Linear Systems. Control and design,* Springer-Verlag, ISBN 1-852333-893-8, London.

Tan, S.C., Lai, Y.M. & Tse C.K. (2008). General design issues fo sliding mode controllers in DC-DC converters. *IEEE Trans. Ind. Electronics*. Vol. 55, No. 3, pp. 1160, 1174. ISSN: 0278-0046.

Tao, H., Duarte, J.L. & Hendrix, M.A.M. (2006). Novel Zero-Voltage Switching Control Methods for a Multiple-Input Converter Interfacing a Fuel Cell and Supercapacitor. *Proc. of Industrial Electronics Conference IEEE*, No. 6-10, Nov. 2006 Page(s): 2341 - 2346.

Ting-Ting Song & Chung, H.S.-h. (2008).Boundary Control of Boost Converters Using State-Energy Plane; *IEEE Trans. Power Electronics*, Vol. 23, No. 2, Page numbers (551 - 563), ISSN: 0885-8993.

Tokat, S., Eksin, I., & Guzelkaya M. (2002). Sliding mode control using a nonlinear time-varying sliding surface. *Proc. of the 10th Mediterranean Conference on Control and Automation - MED2002* Lisbon, Portugal, July 9-12, 2002.

Tsinias, J (2007). Sufficient Conditions for Asymptotic Controllability and Hybrid Feedback Stabilization for a Class of Time-Varying Systems. *IEEE Trans. Automatic Control*, Vol. 52, No. 1, pp: 76 - 83, ISSN 0018-9286.

Verghese,G.C., Kassakian, J.G. , Schlecht M-F. (1991) Principles of Power Electronics. *Addison-Wesley*, ISBN-10: 0201096897. NY.

Xu, X. (2004). Practical stabilizability of a Class of Switched Systems. *Proc. of American Control Conference* , pp. 4537-4542, ISBN 0-7083-8335-4, Boston Massachuttes, July.

Xu, X., He S. & Zhai, G. (2007). Practical uniform asymptotic stabilizability of switched systems with time-varying subsystems. *Proc. of American Control Conference* , pp. 681-686, ISBN 1-4244-0889-6, New York, July.

Yang. T., (2001). *Impulsive Control Theory. Lectures Notes in Control and Information Science.* Springer Verlag ISBN: 3-540-422296-X

Yang. T., (2001). *Impulsive Systems and Control. Theory and Applications.* Nova Science Publisher ISBN: 1-59033-058-7

You, Y.C. & Lee, E.B. (1989). Dynamical boundary control of two-dimensional wave equations: vibrating membrane on general domain. *IEEE Trans. on Automatic Control*, Vol. 34, No. 11, pp. 1181 - 1185, ISSN 0018-9286.

Zhai, G. & Michel A.N. (2002) On practical of stability analysis of switched systems. *Proc of IEEE Conf Decision and Control*. pp. 3488-3493, ISBN 0-7803-7516-5, Vegas, Nevada.

Zhai, G. & Michel A.N. (2003) Generalized practical stability analysis of discontinuous dynamical systems. *Proc of IEEE Conf Decision and Control*. pp. 1663-1668, ISBN 0-7803-7924-1, Maui, Hawaii.

11

Interaction of Renewable Energy Source and Power Supply Network

Branislav Dobrucký, Michal Pokorný and Mariana Beňová
University of Zilina
Slovak Republic

1. Introduction to interaction of RES and network

Renewable energy systems (RES) become more and more important as the serious sources, first of all from the point of view of improvement of power generating efficiency and effectiveness of operation. RES systems (in this sense) consist of renewable energy source, (e.g. photovoltaics, fuel cells, wind power,..), pre-conditioning unit, DC link, inverter, transformer (if necessary) and inductive coupling with electronic switch or direct connection to power supply network.

The loads and supply system can be operated in three modes of operation: autonomous supply from the network or autonomous supply from RES system, and parallel operation of power supply network and RES(s). The chapter thereinafter deals with parallel operation of the both sources at steady-state and transient dynamic states.

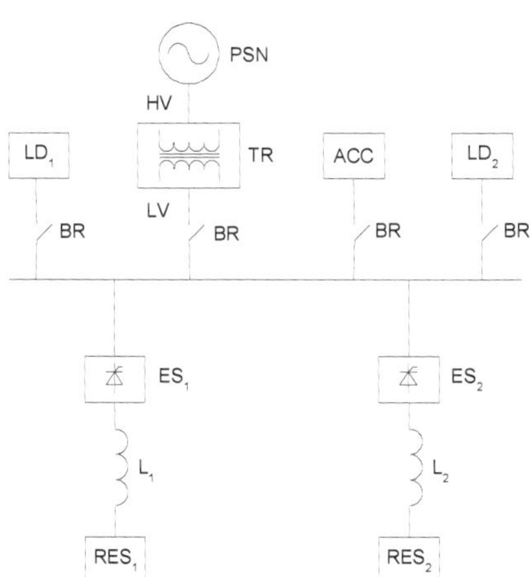

Legend:

PSN - power supply network
HV - high voltage
LV - low voltage
LD - loads
ACC - energy accumulation
BR - breakers
ES - electronic switches
L - separating inductors
RES - renewable energy sources

Fig. 1. Block diagram of power supply network and renewable energy system(s) connection

Interconnection between renewable energy systems RES and power electric network depends on type and power of RES:

- RES with nearly harmonic output voltage, synchronized by grid: direct connection,
- RES with non-harmonic output voltage, synchronized by grid: inductive inter-connection,
- remote RES non-synchronized with grid: HVDC energy transfer.

Single-phase inverters are commonly used to obtain utility grade ac power in small distributed generation systems such as photovoltaics, wind power generators, fuel cells chip systems. When such single-phase systems are aggregated to form microgrids integrable with a three-phase inverter system or a common dc bus, it is desirable to maintain the aggregated power to be constant [Bala & Venkataramanan, 2007]. Various transfigurations of single-phase converter topologies brings [Xue et al, 2004]; [Koutroulis et al, 2001]; [Rajeshekara, 2005]. Large complex wind plants are not explicitly described in this chapter. Their behaviour and back influence on power supply network can be found in [CIGRE, 2007; Spacil, 2006].

The main characteristics of electric power quantities under sinusoidal, non-sinusoidal, balanced, or unbalanced conditions are described in the Standards [IEEE, 1999; IEEE, 2000]. Modelling and simulation of electric power quality parameters differs somewhat from the ordinary power system modelling as far as short -circuit, load flow and transient stability studies are concerned. The reason is that the behaviour of the system equipment must be predicted for frequencies well above the fundamental one [Dumitrescu, 2009; Ghartemani et al, 2004].

All above mentioned problems have to be taken in account during investigation of RES and power network, and proper methods should be used to fulfill the overall goals.

2. Theoretical background for single- and three-phase transients

In general, transient phenomena of RES and power supply network can be investigated by different ways as follows:

- as complex non-linear system, using high volume SW packages (PSCAD/EMTDC, EMTP, EDSA, OrCAD..)
- as linearised system making possible of superposition:
  - in time domain (piece-wise linearization and state-variable method, z-transform and switching function method),
  - in frequency domain (decomposition into single harmonic linear subsystems)

Simulation software packages are a powerful electromagnetic time domain transient simulation environment and study tools. Package usually includes static or dynamic characteristic of the electronic switches, and it works with ***invariant circuit topologic scheme*** and with parametric changes of the system during simulation.

Above mentioned linearisation, superposition method and z-transformation can be conveniently used for investigated electric circuit's analysis.

Piece-wise linearisation in time domain. This method works during simulation with ***periodically variable circuit topologic structure*** of the system, and it uses electronic switches as separators between successive time intervals of operation of the system. Within a single time intervals of operation the topologic structure of the system is to be foreseen as constant one. Therefore one can use state variable method to describe RES system [Mohan et al, 2003; Dobrucky at al, 2007]:

$$\frac{\mathrm{d}}{\mathrm{d}t}(\boldsymbol{x}(t)) = \boldsymbol{A} \cdot \boldsymbol{x}(t) + \boldsymbol{B} \cdot \boldsymbol{u}(t) \tag{1}$$

where: $\boldsymbol{x}(t)$ is the vector of state variables, $\boldsymbol{A}$, $\boldsymbol{B}$ matrices of system elements,
$\boldsymbol{u}(t)$ input vector of exciting functions

$$y(t) = \boldsymbol{C} \cdot \boldsymbol{x}(t) + \sum_{i=0}^{r} \left[\boldsymbol{D} \cdot \boldsymbol{u}^{(i)}(t) \right] \tag{2}$$

where: $y(t)$ is the vector of output variables, $\boldsymbol{C}$, $\boldsymbol{D}$ system matrices,
r highest order of derivatives of the input vector (providing the derivatives exist).
In the next time interval the state variables end-values of the previous time-interval will be considerate as the initial values.
Note: Method of fictitious exciting function could be used in case of non-stationary elements of $\boldsymbol{A}$ matrix of the system.

Using Park-Clarke orthogonal transform and subsequent two orthogonal Fourier series. Any m-phase system (symmetrical or non-symmetrical) can be transformed into equivalent 2-phase orthogonal system using Park-Clarke transform. Transform into stationary α, β-coordinate system is exclusively used in power electronics. That transform is defined by transformation equation

$$x^*(t) = \frac{2}{3}\left[x_1(t) + \mathrm{a} \cdot x_2(t) + \mathrm{a}^2 \cdot x_3(t)\right] \tag{3}$$

where $\mathrm{a} = 1 \cdot \exp\left(\mathrm{j}\frac{2 \cdot \pi}{3}\right)$ and real- and imaginary parts of function (3) for symmetrical system are

$$x_\alpha(t) = \frac{1}{3}\left[2 \cdot x_1(t) - x_2(t) - x_3(t)\right], \tag{4}$$

$$x_\beta(t) = \frac{1}{\sqrt{3}}\left[x_2(t) - x_3(t)\right]. \tag{5}$$

Since $\boldsymbol{x}^*(t)$ is complex time function (= $\boldsymbol{x}_\alpha(t)$ +j$\boldsymbol{x}_\beta(t)$) the time-waveforms of $\boldsymbol{x}_\alpha(t)$ and $\boldsymbol{x}_\beta(t)$ can be expressed by complex Fourier series [Takeuchi, 1973; Bartsch, 1994]

$$x^*(t) \cong \sum_{k=-\infty}^{\infty} x_k\, e^{\mathrm{j}k\omega t} = x_0 + \sum_{k=1}^{\infty}\left(x_k\, e^{\mathrm{j}k\omega t} + x_{-k}\, e^{-\mathrm{j}k\omega t}\right) \tag{6}$$

Properties of orthogonal Fourier series and convergence is described by [Marcokova, 1995; 2009]; and some application of them are given in [Takeuchi, 1973]; [Zaskalicky&Zaskalicka, 2008].

Using z-transform and switching function method. The renewables, as sources of pulse output voltage series, can be described by system of difference equations.

For three-phase system the currents in α-β coordinates are given as

$$i_\alpha(n+1) = \mathrm{f}_{T/6} \cdot i_\alpha(n) + \mathrm{g}_{T/6} \cdot u_\alpha(n) \tag{7}$$
$$i_\beta(n+1) = \mathrm{f}_{T/6} \cdot i_\beta(n) + \mathrm{g}_{T/6} \cdot u_\beta(n) \tag{8}$$

where $u(n)$ are the voltages in α and β coordinates, respectively

$$u_\alpha(n) = \frac{2}{3} \cdot \sin\left(\frac{n \cdot \pi}{3} + \frac{\pi}{6}\right) \cdot U \tag{9}$$

and

$$u_\beta(n) = \frac{-2}{3} \cdot \cos\left(\frac{n \cdot \pi}{3} + \frac{\pi}{6}\right) \cdot U \tag{10}$$

Waveform of exciting impulse function (= switching function) is shown in Figure 2 for 6- and 12-pulses voltages

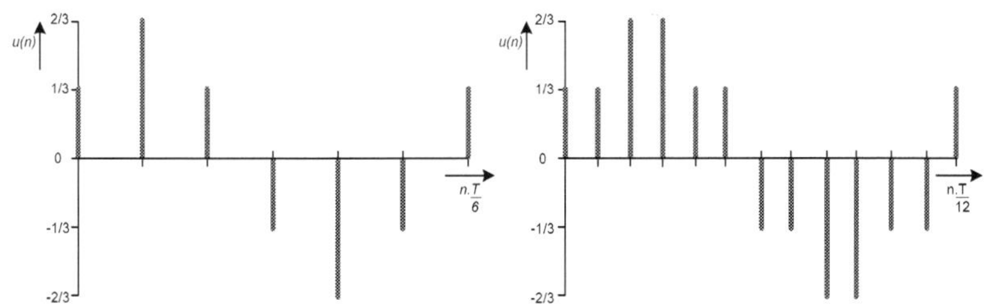

Fig. 2. Switching function of three-phase converter voltage with full width pulses (left)- and PWM modulation 12 pulses (right)

The image of α-component of 6-pulse voltage in z-plain is [Dobrucky et al, 2007]

$$U(z) = \frac{U}{3} \cdot \frac{z^3 + z^2 + z}{z^3 + 1} = \frac{U}{3} \cdot \frac{z \cdot (z+1)}{z^2 - z + 1} \tag{11}$$

Then, the image of α-component of output current in z-plain is

$$I(z) = \frac{U}{3} \cdot g_{T/6} \cdot \frac{z \cdot (z+1)}{(z - f_{T/6}) \cdot (z^2 - z + 1)} \tag{12}$$

Using inverse z-transform the discrete current series in time-domain will be obtained [Moravcik, 1992].

System decomposition into single harmonic linearised subsystems. Method of investigation assumes decomposition of real electric circuit into ν-harmonic separated equivalent schemes for each harmonic component [Dumitrescu, 2009]; [Benova, 2007]. Then transient analysis can be done for each scheme separately using 'impedance harmonic matrices', and each equivalent scheme is now linearised and therefore easily calculated. After finishing of calculation of each harmonic scheme, the effects of each investigated schemes are summed into resulting quantities of real non-linear electric circuit. That means that cumulative effect of sum of v-harmonic circuits is superimposed on basic harmonic waveform with voltage source.

The equivalent scheme for calculation of state quantities of the network with one appliance is drawn in Figure 3.

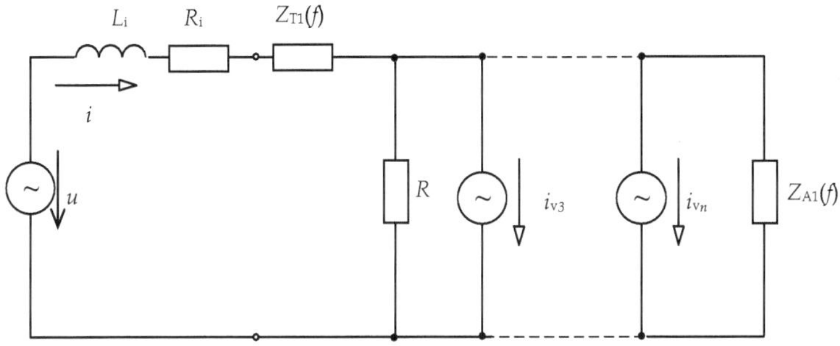

Fig. 3. Equivalent scheme for calculation and state quantities of the network with one R-L appliance for all harmonic components, where $Z_{T1}(f) = f(\nu f_1)$ and $Z_{A1}(f) = f(\nu f_1) = R_1 + j\nu 2\pi f L_1$, f_1 is frequency of fundamental harmonic component, i_v are current sources representing nonlinearity of appliance, $R = \sum R_v$ is the sum of source resistances

This overall scheme is now decomposed into ν-separate schemes for each harmonic component. These schemes for fundamental- and ν- harmonic components of complex magnitudes will be as in Figure 4a and 4b:

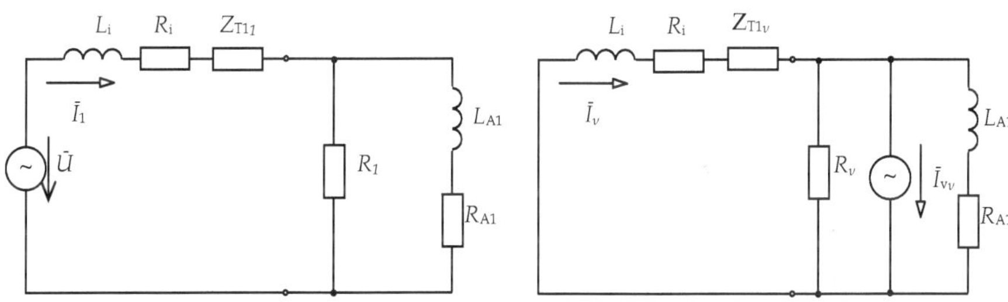

Fig. 4. a,b Equivalent scheme for fundamental (a) and ν - harmonic (b) components of one appliance, where $\bar{U}$ is complex magnitude of network voltage source with voltage of fundamental current source representing nonlinearity of appliance, $\bar{I}_1$ and $\bar{I}_v$ are complex magnitudes of network current for fundamental and ν- harmonic component, $\bar{I}_{vv}$ is complex magnitude of current source representing nonlinearity of appliance

Complex magnitudes of ν-harmonic component can be obtained by harmonic analysis of its current using 'impedance (admittance) harmonic matrices' by nodal voltages for rated power of the appliance. For example, in case of v - harmonic component of one appliance (equivalent scheme on Figure 3b) will be

$$\left(\frac{1}{\overline{Z}_{T1\upsilon}+R_i+j\nu 2\pi f L_i}+\frac{1}{R_\nu}+\frac{1}{R_{A1}+j\nu 2\pi f L_{A1}}\right)\cdot\left(\overline{U}_\nu\right)=\left(\overline{I}_{\nu\nu}\right) \tag{13}$$

where $\bar{U}_\nu$ is nodal voltage (equal to appliance voltage in this case). Complex magnitude of network current for ν- harmonic component can be obtained using

$$\bar{I}_{\nu} = \frac{\bar{U}_{\nu}}{\bar{Z}_{T1\upsilon} + R_i + j\nu 2\pi f L_i} \tag{14}$$

It is possible (after above decomposition) to calculate state quantities of each ν-harmonic.

This overall scheme is now decomposed into ν-separate schemes for each harmonic component. Corresponding harmonic quantities from each scheme are synthesized (summarized) into final resulting waveform of investigated state quantity:

$$i_{total}(t) = i_1(t) + ...+ i_n(t) = \Sigma i_n(t) \tag{15}$$

After synthesis of all harmonic components the total transient current is obtained.

3. Renewables and non-linear and linear passive and active loads

3.1 Harmonic output voltage RES with non-linear load during transients (island operation)

Single-phase bridge rectifier supplied by stiff voltage (from network or RES with small inner impedance) has positive half-vawes on its DC side as depicted in Figure 5.

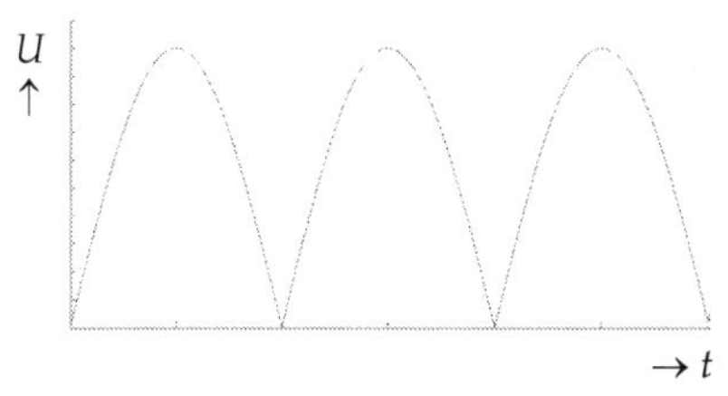

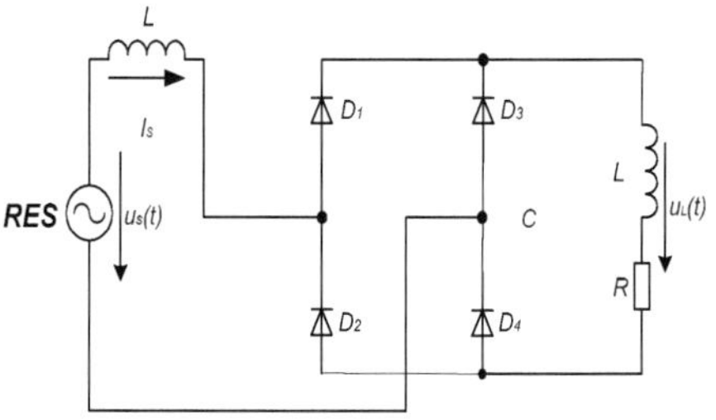

Fig. 5. Scheme of single-phase RES with rectifying load and its DC side voltage

DC voltage can be decomposed into Fourier series as [Bartsch, 1994]

$$u = \frac{4U_m}{\pi}\left[\frac{1}{2} - \frac{1}{1\cdot 3}\cos(2\omega t) - \frac{1}{3\cdot 5}\cos(4\omega t) - ...\right] \tag{16}$$

with DC component of $\frac{2U_m}{\pi}$. Supposing R-L load the corresponding current for single harmonic component can be calculated:.

$$i_{\nu} = \frac{U_{\nu}}{|\bar{Z}_{\nu}|}\left[\cos(\nu\omega t - \varphi_{\nu})\right], \tag{17}$$

where

$$|\bar{Z}_{\nu}| = \sqrt{R^2 + (\nu\omega L)^2} \tag{18}$$

$$\varphi_{\nu} = \arctan\frac{\nu\,\omega L}{R} \tag{19}$$

Using synthesis of all harmonic components the total transient current is obtained

$$i_\nu = \frac{U_\nu}{\left|\overline{Z}_\nu\right|}\left[\cos(\nu\omega t - \varphi_\nu) - \cos\varphi_\nu \cdot e^{t/\tau}\right] \tag{20}$$

where $\tau = L / R$.

Note (nota bene): similarly for three-phase 6- pulse middle point rectifier

$$u = \frac{3\sqrt{3}\cdot U_m}{\pi}\left[\frac{1}{2} - \frac{1}{2\cdot 4}\cos(3\omega t) - \frac{1}{5\cdot 7}\cos(6\omega t) - \ldots\right] \tag{21}$$

with DC component of $\frac{3\sqrt{3}\,U_m}{2\pi}$, and also for other types of rectifiers.

Simulation results are given in Figure 6 and Figure 7 for transient and steady-state considering 999 harmonics.

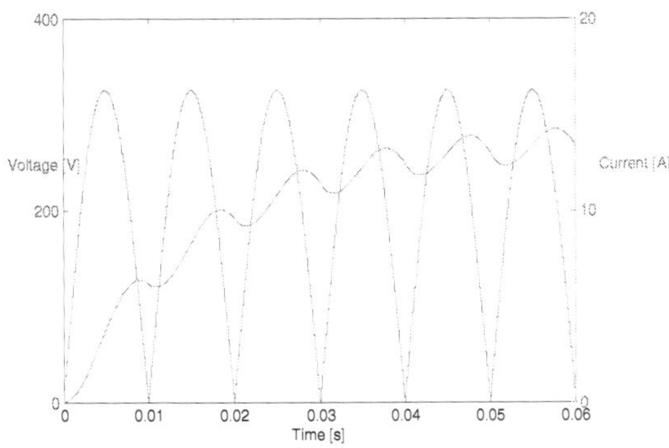

Fig. 6. Dynamic state of connecting nonlinear rectifying load to RES: voltage on DC side (blue) and DC current (green)

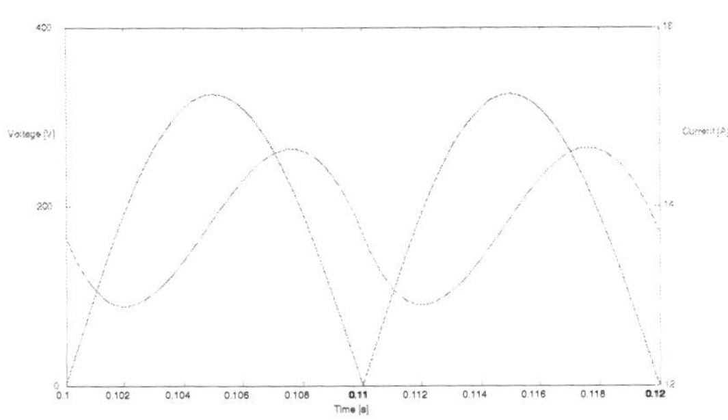

Fig. 7. Steady-state time-vaweforms of DC voltage (blue) and current (green)

3.2 Non-Harmonic output voltage RES with *R-L* load and active voltage source (parallel operation)

Today's converters provide very good quality output quantities regarding to their average and RMS values. Anyway, the output voltage is non-harmonic, switched by high frequency, Figure 8.

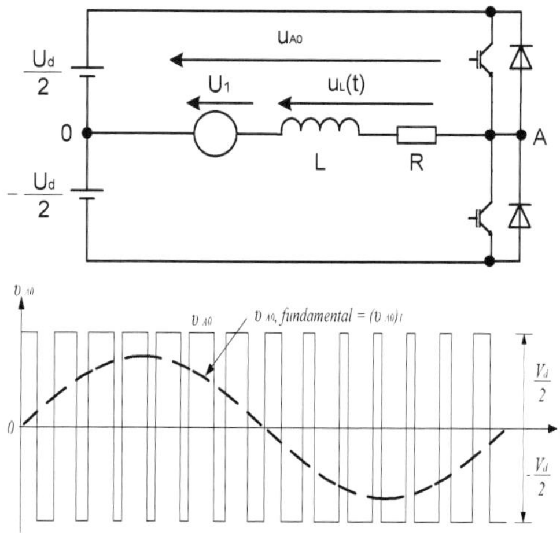

Fig. 8. Scheme of single-phase converter in half-bridge connection (top), and its output voltage

Simulation results for generating and regenerating regimes show Figure 9.

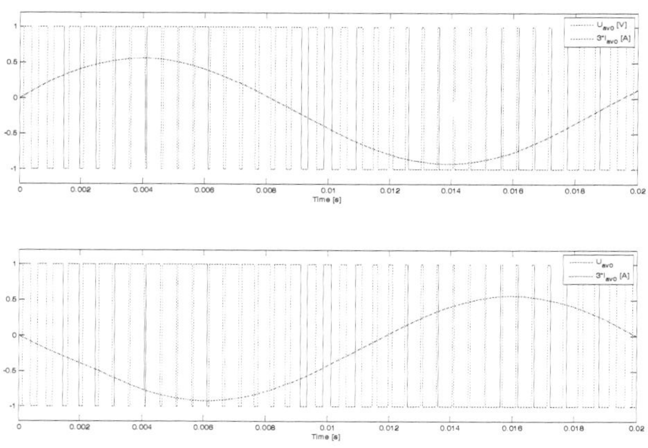

Fig. 9. Time-waveforms of output voltage (blue) and current (red) for generating and regenerating regimes (considering 999 harmonics)

3.3 Three-phase non-harmonic output voltage RES with *R-L* load – behaviour prediction

Three-phase 6-pulse inverter switching function is given in Figure 10.

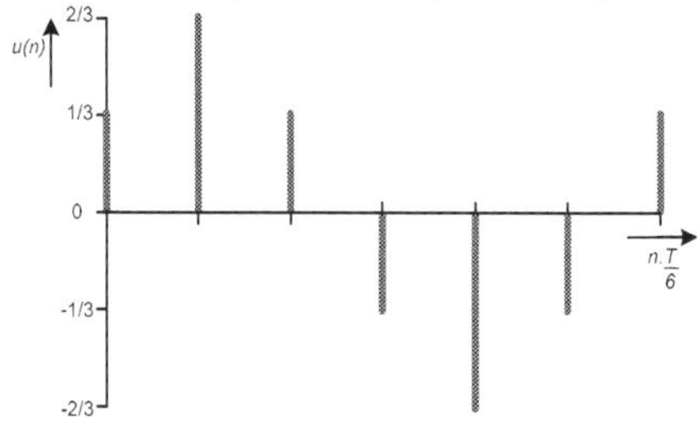

Fig. 10. Switching function for three-phase 6-pulse RES inverter

Corresponding simulation results for complex and time domain are depicted in Figure 11.

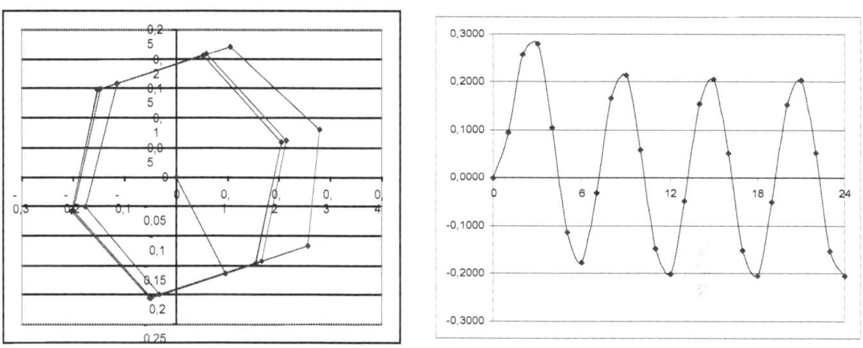

Fig. 11. Trajectories of output voltage in complex (left) and time domain (right)

3.4 Single-phase non-harmonic output voltage RES with *R-L* load – behaviour prediction

Single-phase 2-pulse inverter switching function is given in Figure 12.

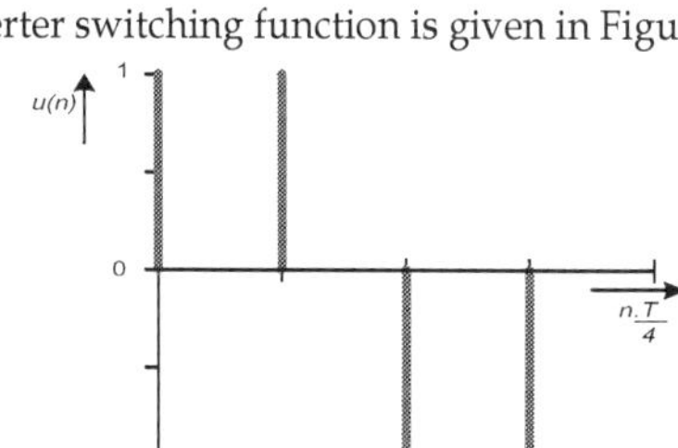

Fig. 12. Switching function for single-phase 2-pulse RES inverter

Corresponding simulation results for complex and time domain are depicted in Figure 13.

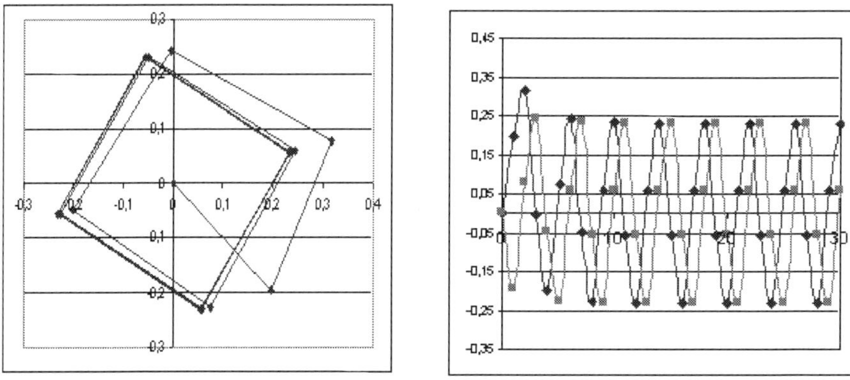

Fig. 13. Trajectories of output voltage in complex (left) and time domain (right)

4. Example of a practical implementation of the introduced methods of solution for different types of operation of RES and power network

The main task of RES system is to deliver power of different composition into power supply network:

- active power delivery (as a priority mission),
- and reactive or/and distortion powers delivery.

Later possibility means that RES provides function of static compensator or/and power active filter. It depends on requests of power supply network what kind of power will be delivered by RES.

The block schemes of connection of both sources and basic scheme of circuit configuration of single-phase voltage inverter are shown in Figure 14.

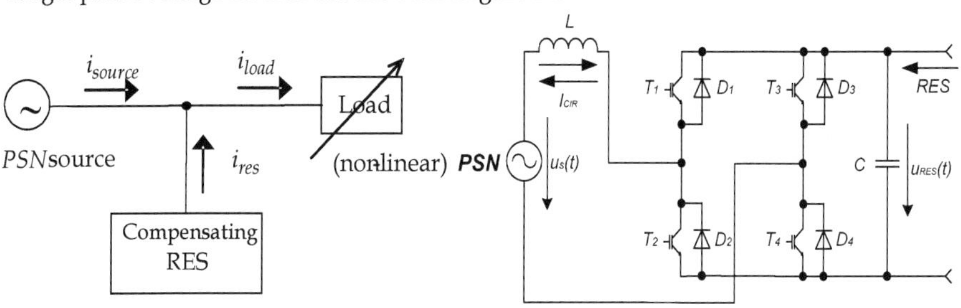

Fig. 14. Basic connection of RES and power network, and circuit configuration of single-phase voltage inverter

Simulation experiments have been carried out by MatLab simulation SW 2008b environment using theory of power active filters [Singh et al, 1999; Dobrucky et al, 2006; Pavlanin et al, 2008] and approach in subsection 3.2 above. Figure 15 shows simulation experiments of active compensation regime depending of time-instant of full load switching-on.

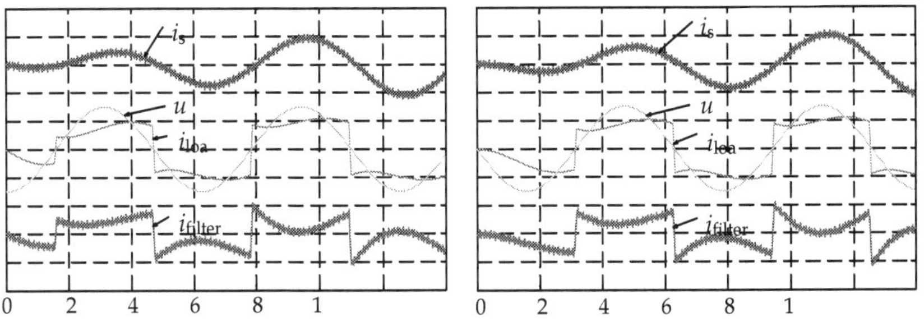

Fig. 15. Simulation of RES in active compensation regime: Load switched-on at negative maximum of voltage (left) and in optimal instant of time (right)

During compensation regime of RES the complementary compensating current i_{res} is generated by such a way, that after its addition with non- sinusoidal and phase shifted load current i_{load}, the only active power (e.g. active and harmonic current i_{source}) will be delivered by supplying network:

$$i_{load} = i_{res} + i_{source} \tag{22}$$

The calculation of this compensating current is the most important activity of the active filter's control circuit. The calculations can be carried out by different ways [Singh et al, 1999; Dobrucky et al, 2006; Pavlanin et al, 2008].
Simulation of RES in regime delivering of active power can be observed in subsection 3.2. Active power can be delivered to- or taken from power supply network.

5. The results of laboratory experiments

There are oscilloscopic records from experimental verification of RES operation in Figs. 16, 17, 18, 19 and Figure 20 for both compensation and active power delivering.

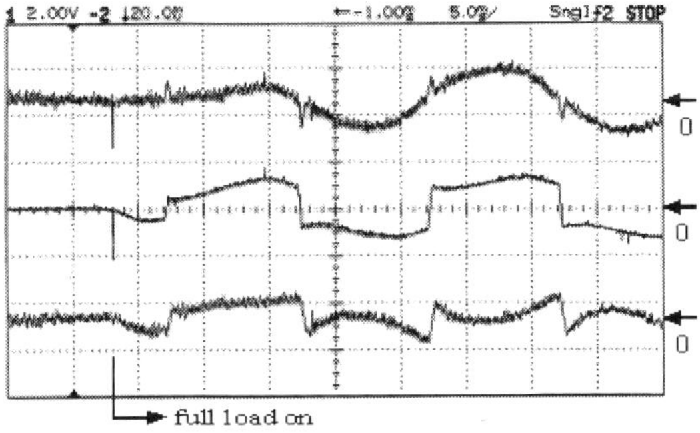

Fig. 16. Transient response of RES for the full load switched on: source current (top), load current (middle) and RES current (bottom) [Dobrucky et al, 2006]

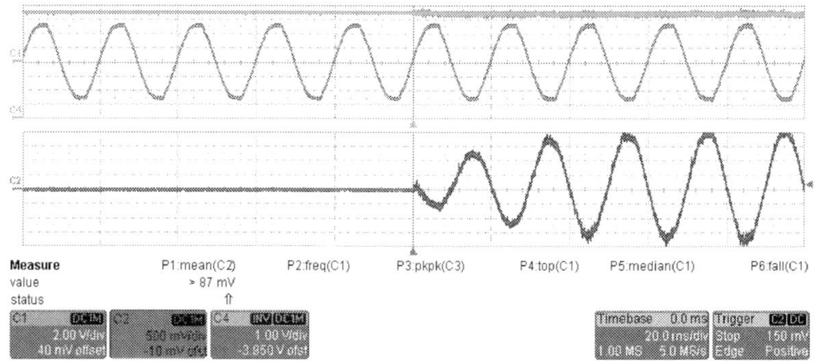

Fig. 17. Switching-on of the RES delivering pure active current [bottom, red] into power supply network (top, yellow)

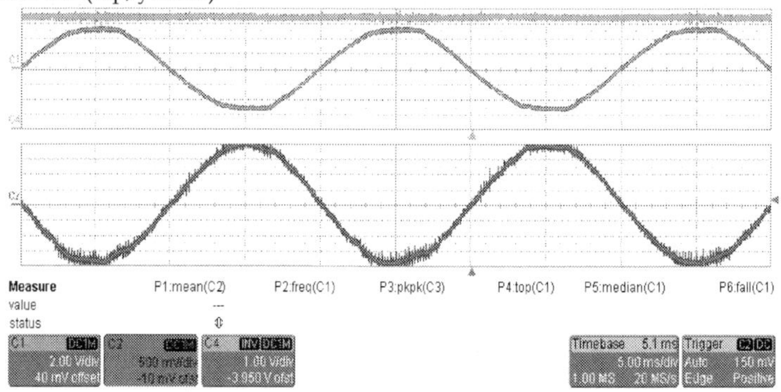

Fig. 18. Steady-state operation of RES in regime of active power delivering: voltage (top) and current (bellow)

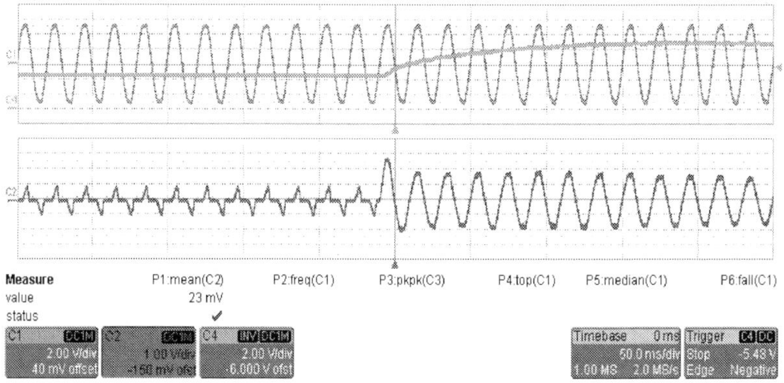

Fig. 19. Transient record of RES in regime of power acceptor from network: network voltage (top, yellow), voltage of DC bus (top, green) of RES and current (bellow)

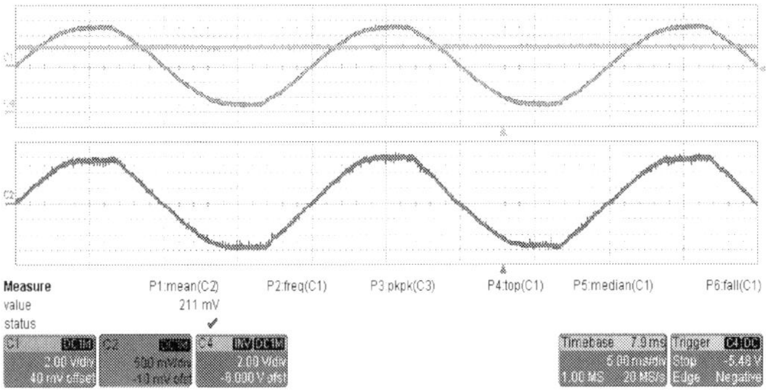

Fig. 20. Steady-state operation of RES in regime of power acceptor from network: voltage (top) and current (bellow)

6. Evaluation and conclusion

It has been shown that renewable energy source can work in two regimes of operation: as power generating or power consumption unit. Simulation and experimental results have proved **excellent transient properties** of the RES in both operations. As can be seen in Figure 16 and 17, respectively, the waveforms indicate an instantaneous reaction of compensation in generic regimes of RES. RES reacts already after the first calculation step Δt, so it is possible to use it as dynamic voltage restorer for small changes of the supply voltage.

7. References

Bala, S. & Venkataramanan, G. (2007). Balanced Power Aggregation of Asymmetric Single-phase Systems. *IEEE Power Electronics Specialists Conference (PESC'07)*, pp. 2772-2778, ISBN: 978-1-4244-0655-5, Orlando, FL, USA, June 2007

Bartsch, H.J. (1994). *Mathematic Formulae*, 4th revised edition (in German), Carl Hanser Publisher, ISBN 80-200-1448-9, Munich (DE)

Benova, M. (2007): *Non-sinusoidal Current and Voltage Analysis with regard to Power Condition of Electrical Machines* (in Slovak), PhD thesis, Faculty of EE, University of Zilina (SK), pp. 115

Burger, B. & Engler, A. (2001). Fast signal conditioning in single phase systems, *9th European Conference on Power Electronics and Applications*, pp. CD-ROM, ISBN, conference Graz (AT),1991

Dobrucky, B. Drozdy, S., Frivaldsky, M. & Spanik P. (2007). Interaction of Renewable Energy Source and Power Supply Network in Transient State, *Proc. of ICCEP'07 Int'l Conf. on Clean Energy Power*, pp. CD-ROM, *Capri* (IT), May 2007

Dobrucky, B., Marcokova, M., Pokorny, M. & Sul, R. (2007). Prediction Of Periodical Variable Structure System Behaviour Using Minimum Data Acquisition Time, *Proc. of IASTED MIC'07 Int'l Conf. on Modelling, Identification, and Control*, pp. CD-ROM, , Innsbruck (AT), Feb. 2007

Dobrucky, B. & Pokorny, M. (2006). Highest Dynamics and Ultra Fast Start-Up of Power Active Filter. *International Review of Electrical Engineering*, No. 8, Aug. 2006, pp. 391-399, ISSN 1827-6660

Dumitrescu, M.: Harmonic Digital Simulation Technique on an Industrial Environment. Electrical Review (Przeglad elektrotechniczny, PL), No. 1 (85), 2009, pp. 217-221, ISSN 0033-2097

Ghartemani, M. K., H. Mokhtari, M. R. Iravani, A Signal Processing System for Extraction of Harmonics and Reactive Current of Single-Phase Systems, *IEEE Trans. on Power Delivery, Vol. 19,* July 2004, pp. 979-986

IEEE, (1999). *Standard 519-1992 Power Grids Distortion Limits. Voltage characteristics of electricity supplied by public distribution systems,* European Norm EN 50160

IEEE (2000). *Trial-Use Standard Definitions for the Measurement of Electric Power Quan-tities Under Sinusoidal, Nonsinusoidal, Balanced, or Unbalanced Conditions*. IEEE Standard 1459

Koutroulis, E., K. Kalaitzakis, N.C. Voulgaris: Development of a Microcontroller-Based Photovoltaic Maximum power Point Tracking Control System. *IEEE Trans, on Power Electronics*, Vol. 16, No. 1, Jan. 2001, pp. 46-54

Marcokova, M. & Guldan V. (2009). On one orthogonal transform applied on a system of orthogonal polynomials in two variables. *Journal of Applied Mathematics*, Vol. 2, No. 2, pp. 239 - 245

Marcokova, M. (1995): Equiconvergence of Two Fourier Series. Journal of Approximation Theory, Vol. 80, No. 2, , pp. 151-16

Mohan, N., Undeland, T.M. & Robbins, W.P. (2003). *Power Electronics : Converters, Applications, and Design 3rd Edition,* John Wiley & Sons, Inc., ISBN 0-471-42908-2, New York

Pavlanin, R. & Spanik, P. (2008). Effect of PCC Voltage Distortion on Shunt active Power filter Function (in Slovak), *Proc. of 8th ERU08 Conference, on Electromagnetic Disturbation in Distributed- and Industrial Networks*, pp. CD-ROM, ISBN 978-80-254-3821-3, Brno (CZ), Nov. 2008

Rajeshekara, K. (2005). Hybrid Fuel-Cell Strategies for Clean Power Generation. *IEEE Trans, on Industry Application,* Vol. 41, No. 3, May-June 2005, pp. 682-689,

Singh, B., Al-Haddad, K., & Chandra, A. (1999). A review of active filters for power quality improvement. *IEEE Transactions on Industrial Electronics,* Vol. 46, pp. 960–971, Oct. 1999

Takeuchi, T.J. (1973). *Theory of SCR Circuit and Application to Motor Control.* Electrical Engineering College Press, 1968 - Russian Edition, Energia Publisher, Leningrad-St. Petersburg (RF), 248 pp., T 051(01)-73 130-73, Tokyo (JP)

CIGRE Working Group (2007). *C4.601 Modeling and Dynamic Behavior of Wind Generation Relates to Power System Control and Dynamic Performance,* CIGRE, August 2007,

Xue, Y., Chang, L., Kjaer, S. B., Bordonau, J. & Shimizu, T. (2004). Topologies of Single-phase Inverters for Small Distributed Power Generators: An Overview. *IEEE Transactions on Power Electronics,* Vol. 19, Sept. 2004, pp. 1305– 1314

Zaskalicky, P. & Zaskalicka M. (2009). Complex Fourier Series Model of a Three-Phase Voltage Source Inverter, *Acta Electrotechnica et Informatica,* Vol. N, No. X, 2009 (accepted)

12

Marine Tidal Current Electric Power Generation: State of Art and Current Status

Yun Seng. Lim and Siong Lee. Koh
University of Tunku Abdul Rahman
Malaysia

1. Introduction

Tidal power is a form of renewable energy generated by the periodic change in the ocean envelope of the Earth while interacting with Sun and the Moon via gravitational forces. The gravitational force of the Earth and the Moon attracts the ocean towards it. The motion of Earth around the centre of mass of the Earth-Moon system develops a bulge on the side of Earth opposite the Moon. The net effect of the two phenomena is two tidal bulges. The rotation of the Earth-Moon system cause the two bulges to rotate, making the sea level to rise and fall periodical to coastal observers. The tides created by the Earth-Moon system are known as lunar tides (Tom, 2007).

The gravitational force of the Earth-Sun system has the same influence on the ocean. It creates bulges that tend to follow the Sun through the day. The influence of the Sun is about 46 % of that from the Moon. The positions of the solar bulges change much more slowly than the position of the lunar bulges. The lunar and solar tides will be additive, resulting in higher high tides and lower low tides. However, if the Moon, Earth and Sun form a right angle, the solar tide will tend to diminish the lunar tide, which is known as neap tides. This is because the moon's contribution is more than twice that of the sun, the solar tide will not completely cancel the lunar tide. During neap tides, the high tides are not very high and low tides not very low. Neap tides occur at two week intervals.

The large tides are caused by the linear alignment of the sun, Earth and moon which is also called spring tides. During spring tides, high tides are very high and low tides very low. Spring tides occur in two week intervals. Neap tides arrive a week after the spring tide.

As Earth turns, landmasses obstruct the tidal crests, diverting, slowing and otherwise complicating their movements. This interference produces different patterns in the arrival of tidal crests at different places. The shape of the basin itself has a significant influence on the patterns and heights of tides. For these and other reasons, some coastlines experience semidiurnal tides: two high tides and two low tides of nearly equal level each day. Others have diurnal tides: one high and one low daily. The tidal pattern is called mixed if successive high tides or low tides are of significantly different heights throughout the cycle. This pattern is caused by blending diurnal and semidiurnal tides.

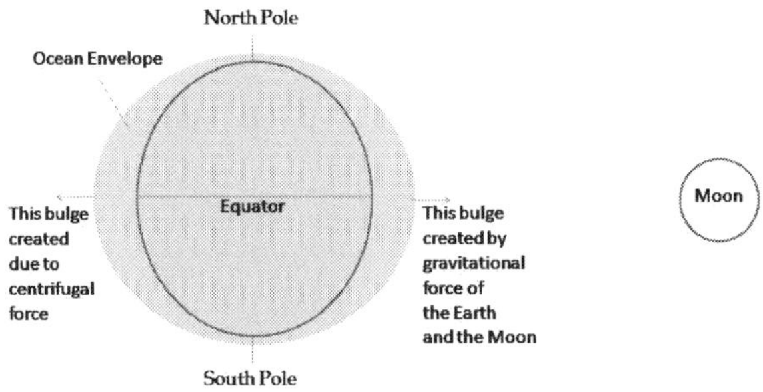

Fig. 1. Lunar tide created by the Earth-Moon System

Humans have found ways to use the tides. As early as the eight century, the Spanish, French and British built tidal storage ponds behind dams that were filled by the incoming tidal through sluice gates. These gates were closed at high tide and the trapped water then released to the sea through a water wheel to mill grain. The Eling Tide Mill in the United Kingdom has been producing flour with tidal power since 1418 (Eling, 2009).

A few tidal power electrical generation plants have been built that operate on the similar principle, in that a dam or barrage is constructed to impound the water at high tide and then releasing the water through openings with conventional hydroelectric turbine-generators when the water in the dam has sufficient potential energy. Although tidal power has not been widely used, tidal power is more reliable and predictable than wind energy and solar power. The Earth's tides are caused by the tidal forces due to gravitational interaction with the Moon and Sun, and the Earth's rotation, tidal power is practically inexhaustible and classified as a renewable energy source.

Numerous techniques have been proposed to extract energy from tidal current energy. The kinetic energy present in marine and tidal currents can be converted to electricity using relatively conventional turbine technology. There are several sites of the global ocean have the potential tidal energy that can be harnessed economically. Some of those sites are Pembrokeshire in Wales, River Severn between Wales and England, Cook Strait in New Zealand, Bay of Fundy in Canada, East River in New York City, Golden Gate in the San Francisco Bay, Piscataqua River in New Hampshire, and The Race of Alderney and The Swinge in the Channel Islands. Those sites can make a significant contribution to electricity supply. This is the reason why the marine renewable sector is currently the focus of much industrial and academic research around the world. There are generally three categories of tidal energy technologies as described in the following.

2. Categories of tidal energy technologies

Tidal power can be classified into three main types:
Barrage tidal power: A physical barrier, namely the Barrage, is constructed within the sea with Sluice Gates to control the flow of sea water. The Sluice Gates are to be closed at high tide so that the water level inside the barrage is held at its highest level. As the tide recedes, a difference in water level in between the barrage and the sea is created. The potential energy from the water level difference can then drive turbines to generate electricity.

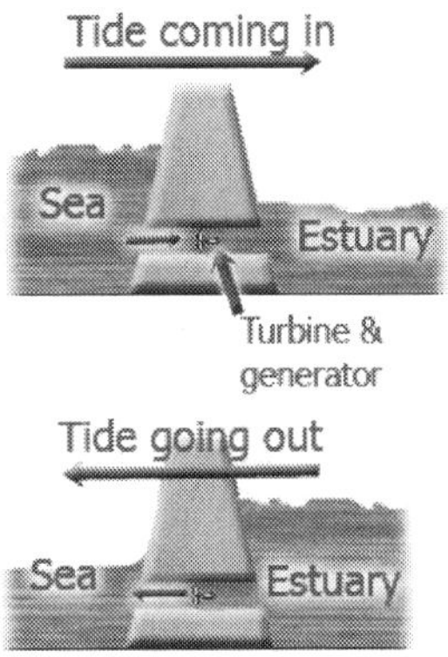

Fig. 2. Barrage tidal power for electricity generation (Source: http://home.clara.net/darvill/altenerg/tidal.htm)

Tidal stream system: A horizontal axis turbines are placed in the path of tidal currents to generate electricity, similar to the operation of wind turbine. This method is getting popular because the cost and ecological impact of tidal turbines is much lower than that of the barrage system.

Fig. 3. Tidal stream turbine (Source: http://www.altdotenergy.com/2008/12/seagen-achieves-maximum-capacity-of-12-mw/)

Tidal lagoons: These approaches are similar to barrages, but can be constructed as self contained structures, not fully across an estuary, and are claimed to incur much lower cost and impact overall. Furthermore they can be configured to generate electricity continuously which is not the case with barrages.

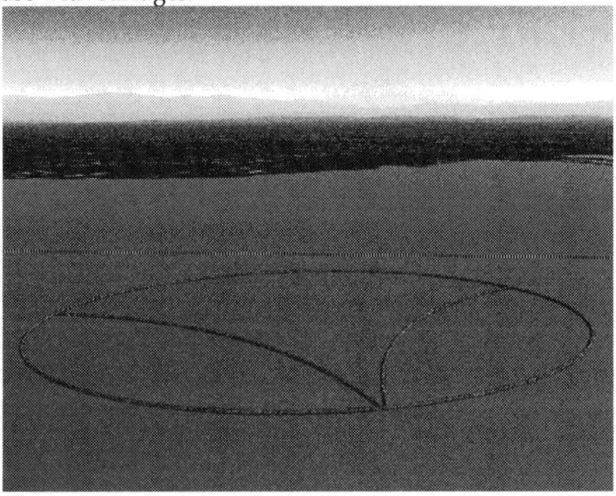

Fig. 4. Tidal lagoon with 3 pools configuration (Source: http://www.tidalelectric.com/technology.html)

3. Barrage tidal power

There are only a few barrage tidal power plants operating globally. The largest one is 240 MW plant on the Rance River in France, and two small plants on the Bay of Fundy and across a tiny inlet in Kislaya Guba Russia. The Rance tidal power plant has been operating since 1966 with the annual production of 600 GWh (Charlie, 2007).

The barrage system consists of caissons, embankments, sluices, turbines, and ship locks. Caisson is a very large concrete blocks used to house the sluices, turbines, and ship locks. Embankments seal a basin where it is not sealed by caissons. The sluice gates used in tidal power are the flap gate, vertical rising gate, radial gate, and rising sector.

Fig. 5. Barrage tidal power plant on Rance River (Source: http://www.britannica.com/EBchecked/topic-art/595132/118418/Tidal-power-generation-station-on-the-Rance-River-in-Saint)

The barrage system generates electricity using the difference in water level in between the barrage and the sea. The difference in water level is created by a process known as outflow or ebb generation where the basin is filled through the sluice gates during high tide while the turbine gates are closed as shown in Figure 6. Then the sluice gates are closed when a significant difference in water level is achieved. The turbine gates are opened to let the water to flow out, hence generating electricity. This cycle repeats itself. If the difference in water level is not great enough, pumping may operate where turbines are powered by the grid electricity to pump water from the sea into the basin in order to raise the height by, say, another 2 ft (61 cm). The cost for creating the extra 2 ft rise can be returned by the total electricity generation as a result of the pumping.

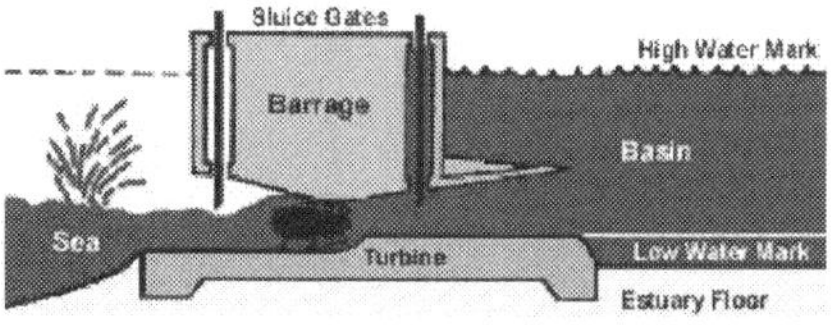

Fig. 6. Ebb generation scheme

Electricity can also be generated through inflow generation where the basin is filled through the turbines instead of sluice gates during high tide. However, this approach is generally less efficient than outflow generation, because the volume contained in the basin is less than the volume contained through outflow generation. Therefore the amount of electricity generated using this method is less than that of outflow generation.

3.1 Environmental impact

Barrage systems have not been widely used because it may create several environmental issues. The placement of a barrage into an estuary alters the flow of saltwater in and out of estuary which eventually change the hydrology and salinity, hence creating negative effects on the marine mammals that use the estuaries as their habitats (Pelc et al., 2002). During the construction of the tidal barrage, the estuary was isolated from the sea water, hence damaging flora and fauna. Some species lost their habitat due to La Rance's construction. Also as a result of the construction, sandbanks disappeared, the beach of St. Servan was badly damaged. (Charlie, 2007).

Estuaries often have high volume of sediments moving through them, from the rivers to the sea. The introduction of a barrage into an estuary may result in sediment accumulation within the barrage, affecting the ecosystem and also the operation of the barrage.

Another environmental impact of barrage system is fish kill. Fish may move through sluices safely, but when these are closed, fish will seek out turbines and attempt to swim through them. If the turbines are moving slowly enough, such as low velocities of 25-50 rpm, fish kill can be minimised. However, some fish will be unable to escape the water speed near a turbine and will be sucked through. Fish can be killed by pressure drop, contact with blades, cavitation, etc. The most fish-friendly turbine design has fish mortality per pass of approximately 15%. Various passage technologies (fish ladders, fish lifts, etc.) have so far failed to overcome this problem for tidal barrages. Research in sonic guidance of fish is ongoing (Peltier, 2003). The Open-Centre turbine may minimise this problem by allowing fish to pass through the open centre of the turbine.

Fig. 7. Open centre tidal turbine (Source: http://www.greentechmedia.com/articles/photos-clean-power-at-the-edge-of-the-world-5723.html)

3.2 Economics

Tidal barrage power schemes have a high capital cost and a very low running cost. Governments may be able to finance tidal barrage power, but many are not willing to do so because they need to wait for many years before investment return and may face high irreversible environmental and social issues. For example, the energy policy of the United Kingdom (European Commission, 2008) realises the potential role of tidal energy and expresses the need for local councils to understand the possible contribution of tidal projects to meeting the national goal of renewable energy. The UK government itself appreciates the technical viability, but has failed to provide meaningful incentives to move these goals forward.

4. Tidal stream generators

Tidal stream generators draw energy from currents, which is the same way as wind turbines. The higher density of water, 832 times the density of air, means that a single generator can provide significant power at low tidal flow velocities as compared with wind speed. In another words, water speeds of nearly one-tenth of the speed of wind provide the same power for the same size of turbine system. The total kinetic power in a marine current can be expressed by the following equation.

$$P = \frac{1}{2}\rho A v^3 \tag{1}$$

Where ρ is the seawater density (kgm^{-3}), A is the swept area of turbine blades (m^2) and v is the velocity of seawater (m/s).

However, a marine energy converter or turbine can only harness a fraction of this power due to losses and Eq. (1) is modified as follows.

$$P = \frac{1}{2}\rho C_p A v^3$$

Where C_p is the power coefficient and the percentage of power that can be extracted from the fluid stream taking into account losses due to Betz's law and the internal mechanisms within the converter or turbine. For wind generators, C_p has typical values in the range 0.25-0.3. The upper limit is for highly efficient machines with low mechanical losses. For marine turbines, C_p is in the range of 0.35-0.5.

Tidal stream generators are not fully mature because, at present, there is no commercial scale production of tidal generators and no standard technology being recognised as the clear winner. A variety of designs are being experimented. Several prototypes have been developed. Some are very close to large scale deployment. However, their efficiencies and feasibility have yet to be independently verified.

At present, there are several prototypes, namely horizontal axis turbine, vertical axis turbine, oscillating devices and tidal turbine with venturi shroud as described in the following.

4.1 Horizontal axis turbines

These horizontal axis turbines have the similar concept to traditional windmills operating under the sea and have the most prototypes currently operating.

A prototype of a 300 kW horizontal axis turbine was installed in Kvalsund, south of Hammerfest, Norway (Penman, 2009) and connected to the grid on 13 November 2003.

Fig. 8. Tidal turbine in comparison with offshore wind turbine

A 300 kW horizontal axis turbine, also known as Seaflow, was installed by Marine Current Turbines off the coast of Lynmouth, Devon, England, in 2003 (Fraenkel, 2004). The turbine has a diameter of 11m being fitted to a steel pile which was driven into the seabed. It was connected to a dump load.

Fig. 9. Seaflow Turbine

A prototype project was installed in the East River between Queens and Roosevelt Island in New York City in the United States in April 2007 (Verdant Power, 2009). However, the blades broke off due to the strong tidal currents. Therefore, new reinforced turbines were installed in September 2008. (Urbina, 2004) (Galbraith, 2008).

A prototype, called SeaGen, was installed by Marine Current Turbines in Strangford Lough in Northern Ireland in April 2008. The turbine began to generate at full power of about 1.2 MW in December 2008 and was reported to feed 150kW into the grid for the first time on 17 July 2008. It is currently the only commercial scale device to have been installed anywhere in the world (MCT, 2008).

Fig. 10. Seagen Farm

A prototype has been developed by OpenHydro, an Irish company (The Renewable Energy Centre, 2008). Its performance is being studied at the European Marine Energy Centre (EMEC), in Orkney, Scotland.

Fig. 11. Seagen installed by Marine Current Turbine in Strangford Lough in Northern Ireland (Source: http://www.iconocast.com/00001/G1/News4.htm)

4.2 Vertical axis turbines

The Gorlov helical turbine (GHT), evolved from the Darrieus turbine design by altering its blade to become helical blades. This cross-flow turbine was developed in 1994. The turbine consists of one or more long helical blades that run along a cylindrical surface like a screw thread that has 'airplane wing' profile. The blades provide a reaction thrust that can rotate the turbine. The turbine shaft (axis of rotation) should be perpendicular to the water current, and the turbine can be positioned either horizontally or vertically. Due to its axial symmetry, the turbine always develops unidirectional rotation, even in reversible tidal currents. This project has been proposed for the Uldolmok Strait in Korea, where a very strong reversible tidal current up to $6ms^{-1}$ (Cascio, 2005).

Fig. 12. Gorlov turbine (Source: http://www.oce.uri.edu/oce311/)

Neptune Renewable Energy has developed Proteus (Neptune, 2009). The Neptune Proteus is a vertical axis turbine mounted within a symmetrical Venturi diffuser duct and beneath a very simple steel deck and buoyancy chambers. The Neptune Proteus is designed for estuarine sites that have high tidal currents, yet have the advantages of lower access, cabling and maintenance costs than in offshore environments. The vertical shaft connects to the 1:200 gearbox and generator in the deck housing. The device is moored in the free stream, minimising environmental impact and operates equally efficiently for both flood and ebb currents. The rotor is maintained at optimal power outputs by sets of computer controlled shutters within the duct and by the variable electrical load. The overall efficiency of the system suggests to be greater than 45%.

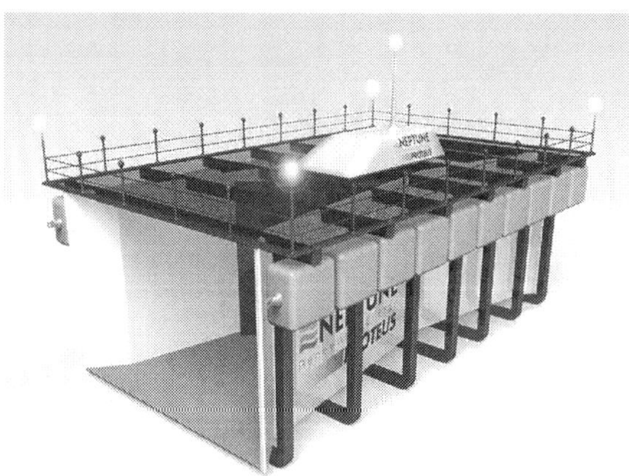

Fig. 13. The Neptune Proteus

The Enermar Project developed Kobold turbine (Ponte, 2009) that can provide very high starting torque, hence providing instant power once loaded. A pilot plant is moored in the Strait of Messina, Sicilian shore in Italy with the average tidal current of 2 m/s. A floating platform is above the surface of the water and is readily accessible for maintenance and repair of the turbine. The system can produce power of 20 kW.

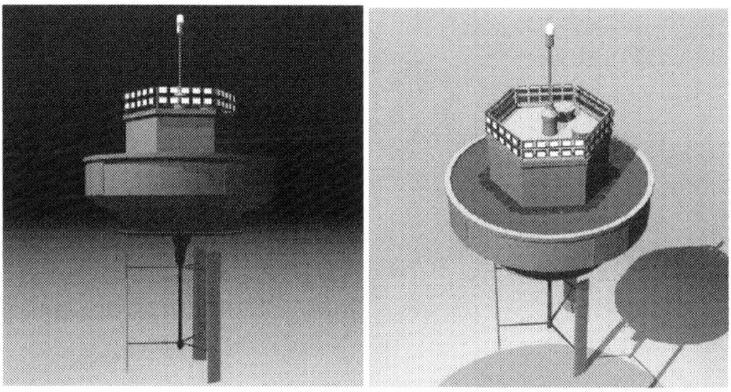

Fig. 14. Kobold turbine (left side) and floating platform (right side)

Blue Energy tidal turbine consists of four fixed hydrofoil blades connected to a rotor that drives an integrated gearbox and electrical generator as shown in Fig. 15. The turbine is mounted in a durable concrete marine caisson which anchors the unit to the seabed. The platform is above the surface of the water and is readily accessible for maintenance and repair. The rotation of the turbine is unidirectional on both the ebb and the flow of the tide. A unit turbine is expected to be about 200 kW output power. For large scale power production, multiple turbines are linked in series to create a tidal fence across an ocean passage or inlet.

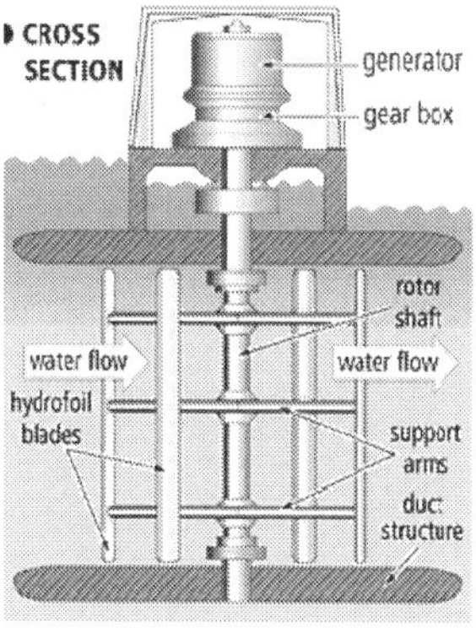

Fig. 15. Blue Energy tidal turbine

4.3 Oscillating devices – Stingray tidal turbine

Stingray uses the flow of the tidal stream over a hydroplane to create an oscillating motion that operates hydraulic cylinders to drive a motor that, in turn, drives an electrical generator. This device is a seabed mounted machine, to be situated typically in any water depth up to 100m. During 2003, a 150kW Stingray was tested off the Scottish coast, including a flexible control system to allow the performance of the generator to be accurately controlled and recorded over a longer period (IHC, 2009).

Fig. 16. Stingray tidal energy conversion device using oscillating hydroplane (Source: http://www.engb.com/services_09a.php)

4.4 Shrouded Tidal Turbines

A Patented tidal stream turbine invented by Aaron Davidson and Craig Hill of Tidal Energy Pty Ltd, Australian company, uses a venturi shaped shroud to increase the flow rate through the turbine, hence increasing the turbine efficiency as much as 3.84 times compared to the same turbine without the shroud. The Tidal Energy installed and tested such shrouded tidal turbines on the Gold Coast, Queensland in 2002. Tidal Energy has installed another shrouded turbine for a remote Australian community in northern Australia, providing 3.5 MW of power to the community. Another larger 5 meter diameter turbine, capable of 800 kW in 4 m/s of tidal flow, is planned for deployment as a tidal powered desalination showcase near Brisbane Australia in October 2008 (Hirch, 2009). Another device, the Hydro Venturi, is to be tested in San Francisco Bay (Hammons, 2008).

Fig. 17. Tidal turbine with venturi shroud (Source: http://www.reuk.co.uk/Lunar-Energy-Tidal-Power.htm)

In late April 2008, Ocean Renewable Power Company, LLC (ORPC) (EMEC, 2009) successfully completed the testing of its turbine-generator unit (TGU) prototype at ORPC's Cobscook Bay and Western Passage tidal sites near Eastport, Maine (Viscarolasaga, 2008). The TGU makes use of cross-flow (ADCF) turbines to drive a permanent magnet generator located between the turbines and mounted on the same shaft. This technology can be used for generating power from river, tidal and deep water ocean currents.

4.5 Current Activities for Commercial Developments of Tidal Technologies

RWE's npower announced that it is in partnership with Marine Current Turbines to build a tidal farm of SeaGen turbines off the coast of Anglesey in Wales (MCT, 2008).

In November 2007, British company Lunar Energy announced that, in conjunction with E.ON, they would be building the world's first tidal energy farm off the coast of Pembrokshire in Wales. It will be the world's first deep-sea tidal-energy farm and will provide electricity for 5,000 homes. Eight underwater turbines, each 25 metres long and 15 metres high, are to be installed on the sea bottom off St David's peninsula. Construction is due to start in the summer of 2008 and the proposed tidal energy turbines, described as "a wind farm under the sea", should be operational by 2010.

British Columbia Tidal Energy Corp. plans to deploy at least three 1.2 MW turbines in the Campbell River or in the surrounding coastline of British Columbia by 2009 (Alternative Energy Press, 2007).

An organisation, named Alderney Renewable Energy Ltd, is planning to use tidal turbines to extract power from the notoriously strong tidal races around Alderney in the Channel Islands. It is estimated that up to 3GW could be extracted. This would not only supply the island's needs but also leave a considerable surplus for export (Alderney Renewable Energy, 2009).

Nova Scotia Power has selected OpenHydro's turbine for a tidal energy demonstration project in the Bay of Fundy, Nova Scotia, Canada and Alderney Renewable Energy Ltd for the supply of tidal turbines in the Channel Islands. Open Hydro.

5. Conclusion

Tidal energy is a promising renewable energy source available to the world. Since the past one decade, numerous research and development efforts have been carried out hoping that tidal energy can become a realistic renewable energy source one day. At present, tidal current technologies are still in the developing stage. A large variety of tidal designs have been developed and experimented. Several prototypes have been claimed to be promising. Some companies are planning for large scale manufacturing and deployment of tidal turbine. Any commercial tidal plants are still in the planning stage.

6. References

Alderney Renewable Energy. (2009). Technology Overview. Source: http://www.are.gb.com/technology-overview.php. Access: 30th March 2009

Alternative Energy Press. (2007). Tidal power coming to west coast of Canada. Source: http://www.alternative-energy-news.info/press/tidal-power-west-coast-canada/. (8th November 2007)

Bahaj, A. S.; & Myers, L. (2004). Analytical estimates of the energy yield potential from the Alderney Race (Channel Islands) using current energy converters. *Renewable Energy*, 29, 12, pp. 1931-1945 (2004), ISSN: 0960-1481

Bellve, Anthony. (2005). NZ: Chance to turn the tide of power supply. New Zealand Herald, (9th May 2005). Available: http://www.nzherald.co.nz/nz/news/article.cfm?c_id=1&objectid=10124626

Cascio, Jamais. (2005). Gorlov's Helical Turbine. *WorldChanging*, Available: http://www.worldchanging.com/archives/002383.html, (21st March 2005)

CBC News. (2007). The resurgence of tidal power. Source: http://www.cbc.ca/news/background/energy/tidal-power.html. 26th February 2007

Charlie, R. H. (2007). Forty Candles for the Rance River TPP tides provide renewable and sustainable power generation. *Renewable and Sustainable Energy Reviews*, 11, 9, (December 2007) pp. 2032-2057, ISSN: 1364-0321

Deng, Zhiqun; Carlson, Thomas J.; Ploskey, Gene R.; Richmond, Marshall C.; & Dauble, Dennis D. (2007). Evaluation of blade-strike models for estimating the biological performance of Kaplan turbines. *Ecological Modelling*, 208, 2-4, (November 2007) pp. 165-176, ISSN: 0304-3800

Eling (2009). Eling Tide Mill. Source : www.elingtidemill.org.uk, Date of Access : 25th March 2009

EMEC. (2008). Tidal Devices. Source : European Marine Energy Centre (EMEC) Ltd (http://www.emec.org.uk/tidal_devices.asp), Date of Access : 25th March 2009

European Commission. (2008). United Kingdom renewable energy fact sheet. Source : http://www.energy.eu/renewables/factsheets/2008_res_sheet_united_kingdom_en.pdf. January 2008

Fraenkel, Peter. (2004). Windmills below the sea: A commercial reality soon?. *Refocus*, 5, 2 (March-April 2004), pp. 46-48, ISSN : 1471-0846

Galbraith, Kate. (2008). Power From the Restless Sea Stirs the Imagination. *The New York Times*, (23 September 2008), Available : http://www.nytimes.com/2008/09/23/business/23tidal.html?_r=1&em

Gevorkian, Peter. (1999). Ocean Energy Technology, In: *Sustainable energy systems engineering: the complete green building design resource*, Peter Gevorkian, (1st Ed.), pp. 255-272, McGraw-Hill Professional, ISBN 0071473599, 9780071473590, New York, USA

Hagerman, G. (2006). Tidal stream energy in the Bay of Fundy. Proceedings of Energy Research and Development Forum 2006 Antigonish, Nova Scotia, Canada, 2006.

Hammons, T. J. (2008). Energy potential of the oceans in Europe and North America: tidal, wave, currents, OTEC and offshore wind. *International Journal of Power & Energy Systems*, 28, pp. 416-428 (2008), ISSN: 1078-3466

Hardisty, Jack. (2007). Power intermittency, redundancy and tidal phasing around the United Kingdom. *Geographical Journal*, 174, 1, pp. 76-84 (December 2007), ISSN: 0373-6245

Hirch, Matthew. (2009). Low tide - Without state or federal funding, S.F.'s tidal power project might be dead in the water. *Sanfrancisco Bay Guardian Online*. Available: http://www.sfbg.com/38/43/news_tidal.html. Access: 30th March 2009.

IHC. (2009). Stingray tidal stream generator. Source: http://www.engb.com/services_09a.php. Access: 30th March 2009

Livingstone, Tomos. (2007). Severn balancing act - Hain. *WalesOnline.co.uk*, (1st September 2007). Available: http://www.walesonline.co.uk/news/welsh-politics/welsh-politics-news/tm_headline=severn-balancing-act-hain&method=full&objectid=19718602&siteid=50082-name_page.html

McDermott, Matthew. (2009). 1.2MW DeltaStream tidal power turbines to be tested in Wales. *Science & Technology*, (5th January 2009). Available: http://www.treehugger.com/files/2009/01/delta-stream-tidal-power-turbine-wales.php

MCT. (2008). World's first commercial-scale tidal power system feeds electricity to the National Grid. Source: http://www.marineturbines.com/3/news/article/10/world_s_first_commercial_scale_tidal_power_system_feeds_electricity_to_the_national_grid__/. 17th July 2008

Neptune. (2009). Tidal Technology - The Neptune Proteus Mark III. Source: http://www.neptunerenewableenergy.com/tidal_technology.php. Access: 25th March 2009

Pelc, Robin.; & Fujita, Rod M. (2002). Renewable energy from the ocean. *Marine Policy, 26*, 6, (November 2002) pp. 471-479, ISSN: 0308-597x

Peltier, Robert. (2003). Fish-friendly hydro turbines move center stage. Energy Industry, April 2003. Available: http://findarticles.com/p/articles/mi_qa5392/is_200304/ai_n21329129/

Penman, Danny. (2003). First power station to harness Moon opens. Source : http://www.newscientist.com/article/dn4188, November 2003

Ponte (2009). Ponte di Archimede International Company Source: http://www.pontediarchimede.coi/ (April 2009).

Retiere, C. (1994). Tidal power and aquatic environment of La Rance. *Biological Journal of the Linnean Society, 51*, 1-2, (January 1993) pp. 25-36, ISSN: 0024-4066

Schmitt, Catherine. (2009). Tidal power proposals proliferate along Maine's coast. *The Working Waterfront*, 2nd April 2009. Available: http://www.workingwaterfront.com/articles/Tidal-power-proposals-proliferate-along-Maines-coast/11432/

The Renewable Energy Centre. (2008). The Renewable Energy Centre supports use of Tidal power to supply the National Grid. Source: http://www.prlog.org/10076285-the-renewable-energy-centre-supports-use-of-tidal-power-to-supply-the-national-grid.html. 30th May 2008

Tom, G. (2007), Oceanography : An Invitation to Marine Science, Thomson Brooks/Cole, Orange Coast College, University of Southern California

Urbina, Ian. (2004). In Search of New Power Source, City Looks Underwater. *The New York Times*, pp B3 (10th July 2004)

Vega, Cecilia M. (2007). PG&E backs new study of bay's tidal power. Source : http://deanzaemtp.googlepages.com/PGEbacksnewstudyofbaystidalpower.pdf. (20th June 2007)

Verdant Power. (2009). The RITE project. Source : http://www.verdantpower.com/what-initiative, Date of Access : 25th March 2009

Viscarolasaga, Efrain. (2008). Tide is slowly rising in interest in ocean power. *Mass High Tech: The Journal of New England Technology,* (1st August 2008) Available : http://www.masshightech.com/stories/2008/07/28/weekly9-Tide-is-slowly-rising-in-interest-in-ocean-power.html/

VLH (2009). Very low head turbine. Source : http://www.vlh-turbine.com/, Date of Access : 25th March 2009

13

Modeling and Simulation of an Induction Drive with Application to a Small Wind Turbine Generator

Levente TAMAS[1] and Zoltan SZEKELY[2]
[1]*Technical University of Cluj-Napoca, Romania*
[2]*Purdue University Calumet, USA*

1. Introduction

In recent years the rapid development of digital signal processors (DSP) based systems and the decreasing cost of power electronics allowed the complex models and control algorithms of AC machines to become popular for a large range of applications including the ones for renewable energies. Among renewable energies the solution of utilizing induction generators for wind energy conversion systems is now in a growing trend. Currently, a wide spread control concept is that of a variable speed rotor with pitch regulation, and this concept is combined with both direct drive and geared drive trains the latter dominating the wind market (Tamas & Szekely, 2008).

A valid choice for operation at variable speed may be the use of induction machine. An induction motor can operate as a generator in super synchronous speed raised by an overhauling type of load, or by lowering the inverter frequency below the machine speed, when there is a converter-fed machine drive. Continuous regenerative operation of a drive is possible if the load machine is a source of power, such as in a wind generation system (Tamas & Szekely, 2008).

In this paper the model of the induction machine is written in terms of fluxes as state variables. This model is then used to design an Extended Kalman Filter (EKF) based estimator that can provide the correct feedback signals for the vector control (VC) of a wind energy conversion system.

Also three other observers are shown with good simulation results, one based on stator voltages and stator currents (Us-Is), one based on stator voltages and rotor speed (Us-ω) and one based on stator currents and rotor speed (Is - ω). A neural network based speed estimator as a forth type of estimator is developed which can be used for sensorless vector control of the induction drive or as a part of a sensor fusion based feedback signal estimator combined with one of the above mentioned estimators.

Finally, a real-time simulation setup is presented using dSpace CLP1104 embedded system (Tamas & Szekely, 2008) together with the possible further research steps in this project.

2. Wind Turbine System Model

Fig. 1 shows a basic wind turbine system that converts the wind energy into mechanical energy which at its turn is used by an electric machine to generate power. The basic configuration consists of the wind turbine blades, the drive train, the induction generator and the AC-AC converter (not shown in Fig. 1 to connect the system to the grid).

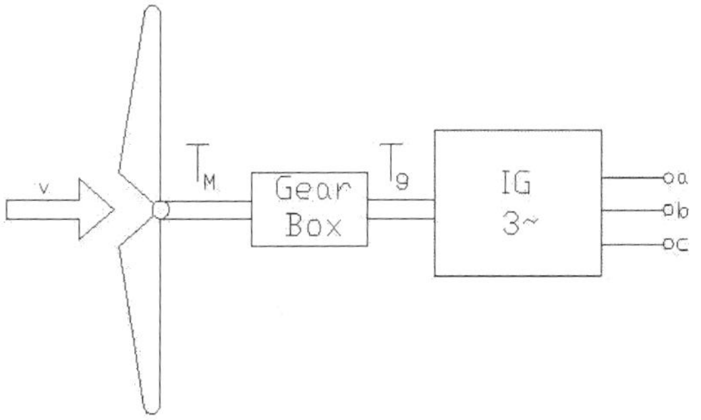

Fig. 1. Basic wind turbine system model

2.1 Wind turbine model

The models of wind turbines take into account several characteristics such as the size, blade radius, nominal power, shafts stiffness, losses, gear box ratio, etc. The mechanical power produced by the wind turbine is:

$$P_M = \frac{1}{2}\rho_{air} c_p \pi R^2 v^3 \tag{1}$$

where ρ is the air density, πR^2 is the turbine swept area, v is the speed of the wind, and cp(λ) is the power efficiency coefficient of the wind turbine which depends on the tip-speed ratio, λ. This is defined as:

$$\lambda = \frac{\omega_M R}{v} \tag{2}$$

where, $\omega_M = \frac{d\theta_M}{dt}$ is the angular speed of the turbine blades, and R is the length of the turbine blade.

There is an optimal value of the tip speed ratio, λopt, which allows a maximum capture of power from the wind. This value is found from the typical characteristic of the wind turbine power coefficient cp = f(λ) provided by the manufacturers.

When the wind turbine is operating with λopt, the nonlinear power expression from (1) may be recalculated as:

$$P_{Mopt} = c_{opt}\omega_M^3 \tag{3}$$

and the corresponding mechanical torque produced by the blades is:

$$T_{M\,opt} = c_{opt}\,\omega_M^2 \tag{4}$$

where, Copt is a constant which depends on the turbine characteristics and air density (Kelemen & Imecs, 1991).

2.2 Drive train model

The drive train dynamics consists of the dynamics of the wind turbine rotor, low-speed shaft, gear box, high-speed shaft, and the induction generator Fig. 2 shows the basic diagram of the drive train that was used for the derivation of the mathematical model (Kelemen & Imecs, 1991).

Thus, for the wind turbine rotor the following torques equation can be written:

$$J_M \frac{d^2\theta_M}{dt^2} + T_G = T_M(t) \tag{5}$$

where, $T_G = K_1 \Delta\theta_{K1}$ is the reaction torque exerted by the flexible low-speed shaft on the rotor of the wind turbine.

Similarly, for the rotor of the induction generator the following torques equation is valid:

$$J_g \frac{d^2\theta_r}{dt^2} - T_g + T_e = 0 \tag{6}$$

where $T_g = K_2 \Delta\theta_{K2}$ is the reaction torque exerted by the high-speed shaft on the induction generator rotor, and T_e is the electromagnetic torque of the induction machine in generator operation mode.

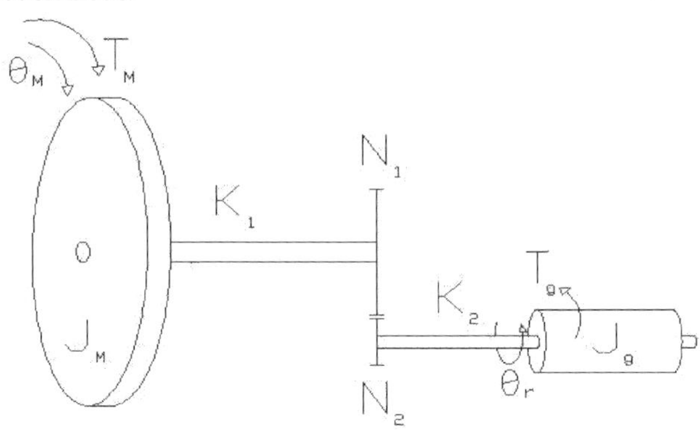

Fig. 2. Drive train model

The reference values for the above mechanical parameters where reported in (Scutaru & Apostoaia, 2004).

2.3 Induction generator model

In this paper, the model of the induction generator is derived using as state parameters the stator and rotor flux d-q axis components and the angular speed of the induction generator. The model is a fifth order dynamic model. It is also assumed that the induction drive is a standalone generator so that the excitation capacitor and load impedance are added to the model.

The mathematical model of the induction machine having fluxes as state variables, and written in terms of inductances, is represented by the set of (7) - (11):

$$\frac{d\psi_{sd}}{dt} = -\frac{R_s}{\sigma L_s}\left(\psi_{sd} - \frac{L_m}{L_r}\psi_{rd}\right) + \omega_\lambda \psi_{sq} + u_{sd} \tag{7}$$

$$\frac{d\psi_{sq}}{dt} = -\frac{R_s}{\sigma L_s}\left(\psi_{sq} - \frac{L_m}{L_r}\omega_{rq}\right) - \omega_\lambda \psi_{sd} + u_{sq} \tag{8}$$

$$\frac{d\psi_{rd}}{dt} = -\frac{R_r}{\sigma L_s}\left(\psi_{sd} - \frac{L_m}{L_r}\psi_{sd}\right) + (\omega_\lambda - \omega)\psi_{rq} + u_{rd} \tag{9}$$

$$\frac{d\psi_{rq}}{dt} = -\frac{R_r}{\sigma L_r}\left(\psi_{rq} - \frac{L_m}{L_s}\psi_{sq}\right) - (\omega_\lambda - \omega)\psi_{rd} + u_{rq} \tag{10}$$

$$\frac{d\omega}{dt} = \frac{3}{2}\frac{z_p^2}{J}\frac{L_m}{L_s}\left(\psi_{sq} i_{rd} - \psi_{sd} i_{rq}\right) - \frac{z_p}{J}T_g \tag{11}$$

An induction generator may be self-excited by providing the magnetizing reactive power by a capacitor bank (Scutaru & Apostoaia, 2004). In Fig. 3 a stand-alone induction generator under an R-L resistive-inductive load is shown.

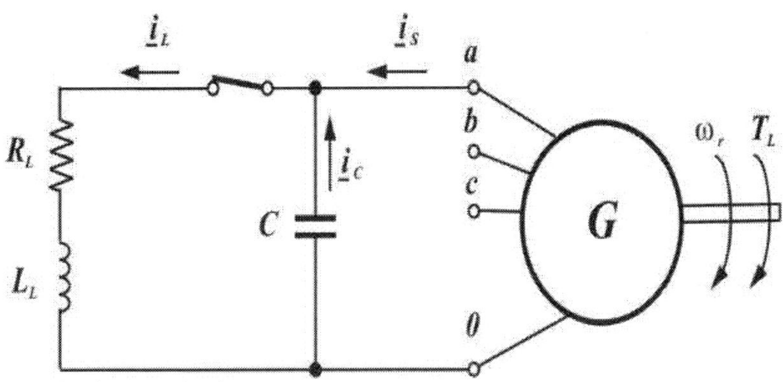

Fig. 3. Single-phase R-L load circuit

The stator voltages usd and usq are computed as state variables in the generator's load impedance model as follows:

$$\frac{du_{sd}}{dt} = -\frac{1}{C} i_{sd} + \frac{1}{C} i_{Ld} \tag{12}$$

$$\frac{du_{sq}}{dt} = -\frac{1}{C} i_{sq} + \frac{1}{C} i_{Lq} \tag{13}$$

$$\frac{di_{Ld}}{dt} = \frac{1}{L_L}(u_{sd} - R_L i_{Ld}) + \omega_\lambda i_{Lq} \tag{14}$$

$$\frac{di_{Lq}}{dt} = \frac{1}{L_L}(u_{sq} - R_L i_{Lq}) - \omega_\lambda i_{Ld} \tag{15}$$

The electromagnetic torque expression can be written in terms of the state variables:

$$T_e = \frac{3}{2} z_p \frac{L_m}{\sigma L_s L_r} \left(\psi_{sq} \psi_{rd} - \psi_{sd} \psi_{rq} \right) \tag{16}$$

Fig. 4 shows the complete Simulink model of the wind conversion system described by (1)-(16).

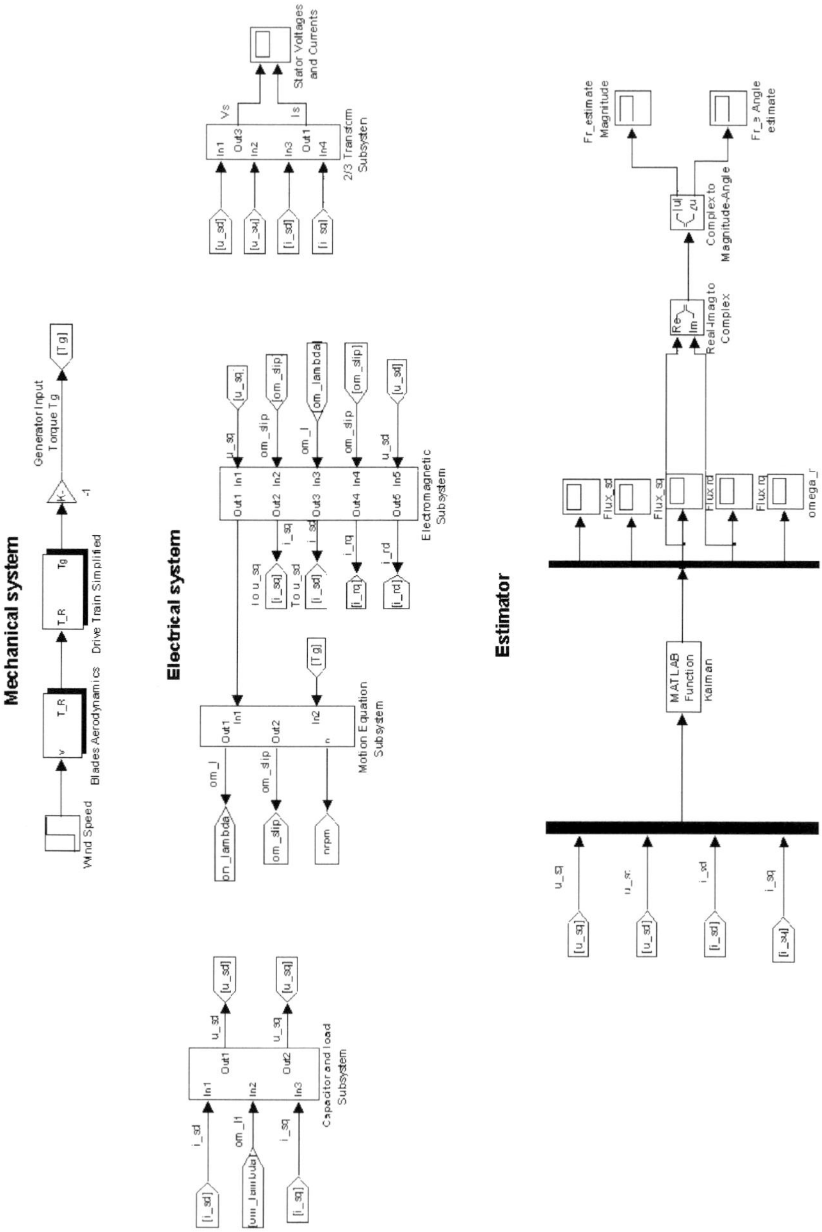

Fig. 4. Simulink block diagram of the dynamic system with EKF estimator

2.4 State space model of the induction generator

The state space equations (7)-(11) of the induction machine can be rewritten in matrix form as in (17):

$$\frac{dx}{dt} = A \cdot x + B \cdot u$$
$$y = C \cdot x \tag{17}$$

In (17) the state variables input and output vectors are as follows:

$$x = \begin{bmatrix} \psi_{sd} & \psi_{sq} & \psi_{rd} & \psi_{rq} & \omega \end{bmatrix}^T \tag{18}$$

$$u = \begin{bmatrix} u_{sd} & u_{sq} & 0 & 0 & 0 \end{bmatrix}^T \tag{19}$$

$$y = \begin{bmatrix} i_{sd} & i_{sq} & i_{rd} & i_{rq} & 0 \end{bmatrix}^T \tag{20}$$

where, urd and urq are considered zero because of the squirrel cage type of the induction generator.

The system matrix A, the input matrix B and the output matrix C are:

$$A = \begin{bmatrix} -\frac{R_s}{\sigma L_s} & \omega_\lambda & \frac{R_s L_m}{\sigma L_s L_r} & 0 & 0 \\ -\omega_\lambda & -\frac{R_s}{\sigma L_s} & 0 & \frac{R_s L_m}{\sigma L_s L_r} & 0 \\ \frac{R_r L_m}{\sigma L_s L_r} & 0 & -\frac{R_r}{\sigma L_r} & \omega_\lambda - \omega & 0 \\ 0 & \frac{R_r L_m}{\sigma L_s L_r} & \omega - \omega_\lambda & -\frac{R_r}{\sigma L_r} & 0 \\ 0 & 0 & 0 & 0 & 0 \end{bmatrix} \tag{21}$$

$$B = \begin{bmatrix} 1 & 0 & 0 & 0 & 0 \\ 0 & 1 & 0 & 0 & 0 \\ 0 & 0 & 0 & 0 & 0 \\ 0 & 0 & 0 & 0 & 0 \\ 0 & 0 & 0 & 0 & 0 \end{bmatrix} \tag{22}$$

$$C = \begin{bmatrix} \frac{1}{\sigma L_s} & 0 & -\frac{L_m}{\sigma L_s L_r} & 0 & 0 \\ 0 & \frac{1}{\sigma L_s} & 0 & -\frac{L_m}{\sigma L_s L_r} & 0 \\ -\frac{L_m}{\sigma L_s L_r} & 0 & \frac{1}{\sigma L_r} & 0 & 0 \\ 0 & -\frac{L_m}{\sigma L_s L_r} & 0 & \frac{1}{\sigma L_r} & 0 \\ 0 & 0 & 0 & 0 & 0 \end{bmatrix} \tag{23}$$

The numerical data for the induction generator parameters found in (21) and (23) can be found in the Appendix.

A successful build-up process, is predicted as in Fig. 5 and in Fig. 6 shows for a slip frequency of -0.5ω, an excitation capacitance value of 700μF, and a resistive inductive load with R=7kΩ and L=10H. The simulation was carried out for a speed of 400, see Fig. 7.

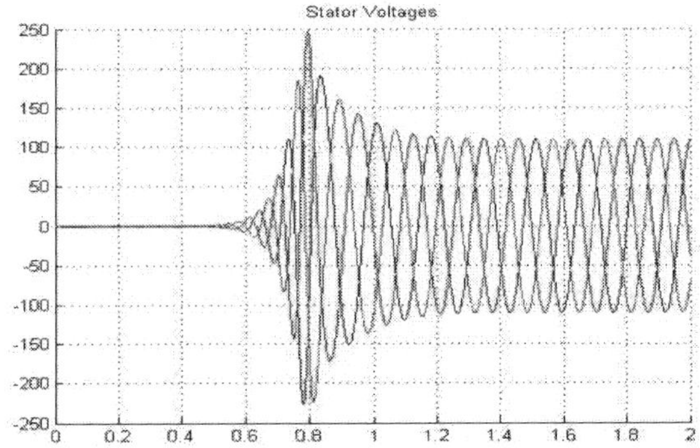

Fig. 5. Stator voltages during a start up of the loaded induction generator

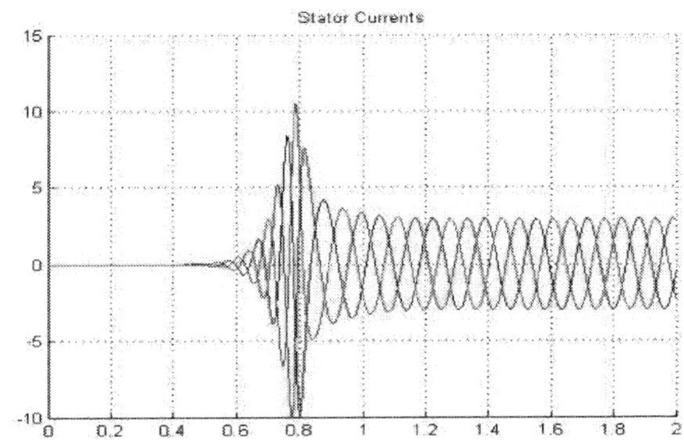

Fig. 6. Stator currents during a start up of the loaded induction generator

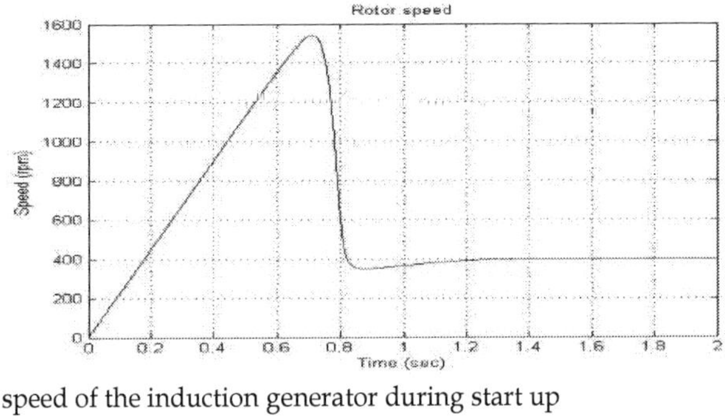

Fig. 7. Rotor speed of the induction generator during start up

3. State observers

3.1 Extended Kalman filter estimator (EKF)

The filtering problem involved in this paper is to find the best estimate of the state vector xk of the induction machine which evolves according to the discrete- time nonlinear state transition equation (Kalman, 1960):

$$x_k = f(x_{k-1}, u_{k-1}) + w_{k-1} \tag{24}$$

where, $f(\bullet,\bullet)$ is the machine dynamics, x_{k-1} is the state observation of the machine at sampling time $k-1$, u_{k-1} is the known input at time step $k-1$ and w_{k-1} is the system noise, which is white noise with a Gaussian distribution with mean zero and covariance Q. Also we admit that we possess a set of noisy measurements, noted as z :

$$z_k = h(x_k) + v_k \tag{25}$$

where, $h(\bullet)$ is a function of the state parameters and v_k is the measurement noise, with a Gaussian distribution with mean zero and covariance R.

First the estimated states are calculated (prediction step) then using the measurements these states are updated in function of the Kalman gain (K).

The estimation (prediction) equations are:

$$\hat{x}_k^- = f(\hat{x}_{k-1}^+, \hat{u}_{k-1}) + \hat{w}_{k-1} \tag{26}$$

$$P_k^- = A_k P_{k-1} A_k^T + B_k U_{k-1} B_k^T + Q_{k-1} \tag{27}$$

where, $\hat{x}_{k-1}^+$ the previous state estimate, with covariance matrix P_{k-1}, $\hat{u}_{k-1}$ is the control input with covariance matrix U_{k-1}, $f(\cdot,\cdot)$ is the system dynamics function and $\hat{w}_{k-1}$ is the system noise with covariance Q_{k-1}.

The update equations following a measurement:

$$\hat{x}_k^+ = \hat{x}_k^- + K_k (z_k - h(\hat{x}_k^-)) \tag{28}$$

$$P_K^+ = (I - K_k H_k) P_k^- \tag{29}$$

$$K_k = P_k^- H_k^T (H_k P_k^- H_k^T + R_k)^{-1} \tag{30}$$

where, K_k is the Kalman gain, z_k is the measurement at time step k, $h(\bullet)$ is a function of the state vector, H is the observation matrix, R_k is the covariance of the measurement noise v_k.

At the first step of the algorithm the values of x and P are initialized with the prior knowledge about the system. It is not a trivial task to tune the values of the covariance R and Q. These values influence the performance of the filter although there is no direct

method of choosing them in real time applications and it is often the case that they are selected in a trial and error method.
The Extended Kalman Filter estimator is implemented using Matlab function. The inputs for the function are the stator voltages, and the stator currents are the noise measurements.
Fig. 8 shows the estimated state variables, the stator and rotor d-q axis fluxes. Fig. 9 shows the rotor flux magnitude and the rotor flux vector angle can be seen from Fig. 10.

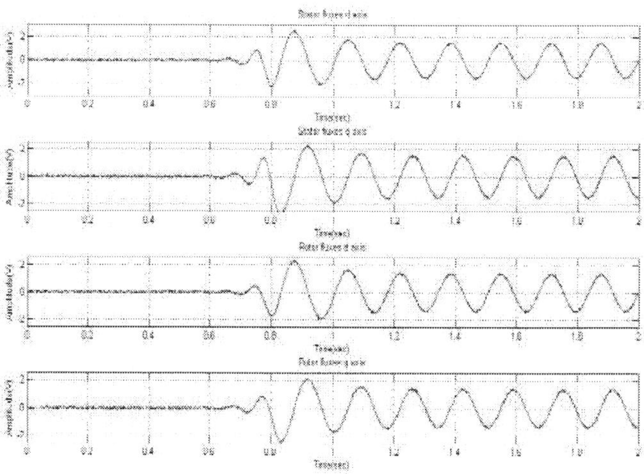

Fig. 8. Estimated stator and rotor d-q fixed axis fluxes using EKF

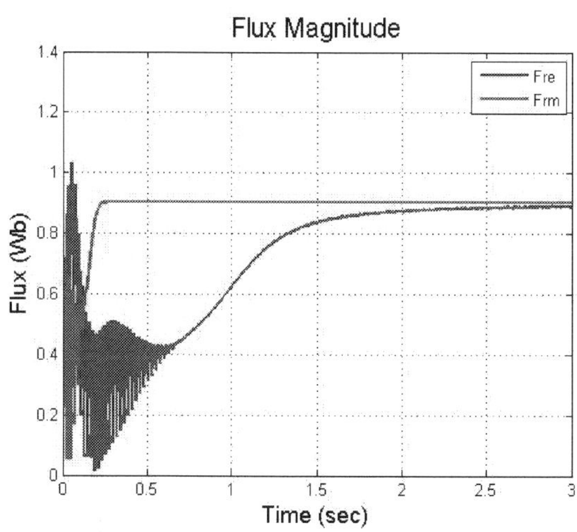

Fig. 9. Estimated rotor flux magnitude in d-q rotating frame using EKF

The estimated electromagnetic torque and the rotor speed of the induction drive can be observed in Fig. 11 and Fig. 12.

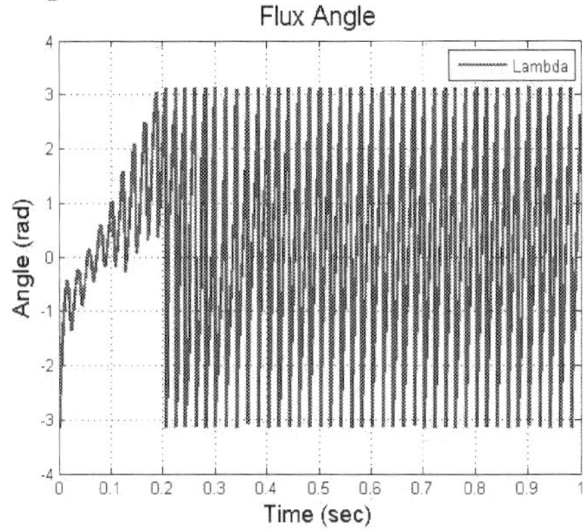

Fig. 10. Estimated rotor flux angle in d-q rotating frame using EKF

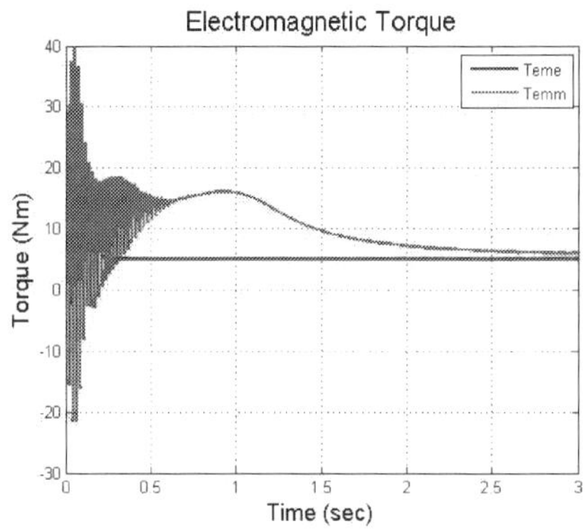

Fig. 11. Estimated electromagnetic torque using EKF

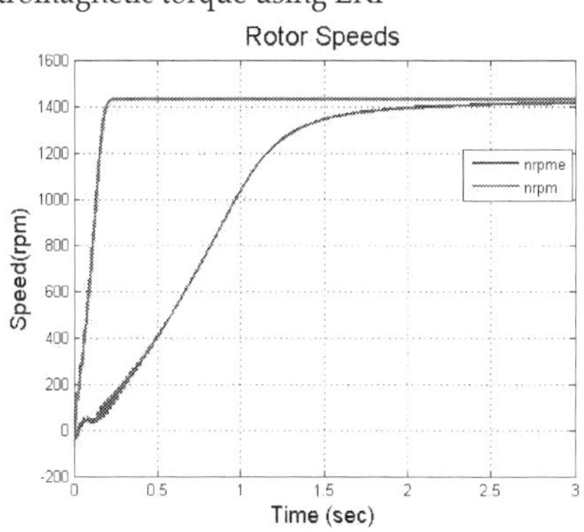

Fig. 12. Estimated rotor speed using EKF

3.2 U_s-I_s estimator

In order to integrate this estimator model in the control system of an induction machine, the stator voltages and currents are considered here as inputs, and the estimated outputs are the magnitude and the angle of the rotor flux and the electromagnetic torque.

This estimator is derived by direct synthesis from the machine state equations which are written in terms of the d-q axis components of the stator and rotor flux as state variables. This choice is justified by the fact that the system matrix is simpler than the d-q axis current state space model

The machine equations are derived from its general model where the speed of the rotational d-q system of axes $\omega\lambda = 0$, since a fixed reference frame is considered here.

The equations can be written in state space form. The outputs will be the stator and rotor currents.

$$
\begin{aligned}
x &= \begin{bmatrix}\psi_{sd} & \psi_{sq} & \psi_{rd} & \psi_{rq}\end{bmatrix}^T \\
u &= \begin{bmatrix}u_{sd} & u_{sq} & 0 & 0\end{bmatrix}^T \\
y &= \begin{bmatrix}i_{sd} & i_{sq} & i_{rd} & i_{rq}\end{bmatrix}^T
\end{aligned} \tag{31}
$$

The estimator mathematical model is based on the inverse model of the system:

$$
\begin{aligned}
\frac{d\psi_{sd}}{dt} &= \frac{L_r}{L_m}\left[u_{sd} - R_s i_{sd} - \sigma L_s \frac{di_{sd}}{dt}\right] \\
\frac{d\psi_{sq}}{dt} &= \frac{L_r}{L_m}\left[u_{sq} - R_s i_{sq} - \sigma L_s \frac{di_{sq}}{dt}\right]
\end{aligned} \tag{32}
$$

The Simulink model of the estimator can be observed in Fig. 13.

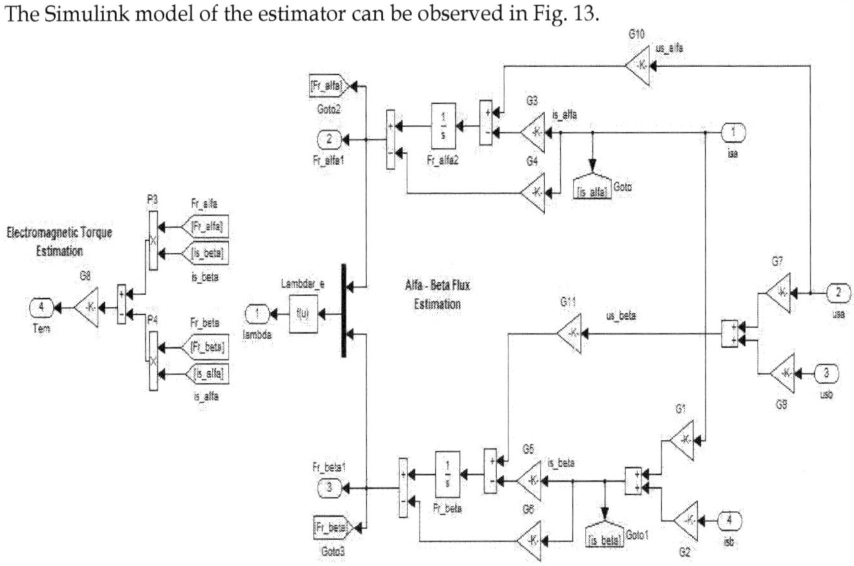

Fig. 13. Us-Is estimator Simulink model

The electromagnetic torque is calculated like:

$$T_e = \frac{3}{2} z_p \frac{L_m}{\sigma L_s L_r} \left(\psi_{sq} \psi_{rd} - \psi_{sd} \psi_{rq} \right) \tag{33}$$

The rotor flux magnitude and angle can be written as:

$$\left| \underline{\psi_r} \right| = \sqrt{\psi_{r\alpha}^{2} + \psi_{r\beta}^{2}} \tag{34}$$

$$\lambda_r = arctg \frac{\psi_{r\beta}}{\psi_{r\alpha}} \tag{35}$$

The magnitude and the angle of the rotor flux estimates using Us-Is observer are shown in Fig. 14 and Fig. 15 and the electromagnetic torque estimation is seen in Fig. 16.
With the estimator Us–Is good output estimates were obtained when the machine was supplied with voltages in the range of rated value. At low speed range the acquisition of the voltages is difficult so the speed estimation is not accurate.

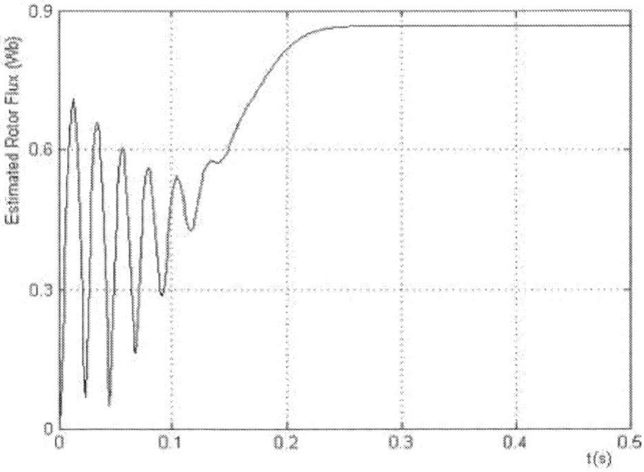

Fig. 14. Estimated rotor flux magnitude in d-q rotating frame using Us-Is estimator

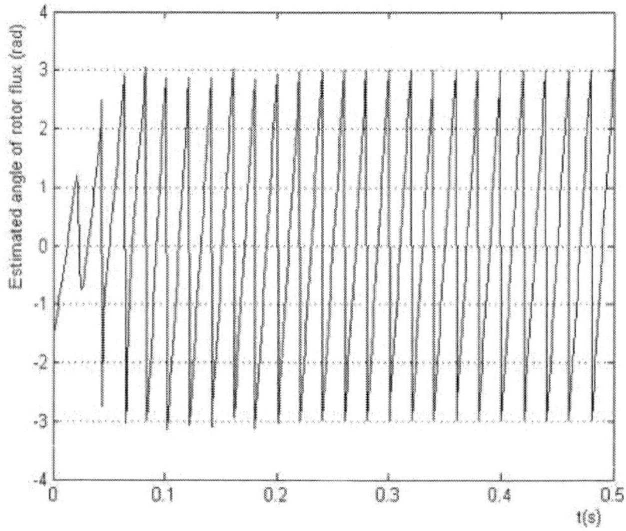

Fig. 15. Estimated rotor flux angle in d-q rotating frame using Us-Is estimator

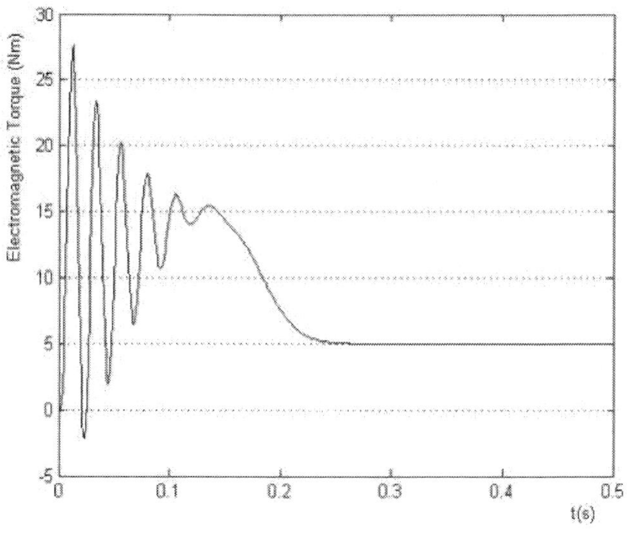

Fig. 16. Estimated electromagnetic torque using Us-Is estimator

3.3 Us-ω estimator

A rotor flux estimator which can operate in the range of low rotational speeds can be designed if one considers as measured inputs the stator voltages and the rotor angular velocity (Apostoaia & Scutaru, 2006). This estimator is derived by direct synthesis from the machine state equations which are written in terms of the d-q axis components of the stator and rotor flux as state variables. The same assumptions were made as in the previous section for the estimator Us-Is.

We will denote the parameters used in the observer, as well as the estimated variables, with the same symbols like in the machine model but having the superscript "e" in addition. Thus, the system of equations of the flux observer is derived as follows:

$$\frac{d\psi_{sd}^e}{dt} = \frac{1}{T_s^e}\left[-\psi_{sd}^e + \omega_\lambda T_s^e \psi_{sq}^e + \frac{L_m^e}{L_r^e}\psi_{rd}^e + T_s^e u_{sd}\right] \tag{36}$$

$$\frac{d\psi_{sq}^e}{dt} = \frac{1}{T_s^e}\left[-\omega_\lambda T_s^e \psi_{sd}^e - \psi_{sq}^e + \frac{L_m^e}{L_r^e}\psi_{rq}^e + T_s^e u_{sq}\right] \tag{37}$$

$$\frac{d\psi_{rd}^e}{dt} = \frac{1}{T_r^e}\left[\frac{L_m^e}{L_s^e}\psi_{sd}^e - \psi_{rd}^e + T_r^e(\omega_\lambda - \omega)\psi_{rq}^e\right] \tag{38}$$

$$\frac{d\psi_{rq}^e}{dt} = \frac{1}{T_r^e}\left[\frac{L_m^e}{L_s^e}\psi_{sq}^e - T_r^e(\omega_\lambda - \omega)\psi_{rd}^e - \psi_{rq}^e\right] \tag{39}$$

The dynamics of the flux estimator described by (36)-(39) is influenced by the time constants:

$$T_s^e = \sigma^e \frac{L_s^e}{R_s^e}, \quad \text{and} \quad T_r^e = \sigma^e \frac{L_r^e}{R_r^e}. \tag{40}$$

The electromagnetic torque is reconstructed in terms of the estimated state variables similarly to equation (33). The rotor flux magnitude and rotor flux angle are calculated like in (34) and (35).

The Simulink block diagram showing the Us-ω estimator can be seen in Fig. 17.

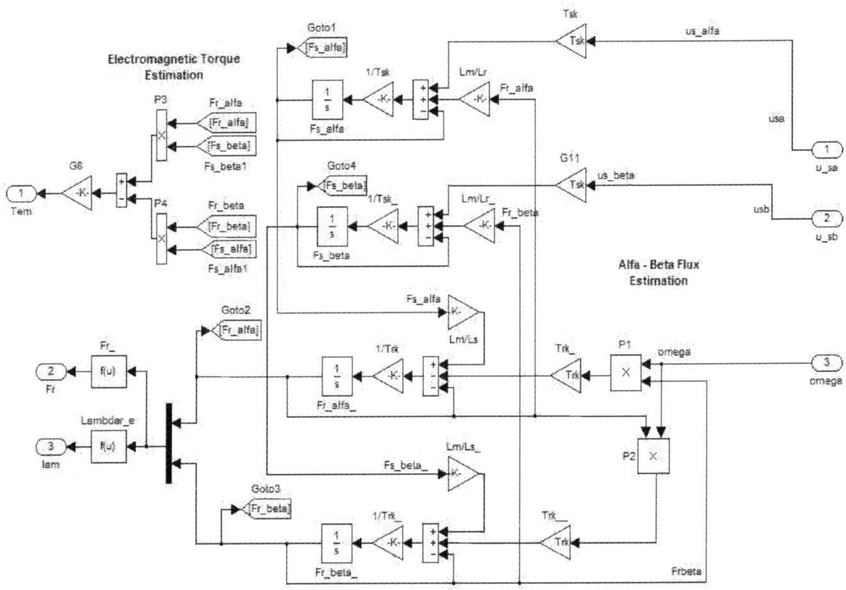

Fig. 17. Us-ω estimator Simulink model

In Fig. 18, Fig. 19, and Fig. 20, simulation results are shown for the Us-ω estimator, during a start up of the squirrel cage induction motor. A rated speed command under full torque load was used in the simulations.
The magnitude and the angle of the rotor flux estimates are shown in Fig. 18 and Fig. 19, and the electromagnetic torque estimation is seen in Fig. 20.

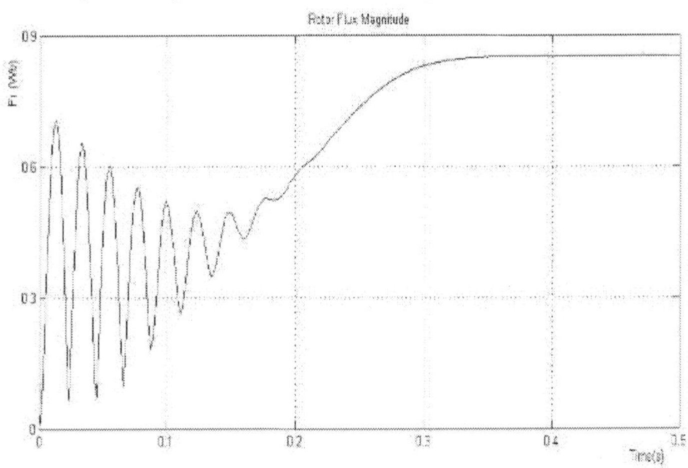

Fig. 18. Estimated rotor flux magnitude in d-q rotating frame using Us -ω estimator

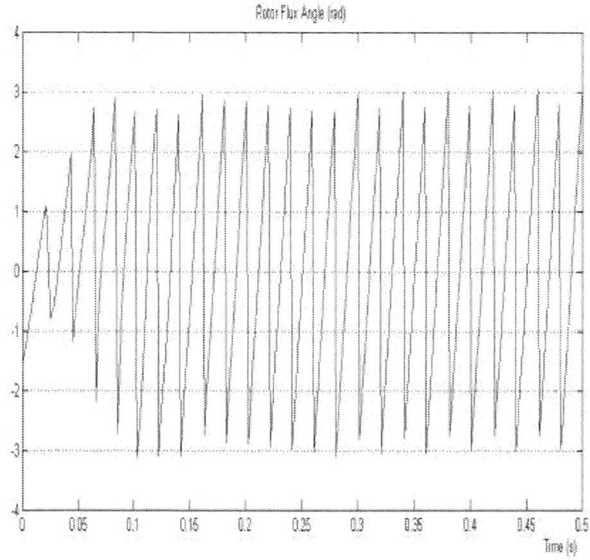

Fig. 19. Estimated rotor flux angle in d-q rotating frame using Us--ω estimator

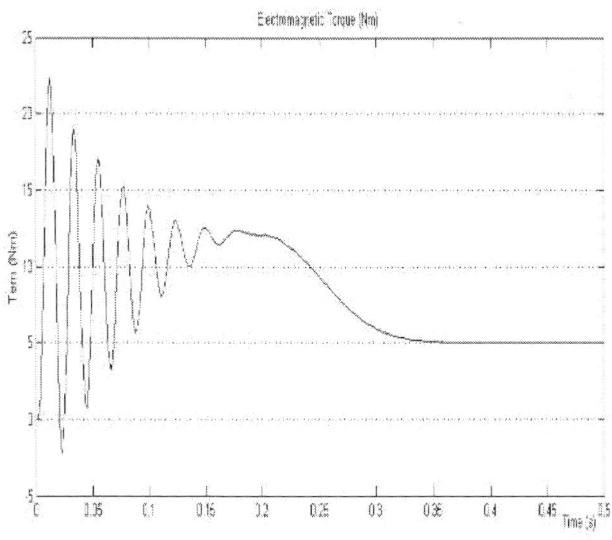

Fig. 20. Estimated electromagnetic torque using Us--ω estimator

3.4 Neural network based speed estimator

The inputs to the neural networks are the stator voltages and stator currents at time step k and $k-1$ $(u_{sd}, u_{sq}, i_{sd}, i_{sq})$.The target is the rotor speed in revolutions per minute at time step k.

The network is a feedforward network with backpropagation algorithm. The training method is Levenberg-Marquardt (Caudill & Butler, 1999).

The neural network has 2 layers with 30 neurons on the hidden layer, and with 'tansig' activation function, and one neuron on the output layer, with 'purelin' activation function.

The derivative of the speed is estimated because the feedforward, backpropagation method is not appropriate of estimating integral components. The estimation of an integral component require some knowledge of the previous states and it would mean that the output of the network would need to be feed back as another input to the network. After the derivative component of the speed is estimated the simple integral method is used to extract the speed value, using 'cumsum' from MATLAB.

The derivative component of the speed can be seen in Fig. 21, while the estimated rotor speed after the integration can be found in Fig. 22.

As it can be seen from the below figures very good simulation results can be obtained using neural network for speed estimation, which can be used in combination with any other estimator in a sensorless vector control scheme or in sensor fusion control scheme(Simon, 2006).

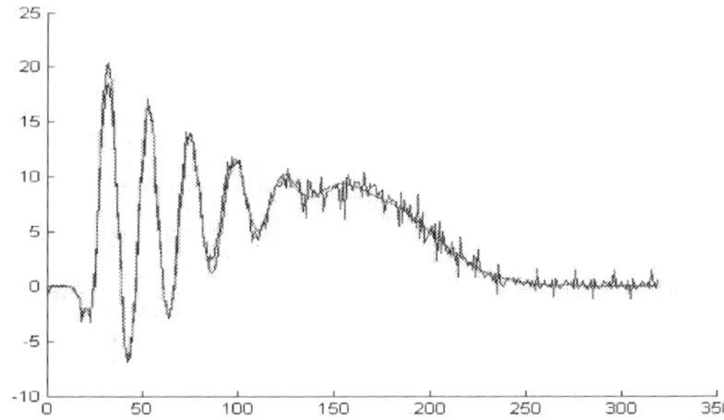

Fig. 21. The derivative component of the rotor speed using neural network

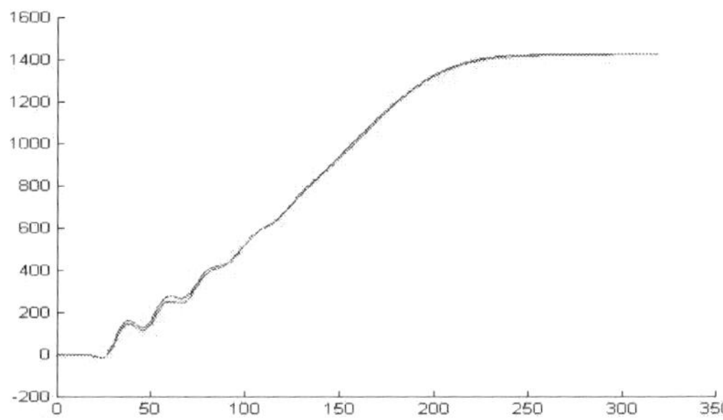

Fig. 22. The estimated rotor speed using neural network

4. Real-time simulation

For the real-time implementation of the wind turbine system a dSPACE CLP1104 embedded system was used (Szekely, 2008).

For the real-time estimation of the rotor flux magnitude and angle, and of the electromagnetic torque, two stator current measurements and a speed measurement are required. Using the analog to digital (ADC 1 and 2) channels of the dSpace board, the two current measurement can be transferred to the computer together with the speed measured by the incremental encoder mounted on the induction generator shaft which is coupled with the wind turbine. The information from and to the computer is transferred through the dSpace board using the Digital I/O connector.

The Real-Time Simulink model of the estimator and measurement channels can be seen in Fig 23. In the feedback loop of the system two stator currents are measured together with the rotor speed. The currents are measured with LEM LA-NP 1752 currents transducers and the user has access to these measurements through BNC connectors. These signals are fed to the dSpace board, and from the board to the computer. The DS1104ADC_C5 and DS1104ADC_C6 blocks are used for the current measurements. The scaling factor for these measurement blocks is 1:20. For smoother measurement and better wave visualization a first order filter is also used.

The speed is measured using digital encoder and with the help of the blocks DS1104ENC_POS_C1 and DS1104ENC_SETUP can be used as an input for the observer. In order to receive the radian angle the DS1104ENC_POS_C1 block needs to be multiplied by $\frac{2\pi}{encoder}$. In this experiment the encoder lines=1000. To obtain the desired speed the delta position scaled has to be divided by the sampling time $\omega = \frac{d\theta}{dt} = \frac{\Delta\theta}{T_s}$.

Using the mentioned measurements the rotor flux components can be estimated together with the rotor flux angle. Also the electromagnetic torque can be computed knowing the rotor flux components and having the two stator current measurements. The TAalfabeta2dq block transforms the rotor flux components from fixed reference frame to d-q rotating reference frame using the estimated rotor flux angle. This transformation is necessary for the field oriented vector control of the induction motor drive. Having the rotor flux components in rotating reference frame the rotor flux magnitude is calculated using equation (33).

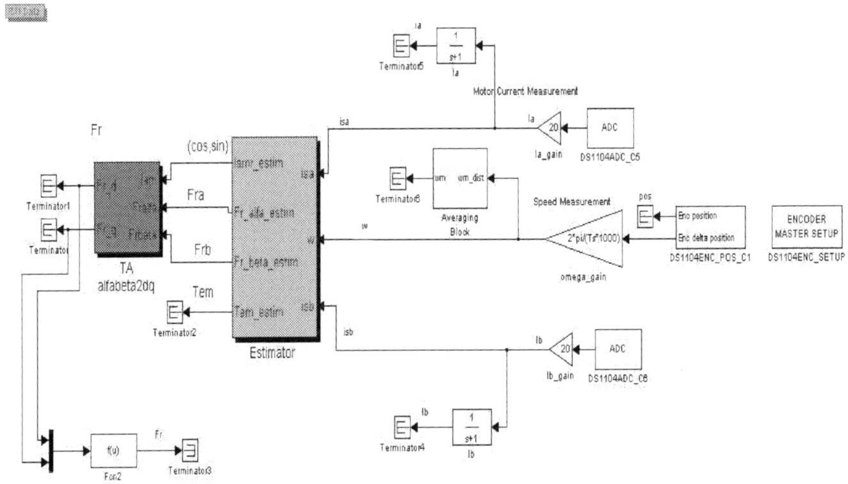

Fig 23. Real-time Simulink model of the measurement channels and the estimator

The real-time estimated values can be followed on the dSpace Control Desk as seen in Fig 24.

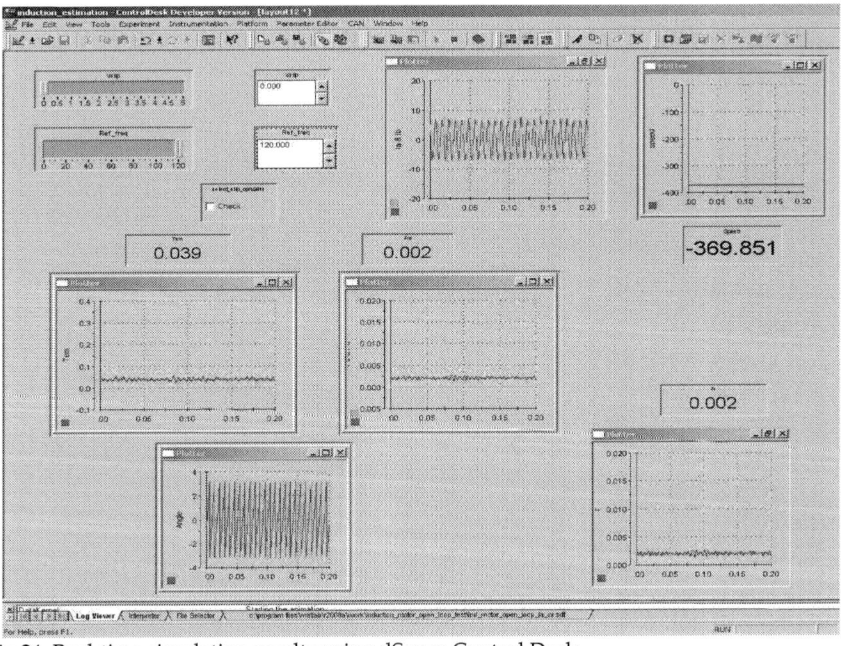

Fig 24. Real-time simulation results using dSpace Control Desk

5. Conclusions

In this paper simulation of a wind energy conversion system based on the induction generator was carried out using Matlab-Simulink programming environment.
Good simulation results were obtained based on real data of the wind turbine and the induction generator to validate them.
Also, a stator and rotor flux estimator model based on EKF was developed to simulate the correct achievement of the field orientation. The developed EKF estimator can be successfully used in sensorless vector control systems.
The Us-Is estimator has a big advantage that it can be used in variable speed applications and a big advantage is the fact that the estimator equations are not containing the rotor resistance as parameter, which eliminates the problems caused by the temperature.
The estimator based on Us-ω has the advantage that is simple to implement but meanwhile is not taking into account any real system noise. Also uses open mathematical integration for the parameter estimation which is hard to implement in real applications.
The block diagrams were used to simulate the system in real time using an existing dSPACE DS 1104 control board. This board is based on a floating point DSP with high speed ADC converters which makes suitable for the cross compilation of the Simulink models into the dedicated platform.
Further steps of this research would involve the validation of the presented models and estimators for a real life small wind turbine. All the necessary software and hardware design is available through the use of the modern HIL dSpace cards.

6. References

L. Tamas & Z. Szekely (2008): "Feedback Signals Estimation of an Induction Drive with Application to a Small Wind Turbine Generator, Automation Computers and Applied Mathematics, Volume 17, Number 4, 2008, p.642-651

Szekely, Z. (2008) "Extended Speed Control of an Induction Motor Drive utilizing Rotor Flux Orientation Technique in Real-Time", Masters Thesis, Purdue University Calumet, Hammond, Indiana, USA.

Scutaru, Gh. & Apostoaia, C. (2004) "MATLAB-Simulink Model of a Stand- Alone Induction Generator", in Proc. OPTIM 2004, "Transilvania" University of Brasov, Romania, May 20-21, 2004, vol. II, pp.155-162

Apostoaia, C. & Scutaru, Gh. (2006) "A Dynamic Model of a Wind Turbine System", in Proceedings OPTIM 2006, "Transilvania" University of Brasov, Romania, May 18-19, 2006, vol. II, pp.261-266

A. Kelemen, M. Imecs (1991): Vector Control of AC Drives, Volume 1; Vector Control of Induction Machine Drives, OMIKK-Publisher, Budapest, Hungary

Kalman, R.E. (1960): A new approach to linear filtering and prediction problems. Transactions of the ASME-Journal of Basic Engineering, Vol. 82

Simon, D. (2006): Optimal State Estimation. v. l., Willey Interscience.

Caudill, M. & C. Butler (1992) Understanding Neural Networks: Computer Explorations, Vols. 1 and 2, Cambridge, MA: The MIT Press.

Photovoltaic/Wind Energy System with Hydrogen Storage

Mamadou Lamine Doumbia and Kodjo Agbossou
Hydrogen Research Institute
Department of Electrical and Computer Engineering
Université du Québec à Trois-Rivières
C.P. 500, Trois-Rivières (Québec) G9A 5H7
Canada

1. Introduction

Renewable energy systems (RES) such as photovoltaic and wind generators are increasingly used as a means to satisfy the growing need for electric energy around the world. For many years, the Hydrogen Research Institute (HRI) has developed a renewable photovoltaic/wind energy system based on hydrogen storage. The system consists of a wind turbine generator (WTG) and a solar photovoltaic (PV) array as primary energy sources, a battery bank, an electrolyzer, a fuel cell stack, different power electronics interfaces for control and voltage adaptation purposes, a measurement and monitoring system. The renewable energy system can operate in stand-alone or grid-connected mode and different control strategies can be developed.

This paper presents the HRI's grid-connected renewable energy system (RES). The system's main components i.e. photovoltaic arrays, wind turbine, batteries, electrolyzer and fuel cell, are described individually and their modelling and simulation methodologies are presented. The complete system model is developed by integrating individual sub-units. Matlab/Simulink and LabVIEW softwares are used for modelling, programming and analyzing the behavior of each system sub-unit. The state of charge control method was used to validate the developed simulation models. The results obtained with the two modelling and simulation softwares were compared. Stand-alone and grid-connected operating conditions are investigated and experimental data are provided to support theoretical and simulation analyses. The power transfer study in the interconnected system is also presented. Such a global model is useful for understanding the system's operation, and optimal dimensioning and effective control of the renewable energy system with hydrogen storage (RESHS) (Kim S-K et al., 2008).

2. System components modelling

Figure 1 shows the block diagram of the HRI renewable energy system (Doumbia et al., 2007). The system consists of a 10 kW permanent magnet wind turbine generator and a 1 kW solar photovoltaic (PV) array as primary energy sources, a battery bank with 48V voltage, a 5 kW electrolyzer, a 1.2 kW proton exchange membrane fuel cell (PEMFC) stack. A 5 kW reversible inverter is used to convert 48V DC bus voltage into alternating current (AC) with 115V. The inverter output can be connected to the utility grid or to power a local AC load. A buck converter is used to control the electrolyzer and a boost converter is used to convert the 24V PEMFC output voltage into 48V DC bus voltage.

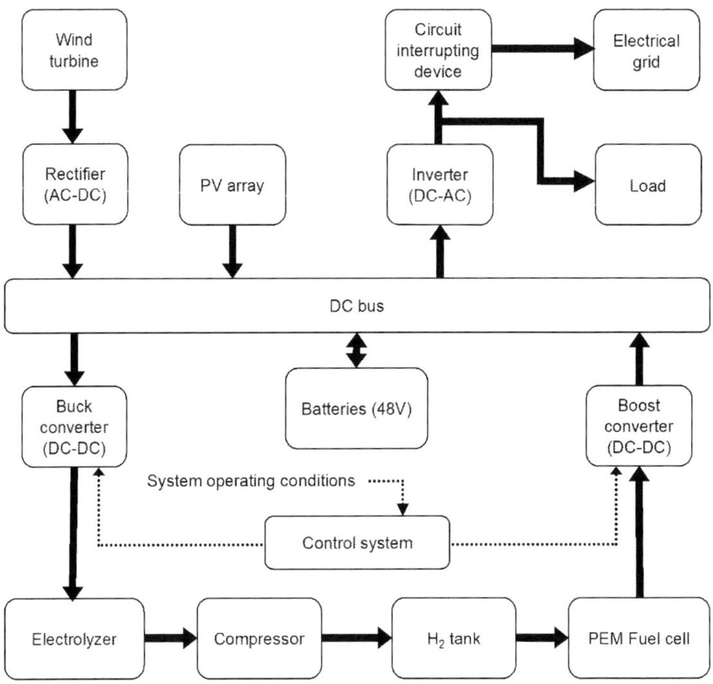

Fig. 1. Block diagram of the HRI's renewable energy system with hydrogen storage

2.1 Photovoltaic Array

The solar array is a group of several modules electrically connected in series-parallel combinations to generate the required current and voltage. The photovoltaic (PV) module output current I is equal to the photo-generated current I_L, less the diode-current I_d and the shunt-leakage current I_{sh}. The series resistance R_s represents the internal resistance to the current flow, and depends on the p-n junction depth, the impurities and the contact resistance. The shunt resistance R_{sh} is inversely related with leakage current to the ground. In an ideal PV cell, $R_s = 0$ (no series loss), and $R_{sh} = \infty$ (no leakage to ground).

In the equivalent circuit, the current delivered to the external load equals the current I_L generated by the illumination, less the diode current I_d and the ground-shunt current I_{sh}. The open circuit voltage V_{oc} of the cell is obtained when the load current is zero, and is given by the equation (1):

$$V_{oc} = V + IR_s \tag{1}$$

The mathematical description of the current-voltage (I-V) characteristic for photovoltaic (PV) cells has been available for some time. The simplest equivalent circuit of a solar cell is a current source in parallel with a diode (Figure 2). The output of the current source is directly proportional to the light falling on the cell. The diode determines the I-V characteristics of the cell (Walker, 2000), (Gow & Manning, 1999). The model includes temperature dependence of the photo-current I_L and the saturation current of the diode I_0.
The equations which describe the I-V characteristics of the cell are:

$$I = I_L - I_0(e^{\frac{q(V+IR_S)}{nkT}} - 1) \tag{2}$$

where :

I_L = photo-generated current (A)
I_0 = diode saturation current (A)
q = electronic charge (C)
V = solar cell terminal voltage (V)
R_S = cell series resistance (Ω)
n = diode quality factor
k = Boltzmann's constant (J/K)
T = ambient temperature (K)

$$I_L = I_{L(T1)}(1 + K_0(T - T_1)) \tag{3}$$

$$I_{L(T1)} = \frac{G * I_{SC(T1)}}{G_{nom}} \tag{4}$$

$$K_0 = \frac{I_{SC(T2)} - I_{SC(T1)}}{I_{SC(T1)}(T_2 - T_1)} \tag{5}$$

where :
G = cell irradiance (W/m^2)
G_{nom} = rated cell irradiance (W/m^2)
T = solar cell temperature (K)
T_1, T_2 = two reference temperatures (K)
$I_{SC(T1)}$ = short circuit current at temperature T_1 (A)
$I_{SC(T2)}$ = short circuit current at temperature T_2 (A)

The diode saturation current I_0 can be determined by the equation (6) :

$$I_0 = I_{0(T_1)} \left(\frac{T}{T_1}\right)^{\frac{3}{n}} e^{\frac{-qVg}{nk}\left(\frac{1}{T}-\frac{1}{T_1}\right)} \quad (6)$$

$$I_{0(T_1)} = \frac{I_{SC(T_1)}}{e^{\frac{qVoc(T_1)}{nkT_1}\left(\frac{1}{T}-\frac{1}{T_1}\right)}-1} \quad (7)$$

where:
V_g = Band gap voltage (V)
$V_{OC(T1)}$ = Open circuit voltage at the temperature T_1 (V)
The cell series resistance R_S is

$$R_S = -\frac{dV}{dI_{Voc}} - \frac{1}{X_V} \quad (8)$$

$$X_V = I_{0(T_1)} \frac{q}{nkT_1} e^{\frac{qVoc(T_1)}{nkT_1}} \quad (9)$$

$\frac{dV}{dI_{Voc}}$ is the voltage-current derivative when the voltage has reached open-circuit value.

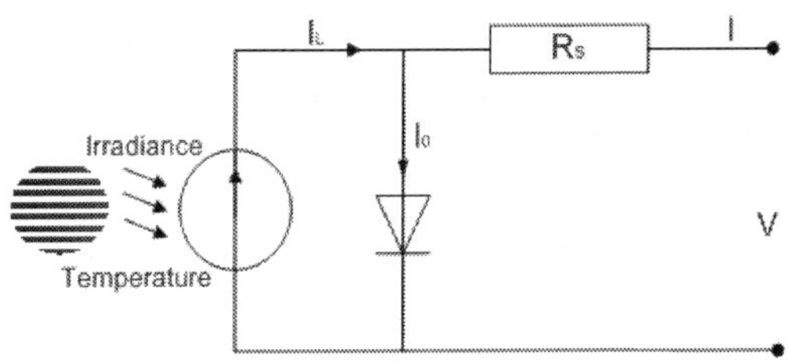

Fig. 2. Circuit diagram of PV model

Solar panels installed at IRH are composed of 16 modules, i.e. four rows of four serial connected modules. The electrical performance and the characteristic curves of the PV modules are dependent on temperature and illumination. From the preview equations, the I-V characteristics of the PV modules are plotted for different temperature (Figure 3) and illumination (Figure 4) conditions.

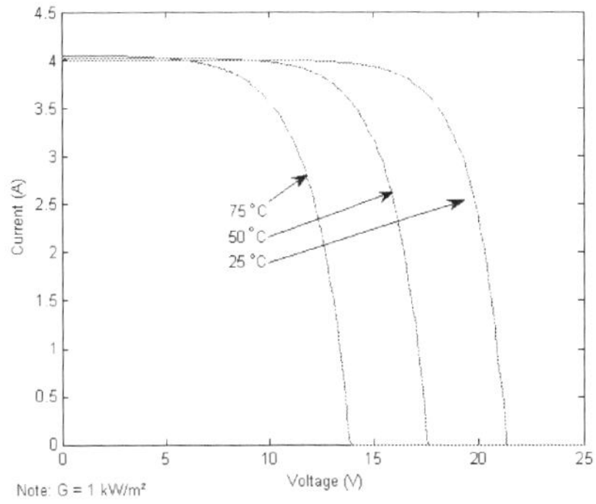

Fig. 3. PV module I-V characteristics for different temperature

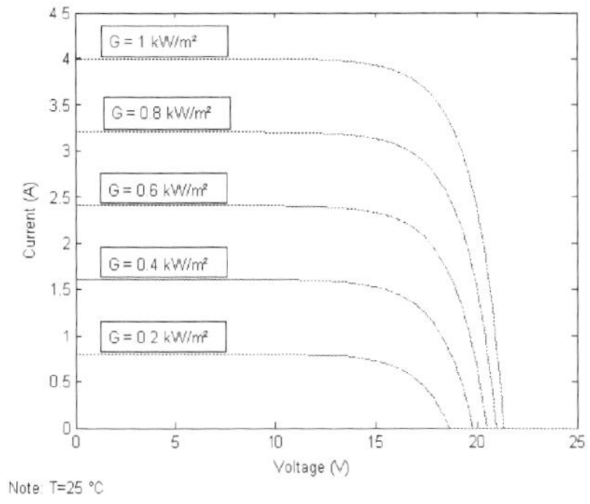

Fig. 4. PV module I-V characteristics for different illumination

2.2 Wind turbine

Some of the available power in the wind is converted by the rotor blades to mechanical power acting on the rotor shaft of the WT. The wind turbine rotor that extracts the energy from the wind and converts it into mechanical power is a complex aerodynamic system.
For state-of-the-art modelling of the rotor, blade element theory must be used. Modelling the rotor using blade element theory has, however, a number of drawbacks (Slootweg et al., 2003).

- Instead of only one wind speed signal, an array of wind speed signals has to be applied.
- Detailed information about the rotor geometry should be available.
- Computations become complicated and lengthy.

To overcome these difficulties, a simplified way of modelling the wind turbine rotor is normally used when the electrical behaviour of the system is the main point of interest. For steady-state calculations of the mechanical power from a wind turbine, the so called $C_p(\lambda, \beta)$-curve can be used. An algebraic relation between wind speed and mechanical power extracted is assumed, which is described by the well-known expression (Slootweg et al., 2003), (Cardenas & Pena, 2004):

$$P_w = \frac{1}{2}\rho\pi R^2 C_p(\lambda,\beta)v^3 \tag{10}$$

where:

P_w = power extracted from the wind (W)
ρ = air density (kg/m^3)
R = blades radius (m)
C_p = power coefficient
λ = tip speed ratio
β = pitch angle of the rotor blades (degrees)
v = wind speed (m/s)

The density ρ of air is quite low less than that of water which powers hydro plant, and this leads directly to the large size of a wind turbine. The power coefficient describes that

fraction of the power in the wind that may be converted by the turbine into mechanical work. It has a theoretical maximum value of 0.593 (the Betz limit). The power coefficient C_p of a rotor varies with the tip speed ratio λ (the ratio of rotor tip speed to free wind speed) and is only a maximum for a unique tip speed ratio. Improvements are continually being sought in the power coefficient by detailed design changes of the rotor and, by operating at variable speed; it is possible to maintain the maximum power coefficient over a range of wind speeds. However, these measures will give only a modest increase in the power output. Major increases in the output power can only be achieved by increasing the swept area of the rotor or by locating the wind turbines on sites with higher wind speeds (Burton et al., 2001).
Numerical approximations have been developed to calculate C_p for given values of λ and β (Slootweg et al., 2003), (Lei et al., 2006). For the Bergey BWC Excel 10 kVA wind turbine used at the HRI, the following equation was used to approximate the power coefficient C_p.

$$C_p = (0.0007391v^3 + 0.023649v^2 - 0.26584v + 1.2934)*(1 - 44.292\exp(-1.0762*v)) \quad (11)$$

Figure 5 shows the wind turbine output electrical power variation with respect to wind speed.

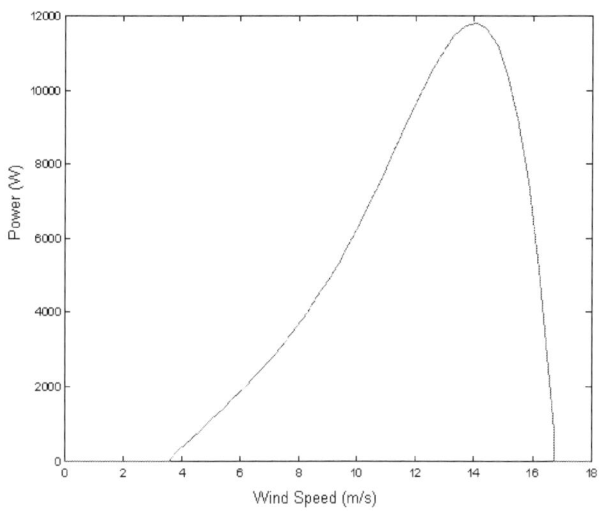

Fig. 5. Wind turbine output electrical power variation with respect to wind speed

2.3 Battery model

The battery plays the role of an energy buffer for short-term energy storage. Different models for batteries are available, in particular those suitable for electrical vehicle applications (Kélouwani et al., 2005). For stationary applications such as the renewable sources, the models described in (Vosen & Keller, 1999) use many experimental parameters that cannot be estimated easily, such as the overcharge effect (though in a properly-controlled RESHS, this effect does not happen, and hence is not included in the present

model). The main parameters which determine the battery's performance are its internal resistance, the polarization effect, and the long-term self-discharge rate. The self-discharge rate is difficult to estimate, and is itself subject to a number of factors such as the operating temperature, the number of operation cycles, and the materials and technology used in its manufacture. The battery model used in this paper presents the relation between voltage, current and the battery state of charge Q as follows (Chérif et al., 2002):
In discharge mode (I<0):

$$V(t) = V_d - g_d \frac{It}{C} + R_d I \left(1 + \frac{M_d It}{C(1 + C_d) - It} \right) \tag{12}$$

In charge mode (I>0):

$$V(t) = V_c - g_c \left(1 - \frac{It}{C} \right) + R_c I \left(1 + \frac{M_c It}{CC_c - It} \right) \tag{13}$$

where :

I	= battery current (A)
V	= battery voltage (V)
C	= battery capacity (Ah)
T	= time (h)
R	= internal resistance (Ω)
g	= coefficient that characterises the voltage variation $\Delta V = f(Q)$ (V)
M	= slope of voltage characteristic $V = f(t,I,Q)$

In (12), (13), (15) and (16) the subscripts *d* and **c** indicate the discharging and charging modes.
The state of charge Q of the battery can be calculated through the current integral.

$$Q = \int Idt + Q_0 \tag{14}$$

where Q_0 is the initial battery charge.
In the previous equations, the term I*t can be replaced by ($C-Q=Q_d$), so equations (12) and (13) become

$$V(Q_d) = V_d - g_d \frac{Q_d}{C} + R_d I \left(1 + \frac{M_d Q_d}{C(1 + C_d) - Q_d} \right) \tag{15}$$

$$V(Q_c) = V_c - g_c \left(1 - \frac{Q_c}{C} \right) + R_c I \left(1 + \frac{M_c Q_c}{CC_c - Q_c} \right) \tag{16}$$

where :

$Q_c = Q$	= cumulated charge (A.s)
Q_d	= transmitted charge (during discharge) (A.s)

2.4 Electrolyzer model

Electrolyzers operate in either current mode or voltage mode. When they are run in voltage mode, voltage is applied to the electrolyzer and, depending on the operating conditions, the electrolyzer draws the current from the source. After a couple of transient cycles it has reached its steady state value. However, most of the commercially available electrolyzers run in current mode, with a polarization characteristic as shown in Figure 6. This characteristic can be represented as a sum of linear, logarithmic and exponential functions.

$$V = E_0(T) + R(T) * I + b(T) * ln(I) + m(T) * exp(n * I) \tag{17}$$

where :

E_0 = Reversible potential

I = current (A)

T = temperature (°C)

b, m, R = coefficients that depend on temperature

n = constant

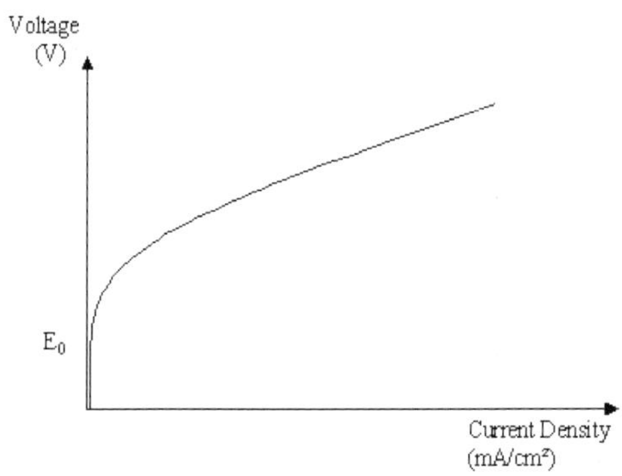

Fig. 6. Electrolyzer polarisation characteristic

Real coefficients were obtained from experimental data for a Stuart electrolyzer. Two series of data at two temperatures (23.5°C and 53°C) and one series with fixed voltage were used. The curves are extrapolated in temperature by modifying eq. 17 (assuming its exponential term can be ignored):

$$V = E_0 + RI + b\,ln(I) \tag{18}$$

By varying the resistance R one can optimize the equation $E_0 + b\,ln(I)$ for the curves at temperatures 23.67°C and 53.51°C via the correlation coefficient. Then the coefficients thus determined: E_0, b and R are extrapolated linearly in temperature. Figure 7 shows several of these polarisation curves, as well as certain points taken from the series of data with fixed voltage.

$$V = E_0 + RI + b\,ln(I) + (0.1716\,exp(-0.0612T)) * exp(0.0551) \tag{19}$$

where :

$E_0(T) = 32.5628 - 0.00677*T$
$R(T) = 0.0002089*T + 0.00955$
$b(T) = 3.374 - 0.0194*T$

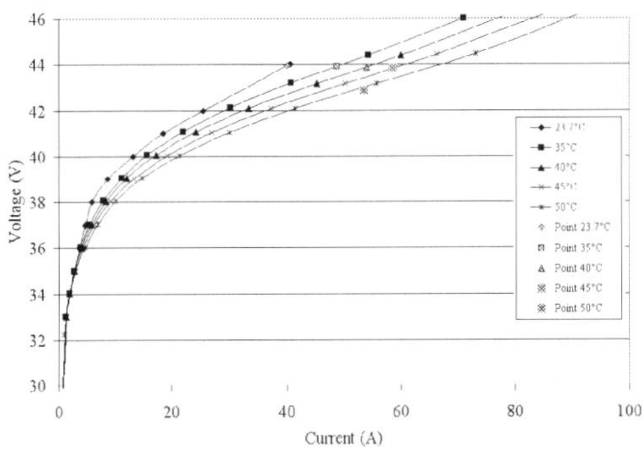

Fig. 7. Stuart electrolyzer polarisation curve

2.5 Fuel cell model

Depending on the current drawn, the fuel cell converts hydrogen and thereby produces voltage as a function of current. The equations used depend on the empirical form of the fuel cell model (Doumbia et al., 2007). The polarisation curve of the fuel cell (Figure 8) can be divided into three regions: the activation overvoltage region, the ohmic overvoltage region, and the thermodynamic overvoltage region. This curve can be represented (again, as in the electrolyzer model) by a sum of linear, logarithmic and exponential functions:

$$V = E_0 - b(T) * ln(I) - R(T) * I - m(T) * exp(n * I) \qquad (20)$$

where :

I = current (A)

T = temperature (°C)

b, R, m = coefficients that depend on the temperature

n = constant

Table 1 summarizes the coefficients values of the Ballard MK5-E fuel cell used at the IRH.

Coefficients	Values
E_0 (V)	1.05
b (V)	$4.01 \times 10^{-2} - 1.40 \times 10^{-4}T$
R (kΩ cm²)	$4.77 \times 10^{-4} - 3.32 \times 10^{-6}T$
m (V) : T ≥ 39°C	$1.1 \times 10^{-4} - 1.2 \times 10^{-6}T$
m (V) : T < 39°C	$3.3 \times 10^{-3} - 8.2 \times 10^{-5}(T-39)$
n	8.0×10^{-3}

Table 1. Ballard MK5-E fuel cell coefficients values

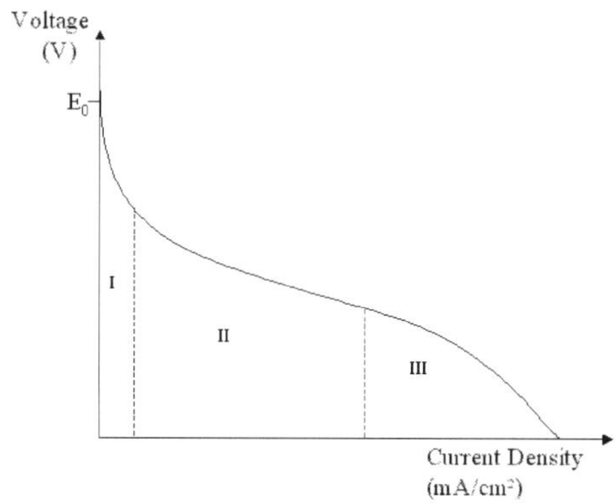

Fig. 8. Fuel cell polarisation curve

The commands for the electrolyzer and for the fuel cell are taken from the equation for the reference current. The form of this equation depends on the power demand and on the temperature of the electrolyzer. To validate system simulation model, the control system is designed in such a way that if the state-of-charge (SOC) exceeds 55%, then the electrolyzer kicks in (i.e. the excess energy is converted to stocked hydrogen), while if the OSC is less than 45%, the extra energy needed is drawn from the hydrogen by activating the fuel cell.

3. Integrated system simulation results

The global renewable energy system with hydrogen storage (RESHS) was modelled and simulated using Matlab/Simulink™ and Labview softwares. The simulation model includes sub-systems for the photovoltaic array, wind power generator, the electrolyzer and the fuel cell. The currents from the photovoltaic array and wind generator are sent to the DC bus which feeds the batteries. The current consumed by the load and the electrolyzer is withdrawn from the DC bus.
For validation purpose, the HRI RESHS was simulated using the state-of-charge (SOC) control strategy. The control system verifies the state of charge of the batteries and sends control signals to the electrolyzer or the fuel cell via DC/DC converters. When the state-of-charge of the batteries is high, the control unit activates the electrolyzer by sending a control signal to the buck converter (to decrease the energy level). The excess energy is thus converted into hydrogen which is stored for latter use. On the other hand, when the state of charge of the batteries is low, the control unit activates the fuel cell via the boost converter. The stored hydrogen is then converted to electrical energy which is sent to the DC bus to supply local loads.

The Figure9 shows the Matlab/Simulink model of the RESHS. This global model integrates the model of all the system sub-units (PV, wind, electrolyzer, fuel cell, batteries). The behaviour of the system is simulated for six-day periods in the months of January (winter time) and July (summer time). The voltage, currents, power of all sub-units can be monitored and plotted. For simulation test purpose, the system is controlled such as if the state of charge of the batteries is too high or too low, respectively the electrolyzer or the fuel cell is started to reduce or increase the energy stocked in the batteries. Input data such as illumination, temperature, wind's speed and load consumption are imported from corresponding sub-systems.

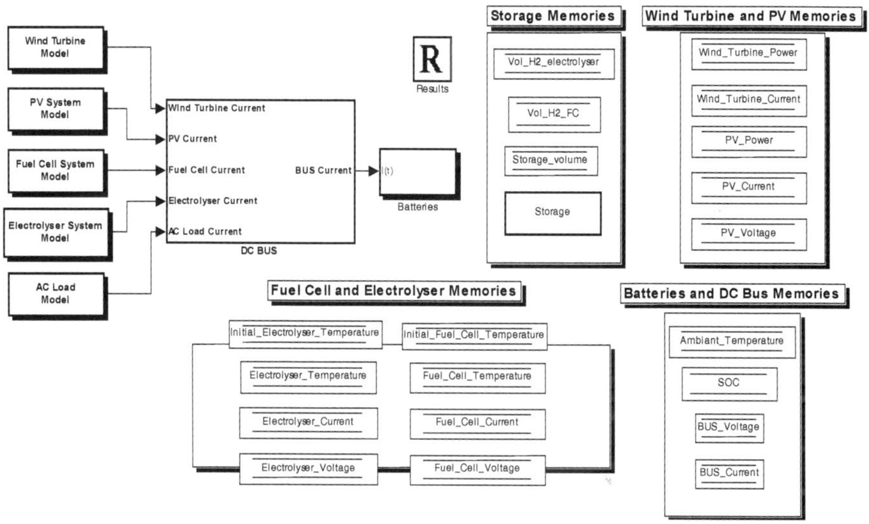

Fig. 9. Matlab/Simulink simulation model of the renewable energy system

Figure 10 and Figure 14 show the input data of temperature, illuminance and wind speed respectively for January and July. With these meteorological input conditions, the resulting photovoltaic and wind powers are shown in Figure 11 and Figure 15. The peaks of power produced by the wind generator can be seen reflected in the total DC bus currents (Figure 12 and Figure 16). The state of charge and the powers of the electrolyzer and of the fuel cell are shown in Figure 13 and Figure 17. These last two figures clearly show the correlation between the state of charge of the batteries and the activation of the electrolyzer (when SOC > 55%) or of the fuel cell (when SOC < 45%).

These simulation results are confirmed using both modelling and simulation softwares Matlab/ Simulink™ (Doumbia et al., 2009) and Labview (Doumbia et al., 2009).

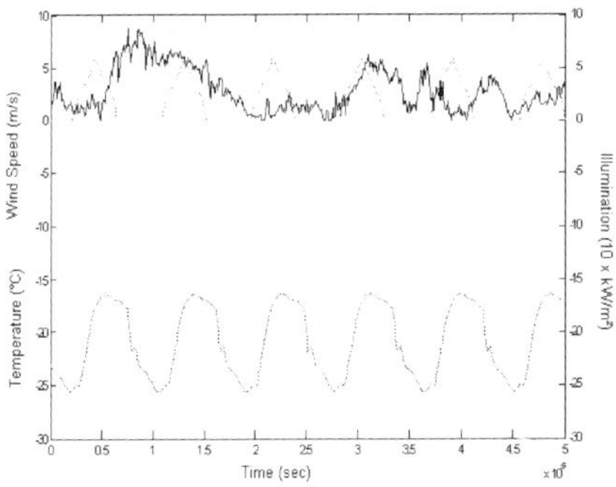

Fig. 10. Wind speed, temperature and illumination in January

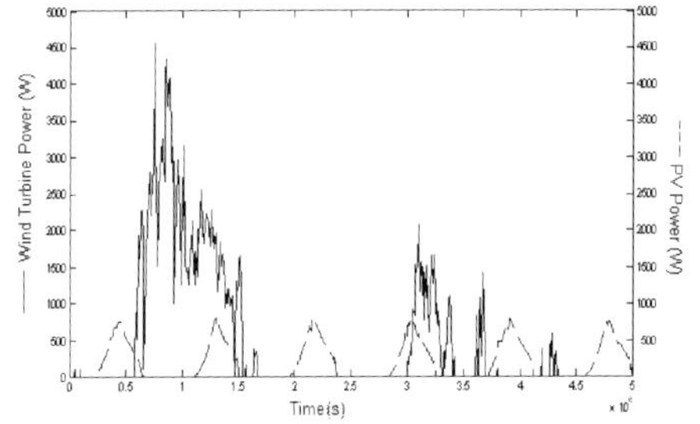

Fig. 11. Wind turbine and photovoltaic array power in January

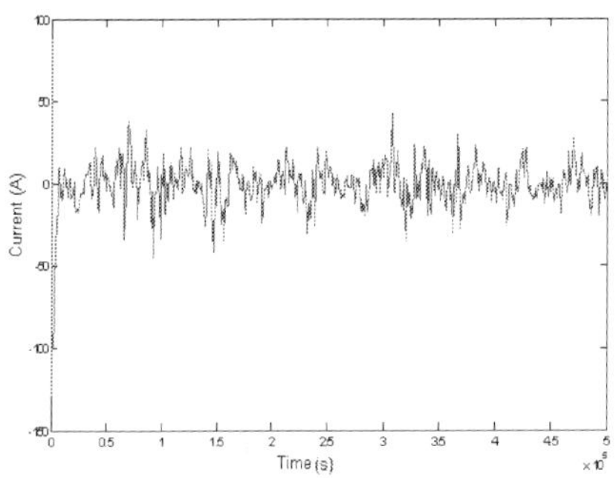

Fig. 12. DC bus current in January

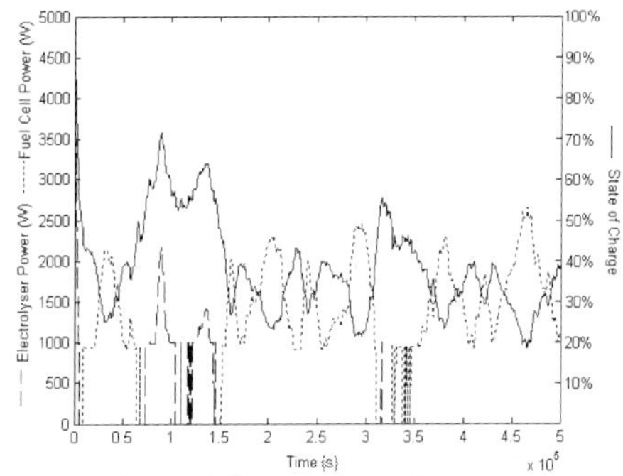

Fig. 13. State of charge, electrolyzer power and fuel cell power in January

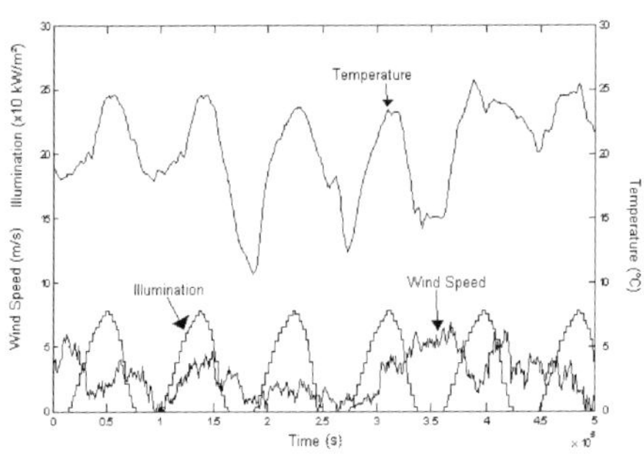

Fig. 14. Wind speed, temperature and illumination in July

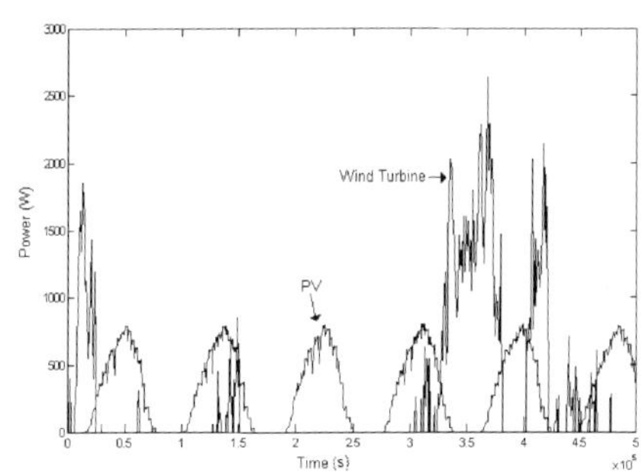

Fig. 15. Wind turbine and photovoltaic array power in July

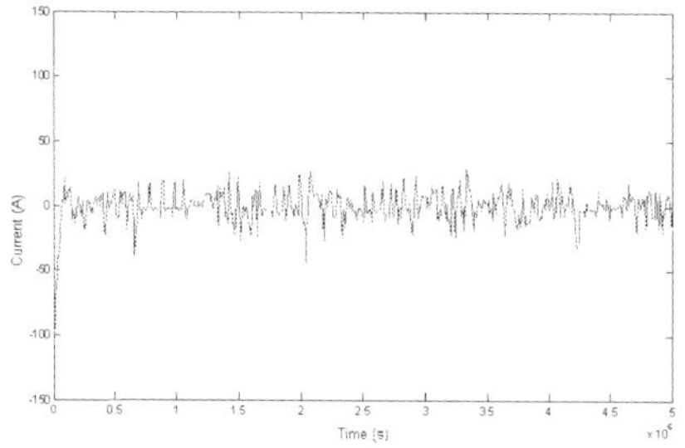

Fig. 16. DC bus current in July

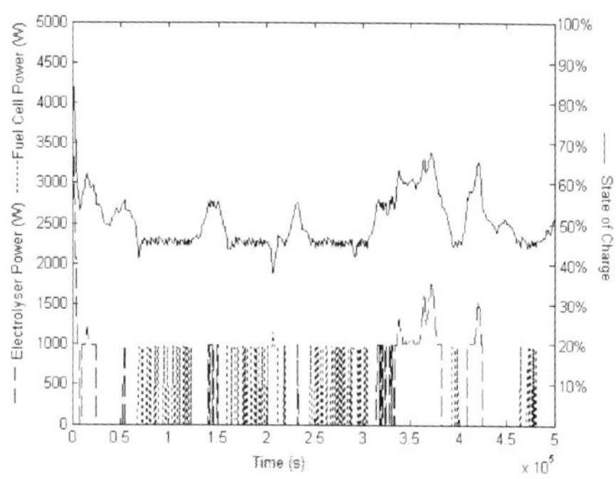

Fig. 17. State of charge, electrolyzer power and fuel cell power in July

4. Experimental study

4.1 Operation in stand-alone mode

The Hydrogen Research Institute's RESHS was investigated experimentally for stand-alone operation (Kodjo et al., 2004). The energy available from the WTG and the PV array are shown respectively in Figure 18 and Figure 19. The limits of energy levels in the control algorithm were set to start the electrolyzer at 99% of energy level at the DC bus or above and to stop the electrolyser at 84%. The fuel cell on and off operations was set to 83% and 85%, respectively. During this operation, the electrolyzer input power and the fuel cell output power are shown in Figure 20, the load profile and the batteries charging/discharging power are shown in Figure 21. The system started its operation at 9:20 AM, when the batteries energy level was at 100% SOC. It has been observed that the power supplied to the electrolyzer is mainly come from the short-term energy storage (batteries) due to the non-availability of sufficient energy from the RE sources during the operation. The power flow of the batteries (charging/discharging) was also carefully monitored. The electrolyzer and the fuel cell operations were started and stopped automatically, when the energy levels at DC bus have reached to the pre-defined levels of the control algorithm.

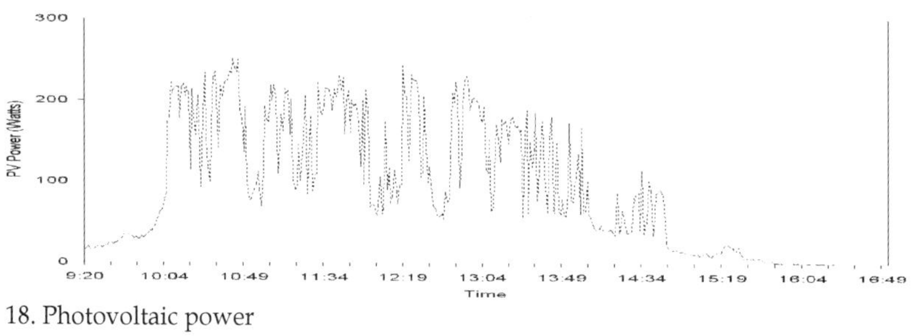

Fig. 18. Photovoltaic power

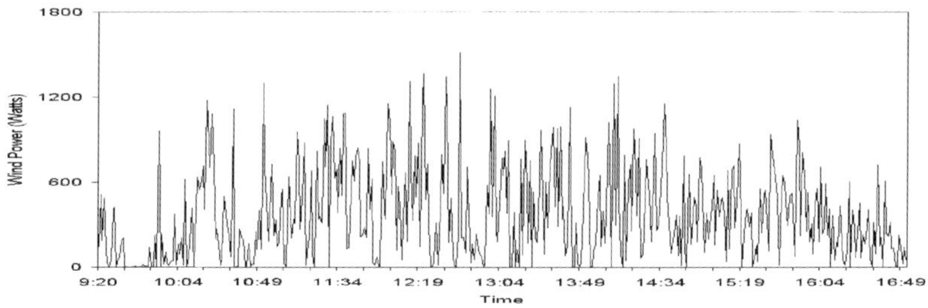

Fig. 19. Wind power

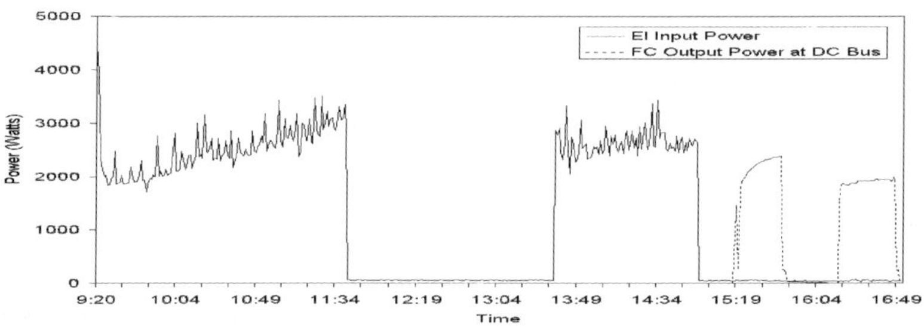

Fig. 20. Electrolyzer and fuel cell powers

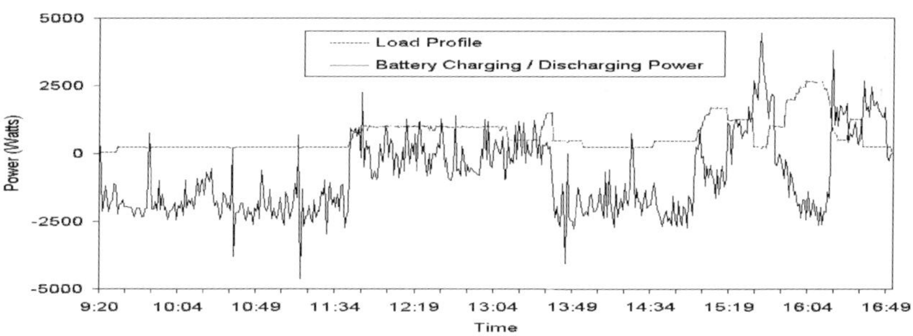

Fig. 21. Load and batteries powers

4.2 Operation in grid-connected mode

The HRI renewable energy system was investigated also in grid-connected operation mode (Figure 22). When connected to the utility grid, the interconnecting power electronic interface (a Trace Engineering SW4458 inverter) can be operated in four modes: Float, Silent, Sell and Low Battery Transfer.

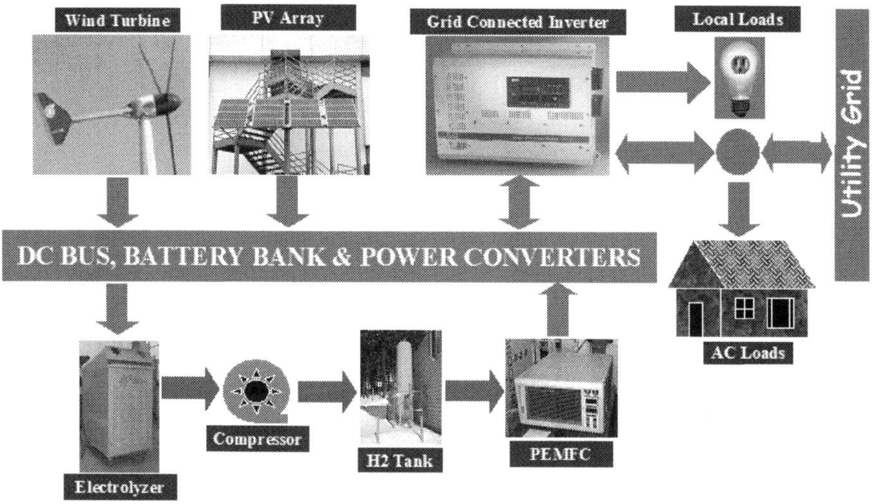

Fig. 22. Grid-connected renewable energy system with hydrogen storage

Operation in Float mode

The Float mode maintains the batteries at the float voltage level. This can be used when the source of power is a utility grid or a generator. When alternating current (AC) power is available, the inverter completes a full three-stage charge cycle and then holds the battery at the float level until the source of utility power is no longer available. This setting is appropriate for use with stand-alone systems with back-up generators or utility back-up systems. If photovoltaic or wind power is available and the battery is full, the excess power will be used to directly power the AC loads connected to the inverter output (Trace Engineering, 1999).

The Float mode was used to operate the RES from 11:55 to 16:38 (April 30, 2009). Figure 23 shows the system's operating characteristics. In the Float mode, the load (Figure 20a) is feed by batteries (Figure 23d) as long as the DC bus voltage (Figure 23f) is greater than a pre-defined threshold minimum (Low Battery Transfer Voltage) set to 47V in our case. When the DC bus voltage falls under 47V, the utility grid begins to feed the loads and batteries are recharged by the energy (if available) coming from solar panels (Figure 23g) and from the wind turbine generator (Figure 23h). The utility grid continues to power the loads as long as the DC bus voltage does not reach another pre-defined voltage level (Low Battery Cut In voltage), set to 52V (Figure 23c) in during this test. Figure 23b and Figure 23e show respectively the batteries current and state of charge variation.

Operation in Silent mode

This mode does not maintain the battery at float voltage all the time. The inverter performs a bulk charge (i.e. charging at a preset voltage level) of the batteries once per day, from the utility grid. The inverter then goes totally silent and will wait for the utility grid to fail, or until the next day when it performs another bulk charge. This mode is typically used in utility back-up applications (Trace Engineering, 1999).

The SER was tested in this mode on April 20th, 2009 from 12:00 to 16:15. The load was composed of the electrolyzer (400W) and lighting (380W) (Figure 24a). At the beginning of the test, the inverter was in bulk charge mode until 12:45 (Figure 24c). During this period, the utility network recharges batteries (Figure 24b, Figure 24d) by maintaining the batteries voltage at 53.6V (Figure 24f). After the Bulk charge period, the inverter works in Silent mode. Then, the DC bus voltage depends on the power balance between the input powers (photovoltaic power Figure 24g; and wind power Figure 24h) and output powers (electrolyzer and lighting).

Operation in Sell mode

SELL mode enables the inverter to send the excess power to the grid. This mode must have the approval of the local power utility prior to its use. In this mode, when power from photovoltaic array or wind generator is available, it will be used to power any AC load connected to the AC output of the inverter. Any excess power available from the system will be sold "into" the utility grid (Trace Engineering, 1999).

The SER was tested in this mode on May 5th, 2009 from 12:00 to 16:30 (Figure 25). A 460W local AC load (Figure 25a) was connected to the inverter output. The inverter was programmed to send up to 7A (800VA supplementary power) to the utility grid if the DC bus voltage is higher than a pre-defined threshold voltage (48V). At the beginning of the test, as the DC bus voltage (Figure 25f) is higher than the pre-defined sell voltage, the loads are powered by batteries (Figure 25d) through the inverter; this leads to the batteries discharging (Figure 25e). When renewable PV (Figure 25g) and wind (Figure 25h) energy sources are available, the power provided by the batteries (inverter) is reduced. When the DC bus voltage reaches the pre-defined 48V, the energy from the batteries is reduced (Figure 25b, Figure 25c) and loads are mainly powered by the utility grid and renewable energy sources.

Operation in Low battery transfer mode

The Low battery transfer mode allows a system to switch automatically between utility connected and stand alone battery operation. In this mode, the inverter will power the loads from the battery and other energy sources until the battery voltage drops to the pre-defined "low battery transfer voltage" value. It will then connect to the utility grid and charge the battery. The loads will be powered by the utility until the battery voltage reaches the "Low battery cut in voltage" value. The inverter will then disconnect the utility and power the loads from the battery or any other connected DC power source. This mode is often used instead of the Sell mode because approval from the utility is not required - no power will be sent into the utility distribution system when this mode is selected (Trace Engineering, 1999).

The SER was tested in Low battery transfer mode on May 11th, 2009 from 11:45 to 16:33 (Figure 26). Around 1200W local AC load (Figure 26a) was connected to the inverter output. At the beginning of the test, as the DC bus voltage (Figure 26f) is higher than the pre-defined threshold voltage (47V), the loads are powered by batteries (Figure 23b, Figure 26d) through the inverter; this leads to the batteries discharging (Figure 26e). When the DC bus voltage reaches the pre-defined 47V, the utility grid begins to power the local load (Figure 26c). At all time, if renewable PV (Figure 26g) or wind (Figure 26h) energy source is available, the power provided by the batteries or by the utility grid is reduced.

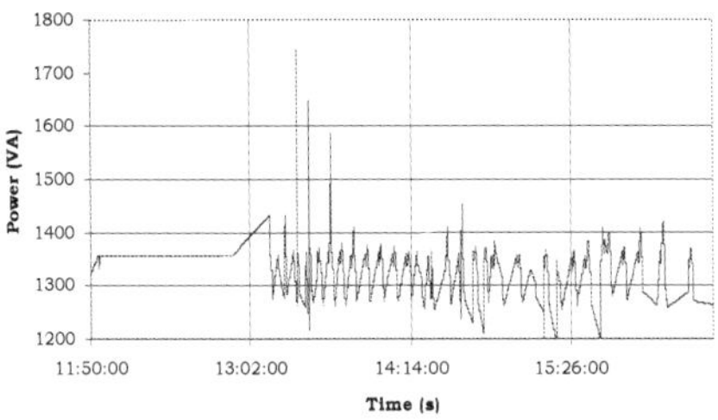

a) Local Load power

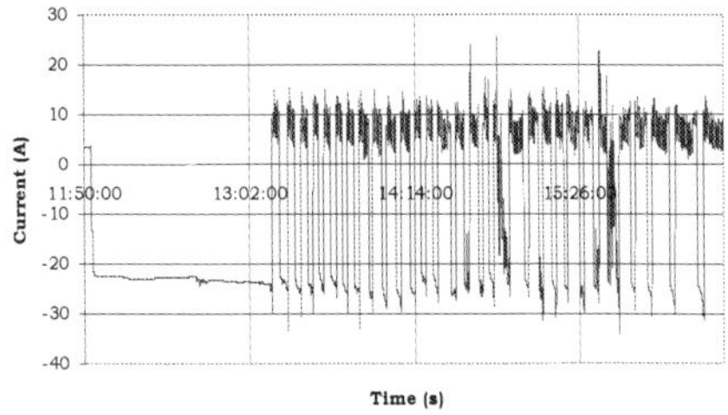

b) Batteries current

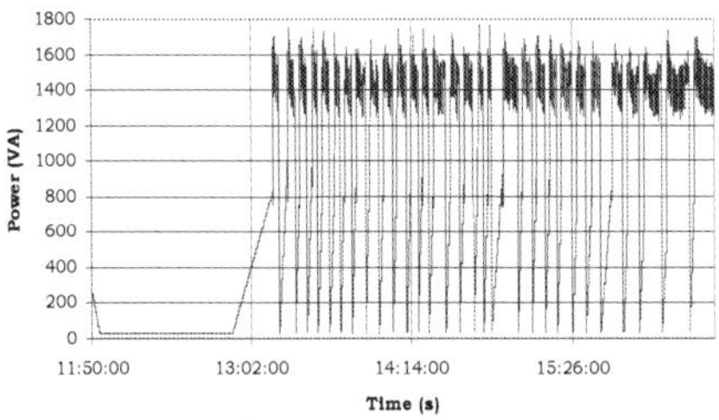

c) Inverter power

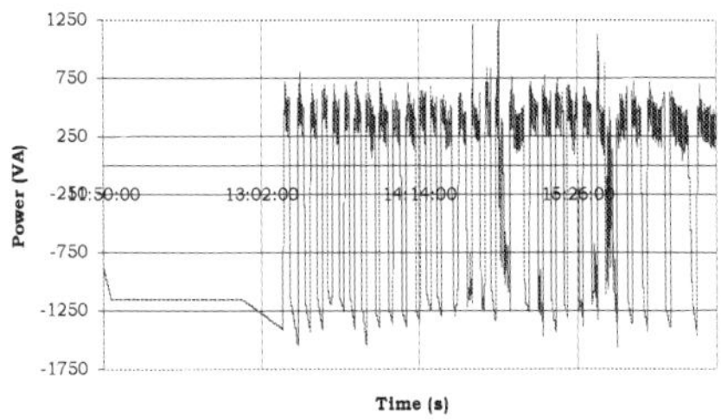

d) Batteries power

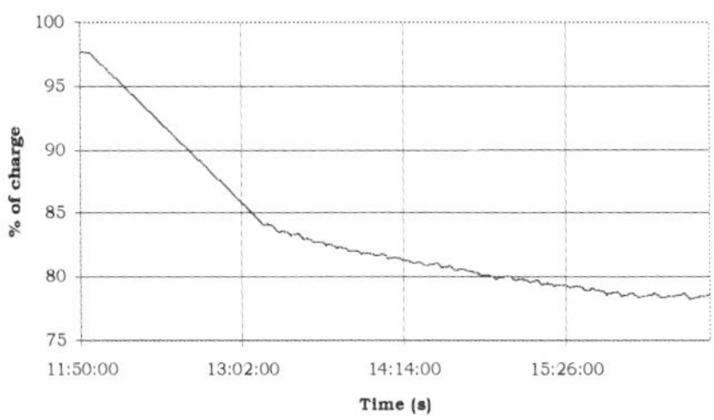

e) State of charge

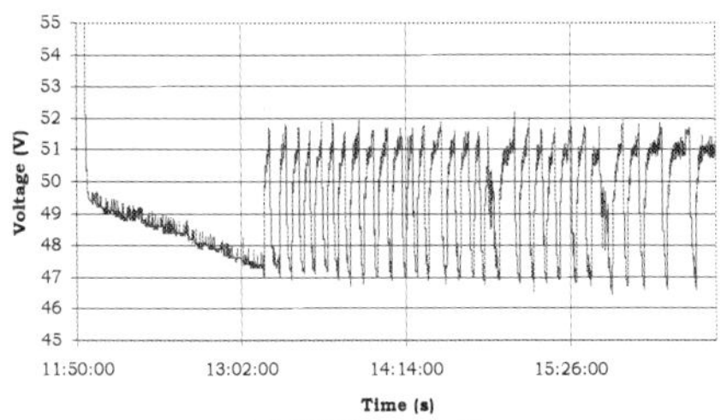

f) DC bus voltage

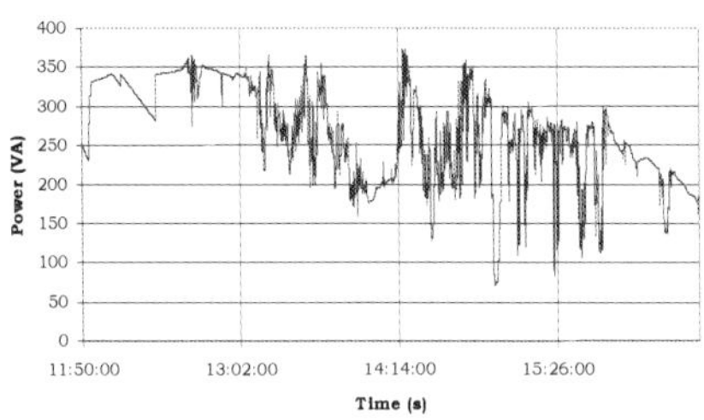

g) Photovoltaic power

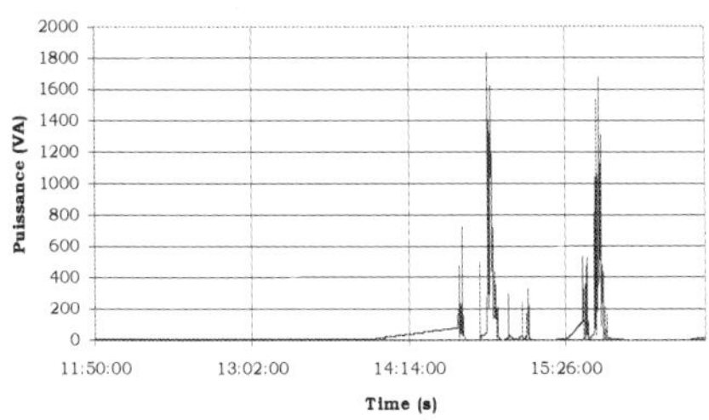

h) Wind power

Fig. 23. Grid-Connected RES operation in Float mode

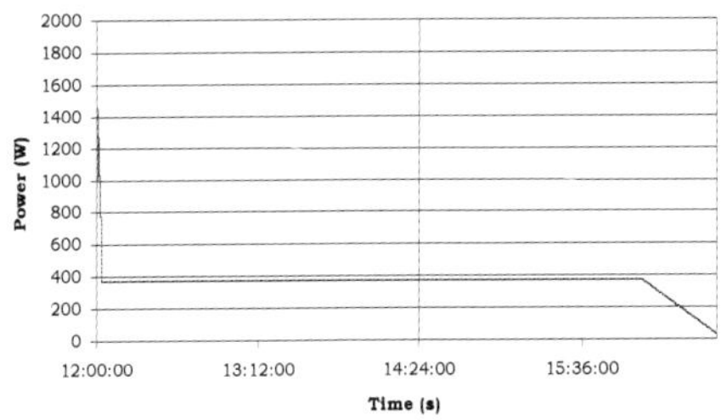

a) Electrolyzer power

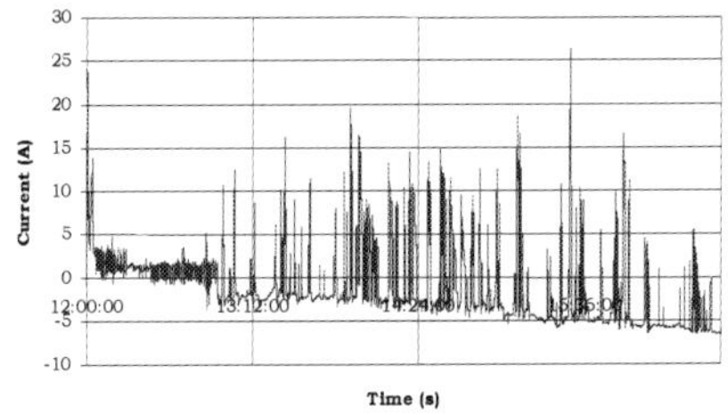

b) Batteries current

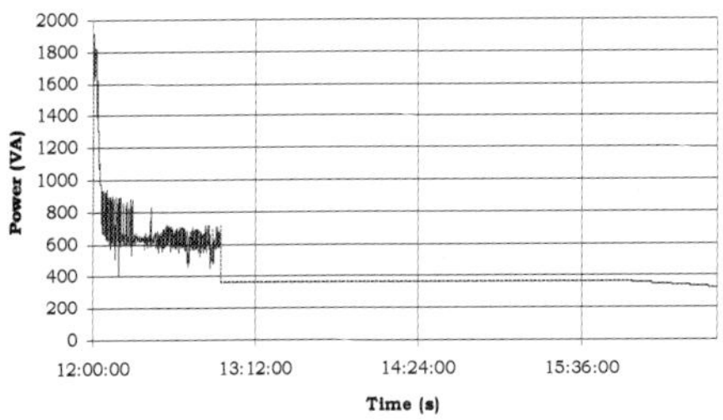

c) Inverter power

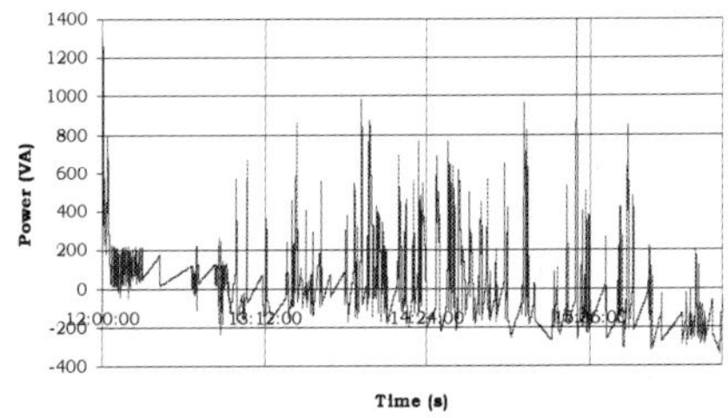

d) Batteries power

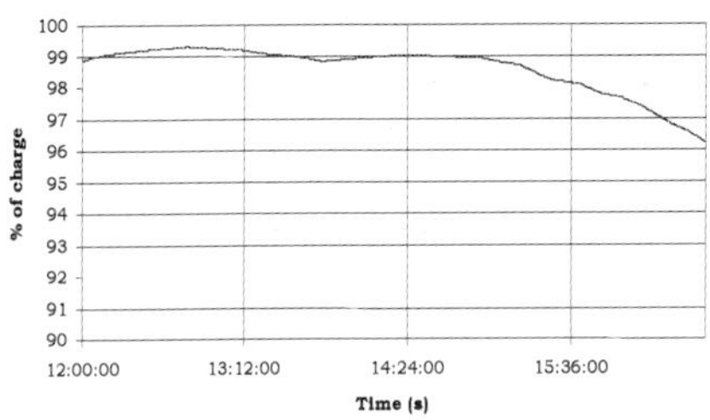

e) State of charge

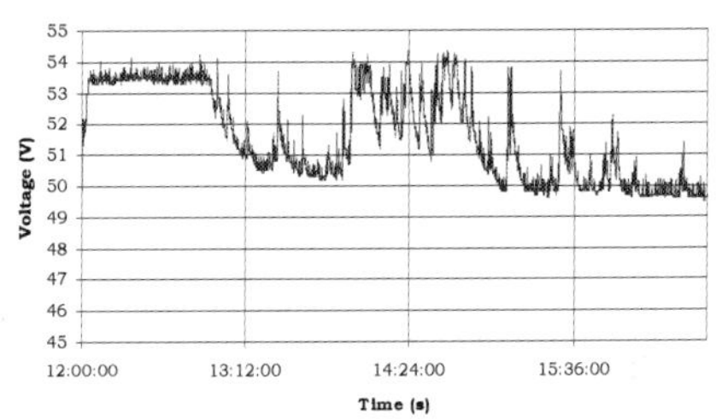

f) DC bus voltage

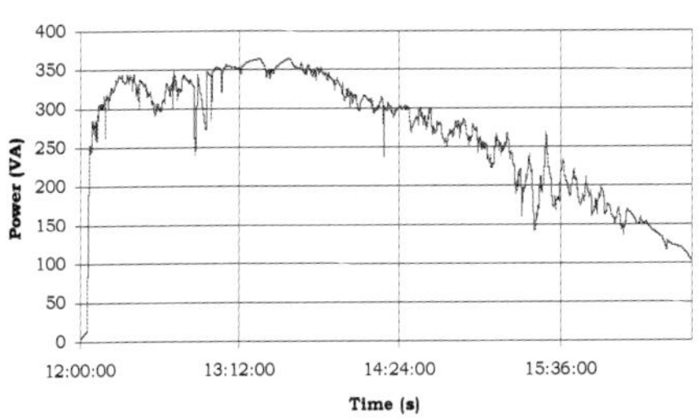

g) Photovoltaic power

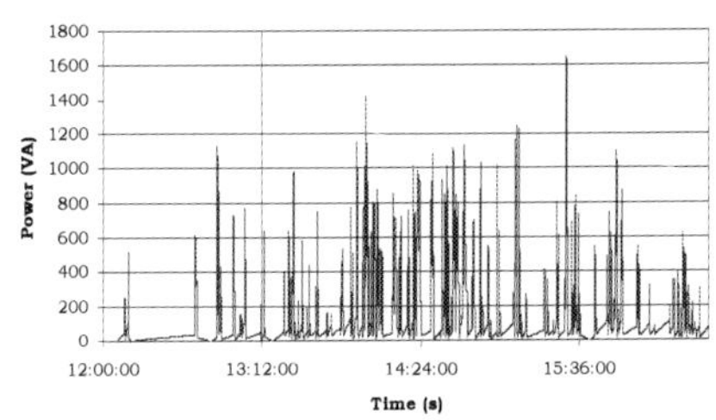

h) Wind power

Fig. 24. Grid-Connected RESHS operation in Silent mode

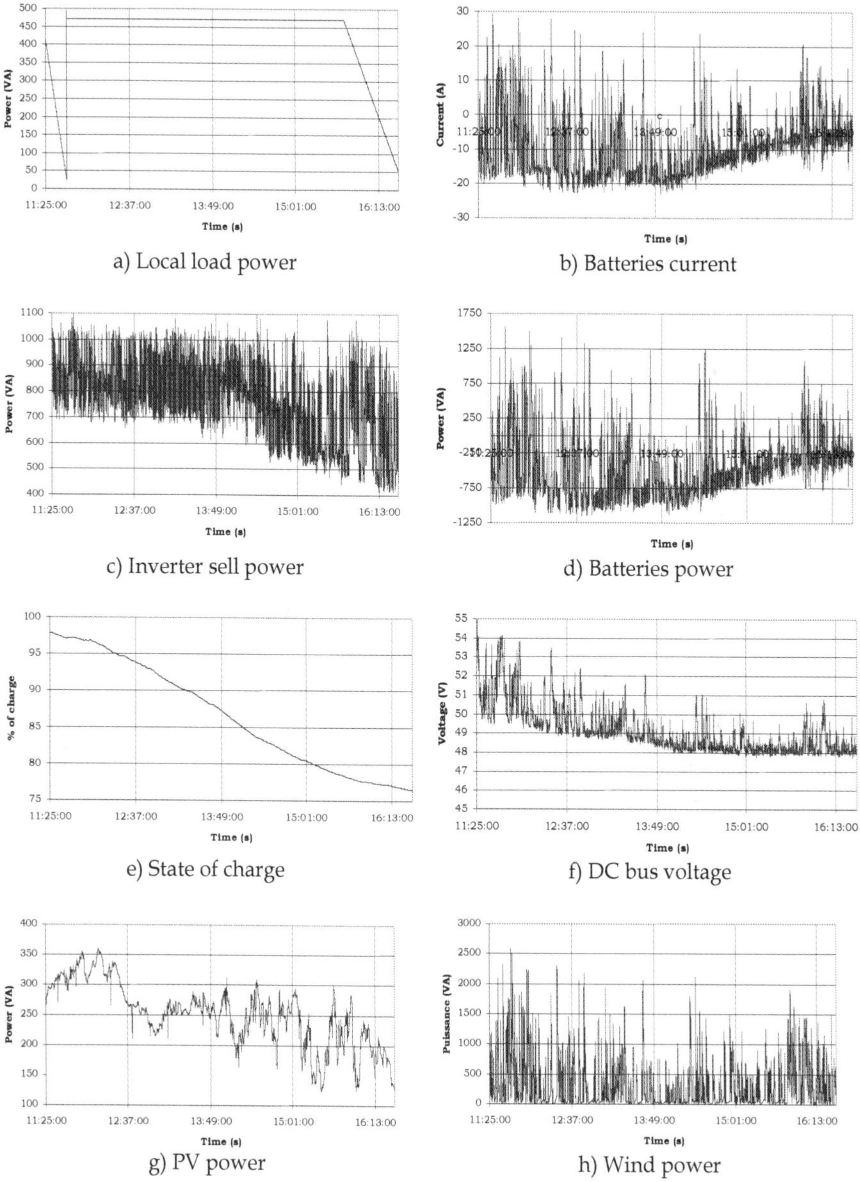

a) Local load power

b) Batteries current

c) Inverter sell power

d) Batteries power

e) State of charge

f) DC bus voltage

g) PV power

h) Wind power

Fig. 25. Grid-Connected RESHS operation in Sell mode

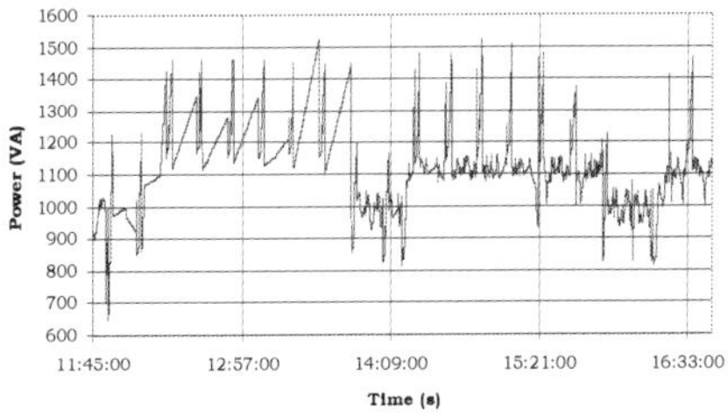

a) Local load power

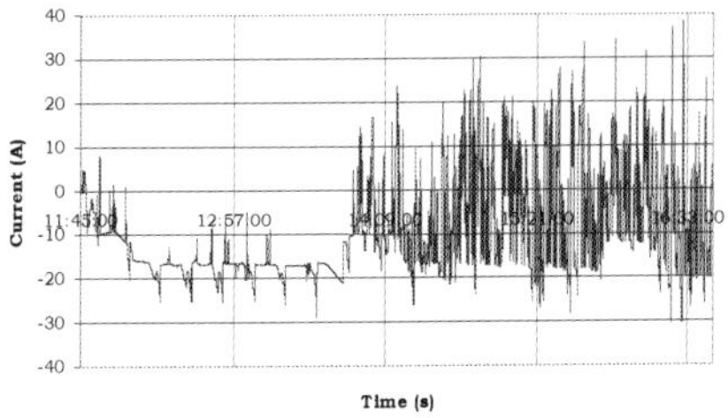

b) DC bus current

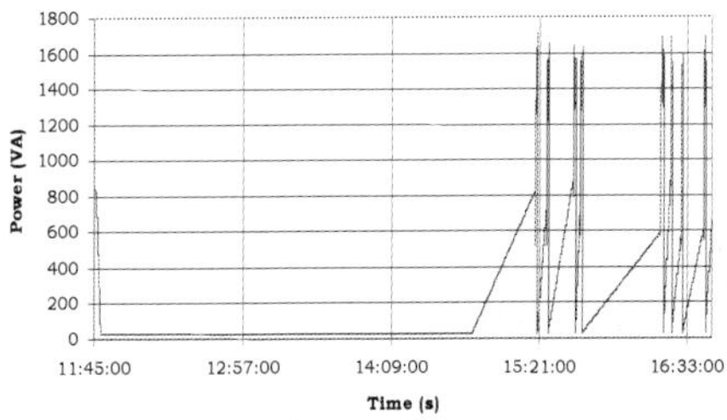

c) Utility grid power

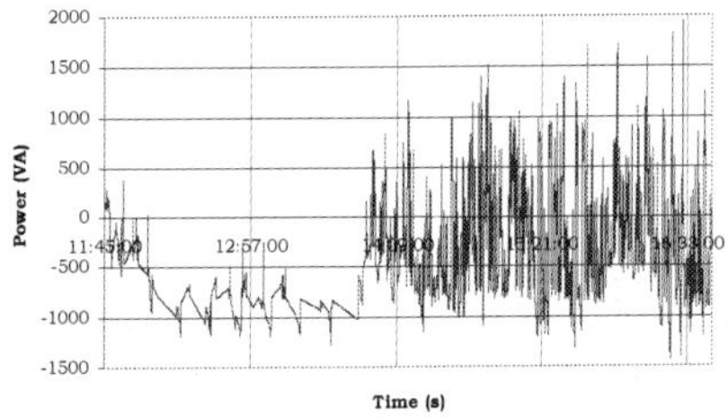

d) Batteries power

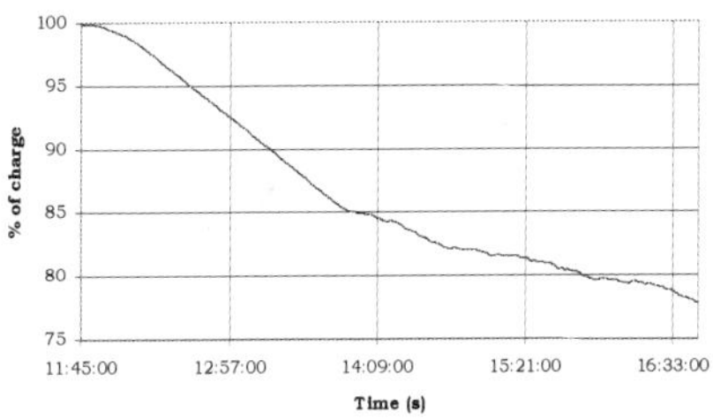

e) State of charge

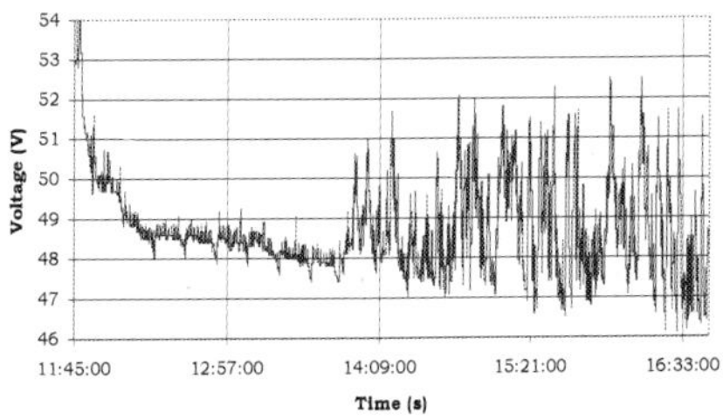

f) DC bus voltage

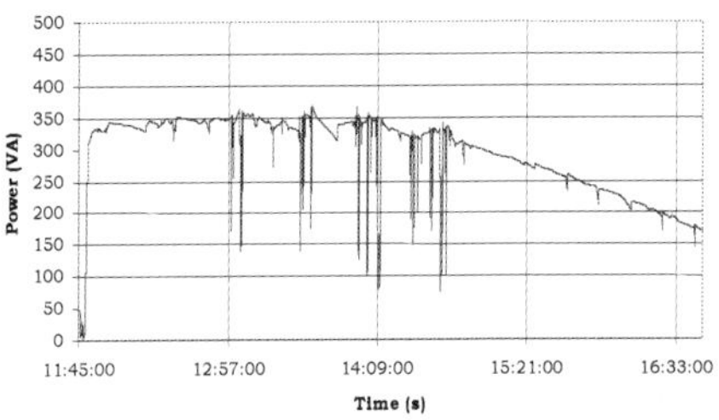

g) PV power

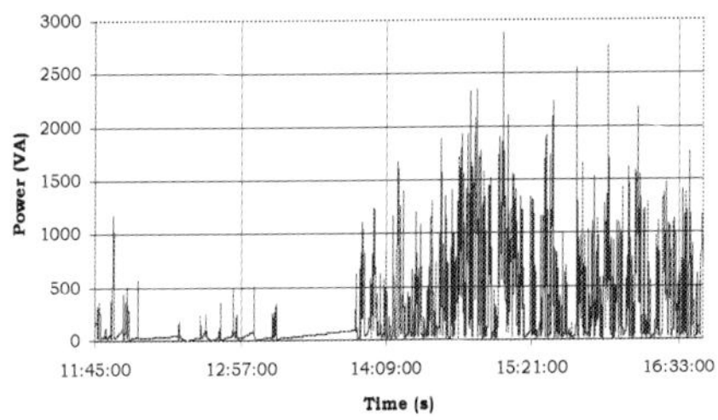

h) Wind power

Fig. 26. Grid-Connected RESHS operation in Low battery transfer mode

5. Conclusion

This paper presents the modelling, simulation and experimental study of a renewable energy system with hydrogen storage (RESHS). Solar photovoltaic array, wind generator, electrolyzer, fuel cell and batteries modelling principles are presented. Matlab/Simulink and Labview programming tools were used for modelling and simulation purpose. The models of the main components have been developed and integrated into the whole system in order to study its general performance. The RESHS was investigated experimentally in stand-alone and grid-connected operation modes. For different grid-connected operating modes (Float, Silent, Sell, Low battery transfer), the system's operating conditions were monitored and power transfer was analyzed. The renewable energy system has shown satisfactory performances for operation with short term (battery) and long-term (hydrogen) energy storage. However, a control strategy based on the state of charge (SOC) of the batteries needs to be developed to improve the performances of the RESHS in grid-connected operation mode.

6. Acknowledgement

This work has been supported by the LTE Hydro-Québec and the Natural Sciences and Engineering Research Council of Canada.

7. References

Agbossou K., Kolhe M., Hamelin J., and Bose T. K. (2004). Performance of a Stand-Alone Renewable Energy System Based on Energy Storage as Hydrogen, *IEEE Transactions on Energy Conversion,* Vol. 19, No. 3, Page(s): 633-640, September 2004.

Burton T., Sharpe D., Jenkins N. and Bossanyi E. (2001). *Wind Energy Handbook,* John Wiley & Sons, LT, 2001.

Cardenas R., R. Pena (2004). Sensorless Vector Control of Induction Machines for Variable-Speed Wind Energy Applications, *IEEE Trans. On Energy Conversion,* Vol. 19, No. 1, Page(s): 196 - 205, March 2004.

Chérif A., Jraidi M. and Dhouib A (2002). A battery ageing model used in stand alone PV systems, *Journal of Power Sources,* vol.112, Issue 1, Page(s): 49-53, 2002.

Doumbia, M.L., Agbossou, A., Granger, E. (2007). Simulink Modelling and Simulation of a Hydrogen Based Photovoltaic/Wind Energy System, *Eurocon* The *International Conference on "Computer as Tools"*, Page(s): 2067 - 2072, Warsaw, September 2007.

Doumbia, M.L., Agbossou, A., Proulx, C-L. (2009). Labview Modelling and Simulation of a Hydrogen Based Photovoltaic/Wind Energy System, *Electromotion,* 8th International Symposium on Advanced Electromechanical Motion Systems, Lille, July 2009, France.

Gow J. A., Manning C.D. (1999). Development of a photovoltaic array model for use in power-electronics simulation studies, *IEE Proc.-Electr. Power Appl.,* Vol. 146, No. 2, March 1999.

Kélouwani S., Agbossou K., and Chahine R. (2005). Model for Energy Conversion in Renewable Energy System with Hydrogen Storage, *Journal of Power Sources,* vol. 140, no 2, p. 392-399, 2005.

Lei Y., Mullane A., Lightbody G., and Yacamini R. (2006). Modelling of the Wind Turbine with a Doubly Fed Induction Generator for Grid Integration Studies, *IEEE Trans. On Energy Conversion*, Vol. 21, No. 1, Page(s): 257 - 264, March 2006.

Kim S-L., Jeon J-H., Cho C-H., Ahn J-B. and Kwon S-H. , (2008). Dynamic Modeling and Control of a Grid-Connected Hybrid Generation System With Versatile Power Transfer. *IEEE Trans. On Industrial Electronics*, Vol. 55 No. 4, April 2008.

Slootweg J. G., De Haan S., Polinder H., and Kling W. (2003). General Model for Representing Variable Speed Wind Turbines in Power System Dynamics Simulations. *IEEE Trans. On Power Systems*, Vol. 18, No. 1, February 2003.

Trace Engineering (Xantrex), *SW Series Inverter/Chargers Owner's Manual*, 1999.

Vosen S.R., Keller J.O. (1999). Hybrid energy storage systems for stand-alone electric power tems: optimization of system performance and cost through control strategies. *International Journal of Hydrogen Energy*, Vol. 24, Page(s): 1139-1156, 1999.

Walker, G. (2000). Evaluating MPPT Converter Topologies Using a MatLab PV Model, *Journal of Electrical & Electronics Engineering*, vol. 21, no 1, pp. 49 - 56, Australia, 2000.

15

Multilevel Converters in Renewable Energy Systems

Alireza Nami and Firuz Zare
Queensland University of Technology, School of Engineering Systems
Australia

1. Introduction

In the current global climate, demand for a renewable energy system has increased due to environmental issues and limited fossil resources. Along with this demand, Photovoltaic (PV) and Wind Turbine (WT) systems have become the most common type of the grid connected renewable energy systems (Carrasco, 2006). However, to connect these systems to the grid, output voltage and frequency adjustment are the challenging issues. Various types of converters have been utilized to provide grid connected renewable energy systems. In PV or Fuel Cell (FC) applications, DC-DC converters are required to adjust the variable and low quality output voltage of the PV panels or fuel cells. A DC-AC converter is employed to generate desired voltage and frequency for the grid connection. As well, an AC-DC-AC converter is necessary for the WT systems as wind energy is variable during the system operation.

In response to the growing demand for medium and high power applications, multilevel inverters have been attracting growing consideration in variable speed WT and PV systems recently (Tolbert & Peng, 2000; Clasis & Agelidis, 1998). Multilevel converters enable the output voltage to be increased without increasing the voltage rating of switching components, so that they offer the direct connection of renewable energy systems to the grid voltage without using the expensive, bulky, and heavy transformers. In addition, multilevel inverters synthesis stair case output voltage which is closer to sinusoidal voltage using DC link voltages compared with two-level inverter. Synthesising a stepped output voltage allows reduction in harmonic content of voltage and current waveforms and eventually size of the output filter. Among different types of the multilevel converters, cascade converters is usually used in PV applications due to its modularity and structure (Alonso, 2003). However, the number of switches is more than the other types of multilevel converters and needs several separated DC sources. The diode-clamped converter is another type of multilevel converters which is widely used in transformerless grid connected systems due to its minimum number of active power components and shared DC link voltage (Busquets-Monge, 2008). Due to the structure of the diode-clamped converter, it suffers from neutral point voltage balancing. Although a solution for capacitor voltage balancing has been addressed in literature (Yazdani, 2005), this increases the complexity of the control strategy. Also, existing methods are not applicable for all numbers of levels and modulation indexes. Furthermore, the use of auxiliary devices or active rectifiers has been proposed to balance the DC link capacitors but these can increase the cost and complexity of the system (Marchesoni, 2002).

In photovoltaic, fuel cells and storage batteries, the low output DC voltage should be boosted. Therefore, a step-up converter is necessary to boost the low DC voltage for the DC link voltage of the inverter. The main contribution of this chapter is to electrical energy conversion in renewable energy systems based on multilevel inverters. Different configuration of renewable energy systems based on power converters will be discussed in detail. Finally, a new single inductor Multi-Output Boost (MOB) converter is proposed, which is compatible with the diode-clamped configuration. Steady state and dynamic analyses have been carried out in order to show the validity of the proposed topology. Then the joint circuit of the proposed DC-DC converter with a three-level diode-clamped converter is presented in order to have a series regulated voltage at the DC link voltage of the diode-clamped inverter. MOB converter can boost the low input DC voltage of the renewable energy sources and at the same time adjust the voltage across each capacitor to the desired voltage levels, thereby solving the main problem associated with capacitor voltage imbalance in this type of multilevel converter.

2. Power Conversion Systems

Power electronic converters are a family of electrical circuits which convert electrical energy from one level of voltage, current, or frequency to another using power switching components (Zare, 2008). In all power converter families, energy conversion is a function of different switching states. The process of switching the power devices in power converter topologies from one state to another is called modulation. Regarding different applications, various families of power converters with optimum modulation technique should be used to deliver the desired electrical energy to the load with maximum efficiency and minimum cost. Three main families of power converters which are usually used in renewable energy systems are:

- AC-DC converters
- DC-DC converters
- DC-AC converters

Fig. 1 shows a scheme of electrical conversion according to the different family of power converters used in renewable energy systems.

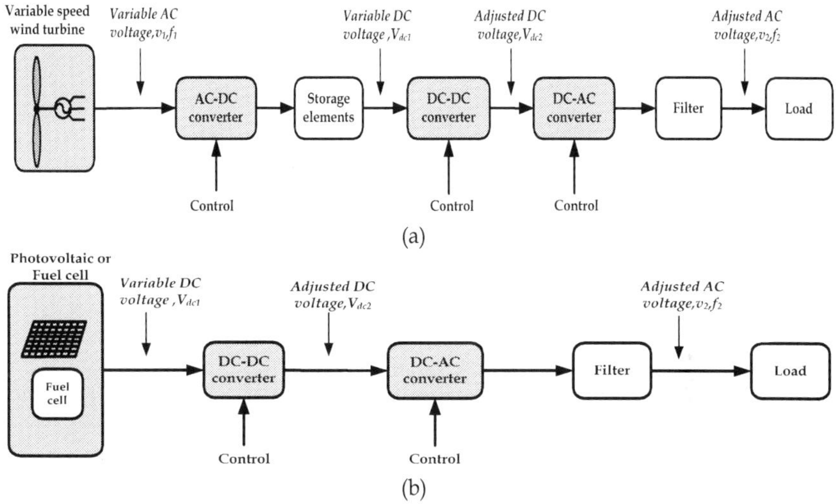

Fig. 1. Families of power converters categorized according to their energy conversion in renewable energy systems

In renewable energy systems, sources can be either AC or DC such as WT or PV systems, respectively. However, due to the load requirement, the power may be changed to DC or AC. Therefore, based on different applications, proper combination and control of above power converters can supply a load. As shown in Fig.1 (a), in residential applications or grid connected systems where the variable voltage of renewable energy systems should be converted to achieve desirable AC voltage and frequency, AC-DC, DC-DC and DC-AC converters may be needed. On the other hand, when the input voltage is variable DC source (Fig.1(b)) such as PV or FC systems, DC-DC converter combined with DC-AC converter may be used to have a regulated AC waveform for residential or grid connected systems.

2.1 AC-DC Converter

A three-phase low frequency rectifier is depicted in Fig.2 (a). This converter operates at line frequency, so that the switching happens at 50Hz or 60Hz. Low frequency rectifiers are made up by diode or thyristor to change the AC voltage to DC. However, rectifiers based on thyristors have a freedom to change the firing angle to switch the thyristor, so that the amount of output DC voltage is controlled compared with diode rectifier. Using a capacitor at output can increase the quality of output voltage.

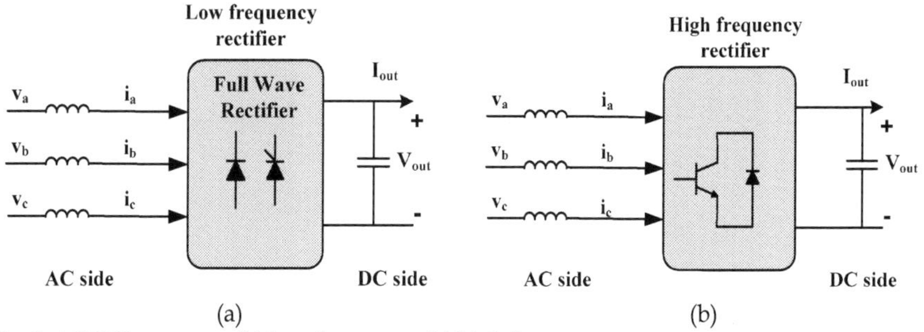

Fig. 2. AC-DC converter (a) low frequency (b) high frequency

A three-phase controlled high frequency converter based on IGBT is shown in Fig.2 (b). This circuit can change the input AC voltage into DC voltage. By controlling the duty cycle of the switches based on different modulation technique, the amount of output DC voltage can be controlled. Since the switches drive in high frequency, the amount of harmonic content in current waveform is decreased. In addition, this configuration can provide bidirectional power flow between load and source.

2.2 DC-DC Converter

DC-DC converters are a kind of high frequency converters which convert unregulated DC power to regulated DC power. Since the output voltage of renewable energy systems or rectifier converter is basically unregulated DC voltage, as shown in Fig. 3, DC-DC converters are necessary to adjust the DC voltage for different applications. Three basic configurations of DC-DC converters are buck, boost and buck-boost converters.

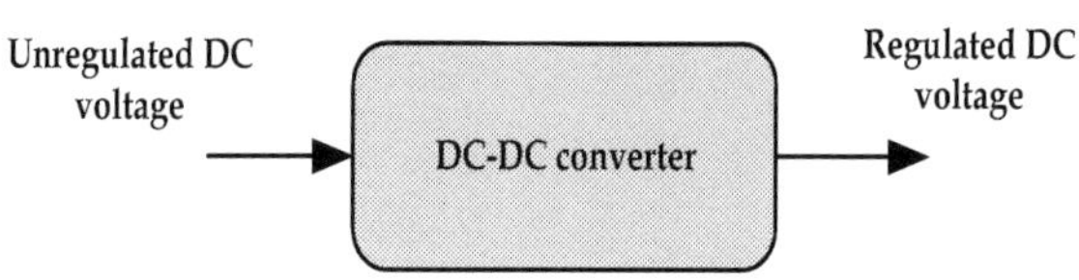

Fig. 3. DC-DC converter

In a buck converter the output voltage is normally less than input voltage. However, a boost converter has the ability to increase the input voltage based on duty cycle of the switch. A buck-boost converter can either buck or boost the input voltage. A boost converter is usually applied in renewable energy systems as the output voltage of these systems is low and unregulated. Configuration of the boost converter is illustrated in Fig. 4. In this converter, output voltage is a function of the duty cycle of switch (S) which can be defined by a proper modulation technique. When the switch is on, the inductor can be charged by the current flowing through it. However, in the next subinterval when the switch is turned off, the capacitor will be charged by the inductor current. Second order LC filter in this configuration can regulate the output voltage and remove the high frequency harmonics.

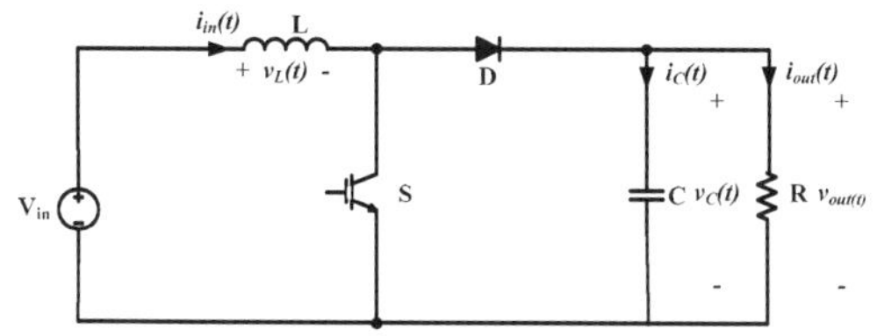

Fig. 4. Schematic configuration of a DC-DC boost converter

2.3 DC-AC Converter

A block diagram of a voltage source converter is shown in Fig. 5. In this configuration the input source is a voltage which is stored in DC link capacitor. This converter chops the input DC voltage and generates an AC voltage with desired magnitude and frequency with respect to the pulse patterns and modulation techniques. Different current and voltage control methods have been proposed to generate a high voltage high current rectangular waveform based on reference voltage characteristics (Nami & Zare, 2008).

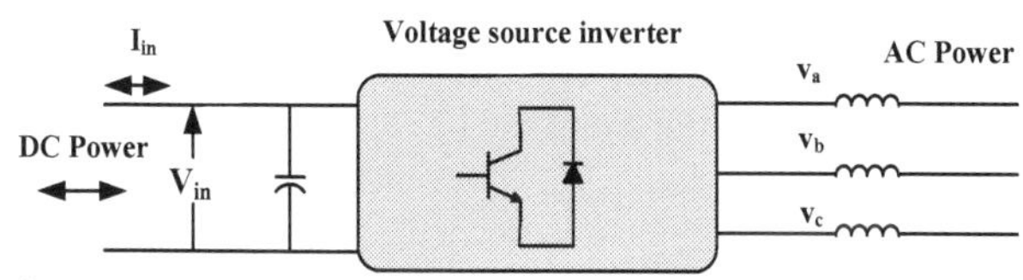

Fig. 5. DC-AC voltage source converter

One leg of a classical voltage source inverter is shown in Fig. 6 (a). Based on this configuration, when S_1 turns on, leg voltage (V_{an}) is V_{dc} and when S_1 turns off, V_{an}=0. Therefore, two different voltage levels appear at the leg voltage. Single-phase and three-phase structure can be constituted by the connection of one and two legs to this structure, respectively. Fig. 6 (b) shows a single-phase inverter. Output voltage levels based on different switching states are given in Table. 1. As shown, three voltage levels appear in output voltage. Different high and low frequency PWM techniques can be applied to generate the reference voltage. Leg voltages and output voltage waveforms of a single-phase inverter are shown in Fig. 7.

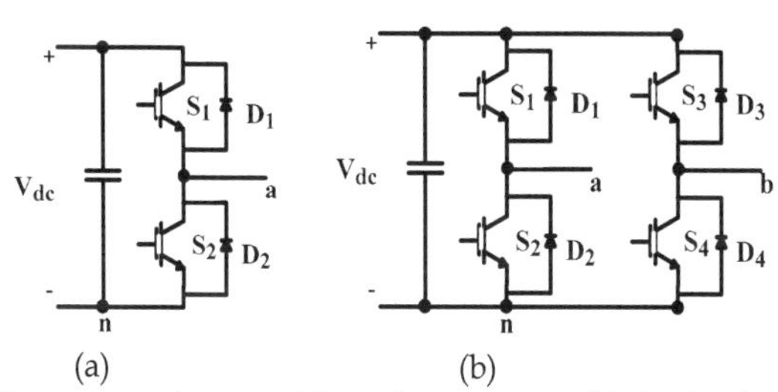

Fig. 6. Classical voltage source inverter (a) one leg structure (b) single-phase inverter

S_1	S_3	V_{an}	V_{bn}	V_{ab}
off	off	0	0	0
off	on	0	V_{dc}	$-V_{dc}$
on	off	V_{dc}	0	V_{dc}
on	on	V_{dc}	V_{dc}	0

Table 1. Switching states of classical voltage source inverter

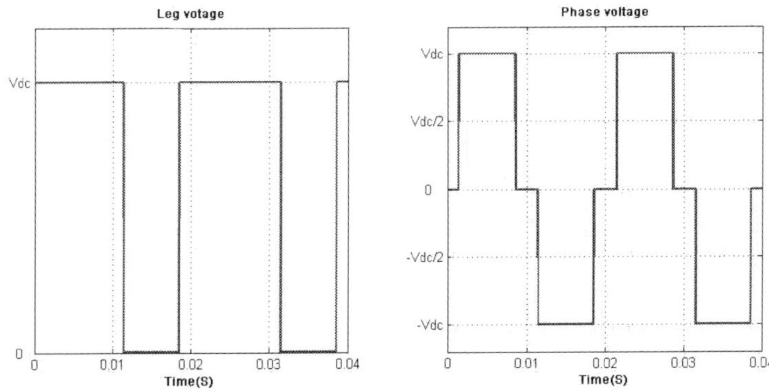

Fig. 7. Leg voltages and output voltage of single-phase inverter

3. Multilevel Topologies

Multilevel converters have many advantages for medium and high power systems as they synthesise a higher output voltage than the voltage rating of each switching device. Stepped output voltage allows a reduction in harmonic content of voltage and current waveforms, switching frequency, and semiconductors voltage level. The presented merits make multilevel converters appropriate for medium and high voltage renewable energy applications. The best known multilevel topologies (Rodriguez et al., 2002) are diode-clamped, flying capacitor, and cascade inverter, the latter being normally considered for PV application. However, the diode-clamped converter is the most popular converter in renewable energy systems due to its structure. Different current (Zare & Ledwich, 2002; Zare & Ledwich, 2008) and voltage control (Leon et al., 2008) techniques have been proposed for multilevel converters to have an optimum efficiency. Although each type of multilevel converters share the advantages of multilevel voltage source inverters, they may be suitable for specific application due to their structures and drawbacks. Operation and structure of the three main types of multilevel converters are discussed in the following sections.

3.1 Diode-clamped Inverter

Concept of the diode clamped topology was proposed by Nabae (Nabae et al., 1981). This topology has found wide acceptance for its capability of high voltage, and high efficiency operation. A phase leg of a three-level diode-clamped inverter is shown in Fig. 8 (a). It consists of two pairs of switches and two diodes. Each switch pairs works in complimentary mode and the diodes are used to provide access to mid-point voltage. The DC bus voltage is split into three voltage levels by using two series connection of DC capacitors, C_1 and C_2. Each capacitor is supposed to have an equal DC voltage and each voltage stress will be limited to one capacitor level through clamping diodes (D_{c1} and D_{c2}). If assumed that total DC link voltage is V_{dc} and mid point is regulated at half of the DC link voltage, the voltage across each capacitor is $V_{dc}/2$ ($V_{c1}=V_{c2}=V_{dc}/2$). Based on the structure of the diode-clamped converter, there are three different possible switching states which apply the staircase voltage on output voltage relating to DC link capacitor voltage rate. Switching states of the three-level converter are summarized in Table 2. To study the effect of the number of output voltage levels in diode-clamped topology, Fig. 8 (b) shows a phase leg of a four-level inverter. If converter works in balance condition, DC link voltage is split in three equal values by the series capacitors ($V_{c1}=V_{c2}=V_{c3}=V_{dc}/3$). There are four different switching combinations shown in Table 3 which can generate four different voltage levels in output leg voltage (V_{an}).

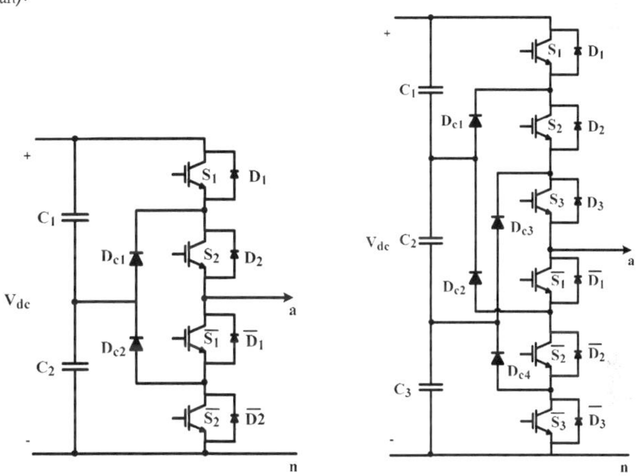

Fig. 8. One leg of diode-clamped converter (a) three-level (b) four-level

According to Fig.6, by connecting another leg named as leg (b) to the above configurations, five and seven voltage levels can be achieved in the output voltage (V_{ab}) of three and four level inverters. Fig. 9 and Fig. 10 show the leg voltage and phase voltage of the three and four level inverters in the balance condition, respectively.

S_1	S_2	Leg voltage (V_{an})
on	on	V_{dc}
off	on	$V_{dc}/2$
off	off	0

Table 2. Switching states in one leg of the three-level diode-clamped inverter

S_1	S_2	S_3	Leg voltage (V_{an})
on	on	on	V_{dc}
off	on	on	$2V_{dc}/3$
off	off	off	$V_{dc}/3$
off	off	off	0

Table 3. Switching states in one leg of the four-level diode-clamped inverter

Improvement in quality of the output voltage is obvious by increasing the number of voltage levels as the voltage waveform becomes closer to sinusoidal waveform (Nami et al., 2008). However, capacitor voltage balancing will be the critical issue in high level converters due to the existence of DC currents in the middle points of the DC link. Thus, capacitor are either charged or discharged for some intervals which limit the practical operation of the high-level diode-clamped converters in some conditions. This issue should be taken into account in this inverter. MOB joint with diode-clamped inverter in the following sections address this problem for renewable energy systems. In general in a converter with n series DC link capacitors, $m= (n+1)$ leg voltage levels and $l=2\times m-1$ phase voltage levels can be achieved in output waveforms.

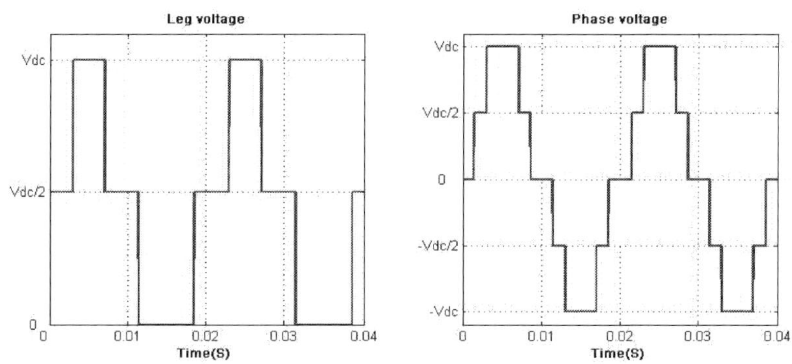

Fig. 9. Output voltage in three-level diode-clamped inverter (a) leg voltage (b) output voltage

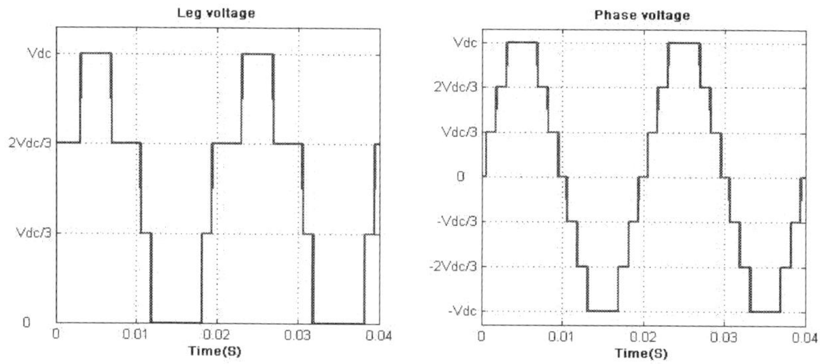

Fig. 10. Output voltage in four-level diode-clamped inverter (a) leg voltage (b) output voltage

3.2 Flying Capacitor Inverter

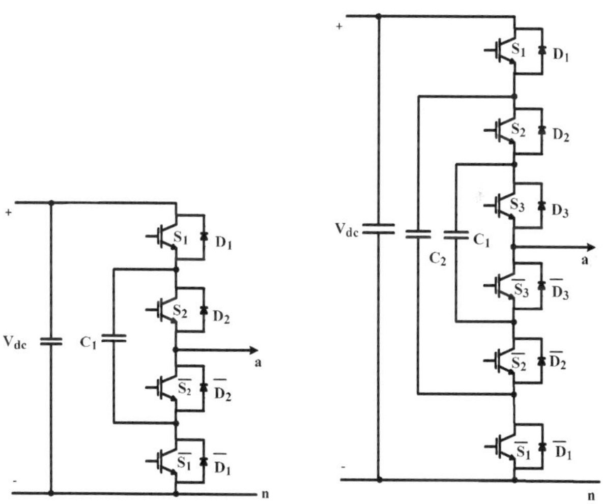

(a) (b)

Fig. 11. A leg of flying capacitor converter (a) three-level (b) four-level

Fig. 11 shows the leg structure of a flying capacitor inverter. This configuration is an alternative to the diode-clamped converter, however, voltage across an open switch is constrained by clamping capacitors instead of clamping diode in diode-clamped topology (Meynard & Foch, 1992). Therefore, it can avoid the use of multiple diodes at the higher voltage levels. Although this type of converter shares the advantages of all multilevel inverters, it faces with some problems. One of the main problems is the requirement of complicated control strategy due to regulation of floating capacitor voltages (Zare & Ledwich, 2002). Another problem is associated with converter initialization that means before the flying capacitor inverter can be modulated, the clamping capacitors must be set up with the required voltage level. By increasing the number of levels, more capacitors are needed (Bum-Seok et al. 1998). If the input DC link is V_{dc} and the flying capacitor works in balance condition mode, in order to have an equal step voltages at output voltage, clamped capacitor should regulated at $V_{c1}=V_{dc}/2$ in the three-level inverter and $V_{c2}= 2V_{c1}=2V_{dc}/3$ in the four-level inverter. It is note that along with increasing the output voltage quality in four-level structure, voltage stress on switching components reduces by $V_{dc}/6$.

Different leg voltage levels associated with different switching states in three and four level flying capacitor inverters are given in Table 4 and 5, respectively. It is clear that one more voltage level is available in four-level inverter. Although the output voltage levels in the flying capacitor inverter is similar to the diode-clamped converter, there are more than one switching state available to achieve the specific level which is called redundant switching states. These redundant switching vectors give freedom to balance the clamped capacitor voltages as they may provide different current loops through the capacitors.

S_1	S_2	Leg voltage (V_{an})
off	off	0
off	on	$V_{dc}/2$
on	off	$V_{dc}/2$
on	off	V_{dc}

Table 4. Switching states of three-level flying capacitor inverter

S_1	S_2	S_3	Leg voltage(V_{an})
off	off	off	0
off	off	on	$V_{dc}/3$
off	on	off	$V_{dc}/3$
off	on	on	$2V_{dc}/3$
on	off	off	$V_{dc}/3$
on	off	on	$2V_{dc}/3$
on	on	off	$2V_{dc}/3$
on	on	on	V_{dc}

Table 5. Switching states of four-level flying capacitor inverter

3.3 Cascade Inverter

The third topology for a multilevel converter is the cascade inverter, which can be synthesised by a series of single-phase full-bridge inverters. Assuming that the DC voltage of each full-bridge cell is same and equal to $V_{dc}/2$, each full-bridge inverter can switch between $-V_{dc}/2$,0, and $V_{dc}/2$. Therefore, by adding and subtracting of the output voltage levels of the two cascaded full-bridge inverter cells, five different voltage levels can be achieved in the output voltage in Fig. 12 (a). Switching states associated with different output voltage levels for the cascade inverter with two full-bridge inverter cells are demonstrated in Table 6. Three-phase configuration can be easily implemented by three single-phase structures. Cascade configuration has been attracted for medium and high voltage renewable energy systems such as photovoltaic, due to its modular and simple structure (Alonso et al., 2003). A higher level can easily be implemented by adding classical H-bridge cells in this configuration. However, it needs additional DC voltage sources and switching devices which can increase the cost of the system.

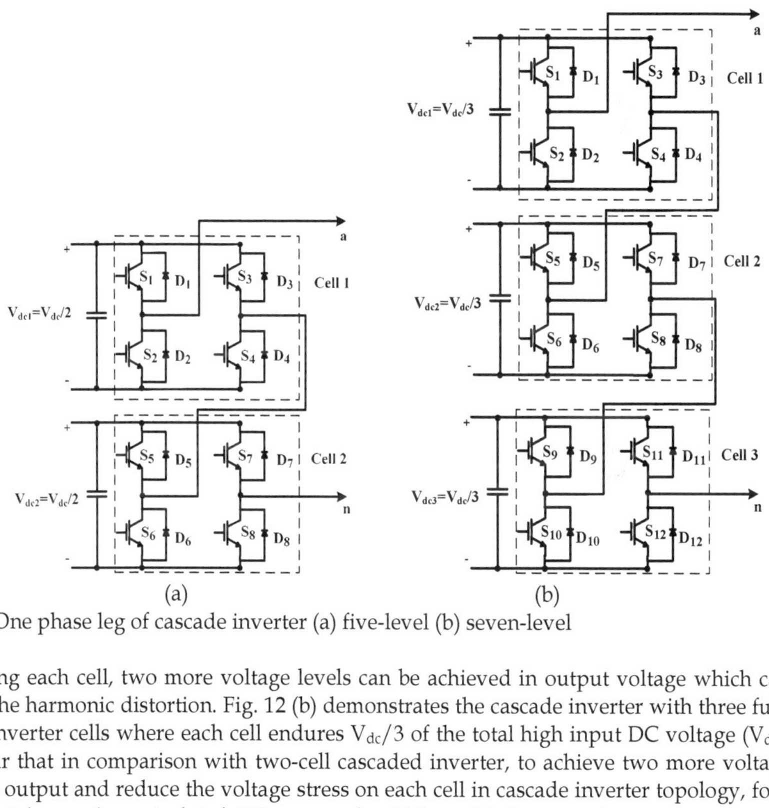

Fig. 12. One phase leg of cascade inverter (a) five-level (b) seven-level

By adding each cell, two more voltage levels can be achieved in output voltage which can reduce the harmonic distortion. Fig. 12 (b) demonstrates the cascade inverter with three full-bridge inverter cells where each cell endures $V_{dc}/3$ of the total high input DC voltage (V_{dc}). It is clear that in comparison with two-cell cascaded inverter, to achieve two more voltage levels at output and reduce the voltage stress on each cell in cascade inverter topology, four extra switches and one isolated DC source should be paid. In general, cascade converter with n-full-bridge inverter cells can synthesis $l=(2n+1)$ voltage levels at the output voltage of each phase structure.

S_1	S_3	S_5	S_7	Leg voltage (V_{an})
off	off	off	off	0
off	off	off	on	$-V_{dc}/2$
off	off	on	off	$V_{dc}/2$
off	off	on	on	0
off	on	off	off	$-V_{dc}/2$
off	on	off	on	$-V_{dc}$
off	on	on	off	0
off	on	on	on	$-V_{dc}/2$
on	off	off	off	$V_{dc}/2$
on	off	off	on	0
on	off	on	off	V_{dc}
on	off	on	on	$V_{dc}/2$
on	on	off	off	0
on	on	off	on	$-V_{dc}/2$
on	on	on	off	$V_{dc}/2$
on	on	on	on	0

Table 6. Switching states of five-level cascade inverter

The hybrid converter proposed by Manjrekar (Manjrekar et al., 2000) is a cascaded structure that has been modified, such that the DC link of full-bridge inverters has unequal DC source voltages. Therefore, based on different switching states it is possible to achieve more voltage levels in output voltage by adding and subtracting DC link voltages compared with conventional multilevel inverters with the same number of components. Diverse topologies have been studied based on a variety of H-bridge cascaded cells and DC voltage ratio to enhance the output voltage resolution compared with the same DC voltage ratio of the cells.

4. Application of Power Electronics in Renewable Energy Systems

Nowadays, the electrical power generation from renewable energy sources has become a focal point in research because of environmental problems and a perceived of traditional energy sources in the near future. Since last decade, researchers have been working on electrical systems for variable speed wind turbines. The main advantages of variable speed are noise reduction, maximum power tracking, and proper controlled torque and in this manner, the possibility to damp resonance and avoid speeds causing resonance. Several electrical systems have been presented to connect the wind turbine with variable speed and frequency to the constant voltage and frequency of the network. The main aspects of these topologies are increased efficiency and robustness, a decrease in the size and maintenance of the system and eventually reduction of whole system expense.

On the other hand, grid connected photovoltaic (PV) systems, mostly single-phase PV systems and their contribution to clean power generation, is recognized more and more worldwide. The main advantages of PV system are long life time, high efficiency and good environmental condition. The most important issues for grid connected PV to gain wide acceptance are reliability and low cost. There are two approaches to achieve high voltage and high efficiency, one is to connect the cells in series to generate high voltage DC and use high voltage DC to an AC inverter circuit. However, this configuration needs high voltage rate devices for the inverter. Another approach is to use low voltage devices for the inverter and then step up the voltage using transformers. This can increase losses and cost of system. Using transformerless concepts are advantageous with regard to their high efficiency and the resulting benefits of reduction in cost, size, weight and complexity of the inverter.

Another renewable energy source is fuel cells which are considered attractive for Distributed Generation (DG) applications. Fuel cells are electrochemical devices that convert the chemical energy of fuel and oxidant directly to electrical energy and heat. In fuel cell powered applications, a fuel cell (low power) will supply the system, then a DC-DC converter is used to boost the low voltage of the fuel cell to make a high voltage DC link. A DC-AC inverter is used to obtain AC voltage to feed the load.

In some high voltage, high power wind turbine or photovoltaic systems, if protection is not a big issue, multilevel inverters are a suitable configuration in transformerless grid connected systems (Lopez et al., 2006). Multilevel converters synthesise a higher output voltage than the voltage rating of each switching device so that they can provide a direct connection of renewable energy systems to the grid. Even in grid connected systems with a transformer, multilevel inverters are suitable topologies due to low Total Harmonic Distortion (THD) and low voltage stress (dv/dt) which minimise EMI, and also switching losses compared to traditional converters. In addition, the above advantages will lead to reduction of the cost and size of the output filter in the systems based on multilevel converters.

4.1 Variable Speed Wind Turbine Systems

This topology basically employs induction or synchronous machine as a generator. The power electronics converter in this topology might be created by either a diode rectifier with boost chopper converter connected to the PWM inverter or two bidirectional PWM-VSI connected back-to-back (Carrasco et al., 2006). Extracting as much power as possible from wind energy and feeding high quality electricity to the grid are the two main objectives of these systems. As shown in Fig. 13, power converters comprise rectifier in generator side and inverter in grid side which are connected together through DC link. This scheme allows, on one hand control of the active and reactive powers of the generator, and on the other hand, a reduction of the harmonics of current waveform by the power converter. In order to benefit from advantages of multilevel inverters in such a system, generator and grid side inverters can be utilized based on the diode-clamped converter as it has shared DC link in its structure (Bueno et al., 2008).

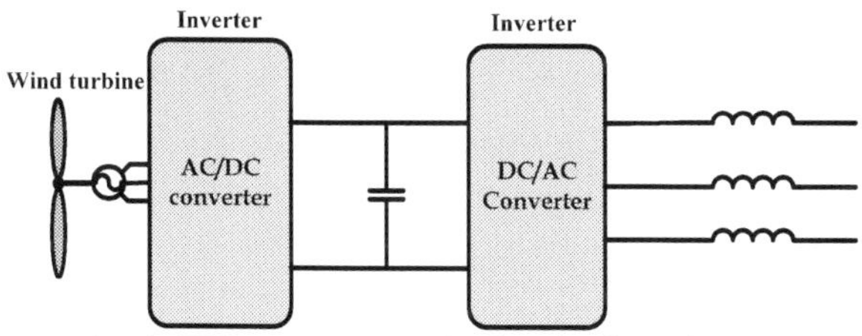

Fig. 13. Power conversion in WT systems using back-to-back configuration

An alternative to the power conversion system of a wind turbine is to use synchronous or permanent magnet generator instead of the induction machine as shown in Fig. 14. The power converter in generator side is replaced by an AC-DC rectifier with step-up DC-DC converter. This is a low cost configuration when compared with back-to-back topology. As the wind energy is variable, the step-up converter is responsible to adapt the rectifier voltage to the DC link voltage of the inverter. Also, this structure may provide transformerless connection systems due to the DC level voltage regulation using boost converter. Using multilevel converters for medium and high voltage applications is advantageous based on this structure as they can increase the voltage without increasing the switching components voltage rate (Alepuz et al., 2006).

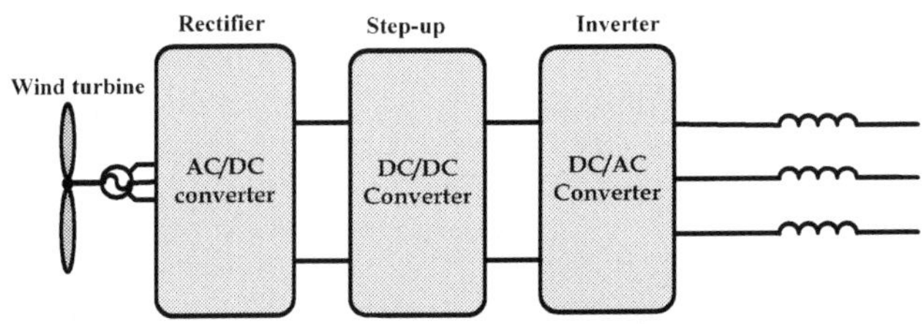

Fig. 14. Power conversion in WT systems using rectifier and step-up converter

4.2 Photovoltaic and Fuel Cell Systems

Photovoltaic and fuel cell systems are mostly attracted in single-phase residential applications with or without grid connection systems. The main advantages of these systems are long life time, high efficiency and good environmental condition. As output voltage of the PV and FC are low DC voltage, one approach for power conversion in this type of system is to use a low voltage inverter and then increase the AC voltage using transformers. However, it can increase losses and also the cost of the system. Alternative power connection topology demonstrated in Fig. 15, utilizes the DC-DC boost converter to generate high DC voltage for inverter DC link. Therefore, high voltage DC to AC inverter is necessary which can impose a high voltage rate for switching devices. To address this problem, multilevel converters such as diode-clamped or flying capacitor are a good candidate for this configuration that can increase the number of voltage levels without increasing the voltage rate of power components in DC-AC inverter (Myrzik, 2003; Sharma & Hongwei, 2006). A configuration of single-phase PV system with cascade converter is shown in Fig. 16. DC-DC converters are responsible for boosting the low input voltage and a cascade converter can synthesise a high voltage AC voltage by adding inverter cells output voltage. As mentioned before, cascade converter is a suitable topology for this kind of application as it needs separated DC voltage. However, a diode-clamped structure can be utilized in this configuration if DC link voltage capacitor voltage imbalance can be solved (Nami et al., 2008).

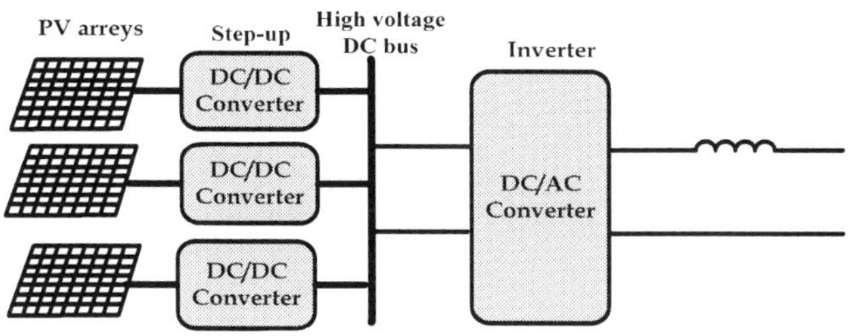

Fig. 15. Power conversion in transformerless PV systems

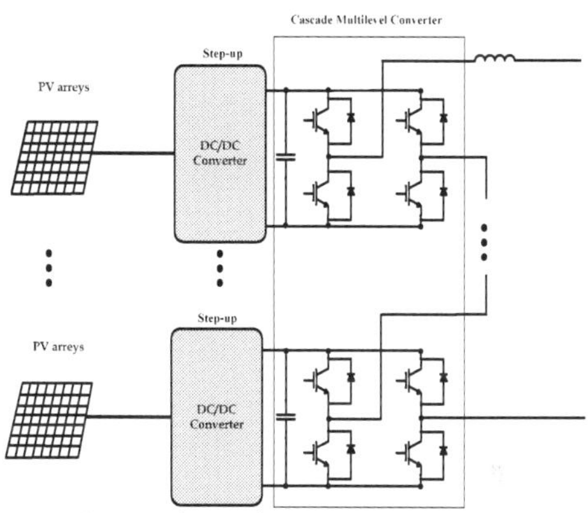

Fig. 16. Multilevel converter in transformerless PV systems

5. A New Topology for Multilevel Diode-clamped Converters in Renewable Energy Systems

DC link capacitor balancing is a challenging issue in the diode-clamped topology due to its series capacitors' connection. To address this limitation, isolated DC sources or alternatively, auxiliary converters can be used for capacitor voltage balancing. In this section, a new DC-DC boost converter with double outputs can be used as a front-end converter to boost the low output voltage of grid connected renewable systems based on the diode-clamped converter.

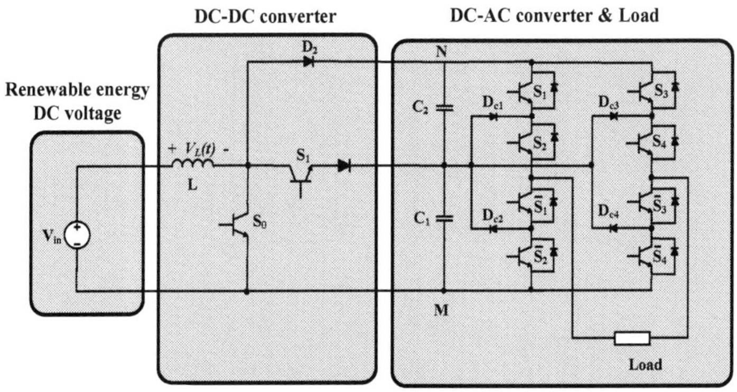

Fig. 17. Three-level diode-clamped topology joint with double-output MOB converter

Fig. 17 shows the proposed DC-DC converter connected to a three-level diode-clamped inverter. By controlling the proposed DC-DC converter, the DC link capacitors of the inverter can be regulated to the desired voltage level. Therefore, the MOB converter can address the capacitor voltage balancing in diode-clamped converters, which then will decrease the complexity of the inverter control strategy. Topology and control strategy of the MOB DC-DC converter is described in the next section.

6. Multi-output Boost Converter (MOB) Topology

Multi output DC-DC converters are efficient and economical devices which are used instead of several separate single output converters to make up a multi output power supply (Yunxiang & Jiuchao, 2004). Recently, several types of multi output DC-DC converters such as switched-capacitor, LLC resonant topology, cross regulation and parallel regulation techniques widely use in telecommunication, computers and industrial fields, are addressed in (Gu et al., 2005). A novel DC-DC converter with series capacitors in order to generate two different voltage levels for diode-clamped inverter is proposed by (Nami et al., 2007). It can also be applied for more voltage levels. This converter basically operates as a boost or buck converter based on duty cycles of the switches in each subinterval of the switching period. In this converter, by controlling the duty cycles of the switches in each subinterval, the output voltage can be controlled to provide the appropriate input DC voltage for the diode-clamped converter. This can avoid the capacitors voltage imbalance in the diode clamped topology. Moreover, by applying the presented topology in renewable energy systems, low rectified output-voltage can be boosted to a desired level.

6.1 Basic Circuit Configuration

A circuit diagram of the *N*-output boost converter is shown in Fig. 18 (a). This circuit consists of a boost switch S_0; N-1 sharing switches S_1 to S_{N-1}, N diodes (D_1 to D_N), an inductor, and N capacitors (C_1 to C_N) with different loads (R_1 to R_N).

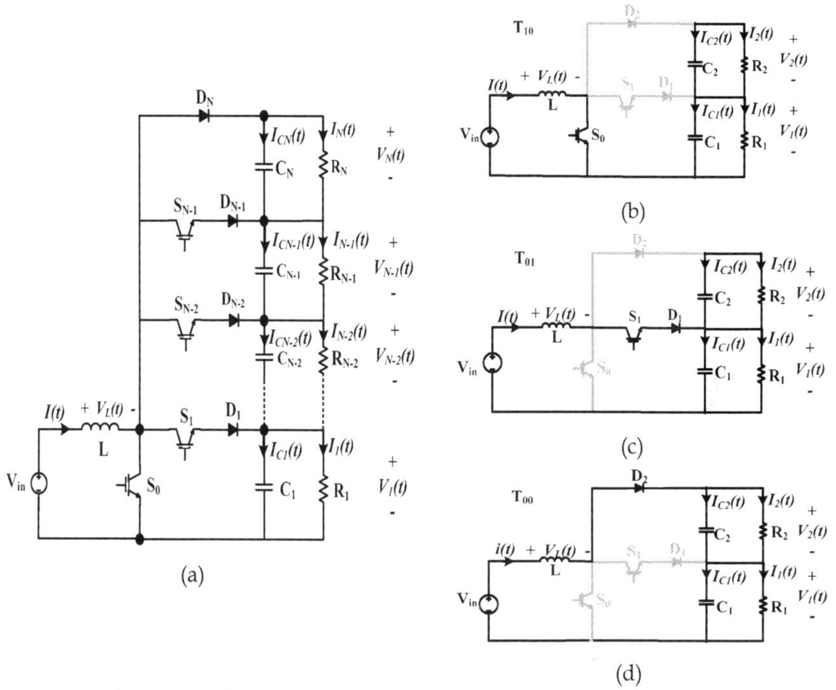

Fig. 18. Configuration of MOB converter (a) schematic; (b)-(d) equivalent circuit in different switching intervals for double-output (N=2)

Switching states	S_0	S_1	C_1	C_2
10	on	off	Discharge	Discharge
00	off	off	Charge	Charge
01	off	on	Charge	Discharge

Table 7. Switching states of double-output boost converter

In the subinterval zero, S_0 is turned "on" and the inductor can be charged by the current flowing through it. In the next N subintervals, S_0 remains "off" and the S_1 to S_{N-1} are switched to charge $N-1$ capacitors into the desired value. When S_1 to S_{N-1} are "off", the diode (D_N) directs the inductor current to charge all C_1 to C_N to generate V_1 to V_N, respectively. D_1 to D_{N-1} are used to block the negative voltage and provide two quadrant operation of S_1 to S_{N-1}. In a double-output converter there are three possible switching states as S_1 can not be turned "on" while S_0 is "on". The operation of the circuit in three different switching states has been summarised in Table 7 for N=2. The equivalent circuits of all switching states have been demonstrated in Fig. 18(b) to 18 (d).

6.2. Steady State and Dynamic Equations

To analyse the steady state performance of the MOB converter, time intervals of each switching are considered as T_{10}, T_{00}, and T_{01}. Then, switching period can be expressed as follows:

$$T_{10} + T_{00} + T_{01} = T \tag{1}$$

where, T is switching period. The average inductor voltage over one cycle should be zero in the steady state:

$$T_{10}(V_{in}) + T_{00}(V_{in} - V_1 - V_2) + T_{01}(V_{in} - V_1) = 0 \tag{2}$$

By definition of duty cycles in Eq.3 and substitution in Eq.2, it can be rewritten as follows:

$$\begin{cases} (D_0 + D_1)' = \dfrac{(T_{00})}{T} \\ D_0' = \dfrac{(T_{00} + T_{01})}{T} \end{cases} \tag{3}$$

$$V_{in} = (D_0')V_1 + (D_0 + D_1)'V_2 \tag{4}$$

where, V_{in} is input voltage and V_1 and V_2 are the bottom capacitor (mid point) and top capacitor voltages, respectively. Also, the average capacitor current over one cycle should be zero in the steady state.

$$\begin{cases} T_{10}(-\dfrac{V_1}{R_1}) + T_{00}(I - \dfrac{V_1}{R_1}) + T_{01}(I - \dfrac{V_1}{R_1}) = 0 \\ I = \dfrac{V_1}{R_1(D_0')} \end{cases} \tag{5}$$

$$\begin{cases} T_{10}(-\dfrac{V_2}{R_2}) + T_{00}(I - \dfrac{V_2}{R_2}) + T_{01}(-\dfrac{V_2}{R_2}) = 0 \\ I = \dfrac{V_2}{R_2(D_0 + D_1)'} \end{cases} \tag{6}$$

From Eq.4 to Eq.6, the steady state equation can be derived as follows:

$$\begin{cases} V_1 = \dfrac{n(D_0')V_{in}}{n(D_0')^2 + (D_0 + D_1)'^2} \\ V_2 = \dfrac{(D_0 + D_1)'V_{in}}{n(D_0')^2 + (D_0 + D_1)'^2} \\ I = \dfrac{V_{in}}{R_1(D_0')^2 + R_2(D_0 + D_1)'^2} \end{cases} \tag{7}$$

where, $n = \dfrac{R_1}{R_2}$. According to Eq.7, two different voltages can be obtained on output.

Comparing this situation with the basic single output (V_1) Boost converter, although the total output voltage (V_T) in both converters is increased, in the MOB converter different output voltages can be obtained based on different duty cycles of the boost switch (S_0) and sharing switch (S_1) in steady state operations. Since the output voltage in Eq.7 is related to the load ratio, Table 8 shows some limitations to achieve diverse voltages that should be considered due to the fundamental nature of the circuit.

$V_1=V_2$	$n=\dfrac{(D_0+D_1)'}{D_0'}$
$V_1>V_2$	$n>\dfrac{(D_0+D_1)'}{D_0'}$
$V_1<V_2$	$n<\dfrac{(D_0+D_1)'}{D_0'}$

Table 8. Nature limitation of load ratio in different output voltage

Using an averaging method in each switching cycle, state equations can be developed for the dynamic analysis of inductor current or capacitor voltages in terms of systems variables. To construct a small-signal ac model at the quiescent operation (*I, V1, V2*), we can assume a small perturbation at the operating point. Thus, all input and output variables are defined as $x(t)=X+\hat{x}(t)$ where X is a DC amount and x (t) is an AC small-signal.

Regarding linearization method, Eq.8 shows the dynamic model space state of the proposed configuration.

$$\begin{bmatrix} L & 0 & 0 \\ 0 & C_1 & 0 \\ 0 & 0 & C_2 \end{bmatrix}\begin{bmatrix} \dot{i} \\ \dot{v}_1 \\ \dot{v}_2 \end{bmatrix}=\begin{bmatrix} 0 & -(D_0') & -(D_0-D_1)' \\ (D_0') & -\dfrac{1}{R_1} & 0 \\ (D_0-D_1)' & 0 & -\dfrac{1}{R_2} \end{bmatrix}\begin{bmatrix} i \\ v_1 \\ v_2 \end{bmatrix}+\begin{bmatrix} 1 & V_1+V_2 & V_1 \\ 0 & -I & 0 \\ 0 & -I & -I \end{bmatrix}\begin{bmatrix} \hat{v}_{in} \\ \hat{d}_0 \\ \hat{d}_1 \end{bmatrix} \tag{8}$$

According to Eq.8, changing $(D_0+D_1)'$ and D_0' with a different ratio in a closed loop control system can keep the output voltages constant while the inductor current value has been modified. Developing this concept will lead to a current control strategy combined with voltage control to achieve the desired output voltage with a proper dynamic response. According to the steady state statements and dynamic model, the proposed MOB converter acts as a boost converter topology with a series of multiple output voltages.

6.3 MOB Converter with Multiple Outputs Configuration

Steady state and dynamic analysis can be extended for the proposed topology with multiple outputs (see Fig. 18 (a)). Rewriting the state space variables for N outputs:

$$\begin{bmatrix} L & 0 & 0 & 0 & 0 & 0 & 0 & 0 \\ 0 & C_1 & 0 & 0 & 0 & 0 & 0 & 0 \\ 0 & 0 & \ddots & 0 & 0 & 0 & 0 & 0 \\ 0 & 0 & 0 & C_i & 0 & 0 & 0 & 0 \\ 0 & 0 & 0 & 0 & \ddots & 0 & 0 & 0 \\ 0 & 0 & 0 & 0 & 0 & C_j & 0 & 0 \\ 0 & 0 & 0 & 0 & 0 & 0 & \ddots & 0 \\ 0 & 0 & 0 & 0 & 0 & 0 & 0 & C_k \end{bmatrix} \begin{bmatrix} \dot{i} \\ \dot{v}_1 \\ \vdots \\ \dot{v}_i \\ \vdots \\ \dot{v}_j \\ \vdots \\ \dot{v}_K \end{bmatrix} = \begin{bmatrix} 0 & -(1-D_0) & \cdots & -(1-\sum_{k=1}^{i} D_{K-1}) & \cdots & -(1-\sum_{k=1}^{j} D_{K-1}) & \cdots & -(1-\sum_{k=1}^{n} D_{K-1}) \\ (1-D_0) & \frac{1}{R_1} & 0 & 0 & 0 & 0 & 0 & 0 \\ \vdots & 0 & \ddots & 0 & 0 & 0 & 0 & 0 \\ (1-\sum_{k=1}^{i} D_{K-1}) & 0 & 0 & \frac{1}{R_i} & 0 & 0 & 0 & 0 \\ \vdots & 0 & 0 & 0 & \ddots & 0 & 0 & 0 \\ (1-\sum_{k=1}^{j} D_{K-1}) & 0 & 0 & 0 & 0 & \frac{1}{R_j} & 0 & 0 \\ \vdots & 0 & 0 & 0 & 0 & 0 & \ddots & 0 \\ (1-\sum_{k=1}^{n} D_{K-1}) & 0 & 0 & 0 & 0 & 0 & 0 & \frac{1}{R_k} \end{bmatrix} \begin{bmatrix} i \\ v_1 \\ \vdots \\ v_i \\ \vdots \\ v_j \\ \vdots \\ v_K \end{bmatrix} + \begin{bmatrix} 1 & \sum_{k=1}^{n} V_K & \cdots & \sum_{k=1}^{n-i} V_K & \cdots & \sum_{k=1}^{n-j} V_K & \cdots & V_1 \\ 0 & -I & 0 & 0 & \cdots & 0 & \cdots & 0 \\ 0 & -I & -I & 0 & \cdots & 0 & \cdots & 0 \\ 0 & -I & -I & 0 & \cdots & 0 & \cdots & 0 \\ 0 & -I & -I & -I & \cdots & 0 & \cdots & 0 \\ 0 & -I & -I & -I & \cdots & 0 & \cdots & 0 \\ 0 & -I & -I & -I & \cdots & -I & \cdots & 0 \\ 0 & -I & -I & -I & \cdots & -I & \cdots & -I \end{bmatrix} \begin{bmatrix} \hat{v}_{in} \\ \hat{d}_0 \\ \vdots \\ \hat{d}_i \\ \vdots \\ \hat{d}_j \\ \vdots \\ \hat{d}_{K-1} \end{bmatrix} \quad (9)$$

Extracting transfer function from these equations:

$$\begin{cases} I(s) = \dfrac{1}{Ls + \sum_{k=1}^{N} \left(\dfrac{R_k (1 - \sum_{j=1}^{k} D_{j-1})^2}{1 + R_k C_k s} \right)} \\ \\ V_k(s) = \dfrac{R_k (1 - \sum_{j=1}^{k} D_{j-1})}{1 + R_k C_k s} I(s) \end{cases} \quad (10)$$

Finally, the steady state equations for N output voltages would be:

$$V_k = \frac{R_K (1 - \sum D_{K-1}) V_{in}}{\sum R_K (1 - \sum D_{K-1})^2}, \quad I = \frac{V_{in}}{\sum R_K (1 - \sum D_{K-1})^2} \quad (11)$$

It is clear that different output voltages can be achieved based on different duty cycles.

6.4 Control System

Several current-mode control strategies have been conducted to improve the dynamic response of boost converters (Mammano, 1999; Siew-Chong et al., 2006). To have the potential of combining the advantages of the logical control and current mode control in a relatively simple controller realization for a MOB converter, a cross voltage control (V_T) with an internal hysteresis current control loop has been performed combined with mid point voltage (V_1) control. Fig. 19 illustrates the block diagram of the control method for a double-output boost converter. As shown, the solid loop is a cross voltage control with a hysteresis current control loop for the inductor current, in which the cross output of the MOB converter is controlled by switching the boost switch (S_0). The dashed loop is the mid point voltages control where the sharing switches (S_1) is forced to balance the capacitors' voltage. The current control loop consists of two cascaded control loops. The outer loop is a voltage control mode through which the reference current is modified based on voltage error to force this error to zero. The inner loop is the hysteresis current control which is the main loop that runs to forcibly constrain the inductor current between the hysteresis bands around the defined reference current.

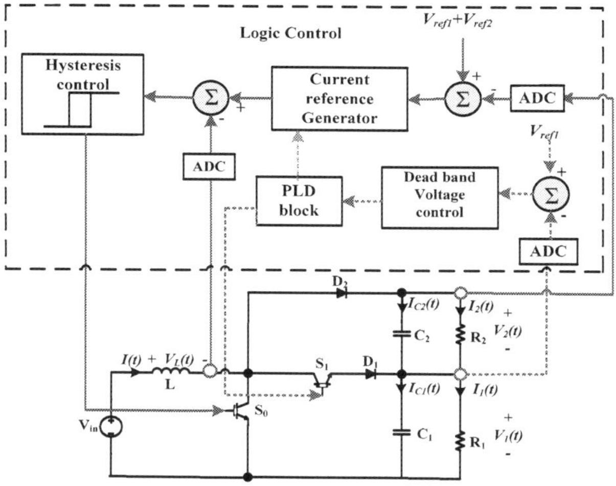

Fig. 19. Block diagram of the proposed control system

In the dead band voltage control block, the amount of the mid point voltage (V_1) error is compared in order to give priority to the voltage with larger error. Then the chosen mid point voltage will be compared with its defined voltage dead band (assumed 1% of reference) , if the voltage is higher (lower) than the upper (lower) dead band output signal of block changes to one (zero). It should be mentioned that defining the dead band can decrease unnecessary switching when the mid point voltage error is negligible; otherwise it can be set to zero to increase accuracy.

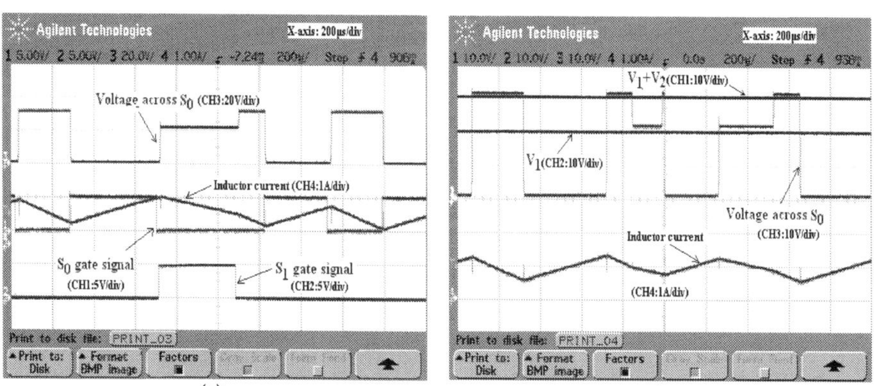

(a) (b)

Fig. 20. (a) Steady-state inductor current and switching pulses for the CCM operation (R_1 = 50 Ω and R_2=50 Ω). (b) Output voltages corresponding to (a).

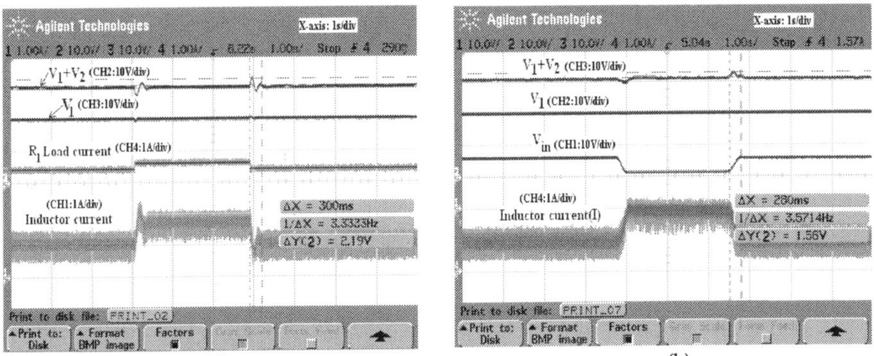

(a) (b)

Fig. 21. Waveforms during transient condition. (a)When R_1 *is* changed. (b) Output voltage during input voltage disturbance (R_1=R_2=*50* Ω).

The steady-state response under the proposed control scheme is shown in Fig. 20 form a laboratory prototype. The switching pulses (control signal) and the inductor current waveforms are shown to verify the control strategy for the Continues Conduction Mode (CCM) operation of the converter with load resistances of R_1=R_2=*50 Ω*. The mid point voltage (V_1) and total output (V_T), are assumed to be kept at *20V* and *30V* respectively, while V_{IN}=*15V*.

The control strategy is tested under disturbances in both load and input voltage, and the corresponding variations in the output voltage and inductor current are demonstrated in Fig. 21. In Fig. 21 (a) R_2=*50* and R_1 is varied from *50* to *35 Ω* and back. It is noted that the operation of the converter still remains in the CCM and the current modification applies in each five hysteresis cycles discontinuously. The output voltages display an undershoot (overshoot) for the load current increase (decrease), but they quickly settle around the reference value. Also, due to the switching options to control V_1 and V_2, the controlling of V_1 surpasses the other output, so that it has less undershoot and overshoot. While the converter is working with a load of R_1=R_2=*50 Ω* (CCM operation), a change in the input voltage from the *15V to 10 V* and back is applied. The change in the input voltage and the corresponding output-voltage waveforms are depicted in Fig. 21 (b). It is observed that the total output voltage shows small undershoot (overshoot) for the input voltage increase (decrease) and settles down after a short time. However, the mid point voltage remains constant, and the transients in voltage are sufficiently diminished.

6.5 Performance of Diode-clamped Joint with MOB Converter

The performance of the new single-phase diode-clamped inverters configuration is simulated in Fig. 22. Herein, on the DC-DC side, input voltage (V_{in}) is assumed as 100V; switching frequency of the DC-DC converter (f_{sw}) is 10 kHz, L=2mH, and C_1=C_2=1mF while on the inverter side fundamental and switching frequencies are f=50Hz, f_{sw}=4 KHz, and the DC link of the three-level diode-clamped inverter (V_{MN}) is boosted to 300 V by using a double-output boost converter. Carrier based PWM strategy has been carried out for the three-level diode-clamped converter to generate reference voltage. As shown in Fig. 22, while the total voltage of an inverter DC link is boosted at 300V, mid point voltages (V_1

$=V_{c1}$) are controlled at 150V to have an equal DC link capacitor voltage arrangement. The double-output boost converter can boost the low input voltage for DC link capacitors as well as balance the capacitors' voltage to the desired level for a high modulation index (Fig. 23), which is impossible without the capacitor voltage balancing algorithm in three-level diode-clamped converters with a passive front-end converter. Output voltage for the proposed configurations for m_a=1 and PF=0.95 is demonstrated in Fig. 22 and 23, respectively.

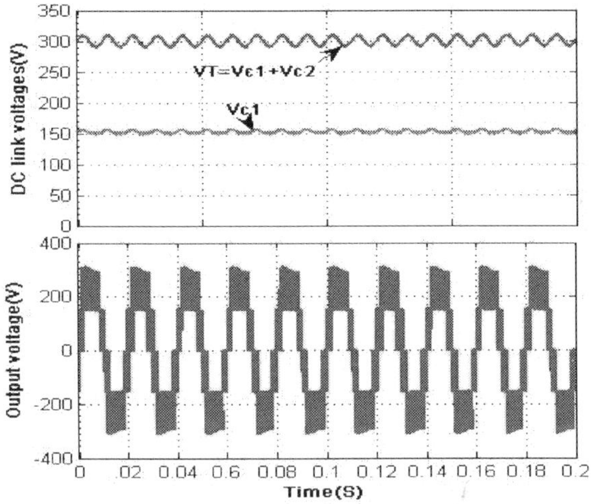

Fig. 22. Simulation results for DC link voltage of three-level diode-clamped Inverter with double-output boost converter

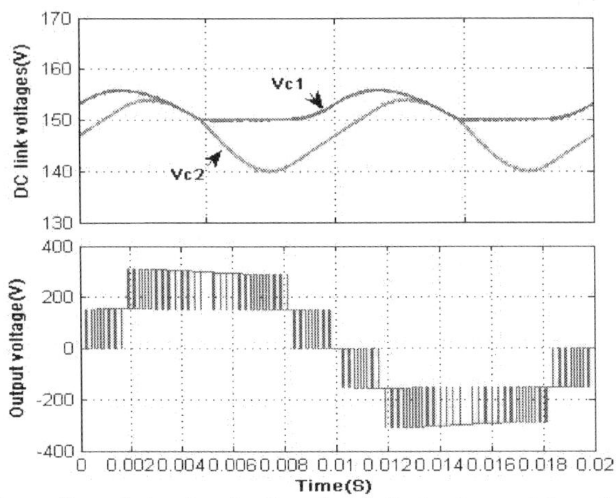

Fig. 23. Capacitor voltage balancing in three-level diode-clamped Inverter with double-output boost converter

7. Conclusion

In this chapter, different topologies for renewable energy systems based on power electronics converters have been studied. A new topology for a multi output DC-DC converter is presented in order to supply input voltages for a diode-clamp multilevel inverter. Using this circuit, the DC link voltages of the diode-clamped can be adjusted to a desired voltage level by the DC-DC converter, thereby solving the main problem associated with balancing the capacitor voltages in the diode-clamped topology. Furthermore, since the DC output voltage of PV or WT systems are not very high, this topology is a suitable candidate for these systems as it can boost the low and unregulated input voltage for a transformer less grid connection based on the multilevel topology. To verify the operation of this topology, both steady-state and small-signal ac models have been compared through simulation and hardware results.

8. References

Alepuz, S.; Busquets-Monge, S.; Bordonau, J.; Gago,J.; Gonzalez, D. & Balcells, J. (2006). Interfacing Renewable Energy Sources to the Utility Grid Using a Three-level Inverter. *IEEE Transaction on Industrial Electronics*, Vol. 53, No. 5, October 2006, 1504-1511, 0278-0046.

Alonso, O.; Sanchis, P.; Gubia, E. & Marroyo,L. (2003). Cascade H-bridge Multilevel converter for Grid-connected Photovoltaic Generators with Independent Maximum power point Power tracking of Each Solar Array. *Proceeding of IEEE Power Electronics Specialist Conference*, pp. 731-735, 0-7803-7754-0/03, 2003.

Bueno, E.J.; Cobreces, S.; Rodriguez, F.J.; Hernandez, A. & Espinaso, F. (2008). Design of NPC Back-to-Back Converter Interface for Wind Turbines with Squirrel-cage Induction Generator. *IEEE Transaction on Energy Conversion*, Vol. 23, No. 3, September 2008, 932-945.

Bum-Seok, S.; Sinha, G.; Manjrekar, M.D. & Lipo, T.A. (1998). Multilevel Power Conversion - An Overview of Topologies And Modulation Strategies, *Proceedings of the 6th International Conference on Optimization of Electrical and Electronic Equipments*, pp. AD-11-AD-24, 1998.

Busquets-Monge, S.; Rocabert, J.; Rodriguez, P.; Alepuz, S. & Bordonau, J. (2008). Multilevel Diode-clamped Converter for Photovoltaic Generators with Independent Voltage control of Each solar Array. *IEEE Transaction on Industrial Electronics*, Vol. 55, No. 7, July 2008, 2713-2723, 0278-0046.

Carrasco, J.M.; Franquelo, L.G.; Bialasiewicz, J.T.; Galvan, E.; Portillo Guisado, R.C.; Prats, Ma.A.; Leon, J.I. & Moreno-Alfonso, N. (2006). Power-Electronics Systems for the Grid Integration of Renewable Energy Sources: A Survey. *IEEE Transaction on Industrial Electronics*, Vol. 53, No. 4, August 2006, 1002-1016, 0278-0046.

Clasis, M. & Agelidis, V.G. (1998). Multilevel converters for Single-phase Grid connected Photovoltaic systems-An Overview, *Proceeding of IEEE International symposium on Industrial Electronics*, pp. 224-229, 0-7803-4756-0, South Africa,1998.

Gu, Y.; Hang, L.; Chen, H.; Lu, Zh.; Qian,Zh. and Li, J. (2005). A simple structure of LLC resonant DC-DC Converter for Multi-output Application. *Proceeding of IEEE Applied Power Electronics Conference*, pp. 1485-1490, 0-7803-8975-1/05 , March 2005.

Leon, J. I.; Portillo, R.; Vazquez, S.; Padilla, J.J; Franquelo,L.J & Carrasco, J.M. (2008). Simple Unified Approach to Develop a Time-Domain Modulation Strategy for Single-Phase Multilevel Converters. IEEE Transactions on Industrial Electronics, Vol. 55, No. 9, Spetember 2008 3239 - 3248.

Lopez, O.; Teodorescu, R. & Doval-Gandoy, J. (2006). Multilevel Transformerless Topologies for Single-Phase Grid-Connected Converters. *Proceeding of IEEE Industrial Electronics Conference,* pp. 5191-5196, 1-4244-0390-1, 2006.

Mammano, R. (1999). Switching Power Supply Topology Voltage Mode vs. Current Mode. *Texas Instruments Incorporated,* 1999.

Manjrekar, M.D.; Lund, R.; Steimer, P.; and Lipo, T.A. (2000). Hybrid Multilevel Power conversion system: A Competitive Solution for High Power Applications. *IEEE Transaction on Industrial Applications,* Vol. 36, No. 3, May/June 2000, 834-841.

Marchesoni, M. & Tenca, P. (2002). Diode-clamped Multilevel Converters: A Practicable Way to Balance DC-Link Voltages. *IEEE Transactions on Industrial Electronics,* Vol. 49, No. 4, August 2002, 752-765, 0278-0046.

Meynard, T.A. & Foch, H. (1992). Multi-level Conversion: High Voltage chopper and Voltage-source Inverter, *Proceeding of IEEE Power Electronics Specialist Conference,* pp. 397-403, Toledo Spain, 1992.

Myrzik, J.M.A. & Calais, M. (2003). String Module Integrated Inverters for Single-phase Grid Connected Photovoltaic Systems-A Review. *Proceeding of IEEE PowerTech Conference,* 8 pp, 0-7803-7967-5/03, Bologna Italy, 2003.

Nabae, A.; Takahashi, I, & Akagi, H. (1981). A new neutral-point-clamped PWM inverter. *IEEE Transaction on Industrial Electronic,* Vol. 1A-17, No. 5, 518–523.

Nami, A. & Zare, F. (2008). A New Random Current Control Technique for a Single-Phase Inverter with Bipolar and Unipolar Modulations. *IEEJ Transactions on Industry Applications,* Vol.128-D, No.4, 2008.

Nami, A.; Zare, F.; Ledwich, G.; Ghosh, A. & Blaabjerg, F. (2007). A new configuration for multilevel converters with diode clamped topology, *proceeding of IEEE International Power Engineering Conference,* pp. 661-665, 978-981-05-9423-7, Singapore, 2007.

Nami, A.; Zare, F.; Ledwich, G.; Ghosh, A. & Blaabjerg, F. (2008). Comparison between Symmetrical and Asymmetrical Single Phase Multilevel Inverter with Diode-Clamped Topology. *Proceeding in IEEE Power Electronics Specialists Conference,* pp.2921-2926, 978-1-4244-1668/08, Rhodes Island Greece, June 2008.

Rodriguez, J.; Jai, J.Sh. & Peng, F.Zh. (2002). Multilevel Inverters: A Survey of Topologies, Controls, and Applications. *IEEE Transaction on Industrial Electronics,* Vol. 49, No. 4, August 2002, 724-738, 0278-0046.

Sharma, R. & Hongwei.G. (2006). A new DC-DC converter for fuel cell powered distributed residential power generation systems, Proceeding of IEEE *Applied Power Electronics Conference and Exposition,* p. 1014-1018, 0-7803-9547-6 ,March 2006.

Siew-Chong, T.; Lai, Y.M. & Tse,C.K. (2006). Implementation of pulse-width-modulation based sliding mode controller for boost converters. *IEEE Power Electronics Letters,* Vol. 3, 130-135, December 2006, 1540-7985.

Tolbert, L.M. & Peng, F.Z. (2000). Multilevel Converters as a Utility Interface for Renewable Energy Systems, *Proceeding of IEEE Power Engineering Society,* pp. 1271-1274, 0-7803-6420-1, USA, 2000.

Yazdani, A. & Iravani, R. (2005). A Generalized State-Space Averaged Model of the Three-level NPC Converter for Systematic DC-Voltage-Balancer and Current-Controller Design. *IEEE Transactions on Power Delivery,* Vol. 20, No. 2, April 2005, 1105-1114, 0885-8977.

Yunxiang, X. and Jiuchao, G. (2004). Study on the Voltage Stability of Multi-output Converters. *Proceeding of IEEE The International Power Electronics and Motion Control Conference,* pp. 482-486, 7-5605-1869-9, August 2004.

Zare, F. & Ledwich, G. (2002). A hysteresis current control for single-phase multilevel voltage source inverters: PLD implementation. *IEEE Transactions on Power Electronics,* Vol. 17, No. 5, Sep 2002, 731 -738.

Zare, F. & Ledwich, G. (2008). A new predictive current control technique for multilevel converters. *Australian Journal of Electrical & Electronics Engineering (AJEEE),* 2008, Vol. 4, No. 1, 2008, 25-35.

Zare, F. (2008). Power Electronics Education E-Book. www.Peeeb.com, ISBN: 978-0-646-49442-5.

16

Isolated hybrid solar-wind-hydro renewable energy systems

Dorin Bică, Cristian Dragoş Dumitru, Adrian Gligor, Adrian-Vasile Duka
"Petru Maior" University of Tg. Mureş, Electrical Engineering Department
Romania

1. Introduction

Renewable energy technologies offer the promise of clean, abundant energy gathered from self-renewing resources such as the sun, wind, water, earth, and plants. Virtually all regions of the world have renewable resources of one type or another.

Renewable energy technologies offer important benefits compared to those of conventional energy sources. Worldwide, 1000 times more energy reaches the surface of the earth from the sun than is released today by all fossil fuels consumed. Photovoltaics and wind generation are also an attractive source of energy because of their benign effect on the environment.

Increased population growth and economic development are accelerating the rate at which energy, and in particular electrical energy is being demanded. All methods of electricity generation have consequences for the environment, so meeting this growth in demand, while safeguarding the environment poses a growing challenge

Each of the renewable energy technologies is in a different stage of research, development, and commercialization and all have differences in current and future expected costs, current industrial base, resource availability, and potential impact on greenhouse gas emissions.

Hybrid power systems consist of a combination of renewable energy sources such as: photovoltaic (PV), wind generators, hydro, etc., to charge batteries and provide power to meet the energy demand, considering the local geography and other details of the place of installation.

These types of systems, which are not connected to the main utility grid, are also used in stand-alone applications and operate independently and reliably. The best applications for these systems are in remote places, such as rural villages, in telecommunications, etc.

The importance of hybrid systems has grown as they appeared to be the right solution for a clean and distributed energy production. It has to be mentioned that new implementations of hybrid systems require special attention on analysis and modelling. One issue is determined by the variable and unpredictable character of energy supply from renewable sources.

A major importance for the theoretical study of hybrid systems, based on renewable energy (photovoltaic, wind, hydroelectric systems), is the availability of models, which can be used to study the behaviour of hybrid systems, and most important, software simulation environments.

As available tools are quite limited, this chapter intends to present several models which can be used for the simulation purposes of hybrid power systems as well as in educational purposes.
The modelling of renewable energy hybrid systems has to be made by knowing all types of renewable energy used in the model. For a good understanding of the system, equivalent models, based on large scale used components, should be considered.

2. Modelling the components of a hybrid power system

2.1 Modelling the Solar Photovoltaic System

A photovoltaic PV generator consists of an assembly of solar cells, connections, protective parts, supports etc.
Solar cells are made of semiconductor materials (usually silicon), which are specially treated to form an electric field, positive on one side (backside) and negative on the other (towards the sun). When solar energy (photons) hits the solar cell, electrons are knocked loose from the atoms in the semiconductor material, creating electron-hole pairs (Lorenzo, 1994). If electrical conductors are then attached to the positive and negative sides, forming an electrical circuit, the electrons are captured in the form of electric current (photocurrent).
The model of the solar cell can be realised by an equivalent circuit that consists of a current source in parallel with a diode (Fig. 1) (Kaltschmitt et al., 2007) (Markvart & Castaner, 2003). In Fig. 1 R_S, R_P and C components can be neglected for the ideal model.

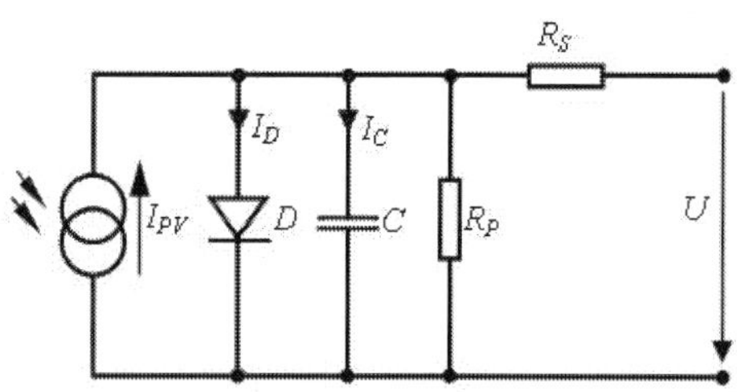

Fig. 1. Equivalent circuit diagram of a solar cell

The p-n junction has a certain depletion layer capacitance, which is typically neglected for modelling solar cells.
At increased inverse voltage the depletion layer becomes wider so that the capacitance is reduced similar to stretching the electrodes of a plate capacitor. Thus solar cells represent variable capacitance whose magnitude depends on the present voltage. This effect is considered by the capacitor C located in parallel to the diode.
Series resistance R_S consists of the contact resistance of the cables as well as of the resistance of the semiconductor material itself.
Parallel or shunt resistance R_P includes the "leakage currents" at the photovoltaic cell edges at which the ideal shunt reaction of the p-n junction may be reduced. This is usually within the kΩ region and consequently has almost no effect on the current-voltage characteristic (Kaltschmitt et al., 2007).
The diode is the one which determines the current-voltage characteristic of the cell. The output of the current source is directly proportional to the light falling on the cell. The open

circuit voltage increases logarithmically according to the Shockley equation which describes the interdependence of current and voltage in a solar cell (Kaltschmitt et al., 2007), (Patel, 1999).

$$I = I_{PV} - I_0\left(e^{\frac{qU}{kT}} - 1\right) \quad (1)$$

$$U = \frac{kT}{q}\ln\left(1 - \frac{I - I_{PV}}{I_0}\right) \quad (2)$$

where:
- k - Boltzmann constant (1.3806 10^{-23} J/K);
- T - reference temperature of solar cell;
- q - elementary charge (1.6021 10^{-19} As);
- U - solar cell voltage (V);
- I_0 - saturation current of the diode (A);
- I_{PV} - photovoltaic current (A).

Equations (1) and (2) lead to the development of a Matlab Simulink model for the PV module presented in Fig. 2.

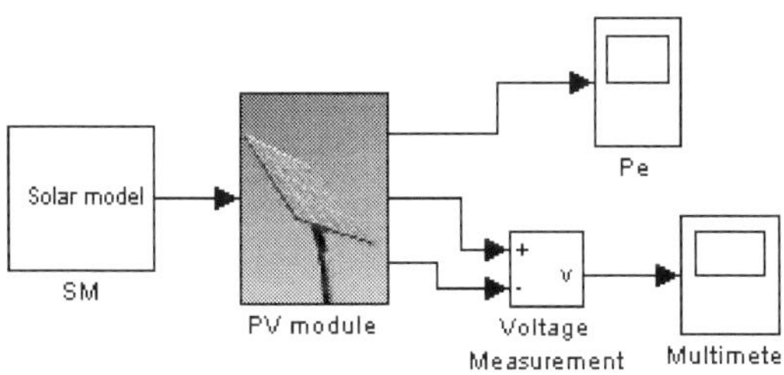

Fig. 2. Matlab Simulink Library PV module

The solar system model consists of three Simulink blocks: the solar model block, the PV model block and energy conversion modules.
The solar model block implements the mathematical model of the solar radiation. This is done by using standard Simulink and Matlab modules and functions. This block allows selecting different type of patterns for the solar radiation (Dumitru & Gligor, 2008a).
The PV module implements the equivalent circuit of a solar cell, shown in Fig. 1. Standard functions and blocks of Matlab and Simulink were used to obtain this model. Its structure is presented in Fig 3.
The output of the PV module is processed by an energy conversion block implemented with an PWM IGBT inverter block from standard Simulink/SimPowerSystems library.

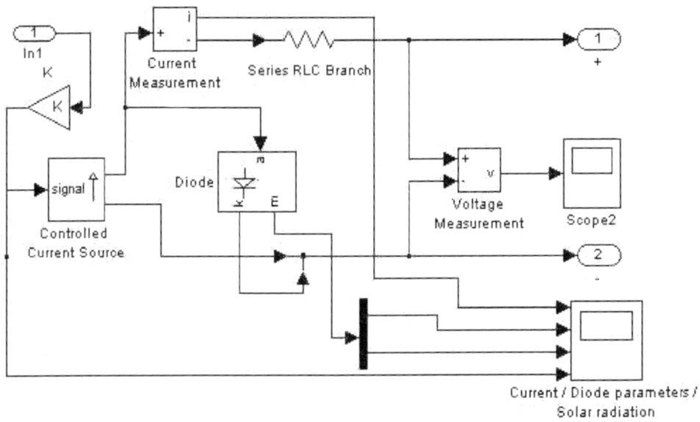

Fig. 3. Matlab Simulink implementation of the PV module.

2.2 Modelling the Wind Energy System

Modelling the wind energy converter is made considering the following assumptions

- friction is neglected;
- stationary wind flow;
- constant, shear-free wind flow;
- rotation-free flow;
- incompressible flow (ρ=1.22 kg/m^3);
- free wind flow around the wind energy converter.

On the above condition the maximum physical achievable wind energy conversion can be derived using a theoretical model that is independent of the technical construction of a wind energy converter.

The flow air mass has certain energy. This energy is obtained from the air movement on the earth's surface determined by the difference in speed and pressure. This is the main source of energy used by the wind turbines to obtain electric power. The kinetic energy W taken from the air mass flow m at speed v_1 in front of the wind turbine's pales and at the back of the pales at speed v_2 is illustrated by equation (3):

$$W = \frac{1}{2} m \left(v_1^2 - v_2^2 \right) \tag{3}$$

The resulted theoretical medium power P is determined as the ratio between the kinetic energy and the unit of time and is expressed by equation (4):

$$P = \frac{W}{t} = \frac{1}{2} \frac{m}{t} \left(v_1^2 - v_2^2 \right) = \frac{1}{2} \frac{V\rho}{t} \left(v_1^2 - v_2^2 \right) \tag{4}$$

where:

- V air mass volume;
- t time;
- ρ air density.

Assuming the expression of the mean air speed $v_{med} = \frac{1}{2}(v_1 + v_2)$ the mean air volume transferred per unit time can be determined as follows:

$$V_{med} = \frac{V}{t} = Av_{med} \tag{5}$$

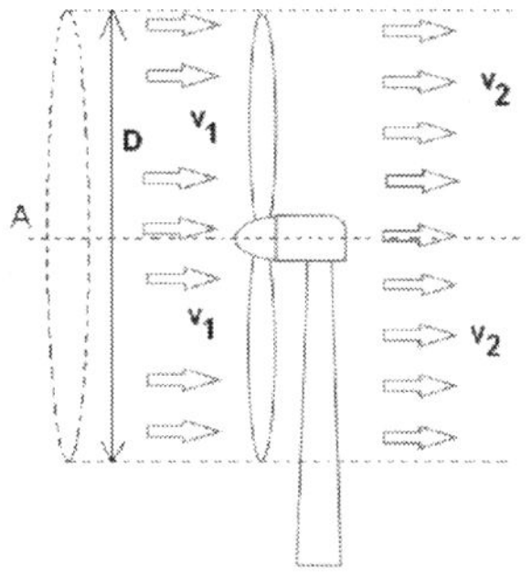

Fig. 4. Flow through a wind energy converter

The equation for the mean theoretical power is determined using equation (5):

$$P = \frac{1}{4} A\rho \left(v_1^2 - v_2^2\right)\left(v_1 + v_2\right) = \frac{A\rho v_1^3}{4}\left(1 - \frac{v_2^2}{v_1^2}\right)\left(1 + \frac{v_2}{v_1}\right) \tag{6}$$

We can conclude that an adequate choice of v_2 / v_1 ratio leads to a maximum power value taken by the wind converter from the kinetic energy of the air masses, as shown by equation (7):

$$P_{max} = \frac{8}{27} A\rho v_1^3 \tag{7}$$

This power represents only a fraction of the incident air flow theoretical power given by:

$$P_{wind} = \frac{1}{2} \rho A v_1^3 \tag{8}$$

Equations (7) and (8) lead to:

$$P_{max} = \frac{8}{27} A\rho v_1^3 = \frac{1}{2} A\rho v_1^3 \cdot 0{,}59 = P_{wind} \cdot C_p \tag{9}$$

where: C_P represents the mechanical power coefficient which expresses that the wind kinetic energy cannot be totally converted in useful energy. This coefficient, meaning the maximum theoretical efficiency of wind power, was introduced by Betz (Burton et al., 2004).

The electrical power obtained under the assumptions of a wind generator's electrical and mechanical part efficiency is given by:

$$P_{el} = \frac{1}{2} C_e \rho A v_1^3 \tag{10}$$

where: C_e represents the total net efficiency coefficient at the transformer terminals (Golovanov et al., 2007).

A Matlab Simulink model, based on the equations mentioned above, was developed for the wind generator module. This model is shown in Figure 5.

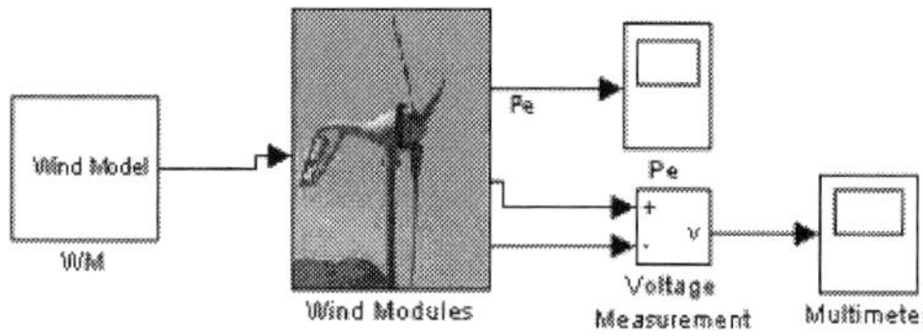

Fig. 5. The Matlab Simulink model of the wind generator module.

The wind system model consists of three Simulink blocks: the wind model block, the wind generator model block and energy conversion modules.
The wind model block implements the mathematical model of the air mass flow. This is done by using standard Simulink and Matlab modules and functions. This block allows the selection of different patterns for the air mass flow and the equations mentioned above were used in the design of this model.
The wind energy generator model was implemented by a module having configurable parameters based on equation (10) and using the equivalent model of a generator. This model takes the following form and is shown in Figure 6.

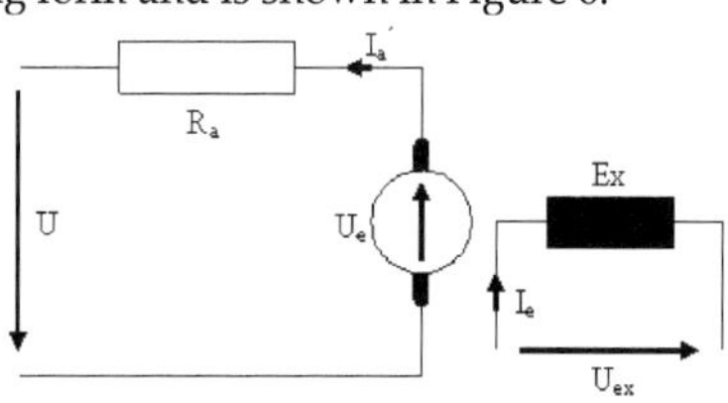

Fig. 6. Equivalent circuit diagram of a small wind generator

In the equivalent circuit diagram of a small wind generator the notations are:

R_a - rotor winding resistance

Ex - generator separate excitation winding; current I_e through this winding generates the main field

U_e - induced voltage in the rotor (armature)

U - terminal voltage $U = U_e - R_a I_a$

The resulted Matlab-Simulink model for the wind generator is a particular case of the more general model of an electrical generator, which is presented in figure 7.

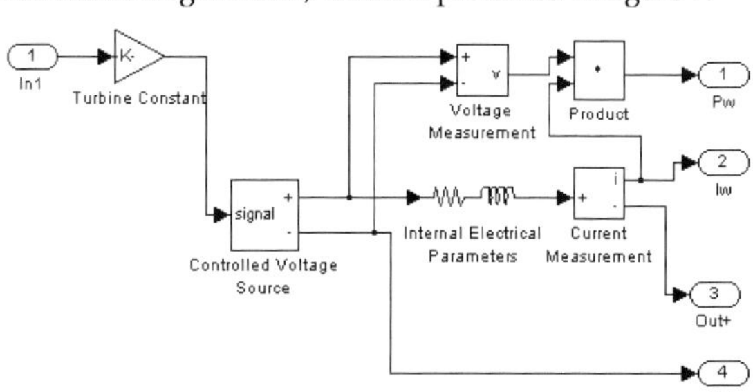

Fig. 7. Matlab-Simulink model of the generator

The output of the wind energy generator module is processed by an energy conversion block implemented with a PWM IGBT inverter block from the standard Simulink/ SimPowerSystems library.

2.3 Modelling the Hydroelectric System

Small hydroelectric power plants harness the falling water kinetic energy to generate electricity. Turbines transform falling water kinetic energy into mechanical rotation energy and then, the alternator transforms the mechanical energy into electricity. Water flows within a river from a higher geodesic site to a lower geodesic site due to gravitation. This is characterized by different particular kinetic and potential energy at both sites. The correct identification of the resulting energy differences of the out-flowing water can be assumed by considering a stationary and friction-free flow with incompressibility. The hydrodynamic Bernoulli pressure equation applied in such conditions is written according to equation (11)

$$p + \rho_{water} gh + \frac{1}{2} \rho_{water} v_{water}^2 = const. \tag{11}$$

where:

- p - hydrostatic pressure;
- ρ_{water} - water density;
- g - acceleration of gravity;
- h - the water height;
- v_{water} - velocity of the water flow.

Equation (11) can be transformed so that the first term expresses the pressure level, the second term the level of the site and the third term the water velocity level.

$$\frac{p}{\rho_{water} g} + h + \frac{1}{2} \frac{v_{water}^2}{g} = const. \tag{12}$$

The term $\frac{1}{2} \frac{v_{water}^2}{g}$ refers to the dynamic height and is defined as the height due to the speed of water flow and can be identified by the term of kinetic water energy.

The usable head h_{util} of a particular section of river can be determined by considering: the difference in pressure, the geodesic difference in height and the different flow velocities of the water, using equation (13). It must be mentioned that the equation is used to analyze an ideal case and does not consider the losses due to the friction of the individual water molecules among each other and the surrounding matter.

$$h_{util} = \frac{p_{up} - p_{down}}{\rho_{water} g} + (h_{up} - h_{down}) + \frac{v_{water,up}^2 - v_{water,down}^2}{2g} \tag{13}$$

where:

- p_{up} - upstream hydrostatic pressure;
- p_{down} - downstream hydrostatic pressure;
- h_{up} - upstream geodesic water height (headwater);
- h_{down} - downstream geodesic water height (tailwater);
- $v_{water,up}$ - upstream water velocity;
- $v_{water,down}$ - downstream water velocity;

Considering equation (13), the power of a water supply P_{water} can be determined using (14):

$$P_{water} = \rho_{water} g q_{water} h_{util} \tag{14}$$

where q_{water} is the volume-related flow rate.

According to equation (14), the power of a water supply is determined by the volume-related flow rate and usable head. The water flow assumes high values in lowland areas, while large heads can be achieved in mountain areas.

Considering two specific points of a river, the theoretical power of the water $P_{water,\ th}$, can be calculated based on:

$$P_{water,th} = \rho_{water} g \dot{q}_{water} (h_{up} - h_{down}) \tag{15}$$

where $\dot{q}_{water}$ is the volumetric flow rate through a hydroelectric power plant.

In the real case, considering the energy balance between two specific points of a river, and also the energy losses, the hydrodynamic Bernoulli pressure equation can be written according to equation (16).

$$\frac{p_{up}}{\rho_{water,up} g} + h_{up} + \frac{v_{water,up}^2}{2g} = \frac{p_{down}}{\rho_{water,down} g} + h_{av} + \frac{v_{water,down}^2}{2g} + \xi \frac{v_{water,down}^2}{2g} = const. \tag{16}$$

where:

- $\frac{p}{\rho_{water} g}$ - hydrodynamic pressure energy;
- h - potential energy of the water;
- $\frac{v_{water}^2}{2g}$ - kinetic energy of the water;
- $\xi \frac{v_{water}^2}{2g}$ - energy losses;
- ξ - loss coefficient.

The energy losses are represented by the part of the rated power which is converted into ambient heat by friction and cannot be used technically.

In the turbine, pressure energy is converted into mechanical energy. The conversion losses are described by the turbine efficiency $\eta_{turbine}$. Equation (17) describes the part of the usable water power that can be converted into mechanical energy at the turbine shaft $P_{turbine}$.

$$P_{turbine} = \eta_{turbine} \rho_{water} g \dot{q}_{water} h_{util} \tag{17}$$

h_{util} is the usable head at the turbine, and the term $(\rho_{water} g \dot{q}_{water} h_{util})$ represents the actual usable water power (Kaltschmitt et al., 2007).

The water model described by the equations mentioned above was introduced in a Matlab-Simulink model of the hydroelectric system. This model is shown in figure 8 and it encapsulates the model of the hydroelectric plant connected to the water model. Measurement of power and voltage is also provided by this model.

The model of the hydroelectric plant (generator) has the same form as the one presented in figure 7 and also an equivalent diagram as the one we considered for the wind generator can be assumed (Figure 6)

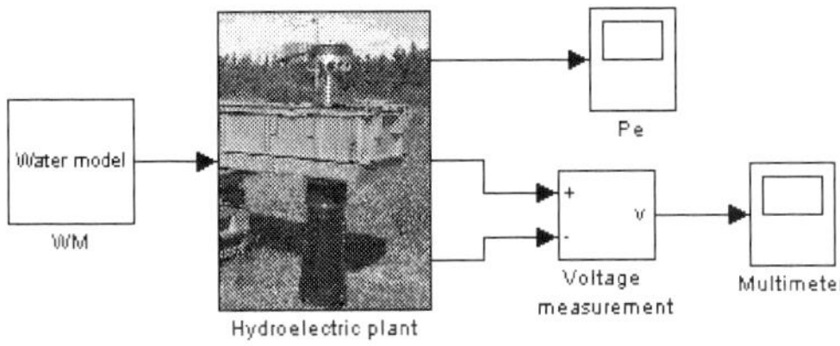

Fig. 8. The Matlab Simulink model of the hydroelectric system.

2.4 Modelling the Storage Device

The energy storage devices/equipments are used basically for three purposes: energy stabilization, ride through capability and dispatchability.

The energy stabilization permits the hybrid system to run at a constant stable output level with the help of the energy storage devices, even if the load fluctuates rapidly.

The ride through capability is the capability of the energy storage device which provides the proper amounts of energy to loads, when the hybrid system generator units are unavailable, (e.g. the solar supply system during the night time or when components of any type are being maintained or repaired). The hybrid system owner who needs power has two options during such periods. The first one is to use another backup or use the utility grid. The second way is to meet out the needs with energy stored when the source is unavailable (Willis & Scott, 2000).

For energy sources like photovoltaic or wind energy systems, the power production depends upon the availability of the resources like sunlight or wind. This makes the nature of power available to loads intermittent, thus making them non-dispatchable sources. However, the energy storage systems with non-dispatchable energy can be deployed as dispatchable energy sources. This only needs a proper design of the energy storage system, by looking into the load curve (Sahay & Dwivedi, 2009).

Batteries are the basic component of an energy storage system. A battery consists of one or more electrochemical cells that are electrically connected. The basic components of an electrolytic cell like a lead-acid cell are a positive electrode, a negative electrode, a porous separator and an electrolyte. During cell operation, ions are created and consumed at the two electrode/electrolyte interfaces by oxidation/reductions reactions. The electrolyte, which can either be a solid or liquid chemical, has high conductivity for ions but not for electrons, because if the electrolyte conducts electrons then the battery will self-discharge. The electrolyte completes the internal circuit between the electrodes.

The parameters associated with battery modelling are (Chan & Sutanto, 2000):

- internal resistance:
    - self discharge resistance: the resistance associated with the electrolysis of water at high voltage levels and slow leakage across the battery terminal at low voltage, inversely proportional to the temperature and very sensitive to it;
    - charge and discharge resistance (R_c / R_d): the resistances associated with the electrolyte resistance, plate resistance and fluid resistance, variable during charging and discharging;

  - overcharge and over discharge resistance: the resistances attributed to the electrolyte diffusion during over charging and over discharging.
- polarization capacitance: the capacitance due to the chemical diffusion within the battery which does not necessarily represent a purely electrical capacitance.
- discharge type:
  - continuous discharging: the battery continuously delivers energy to the load which leads to a continuous drop in the battery capacity;
  - intermittent discharging: the battery delivers energy to the load at regular or irregular intervals of time.
- discharge mode:
  - constant load mode: the battery delivers energy to a constant load and the load current decreases proportional to the decrease in the battery terminal voltage;
  - constant current mode: the battery supplies constant current to the load; this is achieved by continuously reducing the load resistance to match with the decreasing battery terminal voltage in order to maintain a constant current to the load;
  - constant power mode: the battery supplies constant electrical power to the load; the load current increases to compensate for the drop in voltage in order to maintain constant power to the load.
- rate of charge and discharge: the rate of charge and discharge should not be too high in order to extend service life of the battery; the frequency of charging and discharging cycles affects the battery life significantly.

In figure 9 the Thevenin equivalent battery model is presented.

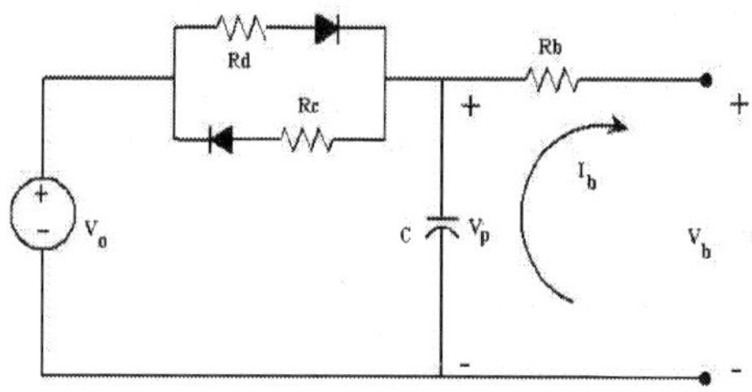

Fig. 9. Thevenin equivalent battery model.

The open circuit voltage, internal capacitor voltage and the terminal voltage are represented by V_0, V_p and V_b. The charging, discharging and the internal resistance of the battery are represented by R_c, R_d and R_b and the polarization capacitance of the battery is represented by C. The current I_b is taken as positive if discharging and negative otherwise (Vairamohan, 2002).

The equations for the circuit model are:

$$\dot{V}_p = \frac{1}{C}\left(\left(\frac{V_0 - V_p}{R_d}\right) - I_b\right) \tag{18}$$

$$V_b = V_p - I_b R_b \tag{19}$$

Based on this model and the equations above, a Matlab-Simulink model was developed for the battery storage device. This model is shown in Figure 10.

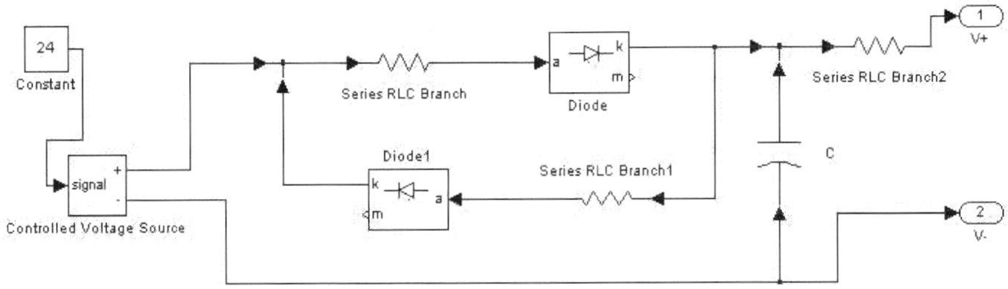

Fig. 10. The Matlab Simulink model of the battery storage device.

3. Distributed Generation in Hybrid Systems

3.1 Issues Regarding Distributed Generation

The problem of generating and using electrical energy by consumers is strictly connected to the study and modelling of local distribution electrical networks.

The distribution of electrical energy in local distribution systems is done by modular generation, using relatively small generation systems, ranging from less than 1kW to approx. 20 MW. These systems are located close to consumer areas. This type of energy production is called distributed generation (DG).

There are two types of distribution systems: interconnected and independent. The main purpose is to redefine the electrical networks, so that, instead of producing electrical energy in large plants and sending it in a single direction, the consumers should have a degree of energetic independence, and the system should be made available to large numbers of small producers in order to assure economic functioning for the electrical distribution network. This way, the electrical energy could be produced by photovoltaic solar systems, wind generation systems, biomass systems etc. The energy surplus could be sold / supplied to the distribution network, and the consumers could use the energy whenever they need it.

The reliability of the distribution system as well as the quality of the electrical energy can be improved by placing the sources close to the consumers, and the efficiency is improved by locally generating electrical and thermal energy. Thermal energy, resulting as side effect in the production of electrical energy, can also be used, and therefore a more efficient plant is obtained.

DG helps complete the traditional centralized production and distribution of electrical energy. Some of the characteristics of DG include the relatively low costs reported to the increase in energy demand meaning that the installation of newer transportation and distribution systems is no longer necessary. Also, energy generation can be placed where it is more needed and has the flexibility of delivering it in a network near the consumer. Another important aspect is represented by the demand of cheaper and less pollutant energy that is safe and reliable for all, including consumers, suppliers, producers and political partners (Bloem, 2006).

Several major influences that can be identified in the operation of distribution systems which are also reflected in DG are:

- according to the produced energy and the level of consumption, the voltage profile is changing across the network, leading to different behavior than usual;
- transitory signals appearing as result of connecting and disconnecting generators or during their operation;
- the increase of short circuit power;
- the variation of losses according to production and consumption levels;
- change in the quality of energy and the system's reliability;
- the distributor's protection systems need to be coordinated with those installed at the generator (Chindriş et al., 2008).

3.2 Case Study: Development of a Hybrid DG System in Mureş County

3.2.1 Solar Energy

The potential of solar radiation hitting the earth's surface is represented by the average energy density of horizontally direct solar radiation, which overcomes 1000 kWh/m^2/year. Romania has five identified geographical areas with differences in solar energy flows, as can be revealed in the map provided by ANM (National Meteorology Agency) shown in Figure 11. The geographical distribution shows that Romania's surface benefits of an average solar energy flow higher than 1000 kWh/m^2/year.

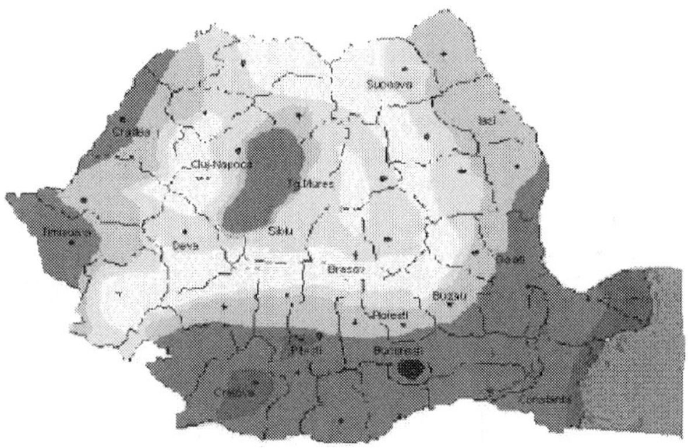

SOLAR RADIATION AREA	SOLAR RADIATION INTENSITY [kWh/m2/year]
I	>1350
II	1300-1350
III	1250-1300
IV	1200-1250
V	<1200

Fig. 11. Solar radiation intensity in Romania

The radiation hitting the earth's surface on a bright summer day contains of levels of approximate 1000 W/m^2 during the year, on average approximate 1000 kWh, which equates to approximate 100 liters of heating oil reaching every m^2 surface of our country's territory.
Considering the geographical position of Mureş County and analyzing its solar potential, we may affirm that two large areas with different incident solar radiation values are available. The first area, with incident solar radiation values between 1300 and 1350

kWh/m^2/year presents a good solar potential, and is situated in the west of Mureş County (an area which includes the Transylvanian Plain bounded in the North by Bistriţa Năsăud County, in the West by Cluj County, in the South by Sibiu County and at the East by an imaginary line connecting the towns Târnăveni, Târgu Mureş and Reghin).
The second, more narrow, area with a moderate solar potential having incident solar radiation values between 1250 and 1300 kWh/m^2/year is situated in the hill area of Mureş County which includes the Târnave Plateau and the Carpathian lower mountains.
There are also areas with different particularities such as the Mureş County mountain area, where the direct solar radiation has large variations, the negative relief forms (e.g. valleys) favoring the persistence of fog and even diminishing the sun light, while the positive relief forms (e.g. hills, mountains) favor the growth or the decrease of the value for the direct solar radiation, considering their orientation in relation to the sun and the dominant air mass circulation.
The study is based on data provided by ANM from weather stations in Târgu Mureş and Târnăveni. In 2006, ANM in cooperation with NASA, JRC and Meteotest produced a detailed map with solar radiation areas of Romania, which was presented in Figure 11.
A local detailed analysis can be made by accessing JRC website: http://re.jrc.ec.europa.eu (JRC Website, 2009), which offers data regarding solar irradiation on daily, monthly or yearly time base as values or as histogram. For example, for Târgu Mureş (46°32'59" North and 24°33'59" East), the recorded data is presented in Figure 12. JRC's website provides also other important information, such as: the estimated amount of daily, monthly or yearly power generated by a solar panel placed in a specified location. It also takes in consideration the panel type, its installed power and its position relative to the sun. This way, the optimum place for a solar photovoltaic panel in a certain area can be determined technically and economically based on this analysis.

Month	Irradiation at inclination: (Wh/m2/day) 0 deg
Jan	1166
Feb	2024
Mar	3413
Apr	4553
May	5797
Jun	6068
Jul	6224
Aug	5563
Sep	3933
Oct	2738
Nov	1435
Dec	922
Year	3663

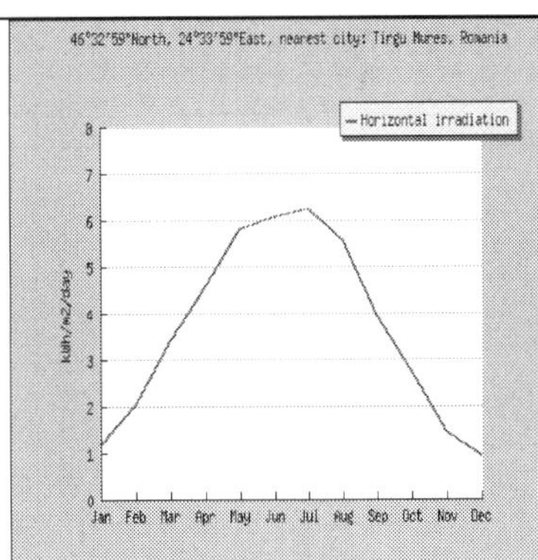

Fig. 12. Solar data for Târgu Mureş according to JRC's website

An alternative to JRC's website is represented by the "Calculator for a photovoltaic system estimation" provided by LP Electric (Alba Iulia, Romania) on their website: http://www.lpelectric.ro/ro/support/aplicatie_solar_ro.html. Depending on the consumer's required power, on the type of system and on the available solar data for a certain location, using the LP Electric application different photovoltaic panels can be chosen, an estimated cost of investment cost and its depreciation period can be calculated. Such an example is shown in Figure 13.

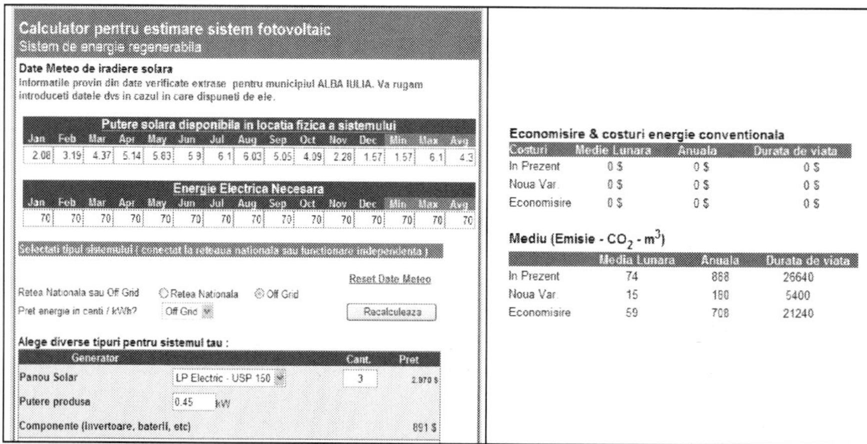

Fig. 13. Calculator for a photovoltaic system estimation provided by LP Electric (LP Electric Website, 2009).

3.2.2 Wind Energy

Assessing the wind potential in a certain areas encounters difficulties in measuring the wind components, especially the wind speed and direction. According to the recommendations, of the World Meteorological Organization, the wind speed and direction have to be measured at 10m above ground. However, the wind speed at which wind exploitation becomes feasible as a renewable energy resource has to be considered at the wind rotor's height, which is situated at 50 to 90 m above ground level for wind power turbines. Consequently, a reevaluation of Romania's wind power potential, by using dedicated software and devices, is necessary. In order to meet this necessity a wind map of Romania, containing the annual average wind speed measured at 50 m above ground, was elaborated by ANM (Figure 14).

According to a study performed by ANM based on the information provided by Weather Stations of Târgu Mureş and Târnăveni, three areas of different wind potential were identified in Mureş County.

The Transylvanian Plain is a region of low wind potential, with average wind speeds of 3-4 m/s and it is not of a great interest from the wind power turbines point of view. Still, there are areas where the relief allows the location of low and medium wind turbines that can give relatively high yield (e.g. Ernei village is suitable for wind turbines of 0,5 kW power).

An area having higher potential with an average wind speed of 4-6 m/s is represented by the hill area of Mureş County which includes the Târnave Plateau (the hill area between Târnava Mică and Târnava Mare rivers), the Carpathian lower mountains and the Gurghiu Mountains. This area is favorable for developing applications with medium and high power wind systems according to the particular structure of the relief.

However, the area with the highest wind potential is the region of the Călimani Mountains with an annual average wind speed of 8-10 m/s or more. Unfortunatelly, exploitation in this part is quite difficult because of the Călimani National Park and of some protected environmental areas. The installation of wind turbines in national parks or reservations is restricted. Presently, some studies are performed by an Austrian company to locate these turbines in the limitary areas of the Călimani National Park.

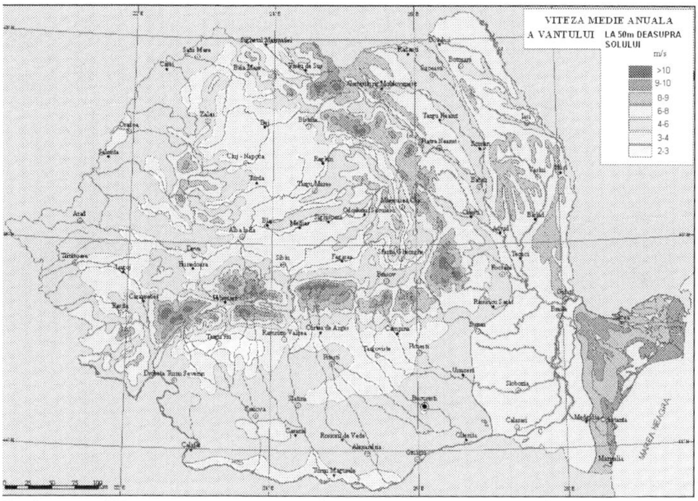

Fig. 14. Wind map of Romania containing the annual average wind speed measured at 50 m above ground [kWh/m^2] (Electrica, 2007).

3.2.3 Hydroelectric Energy

Water represents the most important renewable energy resource of Romania. An inventory of micro-hydropotential is required for the studies concerning the location of future micro-hydro power plants.

This can be done in areas where the watercourse is present permanently all year long with a volume of at least 50 l/s and a slope greater than 10 m/km on rivers that are not included in the great hydroelectric energy locations.

The technical feasible micro-hydropotential is represented by the power or electric energy that can be produced using the water courses mentioned above.

The linear theoretical micro-hydropotential can be calculated to highlight the river sectors with the highest hydroelectrical potential. These sectors can be determined based on the following hypothesis:

- the use of all drained water volume;
- the use of all available fall;
- the total yield of hydraulic energy conversion into electric power equals 1 (losses are ignored).

The hydrographic basin of River Mureş is located in the center and west of Romania, between the Eastern, Southern and Western Carpathians and its lower sector is located in the middle of Tisa Plain. River Mureş is an affluent of River Tisa and collects its main waters from the volcanic chain of the Eastern Carpathians, from the Transylvanian Plateau, from east and south of the Western Carpathians and from the northern slopes of the Southern Carpathians located at the west of River Olt. The dividing crest of River Mureş crosses over different units and sub-units of relief with specific physic-geographical characteristics that separates it from other important hydrographic basins. River Mureş collects a large number of affluents with hydrographic basins between 100 and 1900 km^2. These affluent rivers are: Topliţa, Răstoliţa, Gurghiu, Arieş, Pârâul de Câmpie, Târnava, Sebeş, Cugir, Râul Mare, Strei, Cerna. The surface of the hydrographic basins inventoried in the River Mureş hydrographic basin sums to 15340 km^2 and the proper lenght of 3050 km.

The linear theoretical microhydropotential of river sectors studied in the River Mureş hydrographic basin sums 300994 kW and has a share of 31% with an average specific potential p = 74,6 kW/km. Technical feasible micro-hydropotential studied for rivers with p > 150 kW/km sums to P_i = 76.340 kW with a calculated share of 32% and E_m = 443 GWh/an situates the hydrographic basin of River Mureş among the hydrographic basins of Romania with a good hydroelectric potential to be asserted.

The River Mureş hydrographic basin includes the counties Mureş and Alba and part of Sibiu, Harghita, Cluj, Arad and Hunedoara counties and a very small part of Bistriţa Năsăud country. According to a study performed by ISPH based on data provided by A.N. Apele Române Mureş the location for future micro-hydro power plants is synthesized in Table 1.

Nr.	Name of the power plant	Water course	County	Basin	Brute fall [m]	Installed flow [m^3/s]	Installed power [MW]	Mean energy [GWh/yr.]
1	Duşa	Mureş	Mureş	Mureş	10	11.62	0.8	4.2
2	Zebrac	Mureş	Mureş	Mureş	10	12.19	0.8	4.4
3	Stânceni	Mureş	Mureş	Mureş	10	12.19	0.8	4.4
4	L. Bradului	Mureş	Mureş	Mureş	20	14.05	1.9	10.2
5	Jirca	Mureş	Mureş	Mureş	20	14.17	1.9	10.2
6	Beleiu	Mureş	Mureş	Mureş	10	16.12	1.1	5.8
7	Ferigilor	Mureş	Mureş	Mureş	10	16.24	1.1	5.9
8	Bradului	Mureş	Mureş	Mureş	15	16.35	1.7	8.9
9	Iodului	Mureş	Mureş	Mureş	7	16.41	0.8	4.2
10	Sărăţeni	Târnava Mică	Mureş	Mureş	11	4.5	0.3	2.1
11	Sălard	Sălard	Mureş	Mureş	78	1.65	0.9	5.4

Table 1. Future hydroelectric developments

Among the main affluents of River Mureş are Topliţa, Răstoliţa, Gurghiu, the lower course of River Arieş which crosses Mureş county, and also Rivers Târnava Mică and part of Târnava Mare. All of these affluents have a good hydroelectric potential.

At the moment, there are five functional micro-hydro power plants in Mureş County: three are placed on the Iod Valley in Răstoliţa and two near the Town of Sovata. A development project planned for the upper sector of River Mureş intends to locate a series of micro-hydro power plants to regulate the water-course starting with Răstoliţa dam.

3.2.4 Hybrid DG System

In order to implement a real hybrid system a theoretical preliminary study is required. Such study can be performed on simulation models. A simulation model is presented in Fig. 15.

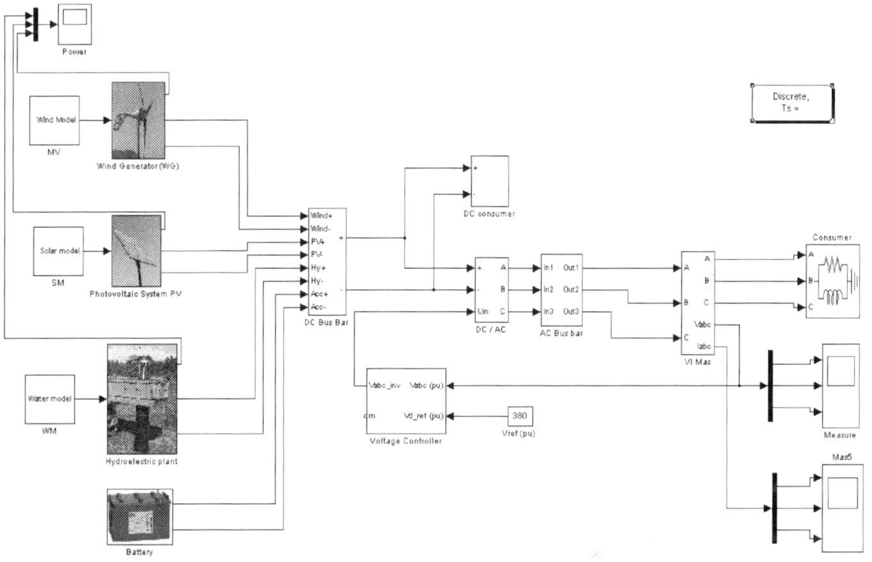

Fig. 15. Simulation model of a hybrid renewable energy system

The simulation model basically consists of the models presented above connected together to form an isolated hybrid system. The proposed model allows studies of modelled DC and AC consumers.

The simulation model allows studies such as:

- renewable energy sources electrical parameters (powers, voltages, currents, etc.);
- renewable energy sources constructive parameters (blades length and number of wind turbine, PV panels number and dimensions, number of hydroelectric turbines, batteries number, etc.);
- voltage and frequency control (control algorithms);
- electrical energy conversion (type of DC/AC conversion methods);
- consumer modelling and control
- power quality distortion phenomena and analysis
- renewable energy availability

Some examples of simulation results are presented below.

Fig. 16 illustrates the voltage waveform measured at the AC bus bar. It can be seen a voltage waveform distortion caused by electronic devices - inverters - used for energy conversion in DC/AC module.

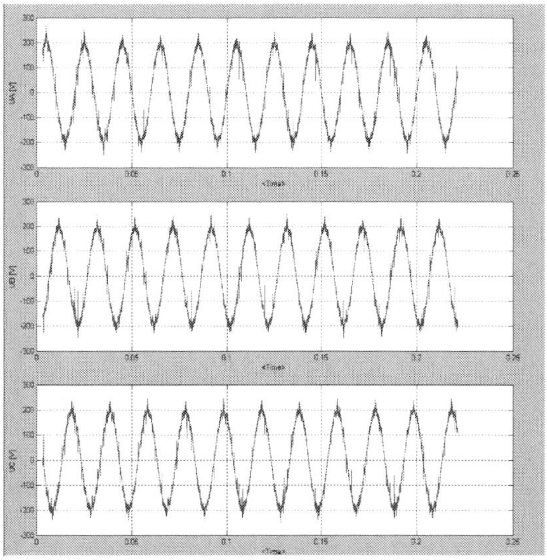

Fig. 16. Voltage waveform at the AC three-phased bus bar

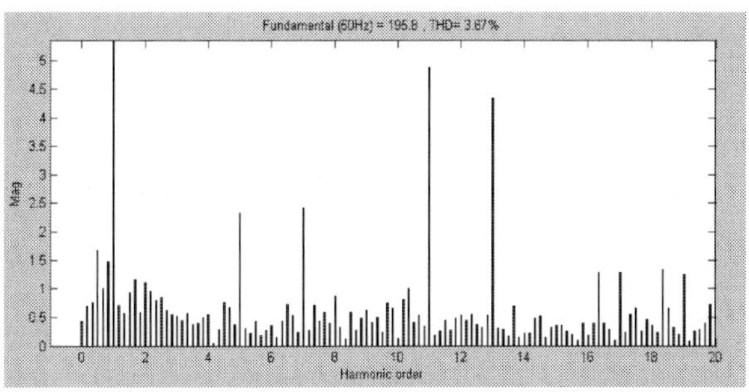

Fig. 17a. Harmonic analysis of the voltage waveform corresponding to phase A

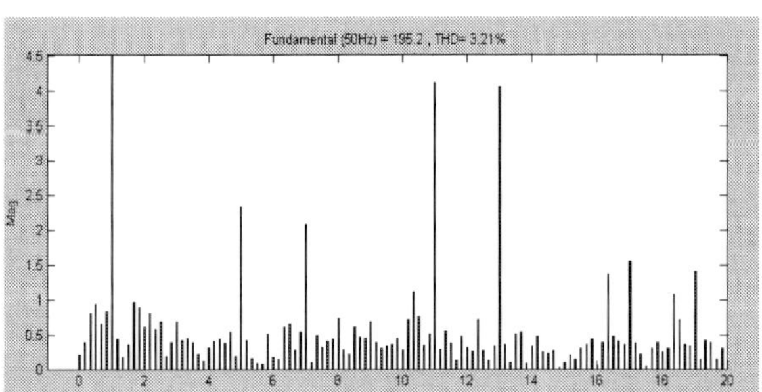

Fig. 17b. Harmonic analysis of the voltage waveform corresponding to phase B

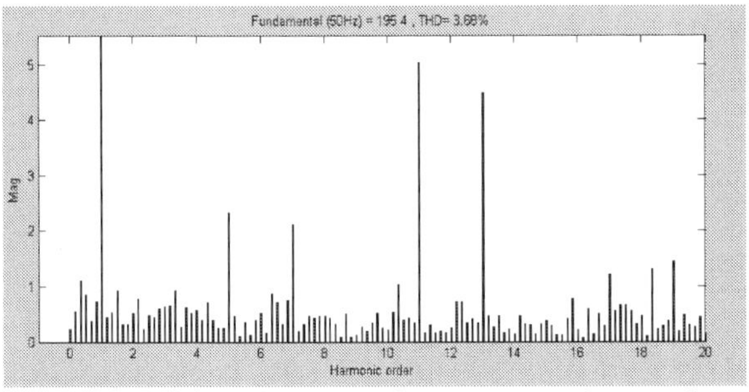

Fig. 17c. Harmonic analysis of the voltage waveform corresponding to phase C

The above simulation results are provided by a hybrid system with installed PV power of 1kW, wind power of 1kW and a hydroelectric turbine of 1kW. It has to be mentioned that the output voltage of hybrid power system generators is 12V and the consumer uses 60% of the available power the rest being used for battery charging. It can be also notified the presence of harmonics caused by consumer but also by the power electronics from electric energy conversion modules. If necessary, on AC side, the power quality can be raised up by using power active filtering devices (Dumitru & Gligor, 2008b).

4. Conclusion

In the conditions of an accelerated economic development in Romania, by a good energy policy it might be assured an increasing level of power supply safety and the reduction of energetic resources import. This demand can be accomplished by implementing a sustained policy of energy conservation, by increasing energy efficiency concurrent with a better capitalization of renewable resources.

The capitalization of renewable resources potential confers real premises to achieve some strategic aims c, but also the durable development of energy sector and the protection of the environment. In order to exploit the economic potential of renewable resources in competitive conditions on the energy market, it is necessary to adopt and implement some energy policies and specific resources. The promotion of energy production from renewable resources (E-SRE) represents an imperative objective in present times justified by environment protection, the increase of energetic independence by supplying sources diversity and, of course, economic and social cohesion reasons. The reasons to promote energy production from renewable resources (E-SRE) were accumulated in time, from economic nature reasons which appeared after the oil crisis in the 70's, to environmental, reasons, especially after signing the Kyoto Protocol in December 1997 and nowadays social cohesion reasons.

Depending on the methods or technologies, rational use of energy can have a positive or a negative impact on power quality. The cause of power quality problems is mainly the improper use of local loads and equipments. But, even if efficient loads are connected to a local grid or bus by a power electronics interface, the power quality might deteriorate. There are also devices with malfunction when the power quality becomes too poor.

5. References

Bloem, J., (2006). *Distributed Generation and Renewables - Integration & Interconnection*, Power Quality and Utilization Guide - LPQI Program, Issue Winter 2006

Burton, T., Sharpe, D., Jenkins, N., Bossanyi, E. (2004). *Handbook of Wind Energy*, John Wiley & Sons Ltd, ISBN: ISBN 0-471-48997-2, The Atrium, Southern Gate, Chichester, England

Chan, H.L., Sutanto, L., (2000). A New Battery Model for Use with Battery Energy Storage Systems and Electric Vehicles Power Systems, In: *IEEE Power Engineering Society Winter Meeting Conference*, Vol.1, pp. 470-475, NJ, USA

Chindriş, M., Cziker, A., Miron, A., (2008). Conectarea la reţea a generatoarelor distribuite. Legislaţia din România, In: *Forumul Naţional al Energiei - Foren 2008*, 15-19 iun. 2008, Neptun, România, on CD

Dumitru, C.D., Gligor, A. (2008a). Power Quality Analysis Of A System Based On Renewable Energy Supplying A Local Distribution Network, *Acta Electrotehnica*, Special Issue: Proceedings of the 2nd International Conference on Modern Power Systems MPS 2008, 12-14 nov. 2008, pag. 224-226, ISSN 1841-3323

Dumitru, C.D., Gligor, A. (2008b). Software Development For Analysis Of Solar-Wind Hybrid Systems Supplying Local Distribution Networks, *Acta Electrotehnica*, Special Issue: Proceedings of the 2nd International Conference on Modern Power Systems MPS 2008, 12-14 nov. 2008, pp. 220-223, ISSN 1841-3323

ELECTRICA S.A., (2007). Preocupările SC. ELECTRICA S.A în domeniul energiei verzi, In: Conferinţa Energiile regenerabile şi actorii principali. O provocare pe termen lung, 30 august 2007, Bucureşti, România

Golovanov, N., Postolache, P., Toader, C. (2007). *Eficienţa şi calitatea energiei electrice*, Ed. AGIR, ISBN: 978-973-720-151-5, Bucureşti, Romania

JRC Website (2009). http://re.jrc.ec.europa.eu/pvgis/apps/pvest.php?lang=en&map

Kaltschmitt, M., Streicher, W., Wise A. (Eds.) (2007). *Renewable energy. Technology, economics and environment*, Springer Berlin Heidelberg New York, ISBN 978-3-540-70947-3

Lorenzo, E. (1994). *Solar Electricity Engineering of Photovoltaic Systems*. Artes Graficas Gala, S.L., ISBN: 84-86505-55-0, Spain

LP Electric Website (2009). http://www.lpelectric.ro/ro/support/aplicatie_solar_ro.html

Markvart, T., Castaner, L. (2003). *Practical Handbook of Photovoltaics, Fundamentals and Applications*, Ed. Elsevier, ISBN 1-85617-390-9, Oxford, UK

Patel, M. R. (1999). *Wind and solar power systems*, CRC Press LLC, ISBN 0-8493-1605-7, Boca Raton, Florida

Sahay, K., Dwivedi, B. (2009). Energy Storage Technology for Performance Enhancement of Power Systems, *Electrical Power Quality & Utilization Magazine*, Volume 4, Issue 1, Available online March 2009

Vairamohan, B., (2002). *State of Charge Estimation of Batteries*, A Thesis Presented for the Master of Science Degree, The University of Tennessee, Knoxville

Willis, H. L., Scott, W. G. (2000). *Distributed Generation Planning and Evaluation*, Marcel Dekker Inc., ISBN: 0-8247-0336-7, New York

17

Planning of Distributed Energy Systems with Parallel Infrastructures: A Case study

Bjorn H. Bakken
SINTEF Energy Research
Norway

1. Introduction

Traditionally, energy supply systems have been designed from a centralized perspective where the prime concern has been to cover the customers' demand for energy at minimum cost. In the increasing drive for more renewable energy and a sustainable society, changes are needed both on the supply and demand side. New technologies and advanced solutions for distributed energy systems are emerging, removing the previous clear distinction between centralised supply options and distributed passive loads. These new technologies yield better possibilities to design sustainable energy systems for the future, but also introduce more complex energy systems to design, operate and maintain. Different types of micro-cogeneration, heat pumps and possibly fuel cells create mutual influence and dependency between different infrastructures.

Furthermore, the ongoing liberalization is causing a change from previously vertically integrated (mono-energy) utilities to horizontally integrated "multi-utilities" supplying electricity, heat and gas to their customers. In a liberalized market regime, also independent energy suppliers can enter the traditional supply area of a local utility and offer their products to customers at competitive conditions. Strategic planners in energy companies thus need to consider both complementarities within their own multi-utility company and competition from others in their "home" market. Public planners, on the other hand, need to be able to give a fair and neutral evaluation of different projects across the traditional energy supply systems of electricity, heat and gas. To meet this development, more comprehensive and flexible planning tools are needed, in particular at distribution system level.

Several approaches have appeared the last years that integrate two or more energy infrastructures in the analysis. Many of these focus on the integrated operation of large scale gas (fuel) and electricity networks for optimal dispatch of generating units and/or pricing of transmission capacity (An et al., 2003; Gil et al., 2003; de Mello & Ohishi, 2005; Morais & Marangon Lima, 2003; Quelhas et al., 2006; Shahidehpour et al., 2005). Others consider downstream optimization of electricity and heat demand from cogeneration units (Henning, 1997; Sandou et al., 2005). Some attack the optimization of multiple energy carriers more generalised, incorporating electricity, gas, heat and hydrogen on the supply side as well as electricity, heating and cooling on the demand side (Aki et al., 2006; Bruckner et al., 1997; Geidl & Andersson, 2005; Lindenberger, 2000; Lindenberger, 2004). However, few, if any, of these approaches consider the issue of expansion/investment planning of such multiple infrastructures.

The area of optimal expansion planning in energy systems with multiple energy carriers is currently dominated by large scale optimisation tools for regional or global system studies like MARKAL/TIMES, EFOM, MESSAGE and similar models (Beller, 1979; Henning, 1999; Goldstein et al., 2003; Messner & Strubegger, 1995; Seebregts et al., 2001). In such large scale studies the energy system is typically represented with an aggregated type of modelling with one energy balance per energy carrier, and with resources deployed on one side and end use extracted on the other side. Various technologies are modelled with emissions and energy losses. This approach is usually sufficient for energy system studies on a national or international level. In an improved optimisation approach for expansion planning in local energy supply systems, however, different infrastructures within the geographical area of concern have to be identified. Geography, topology and timing are all key elements in this approach. It is thus not only a question of which resources and which amounts to use, but also where in the system the necessary investments should take place and when investments should be carried out.

Over the last decade, SINTEF Energy Research has been developing a new optimization model called 'eTransport' that takes into account both the topology and geographic distance of multiple energy infrastructures, and the technical and economic properties of different investment alternatives. The model employs nested optimization algorithms of linear, mixed integer and dynamic programming, calculating both the optimal diurnal operation of the given energy system and the optimal expansion plan typically 15-20 years into the future. The model offers a systematic approach to meet the challenges of planning future energy supply systems with multiple energy carriers.

In this paper, a brief documentation of the eTransport model will be given, and a case study will be presented to demonstrate the use of the model. In the case study, a supply area with variable heat and electricity demand during winter and summer periods considers investments in both electric boilers and a local CHP unit to cover increasing energy demand.

2. The eTransport model

The optimization model "eTransport" is developed for expansion planning in energy systems where several alternative energy carriers and technologies are considered simultaneously (Bakken & Holen, 2004; Bakken et al., 2007). The model uses a detailed network representation of technologies and infrastructure to enable identification of single components, cables and pipelines. The current version optimizes investments in infrastructure over a planning horizon of typically 15 to 20 years for most relevant energy carriers and conversion between these. It is not limited to continuous transport like lines, cables and pipelines, but can also include discrete transport by ship, road or rail.

The model is separated into an operational model (energy system model) and an investment model, as illustrated in Figure 1 (Bakken et al., 2007). In the operational model there are sub-models for each energy carrier and for conversion components. The operational planning horizon is relatively short (1-3 days) with a typical time-step of one hour. The operational model finds the cost-minimising diurnal operation for a given infrastructure and for given energy loads. Annual operating costs for different energy system designs are calculated by solving the operational model repeatedly for different seasons (e.g. peak load, low load, intermediate etc), investment periods (e.g. 2-5 year intervals) and relevant system designs. Annual operating and environmental costs for all different periods and energy system designs are then used by the investment model to find the investment plan that minimises the present value of all costs over the planning horizon.

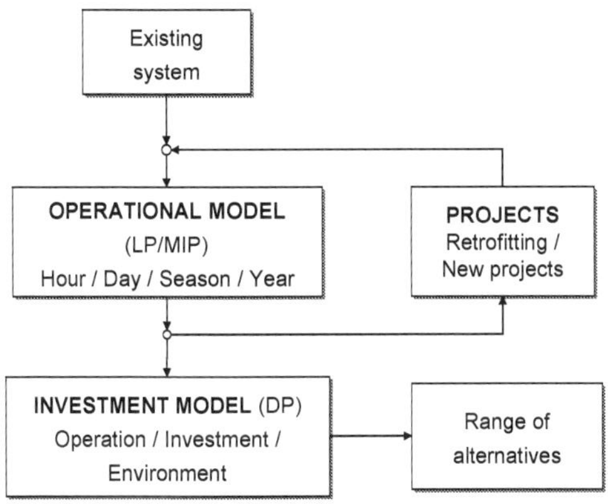

Fig. 1. Combination of operation and investment optimization in eTransport

Mathematically, the model uses a combination of linear programming (LP) and mixed integer programming (MIP) for the operational model, and dynamic programming (DP) for the investment model. The operational model is implemented in the AMPL programming language with CPLEX as solver, while the investment model is implemented in C++. A modular design ensures that new technology modules developed in AMPL for the operational model are automatically embedded in the investment model. A full-graphical drag-and-drop Windows interface is developed for the model in MS Visio. All data for a given case are stored in a database.

In the *operational model* the sub-models for different components are connected by general energy flow variables that identify the flow between energy sources, network components for transport, conversion and storage, and energy sinks like loads and markets. These general variables are included in and restricted by the various models and they are the link between the different models. The different technology models are added together to form a single linear optimisation problem where the object function is the sum of the contributions from the different models, and the restrictions of the problem include all the restrictions defined in the models. Emissions are caused by a subset of components (power plants/CHP, boilers, road/ship transport etc) that are defined as emitting CO_2, NOx, CO and SOx.

Further environmental consequences can be defined. Emissions are calculated for each module and accounted for as separate results. When emission penalties are introduced by the user (e.g. a CO_2 tax), the resulting costs are included in the objective function and thus added to operating costs.

The task for the *investment model* is to find the optimal set of investments during the period of analysis, based on investment costs for different projects and the pre-calculated annual operating costs for different periods and states. The optimal investment plan is defined as the plan that minimises the discounted present value of all costs in the planning period, i.e. operating costs plus investment costs minus the rest value of investments. The optimal plan will therefore identify the optimal design of the energy system (i.e. the optimal state) in different periods. The user defines a set of investment alternatives where each alternative typically consists of several physical components with predefined connections to the rest of the energy system. The same components can be included in several competing investment alternatives, making the different alternatives mutually exclusive from an economic point of view. Mutually exclusive alternatives will be identified by the model in the search for the best expansion plan. More details of the investment algorithm in eTransport can be found in (Bakken et al., 2007).

The combined operational and investment analysis enables a very flexible time resolution as illustrated in Figure 2. The user specifies *Hourly profiles* of prices and loads for one or more days, which are aggregated into one or more seasonal *Segments* (e.g. winter, summer, spring and autumn). The sum of the Segments equals one *Year*, which is the base for the results from the operational analysis. Yearly values of costs and emissions are input to the investment analysis, where one or more years define an *Investment period* (where the model is allowed to make investments). The sum of investment periods, which do not have to be of equal length, is the *Planning horizon* of the case.

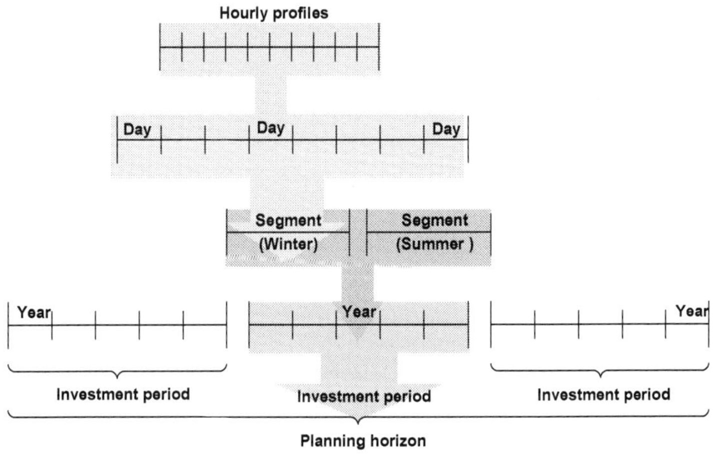

Fig. 2. Time resolution in eTransport

3. Case study

In this paper a case study is used to demonstrate the use of the eTransport model for planning of a distributed energy system with multiple infrastructures and several alternative energy sources. It is worth noting that the example is based on a real case, but the data are slightly tuned to emphasize the different aspects discussed in the paper.

3.1 Case input parameters and assumptions

The case is based on a simplified model of a small municipal/suburban area as shown in Figure 3. There is a mix of residential, service and industrial loads in the area, aggregated into 6 load centres of electricity (blue circles) and heat (red circles). The loads are supplied by a 22 kV distribution grid of underground cables (blue) and a district heating network (red). The main electricity source is the Elspot market (blue square). The old oil fired heat central is being decommissioned, and the challenge in the case is to decide whether to base the future heat supply in the area on:

i) a gas fired CHP (CHP 1),
ii) a waste fired CHP with low quality domestic waste (CHP 2),
iii) a waste/biomass fired CHP with higher quality recycled waste and wood construction materials (CHP 3), or
iv) an electric boiler.

In the case of the three CHP's, surplus electricity may be sold back to the local utility at a fixed price. Electricity demand exceeding the capacity of the CHP's has to be purchased from the Elspot market to market prices.

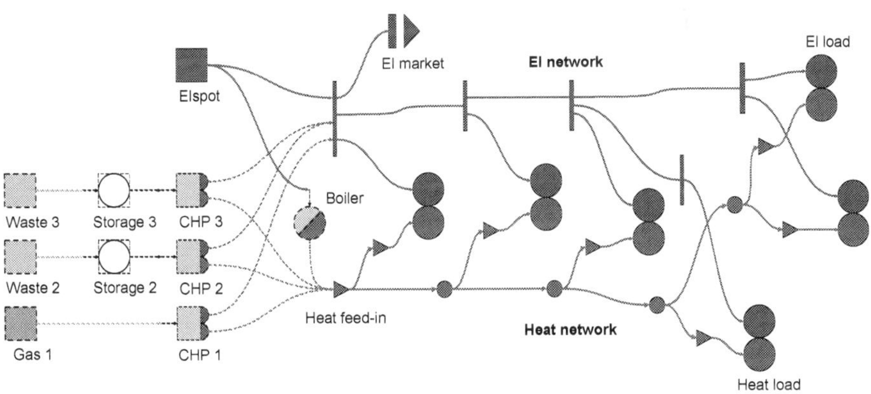

Fig. 3. Energy supply system layout

When a case model is set up as in Figure 3, special consideration has to be given to the supply of heat to the customers. Heat demand can be covered by either electricity through local boilers or direct electric heating, or by hot water from the district heating network. If both alternatives are modelled as in the left hand illustration in Figure 4, a "flip-flop" solution may occur where the model changes from electricity to district heating and back every hour depending on which supply is cheapest at any time. This would require that customers have installed multiple heat systems and is not a feasible solution, so the modeller should take care to separate electricity and heat supplies as shown in the right hand illustration of Figure 4. Physically, this implies a policy/regulatory decision that prohibit users to install electric heating equipment once the district heating system is in operation.

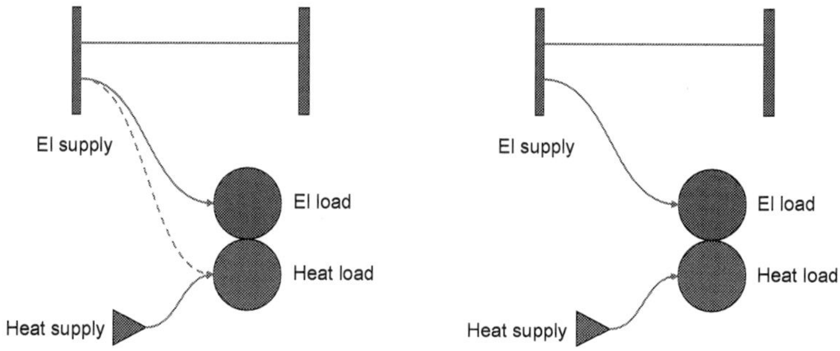

Fig. 4. Heat supply alternatives; *Left*: Multiple heat supply by electricity and heat network; *Right*: Separate electricity and heat supply.

In order to analyse the profitability of a district heating network in a new area, the network itself must be defined as part of the investment. The heat load should then be supplied from the heating network (Figure 4, Right) in the states where the network existed, and from the electricity network in the states with only electricity distribution. Such a case is however not treated in this paper, where we presume the district heating network already exists.

The planning horizon is 15 years (2010-2025), split into three 5-year investment periods. The case is set up with two annual segments: "Summer" of 265 days and "Winter" of 100 days (see Figure 2). The total load profiles for the area for the Summer segment are shown in Figure 5. For simplicity, the Winter segment has the same load profiles but they are 50% higher. The electricity loads are assumed to increase with 10% for each 5-year period, while the heat loads do not increase.

There are no renewable electricity sources in the area, so electricity has to be purchased at the Elspot market to the price shown in Figure 6 for the Summer period. The Elspot price is assumed to be 50% higher in the Winter period, but the same profile is used. Furthermore, if the municipality decides to invest in a CHP, the local utility offers to purchase surplus electricity at a fixed price of 20 Euro/MWh. The Elspot price is assumed to increase 10% for each 5-year period. The interest rate is set to 5% pa.

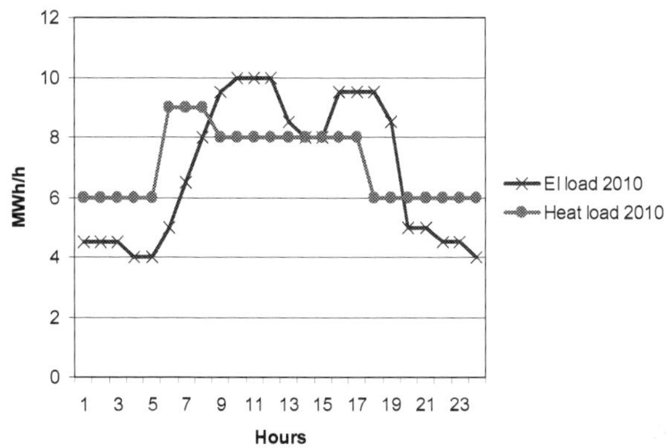

Fig. 5. Total electricity and heat loads in "Summer" period 2010 (MWh/h)

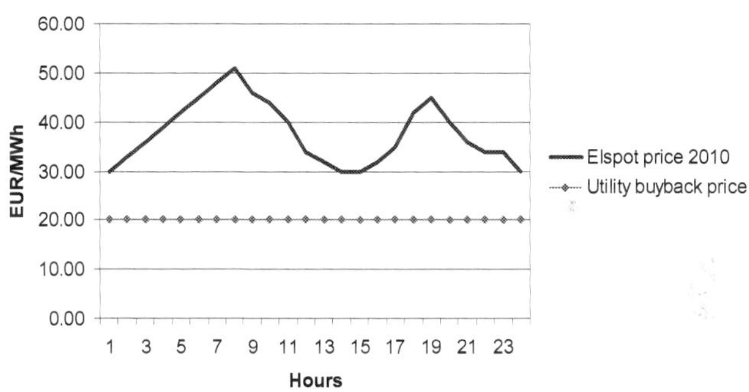

Fig. 6. Elspot market price and utility buyback price in "Summer" period 2010 (EUR/MWh)

Table 1 summarizes the main technical and economic parameters for the four investment alternatives. We assume the same contractor is offering both waste fired units (CHP 2 and CHP 3) as alternative designs for the same investment cost. However, the domestic waste has lower quality than the recycled biomass, so both fuel cost and unit efficiency is lower.

Unit	**Fuel**	**Fuel price**	**Unit rating (MW)**	**Unit efficiency (El / Heat)**	**CAPEX (1000 EUR)**	**OPEX (EUR/year)**	**Service life (years)**
Boiler	Electricity	30-51 EUR/MWh	15	- / 0.9	500	0	30
CHP 1	Gas	0.3 EUR/Sm^3	35	0.4 / 0.45	5 000	50 000	30
CHP 2	Waste	10 EUR/MWh	20	0.3 / 0.4	10 000	70 000	25
CHP 3	Waste	25 EUR/MWh	20	0.35 / 0.4	10 000	60 000	25

Table 1. Basic data for investment alternatives

In some cases the cost of domestic waste can even be negative. The municipality can be obliged to collect and dispose of waste in a safe way, and the alternative cost to incineration can be high. In this case, however, we set a positive but low cost for the waste fuel.

3.2 Results based on initial assumptions

Due to limitations in the waste supply, the rating of the waste units is lower than the gas fired CHP. The model is therefore allowed to combine the electric boiler with the two waste fired CHP's, while the gas fired CHP is large enough to supply the whole heat load alone. Note that the case only deals with the total unit rating, and does not consider whether the heat central consists of one large or several smaller units for backup/redundancy.

With initial assumptions of investments and fuel prices as given above Figure 7 shows the resulting ranking of the alternatives. The CHP 2 unit with cheap domestic waste in combination with the electric boiler turns out to be the cheapest alternative with an annuity of 4.4 mill. Euro. The gas fired CHP 1 unit comes in second place with an annuity of 5.4 mill. Euro. The waste fired CHP 3 comes in third place while the electric boiler is last. The boiler itself has lowest investment cost, but the purchase of electricity for heat production causes the operational costs to exceed the other alternatives. Figure 8 shows that all investments are made in the first investment period.

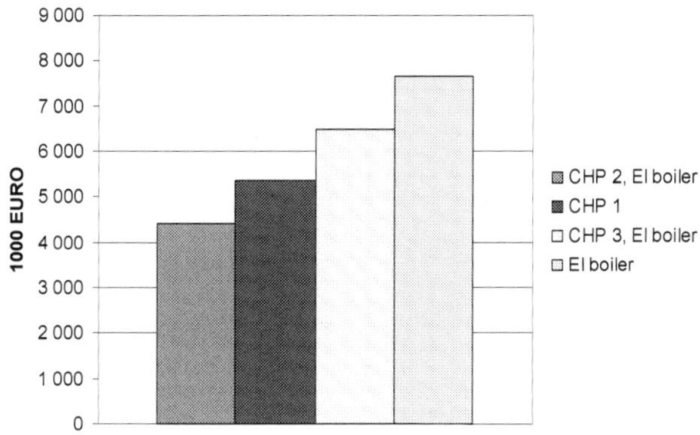

Fig. 7. Ranking of investment alternatives based on annuity of total system cost over planning horizon

The operation of the system can be examined by inspecting the energy profiles of different components. All variables for all components in all segments and investment periods are available to the user in the eTransport user interface, and can also be exported to Excel for further documentation as in this paper. Figure 9 shows the operating profiles of the CHP 2 and the boiler in the "Summer" period in 2010, i.e. the low load period in the first year (corresponding to the load situation shown in Figure 5). Already in this period the CHP is rather heavily loaded, while the boiler is used during peak load hours only. The corresponding electricity generation is shown in Figure 10. As small amount of surplus electricity is sold back to the local utility during the day, but most of the time electricity has to be purchased from the Elspot market.

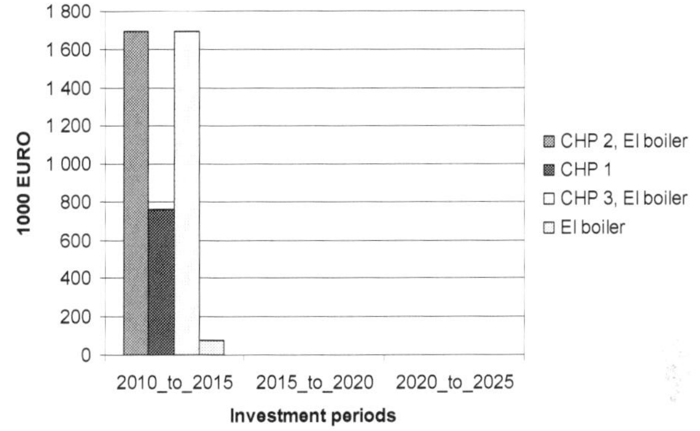

Fig. 8. Investments (costs shown as annuities over the 5-year investment periods)

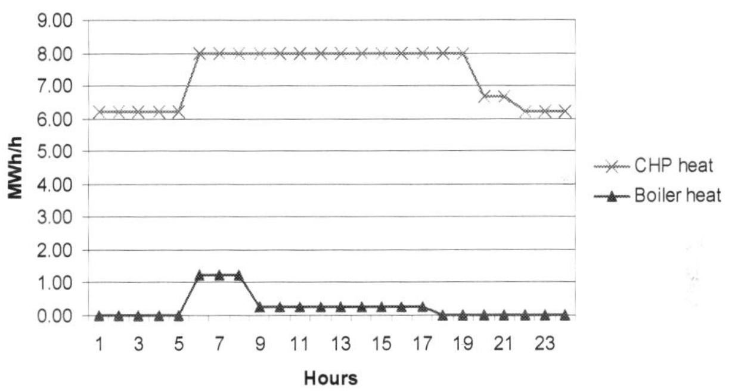

Fig. 9. Heat production from boiler and CHP 2 in "Summer" period 2010 (MWh/h)

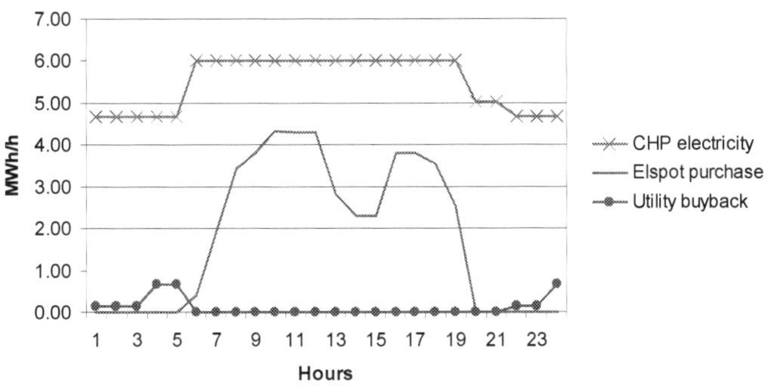

Fig. 10. Electricity production from CHP 2, Elspot purchase and sales of surplus electricity in "Summer" period 2010 (MWh/h)

For comparison, we can show the same variables for the "Winter" period in 2020, i.e. the period of heaviest load in the case. Figure 11 shows that the CHP is running full load all day, and also the boiler has to be in operation the whole day to cover the rest of the load. There is still sufficient capacity in the system to supply the needed heat, but the CHP is obviously due for an upgrade. Similarly, Figure 12 shows that a large amount of electricity has to be purchased from the Elspot market. There is no surplus electricity to sell back to the local utility at this stage.

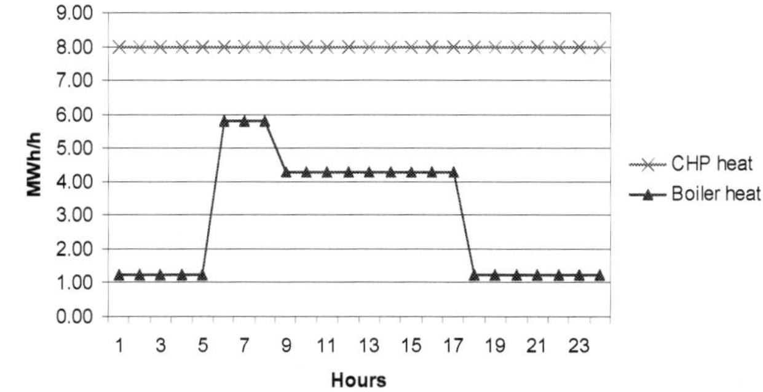

Fig. 11. Heat production from boiler and CHP 2 in "Winter" period 2020 (MWh/h)

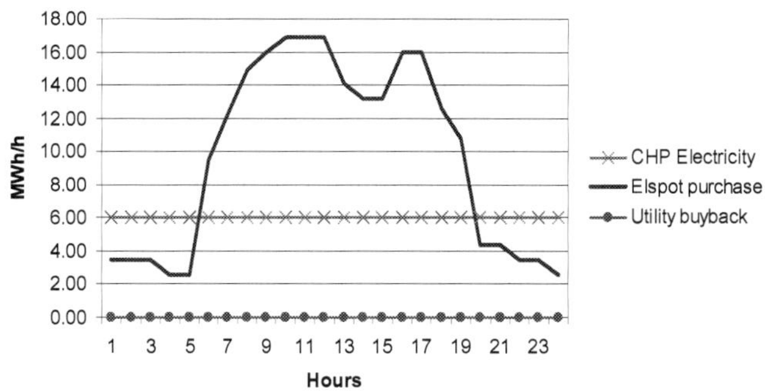

Fig. 12. Electricity production from CHP 2, Elspot purchase and sales of surplus electricity in "Winter" period 2020 (MWh/h)

3.3 Re-negotiation of gas price

As a second step, let us assume that the gas supplier is not satisfied with second place in the investment analysis. He therefore returns to the municipality with an offer for a long-term gas contract at the reduced price of 0.2 EUR/Sm^3.

When all other parameters are kept at initial values, a re-run of the analysis with reduced gas price allows the gas fired CHP 1 to move up to first place with the reduced annuity of 3.8 mill. Euro, as shown in Figure 13. Figure 14 shows that all investments are again made in the first period

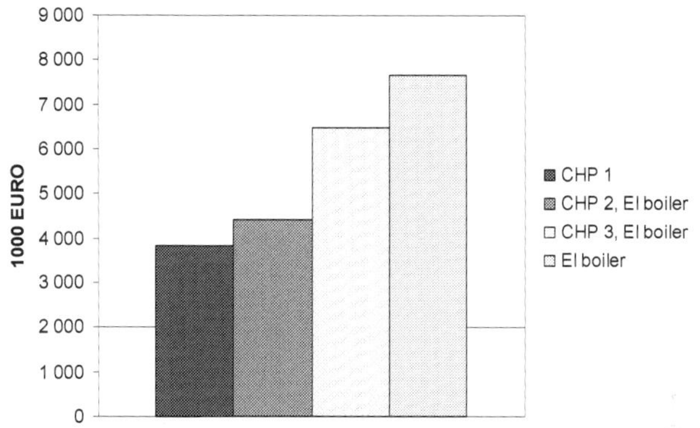

Fig. 13. Ranking of investment alternatives based on annuity of total system cost over period of analysis

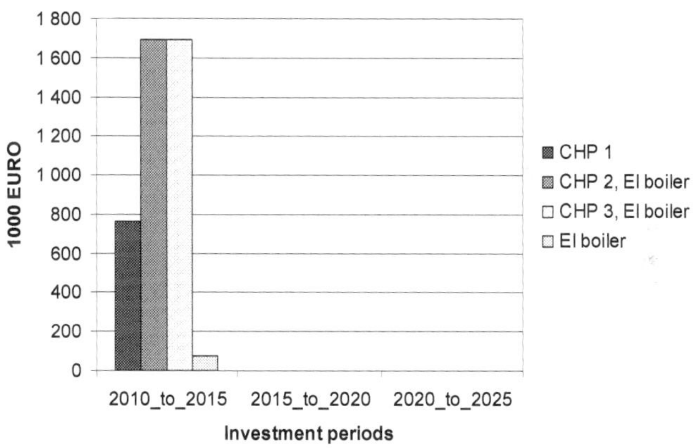

Fig. 14. Investments (costs shown as annuities over the 5-year investment periods)

In this case, the gas fired CHP 1 is large enough to cover the full heat load, but it is more interesting to examine the electricity balance. Figure 15 shows the electricity production, the purchase from the Elspot market and the electricity sold back to the local utility in the "Summer" period of 2010. The hourly distribution between CHP production, Elspot purchase and electricity sales is defined by the relation between the energy prices, so the CHP is not running full load all day to minimize purchase from the Elspot market.

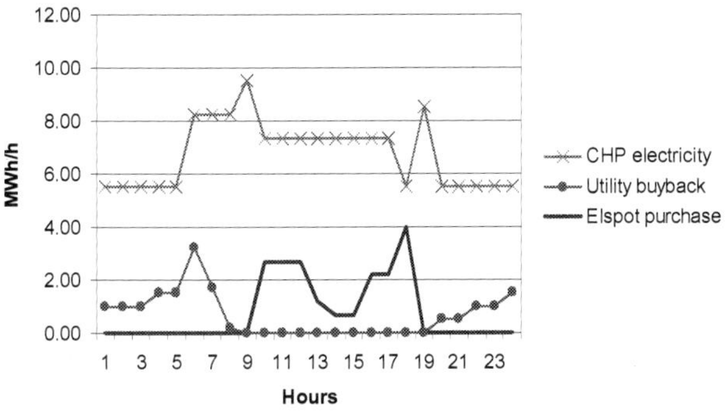

Fig. 15. Electricity production from CHP 1, Elspot purchase and sales of surplus electricity in "Summer" period 2010 (MWh/h)

Moving to the heavy load period of 2020 in Figure 16, the picture is similar. However, now the load is so high that the CHP has to run full load during peak hours and still electricity has to be purchased to cover the load. During off-peak hours some electricity is still sold back to the utility.

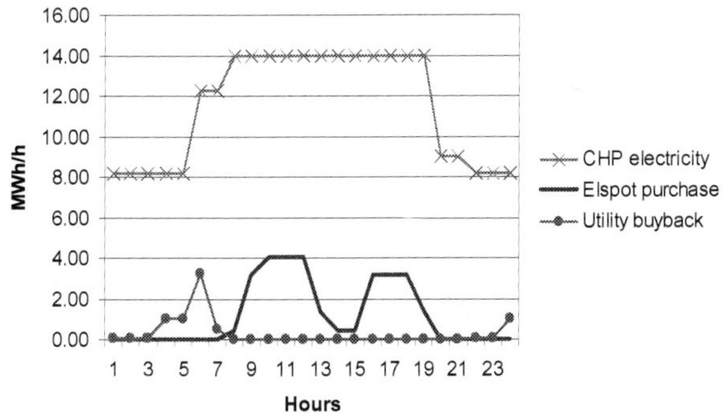

Fig. 16. Electricity production from CHP 1, Elspot purchase and sales of surplus electricity in "Winter" period 2020 (MWh/h)

The gas supplier is naturally satisfied with the changed outcome. However, let us further assume the "Green Party" in the municipality raises the question of CO_2 emissions. The gas fired unit causes average CO_2 emissions of 37.78 tonnes/year over the period of analysis. The domestic waste of CHP 2, on the other hand, is estimated to consist of 40% fossil waste and 60% organic waste, the latter considered as CO_2 neutral. The recycled wood waste of CHP 3 is also CO_2 neutral. Assuming the Elspot electricity is produced with the Norwegian generation portfolio of 99% hydropower, also the electricity can be considered as CO_2 free.

A sensitivity analysis shows that the waste fired CHP 2 moves back as cheaper than the gas fired CHP 1 for a CO_2 tax above 25 Euro/tonne. Although higher than current CO_2 prices, this causes some concern in the municipality. The final decision is to build the waste fired unit CHP 2 in combination with the electric boiler as peak load/backup unit.

3.4 Delayed commissioning

The municipality has decided to sign a contract for the waste fuelled CHP 2. However, such technologies are still quite young, so in a final sensitivity analysis we assume there will be problems with the delivery and/or commissioning of the unit. This is modelled as a delay in the CHP 2 investment until the second investment period (2015), changing also the cash flows accordingly. Figure 17 shows that in this case the gas fired unit CHP 1 again moves up as cheapest alternative even with the original gas price of 0.3 Euro/Sm^3.

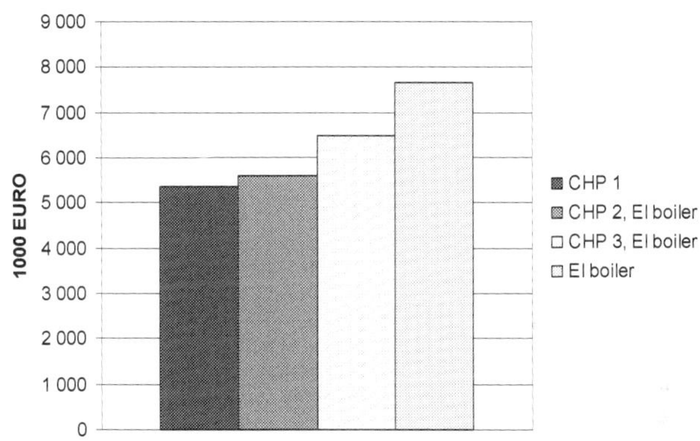

Fig. 17. Ranking of investment alternatives based on annuity of total system cost over period of analysis

Another interesting observation from Figure 18 is that the electric boiler is put into operation in 2010, while the CHP 2 does not come online until 2015. The added operational expenses of supplying heat only from electricity for the first 5 years cause the waste alternative to come in second place.

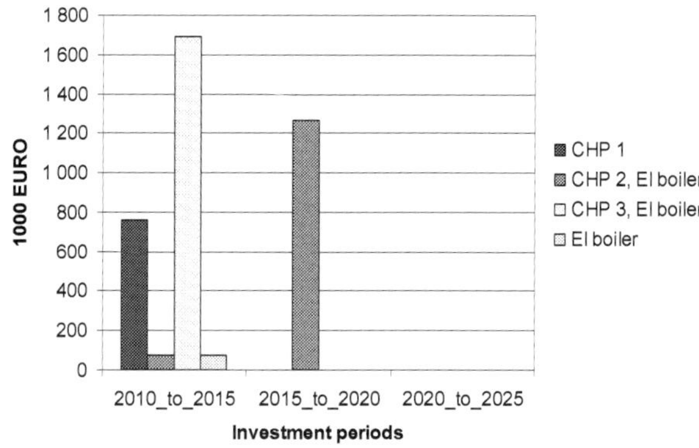

Fig. 18. Investments (costs shown as annuities over the 5-year investment periods)

4. Summary

This paper has presented a novel optimization model 'eTransport' for planning of distributed energy supply systems with parallel infrastructures. The model minimises total energy system costs (investments, operation and emissions) of meeting predefined energy demands of electricity, space heating and tap water heating within a geographical area over a given planning horizon. Many topographical details can be included for the different supply infrastructures (electricity, natural gas, biomass, waste, district heating etc), and this makes the model especially appropriate for local energy planning e.g. in municipalities or cities.

The main user group for the eTransport model are decision makers involved in planning of local energy service systems including new DG plants. Local authorities (e.g. municipalities) need to analyze the local energy system with respect to concessions and/or energy planning. It is also useful for utilities that do regional energy studies, including large energy suppliers that must find the least-cost option for their supply. Governmental agencies that give investment subsidies on basis of socio-economic efficiency can also use the model to analyse the effects of the support.

The model currently employs a nested optimization of mixed integer programming and dynamic programming, calculating both the optimal diurnal operation of the complete energy system and the optimal expansion plan typically 20-30 years into the future. A full graphical user interface is also developed to increase the user-friendliness of the model. In the next version stochastic optimization is implemented to handle uncertainties in energy prices, demand and investments.

The main focus of this paper has not been the optimisation algorithm itself; rather the usefulness of such an optimisation model is demonstrated though a case study. Initially, the model is used to find the least-cost alternative for future energy supply in a suburb or municipality. Then various assumptions are changed to see how the results are affected. The case study uses fictive input data that have been tuned to show the effects of various changes. Usually, the process of analysing the sensitivity of different solutions to changes in parameters is done offline by an analyst and presented afterwards. However, due to the full graphic user interface developed for the eTransport model, sensitivity analyses can also be performed "live" with municipal or corporate decision makers present. This will enhance the involvement of the decision makers and thus increase the value of the analysis. This is especially important in the case of municipal planning, where political decision makers may have low technical competence.

5. Acknowledgement

The author gratefully acknowledges the support from the Research Council of Norway and from the sponsors of the "eTransport" development.

6. References

Aki, H.; Yamamoto, S.; Kondoh, J.; Maeda, T.; Yamaguchi, H.; Murata, A. & Ishii, I. (2006). Fuel cells and energy networks of electricity, heat and hydrogen in residential areas, *Hydrogen Energy*, Vol. 31, 2006, 967-980

An, S.; Li, Q. & Gedra, T.W. (2003). Natural gas and electricity optimal power flow, *Proceedings of IEEE PES 2003 Transmission and Distribution Conference*, pp. 138-143, Dallas, USA

Bakken, B.H. & Holen, A.T. (2004). Energy service systems: Integrated planning case studies, *Proceedings of IEEE PES 2004 General Meeting*, Denver, USA, 2004

Bakken, B.H.; Skjelbred, H.I. & Wolfgang, O (2007). eTransport: Investment Planning in Energy Supply Systems with Multiple Energy Carriers, *Energy*, Vol. 32, 2007, 1676-1689

Beller, M. (1979). The Applications of Energy Systems Analysis to Policy and Technology Studies, *Proceedings of Int. Conf. of Energy Systems Analysis*, Dublin, Ireland, 1979

Bruckner, T.; Groscurth, H.M. & Kümmel, R. (1997). Competition and synergy between energy technologies in municipal energy systems, *Energy*, Vol. 22, No. 10, 1997, 1005-1014

Geidl, M. & Andersson, G. (2005). A modeling and optimization approach for multiple energy carrier power flow. *Proceedings of IEEE PES 2005 PowerTech Conference*, St. Petersburg, Russian Federation, 2005.

Gil, E.M.; Quelhas, A.M.; McCalley, J.D. & Van Voorhis, T. (2003). Modeling integrated energy transportation networks for analysis of economic efficiency and network interdependencies, *Proceedings of 35th North American Power Symposium (NAPS)*, Rolla, USA, 2003

Goldstein, G.A.; Kanudi, A. & Loulou, R. (2003). *MARKAL: An Energy-Environment-Economic Model for Sustainable Development*, International Resources Group, NESCAUM, Boston, USA, 2003. Available from: http://www.nescaum.org

Henning, D. (1997). MODEST - An Energy-System Optimisation Model Applicable to Local Utilities and Countries, *Energy*, Vol. 22, No. 12, 1997, 1135-1150

Henning, D. (1999). *Optimisation of Local and National Energy Systems. Development and Use of the MODEST Model*, PhD Dissertation, Linköping University, ISBN 91-7219-391-1, Linköping, Sweden

Lindenberger, D.; Bruckner, T.; Groscurth, H.M. & Kümmel, R. (2000) Optimization of solar district heating systems: Seasonal storage, heat pumps and cogeneration, *Energy*, Vol. 25, 2000, 591-608

Lindenberger, D.; Bruckner, T.; Morrison, R.; Groscurth, H.M. & Kümmel, R. (2004). Modernization of local energy systems, *Energy*, Vol. 29, 2004, 245-256

de Mello, O.D. & Ohishi, T. (2005). An integrated dispatch model of gas supply and thermoelectric systems. *Proceedings of 15th Power Systems Computation Conference*, Liege, Belgium, 2005

Messner, S. & Strubegger, M. (1995). *User's Guide for MESSAGE III, WP-95-69*, International Institute for Applied Systems Analysis, Laxenburg, Austria, 1995. Available from: http://www.iiasa.ac.at/Research/ECS/docs/MESSAGE_man018.pdf

Morais, M.S. & Marangon Lima, J.W. (2003). Natural gas network pricing and its influence on electricity and gas markets, *Proceedings of IEEE PES 2003 PowerTech Conference*, Bologna, Italy, 2003

Quelhas, A.M.; Gil, E. & McCalley JD (2006). Nodal prices in an integrated energy system. *Int. Journal of Critical Infrastructures*, Vol. 2, No. 1, 2006, 50-69

Sandou, G.; Font, S.; Tebbani, S.; Hiret, A. & Mondon, C. (2005). Short term optimization of cogeneration systems considering heat and electricity demands. *Proceedings of 15th Power Systems Computation Conference*, Liege, Belgium, 2005

Seebregts, J.; Goldstein, G.A. & Smekens, K. (2001). Energy/Environmental Modeling with the MARKAL Family of Models. In: *2001 Int. Conf. on Operations Research (OR 2001)*, Chamoni, P.; Leisten, R.; Martin, A.; Minnemann, J. & Stadtler, H. (Ed.), Duisburg, Germany, 2001

Shahidehpour, M.; Fu, Y. & Wiedman, T. (2005). Impact of natural gas infrastructure on electric power systems. *Proceedings of the IEEE*, Vol. 93, No. 5, 2005, 1042–1056

18

Power Electronics Control of Wind Energy in Distributed Power Systems

Florin Iov and Frede Blaabjerg
Department of Energy Technology, Aalborg University
Denmark

1. Introduction

In classical power systems, large power generation plants located at adequate geographical places produce most of the power, which is then transferred towards large consumption centres over long distance transmission lines. The system control centres monitor and regulate the power system continuously to ensure the quality of the power, namely frequency and voltage. However, now the overall power system is changing, a large number of dispersed generation (DG) units, including both renewable and non-renewable sources such as wind turbines, wave generators, photovoltaic (PV) generators, small hydro, fuel cells and gas/steam powered Combined Heat and Power (CHP) stations, are being developed (Heier, 1998), (Bossany, 2000) and installed. A wide-spread use of renewable energy sources in distribution networks and a high penetration level will be seen in the near future in many places. For instance Denmark has a high power capacity penetration (> 20%) of wind energy in major areas of the country and today around 20% of the whole electrical energy consumption is covered by the wind energy. The main advantages of using renewable energy sources are the elimination of harmful emissions and inexhaustible resources of the primary energy. The availability of renewable energy sources has strong daily and seasonal patterns and the power demand by the consumers could have a very different characteristic. Therefore, it is difficult to operate a power system installed with only renewable generation units due to the characteristic differences and the high uncertainty in the availability of the renewable energy sources. This is further strengthened as no real large energy storage systems exist.

The wind turbine technology is one of the most emerging renewable energy technologies. It started in the 1980'es with a few tens of kW power capacity to today with multi-MW size wind turbines that are being installed. It also means that wind power production in the beginning did not have any impact on the power system control but now due to their size they have to play an active role in the grid. The technology used in wind turbines was in the beginning based on a squirrel-cage induction generator connected directly to the grid. By that power pulsations in the wind are almost directly transferred to the electrical grid. Furthermore, there is no control of the active and reactive power, which typically are important control parameters to regulate the frequency and the voltage. As the power range of the turbines increases these control parameters become more important and it is necessary to introduce power electronics (Hansen et al., 2002) as an interface between the wind turbine and the grid. The power electronics is changing the basic characteristic of the wind turbine from being an energy source to be an active power source. The electrical technology used in wind turbines is not new. It has been discussed for several years e.g. (Hansen et al. 2004); (Gertmar, 2003); (Blaabjerg et al., 2004); (Iov & Blaabjerg, 2007) etc. but now the price pr. produced kWh is so low, that solutions with power electronics are very attractive.

Firstly, the basic development in power electronics and power electronic conversion will be discussed. Then, different wind turbine configurations will be explained both aerodynamically and electrically. A short overview of the interconnection requirements will be given. Also, different control methods for power converters within a wind turbine system will be presented. Finally, a general technology status of the wind power is presented demonstrating still more efficient and attractive power source for the future.

2. Modern power electronics

Power electronics has changed rapidly during the last thirty years and the number of applications has been increasing, mainly due to the developments of the semiconductor devices and the microprocessor technology. For both cases higher performance is steadily given for the same area of silicon, and at the same time they are continuously reducing in price. A typical power electronic system, consisting of a power converter, a renewable energy source and a control unit, is shown in Fig. 1.

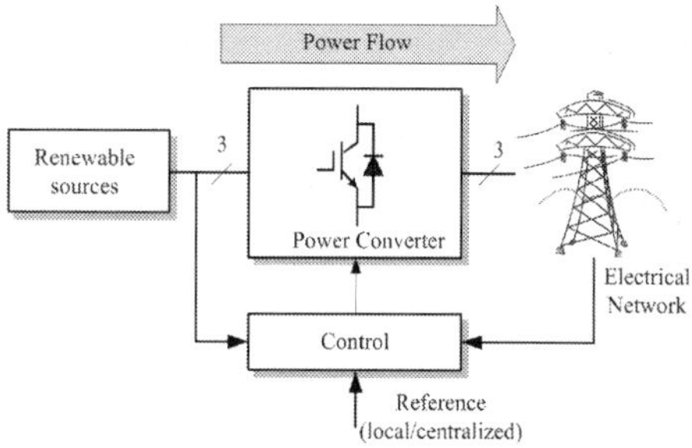

Fig. 1. Power electronic system with the grid, renewable source, power converter and control.

The power converter is the interface between the generator and the grid. Typically, the power flow is uni-directional from the generator to the electrical network. Three important issues are of concern using such a system namely the reliability, the efficiency and last but not least the cost. Currently, the cost of power semiconductor devices is decreasing 1÷5 % every year for the same output performance and the price pr. kW for a power electronic system is also decreasing. An example of a mass-produced and high competitive power electronic system is an adjustable speed drive (ASD). The trend of weight, size, number of components and functions in a standard Danfoss Drives A/S frequency converter can be seen in Fig. 2. It clearly shows that power electronic conversion is shrinking in volume and weight. It also shows that more integration is an important key to be competitive as well as more functions become available in such a product.

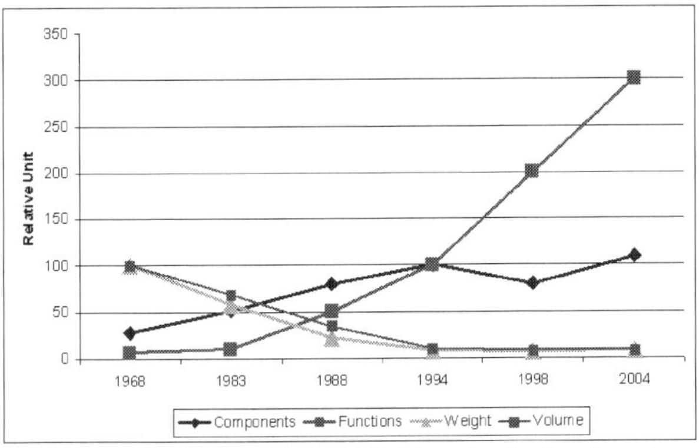

Fig. 2. Development of standard adjustable speed drives for the last four decades.

The key driver of this development is that the power electronic device technology is still undergoing important progress. An overview of different power devices and the areas where the development is still going on is presented in Fig. 3.

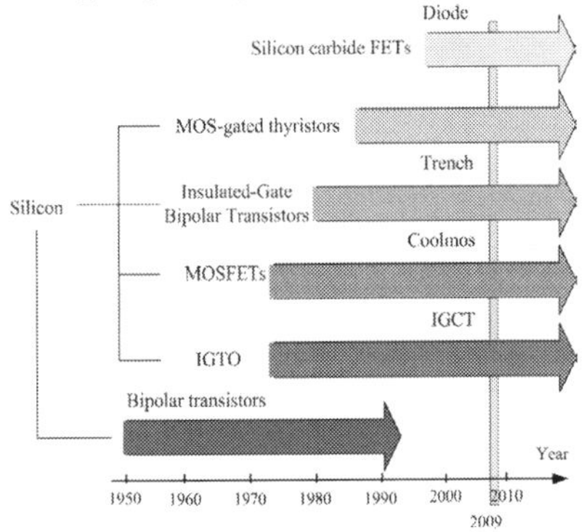

Fig. 3. Development of power semiconductor devices in the past and in the future (based on (Baliga, 1995)).

The only power device which is not under development anymore is the silicon-based power bipolar transistor because MOS-gated devices are preferable in the sense of easy control. The breakdown voltage and/or current carrying capability of the components are also continuously increasing. Important research is going on to change the material from silicon to silicon carbide, which may dramatically increase the power density of power converters as well as their voltage capability. Using such devices a direct connection on the medium voltage networks of the power converters without a step-up transformer may be possible.

3. Wind energy conversion

Wind turbines capture power from the wind by means of aerodynamically designed blades and convert it to rotating mechanical power. The number of blades is three in a modern wind turbine. As the blade tip-speed should be lower than half the speed of sound the rotational speed will decrease as the radius of the blade increases. For multi-MW wind turbines the rotational speed is typically 10-15 rpm. The most weight efficient way to convert the low-speed, high-torque power to electrical power is to use a gear-box and a standard fixed speed generator as illustrated in Fig. 4.

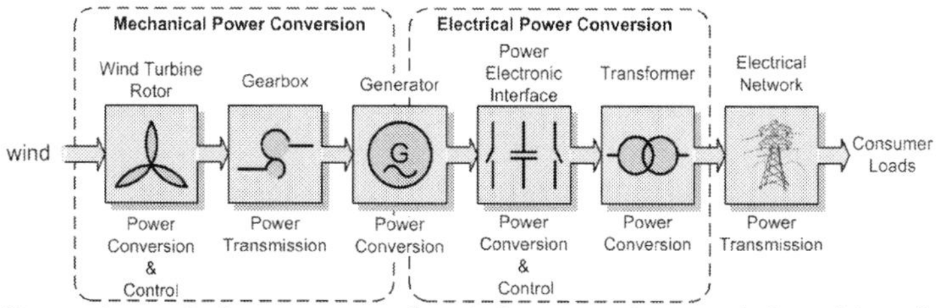

Fig. 4. Converting wind power to electrical power in a wind turbine (based on Kazmierkowski et al., 2002)).

The gear-box is optional as multi-pole generator systems are also possible solutions. Between the grid and the generator a power converter can be inserted. The possible technical solutions are many and a technological roadmap starting with wind energy/power and converting the mechanical power into electrical power is shown in Fig. 5. The electrical output can either be AC or DC. In the last case a power converter will be used as interface to the grid.

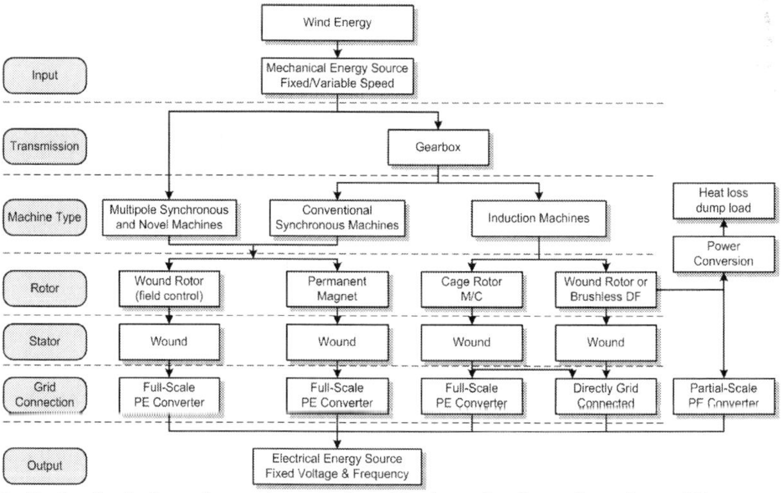

Fig. 5. Technological roadmap for wind turbine's technology (based on (Hansen et al., 2001)).

The development in wind turbine systems has been steady for the last 25 years and four to five generations of wind turbines exist and it is now proven technology. It is important to be able to control and limit the converted mechanical power at higher wind speed, as the power in the wind is a cube of the wind speed. The power limitation may be done either by stall control (the blade position is fixed but stall of the wind appears along the blade at higher wind speed), active stall (the blade angle is adjusted in order to create stall along the blades) or pitch control (the blades are turned out of the wind at higher wind speed) (Hansen et al., 2004), (Iov & Blaabjerg, 2007). The basic output characteristics of these three methods of controlling the power are summarized in Fig. 6. The modern wind turbines use only active stall and pitch control.

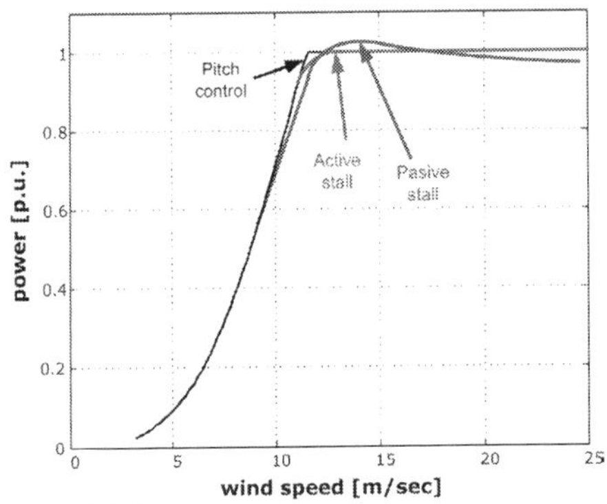

Fig. 6. Power characteristics of different wind turbine systems.

Another control variable in a wind turbine system is the speed. Based on this criterion the wind turbines are classified into two main categories (Hansen et al., 2004), (Iov & Blaabjerg, 2007); namely fixed speed and variable speed wind turbines respectively.
Additionally, the presence of power converters in wind turbines also provides high potential control capabilities for both large modern wind turbines and wind farms to fulfil the high technical demands imposed by the grid operators (Sørensen et al., 2000); (Hansen et al., 2004); (Milborrow, 2005) and (Iov & Blaabjerg, 2007), such as: controllable active and reactive power (frequency and voltage control); quick response under transient and dynamic power system situations, influence on network stability and improved power quality.

4. Wind Turbine Concepts

The most commonly applied wind turbine designs can be categorized into four wind turbine concepts. The main differences between these concepts concern the generating system and the way in which the aerodynamic efficiency of the rotor is limited during above the rated value in order to prevent overloading. These concepts are presented briefly in the following paragraphs.

4.1 Fixed Speed Wind Turbines

This configuration corresponds to the so called Danish concept that was very popular in 80's. This wind turbine is fixed speed controlled machine, with asynchronous squirrel cage induction generator (SCIG) directly connected to the grid via a transformer as shown in Fig. 7. This concept needs a reactive power compensator to reduce (almost eliminate) the reactive power demand from the turbine generators to the grid. It is usually done by continuously switching capacitor banks following the production variation (5-25 steps) Smoother grid connection occurs by incorporating a soft-starter. Regardless the power control principle in a fixed speed wind turbine, the wind fluctuations are converted into mechanical fluctuations and further into electrical power fluctuations. These can yield to voltage fluctuations at the point of connection in the case of a weak grid. Because of these voltage fluctuations, the fixed speed wind turbine draws varying amounts of reactive power from the utility grid (in the case of no capacitor bank), which increases both the voltage fluctuations and the line losses.

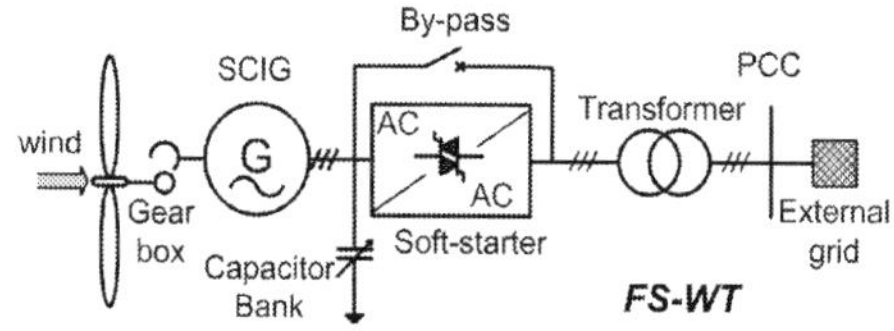

Fig. 7. Fixed speed wind turbine with directly grid connected squirrel-cage induction generator.

Thus, this concept does not support any speed control, requires a stiff grid and its mechanical construction must be able to support high mechanical stress caused by wind gusts.

4.2 Partial Variable Speed Wind Turbine with Variable Rotor Resistance

This configuration corresponds to the limited variable speed controlled wind turbine with variable rotor resistance, known as OptiSlip (VestasTM) as presented in Fig. 8.

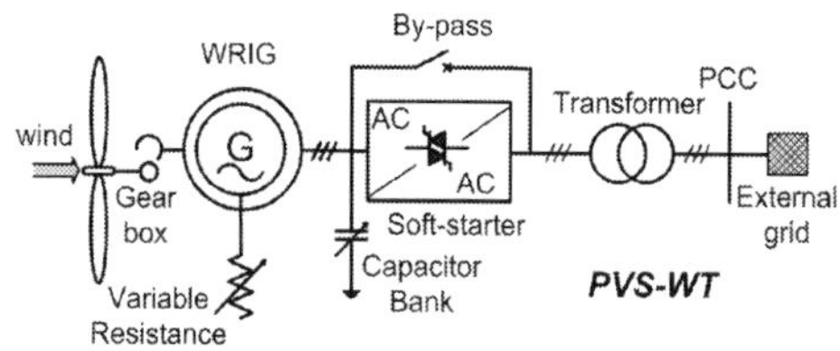

Fig. 8. Partial variable speed wind turbine with variable rotor resistance.

The generator is directly connected to the grid. The rotor winding of the generator is connected in series with a controlled resistance, whose size defines the range of the variable speed (typically 0-10% above synchronous speed). A capacitor bank performs the reactive power compensation and smooth grid connection occurs by means of a soft starter. An extra resistance is added in the rotor circuit, which can be controlled by power electronics. Thus, the total rotor resistance is controllable and the slip and thus the power output in the system are controlled. The dynamic speed control range depends on the size of the variable rotor resistance. Typically the speed range is 0-10% above synchronous speed. The energy coming from the external power conversion unit is dumped as heat loss.

4.3 Variable Speed WT with partial-scale power converter

This configuration, known as the doubly-fed induction generator (DFIG) concept, corresponds to the variable speed controlled wind turbine with a wound rotor induction generator (WRIG) and partial-scale power converter (rated to approx. 30% of nominal generator power) on the rotor circuit as shown in Fig. 9. The stator is directly connected to the grid, while a partial-scale power converter controls the rotor frequency and thus the rotor speed.

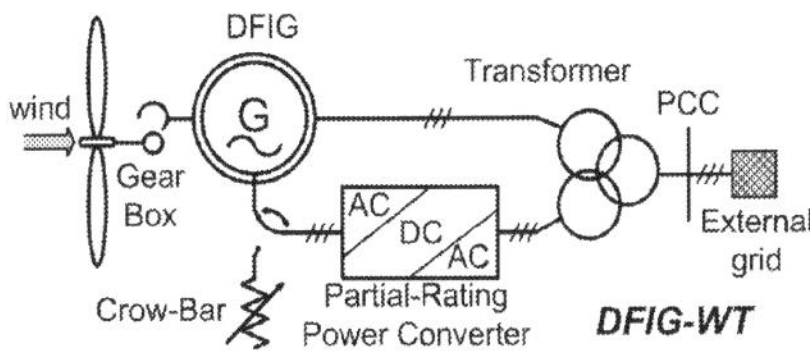

Fig. 9. Variable speed wind turbine with partial scale power converter.

The power rating of this partial-scale frequency converter defines the speed range (typically ±30% around synchronous speed). Moreover, this converter performs the reactive power compensation and a smooth grid connection. The control range of the rotor speed is wide compared to that of OptiSlip. Moreover, it captures the energy, which in the OptiSlip concept is burned off in the controllable rotor resistance. The smaller frequency converter makes this concept attractive from an economical point of view. Moreover, the power electronics is enabling the wind turbine to act as a more dynamic power source to the grid. However, its main drawbacks are the use of slip-rings and the complicated protection schemes in the case of grid faults.

4.4 Variable speed wind turbine with full-rating power converter

This configuration corresponds to the variable speed controlled wind turbine, with the generator connected to the grid through a full-rating power converter as shown in Fig. 10.

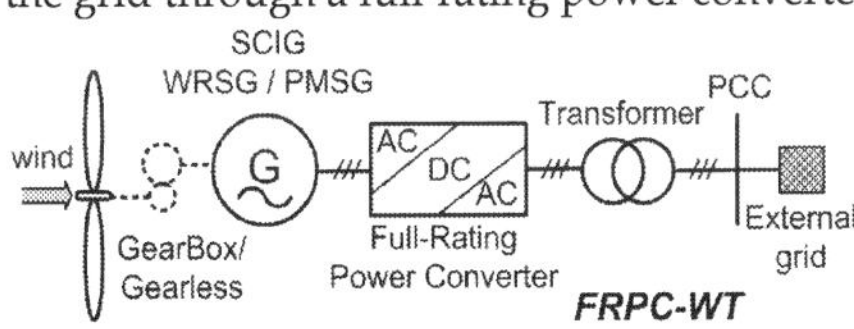

Fig. 10. Variable speed wind turbine with full-rating power converter.

The power converter performs the reactive power compensation and a smooth grid connection for the entire speed range. The generator can be electrically excited (wound rotor synchronous generator WRSG) or permanent magnet excited type (permanent magnet synchronous generator PMSG). The stator windings are connected to the grid through a full-scale power converter. Recently, due to the advancement in the power electronics the squirrel-cage induction generator has also started to be used for this concept.

Some variable speed wind turbines systems are gearless - see dotted gearbox in Fig. 10. In these cases, a direct driven multi-pole generator is used. The wind turbine companies Enercon, Siemens Wind Power, Made and Lagerwey are examples of manufacturers using this configuration.

The variable speed wind turbines are designed to achieve maximum aerodynamic efficiency over a wide range of wind speed. By introducing the variable speed operation, it is possible to continuously adapt (accelerate or decelerate) the rotational speed of the wind turbine to the wind speed, in such a way that tip speed ratio is kept constant to a predefined value corresponding to the maximum power coefficient. Contrary to a fixed speed system, a variable speed system keeps the generator torque nearly constant. Thus, the variations in wind are absorbed by the generator speed changes.

Seen from the wind turbine point of view, the most important advantages of the variable speed operation compared to the conventional fixed speed operation are: reduced mechanical stress on the mechanical components such as shaft and gearbox, increased power capture and reduced acoustical noise.

5. Grid connection requirements for wind power

Some European countries have at this moment dedicated grid codes for wind power. These requirements reflect, in most of the cases, the penetration of wind power into the electrical network or that a future development is prepared.

The requirements for wind power cover a wide range of voltage levels from medium voltage to very high voltage. The grid codes for wind power address issues that make the wind farms to act as a conventional power plant into the electrical network. These requirements have focus on power controllability, power quality, fault ride-through capability and grid support during network disturbances. According to several references e.g. (Hansen et al., 2004) and (Milborrow, 2005) in some of the cases these requirements are very stringent. An overview of the most important requirements of the European grid codes is given in the following based on (Iov & Blaabjerg, 2007).

5.1 Active power control

According to this demand the wind turbines must be able to control the active in the Point-of-Common-Coupling (PCC) in a given power range. The active power is typically controlled based on the system frequency e.g. Denmark, Ireland, Germany so that the power delivered to the grid is decreased when the grid frequency rise above 50 Hz. A typical characteristic for the frequency control in the Danish grid code is shown in Fig. 11.

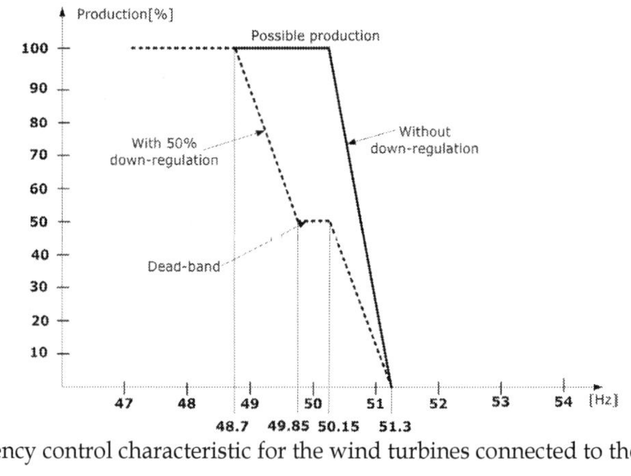

Fig. 11. Frequency control characteristic for the wind turbines connected to the Danish grid.

On the contrary other grid codes, e.g. Great Britain specifies that the active power output must be kept constant for the frequency range 49.5 to 50.5 Hz, and a drop of maximum 5% in the delivered power is allowed when frequency drops to 47 Hz.
Curtailment of produced power based on system operator demands is required in Denmark, Ireland, Germany and Great Britain. Currently, Denmark has the most demanding requirements regarding the controllability of the produced power. Wind farms connected at the transmission level shall act as a conventional power plant providing a wide range of controlling the output power based on Transmission System Operator's (TSO) demands and also participation in primary and secondary control. Seven power regulation functions, each of them prioritized, are required in the wind farm control as shown in Fig. 12.

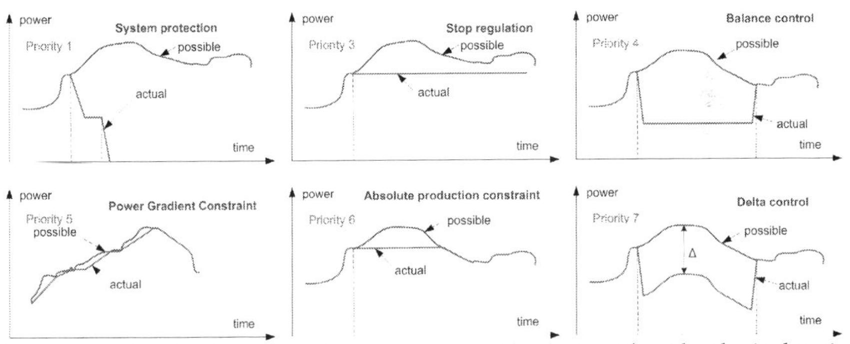

Fig. 12. Regulation functions for active power required by the Danish grid codes in the wind farm controller.

It must be noticed that the frequency control characteristic presented in Fig 11 has the second priority. Using all these power regulation functions a wind farm can be operated as a conventional power plant.

5.2 Reactive power control and voltage stability

Reactive power is typically controlled in a given range. The grid codes specify in different ways this control capability. The Danish grid code gives a band for controlling the reactive power based on the active power output as shown in Fig. 13.

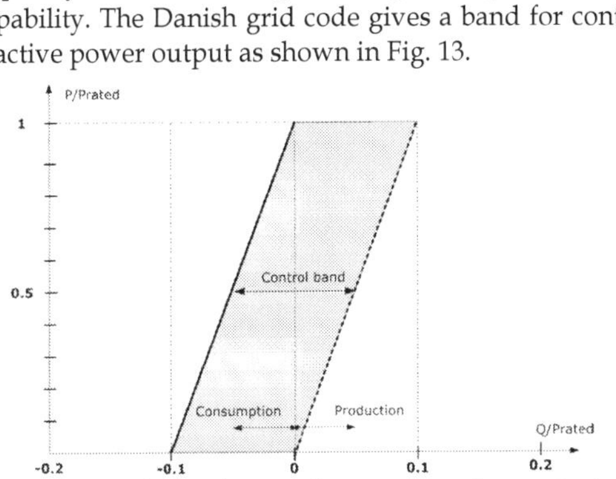

Fig. 13. Danish grid code demands for the reactive power exchange in the PCC.

The Irish grid code specifies e.g. the reactive power capability in terms of power factor and reactive power as shown in Fig. 14 while the German transmission grid code for wind power specifies that the wind power units must provide a reactive power provision in the connection point without limiting the active power output as shown in Fig. 15.

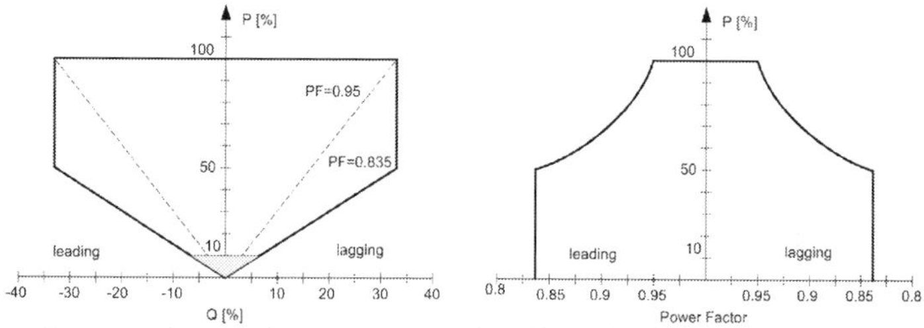

Fig. 14. Requirements for reactive power control in the Irish grid code for wind turbines.

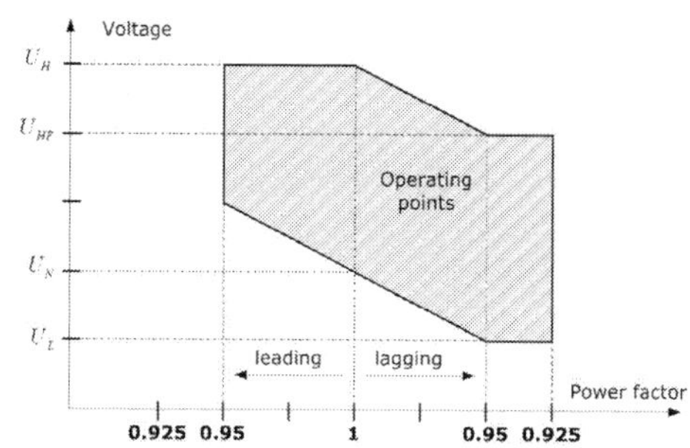

Fig. 15. Requirements for reactive power provision of generating units without limiting the active power output in the German transmission grid code.

5.3. Low-voltage ride-through capability

All considered grid codes requires fault ride-through capabilities for wind turbines. Voltage profiles are given specifying the depth of the voltage dip and the clearance time as well. One of the problems is that the calculation of the voltage during all types of unsymmetrical faults is not very well defined in some grid codes. The voltage profile for ride-through capability can be summarized as shown in Fig. 16.

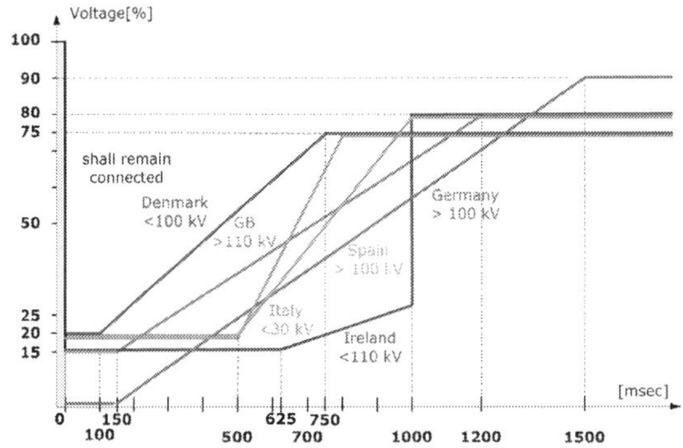

Fig. 16. Voltage profile for fault ride-through capability for wind power in European grid codes.

Ireland's grid code is very demanding in respect with the fault duration while Denmark has the lowest short circuit time duration with only 100 msec. However, Denmark's grid code requires that the wind turbine shall remain connected to the electrical network during successive faults which is a technical challenge. On the other hand Germany and Spain requires grid support during faults by reactive current injection up to 100% from the rated current as shown in Fig. 17. This demand is relative difficult to meet by some of the wind turbine concepts e.g. active stall wind turbine with directly grid connected squirrel cage induction generator.

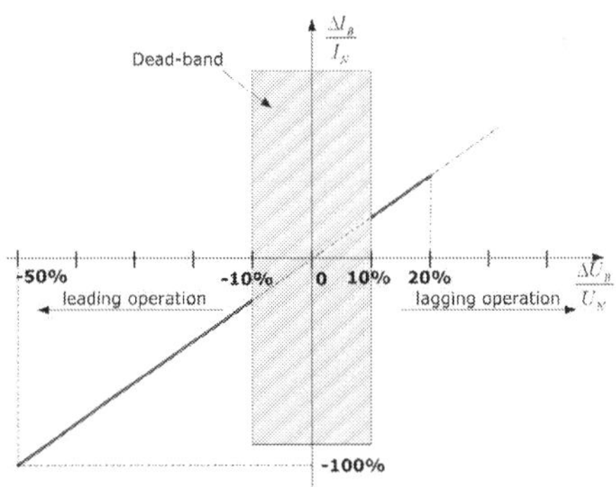

Fig. 17. Reactive current support during faults as specified in the German grid code.

Since more European countries are planning to increase the installed wind power capacity a change in their grid code requirements is expected in the near future. Also, in order to accommodate more wind power into the electrical networks, these connection requirements will be more demanding.

6. Power converter topologies for wind turbines

Basically two power converter topologies with full controllability of the generated voltage are currently used in the commercial wind turbine systems. These power converters are related to the partial-rating power converter wind turbine and the full-rating one. However, other topologies have been proposed in the last years.

6.1 Bi-directional back-to-back two-level power converter

The back-to-back Pulse Width Modulation-Voltage Source Converter is a bi-directional power converter consisting of two conventional PWM-VSCs. The topology is shown in Fig. 18.

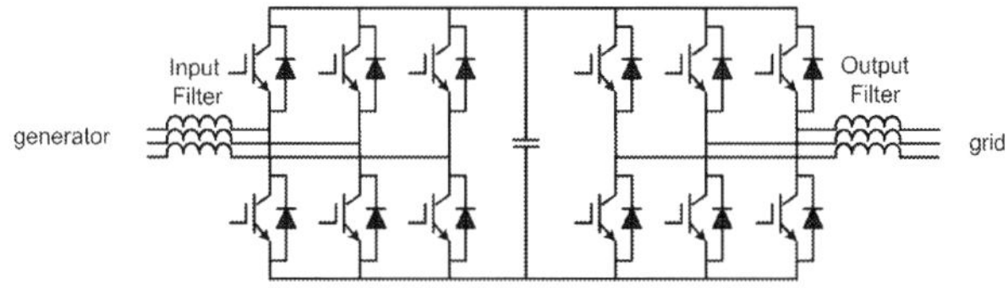

Fig. 18. Structure of the back-to-back voltage source converter.

The PWM-VSC is the most frequently used three-phase frequency converter. As a consequence of this, the knowledge available in the field is extensive and very well established. Furthermore, many manufacturers produce components especially designed for use in this type of converter (e.g., a transistor-pack comprising six bridge coupled transistors and anti-paralleled diodes). Therefore, the component costs can be low compared to converters requiring components designed for a niche production. A technical advantage of the PWM-VSC is the capacitor decoupling between the grid inverter and the generator inverter. Besides affording some protection, this decoupling offers separate control of the two inverters, allowing compensation of asymmetry both on the generator side and on the grid side, independently. The inclusion of a boost inductance in the DC-link circuit increases the component count, but a positive effect is that the boost inductance reduces the demands on the performance of the grid side harmonic filter, and offers some protection of the converter against abnormal conditions on the grid.

However, some disadvantages of the back-to-back PWM-VSC are reported in literature (Hansen et al., 2002) and (Kazmierkowski et al., 2002). In several papers concerning adjustable speed drives, the presence of the DC-link capacitor is mentioned as a drawback, since: it is bulky and heavy; - it increases the costs and maybe of most importance; - it reduces the overall lifetime of the system.

Another important drawback of the back-to-back PWM-VSI is the switching losses. Every commutation in both the grid inverter and the generator inverter between the upper and lower DC-link branch is associated with a hard switching and a natural commutation. Since the back-to-back PWM-VSI consists of two inverters, the switching losses might be even more pronounced. The high switching speed to the grid may also require extra EMI-filters. To prevent high stresses on the generator insulation and to avoid bearing current problems the voltage gradient may have to be limited by applying an output filter.

This topology is state-of-the-art especially in large DFIG based wind turbines e.g. (Carlson et al., 1996); (Bhowmik et al., 1999); (Ekanayake et al., 2003); (Gertmar, 2003) and (Carrasco et al., 2006). However, recently some wind turbine manufacturers use this topology for full-rating power converter wind turbines with squirrel-cage induction generator (e.g. Siemens Wind Power). The topology can also be used for permanent magnet synchronous generators.

6.2 Unidirectional power converter

A wound rotor synchronous generator requires only a simple diode bridge rectifier for the generator side converter as shown in Fig. 19.

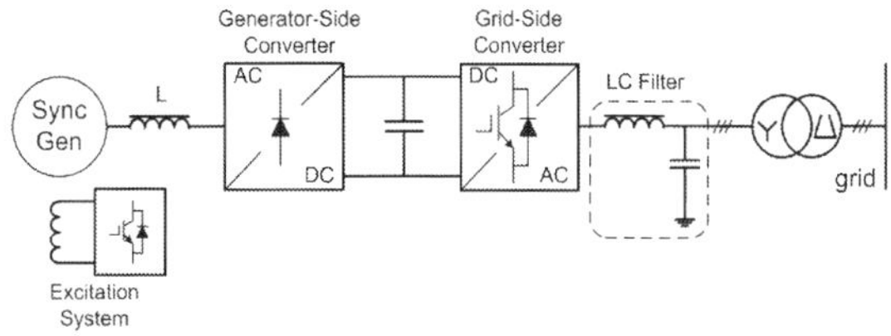

Fig. 19. Variable speed wind turbine with synchronous generator and full-rating power converter.

The diode rectifier is the most common used topology in power electronic applications. For a three-phase system it consists of six diodes. The diode rectifier can only be used in one quadrant, it is simple and it is not possible to control it. It could be used in some applications with a DC-link. The variable speed operation of the wind turbine is achieved by using an extra power converter which feed the excitation winding. The grid side converter will offer a decoupled control of the active and reactive power delivered to the grid and also all the grid support features. These wind turbines can have a gearbox or they can be direct-driven (Dubois et al., 2000). In order to achieve variable speed operation the wind turbines equipped with a permanent magnet synchronous generator (PMSG) will require a boost DC-DC converter inserted in the DC-link.

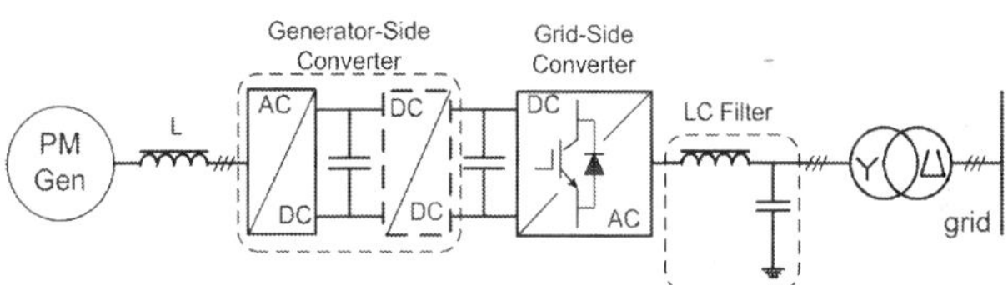

Fig. 20. Full rating power converter wind turbine with permanent magnet generator.

6.3 Multilevel power converter

Currently, there is an increasing interest in multilevel power converters especially for medium to high-power, high-voltage wind turbine applications (Carrasco et al., 2006) and (Portillo et al., 2006).

Since the development of the neutral-point clamped three-level converter (Nabae et al., 1981), several alternative multilevel converter topologies have been reported in the literature. The general idea behind the multilevel converter technology is to create a sinusoidal voltage from several levels of voltages, typically obtained from capacitor voltage sources. The different proposed multilevel converter topologies can be classified in the following five categories (Hansen et al., 2002); (Carrasco et al., 2006) and (Wu, 2006): multilevel configurations with diode clamps, multilevel configurations with bi-directional switch interconnection, multilevel configurations with flying capacitors, multilevel configurations with multiple three-phase inverters and multilevel configurations with cascaded single phase H-bridge inverters. These topologies are shown in Fig. 21 (Hansen et al., 2002).

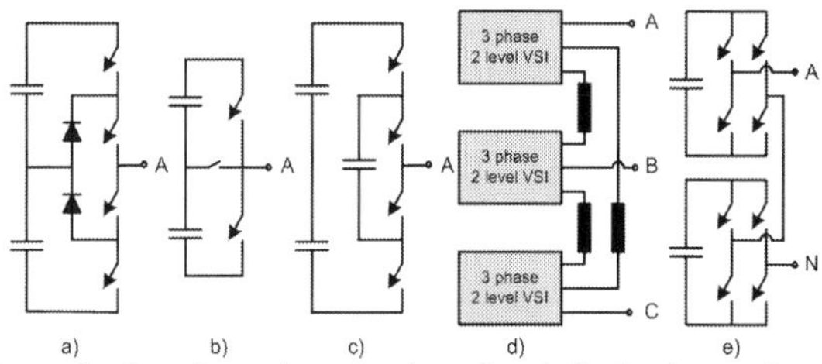

Fig. 21. Multilevel topologies: a) one leg of a three-level diode clamped converter; b) one leg of a three-level converter with bidirectional switch interconnection; c) one leg of a three-level flying capacitor converter; d) three-level converter using three two-level converters and e) one leg of a three-level H-bridge cascaded converter (Hansen et al., 2002).

Initially, the main purpose of the multilevel converter was to achieve a higher voltage capability of the converters. As the ratings of the components increases and the switching- and conducting properties improve, the secondary effects of applying multilevel converters become more and more advantageous. The reduced content of harmonics in the input and output voltage as well as a reduced EMI is reported (Hansen et al., 2002). The switching losses of the multilevel converter are another feature, which is often accentuated in literature. In (Marchesoni & Mazzucchelli, 1993), it is stated, that for the same harmonic performance the switching frequency can be reduced to 25% of the switching frequency of a two-level converter. Even though the conduction losses are higher for the multilevel converter, the overall efficiency for the diode clamped multilevel converter is higher than the efficiency for a comparable two-level converter (Hansen et al., 2002). Of course, the truth in this assertion depends on the ratio between the switching losses and the conduction losses.

However, some disadvantages exist and are reported in literature e.g. (Hansen et al., 2002); (Carrasco et al., 2006); (Portillio et al., 2006) and (Lai & Peng, 1995). The most commonly reported disadvantage of the three level converters with split DC-link is the voltage imbalance between the upper and the lower DC-link capacitor. Nevertheless, for a three-level converter this problem is not very serious, and the problem in the three-level converter is mainly caused by differences in the real capacitance of each capacitor, inaccuracies in the dead-time implementation or an unbalanced load (Shen & Butterworth, 1997) and (Hansen et al., 2001). By a proper modulation control of the switches, the imbalance problem can be solved (Sun-Kyoung Lim et al., 1999). In (Shen & Butterworth, 1997) the voltage balancing problem is solved by hardware, while (Newton & Sumner, 1997) and (Peng et al., 1995) proposed solutions based on modulation control. However, whether the voltage balancing problem is solved by hardware or software, it is necessary to measure the voltage across the capacitors in the DC-link.

The three-level diode clamped multilevel converter (Fig. 21a) and the three-level flying capacitor multilevel converter (Fig. 21c) exhibits an unequal current stress on the semiconductors. It appears that the upper and lower switches in an inverter branch might be de-rated compared to the switches in the middle. For an appropriate design of the converter, different devices are required (Lai & Peng, 1995). The unequal current stress and the unequal voltage stress might constitute a design problem for the multilevel converter with bidirectional switch interconnection presented in Fig. 21b (Hansen et al., 2002).

It is evident for all presented topologies in Fig. 21 that the number of semiconductors in the conducting path is higher than for e.g. a two-level converter. Thus, the conduction losses of the converter might increase. On the other hand, each of the semiconductors need only to block half the total DC-link voltage and for lower voltage ratings, the on-state losses per switch decreases, which to a certain extent might justify the higher number of semiconductors in the conducting path (Hansen et al., 2002).

6.4 Modular power converters

At low wind speeds and hence low level of the produced power, the full-rating power converter concept exhibits low utilization of the power switches and thus increased power losses. Therefore, a concept in which several power converters are running in parallel is used as shown in Fig. 22. The power converter in this case can be one of the structures presented above. This configuration can also be used for standard generators.
By introducing power electronics many of the wind turbine systems get similar performances with the conventional power plants. Modern wind turbines have a fast response in respect with the grid operator demands. However the produced real power depends on the available wind speed. The reactive power can in some solutions, e.g. full scale power converter based wind turbines, be delivered without having any wind producing active power.

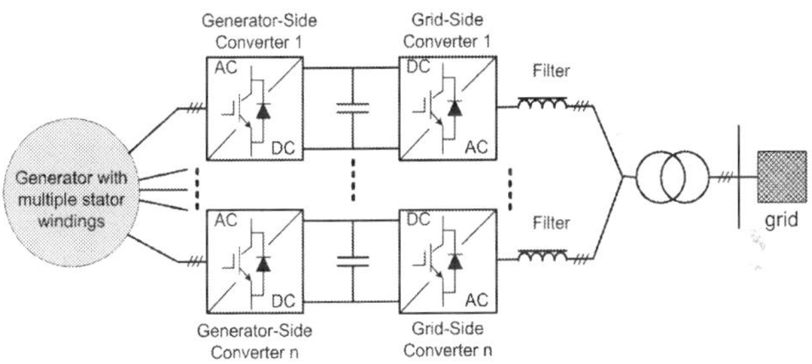

Fig. 22. Full-rating power converter based wind turbine with n-paralleled power converters.

These wind turbines can also be active when a fault appears on the grid and where it is necessary to build the grid voltage up again (Hansen et al., 2004) and (Iov & Blaabjerg, 2007); having the possibility to lower the power production even though more power is available in the wind and thereby act as a rolling capacity for the power system. Finally, some systems are able to work in island operation in the case of a grid collapse.

7. Control of generator-side converter

The control of the generator side-converter is basically determined by the generator type. However, since the wind turbine concepts available on the market are based on AC machines some basic control configurations can be identified. It must be noticed that these control schemes have their origins in the motor drives applications and they have been adapted to generator mode of operation. The general structure of a generator fed by an IGBT based power converter is shown in Fig. 23.

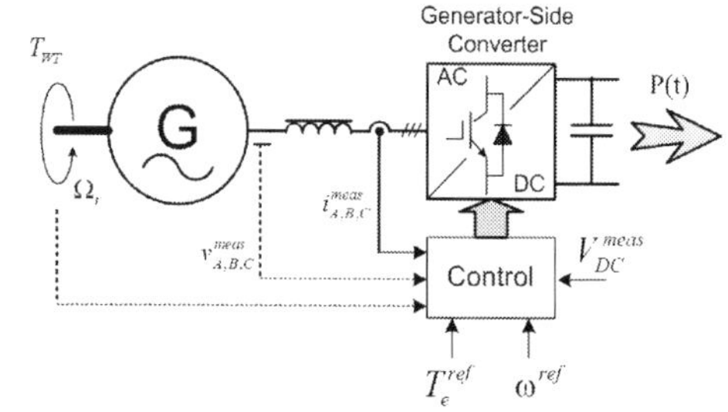

Fig. 23. General layout of a VSC-fed based three-phase AC generator.

7.1 Field oriented control

Field oriented control is one of the most used control methods of modern AC drives systems (Vas, 1998) and (Godoy Simoes & Farrat, 2004). The basic idea behind this control method is to transform the three-phase quantities in the AC machine in an orthogonal dq system aligned to one of the fluxes in the machine. Thus, a decoupling in controlling the flux and electromagnetic torque of the machine is achieved. Two methods of field oriented control for induction machines are used namely: indirect and direct vector control. Each of these methods has advantages and drawbacks. The indirect vector control can operate in four-quadrant down to standstill and it is widely used in both motor drives and generator applications. Typically the orthogonal synchronous reference frame is aligned on the rotor flux. However, this control is highly dependent on machine parameters. The direct vector control oriented along the stator flux does not need information about the rotor speed and is less sensitive to the machine parameters. However, it presents low performances for low speeds near to standstill. A general control structure for field oriented control in synchronous reference frame for induction machines is shown in Fig. 24.

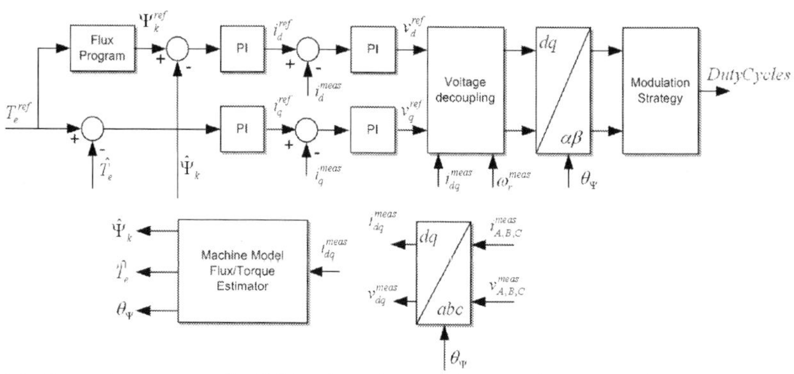

Fig. 24. General structure of a field oriented control in synchronous reference frame for an induction machine.

The electromagnetic torque is controlled in q-axis while the d-axis controls the flux of the machine. The actual flux and torque as well as the flux angle are determined based on the machine equations using the currents

Similar control structure is used for the DFIG systems. Typically, the outer control loops are used to regulate the active and reactive power on the stator side of the machine.

7.2 Direct torque control

The Direct Torque Control proposed by Depenbrock eliminates the inner current loops and the needs of transformations between different references frames (Godoy Simoes & Farrat, 2004). It controls directly the magnitude of the stator flux and the electromagnetic torque of the machine by using hysteresis comparators as shown in Fig. 25.

The outputs of the hysteresis comparators as well as the flux angle are used directly to determine the switching states of the converter.

The performances of all the control schemes used for the generator-side converter must be evaluated in terms of current and hence torque ripple. High torque ripple can cause damages into the gearbox, while important low frequency harmonics can induce resonances with the mechanical structure of the wind turbine.

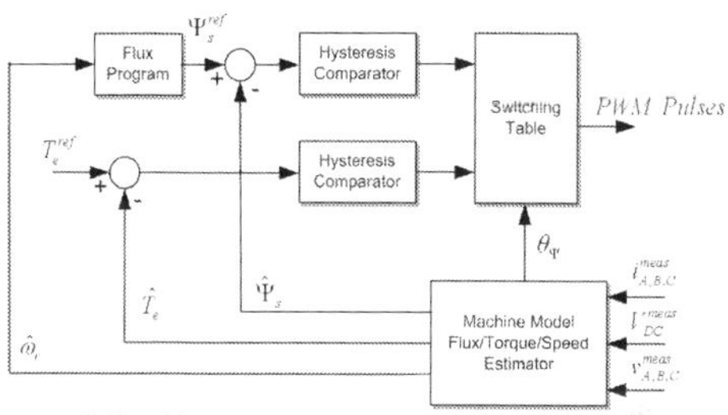

Fig. 25. General structure of the direct torque control for AC machines.

8. Control of grid-side converter

Independently of the generator type and the power converter configuration, the grid side converter is responsible for the quality of the generated power and the grid code compliance. A typical configuration of the grid side converter in wind turbine applications is given in Fig. 26.

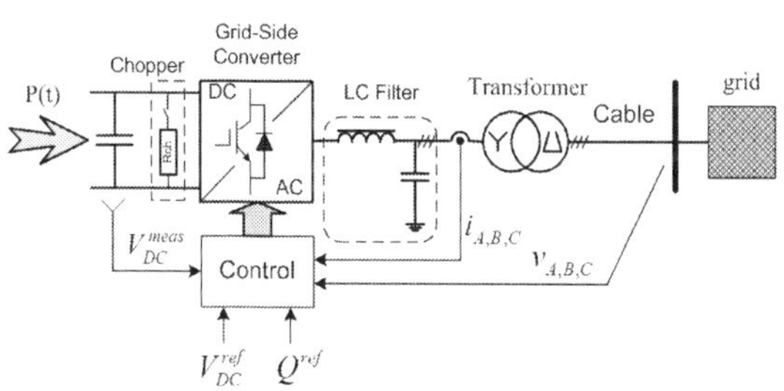

Fig. 26. Structure of the grid-side converter in wind turbine applications.

The system consist of a DC-link circuit, an IGBT based Voltage Source Converter, an LC filter, a Dyn11 transformer and a cable to the Point of Common Coupling. The LC filter is used to minimize the ripple of the output current due to the switching of the power devices

8.1 Grid synchronization

Initially, the synchronization of the delivered current with the utility network voltage was a basic requirement for interconnecting distributed power generators with the power system. In case of wind turbines, reactive power control at the point of common coupling is requested. Consequently, the wind turbine control should accommodate an algorithm capable of detecting the phase angle of grid voltage in order to synchronize the delivered current. Moreover, the phase angle plays an important role in control, being used to transform the feedback variables to a suitable reference frame in which the control structure is implemented. Hence, phase angle detection has a significant role in control of the grid side converter in a wind turbine. Numerous research papers report several algorithms capable of detecting the grid voltage phase angle, i.e. zero crossing detection, the use of *atan* function or Phase-Locked Loop (PLL) technique.

An overview of the grid synchronization and monitoring methods is presented in the following, based on (Iov & Blaabjerg, 2007) and (Iov et al., 2008).

8.1.1 Zero crossing method

A simple method of obtaining the phase and frequency information is to detect the zero-crossing point of the grid voltage (Mur et al., 1998); (Choi et al., 2006). This method has two major drawbacks as described in the following.

Since the zero crossing point can be detected only at every half cycle of the utility frequency, the phase tracking action is impossible between the detecting points and thus the fast dynamic performance can not be obtained (Chung, 2000). Some work has been done in order to alleviate this problem using multiple level crossing detection as presented in (Nguyen & Srinivasan, 1984).

Significant line voltage distortion due to notches caused by power device switching and/or low frequency harmonic content can easily corrupt the output of a conventional zero-crossing detector (McGrath et al., 2005). Therefore, the zero-crossing detection of the grid voltage needs to obtain its fundamental component at the line frequency. This task is usually made by a digital filter. In order to avoid the delay introduced by this filter numerous techniques are used in the technical literature. Methods based on advanced filtering techniques are presented in (Vainio et al., 1995); (Valiviita et al., 1997); (Vainio et al., 2003); (Wall, 2003) and (McGrath et al., 2005). Other methods use Neural Networks for detection of the true zero-crossing of the grid voltage waveform (Valiviita, 1998); (Valiviita, 1999) and (Das et al., 2004). An improved accuracy in the integrity of the zero-crossing can also be obtained by reconstructing a voltage representing the grid voltage (Weidenbrug et al., 1993); (Baker and Agelidis, 1998); (Nedeljkovic et al, 1998) and (Nedeljkovic et al, 1999).

However, starting from its simplicity, when the two major drawbacks are alleviated by using advanced techniques, the zero-crossing method proves to be rather complex and unsuitable for applications which require accurate and fast tracking of the grid voltage.

8.1.2 Arctangent method

Another solution for detecting the phase angle of grid voltage is the use of *arctangent* function applied to voltages transformed into a Cartesian coordinate system such as synchronous or stationary reference frames as shown in Fig. 27a and Fig. 27b respectively.

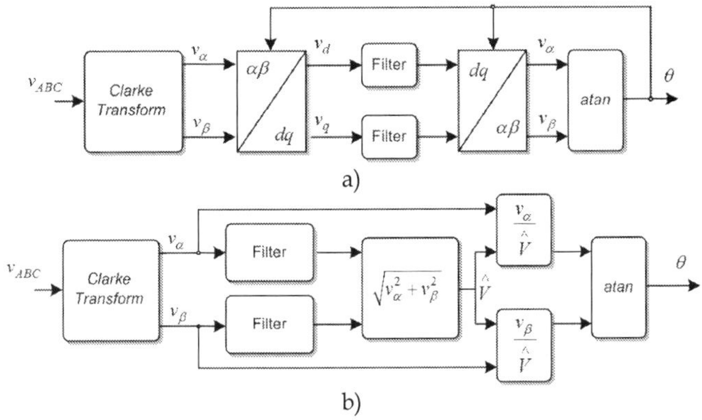

Fig. 27. Synchronization method using (a) filtering on the dq synchronous rotating reference frame and (b) filtering on stationary frame.

This method has been used in drives applications (Kazmierkowski et al., 2002), for transforming feedback variables to a reference frame suitable for control purposes. However, this method has the drawback that requires additional filtering in order to obtain an accurate detection of the phase angle and frequency in the case of a distorted grid voltage. Therefore, this technique is not suitable for grid-connected converter applications.

8.1.3 PLL technique

Phase-Locked Loop (PLL) is a phase tracking algorithm widely applied in communication technology (Gardner, 1979), being able to provide an output signal synchronized with its reference input in both frequency and phase.

Nowadays, the PLL technique is the state of the art method to extract the phase angle of the grid voltages (Nguyen & Srinivasan, 1984); (Kaura & Blasko, 1997); (Chung, 2000a) and (Chung, 2000b). The PLL is implemented in dq synchronous reference frame and its schematic is illustrated in Fig. 28. As it can be noticed, this structure needs the coordinate transformation form abc to dq and the lock is realized by setting the reference to zero. A controller, usually PI, is used to control this variable. This structure can provides both the grid frequency as well as the grid voltage angle.

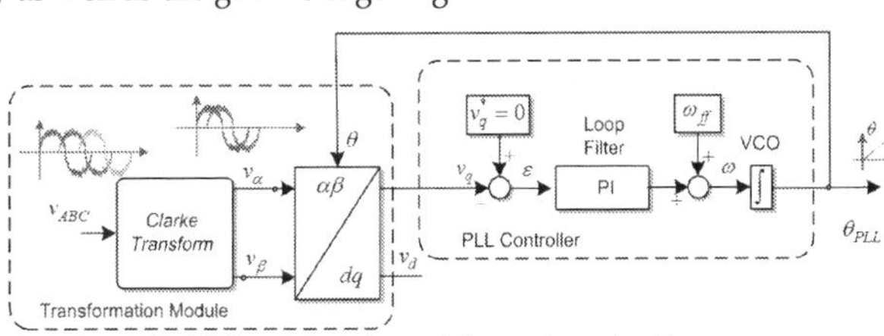

Fig. 28. Basic structure of a PLL system for grid synchronization.

After the integration of the grid frequency, the utility voltage angle is obtained, which is fed back into the Park Transform module in order to transform into the synchronous rotating reference frame.

This algorithm has a better rejection of grid harmonics, notches and any other kind of disturbances but additional improvements have to be done in order to overcome grid unbalance (Lee et al., 1999); (Song et al., 1999); (Karimi-Ghartemani & Iravani, 2004); (Rodriguez et al., 2005) and (Benhabib & Saadate, 2005);. In the case of unsymmetrical voltage faults, the second harmonics produced by the negative sequence will propagate through the PLL system and will be reflected in the extracted phase angle. In order to overcome this, different filtering techniques are necessary such that the negative sequence is filtered out. As a consequence, during unbalance conditions, the three phase dq PLL structure can estimate the phase angle of the positive sequence of the grid voltages.

8.1.4 Grid Monitoring

Grid requirements applying to utility connected power generation units impose the operation conditions in respect to voltage and frequency values. The demands are country specific. A graphical representation of allowed operation area in respect to the grid voltage amplitude and grid frequency as specified in the Danish Grid code for wind turbines connected to the distribution system (Iov & Blaabjerg, 2007) is illustrated in Fig. 29.

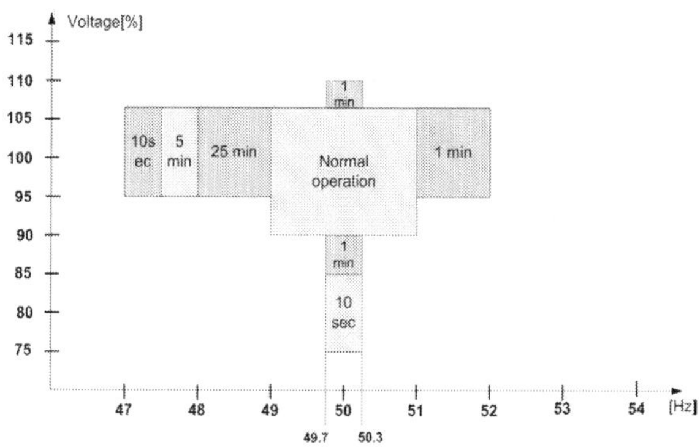

Fig. 29. Voltage and frequency operational ranges for wind turbines connected to the Danish distribution system.

A normal operation area between 95 and 105% of the nominal grid voltage and ±1 Hz around the nominal frequency is defined. Either frequency or voltage exceeds the predefined limits, the wind turbine should disconnect within the specified time interval. Therefore, in order to be able to disconnect in time, the wind turbine should accommodate a fast and reliable grid monitoring unit.

The PLL structures are used in the grid monitoring techniques. In a three-phase system, the grid voltage information can easily be obtained through the Clarke Transform as shown in Fig. 30.

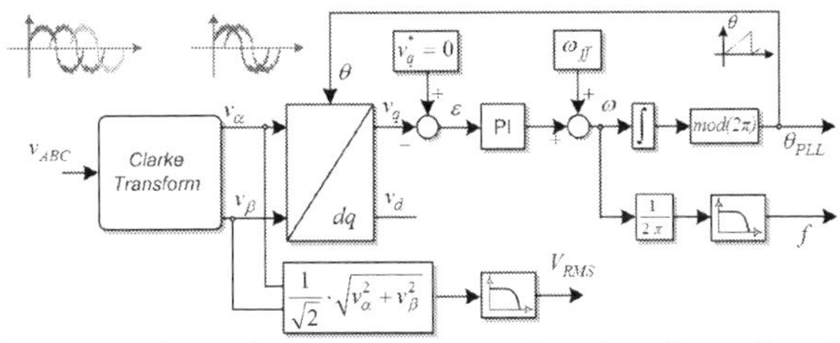

Fig. 30. General structure of a grid monitoring system based on three-phase PLL.

8.2 Converter control

Different control strategies can be used for the power converter such as synchronous Voltage Oriented Control (VOC) with PI controllers, stationary VOC with Proportional Resonant controllers, Synchronous Virtual Flux Oriented Control (VFOC) with PI controller, Adaptive Band Hysteresis (ABH) Current Control, Direct Power Control (DPC) with Space Vector Modulation (SVM), Virtual Flux DPC with SVM (Kazmierkowski et al., 2002), (Iov et al., 2006), (Teodorescu et al., 2006). However, in industrial applications few of these control strategies are used.

The synchronous Voltage Oriented Control with PI controllers is widely used in grid applications. This control strategy is based on the coordinate transformation between the stationary αβ and the synchronous dq reference frames. It assures fast transient response and high static performance due to the internal current control loops (Kazmierkowski et al., 2002) and (Iov et al., 2006). The decomposition of the AC currents in two axes provides a decoupled control for the active and reactive power.

A block diagram of the VOC control with PI controllers is shown in Fig. 31

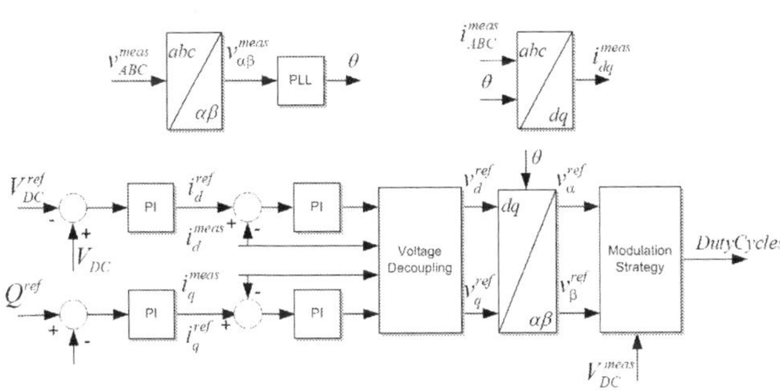

Fig. 31. Block diagram of the VOC in the synchronous reference frame.

A Phase Locked Loop (PLL) is used for the coordinate transformation. The control scheme comprises the DC-link voltage controller and the current controller in the d-axis, while the reactive power and the reactive component of the current are controlled in the q-axis.

In order to achieve a high accuracy current tracking the control algorithm accounts for the output filter inductance. Therefore, the output of the current controllers is compensated with the voltage drop on the output filter. Then the reference voltages are translated to the stationary reference frame and applied to a Space Vector Modulator (SVM).

It should be noticed that the performances of this control strategy relies on the accuracy of the PLL system for the voltage grid angle estimation.
Another control strategy used by some wind turbine manufacturers is the adaptive hysteresis-band current control that provides a very fast-response current-loop (Iov et al., 2006). This control strategy is also based on the synchronous reference frame; therefore a PLL is used for the reference frame transformation of the currents.

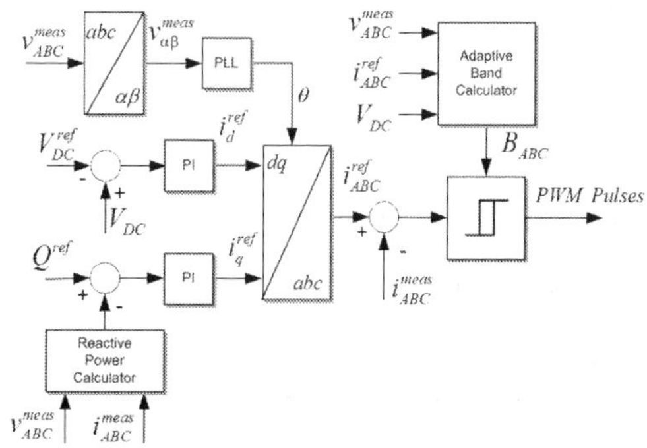

Fig. 32. Block diagram of the Adaptive Band Hysteresis Current Control.

A PI controller provides the control of the DC-link voltage in the d-axis while the reactive power is controlled in the q-axis. The current control is performed by the hysteresis comparators for each phase independently after the transformation of the reference dq currents (Iov et al., 2006) and (Teodorescu et al., 2006).
Both control algorithms require the estimation of the reactive power based on measured voltages and currents. Also, both control algorithms can meet the requirements for harmonic current injection in the PCC (Iov et al., 2006). However, the VOC algorithm is more sensitive to voltage unbalances and asymmetries than the ABH current control due to the double transformation of axes.
A typical control scheme for a cascaded H-bridge multilevel converter is presented in Fig. 33 (Ciobotaru et al., 2008).

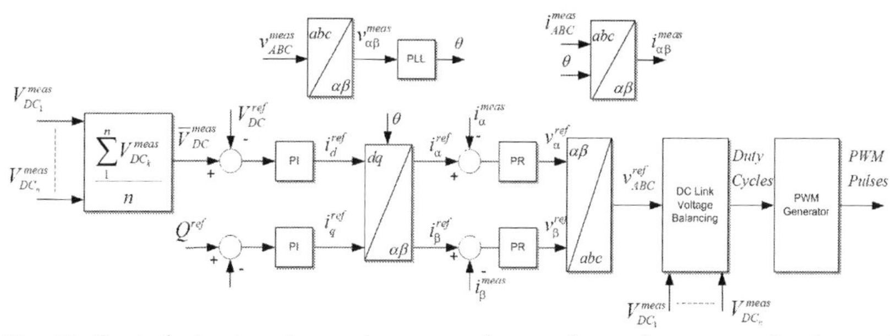

Fig. 33. Control structure in synchronous reference frame for a seven level cascaded H-bridge multilevel converter.

The overall structure of this control is shown in Fig. 4.6 and it is based on PR current controllers. The PR controller has been chosen due to the fact that it gives better performances compare to the classical PI. The two well known drawbacks of the PI controller (steady-state errors and poor harmonics rejection capability) can be easily overcome by the PR controller. The PR controller is able to remove the steady-state error without using voltage feed-forward, which makes it more reliable compared with the "classical" synchronous reference frame control. However, a three-phase PLL (Phase-Locked Loop) system is used to obtain the necessary information about the grid voltage magnitude and its angle.

In order to support the bidirectional power flow through the system, the current reference in d-axis is obtained based on the active power set-point and the average voltage (provided by PLL) and a correction given by the DC-link voltage control loop. This controller controls the average value of the DC link voltages. The reactive power is controlled using a PI controller which provides the current reference in the q-axis. One of the drawbacks of this control is the transformation of the control variables from synchronous reference frame to the stationery one. Based on the grid voltage angle the current references in synchronous reference frame are transformed into the stationary frame and then applied to the PR current controllers. It is also worth to notice that this control does not require a neutral connection.

It is obviously that all these control strategies need information about the grid voltage angle in order to operate properly. The number of transformations between reference frames (*abc* to *dq* and back) makes them more or less sensitive to changes into the grid behaviour. Thus, their capability to handle events into the network such as voltage asymmetries/unbalances, frequency excursions and phase jumps as well as grid faults is very important for fulfilling the grid connection requirements. Bottom-line, the harmonic content injected into the grid as well as their capability of selective harmonic elimination will be a factor key in choosing a particular control structure for the grid-side converter.

9. Wind Farm Connection

In many countries energy planning is going on with a high penetration of wind energy, which will be covered by large offshore wind farms. These wind farms may in the future present a significant power contribution to the national grid, and therefore, play an important role on the power quality and the control of complex power systems. Consequently, very high technical demands are expected to be met by these generation units, such as to perform frequency and voltage control, regulation of active and reactive power, quick responses under power system transient and dynamic situations, for example, to reduce the power from the nominal power to 20 % power within 2 seconds. The power electronic technology is again an important part in both the system configurations and the control of the offshore wind farms in order to fulfil the future demands.

One off-shore wind farm equipped with power electronic converters can perform both active and reactive power control and also operate the wind turbines in variable speed to maximize the energy captured and reduce the mechanical stress and acoustical noise. This solution is shown in Fig. 34 and it is in operation in Denmark as a 160 MW off-shore wind power station.

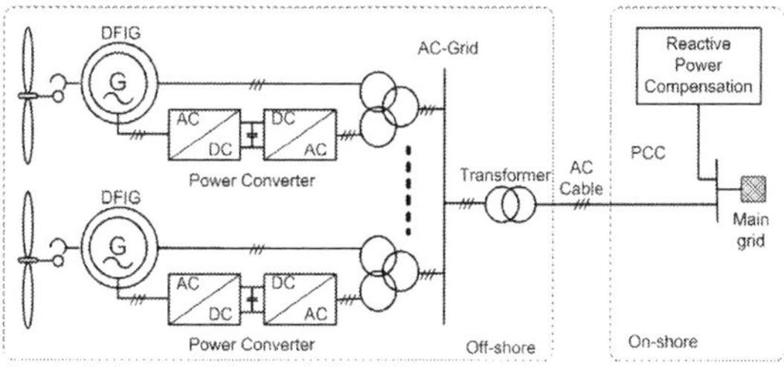

Fig. 34. DFIG based wind farm with an AC grid connection.

For long distance power transmission from off-shore wind farm, HVDC may be an interesting option. In an HVDC transmission system, the low or medium AC voltage at the wind farm is converted into a high dc voltage on the transmission side and the dc power is transferred to the on-shore system where the DC voltage is converted back into AC voltage as shown in Fig. 28.

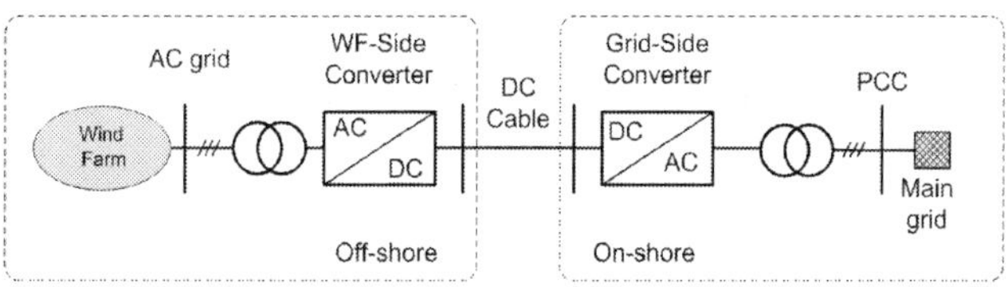

Fig. 35. HVDC system based on voltage source converters for wind farm connection.

This connection type is considered for Borkum wind farm in the Baltic Sea and it will be commissioned in 2009.
Another possible DC transmission system configuration is shown in Fig. 36, where each wind turbine has its own power electronic converter, so it is possible to operate each wind turbine at an individual optimal speed. A common DC grid is present on the wind farm while a power converter terminal realizes the on-shore grid connection. In this case any of the wind turbine configurations based on full-rating power converter can be used.

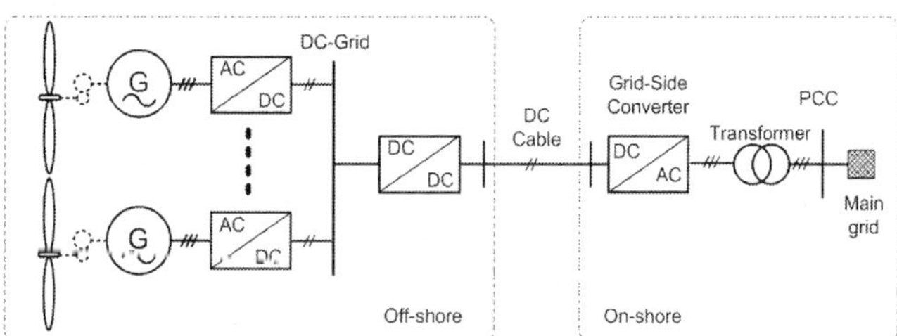

Fig. 36. Wind farm with common DC grid based on variable speed wind turbines with full rating power converter.

This topology can be extended by adding the input of more wind farms in the DC grid and then multiple DC cables and converters for connection to other AC power grids. In this case a multi-terminal DC connection will result.
As it can be seen the wind farms have interesting features in order to act as a power source to the grid. Some have better abilities than others. Bottom-line will always be a total cost

scenario including production, investment, maintenance and reliability. This may be different depending on the planned site.

10. Developments and trends in wind energy systems

Wind turbine's size was in a continuous growth in the last 25 years as shown in Fig. 37 and today prototype turbines of 5-6 MW are seen around the world being tested.

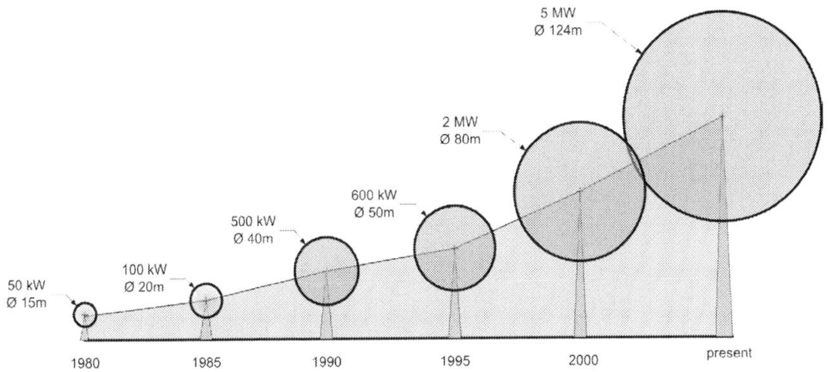

Fig. 37. Development of wind turbines during the last 25 years.

Currently, dispersed single wind turbines are replaced with MW-size wind turbines concentrated in wind power plants. Also, the grid penetration of wind power is increasing dramatically especially in the European countries like Denmark, Germany and Spain. It is expected that more countries worldwide will follow this trend. Due to the unpredictable nature of the wind, the grid connection of these modern MW-size wind turbines and wind power plants has a large impact on grid stability and security of supply. Thus, in order to accommodate more wind power into the grid, one of the major challenges in the future is directed towards the integration of wind power within the existing electrical network.
Today, the grid connection requirements, in countries with a relatively high penetration of wind power, require MW-size wind turbines and wind farms to support the grid actively and to remain connected during grid events. These requirements are mainly fulfilled by using power electronics within the wind turbines and wind farms. The power electronics make also possible the variable speed operation of wind turbines and further improve the performances of the wind turbines by reducing the mechanical stress and acoustical noise, and by increasing the wind power capture. Also, the power electronics increase the controllability of the wind turbines, which is a major concern for their grid integration. All these features are provided at low price share of power electronics compared to other components from a modern MW-size wind turbine as shown in Fig. 38.

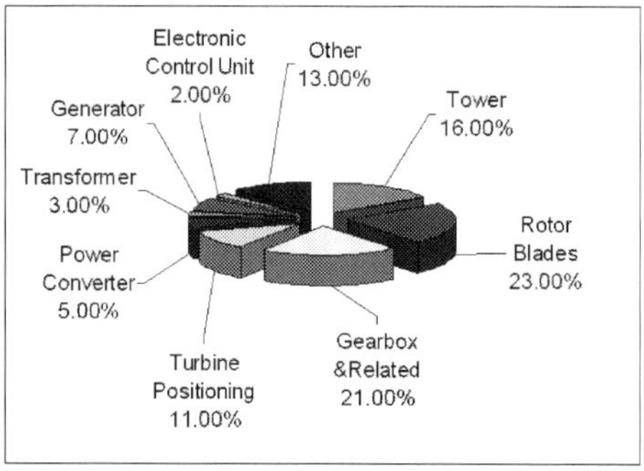

Fig. 38. Cost share of main components in a typical 2 MW variable speed wind turbine [Bernstein Research, 2007].

Currently, the variable speed wind turbines dominate the market. The doubly-fed induction generator based wind turbine has developed into a semi-industry standard for gear-driven wind turbines. On the other hand, more manufacturers are coming with wind turbine prototypes based on the full-rating power converter using different generator types. The main advantage of the doubly-fed wind turbine of using a partial scale power converter may be overcame by its behaviour during grid events. Moreover, the grid support during faults by means of 100% reactive current injection into the grid required by some grid operators cannot be provided by this concept. On the contrary, the full-rating power converters based wind turbines are very attractive because they can provide complete grid support during network events. Thus, further developments of both wind turbine concepts are expected, focusing on more optimised turbines and, thus, towards more cost-effective machines.

Moving to large wind power plants installations these wind turbine concepts may be modified for a better integration into the electrical grid. The HVDC solution for connecting wind farms might create a new class of wind turbines that will deliver only DC power into a common DC-grid. Thus, the present layout of a full-rating power converter based wind turbine can be simplified by removing some of the conversion stages, e.g. DC to AC including the step-up transformer. The receiving-end station of such a system will be responsible with fulfilling the grid connection requirements, while the wind turbine itself will just maximize the wind power conversion.

The future is difficult to predict. Technologically, improvements of the current designs are expected. Concepts borrowed from other fields or other applications might change the future design of the wind turbines. Economically, the cheapest and the most reliable solution will be preferred. Looking at the wind power plants the overall costs against performances will definitely be the main driver. Thus, in order to minimize de costs at high control capabilities required by the system operators, an integrated design of wind power plants will be required. However, power electronics including control will be the key technology for the large scale grid integration of wind power.

11. References (selected)

Arruda, L.N.; Cardoso Filho, B.J.; Silva, S.M.; Silva, S.R. & Diniz, A.S.A.C. (2000). Wide bandwidth single and three-phase PLL structures for grid-tied, *Proceedings of Photovoltaic Specialists Conference*, 2000, pp. 1660-1663.

Arruda, L. N.; Silva, S. M. & Filho, B. (2001). PLL structures for utility connected systems, *Proceedings of IAS'01*, vol. 4, 2001, pp. 2655–2660.

Baker, D.M. & Agelidis, V.G. (1998). Phase-locked loop for microprocessor with reduced complexity voltage controlled oscillator suitable for inverters, *Proceedings of PEDS*, 1998, Vol.1, pp. 464-469.

Baliga, B.J. (1995). Power IC's in the saddle. *IEEE Spectrum*, July 1995, pp. 34-49.

Benhabib, M. C. & Saadate, S. (2005). A new robust experimentally validated Phase-Looked Loop for power electronic control, *EPE Journal*, vol. 15, no. 3, pp. 36–48, August 2005.

Bernstein Research (2007). Technology Sector Strategy: Global Warming Challenges. Information Technology Solutions, October 2007

Blaabjerg, F.; Teodorescu, R.; Liserre, M. & Timbus, A. (2006). Overview of control and grid synchronization for distributed power generation systems, *IEEE Transactions on Industrial Electronics*, Vol. 53, No. 5, 2006, pp. 1398-1409.

Bhowmik, S.; Spee, R. & Enslin, J.H.R. (1999). Performance optimization for doubly fed wind power generation systems, *IEEE Trans. on Industry Applications*, 1999, Vol. 35, No. 4, pp. 949-958.

Bogalecka, E. (1993). Power control of a doubly fed induction generator without speed or position sensor, *Proceedings of EPE*, 1993, Vol.8, pp. 224-228.

Bossanyi, E. (2000). *Wind Energy Handbook*, John Wiley, 2000.

Cameron, A. & De Vries, E. (2006). Top of the list, *Renewable Energy World*, James & James, January-February 2006, Vol. 9, No. 1, pp. 56-66, ISSN 1462-6381.

Carrasco, J.M.; Galvan, E.; Portillo, R.; Franquelo, L.G. & Bialasiewicz, J.T. (2006). Power Electronics System for the Grid Integration of Wind Turbines, *Proceedings of IECON '06 Conference*, November 2006, pp. 4182 - 4188.

Carlson, O.; Hylander, J.& Thorborg, K. (1996). Survey of variable speed operation of wind turbines, *Proceedings of European Union Wind Energy Conference*, 1996, pp. 406-409.

Choi, J. W.; Kim, Y.K. & Kim, H.G. (2006). Digital PLL control for single-phase photovoltaic system, *IEE Trans. on Electric Power Applications*, 2006, Vol. 153, pp. 40-46.

Chung, S.-K. (2000). A phase tracking system for three phase utility interface inverters, *IEEE Trans. on Power Electronics*, 2000, Vol. 15, No. 3, pp. 431–438.

Chung, S.-K. (2000). Phase-Locked Loop for grid-connected three-phase power conversion systems, *IEE Proceedings on Electronic Power Applications*, vol. 147, no. 3, pp. 213–219, 2000.

Ciobotaru, M.; Teodorescu, R. & Blaabjerg, F. (2005). Improved PLL structures for single-phase grid inverters, *Proceedings of PELINCEC*, 2005, pp. 1-6.

Ciobotaru, M.; Teodorescu, R. & Blaabjerg, F. (2006). A New Single-Phase PLL Structure Based on Second Order Generalized Integrator, *Proceedings of PESC*, 2006, pp. 1-6.

Ciobotaru, M.; Iov, F.; Zanchetta, P.; De Novaes, Y. & Clare, J. (2008) A stationary reference frame current control for a multi-level H-bridge power converter for universal and flexible power management in future electricity network. *Proceedings of Power Electronics Specialists Conference*, 2008. PESC 2008. IEEE. 3943-3949, 0275-9306.

Dahlgren, M.; Frank, H.; Leijon M.; Owman, F. & Walfridsson, L. (2000). Wind power goes large scale, *ABB Review*, 2000, Vol.3, pp. 31-37.

Das, S.; Syam, P.; Bandyopadhyay, G. & Chattopadhyay, A.K. (2004). Wavelet transform application for zero-crossing detection of distorted line voltages in weak AC-systems, *Proceedings of INDICON*, 2004, pp. 464-467.

De Brabandere, K.; Loix, T.; Engelen, K.; Bolsens, B.; Van den Keybus, J.; Driesen, J. & Belmans, R. (2006). Design and Operation of a Phase-Locked Loop with Kalman Estimator-Based Filter for Single-Phase Applications, *Proceedings of IECON*, 2006, pp. 525-530.

Dubois, M.R.; Polinder, H. & Ferreira, J.A. (2000). omparison of Generator Topologies for Direct-Drive Wind Turbines, *Proceedings of IEEE Nordic Workshop on Power and Industrial Electronics* (Norpie 2000), Aalborg, Denmark, pp. 22-26.

Ekanayake, J.B.; Holdsworth, L.; XueGuang, W. & Jenkins, N. (2003). Dynamic modelling of doubly fed induction generator wind turbines, *IEEE Trans. on Power Systems*, 2003, Vol. 18 , No. 2, pp. 803-809.

Gardner, F. M. (1979). *Phase Lock Techniques*. New York, Wiley, 1979.

Gertmar, L. (2003). Power Electronics and Wind Power, *Proceedings of EPE 2003*, paper 1205.

Godoy Simoes, M; Farret, F.A. (2004). *Renewable Energy Systems*. New York, CRC Press 2004, ISBN 0-8493-2031-3.

Hansen, A.D.; Iov, F.; Blaabjerg, F. & Hansen, L.H. (2004). Review of contemporary wind turbine concepts and their market penetration, *Journal of Wind Engineering*, 28(3), 2004, 247-263, ISSN 1095-4244

Hansen, L.H.; Helle, L.; Blaabjerg, F.; Ritchie, E.; Munk-Nielsen, S.; Bindner, H.; Sørensen, P. & Bak-Jensen, B. (2002). *Conceptual survey of Generators and Power Electronics for Wind Turbines*, Risø-R-1205(EN), Pitney Bowes Management Services Denmark, 2002, ISBN 87-550-2743-1.

Heier, S. (1998). *Grid integration of wind energy conversion systems"*, John Wiley, 1998, ISBN-10: 0-47-197143X.

Iov, F.; Teodorescu, R.; Blaabjerg, F.; Andresen, B.; Birk, J. & Miranda, J. (2006). Grid code compliance of grid-side converter in wind turbine systems, *Proceedings of PESC 2006 Conference*, 18-22 June 2006, Jeju, South-Korea, 7 p., IEEE Catalog Number: 06CH37819C, ISBN: 1-4244-9717-7, ISSN 02-75-9306.

Iov, F.; Blaabjerg, F. (2007). UNIFLEX-PM. Advanced power converters for universal and flexible power management in future electricity network - Converter applications in future European electricity network, Deliverable D2.1, EC Contract no. 019794(SES6), February 2007, pp. 171, (available on line www.eee.nott.ac.uk/uniflex/Deliverables.htm).

Iov, F.; Hansen, A.D.; Sørensen, P. & Cutululis, N.A. (2007). Mapping of grid faults and grid codes, Risø-R-1617(EN) (2007) 41 p. (available online at www.risoe.dk).

Iov, F.; Ciobotaru, M. & Blaabjerg, F. (2008). Power Electronics Control of Wind Energy in Distributed Power System, keynote paper in *Proceedings of 11th International Conference on Optimization of Electrical and Electronic Equipment Optim'08*, pp. 16, May 24-26, Brasov, Romania, ISBN 1-4244-1545-4

Karimi-Ghartemani, M. & Iravani, M. (2004). A method for synchronization of power electronic converters in polluted and variable-frequency environments, *IEEE Trans. on Power Systems*, vol. 19, no. 3, pp. 1263–1270, 2004.

Kaura, V. & Blasko, V. (1997). Operation of a phase loop system under distorted utility conditions, *IEEE Trans. on Industry Applications*, vol. 33, no. 1, pp. 58–63, 1997.

Kazmierkowski, M.P.; Krishnan, R. & Blaabjerg, F. (2002). *Control in Power Electronics-Selected problems, Academic Press*, 2002. ISBN 0-12-402772-5.

Lai, J.S. & Peng, F.Z. (1995). Multilevel converters - A new breed of power converters, *Proceedings of IEEE Industrial Application Society Annual Meeting*, 8-12 October 1995 Vol. 3. pp. 2348-2356.

Lee, S.; Kang, J. & Sul, S. (1999). A new phase detection method for power conversion systems considering distorted conditions in power system, *Proceedings of IAS'99*, vol. 4, 1999, pp. 2167–2172.

Marchesoni, M. & Mazzucchelli, M. (1993). Multilevel converters for high power ac drives: A review, *Proceedings of IEEE International Symposium on Industrial Electronics*, 1993, pp. 38-43.

Matsuzaka, T.; Trusliga, K.; Yamada, S. & Kitahara, H. (1998). A variable speed wind generating system and its test results, *Proceedings of EWEC*, 1989, Part Two, pp. 608-612.

McGrath, B.P.; Holmes, D.G. & Galloway, J.J.H. (2005). Power converter line synchronization using a discrete Fourier transform (DFT) based on a variable sample rate, *IEEE Trans. on Power Electronics*, 2005, Vol. 20, pp. 877-884.

Milborrow, D. (2005). Going mainstream at the grid face. Examining grid codes for wind, *Windpower Monthly*, September 2005, ISSN 109-7318.

Mur, F.; Cardenas, V.; Vaquero, J. & Martinez, S. (1998). Phase synchronization and measurement digital systems of AC mains for power converters, *Proceedings of CIEP*, 1998, pp. 188-194.

Nabae, A.; Takahashi, I. & Akagi, H. (1991). A new neutral-point-clamped PWM-inverter. *IEEE Trans. on Industry Applications*, IA-17(5) 1981, pp. 518-523.

Nedeljkovic, D.; Nastran, J.; Voncina, D. & Ambrozic, V. (1999). Synchronization of active power filter current reference to the network, *IEEE Trans. on Industrial Electronics*, 1999, vol. 46, pp. 333-339.

Nedeljkovic, D.; Ambrozic, V.; Nastran, J. & Hudnik, D. (1998). Synchronization to the network without voltage zero-cross detection, *Proceedings of MELECON*, 1998, Vol. 2, pp. 1228-1232.

Newton, C. & Sumner, M. (1997). Neutral point control for multi-level inverters: theory, design and operational limitations. *Proceedings of IEEE Industrial Application Society Annual Meeting IAS '97*, 1997, pp.1336-1343.

Nguyen, C.T. & Srinivasan, K. (1984). A New Technique for Rapid Tracking of Frequency Deviations Based on Level Crossings, *IEEE Trans. on Power Apparatus and Systems*, 1984, Vol. PAS-103, pp. 2230-2236.

Pena, R.; Clare, J.C. & Asher, G.M. (1996). Doubly fed induction generator using back-to-back PWM converters and its application to variable speed wind-energy generation, *IEE Trans. on Electronic Power Application*, 1996, pp. 231-241.

Portillo, R.; Prats, M.; Leon, J.I.; Sanchez, J.A.; Carrasco, J.M.; Galvan, E. & Franquelo, L.G. (2006). Modelling Strategy for Back-to-Back Three-Level Converters Applied to High-Power Wind Turbines, *IEEE Trans. on Industrial Electronics*, vol. 53, no. 5, October 2006 pp. 1483-1491.

Peng, F.Z.; Lai, J.S.; McKleever, J. & Van Coevering, J. (1995). A multilevel voltage-source converter system with balanced DC-voltages. *Proceedings of PESC '95*, 1995, Vol. 2, pp. 1144-1150.

Rodriguez, J.; Moran, L.; Gonzalez, A. & Silva, C. (1999). High voltage multilevel converter with regeneration capability, *Proceedings of PESC*, 1999, Vol. 2, pp. 1077-1082.

Rodriguez, P.; Pou, J.; Bergas, J.; Candela, I.; Burgos, R. & Boroyevich, D. (2005). Double synchronous reference frame PLL for power converters, *Proceedings of PESC'05*, 2005, pp. 1415–1421.

Saad-Saoud, Z. & Jenkins, N. (1997). The application of advanced static VAr compensators to wind farms, *IEE Colloquium on Power Electronics for Renewable Energy*, 1997, pp. 6/1 - 6/5.

Saitou, M.; Matsui, N. & Shimizu, T. (2003). A control strategy of single-phase active filter using a novel d-q transformation, *Proceedings of IAS'03*, 2003, Vol. 2, pp. 1222-1227.

Salaet, J.; Alepuz, S.; Gilabert, A. & Bordonau, J. (2004). Comparison between two methods of DQ transformation for single phase converters control. Application to a 3-level boost rectifier, *Proceedings of PESC*, 2004, Vol.1, pp. 214-220.

Shen, J. & Butterworth, N. (1997). Analysis and design of a three-level PWM converter system for railway-traction applications, *IEE Proceedings on Electronic Power Application*, 144(5), 1997, pp. 357-371.

Silva, S. M.; Lopes, B. M.; Filho, B. J. C.; Campana, R. P. & Bosventura, W. C. (2004). Performance evaluation of PLL algorithms for single-phase grid-connected systems, *Proceedings of IAS*, 2004, Vol. 4, pp. 2259-2263.

Song, H.-S.; Park, H.-G. & Nam, K. (1999). An instantaneous phase angle detection algorithm under unbalanced line voltage condition, *Proceedings of PESC'99*, vol. 1, 27 June-1 July 1999, pp. 533–537.

Sun-Kyoung Lim, Jun-Ha Kim, Kwanghee Nam (1999). A DC-link voltage balancing algorithm for 3-level converter using the zero sequence current, *Proceedings of IEEE PESC '99*, 1999, Vol. 2, pp. 1083-108.

Sørensen, P.; Bak-Jensen, B.; Kristian, J.; Hansen, A.D.; Janosi, L. & Bech, J. (2000). Power Plant Characteristics of Wind Farms, *Proceedings of the International Conference in Wind Power for the 21st Century*, 2000, Kassel, Germany.

Teodorescu, R.; Iov, F. & Blaabjerg, F. (2006). *Modelling and Control of Grid converter – Basic Grid Inverter Control*, Aalborg University, February 2006, ISBN - 87-89179-62-5, 67 p.

Vainio, O. & Ovaska, S. J. (1995). Noise reduction in zero crossing detection by predictive digital filtering, *IEEE Trans. on Industrial Electronics*, 1995, vol. 42, pp. 58-62.

Vainio, O.; Ovaska, S.J. & Polla, M. (2003). Adaptive filtering using multiplicative general parameters for zero-crossing detection, *IEEE Trans. on Industrial Electronics*, 2003, vol. 50, pp. 1340-1342.

Valiviita, S.; Ovaska, S. J. & Kyyra, J. (1997). Adaptive signal processing system for accurate zero-crossing detection of cycloconverter phase currents, *Proceedings of PCC'97*, 1997, Vol.1, pp. 467-472.

Valiviita, S. (1998). Neural network for zero-crossing detection of distorted line voltages in weak AC-systems, *Proceedings of IMTC*, 1998, Vol.1, pp. 280-285.

Valiviita, S. (1999). Zero-crossing detection of distorted line voltages using 1-b measurements, *IEEE Trans. on Industrial Electronics*, 1999, Vol. 46, pp. 917-922.

Vas, P. (1998). *Sensorless Vector and Direct Torque Control*. Oxford University Press 1998, ISBN 0-19-856465-1.
Wall, R.W. (2003). Simple methods for detecting zero crossing, *Proceedings of IECON*, 2003, Vol.3, pp. 2477-2481.
Wallace, K. & Oliver, J.A. (1998). Variable-Speed Generation Controlled by Passive Elements, *Proceedings of ICEM*, 1998, pp. 1554-1559.
Yamamoto, M. & Motoyoshi, O. (1990). Active and Reactive Power control for Doubly-Fed Wound Rotor Induction Generator, *Proceedings of PESC*, 1990, Vol. 1, pp. 455-460.
Wu, B. (2006). *High-Power Converters and AC drives*, IEEE Press, Wiley Interscience, 2006, ISBN 13-978-0-471-73171-9.
Weidenbrug, R.; Dawson, F. P. & Bonert, R. (1993). New synchronization method for thyristor power converters to weak, *IEEE Trans. on Industrial Electronics*, 1993, Vol. 40, pp. 505-511.

19

Renewable Energy in Lebanon

Nazih Moubayed*, Ali El-Ali** and Rachid Outbib**
*Lebanese University, Faculty of Engineering 1
Lebanon
**University of Aix-Marseille, LSIS
France

1. Introduction

All along the history, the energies included a strong imaginary content. We mention for example, the discovery of fire, the multiple effects of the sun, and the discovery of oil, petrol and natural gas. Therefore, the energy held an important place in the world.

Nowadays, there is a real need to develop renewable energies and the main reasons are as follows:

- A first reason is, of course, that they are inexhaustible, contrary to the other energies, notably the fossil energies, whose stocks are limited. For example, in the case of petrol, a recent survey has the tendency to prove that the reserve in years would be in reduction what lets foretell a production decreasing from 2010.

- A second reason is the risk that presents the nuclear energy; therefore, many nuclear countries want today to change production of energy to safety methods: United States, Germany, Switzerland, etc.

- A third reason, absolutely vital, and short-term, that imposes us to bet on the renewable energies: it is the notion today unanimous recognized of "lasting development", bound to the pollution in the air. One can expect an increase of the number and the power of the cyclones, desertification of subtropical zones, however flooding of some countries (Low Country, Bangladesh), deterioration of the earths by erosion, deviation of the Gulf Stream allowing the polar air mass to arrive to Europe.

Today, there is no a single renewable source which is capable to fill every requirement of energy. Hence, the renewable energy solution of the future will be necessary hybrid (i.e. combining two or more sources of energy) and it will use the potential of local sources. In the last decades, we remark a great activity in the scientific community, in the first hand, to study the renewable sources and hybrid system of energy (Chedid & Rahman, 1997), (Borowy & Salameh, 1994), (Borowy & Salameh, 1996), (Kellog et al., 1998), (Broe et al., 1999) and, in the second hand, to analyze the potential of renewable sources in some area (Ucar & Balo, 2009).

This chapter is a contribution to renewable energies study and it concerns the case of Lebanon. More precisely, we consider the conversion of photovoltaic and wind energy to electrical one. In this work, annual data are discussed, efficiencies of each conversion source are calculated and their economic costs are compared.

2. Review on renewable energy sources

The human consumption of energy became dangerous for the environment and it is necessary to reconsider our resources notably while making appear the part of the renewable. This section brings us to note that the renewable resources (coming from the sun, the terrestrial core and from the phenomena of tides) are enormous, often comfortably available for our long-term needs.

2.1 Hydraulic energy

Hydraulic power station uses the energy provided by the mass of water in movement to turn a turbine. This one is coupled mechanically to an alternator to produce electricity.

The power of these power stations varies from a few hundreds of Kilowatt to several hundreds of Megawatt. The recoverable energy annually on the planet reaches 40.10^{12} kWh and the technically exploitable value is worth, according to the evaluations, between 15 and 20.10^{12} kWh (Multon et al., 2004). Hydraulic power stations supply 20% of the world electricity and it represents 95% in Norway, 83% in Iceland, 70% in Canada, 67% in Austria, 32.02% in Romania (Ciobanu, 2008) and 10% in Lebanon.

2.2 Solar energy

A photovoltaic cell is a semiconductor component, exposed to light, generates an electric voltage (0.5 V to 1.7 V according to the used matter). The cells are joined between them to constitute a solar panel. The energy received at the surface of the earth (720.10^{15} kWh) varies, by m^2, between 1100 kWh and 2300 kWh/year (Multon et al., 2004).

2.3 Wind energy

The use of the wind to produce electricity from the energy of wind was also exploited for a long time. This energy represents an enormous resource, 32.10^{15} kWh, whose exploitable terrestrial part is estimated to 50.10^{12} kWh/year (Multon et al., 2004). The produced wind energy in 2002 represents 0.4% of the total electricity production.

2.4 Biomass

In the domain of the energy, the term of biomass regroups the set of the organic matters capable to become sources of energy. These organic matters that come from the plants are a shape of storage of the solar energy, captured and used by the plants thanks to the chlorophyll. The biomass is an energy that can be chemically polluting when it is badly used. The annual renewable part represents energy of about 800 to 900.10^{12} kWh of which 60.10^{12} kWh is exploitable (Multon et al., 2004).

2.5 Gravitational interactions Moon-Sun

The tide designates the oscillating movement of the sea level that results from the attraction of the moon and the sun on the liquid particles. The yearly energy of the marine currents represents about 25.10^{12} kWh. The exploitable part is between 270 and 500.10^{9} kWh (Multon et al., 2004).

2.6 Surge energy
It is due to the action of the wind on the surface of the seas and oceans. It is estimated to 8.10^{12} kWh of which 90.10^{9} kWh is technically usable per year (Multon et al., 2004).

2.7 Geothermal energy
The terrestrial core gives a yearly energy of about 300.10^{12} kWh. The exploitable reserves are of about 40.10^{9} kWh (Multon et al., 2004). The most important geothermal plants are: Gheiserele - USA (502 MW), Wairaki - New Zeeland (192 MW), Krafla - Iceland (50 MW). Ocean thermal energy conversion plant extracts the water temperature between 25°C at surface and 5°C at a depth of 1000m (Ciobanu, 2008).

3. Energy in Lebanon

In the Mediterranean region, the encouragement of the big scale use of renewable energy constitutes the alternative solution for power consumption in rural zones that are occupied by 40% to 50% of the total population. However, in urban zones, a variety of applications of renewable energy are put in action such as: the desalination and water treatment, the solar hot water production, the domestic and industrial garbage treatment, the bioclimatic architecture, etc. Thus, answering the needs of energy while reducing the troubles in the cities of high and medium density. The structural demand of the Mediterranean countries indicates realistically that up to 45% of the energy demand in this area can be fulfilled by renewable energy from now till 2015, about 3 times the yearly consumption of Algeria, Morocco and Tunisia together (Frenn, 2003). This can be achieved by the implementation of a common voluntary strategy for the renewable energies in the region. Effectively, we are lucky in Lebanon to have a climate allowing more than 300 sunny days per year. Thus, Lebanon is known for being 'the 300 sun day country'. Also, Lebanon is favoured naturally with a big hydraulic energy potential, keeping in mind the possibilities arising from the use of wind energy in the mountains and the bioclimatic architecture. This serves in the preservation of the environment and saves in the energy resources while helping up with the incremental tendencies. Highly industrialized nations are witnessing an increase in the number of hydraulic and solar stations, despite the fact that these countries already possess numerous nuclear plants. The situation in Lebanon indicates a wide and increasing public awareness towards the use of solar energy especially in the use of solar water heaters as they are topping almost every roof. This shows that the Lebanese are well receptive of this aspect; especially that in Lebanon, human resources are capable of introducing and applying correctly these types of renewable and natural energy techniques in our homes, factories and establishments. As the technical know how is well available. These resources must be supported and assisted in order to use them for the good of our country. Economically, these techniques are cost effective both in the short and long terms, because expenses related to installation and maintenance are reduced and the raw materials are naturally available. For example: the annual petroleum bill in Lebanon amounts to 750 Million Dollars, of which more than the half goes to electricity production. The pollution resulting from our public power stations, and those of privates generators installed almost everywhere in the country in addition to the absence of a coherent policy of urbanization, aggravates the negative impact on the environment. Unluckily, this gloomy does not seem to show a tendency of improving in the near future (Frenn, 2003).

In Lebanon, and according to an investigation done by the ministries of industry and petrol, the hydraulic and electric resources as well as the statistical administration, it takes out again of it that:

- 98% of our needs in primary energy have been imported. The renewable energies (solar, wind, etc), in spite of a geographical and socioeconomic context auspicious to their development, represent even less of 1% in the global energizing balance of this country.
- The yearly consumption in energy per person remained less then the world average and represents 1/5 of the one of the European Community and 1/8 of the one of the USA.
- The invoice of energy increased 20% in 2001 in relation to 2000 and this following the increasing in prices of oil and its derivatives on the international markets.
- The analysis of final electricity in relation to the primary resources permits to note that the efficiencies of the thermal power stations don't pass 33% and that the losses on the networks high voltage and of distribution are estimated to 12%.
- The electric consumption rose, in 2001, to 7650 GWh, either in increase of 37% in relation to 2000, but remains even lower to the efficient needs and especially to the level of the seasonal crests and the reserve powers. The electric production is to 11% hydraulic and 89% thermal.
- The combustion of our based energy on the hydrocarbons (4200 tons) and other primary resources, give out in air harmful substances estimated to more then 15 tons of dusts, 80 tons of SO_2 and as many of organic compounds. It also produced 3.5 millions of tons of CO_2, which is 0.88 ton per person and per year. It is greater then 25% to the average production of the countries of the region (0.7 ton of CO_2 per person and per year).

4. A survey on solar and wind energy

4.1 Solar energy

Ancient Egyptians worshiped God Ra' or the Sun God. They granted Ra' the supreme power above all Gods. Most likely, they were not thinking of solar power at the time, but if they did, they would have every good reason to do so since most power resources in our plant earth come from this same Sun once worshiped. Fossil fuel, the arrogant energy source of the era, for example, descended from the very organics eventually ran on solar power! Wood, hydraulic, etc are the outcome of 'Solar power'. One thing in common for the above resources is that they all naturally and 'indirectly' divert, or store, the sun power for our later utilization. It should be worth investigating, however, innovative methods that could utilize solar power as soon as it arrives from the gigantic nuclear plant we used to call SUN.

Enormous researches were conducted in the past to address such direct conversion. Some of these researches did arrive to novel systems including Photovoltaic, or PV, which in no time, directly and smoothly, convert solar photons to electrical current. PV is one kind of PN junction that is a two-semiconductor plates one positive and the other negatively doped and separated by a thin film. While some scientists were doing experiment on PN junctions, they accidentally discovered that if light hits an exposed PN junction some electron will gain energy to jump from one side to the other side. If the circuit was closed from the other side, the missing electron has to be replaced and electron flow will occur. This flow is what we call electric current (Abou Said, 2003).

Thus, the phenomenon named "photovoltaic effect" consists mainly in converting the solar light in electric energy by means of semiconductors devices named photovoltaic cells (Gergaud et al., 2002). The photovoltaic generator is constituted of a series and parallel association of the number of necessary modules to assure the requisite energy to product (De Soto et al., 2005).

The industry of PV modules is still in progress. Besides, they are produced in many forms or types including amorphous, crystalline, and multi-crystalline. They can be made of some particular semiconductor materials but the most are made of Silicon.

To generate sufficient electrical energy, light source must be intense especially because with the current PV'S technologies and material, efficiency of light to current conversion in all PV types is still relatively low, at best just below 20%. Therefore, the light has to have enough intensity that it has to come from the sun to produce effective power. In addition, the sun light intensity varies extensively during the day or in different times of the year and many obstacles may easily shade the sun light, and sometimes for every long, like in case of cloudy days. As a result, additional component should be combined to PV modules to have reliable power sources. Batteries can solve the frequent absence of sunlight and innovative electronics will do the regulating job. These components, when tuned together along with the PV modules construct a system so-called 'Electrical Solar Power Supply'.

Sunlight is free, but, the electrical power produced from it is far from being cheap. In fact, solar electrical power is still one expensive conventional methods of electricity generation. Nevertheless, in many cases, solar power is quite justifiable and it had a wide range of applications in remote places (Abou Said, 2003).

The Sun is a sphere of diameter 139×10^9 m and at a distance from earth equal to 1.5×10^{11} m is called an astronomical unit. The sun as seen from earth may be considered as an equivalent black body with a temperature equal to 57620°K. The energy is produced in the interior of the solar sphere, at temperature of many millions of degrees. The amount of solar energy received on earth surface is depleted by the atmosphere. Generally speaking, the Earth has two global movements that affect the reception of the solar energy to its surface: the rotation that it makes once on itself per day and the yearly revolution that it makes around the sun.

The combination of these movements explains the daily changes in the reception of the solar light in particular places (Bernard, 2004). The reason for which the energizing flux received to soil does not pass 1000 W/m^2 is that the atmosphere modifies in an important way the direct radiance of the sun due to the following mechanisms (Ikegami et al., 2001):

- absorption of light by the various gases constituent,
- diffusion by their molecules,
- absorption and diffusion by the dusts.

In addition, the solar flux received on a surface depends on (Bernard, 2004):

- the orientation and the slant of the surface,
- the latitude of the place and its degree of pollution,
- the period of the year,
- the time considered in the day,
- the nature of the cloudy layers.

The place of solar panel should be cleared well. The orientation of the panel depends on:

- the impact angle: It is the angle formed by the solar panel and the solar rays. The optimal impact angle is an angle of 90°.
- the slant angle: It is the angle formed by the solar panel and the horizontal.
- the zenith angle: It is the angle formed by the solar rays and the horizontal.

Besides its great ecological benefits many application demands solar energy, starting from remote communication relays, remote houses, winter and summer resorts, bill boards lighting, remote water pumping, etc. Some of these systems are already in operation at the time being; but most of which are found in hybrid systems where wind and solar power are combined to one system for practical and optimized operation.

4.2 Wind energy

Wind is simple air in motion. It is caused by the uneven heating of the earth's surface by the sun. Since the earth's surface is made of very different types of land and water, it absorbs the sun's heat at different rates. During the day, the air above the land heats up more quickly than the air over water. The warm air over the land expands and rises, and the heavier, cooler air rushes in to take its place, creating winds. At night, the winds are reversed because the air cools more rapidly over land than over water. In the same way, the large atmospheric winds that circle the earth are created because the land near the earth's equator is heated more by the sun than the land near the North and South Poles. Today, wind energy is mainly used to generate electricity. Wind is called a renewable energy source because the wind will blow as long as the sun shines (Poitiers, 2003).

Like old fashioned windmills, today's wind machines use blades to collect the wind's kinetic energy. Windmills work because they slow down the speed of the wind. The wind flows over the airfoil shaped blades causing lift, like the effect on airplane wings, causing them to turn. The blades are connected to a drive shaft that turns an electric generator to produce electricity. There are two types of wind machines used today: horizontal-axis wind machines and vertical-axis wind machines (Laverdure et al., 2004).

Wind power plants, or wind farms as they are sometimes called, are clusters of wind machines used to produce electricity. A wind farm usually has dozens of wind machines scattered over a large area. Unlike power plants, many wind plants are not owned by public utility companies. Instead they are owned and operated by business people who sell the electricity produced on the wind farm to electric utilities. Operating a wind power plant is not as simple as just building a windmill in a windy place. Wind plant owners must carefully plan where to locate their machines. One important thing to consider is how fast and how much the wind blows (Iov et al., 2004).

As a rule, wind speed increases with altitude and over open areas with no windbreaks. Good sites for wind plants are the tops of smooth, rounded hills, open plains or shorelines, and mountain gaps that produce wind funneling. Wind speed varies throughout the country. It also varies from season to season (Chowdhury & Chellapilla, 2005).

New technologies have decreased the cost of producing electricity from wind, and growth in wind power has been encouraged by tax breaks for renewable energy.

The major potential disadvantages for wind turbines are noise, effect on birds of prey, and aesthetics. Noise is an issue with wind turbines as the mechanical blade and rotor movements tend to produce some noise. With the development of better equipment, the noise issue is being properly handled and noise level is rapidly decreasing. Since most of the appropriate sites are located in uninhabited wind swept, remote areas and sometimes on mountaintops, it is to be expected that the noise factor will be of minor importance. Research has indicated that many people who had originally opposed the presence of wind turbine, had a much more positive feeling after the turbines were installed. As for the effect on birds of prey, it has been found that the probability of a bird hitting the blades is minor and does not present a credible threat to wild species. Finally, the issue of aesthetics is a highly debatable issue. Some people are strongly opposed to the presence of large wind turbines in landscape while others find it attractive. Despite these concerns, it has been positively assured that the presence of wind turbine is a tourist attraction. Actually, wind energy technology is developed well enough to complete with non-renewable sources of electricity (Houri, 2003).

5. Solar and wind energy in Lebanon

5.1 Balance of solar and wind energy

Our study is achieved within the Laboratory of electricity of the Faculty of Engineering in the Lebanese University (Tripoli - Lebanon). The solar experimental part of the system is constituted of a fixed panel and a mobile one of 50 W each one, two batteries and two load voltage regulators. The position control of the mobile panel took place so that its surface is always perpendicular to the solar rays. The wind experimental part is constituted of a wind converter of 400 W, tow batteries, one rectifier and one load voltage regulator. In the global experimental system, a microcontroller is used in order to take data about the solar and wind generated energy and that in order to archive them. Another goal of this microcontroller is to control the functioning of the electric motor moving the mobile solar panel. To make comparison between these two types of renewable energy which should be of the same rated power, we multiplied the data given by the solar panels by eight. Figure 1 shows the generated power, for each month of the year 2006, by solar panels and wind turbine.

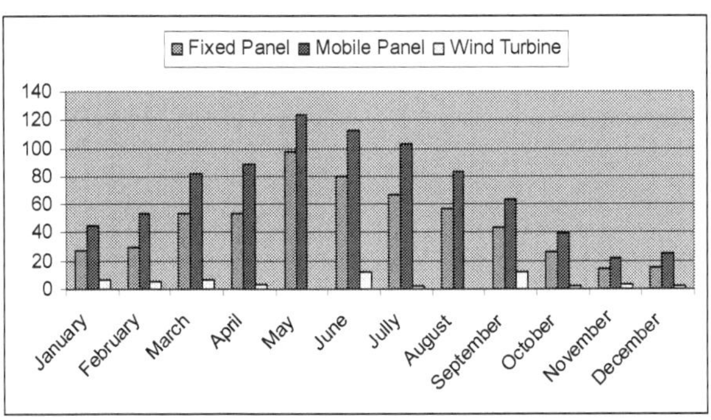

Fig. 1. Comparison between solar and wind energy in kWh

From this figure, one can see that the solar energy is increasing from January to June and decreasing from June to December, which is a logic variation. Regarding to the wind energy, this one takes a normal values in winter and an important values in June and September. These two months are the most important of the year. Figure 2 represents the percentage of the involvement of these two types of energies, solar and wind, in the production of the total renewable energy.

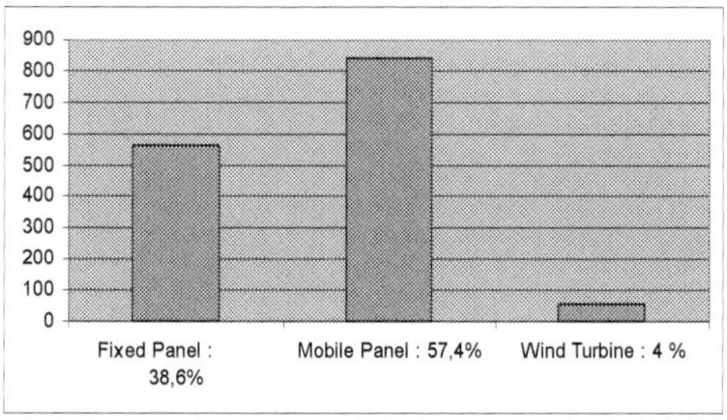

Fig. 2. Percentage value for solar and wind energy

During the year 2006, the energy generated by the wind turbine of 400 W of power is 57 kWh. To do a comparison between the two sources of renewable energy, one multiplies the value obtained by the solar panel by eight. Therefore, the energy produced by eight mobile solar panels of total power 400 W is equal to 840 kWh. Thus, for the same rated power of 400 W, the produced solar energy is very important compared to wind one. It should be noted that the generated solar energy is maximum, because the sky is cleared around the Faculty of Engineering, but, this Faculty is not placed in a windy zone. Consequently, if one places this wind turbine in a windy zone, the produced wind energy will be more important than that obtained in the Faculty of Engineering.

5.2 Efficiency of solar and wind energy conversion

Experimental results indicate that the efficiency of the used solar panel is equal to 10 %. That of the wind turbine is 23 %. Therefore, from these results, one notes that the efficiency of the wind energy conversion is 2.2 times the one of the solar panel.

5.3 Profitability of solar and wind energy conversion

To choose the most profitable type of energy conversion, it should take some hypotheses:

- The 1 kWh is the measurement unit,
- The working duration of each conversion system is estimated to 20 years.

The cost of the eight solar panels of total power equal to 400 W was equal to 6000 USD. This cost includes the installation price. Adding the batteries prices to this cost, the actual value to have this solar station becomes 6500 USD.

The cost of the wind turbine of 400 W was equal to 2000 USD. This cost includes also the installation price. Adding the batteries prices to this cost, the actual value to have this wind station becomes 2500 USD.

Therefore, for the same rated power (400 W), the price of the wind turbine is three times less expensive than that of the solar panels.

6. Conclusion

The Sun is capable of supplying ten thousand times the overall energy needs of humanity. The technological progress is moving at large strips in the development of fire and clean technologies, in particular photovoltaic energy and wind energy. Most developing countries possess renewable sources that ought to be exploited. Wind and solar energy offer a viable, economical alternative to conventional power plants in many areas of the world. Wind and solar are clean fuel; they produce no air or water pollution because no fuel is burned.
This survey consists in creating a data base of the power delivered by the wind machine and solar panels. It permitted during the year 2006 to determine the monthly and yearly balances of these energies in Lebanon.
For the selected zone, experimental results show that the solar panels generated more power than wind turbine. In other hand, the calculation of the efficiency and the profitability of these tow types of conversion are also analyzed. We noted that the wind machine efficiency is 2.2 times than that of a solar panel and its price is three times less expensive than that of the solar panels.

7. References

Abu Said, R. (2003). Solar Energy via photovoltaic, *Solar and renewable energy - Lebanon guide*, pp. 20-21.

Bernard, J. (2004). Energie solaire : calculs et optimisation, Ellipses.

Borowy, B.S. & Salameh, Z.M. (1994). Optimum photovoltaic array size for a hybrid wind/PV system. *IEEE Transactions on energy conversion*, Vol. 9, N°2, September 1994, pp. 482-488.

Borowy, B.S. & Salameh, Z.M. (1996). Methodology for optimally sizing the combination of a battery bank and PV array in a wind/PV hybrid system. *IEEE Transactions on energy conversion*, Vol. 11, N°2, June 1996, pp. 367-375.

Chedid, R. & Rahman, S. (1997). Unit sizing and control of hybrid wind-solar power systems. *IEEE Transactions on energy conversion*, Vol. 12, N°1, March 1997, pp. 79-85.

Chowdhury, B. H. & Chellapilla, S. (2005). Double-fed induction generator control for variable speed wind power generation. *Electric power systems research, ELSEVIER*, pp. 1-15.

Ciobanu, L. (2008). History of renewable electric source and perspectives, *IEEI'08, 2nd International Symposium on the History of the Electrical Engineering and of Tertiary-Level Engineering Education*, IASI - Roumanie, Octobre 2004, Vol. 4, pp. 33-38.

De Broe, A.M.; Drouilhet, S. & Gevorgian, V. (1999). A peak power tracker for small wind turbines in battery charging applications. *IEEE Transactions on energy conversion*, Vol. 14, N°4, December 1999, pp. 1630-1635.

De Soto, W.; Klein, S.A. & Beckman, W.A. (2005). Improvement and validation of a model for photovoltaic array performance, *Solar Energy 80, ELSEVIER*, pp. 78-88.

Frenn, B. (2003). Renewable energy, *Solar and renewable energy - Lebanon guide*, pp. 1.

Gergaud, O.; Multon, B. & Ben Ahmed, H. (2002). Analysis and experimental validation of various photovoltaic system models, *7th International ELECTRIMACS Congress*, Montréal, pp. 1-7.

Houri, A. (2003). Wind energy, *Solar and renewable energy - Lebanon Guide*, pp. 22-23.

Ikegami, T.; Maezono, T.; Nakanishi, F.; Yamagata, Y. & Ebihara, K. (2001). Estimation of equivalent circuit parameters of PV module and its application to optimal operation of PV system, *Solar Energy Materials & Solar Cells 67, ELSEVIER, 2001,* pp. 389-395.

Iov, F.; Hansen, A. D.; Sorensen, P. & Blaabjerg, F. (2004). Wind Turbine Blockset in Matlab / Simulink, General overview and description of the models, *Internal report, Aalborg University.*

Kellog, W.D.; Nehrir, M.H.; Venkataramanan, G. & Gerez, V. (1998). Generation unit sizing and cost analysis for stand-alone wind photovoltaic and hybrid wind/PV systems. *IEEE Transactions on energy conversion.* Vol. 13, N°1, March 1998, pp. 70 -75.

Laverdure, N. ; Roye, D. ; Bacha, S. & Belhomme R. (2004). Technologie des systèmes éoliens Intégration dans les réseaux électriques. *La revue 3EI*, n°39.

Multon, B.; Robin, G.; Ruellan, M. & Ben Ahmed, H. (2004). Situation énergétique mondiale à l'aube du 3ème millénaire. *La Revue 3EI,* n°36, mars 2004, pp. 20-33.

Poitiers, F. (2003). Etude et commande de génératrices asynchrones pour l'utilisation de l'énergie éolienne, *PhD form Ecole polytechnique de Nantes.*

Ucar, A. & Balo, F. (2009). Evaluation of wind energy potential and electricity generation at six locations in Turkey. *Applied Energy.* Article in press.

$RenH_2$ – A Stand-Alone Sustainable Renewable Energy System

João Martins
CTS-UNINOVA and Departamento de Eengenharia Electrotécnica, Faculdade de Ciências e Tecnologia, Universidade Nova de Lisboa Portugal
Carmen M. Rangel
Fuel Cells and Hydrogen Research Unit LNEG (former INETI), Laboratório Nacional de Energia e Geologia Portugal
António Joyce
Unidade de Energia Eólica, Solar e dos Oceanos LNEG (former INETI), Laboratório Nacional de Energia e Geologia Portugal
João Sotomayor
REQUIMTE, Departamento de Química, Faculdade de Ciências e Tecnologia, Universidade Nova de Lisboa Portugal
Armando Pires
CTS-UNINOVA and Polytechnic Institute of Setúbal Portugal
Rui Castro
Cie3 / IST - Technical University of Lisbon Portugal

1. Introduction

Rural and remote sites electrification, where grid connection is almost impossible in terms of cost and geography, is nowadays an important market for renewable energy based electricity production systems. It is becoming largely consensual that autonomous electricity production systems based on renewable energies are the most competitive economical option, when compared with solutions based only on conventional diesel generators.

The main issue, related with renewable energy based autonomous systems, is the intermittent nature of the renewable resources, namely, solar and essentially wind. The variability of these renewable energy sources turns difficult its synchronisation with the load pattern. Therefore, some sort of additional regulation systems capable of performing such matching are required. In the past, conventional diesel generators have been the preferred solution to ensure the perfect balance between generation and load. In most cases, batteries are associated as a low energy storage device, in order to somehow lighten the diesel generator operation. These types of systems fall under the general classification of *hybrid systems*.

It is well-known that diesel generators have several disadvantages, the main ones being that its operation is far from being environmental friendly and the fuel is difficult and expensive to provide at remote sites. Thus, the current trend is to focus on energy storage solutions that are able to dismiss the diesel generators. The basic general idea is to store the energy produced in excess during low demand periods and to recover it in the peak hours.

However, the problem remains as the basic conventional energy storage solution is battery-based. Batteries are being worldwide used for ages and its technology is nowadays maturing, reliable and very well understood. Most batteries used in hybrid systems are of the lead-acid type. There are several other types, such as, nickel-cadmium, lithium-ion, but these are generally either too expensive or too unreliable for practical application. In spite of their numerous advantages, battery technology has experienced little advances in recent years. Their known drawbacks, such as low energy density, limited number of full discharge cycles, self-discharge and environmentally unfriendly characteristics, still persist. Therefore, batteries are only envisaged as short-term storage devices.

In order to minimise these recognised pitfalls, alternative energy storage solutions have come to force. One of the most promising is the production of hydrogen through the electrolysis of water. This hydrogen is to be subsequently used to generate electricity through fuel cell technology or even in a combustion engine linked to an electrical generator. The same basic idea as for batteries applies: electrical energy in excess during off-peak hours is used to produce hydrogen, which is later used to generate electrical energy in high demand periods. In this way, the known limitations of the diesel/battery solutions are overcome and a cleaner and more efficient stand-alone electrical production system is achieved. Moreover, hydrogen is an environmentally benign and sustainable fuel and provides long-term storage facilities.

The main objective of this project is the development of a fully autonomous system, in which every component is based upon renewable energies. In this way, a renewable energy autonomous production system based on the hydrogen technology is proposed. This type of systems can be designated as *Stand-Alone Sustainable Renewable Energy System* (SASRES).

The proposed SASRES is composed by three generators - PhotoVoltaics (PV), wind turbine (WTG) and fuel-cell (FC), two loads - an effective load bank and an electrolyser, two buses - AC and DC, and several advanced power electronics converters. Briefly, the system runs under two basic operating modes: 1) whenever there is enough PV and wind power available, the load is supplied through these generators and hydrogen is produced in the electrolyser; 2) when the load demand is higher than the renewable based generators production, the load is supplied from the operation of the fuel-cell. It is worth to mention that the successful integration of multiple generating sources must rely on complex controls to ensure correct sharing of the intermittent renewable energy and controllable fuel-cell generation to meet the demand of the variable load. A general scheme of the overall system is presented in Fig. 1.

The obtained system is a suitable choice regarding the actual stand-alone systems based upon diesel generators and lead-acid batteries. Several benefits can be listed as far as the proposed hybrid PV/WTG/FC system is concerned: improved reliability and energy services, zero emissions and noise pollution, continuous power, increased operational life and efficient use of energy. Therefore, this hybrid system meets the sustainability and environmental respect criteria regarding the energetic solutions of the future - zero emitting either on production or consumption.

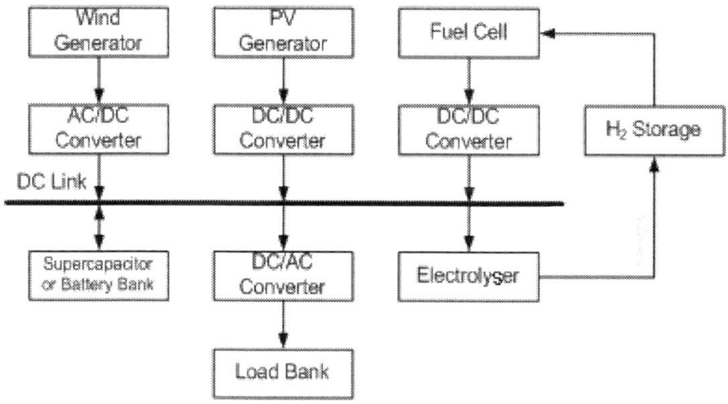

Fig. 1. General scheme of the overall system

This paper presents the proposed PV/WTG/FC stand-alone system supported by totally renewable hydrogen production. The dedicated experimental test prototype is described and the correspondent modelling approach of each component and overall system is introduced. The principles of the implemented control strategy are presented, so that proper behaviour of the system is achieved and its operation can be predicted. Some results obtained both from the experimental test facility and from the implemented software model are presented and discussed. Finally, some conclusions regarding the overall behaviour of the system and the sizing of its components are drawn.

Proposed innovations

Taking the performed state of the art assessment into consideration, several innovations can be found in the proposed SASRES. Although those innovative aspects are identified in the core of the paper, hereafter we list the three most significant ones.

1. No diesel generator is foreseen what makes the proposed system a real SASRES, as it is totally based on renewable resources. This way, energy production is clean and environmentally friendly with no CO_2 emissions.
2. The layout of the SASRES is remarkably elegant. All the generators are connected to a common DC bus and all the loads are connected to an AC bus.
3. Advanced power electronics is used in order to provide efficient power management. This allows dismissing the battery, which is another significant environmental benefit.

2. Brief Review of the Literature

Hybrid renewable based systems with hydrogen storage have been in recent years and are currently still being a highly demanded research topic. A lot of research is being done, making it impossible to take account of all the relevant published research on the topic. To offer the reader some technical guidance, this chapter is divided into sections, each one concerning a particular research domain.

2.1 Demonstration projects

A number of demonstration hybrid renewable based systems are in operation all over the world. Based on (Sovacool et al., 2006), three examples of such systems, located in Europe, are presented in chronological sequence.

In the island of Utsira, Norway, the first large-scale test wind-hydrogen system has been installed in 2004 and is described by (Nakken et al., 2006). The plant produces hydrogen through an electrolyser when there is excess of wind energy available; it provides electricity to domestic customers via a fuel cell and a hydrogen combustion engine when the wind turbine slows or stops. The main components of the system are: i) a 600 kW WTG; ii) a 48 kW (10 Nm^3/h) electrolyser; iii) a 5 kW compressor to increase the pressure of the hydrogen to a maximum of 200 bar; iv) a 12 m^3 H_2-storage tank having enough capacity to cover the customer's demand for 2-3 days with low wind; v) a 10 kW fuel cell and a 55 kW hydrogen combustion engine/generator; vi) a 5 kWh flywheel and a 100-kVA synchronous machine stabilizing the local grid; vii) and a 35 kWh battery providing emergency back-up power. The domestic customers connected to the plant have a peak demand of approximately 50 kW. (Nakken et al., 2006) states that after one year of operation, an average availability of approximately 90% has been achieved and the power quality is reported to be good.

At the northernmost part of the United Kingdom, in the island of Unst, 200 miles north of the Scottish mainland, a hybrid wind-hydrogen energy system is in operation since 2005, in the framework of the PURE (Promoting Unst Renewable Energy) project. (Gazey et al., 2006) explains that on the island, two 15 kW wind generators, whose design is based on the concept of using a permanent magnet generator and a direct drive, provide electrical heating for five office buildings. During times of low electricity demand or high wind speeds, the WTG send their excess electricity to a hydrogen electrolyser. This device demands between 2 and 7 kW to produce a daily average of 2 kg of hydrogen daily. Stored in a high-pressure container, the hydrogen is then dispensed to fill hydrogen canisters that power a hybrid electric car with a 1.2 kW daily plug power fuel cell. The system also contains a 5 kW backup power supply that utilizes another fuel cell and inverter. This system uses the hydrogen to power the office buildings during the times of no wind turbines operation.

Another demonstration project exists on the island of Lolland, Denmark, whose details can be found in (Jensen et al., 2007). The project, known as the "Lolland Hydrogen Community", opened in May 2007 and intends to be the first full-scale hydrogen demonstration facility for residential fuel cells that generate heat and power. Phase I of the project began in early 2007 with the installation in ten homes of combined heat and power (CHP) units that draw on hydrogen-based fuel cells. A centralized electrolyser splits oxygen from hydrogen, with the latter gas stored in low-pressure tanks that are connected to the fuel cells in the houses. As reported in (Jensen et al., 2007), the demonstration plant consists of an energy container with two 4 kW electrolysis installations and two PEM (Proton-Exchange Membrane) fuel cell stacks with an installed capacity of 2 kW and 7.5 kW. This installation is not directly based on local renewable resources, as the electrolyser requires a mains power supply; however, it is indirectly powered by grid integrated wind generators, following a strategy of production of hydrogen in periods with high wind power and low consumption.

2.2 Research oriented projects

In the framework of EU or governments research funded projects, several installations have been raised and are operating. A comprehensive survey of these projects can be found in (Yilanci et al., 2008), (Zoulias et al., 2008) and (Zini et al., 2009). Some of these projects are hereafter reported.

The EU FP5 project "Cluster Pilot Project for the Integration of Renewables into European Energy Sectors using Hydrogen" (RES2H2 in brief) started in January 2002 and is concerned with the design, installation, operation and optimization of two different wind-hydrogen systems. One unit is installed in Gran Canaria, Spain, the second in Keratea, approximately 40 km south of Athens, Greece. The Greek facility has been in operation for over three years now (started 2005). Details of the Greek installation are available in (Varkaraki et al., 2006) and (Varkaraki et al., 2008), for instance. The system is composed of a 25 kW water electrolyser, metal hydride tanks filled with a LaNi5-type alloy and a hydrogen compressor for filling hydrogen cylinders, all powered by a 500 kW WTG.

As described in (Chaparro et al., 2005), a stand-alone system to convert direct solar energy in hydrogen has been studied in Spain within the EU FP5 project FIRST (Fuel cell Innovative Remote System for Telecoms). The system has been designed to supply a power of 200 W, without interruption, from a 1.4 kWp PV field (thin-film technology). The electrical energy produced is stored in batteries (20 kWh total capacity) to smooth PV fluctuations. A PEM electrolyser produces hydrogen at 30 bar and feeds directly seven metal hydrides tanks of 10Nm3 capacity each. Metal hydrides work under pressure control in the temperature range 0-40°C. Hydrogen is converted back to energy in a 275 W PEM fuel cell.

A stand-alone renewable energy system employing hydrogen storage has been built within the Hydrogen and Renewables Integration (HaRI) project at West Beacon Farm, Leicestershire, UK. As remarked by (Gammon et al., 2006) and more recently by (Little et al., 2007), before the start of the HaRI project, the existing renewable energy devices included two 25 kW WTG, a 13 kWp PV array and two micro-hydro turbines with a combined output of 3 kW. Further sustainable energy features include a 10 kW heat pump, and a 15 kWel, 38 kWth CHP unit. The above mentioned authors explain that the addition of a hydrogen energy storage system to the existing supply network was proposed as a means of testing the feasibility of a stand-alone system. The three primary components of the newly installed hydrogen based system are a 36 kW electrolyser, pressurized hydrogen storage cylinders with a capacity of 2856 Nm^3 of hydrogen at 13.7 MPa, and two fuel cells, 2 kW and 5 kW rated power.

Outside Europe there are some demonstration projects underway. The hydrogen production test facility located at Kuala Terengganu, East Coast of Peninsular Malaysia, can be mentioned and is presented in (Sopian et al., 2009). A 1 kWp PV array consisting of 12 amorphous silicon modules to operate nominally at 24 VDC is installed together with a 1 kW WTG, equipped with a permanent magnet generator. The voltage is regulated and converted to 24 V on a power controlled centre. A set of deep-discharge batteries bank with a capacity of 1000 Ah acts as a buffer between the PEM electrolyser and the power sources. The above mentioned (Sopian et al., 2009) presents results of the validation of a complete performance model of the hybrid PV–wind hydrogen production system against experimental results.

In the framework of university research, several projects have also been developed. Some examples are also presented, again following a chronological sequence.

A Multi-component Laboratory for Integrated Energy Systems has been established at the University of Applied Sciences, Stralsund, Germany. As early noted by (Menzl et al., 1999), one feature of this laboratory is a windmill–electrolyser system for carbon dioxide-free hydrogen production. (Menzl et al., 1999) continues by describing the system components: a 100 kW WTG equipped with a two-speed asynchronous generator, a 10 kWp PV installation consisting of mono-crystalline, poly-crystalline and amorphous cells, a 20 kW alkaline pressure electrolyser delivering hydrogen at up to 25 bars without using a compressor, a hydrogen storage tank with a geometrical volume of 8 m^3, a 370-W PEM fuel cell, a catalytic burner with a thermal power output of 21 kW converting hydrogen directly in thermal energy, a cogeneration plant with a power output of 30 kWel, 70 kWth, fed by natural gas and a diesel generator.

Agbossou (Agbossou et al., 2001) reports that at the Hydrogen Research Institute (HRI), Canada, a test facility is installed since May 2001. The system consists of a 10 kW WTG and 1 kWp PV array as primary energy sources. The excess energy with respect to the load requirement is stored as electrolytic hydrogen through a 5 kW electrolyser and utilized to produce electricity as per energy demand through a 5 kW PEM FC system. This facility allows experimental results to be obtained as in (Agbossou et al., 2004), which is always a much appreciated plus.

Recently (beginning of 2006), a new 5 kWp solar-hydrogen system was installed at Pamukkale University, Turkey. This integrated clean energy system consists of some major components such as sun tracking and fixed PV panels, charge controllers, batteries, inverters, a deionizer, a PEM electrolyser, metal hydride tanks and PEM fuel cells. Further information on this research facility can be found in (Ozturk et al., 2007).

2.3 Software tools

With the purpose of simulating the behavior of stand-alone renewable based systems, either dedicated or adapted software programs have been developed. In (Turcotte et al., 2001), a review of the main available tools is presented. A selection of references regarding the main sizing and simulation tools follows.

HOMER (Hybrid Optimization Model for Electric Renewables) is a sizing and optimization tool developed by NREL (National Renewable Energy Laboratory, USA). Alam (Alam et al., 2007) shows that it is able of comparing different system configurations and components sizes, by automatically running many simulations.

Hybrid2 (Hybrid Power System Simulation Model), developed by the University of Massachusetts, USA, and NREL, USA, is perhaps the more used hybrid power systems simulation tool. Mills (Mills et al., 2004) highlights an important feature of Hybrid2: it allows the user to include manufacturer specified parameters, such as the wind power curve for a WTG, or the I–V curve of a PV panel. Additionally, Hybrid2 is based on a quasi-steady probabilistic/time-dependent model as noted by (Mills et al., 2004).

TRNSYS (Transient Energy System Simulation Tool) was developed by a joint team lead by the University of Wisconsin-Madison, USA. It is an open architecture research tool based on a modular architecture of FORTRAN code blocks. Hundreds of simulation blocks are available, thus enabling an extensive use of TRNSYS for hybrid systems simulation, where Ulleberg (1999) and Ulleberg (1997) are good examples. TRNSYS is based on an unsteady time-dependent model.

2.4 System modeling

In the literature, several models able to predict the overall behavior of both SASRES and general hybrid systems are available. From the abundant offer in this domain, some examples are selected.

In (Senjyu et al., 2005) a model of the different components and the resulting overall system model are presented. The modeled hybrid system includes wind and diesel generators but no battery, therefore special emphasis is put on the control system. A large number of simulation results concerning different case-studies are displayed.

A stand alone PV system was modeled by Joyce et al (2001) addressing the PV modules, the inverter and the batteries, and the results were validated against data obtained in a 150 W stand alone system installed in Portugal at INETI.

A study regarding the improvement of the utilization of wind power in a Greek island is presented in (Ntziachristos et al., 2005). A part of the energy produced by the WTG is stored in the form of hydrogen and is then delivered to the consumption at constant power through a fuel cell. The model is used to simulate the operation of the system over a year. The focus of the paper is the grid connected operating mode and not the autonomous running. The conclusion stated by (Ntziachristos et al., 2005) is that the operation is possible with fuel cell sizes that reach almost up to 1/3 of the nominal wind-turbine power and overall efficiencies that may exceed 60%.

A comparative analysis of the behavior of a SASRES whose primary resource is alternatively wind, sun or hydro is performed in (Santarelli et al., 2004). Regarding a specific location in Italy, an overall model has been built. This model has been used to produce simulation results of plant operation over a year.

In (Chedid et al., 2007), an interesting equivalent electric circuit of a hybrid WTG/FC system is derived after a complete mathematical model for each component of the studied system. The complete electric circuit is composed of two parts; the first part models a wind turbine driving a three-phase permanent magnet alternator connected to a rectifier and an electrolyser, and the second part models the fuel cell and the load. (Chedid et al., 2007) provides simulation results to determine the values of the main system variables and therefore to predict its behavior.

Advanced algorithms have been applied to hybrid/hydrogen systems modeling. For instance, (Dufo-López et al., 2007) presents a strategy, optimized by genetic algorithms, to control a stand-alone hybrid PV/diesel/battery system with hydrogen storage. In this paper, genetic algorithms are the tool at hand to optimize the various system control parameters. Another instance is the use of fuzzy logic controllers to achieve maximum power tracking for both PV and wind generators described in (El-Shatter et al., 2006). The paper makes use of fuzzy logic to design the appropriate power management of the flows between the system components in order to satisfy the load requirements.

Power management is another issue that has been subject to the attention of researchers. In (Wang et al., 2008) the power management of a SASRES is assessed. The proposed strategy allows appropriate management of the power flows among the different energy sources and the storage unit in the system. The system performance under different scenarios has been verified after the development of a model of the overall system.
The power management subject is also addressed in (Zhou et al., 2008) where a WTG with hydrogen based long-term storage and super-capacitor based fast dynamic storage is studied.
Power electronics is a key issue to achieve proper and cost-effective power management between the different players. That is why, for instance, (De Battista et al., 2006) proposes an advanced control for a wind-electrolysis system, which match the wind power output to the electrolyser power requirements, thus gaining in system performance. This control strategy is developed using concepts of the reference conditioning technique and the sliding mode control theory.

2.5 Economics

Cost assessment of SASRES systems is a matter of great concern and several studies were published on this subject.
In (Ghosh et al., 2003) a comparison of hydrogen storage with diesel generation in a PV-WTG hybrid system is performed from a cost analysis point of view. The critical fuel cost is calculated depending on the seasonal solar and wind energy difference.
Nelson (Nelson et al., 2006) addresses an economic evaluation of a hybrid PV/WTG/FC generation system for a typical home in the Pacific Northwest; furthermore, a comparison to a traditional hybrid energy system with battery storage is performed.
The above mentioned HOMER software is used as a sizing and optimization tool and a sensitivity analysis with wind speed data, solar radiation level, diesel price and fuel cell cost is done in (Khan et al., 2005) for a remote house in Newfoundland, Canada.
The "discounted cash flow" method, with the "levelized energy cost" as a financial indicator, is used for the economical analysis of an application regarding a stand-alone hybrid PV/WTG/FC system incorporating compressed hydrogen gas storage in Cooma, Australia; the results of this investigation are reported in (Shakya et al., 2005).
All these studies point to the same conclusion: hybrid/hydrogen systems are at the edge of becoming the cost-effective choice for the electrification of remote areas, providing that the cost of the electrolyser and fuel cell decreases a little in the near future.

3. System's Components

The main goal of RenH$_2$ project is the fully study of low power stand-alone totally renewable power supply systems. The project comprises model development and prototype implementation and integration of the different components, namely:

- Photovoltaic cells
- Wind turbine
- Electrolyser system
- Fuel Cell
- Power electronics converters
- Control system

A general scheme of the overall system has already been presented in Fig. 1.

3.1 Photovoltaic System
The Photovoltaic system comprises a 530 Wp PV array with 10monochristaline Isofoton M-55L modules, facing South ant tilted at 45° connected to the 24 V DC bus through a power electronics converter. A pyranometer placed in the plane of the modules measures solar incident radiation. Figure 2 depicts the PV array installed at INETI's campus.

3.2 Wind Generator
The wind generator is a 750 W Aerocraft AC752 with 2.4 m diameter, installed at 10 m, delivering 3 phase AC power at a nominal speed of 9 m/s. A power electronics controller rectifies the 3 phase AC power connecting the wind generator to the DC bus. Figure 2 depicts the wind generator used.

Fig. 2. Wind Generator and PV array of RenH2 project

3.3 Fuel-Cell
A prototype proton exchange membrane (PEM) fuel cell with nominal power of 100 W, built by SRE, Portugal, was used in this work. The main characteristics of the stack are presented in Table 1. A view of the power source with already integrated fuel cell and auxiliary system and the fuel cell stack are shown in Figure 3a) and 3b) respectively. The fuel cell uses hydrogen without previous humidification supplied in dead-end mode and an open cathode fed by ambient air.

Fuel cell stack	4 stacks of 24 membranes
Nominal power	100 W
Power density	350 mW/cm2
Ohmic resistance per MEA*	470 mohm/cm2
Stabilized voltage (back-up model)	6,12 and 24 V (Imax=4A)
Charge-Imax (charger mode)	24 V or 48 V
Voltage decay at 0,5 A	3 mV/hr
Useful functioning lifetime	Expected 1500 hr
Start-up	Lithium ion rechargeable battery
Weight	2 kg
Temperature	5 - 40°C

*MEA- membrane electrode assembly

Table 1. Main characteristics of the PEM fuel cell stack used in this work

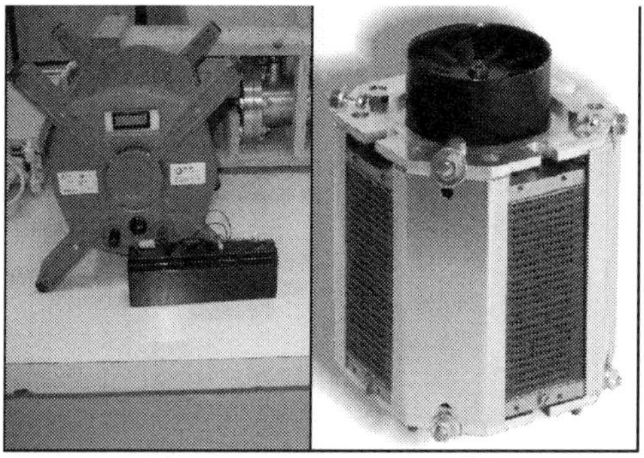

(a) (b)

Fig. 3. General aspects of the power supply (a) and the PEM fuel cell stack (b) used in this work.

3.4 Electrolyser

As the idea of producing hydrogen by reforming fossil fuels, such as methane, methanol, ethanol, ammonia, releasing carbon dioxide as a greenhouse gas, is no more an option, due to environmental considerations, H_2 production by means of renewable energy sources is one of the most promising alternative for the future. Among the renewable energy sources, solar energy is one of the strongest options.

There are three pathways for hydrogen production using solar energy: i) electrochemical, by means water electrolysis, using electric energy to promote a redox reaction in a water molecule and producing reduced and oxidized species, namely hydrogen and oxygen, respectively; ii) thermochemical, using solar thermolysis, based on the use of concentrated solar radiation as the energy source of high temperature process heat for endothermic reactions and producing the water dissociation without the redox reaction or solar thermochemical cycles, in which water splitting can be achieved using heat as energy source to create reactive species that can be recycled; and iii) photochemical, where light sensitive

photoelectrodes immersed in an aqueous electrolyte, make up a photoelectrochemical cell, converting light energy into electricity.
Solar thermolysis and solar thermochemical cycles decompose water into H_2 and O_2 without electrolysis, but high temperatures must be achieved (T>2000°C) and the product gases must be separated to avoid the make up of an explosive mixture. Photoelectrochemical cells are promising but in nowadays only around 18% energy efficiency is attained with TiO_2 as anode. Solar electrochemical seems to be the ideal option, especially for stationary applications. The water electrolysis is made by means an electrolyser working at relatively low temperature using a proton exchange membrane, PEM. These electrolysers show a energy efficiency between 65% and 80%, while the solar photovoltaic energy reach a energy efficiency from 3% to 17%, reflecting a total solar energy-to-hydrogen conversion for solar photovoltaic H_2 production by an electrolyser between 2% and 14%. Nevertheless, although expensive at the moment and presenting reduced energy efficiency, this option is very versatile, can be operate in a large or a small scale, and is totally free of pollution (Licht, 2005).
A 500 W nominal power electrolyser, model LM-1000 from Sandong Institute of Chemical Industry was installed with a maximum hydrogen production rate of 1000 cm^3/min at low pressures going from 1 to 4 bar. The electrolyser has 2 stacks of PEM cells and can work both on AC (230 V) or DC with voltages reaching 25 V which is useful for working directly connected to the 24 V DC bus. Figure 4 depicts the electrolyser. The hydrogen produced by water electrolysis is 99,99% pure and does not need any further purification other than hydrogen dryer before to be fed to the storage system.

Fig. 4. Stand-alone system's electrolyser, fuel-cell, and electrolyser's DC/DC power converter

3.5 Hydrogen Storage

An AB5 type intermetallic alloy, namely LaNi5 containing Al as a substitution element, was chosen as a storage option. This allowed adjustments of the equilibrium pressure and absorption kinetics of the base metallic hydride, making the modified alloy appropriate to be charged using a low pressure PEM electrolyser as well as the fuel cell selected for the present application. Furthermore, since weight is not an issue in stand-alone systems applications, the present storage option represents a more compact solution (by a factor of 3) when compared with high pressure compressed gas. Figure 5 shows the hydrogen storage reactor used in this work, containing alloy LaNi4.7Al0.3 with capacity for 1500 NL H2, which allows more than 20 hours autonomy to the system when running on the fuel cell as sole power supply.

Fig. 5. Hydrogen storage reactor containing LaNi4.7Al0.3 alloy with a capacity for 1500 NL H2.

The process of hydrogen absorption/desorption in metal hydrides is best illustrated by pressure-composition-temperature profiles, denoted as PCT curves - the pressure at a given H-content increases with temperature and is a direct consequence of the thermodynamics associated with the hydriding reaction (1).

$$M + x/2\, H_2 \Leftrightarrow M\text{-}Hx \tag{1}$$

Previous to the selection of the alloy, the thermodynamic properties and the absorption/ desorption capacity of LaNi5 and LaNi5-yAly alloys were studied using a purpose built Sievert-type apparatus to draw the PCT curves. LaNi5-yAly, absorption/desorption cycles were implemented for alloys with different contents of Al ($0 \geq y \leq 0.4$) at different temperatures (Van't Hoff diagrams). The increase in aluminium content in LaNi5 alloy decreases the equilibrium pressure of the hydride. It was evident that the Al content in the alloy increases the enthalpy of formation and therefore the stability of the metal hydride, consequently smaller pressures are needed to charge the alloy (1 bar at 20°C) but higher temperatures are required for desorption, ~60°C. The alloys can absorb hydrogen at sub-atmospheric pressures at ambient temperature. A slight decrease in the hydrogen storage capacity was noticed.

3.6 Power Electronics Converters

As the fuel cell operates in a wide range voltage output, a fuel cell DC/DC converter is used to connect it to the DC busbar. To guarantee a good performance and stability, the fuel cell output current should be continuous. Hence, a DC/DC converter topology with a continuous input current must be selected. For this project a SEPIC (Single-Ended Primary Inductance Converter) was chosen and presented in Fig. 6.

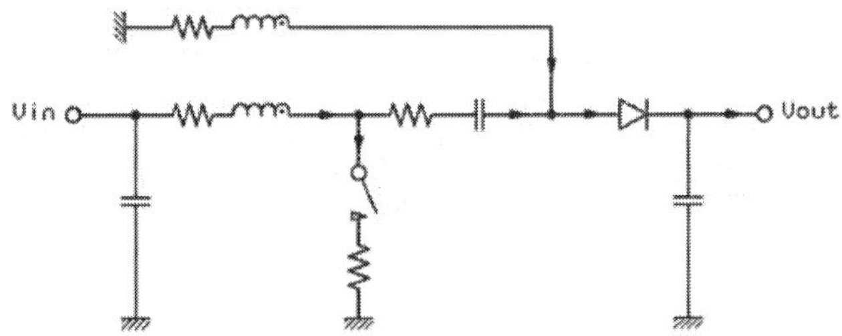

Fig. 6. SEPIC topology

This topology, where its input voltage range can overlap the output voltage, is based on the well-known Boost Converter discussed in (Ismail, 2009). In steady-state conditions the mean voltage across inductance L_1 remains zero, forcing the voltage across this inductance during the time the switch S is on to be equal to the voltage experienced during the switch off time. This leads to relationship (2), where V_o is the output voltage, V_i the input voltage, V_d the diode's forward voltage drop, α the duty cycle and A is called the Amplification Factor (neglecting the parasitic resistances).

$$(V_o - V_d)/V_i = \alpha/(1-\alpha) = A \tag{2}$$

The DC/DC converter control is performed as a current source (adopting a current control scheme). The power value to be outputted is applied as a reference value, and the control acts in order to obtain the current corresponding to the desired power output value. Once the fuel cell power limit is reached, the converter control switches to a voltage control scheme.

Since the DC busbar voltage operates in 24 Vdc and the AC busbar voltage is 230 Vrms it was necessary to consider another power electronics converter toperform DC/AC conversion and also amplifying the voltage. Both of these tasks could either be accomplished by connecting an inverter to the DC busbar followed by a 50 Hz transformer (AC/AC voltage gain) or by connecting a switch mode DC/DC converter to the DC bus bar (DC/DC voltage gain) followed by an inverter. A DC/DC voltage gain (switch mode DC/DC converter) is preferred in order to limit the physical size and the cost of the system. Hence, a system with a voltage source switch mode DC/DC converter followed by an inverter has been selected, as presented in Fig. 7.

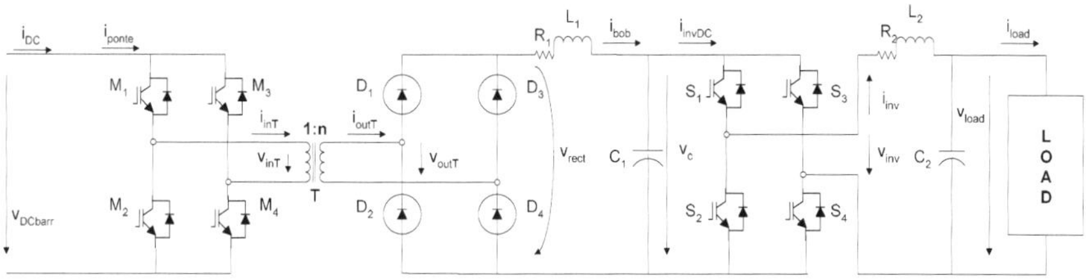

Fig. 7. DC/AC busbars power electronic converter

The wind generator is connected to the DC busbar through a standard diode rectifier followed by a DC/DC Buck converter. The PV array system considers a DC/DC boost conversion, presented in Fig. 8, with MPPT capabilities. Experimental results are presented in Fig. 9. The electrolyser is connected to the DC busbar through a standard controlled DC/DC power converter. This converter was already shown in Fig. 4, along with the electrolyser and the fuel-cell.

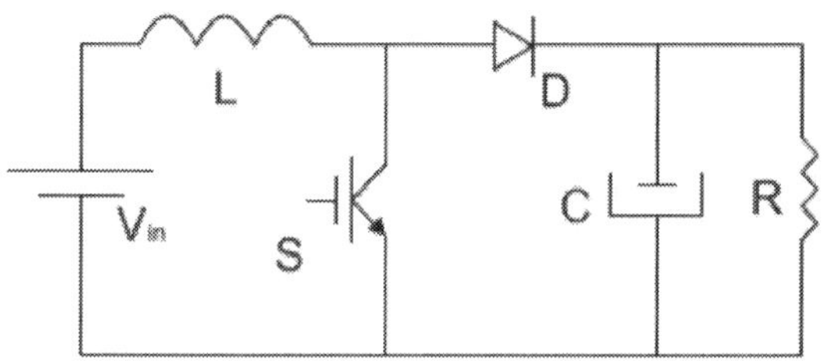

Fig. 8. PV array DC/DC boost converter

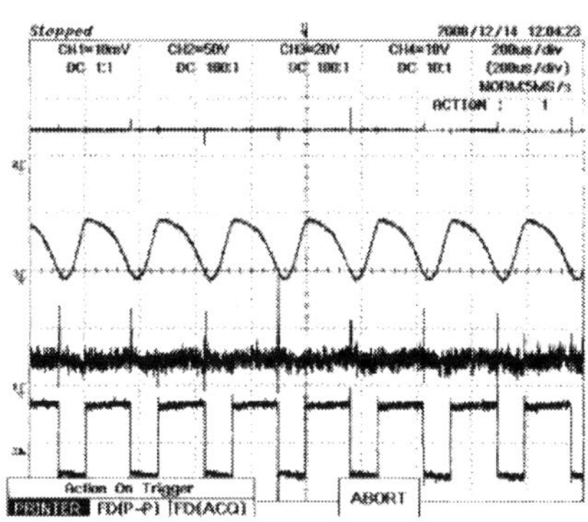

Fig. 9. PV DC/DC converter experimental results: output current, output voltage, input voltage and command signal

3.7 DC Link Busbar

The DC link acts as interface between the production sub-system and the electrical load. In the production sub-system one can consider the Photovoltaic Modules, the Wind Generator and the Fuel Cell. Whenever there are no renewable resources available the control system relies on the fuel cell in order to provide the power demanded by the load. However the fuel cell may not be sufficient to rapidly satisfy load-changing demands. In this way a DC buffer is mandatory. One can establish a set of batteries or supercapacitors to meet sudden load changes. Supercapacitors are a better choice because they are devices that provide higher power densities than conventional batteries. Their charge/discharge times can be extremely fast, reliable, maintenance-free and present long lifespan. A supercapacitor bank with a series of 5 parallel branches of 11 standard 2600F/3.5V supercacitors allows keeping the DC voltage between 23V and 27V, over the 24V DC busbar.

4. Control Strategy

The control of the overall system is a fully automated process that, regarding the sensor array information, establishes a set of controls that will run all of the system's components. The sensor array includes information from all available data, however only some of them are essential for control purposes:

- wind and solar energy;
- H_2 conditions;
- fuel-cell power;
- AC and DC busbar voltages;
- load requirements.

The control considers the following set of main commands:

- electrolyser set-point and command;
- fuel-cell set-point and command (including its DC/DC converter);
- DC/AC converter set-point.

One should note that both the PV generator and the wind generator should generate the maximum energy available. Fig. 10 presents the power conditions, acquired from the sensor array, for two consecutive days. It presents the output electric power of the wind generator and of the PV system, the load power and its difference to the produced power.

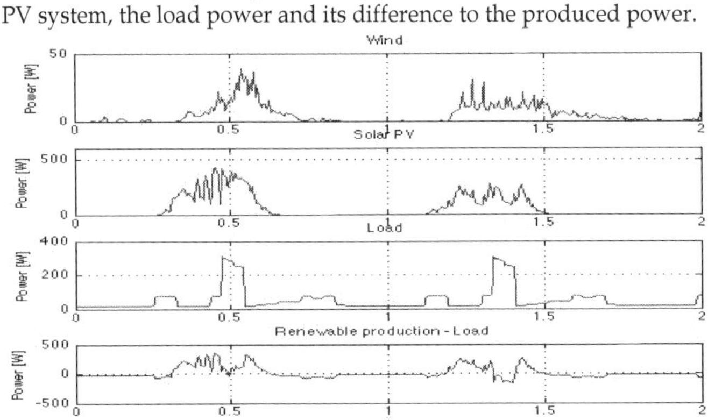

Fig. 10. Renewable power supply and load demand

The basic control idea is that the electrolyser generates hydrogen whenever there is an excess of solar or/and wind energy. This means that if the solar and wind energy are more than enough to demand the load requirements its excess should be used to produce hydrogen. Whenever the wind and solar energy are insufficient to face the load demand the fuel cell uses the stored hydrogen to produce the required lack of energy. The control algorithm flowchart is presented in Fig. 11.

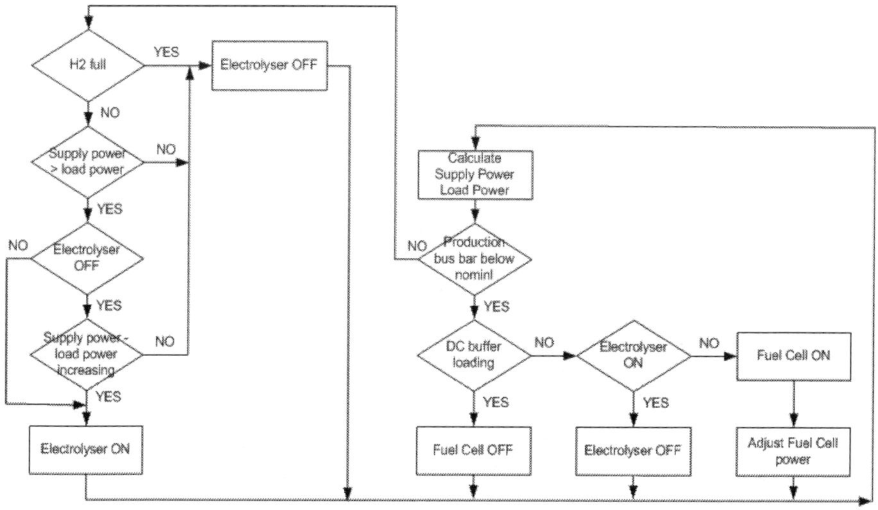

Fig. 11. RenH_2 system control strategy flowchart

At this time it is important to note that a building management system is considered in order to manage the load demand: if the overall power in insufficient to cover the load demand non essential loads are disconnected and only reconnected whenever the generated power is in conditions of supplying the considered load.

Several interlocks are considered to protect the system. A typical example is the one that does not allow the electrolyser and fuel cell to work at the same time, thus the electrolyser should only work when in presence of excess power.

The considered control scheme runs as follows: First the produced and demand power are computed. If the produced power covers the demand one should consider the following issues. If the hydrogen tanks are full then the electrolyser is switched off. If not and the excess produce power can run the electrolyser, then the electrolyser is switched on. If the produced power does not cover the demand the electrolyser should be switched off, if it was running. Then the fuel cell should be switched on and its set-point be adjusted to cover the power demand.

Fig. 12 presents the fuel cell and electrolyser electrical power correlation with the H_2 state-of-charge. One can see that whenever there is enough input power the electrolyser produces hydrogen, increasing its state-of-charge. Lack of renewable input power forces the fuel cell to operate decreasing the amount of stored hydrogen.

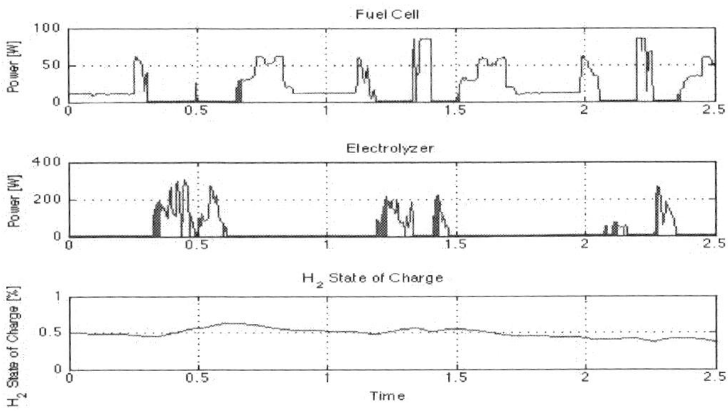

Fig. 12. $RenH_2$ electrolyser and fuel cell electrical power and H_2 state-of-charge

The control system implements the algorithm described above. It was developed in a VEE7.0 programming language implemented in the AGILENT data acquisition system. The program runs a continuous loop, considering a 5 second cycle. At each cycle, reads each channel and converts each signal (for quantities where this procedure is not automatic) for the physical quantity to be measured by means of programmed relations. The data is saved in two files that alternate cyclically, allowing access instantaneous values or 10 minutes average values. Afterwards it will decide how to act on the various systems' components. This is done by imposing conditions that implement the logical algorithm described above, comparing the instantaneous values of the physical system with the programmed references.

5. System Modelling

Obtaining the system model is essential for understanding and predicting systems' behaviour as well for specifying systems' components. In order to completely simulate the global system, sub-models for each system component were obtained. Each physical component was separately modelled as a Simulink block within the Matlab environment. The developed models, based on the fundamental physical, chemical and electromagnetic theories, were designed as general features that can be parameterized to match the real components characteristics. The structure of the simulation system is presented in Fig. 13.

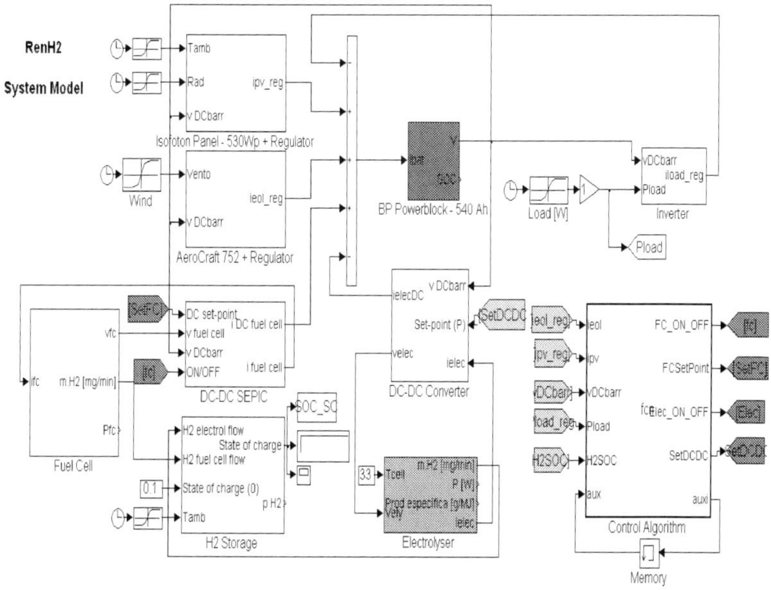

Fig. 13. Structure of the simulation system

The model of the PV modules uses a 5-parameter IV curve of the diode (3), where I_L denotes the current due to solar radiation exposure, I_0 the saturation current, q the electron charge, k the Boltzmann constant, T the temperature, R_s the series resistance, R_{sh} the shunt resistance, N_{cell} the number of cells in series per module and n the ideality factor ($n \in [1,2]$).

$$I = I_L - I_0 \left[\exp\left(\frac{q(V + R_S I)}{nkTN_{cell}} \right) - 1 \right] - \frac{V + R_S I}{R_{sh}} \tag{3}$$

The wind generator's Permanent Magnet Synchronous Machine is described in a classical dq axis model based on the evaluation of the fem and rotor speed, as in (Bose, 1996), being the stator transients neglected (4).

$$\begin{cases} v_q = R_s i_q + L_q \frac{di_q}{dt} + \omega_r L_d \frac{di_d}{dt} + L_{mq} \frac{di_{kq1}}{dt} + \\ \quad + L_{mq} \frac{di_{kq2}}{dt} + \omega_r L_{md} \frac{di_{fd}}{dt} + \omega_r L_{md} \frac{di_{kd}}{dt} \\ v_d = -\omega_r L_q \frac{di_q}{dt} + R_s i_d + L_d \frac{di_d}{dt} - \omega_r L_{mq} \frac{di_{kq1}}{dt} - \\ \quad - \omega_r L_{mq} \frac{di_{kq2}}{dt} + L_{md} \frac{di_{fd}}{dt} + L_{md} \frac{di_{kd}}{dt} \\ v_{kq1} = L_{mq} \frac{di_q}{dt} + R_{kq1} i_{kq1} + L_{kq1} \frac{di_{kq1}}{dt} \\ v_{kq2} = L_{mq} \frac{di_q}{dt} + R_{kq2} i_{kq2} + L_{kq2} \frac{di_{kq2}}{dt} \\ v_{fd} = L_{md} \frac{di_d}{dt} + R_{fd} i_{fd} + L_d \frac{di_{fd}}{dt} + L_{md} \frac{di_{kd}}{dt} \\ v_{kd} = L_{md} \frac{di_d}{dt} + L_{md} \frac{di_{fd}}{dt} + R_{kd} i_{kd} + L_{kd} \frac{di_{kd}}{dt} \end{cases} \tag{4}$$

The detailed model of the fuel-cell considers the fundamental anode/cathode equations with temperature effects. The relationship between the gas molar flow through the valve and its partial pressure inside the channel can be expressed as (5) p_{H2} denotes the hydrogen partial pressure, R the universal gas constant, T the temperature and V_{an} the anode voltage.

$$\frac{d}{dt} p_{H_2} = \frac{RT}{V_{an}} \left(q_{H_2}^{in} - q_{H_2}^{out} - q_{H_2}^{r} \right) \tag{5}$$

Assuming constant temperature and oxygen concentration, the fuel cell output voltage can be expressed as (6). E denotes the Nernst instantaneous voltage (7), V_{act} the activation voltage losses (8), V_{ohmic} the resistive voltage losses (9), $V_{concentration}$ the voltage losses due to the mass transport losses (10), N_0 the number of cells in series, E_0 the open circuit output voltage, T the stack temperature, F the Faraday's constant, $p_{H2}/p_{O2}/p_{H20}$ denote respectively the hydrogen/oxygen/water partial pressure, R_{FC} the electrical resistance, I_{dc} the fuel cell stack current, α the electrodes' charge transfer coefficient, I_{Lim} the limiting current of the fuel stack and I_0 is the exchange current.

$$V_{dc} = E - V_{ohmic} - V_{activation} - V_{concentration} \tag{6}$$

$$E = N_0 \left[E_0 + \frac{RT}{2F} \log\left(\frac{p_{H_2} \sqrt{p_{O_2}}}{p_{H_2O}} \right) \right] \tag{7}$$

$$V_{ohmic} = R_{FC} I_{dc} \tag{8}$$

$$V_{activation} = N_0 \left(\frac{RT}{2\alpha F} \right) \ln\left(\frac{I_{dc}}{I_0}\right) \tag{9}$$

$$V_{concentration} = -\frac{RT}{2F} \ln\left(1 - \frac{I_{dc}}{I_{Lim}} \right) \tag{10}$$

Experimental polarization and power curves for the fuel cell stack are shown in Fig. 14 evidencing how the fuel cell responds to load.

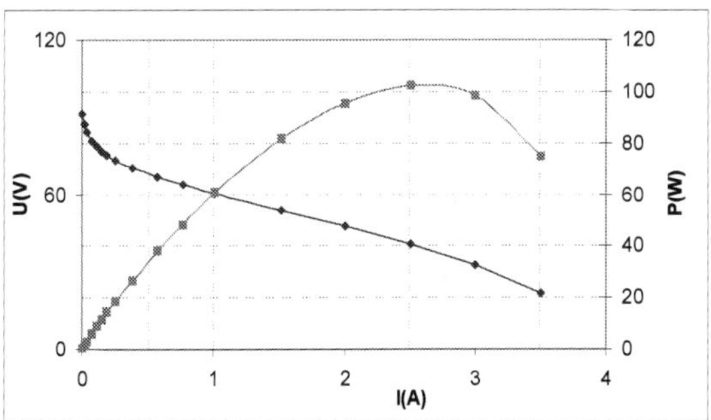

Fig. 14. Polarization and power curves for the 100 W fuel cell stack used in this work.

The electrolyser was modelled with experimentally obtained V(I) curves, presented in Fig. 15, and considering Faraday's law for optimal performance comparison.

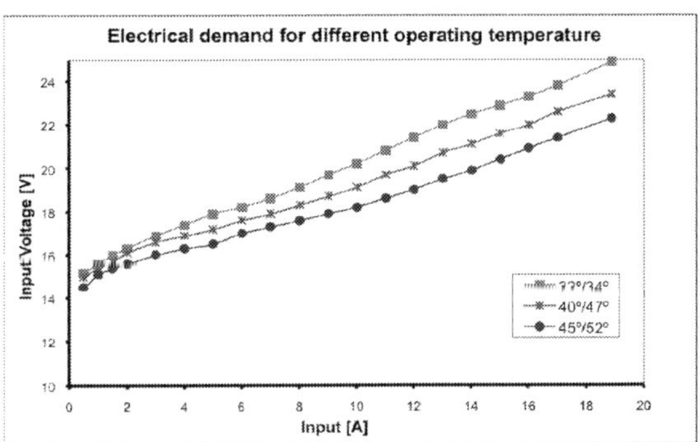

Fig. 15. Electrolyser experimental V(I) curves

Regarding the storage option, it is demonstrated that the metal hydride is able to supply hydrogen at the flows demanded by the fuel cell. A device was built incorporating a water circuit for cooling and heating of the metal hydride reactor during the charge and discharge of hydrogen respectively, allowing the fuel cell to be fed with fuel in a secure, fast and simple way, see Fig. 16. Using a pressure controller, inserted in the hydrogen circuit of the device, allows activation of the water circuit when the pressure of hydrogen is less than 2 Bar.

Fig. 17 presents the hydrogen absorption rate, during the charging process, of alloy LaNi4.7Al0.3. The alloy showed a high absorption kinetics, achieving total loading of the reactor storage in ~ 30 min, for a cooling water temperature of 22 ° C.

The pressure of the reactor containing metal hydrides during discharge for different hydrogen flow-out is presented in Fig. 18. A full loading with partial discharging was used for 1 and 3 $Lmin^{-1}$ and a partial loading (30%) with total discharging was used for 2 $Lmin^{-1}$. It was possible to observe a rapid decrease in pressure in the initial download time, reaching a level close to 2 bar. This rapid decrease in pressure is explained by the fact that, until the pressure of the programmed controller is reached (2 bar), there is no water flow within the reactor. This leads to a decrease in temperature of the reactor and the consequent decrease of the hydrogen desorbed.

It is to be noticed that for a hydrogen demand of 1 $Lmin^{-1}$, it is possible to keep the discharge at 2 bar as required in this case; for 3 $Lmin^{-1}$, the plateau at 2 Bar was only managed in the first 60 minutes of the discharge, being lower for the rest of the discharge, (it is suggested that the rate of desorption of the alloy is lower than the required flow).

Data presented for discharge at 2 $Lmin^{-1}$ correspond to partial loading of the hydride, which may account for a situation where more than 70% of the hydrogen has been consumed. Results show that it is possible to proceed up to the total discharge of the alloy at a constant pressure of 2 bar.

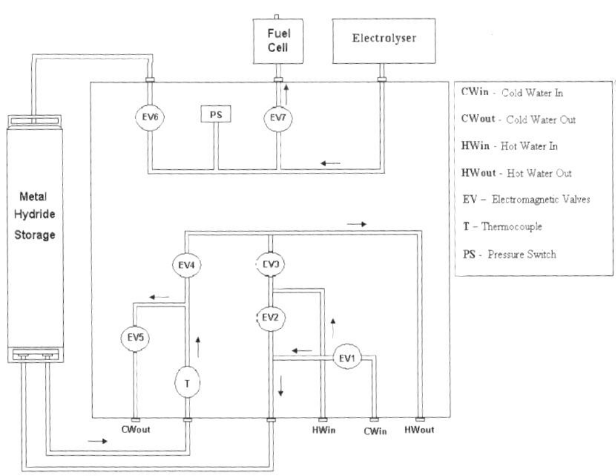

Fig. 16. Schematic drawing of the device implemented for the charging and discharging of metallic hydride hydrogen storage of 1500 NL feed a fuel cell.

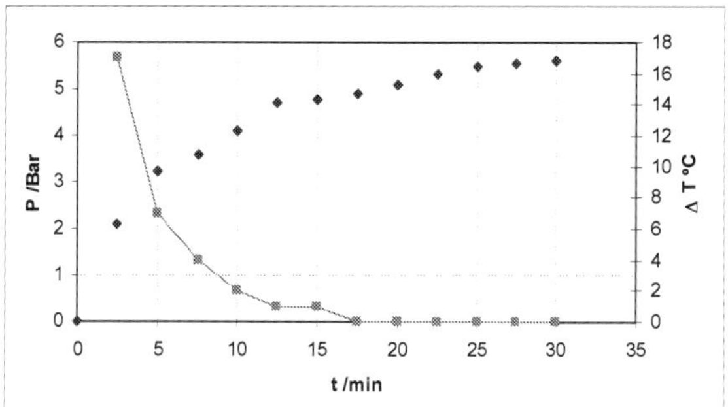

Fig. 17. Pressure of the reactor as a function of charging time, obtained in the process of charging with hydrogen of alloy LaNi4.7Al0.3. Represented also is the temperature variation as function of time during the charging process. The cooling water initial temperature was 22° C.

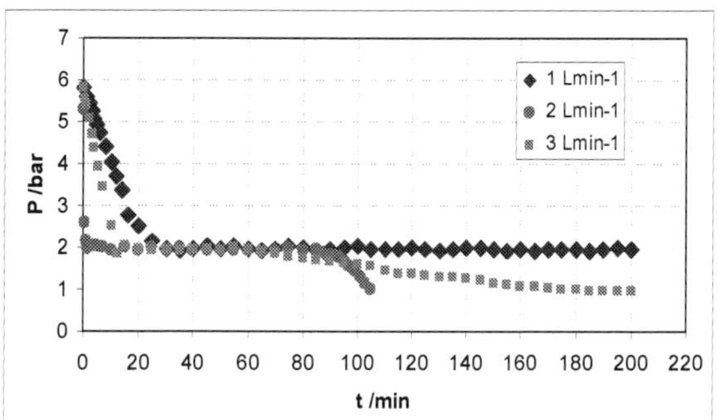

Fig. 18. Discharge curve of alloy LaNi4.7Al0.3 for flow rates between 1 and 3 Lmin-1. Water temperature and delivery pressure were programmed to ~70°C and to 2 bar respectively.

The process of hydrogen absorption/desorption is best illustrated by pressure-composition-temperature profiles, denoted as PCT curves, the pressure at a given H-content increases with temperature and is a direct consequence of the thermodynamics associated with the hydriding reaction (1).

$$M + x/2\, H_2 \Leftrightarrow M\text{-}H_x \tag{1}$$

Since the metal hydride is able to supply hydrogen at the demanded flows, the metal hydride component is modelled using the state of charge (SOCMH) in a simplified manner, described by equation (1), where $N_{MH,initial}$ is the initial hydrogen content; $N_{MH,total}$ is the total capacity of hydrogen in the metal hydride, $m_{EL,H2}$ is the hydrogen flow from the electrolyser and $m_{FC,H2}$ is the hydrogen flow demanded by the fuel cell. In this way the model considers a simple and effective hydrogen mass balance.

$$SOC_{MH} = \frac{N_{MH,\ initial} + \int \dot{m}_{EL,H_2} dt - \int \dot{m}_{FC,H_2} dt}{N_{MH,\ total}} \tag{11}$$

The global system has several topologies of power electronics converters: AC/DC (wind generator), DC/DC (Pv system, fuel cell), DC/AC (load). The complete modelling of all of these components strongly slows down the computation time. Several methods have been proposed in order to reduce computation times, however simple ON/OFF switches and average values models were used in the simplified model. Fig. 19 presents the DC busbar voltage for both models. The gray line refers to the detailed model while the black line refers to the simplified model.

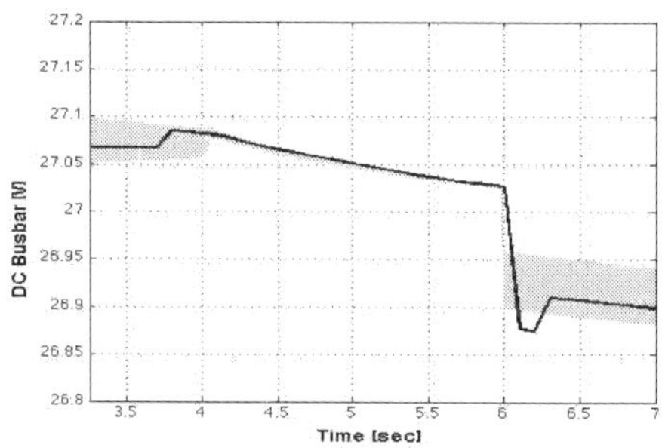

Fig. 19. DC busbar voltage for the detailed and simplified model

6. Conclusions and Remarks

The main goal of $RenH_2$ project is the full study of low power stand-alone totally renewable power supply systems. Thus, the main idea behind this project was the development of a fully autonomous system, where every component is based upon renewable energies. The project comprises model development and prototype implementation with full integration of the different components, namely: Photovoltaic cells, Wind turbine, Electrolyser system, Fuel Cell, Hydrogen storage, Power electronics converters and Control system.

The paper also describes the optimization and integration of production modules, storage and energy conversion within fuel cells, photovoltaic cells and wind turbines to achieve the production of hydrogen. Each component was separately modelled, in order to test the overall control strategy. Key feature in the modelling process was the correct choice of input/output variables, in order to perfectly establish the sensor array that should be used in the experimental prototype. Two distinct models were considered: a detailed one and a simplified one. While each component was accurately detailed modelled for the overall system, due to its complexity, a simplified model was assumed. Each modelling was experimentally validated.

The obtained system is a suitable choice regarding the actual stand-alone systems, based upon diesel generators and lead-acid batteries energy storage. It meets the sustainability and environmental respect criteria regarding the energetic solutions of the future - zero emitting either on production or consumption. Conceived to meet the energetic needs of a rural facility, this system considers the energy storage within hydrogen fuel cells and the connection to several energy sources, which may be placed where the receptive resource is by far more abundant.

7. Acknowledgements

The authors wish to thank Fundação para a Ciência e Tecnologia for financial support (Projects POCI/ENR/59623/2004 and PPCDT/ENR/59623/2004). Grateful appreciation is also due to Susana Viana, João Pedro Simões, Miguel Caldeira Coelho and Vitor R. Fernandes for their fundamental work in implementing the prototype.

8. References

Agbossou, K.; Chahine, R.; Hamelin, J.; Laurencelle, F.; Anouar, A.; St-Arnaud, J.M. &Bose, T.K. (2001). Renewable energy systems based on hydrogen for remote applications. *Journal of Power Sources* 96 (2001) 168-172.

Agbossou, K.; Kolhe, M.; Hamelin, J. &Bose, T.K. (2004). Performance of a Stand-Alone Renewable Energy System Based on Energy Storage as Hydrogen. *IEEE Transactions on Energy Conversion*, Vol. 19, No. 3, September 2004.

Alam, M.S. & Gao, D.W. (2007). Modeling and Analysis of a Wind/PV/Fuel Cell Hybrid Power System in HOMER. *Proceedings of 2nd IEEE Conference on Industrial Electronics and Applications - ICIEA2007*, Harbin, China, May 2007.

Bose, B.K. (1996). Power Electronics and Variable Frequency Drives, IEEE Press (1996).

Chaparro, A.M.; Soler, J.; Escudero, M.J.; de Ceballos, E.M.L.; Wittstadt, U. & Daza, L. (2005). Data results and operational experience with a solar hydrogen system. *Journal of Power Sources* 144 (2005) 165-169.

Chedid, R; Chaaban, F.B. &Shihab, R. (2007). A Simplified Electric Circuit Model for the Analysis of Hybrid Wind-Fuel Cell Systems. *Proceedings of 2007 IEEE Power Engineering Society General Meeting*, June 2007.

De Battista, H.; Mantz, R.J. &Garelli, F. (2006). Power conditioning for a wind-hydrogen energy system. *Journal of Power Sources* 155 (2006) 478-486.

Dufo-López, R.; Bernal-Agustín, J.L. & Contreras, J. (2007). Optimization of control strategies for stand-alone renewable energy systems with hydrogen storage. *Renewable Energy* 32 (2007) 1102-1126.

El-Shatter, T.F.; Eskander, M.N. &El-Hagry, M.T. (2006). Energy flow and management of a hybrid wind/PV/fuel cell generation system. *Energy Conversion and Management* 47 (2006) 1264-1280.

Gammon, R.; Roy, A.; Barton, J. &Little, M. (2006). Hydrogen and Renewables Integration (HaRI). Case Study of International Energy Agency Hydrogen Implementing Agreement (IEAHIA), 2006.

Gazey, R.; Salman, S.K.&Aklil-D'Halluin, D.D. (2006). A field application experience of integrating hydrogen technology with wind power in a remote island location. *Journal of Power Sources* 157 (2006) 841-847.

Ghosh, P.C.; Emonts, B. &Stolten, D. (2003). Comparison of hydrogen storage with diesel-generator system in a PV-WEC hybrid system.*Solar Energy* 75 (2003) 187–198.

Ismail, Esam H. (2009). Bridgeless SEPIC Rectifier With Unity Power Factor and Reduced Conduction Losses, *IEEE Trans. On Industrial Electronics*, Vol. 56, No. 4 (2009) 1147-1157.

Jensen, J.K. &Madsen, J.B. (2007). Utilisation of hydrogen and fuel cell technology for micro combined heat and power production. Internal report of the "Lolland Hydrogen Community", 2007.

Joyce, A.; Rodrigues, C.; Manso, R. (2001).Modelling a PV system. Renewable Energy 22 (2001) 275-280

Khan, M.J. &Iqbal, M.T. (2005). Pre-feasibility study of stand-alone hybrid energy systems for applications in Newfoundland. *Renewable Energy* 30 (2005) 835–854.

Licht, Stuart (2005). Thermochemical solar hydrogen generation. *Chemical Communications* (2005) 4635-4646.

Little, M.; Thomson, M. &Infield, D. (2007). Electrical integration of renewable energy into stand-alone power supplies incorporating hydrogen storage. *International Journal of Hydrogen Energy* 32 (2007) 1582 - 1588.

Menzl F.; Wenske, M. &Lehman, J. (1999). Windmill-electrolyser system for ahydrogen based energy supply. *Proceedings of the 1999 Wind Energy Conference*, pp. 911-914, Nice, France,March 1999.

Mills, A. &Al-Hallaj, S. (2004). Simulation of hydrogen-based hybrid systems using Hybrid2.*International Journal of Hydrogen Energy* 29 (2004) 991 - 999.

Nakken, T.; Strand, L.R.; Frantzen, E.; Rohden, R.& Eide, P.O. (2006). The Utsira wind-hydrogen system - operational experience. *Proceedings of 2006 European Wind Energy Conference & Exhibition – 2006EWEC*, Athens, Greece, March 2006.

Nelson, D.B.; Nehrir, M.H. &Wang, C. (2006). Unit sizing and cost analysis of stand-alone hybrid wind/PV/fuel cell power generation systems. *Renewable Energy* 31 (2006) 1641-1656.

Ntziachristos, L.; Kouridis, C.; Samaras, Z. & Pattas, K. (2005). A wind-power fuel-cell hybrid system study on the non-interconnected Aegean islands grid. *Renewable Energy* 30 (2005) 1471–1487.

Ozturk, H.K.; Yilanci, A.; Atalay, O. & Dincer, I.(2007). A Preliminary Assessment of a 5 kWp Solar Hydrogen System in Denizli, Turkey.*The Third International Energy, Exergy and Environment Symposium*, Évora, Portugal, July 2007.

Santarelli, M.; Cali, M. & Macagno, S. (2004). Design and analysis of stand-alone hydrogen energy systems with different renewable sources. *International Journal of Hydrogen Energy* 29 (2004) 1571 - 1586.

Senjyu, T.; Nakaji, T.; Uezato, K. & Funabashi, T. (2005). A Hybrid Power System Using Alternative Energy Facilities in Isolated Island. *IEEE Transactions on Energy Conversion*, Vol. 20, No. 2, June 2005.

Shakya, B.D.; Aye, L. &Musgrave, P. (2005). Technical feasibility and financial analysis of hybrid wind–photovoltaic system with hydrogen storage for Cooma. *International Journal of Hydrogen Energy* 30 (2005) 9 - 20.

Sopian, K.; Ibrahim, M.Z.; Daud, W.R.W.; Othman, M.Y.; Yatim, B. &Amin, N. (2009). Performance of a PV-wind hybrid system for hydrogen production. *Renewable Energy* (2009), doi: 10.1016/j.renene.2008.12.010

Sovacool, B.K. &Hirsh, R.F. (2008). Island wind-hydrogen energy: A significant potential US resource. *Renewable Energy* 33 (2008) 1928–1935.

Turcotte, D.; Ross, M. &Sheriff, F. (2001). Photovoltaic hybrid systems sizing and simulation tools: status and needs. *Proceedings of PV Horizon: Workshop on Photovoltaic Hybrid Systems*, Montreal, September 2001.

Ulleberg, O. & Morner, S.O. (1997). TRNSYS Simulation Models for Solar-Hydrogen Systems.*Solar Energy* Vol. 59, No. 4-6, pp. 271-279.

Varkaraki, E.; Lymberopoulos, N.; Zoulias, E.; Kalyvas, E.; Christodoulou, C.; Vionis, P. & Chaviaropoulos, P. (2006). Integrated Wind-Hydrogen Systems for Wind Parks. *Proceedings of 2006 European Wind Energy Conference & Exhibition - 2006EWEC*, Athens, Greece, March 2006.

Varkaraki, E.; Tzamalis, G. & Zoulias, E. (2008). Operational experience from the RES2H2 wind-hydrogen plant in Greece. *Proceedings of 17th World Hydrogen Energy Conference - WHEC2008*, Brisbane, Australia, June 2008.

Wang, C. &Nehrir, M.H. (2008). Power Management of a Stand-Alone Wind/Photovoltaic/Fuel Cell Energy System. *IEEE Transactions on Energy Conversion*, Vol. 23, No. 3, September 2008.

Yilanci, A.; Dincer, I. &Ozturk, H.K. (2008). A review on solar-hydrogen/fuel cell hybrid energy systems for stationary applications. *Progress in Energy and Combustion Science* (2008), doi:10.1016/j.pecs.2008.07.004

Zhou, T.; Lu, D.; Fakham, H. & Francois, B. (2008). Power flow control in different time scales for a wind/hydrogen/super-capacitors based active hybrid power system. *Proceedings of 13th International Power Electronics and Motion Control Conference - EPE-PEMC2008*, September, Poznań, Poland.

Zini, G. &Tartarini, P. (2009). Hybrid systems for solar hydrogen: A selection of case-studies. *Appl. Therm. Eng.*(2009), doi:10.1016/j.applthermaleng.2008.12.029

Zoulias, E.I. & Lymberopoulos, N. (2008). *Hydrogen-based Autonomous Power Systems: Techno-economic Analysis of the Integration of Hydrogen in Autonomous Power Systems.* Springer London, ISBN 978-1-84800-246-3 (Print) 978-1-84800-247-0 (Online).

Solar Power Source for autonomous sensors

José Pelegrí-Sebastiá, Rafael Lajara Vizcaíno and Jorge Alberola Lluch
Universitat Politècnica de Valencia
Spain

1. Introduction

This chapter focuses on solar power source for autonomous sensors. Nowadays, low power consumption is assuming the main role in the design of embedded electronic devices or battery powered devices, mainly in wireless sensor networks (WSN) 0. In fact the power consumption of current wireless transceivers IC's is going down until such levels that take us to explore the application of renewable energies for autonomous powered wireless sensor networks, that turned out inefficient or not enough efficient up to now. Several articles discuss about the amount of energy that can be obtained from many environmental energies (Raghunathan et al., 2005; Roundy et al., 2004), (Park et al., 2004; Hande et al., 2006; Chou et al., 2005; Raghunathan et al., 2005; Kansal et al., 2004; Alberola et al., 2008) and it seems sunlight is the most powerful energy available at the probable location of a wireless sensor network outdoors. Currently, small solar panels deliver enough power for both charging batteries and supplying power to these low power wireless nodes (Arms et al., 2005; Rahimi et al., 2003; Hsu et al., 2005). Therefore, the main trade-offs that involve the design of an inexhaustible power source are not only taking care of getting a lot of energy from the solar panel, but mainly taking care of holding-up the stored energy. Thus, a careful selection of the electronic components in the power management circuit is essential to reach high efficiency at really low currents. Furthermore, the designer has to have in mind that batteries suffer the aging problem, by which they can reduce dramatically their capacity in two or three years if they are recharged daily. This chapter focuses on the practical application to automatically manage the sunlight energy and store it efficiently, while minimizing the strain on the storage components to extend the power source lifetime.

2. Background and related work

Most of solar powered devices with backup system are composed by a solar panel, a charging control unit and a single battery backup element (Dreher et al., 2004; Panasonic, 1999). Unfortunately the aging problem is always patent in batteries. Hence, the design of a real inexhaustible power source must take care of it by using storage elements without almost fatigue, like supercapacitors. As it is described in (Jiang et al., 2005), "Prometheus" use a system architecture with two energy buffers, two supercapacitors and a Li+ Battery. There, a electronic system as a microcontroller controls the recharging cycles of the battery and selects the energy path to the load by means of a switch. This architecture is strongly dependent of the microcontroller, these supervises the switching at the right voltage threshold. Working at low rate duty cycles (among sleeping and working time), typically in network sensors, the wake-up of the microcontroller could be too delayed to switch the backup battery on time. This way, the system could lose the voltage supply and do not wake-up anymore. On the other hand, the consumption of an active microcontroller supervising the stored energy in the supercapacitors at high rate duty cycles could be comparable to the consumption of the current micro power supervisors IC's working at full time, but assuring a safe backup switching.

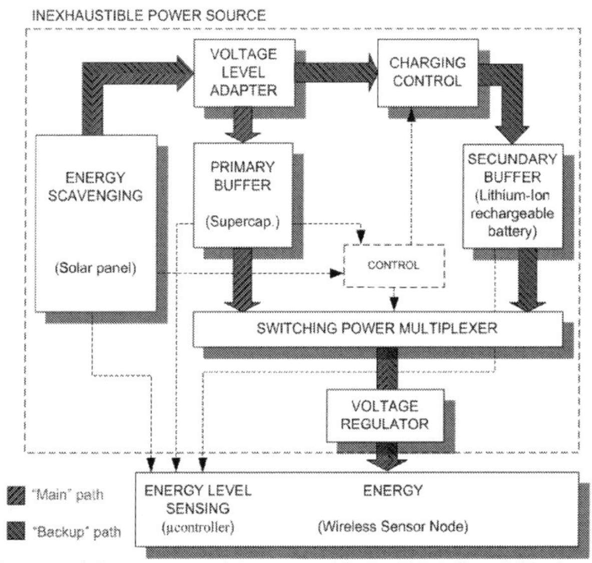

Fig. 1. Block diagram of the system architecture with the "Main" and "Backup" paths.

Therefore we evaluate a new system architecture that is shown in Figure 1, which is entirely independent of the microcontroller. An automatic micro-power control unit manages the energy flow to the load. Besides, the battery is recharged from the solar energy regardless of the state of the supercapacitors. This maintains the energy stored in the supercapacitors at its maximum while the sun hides. In addition an analysis of the current requirements in the two power paths, as much in charge as in discharge, take us to maximize the efficiency within each power path, by choosing the most efficient DC-DC converters for the required currents.

3. Design

Our inexhaustible power source supplies a regulated voltage of 3.3 V to a wireless sensor node with the low power RF transceiver CC2420, which is IEEE 802.15.4 compliant. These kind of wireless devices use to spend most of the time sleeping or within some low power consumption mode in order to save energy, and duty cycles among 1 % and less than 10 % (see Figure 2) are typical in field sensing applications (Roundy et al., 2003; Polastre et al.,

2004). Therefore the design is focused for efficiently managing really low currents (around 6 μA) at sleeping mode, and responding quickly to the relatively large currents (around 22 mA) requested by the load in some active mode (transmitting, receiving or synchronizing). The active time (when the consumption is 22 mA) is fixed arbitrarily to 100 ms and the percentage is referred to this time. The rest of the time (with a load current of 6 μA) corresponds to the sleeping time. The straight lines correspond to the average current. These values of current are only approximated and they try to hardly model the power consumption of a wireless node in a network. Nevertheless these values are useful to choose the most efficient DC-DC converter for the required currents.

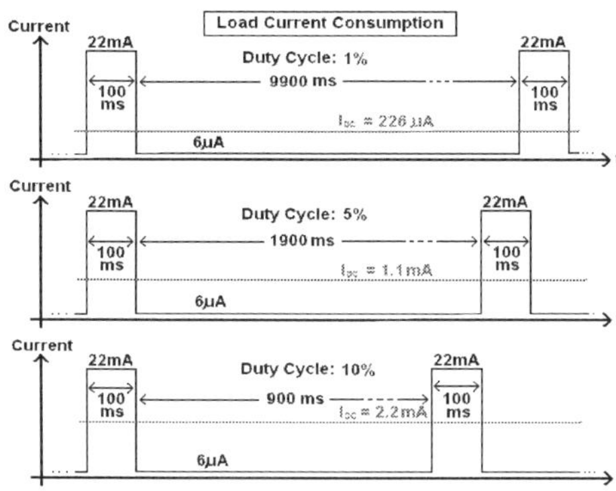

Fig. 2. Load currents tested at different duty cycles to emulate the power consumption of Wireless Sensor Node.

Thus, having in mind the amount of current requested by the storage elements and the load, and also evaluating the step-up and step-down converters in the current market, we propose and evaluate the new system architecture showed in Figure 1. There we differentiate two energy paths with different constraints; "Main", and "Backup". The currents in each path have different levels, so we take their value into account and maximize the efficiency of the "Main" path, that is the most critical path.

3.1. "Main" path

This is the energy path by default because when solar energy is available, block control gives the highest-priority to it. This is the path by where we would like to supply energy to the load permanently; this is, taking energy from the sun for both replenishing the primary buffer and delivering the required current to the load without degrading the primary buffer. This way, the use of the backup system would be minimized and lifetime of the whole system expanded.

The first element in the "Main" path is the solar panel. Since our power source needs to supply a current of at least 22 mA to the load, the selected solar panel has to be able to deliver this minimum value of current directly to the load. Nevertheless, the power source needs to supply more current for charging both storage elements. The larger is the maximum current that the solar panel is able to deliver the quicker is the recharge of the storage elements. Hence, the only upper limit of current is determined by the size of the solar panel that fits in the application. Solar panels behave like voltage-limited current sources (Würfel et al., 2005; Chuck, 2006; Panasonic, 1999) as opposite to batteries which behave like voltage sources. They have a single operating point or well-known as Maximum Power Point (MPP). Since this operating point moves along the solar panel curves depending on the incident solar radiation, the manner of extracting the maximum power could be tracking the MPP. One could think to use some advanced DC-DC converter that adjusts its duty cycle dynamically (Koutroulis et al., 2001; Lim et al., 2000). Unfortunately, at the moment there are no commercial MPP trackers for such a low power devices. Furthermore, implementing it in a microcontroller possibly would waste more power than we earn by following the MPP with such a low voltage levels. Therefore, there are only two feasible options for efficiently coupling the solar panel and the supercapacitors; directly or using a DC-DC converter. We choose to use a DC-DC converter because it steps-up the solar panel voltage even when the sky is cloudy, reaching the appropriate voltage level for charging the supercapacitors near to its maximum every day, regardless of the solar panel voltage that depends on the solar radiation intensity.

The next element in the "Main" path is the primary buffer. This is the critical element in the design. The key design for autonomous operation is that the primary buffer needs to be recharged daily without accusing a remarkable fatigue. At the same time, it requires enough capacity to hold-up the load during the night, which minimizes the use of the secondary buffer. As it is described in "Prometheus" (Jiang et al., 2005), the only current solution is using supercapacitors. One could think about using other energy storage elements for the primary buffer, because there are other commercial elements with larger capacity like batteries. But the fact is they suffer the aging issue, that is, capacity loss manifests itself in increased internal resistance caused by oxidation with the successive recharging cycles (Buchmann, 2005; MPower, 2006). For example, a Lithium battery suffers a 10% - 20% degradation in capacitance and resistance after only 300 - 500 discharge/charge cycles, this is among one and two years supposing one cycle a day. However, nowadays the supercapacitors have reached large capacities with competitive prices with respect to the current high-capacity batteries. Compared to the Lithium battery, a supercapacitor can be deep cycled at high rates for 500.000–1.000.000 cycles for the same change in characteristics (10–20% degradation), this is among 1.500 and 3.000 years supposing one cycle a day again (Burke, 2000; Cooper, 2006). Moreover, the supercapacitors have additional advantages as pulse power devices; they have high power density, high efficiency, short recharging times, and long shelf and cycle life. In contrast, the primary disadvantage of supercapacitors is their relatively low energy density compared to batteries (Stor & Bussmann, 2007). Nevertheless, they turn the best feasible solution within this kind of low power and low duty cycle applications, where the solar power has to replenish the primary buffer daily and delivering pulsating currents to the load.

By the other hand, when there is no solar energy available, the supercapacitors will start to deliver current to the load and their voltage will start to drop. Therefore a voltage regulator is required to maximize the energy extracted from the supercapacitors and deliver it to the load, which requires a constant voltage. That voltage regulator needs to be very efficient in order to maximize the use of the energy stored in the supercapacitors and therefore minimizing the use of the secondary buffer.

3.2. "Backup" path

This energy path should provide energy to the load when the "Main" path fails; this is, when there is no direct sunlight and the supercapacitors drop below the minimum operating voltage. As it has been named earlier, the "Main" path has been designed to work as much time as possible. Therefore, it is expected "Backup" path only takes part during exceptional situations, which minimizes the recharging cycles on the secondary buffer and therefore, the stress and degradation on it.

The "Backup" path is characterized by housing a large rechargeable energy storage element, this is the secondary buffer (Figure 1), that assures the power supply to the load during long time periods without sunlight. The length of such black-out periods depends essentially on weather factors and on the geographic location. Therefore the optimal autonomy for the secondary buffer remains uncertain and one need to choose the largest rechargeable battery that adapts to the cost and size of the application. Currently there are many type of batteries in the market (Ni-Cd, Ni-MH, Lithium) but it is well-known that Lithium batteries have the highest energy density, the lowest self-discharge rates, and the lowest "memory effect" (Buchmann, 2005; MPower, 2006). Furthermore they are becoming less expensive over time. Although they require a more complex charging method, we can use some dedicated charger chip because the battery charge should only be enabled under excess of solar power conditions. This means when the primary buffer has been full replenished and solar panel maintains the appropriate voltage in their terminals.

By the other hand, the capacity of the secondary buffer can be so much larger than the primary buffer one. Hence, the secondary buffer is able to supply power to the load much more time than the primary buffer even assuming a less efficient path. Therefore "Backup" path, powered from the right sized solar panel to charge it quickly, is not such critical with the component selection, in contrast with the "Main" one.

4. Implementation

We have implemented the inexhaustible power source to supply a regulated voltage of 3.3 V to a wireless sensor node. The prototype board contains a solar panel, two supercapacitors, a Lithium-Ion battery and the energy management integrated circuits that can be seen in Figure 4.

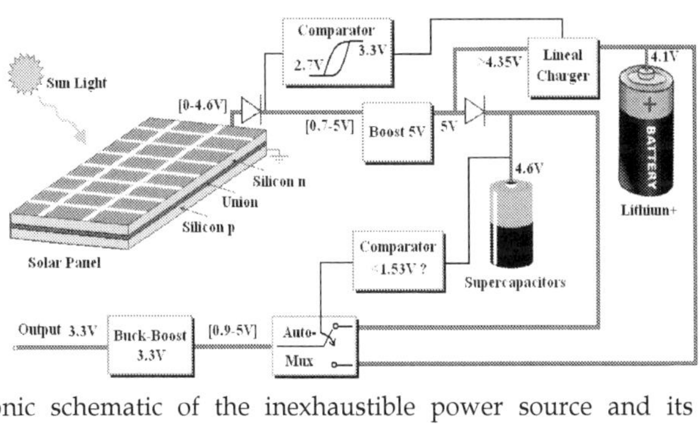

Fig. 3. Electronic schematic of the inexhaustible power source and its voltages while working.

The board also provides a 10 pin straight header that is used for measuring all the important voltages in the circuit, by means of those which an acquisition board is connected through.

4.1 Hardware

This section describes the selection of the components inside the blocks shown in Figure 1 and Figure 3, and the workbench used to test our inexhaustible power source in a house roof.

Fig. 4. Picture of the solar inexhaustible power source prototype.

1) Solar Panel

We use the MSX-005F (114 mm x 66 mm) of 0.5 W from Solarex. It was selected because its maximum power point (3.3 V) is the closest to the voltage range that manages our storage components. Moreover, it delivers 150 mA on the nominal point (at maximum solar radiation), which is much more than the maximum current that our load needs (around 22 mA). This way, the solar panel has a huge current margin to recharge the energy storage elements quickly under sunlight conditions.

2) Voltage Level Adapter

We use a step-up converter (MAX1795) between the solar panel and the supercapacitors (Figure 3). Thus, we reach the required voltage (more than 4.35V) for charging the Lithium battery and also for maximizing the charge stored in the supercapacitors. The step-up converter elevates the voltage to 5V and lets the current flows into the supercapacitors even when the sunlight is weak. This way the converter maximizes the energy extracted from the solar panel.

3) Primary Buffer (Supercapacitors)

Since each capacitor admits only 2.3V we connect two in series, that permits to charge the supercapacitors close to 4.6V and reduces the leakage current. The larger is the capacity of the supercapacitor the lower is its leakage current, but the price increases much more. We use two supercapacitors of 50F from PANASONIC due to its availability and relatively low price.

4) Switching Power Multiplexer
The new automatic power multiplexer from Texas (TPS2113PW) is the one in charge of selecting the power path to the load. It switches automatically between the primary and the secondary buffer depending on the voltage of the supercapacitors.

5) Control Block
Control block is actually embedded in other blocks. That is why it appears with dotted lines in Figure 1. It is composed by a comparator used within the voltage level adapter and a comparator within the switching power multiplexer. The first comparator, which is built in the step-up converter of the voltage level adapter, is used for enabling the charge of the Lithium-Ion battery. This takes place when the voltage measured on the solar panel is above 3.3 V (MPP), and it is disabled below 2.7 V. This comparator has a large hysteresis to prevent a false halt in the charge, due to a voltage drop when the battery starts being recharged.
The second one is a comparator built into the switching power multiplexer. It automatically switches among the inputs, assigning priority to the first one (supercapacitors). This means the second input (the battery) only is selected when the voltage of the supercapacitors drops below the minimum threshold that keeps the switching power multiplexer working on (that in this case is 1.53 V, Figure 7). This comparator also has a small hysteresis (around 60 mV) to prevent from false commutations among the inputs or energy paths.

6) Voltage Regulator
This block is composed by a single chip with a step-up converter and a voltage regulator. Its output is a fixed voltage at 3.3 V. We use the TPS61025 from Texas Instruments because it has a really flat efficiency graph (90 % -94 %) even at low currents (from 1 mA to 40 mA).

7) Secondary Buffer (Battery)
We use a 1Ah Lithium-Ion (Li+) battery due to its availability and low price.

8) Charging control
Since the Lithium-Ion battery is going to be used outdoors and it requires a careful and safety recharge with a limited current, we decided to use a dedicated charge control chip, the MAX1811, which limits the charge current and protects the battery.

4.2 Load (Simulation)
We include two mosfets in the prototype board. Each one has a resistor between its drain node and the output voltage (3.3 V), whose value has been calculated to obtain the load currents shown in Figure 2. This way, we digitally switch them to simulate the load current that a wireless sensor node approximately consumes, but only using two average values (22 mA and 6 μA), which correspond to the working modes previously described (active and sleeping).
The switching of the mosfets is digitally managed by a microcontroller located in an external board that is used exclusively for the load simulation purposes, and it is powered by a portable acquisition system.

4.3 Acquisition System

We use the data acquisition board NI USB-6008 from National Instruments to sample the evolution of the most significant voltages. Since the voltages in our power source circuit do not change quickly it is enough by using this cheap acquisition module to sample eight voltage nodes every second.

The acquisition system is connected to a computer by means of a USB cable that supplies power to it. Then, a program developed using LabVIEW displays and stores the sampled data.

5. Results and discussion

We located the inexhaustible power source on a house roof for several days. Using a laptop computer connected to the portable acquisition system, we tested our inexhaustible power source sampling every second the voltage on the following nodes:

* Solar panel terminals.
* Output of the voltage level adapter.
* Supercapacitors.
* Output of the power source.
* End battery charge signal.
* Enable battery charge signal.
* State multiplexer signal.
* Battery.

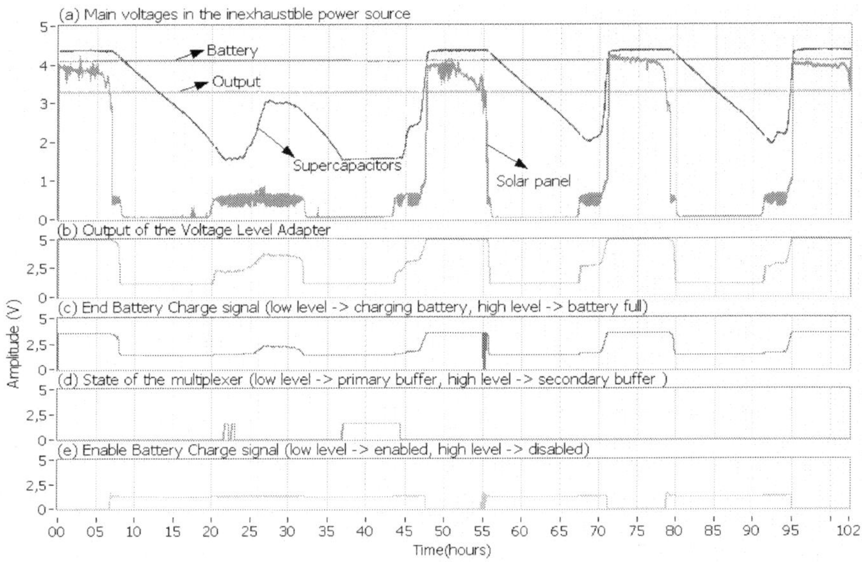

Fig. 5. Evolution of the voltages in the inexhaustible power source.

Figure 5 shows the evolution of the main voltages in the power source. In this case we let the system running autonomously during 5 days. It was supplying 3.3 V to the simulated load, which was working at 4% of duty cycle. We started the measurements on September 22 at midday 12:00 p.m. with full sunlight, the supercapacitors fully replenished at 4.35 V, and the battery fully charged at 4.1 V (Figure 5a). The solar panel was supplying power directly to the load. After seven hours the sun started to hide and the supercapacitors assumed the main role powering the load. The first night came out (follow the solar panel voltage), and the supercapacitors did not hold-up the load current during the whole night because the second day dawned raining. Although the sunrise was at 8 o'clock (20 h after starting), the solar radiation was too weak for replenishing the supercapacitors. Therefore at 10 o'clock (22 h after starting) the supercapacitors dropped below the working threshold 1.53 V, and the system automatically switched to battery for sustaining the load current (Figure 5d). In spite of the weak sunlight on a very rainy day, the supercapacitors got to raise its voltage within the next hours thanks to the output level adaptor (Figure 5b), but clearly with a smaller slope.

This way, with the supercapacitors charged at 3 V, the system did not support the load during the whole second night, but the backup battery sustained the output voltage at 3.3 V again.

The third day came out again at 8 o'clock (44 h after starting) and this time it was a shiny day. The supercapacitors were replenished from its minimum voltage 1.53 V to its maximum voltage 4.35 V in less than three hours. Moreover since the battery was hardly discharged last day, the power source recharged the battery in a while and it is hardly appreciable in the Figure 5c. Although the Figure 5e shows when the battery is enabled to be recharged, that does not implies the battery is being recharged because this depends on whether the battery is full or not, and whether the output voltage of the voltage level adapter is higher than 4.35 V or not.

Next two sunny days, the power source kept on powering the load without using the Lithium battery. Since supercapacitors are hardly deteriorated as time goes by, the power source promises a long lifetime because the battery will be used only during adverse climate conditions, this is without almost sunlight.

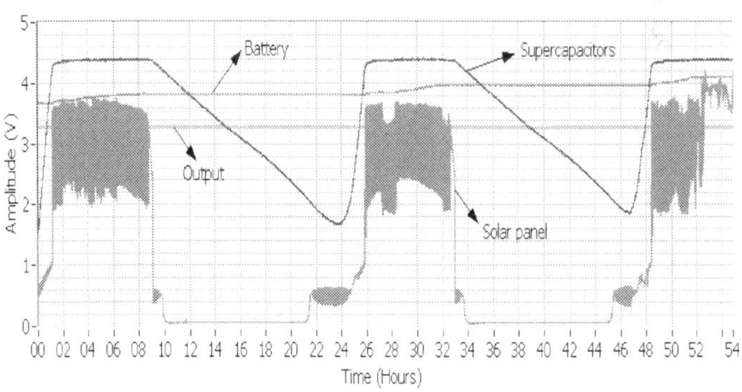

Fig. 6. This figure shows the recovery of the Lithium-Ion battery which was intentionally deeply discharged

A second measurement was carried out to evaluate the recovery time of the Lithium-Ion battery. Initially, we intentionally left the supercapacitors and the battery discharged at 1.57 V and 3.67 V respectively. Then, we exposed the solar panel to the sunlight during three days. Figure 6 shows the evolution of the voltage on the supercapacitors and on the battery, while powering the load at 4 % of duty cycle. Firstly, the energy from the solar panel was used for fully replenishing the supercapacitors from 1.57 V to 4.35 V, which took less than 2 h. After that, although the second day was a bit cloudy, the 3.7/1Ah Lithium-Ion battery started getting energy from the sun and the battery was recharged from 3.67 V to 4.1 V in two and a half days while powering a load at 4 % of duty cycle. This demonstrates the recovery ability of this inexhaustible power source even under a deep battery discharge.

Finally we carried out some more experiments to evaluate the longevity of our power source. For that, we consider that the autonomy can be near perpetual if under normal climate conditions, this is sunny days, the solar panel and the supercapacitors hold-up the load during the whole day and the whole night. This way, depending on the geographic location, the battery is only eventually used during very rainy days.

Therefore we acquired voltage samples of the supercapacitors during many days and nights with different load currents, this is, varying the duty cycle. We centred our attention at nights, when the supercapacitors remain as the unique active energy source. Figure 7 shows the evolution of the voltage in the supercapacitors during four different nights with a load working at 1 %, 3 %, 4 % and 5 % of duty cycle. Starting with the supercapacitors fully charged and overlapping the four curves to the same hour of a day, the slope for each duty cycle can be easily compared. Moreover, Figure 7 determines that 4 % is the maximum continuous duty cycle which our power source is able to maintain in a September night without using the battery. This means 4 % is approximately the continuous duty cycle that avoids a daily discharge and recharge of the Lithium-Ion battery, expanding much more the lifetime of the entire power source.

Furthermore, using a variable duty cycle, this is adapting the load consumption to the energy available, the designed power source can work even at higher duty cycles than 4% during the day and reducing it during the night depending on the tasks of the wireless sensor node powered.

It is a hard work to predict the lifetime of our power source because it depends among other parameters on the unpredictable climate conditions, but we expect a so much longer lifetime than 3 years, which is the typical lifetime of Lithium-Ion batteries if used daily and charged and discharged deeply (Buchmann, 2005).

In addition, it is expected the battery is not deeply discharged when used in our power source, let us to assume a maximum use of 7 days, which is a long period without sunlight. Using a duty cycle of 4 % for the load of Figure 5, the average current is 886 µA, as it is noted in equation 1. With this current consumption the two supercapacitor in series are able to hold-up the system power supply during 1 day ($t_{supercap}$) without any recharge from the sun's energy, according to the equation 2 (Stor, 2007).

$$I_{load} = \frac{(4 \cdot 22\text{mA}) + (96 \cdot 6\mu\text{A})}{100} \approx 886\mu\text{A} \tag{1}$$

$$t_{supercap} = \frac{C \cdot \left(V_{max}^2 - V_{min}^2\right)}{I_{load} \cdot \left(V_{max} + V_{min}\right)} = 1\text{day} \tag{2}$$

$$t_{battery} = \frac{1\text{Ah}}{886\mu\text{A} \cdot 24\text{hours}} \approx 47\text{days} \tag{3}$$

Where C is the nominal capacitance of the supercapacitor in Farads, I_{load} is the average current delivered to the load, V_{max} and V_{min} are the maximum and minimum threshold voltages for proper working and t is the held-up time. The expected time has been calculated with 25F, 886μA, 4.6V and 1.53V. The expected lifetime of the Lithium-Ion battery is 47 days ($t_{battery}$), as it has been calculated in equation 3. Thus, 7 days of discharge correspond to 15% of the total capacity, and this low discharge percentage of charge/discharge means a longer lifetime, close to what we can call near perpetual operation for such a changing technology.

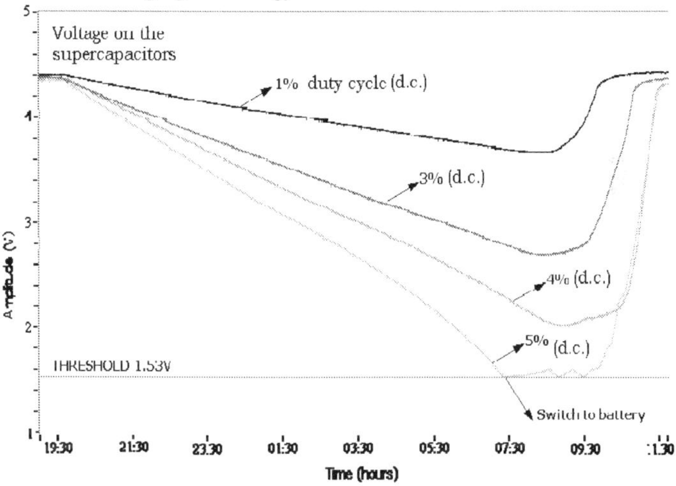

Fig. 7. This figure shows the discharge of the supercapacitors during 4 nights overlapped at different load currents. Each curve belongs to a different night.

6. Conclusion

This chapter discussed a method of design and implementation of an inexhaustible power source that without the human intervention manages and fully recharge the two energy buffers from the solar energy. After several weeks of tests the output of 3.3V has never failed and the battery has been resorted only a very rainy day. We have demonstrated autonomous operation for wireless sensor nodes with an average consumption of 886 μA. This value corresponds to a load with a fixed 4 % of duty cycle, but we could also assure an autonomous power for even higher duty cycles in case the load adjusts its duty cycle dynamically. Since the wireless sensor node can sense the energy available in our power source, it could reduce its activity during the night and increase it during the day. Nevertheless, the duty cycle depends on the specific application and our evaluated power source could widely cover most of field sensor applications where duty cycles of 1 % or less turn out to be enough (Zhang et al., 2004; Werner-Allen et al., 2005; Noda et al., 2006).

At the end, the new system architecture, in which this power source is based, opens new frontiers to experiment with other energy scavenging sources, because an appropriate use of supercapacitors as the single primary source could eliminate the necessity of large and degradable batteries, which would mean full autonomous power.

7. References

Alberola, J.; Pelegri, J. & Lajara, R. & Perez, Juan J. (2008). Solar Inexhaustible Power Source for Wireless Sensor Node, IEEE International Instrumentation and Measurement technology Conference, I2MTC 2008, Victoria-Canada.

Arms, S.W.; Townsend, C.P. & Churchill, D.L. & Galbreath, J.H. & Mundell, S.W. (2005). Power Management for Energy Harvesting Wireless Sensors, International Symposium on Smart Structures & Smart Materials (SPIE), 9 March 2005, San Diego, CA.

Buchmann, I. (2005). How to prolong lithium-based batteries (BU34), Cadex Electronics Inc., Vancouver. http://www.batteryuniversity.com/parttwo-34.htm

Burke, A. (2000). Ultracapacitors: why, how, and where is the technology, Journal of Power Sources, Vol. 91, Issue 1, Nov. 2000, pp. 37-50.

Chou, Pai H. & Park, C. (2005). Energy-Efficient Platform Designs for Real-World Wireless Sensing Applications, IEEE International Conference on Computer Aided Design, (ICCAD), San Jose, CA, 2005.

Chuck Wright Consulting, LLC, (2006). The Solar Sprint PV Panel, website. http://www.chuck-wright.com/SolarSprintPV/SolarSprintPV.html

Cooper Industries, (2006), Aerogel Supercapacitors - b series datasheet. http://www.cooperet.com/products/products.cfm?page=supercapacitors

Dreher, R. (2004). Remote Observation Station, Circuit Cellar: The magazine for computer applications, Issue 162 January 2004.

Dubois-Ferrière, H.; Meier, R. & Fabre L. & Metrailler, P. (2006). TinyNode: A Comprehensive Platform for Wireless Sensor Network Applications, *IEEE International Conference on Information Processing in Sensor Networks (IPSN)*, April 19–21, 2006, Nashville, Tennessee, USA.

Hande, A.; Polk, T. & Walker, W. & Bhatia, D. (2006). Indoor Solar Energy Harvesting for Sensor Network Router Nodes, Journal of Microprocessors and Microsystems Special Issue on Sensor Systems, 2006.

Howell Jones, M. (2004). Energy Scavenging for Sensor Networks. ELG7178F Topics in Communications II Wireless Ad-Hoc Networking November, 2004. www.howelljones.ca/tech/energyscavengefull.pdf

Hsu, J.; Friedman, J. & Raghunathan, V. & Kansal, A. & Srivastava, M. (2005). Heliomote: Enabling self-sustained wireless sensor networks through solar energy harvesting, International Symposium on Low Power Electronics and Design (ISLPED), San Diego, California, August 8-10, 2005.

Jiang, X.; Polastre, J. & Culler, D. (2005). Perpetual Environmentally Powered Sensor Networks. IEEE International Conference on Information Processing in Sensor Networks (IPSN), Los Angeles, CA, USA, April 25-27, 2005.

Kansal, A.; Potter, D. & Srivastava, M.B. (2004). Performance Aware Tasking for Environmentally Powered Sensor Networks, ACM Joint International Conference on Measurement and Modelling of Computer Systems (SIGMETRICS), New York, USA, 2004.

Koutroulis, E.; Kalaitzakis, K. & Voulgaris, N. C. (2001). Development of a Microcontroller-Based, Photovoltaic Maximum Power Point Tracking Control System, IEEE Trans. Power Electronics, vol. 16, no. 1, Jan 2001.

Lim, Y. H. & Hamill, D.C. (2000). Simple maximum power point tracker for photovoltaic arrays. Electronics Letters, IEE 2000, 24 February 2000.

MPower Solutions for All Your Specialist Portable Power Needs, (2006). http://www.mpoweruk.com/performance.htm

Noda, C.; Fernández, J. & Pérez, C. & Altshuler, E. (2006). Measuring Activity in Ant Colonies: The Wireless Perspective, International Centre for Theoretical Physics (ICTP), February 2006, Trieste, Italy.

Panasonic, (1999). Solar Cells Technical Handbook 98/99. http://downloads.solarbotics.com/PDF/sunceramcat.pdf

Park, C. & Chou, P. H. (2004). Power Utility Maximization for Multiple Supply Systems by a Load Matching Switch. IEEE International Symposium on Low Power Electronics and Design (ISLPED), pp. 168–173, Newport Beach, California, USA, 2004.

Park, C.; Liu, J. & Chou, P.H. (2005). Eco: An ultra-compact low-power wireless sensor node for real-time motion monitoring, IEEE International Conference on Information Processing in Sensor Networks (IPSN), Los Angeles, CA, USA, April 25-27, 2005.

Polastre, J.; Szewczyk, R. & Sharp, C. & Culler, D. (2004). The Mote Revolution: Low Power Wireless Sensor Network Devices, Hot Chips 2004, Aug 22-24, 2004.

Rabaey J. M. (2003). Ultra-low Power Computation and Communication enables Ambient Intelligence, Keynote Presentations, ELMO Seminar, Helsinki, Finland, 2003. http://bwrc.eecs.berkeley.edu.

Raghunathan, V.; Kansal, A. & Hsu, J. & Friedman, J. & Srivastava, M. (2005). Design Considerations for Solar Energy Harvesting Wireless Embedded Systems, IEEE International Conference on Information Processing in Sensor Networks (IPSN), pp. 457-462, Los Angeles, CA, USA, April 25-27, 2005.

Raghunathan, V.; Kansal, A. & Hsu, J. & Friedman, J. & Srivastava, M. (2005) Design Considerations for Solar Energy Harvesting Wireless Embedded Systems, Fourth IEEE International Conference on Information Processing in Sensor Networks (IPSN), Special Track on Platform Tools and Design Methods for Network Embedded Sensors (SPOTS), April 2005.

Rahimi, M.; Shah, H. & Sukhatme, G.S. & Heideman, J. & Estrin, D. (2003). Studying the Feasibility of Energy Harvesting in a Mobile Sensor Network, Proc. IEEE International Conference on Robotics and Automation (ICRA), pp. 19-24, Taipei, Taiwan, May 2003.

Roundy, S.; Otis, B. P. & Chee, Y. & Rabaey, J. M. & Wright, P. (2003). A 1.9GHz RF Transmit Beacon using Environmentally Scavenged Energy, IEEE International Symposium on Low Power Electronics And Devices (ISLPED), Seoul, Korea, 2003.

Roundy, S.; Steingart, D. & Frechette, L. & Wright, P. K. & Rabaey, J. M. (2004). Power Sources for Wireless Sensor Network, Proc. 1st European Workshop on Wireless Sensor Networks (EWSN), Berlin, Germany, 2004.

Stor, P.; Bussmann C. (2007) Application Note: Design Considerations In Selecting Aerogel Supercapacitors, http://www.cooperbussmann.com/3/PowerStor.html

Werner-Allen, G.; Johnson, J. & Ruiz, M. & Lees, J. & Welsh, M. & Marcillo, O. & (2005). Monitoring Volcanic Eruptions with a Wireless Sensor Network, Second European Workshop on Wireless Sensor Networks (EWSN), Istanbul, Turkey, January-February, 2005.

Würfel, P. (2005). Chapter 1. Introduction of Physics of Solar Cells: From Principles to New Concepts, 2005 WILEY-VCH Vewrlag gmbh & Co. kgaa, Weinheim. ISBN: 3-527-40428-7.

Zhang, P.; Sadler, C.M. & Lyon, S.A. & Martonosi, M. (2004). Hardware Design Experiences in ZebraNet, ACM Conference on Embedded Networked Sensor Systems (SenSys), Baltimore, Maryland, USA, November 3-5, 2004.

22

The Temperature Dependant Efficiency of Photovoltaic Modules - a long term evaluation of experimental measurements

Jan Machacek, Zdenek Prochazka and Jiri Drapela
Brno University of Technology
Czech Republic

1. Solar energy use, energetic potential of the Sun

1.1 The Sun as energy source

All energy existing on the Earth comes from the Sun. This energy can be present in the chemical form in fossile fuels, as biomass energy in plants and animals, as luminous energy contained in the falling radiation or as thermal energy which affects processes in the atmosphere and the circulation of water in the environment. At the heart of the Sun, nuclear fusion takes place - the fusion of hydrogen nuclei during which helium and other heavier elements are formed. Every second, the Sun changes approximately 600 million tonnes of hydrogen into helium. According to the well-known Einstein's formula $E = m \cdot c^2$, the difference between masses can be converted to energy. That is, 4.26 million tonnes of mass is transformed on the Sun every second, which results in the production of about $3.8 \cdot 10^{26}$ J of energy.

The Sun as a star is now in a stable phase of its development, which lasts some 4.6 billion years and is assumed to continue for about another 5 billion years. From the perspective of human life and human society it is therefore an inexhaustible source of energy.

It is assumed that at the end of this stable phase the Sun will approximately 1.4times enlarge its volume, exhaust the supply of hydrogen in its core, a thin layer of hydrogen enveloping its core ignites, the core shrinks as a result. The inner part of the Sun will expand and become a "red giant", who will engulf most of the planets in the solar system. The helium fusion will start in the helium envelope, with carbon and oxygen being formed. The thin envelope that will surround the burning core, will move out into space as a result of radiation effect. At the end of this process, only a carbon-oxygen core will remain of a 60% mass of the original Sun, wrapped in a thin hydrogen layer. This remainder will start cooling down and change into a "white dwarf", and after finally cooling down, the Sun will become a "black dwarf". The temperature on the Sun today is approx. 5900 K and its energy flows to as electromagnetic radiation.

1.2 Energy consumption of mankind

The civilization of today needs ever more energy for its development and needs to find new and long-term sources of energy. At present, the greater amount of the electric power is produced in thermal and nuclear power stations. The consumption of the technosphere is stated to be approx. 10 TW (all the fossil fuels, biomass energy and energy present in foodstuffs). According to available sources, about 1.4 $kW{\cdot}m^{-2}$ energy (solar constant) falls onto the outer boundary of the Earth's atmosphere. The whole Earth's surface is exposed to a radiation power of 180.000 TW. Theoretically, the complete radiation power falling to the Earth's surface would amply cover the energy consumption of humankind.

1.3 Use of solar radiation on earth

If we want to start using solar energy on the earth surface, we first need to know its usable potential at the place of installation of equipment. There is a difference between the installation of photovoltaic panels outside the earth's atmosphere and right on the earth's surface. Beyond the outer boundary of the atmosphere the conditions of sunlight are fully predictable, the period, time and intensity of solar radiation can be accurately foreseen. Panels may only be screened by the Earth, or by another celestial body, which is predictable, and the screening does not last very long.

Also, it is no problem to provide a perpendicular position for panels in relation to the Sun. On the other hand, installation of panels on the Earth entails a number of other problems. Solar radiation which reaches the Earth through the atmosphere is affected by processes going on within it and by its pollution. Further it is necessary to consider the changing position of the Sun in the sky during the day and year (azimuth and altitude above the horizon). Also we have to consider the screening of objects around the place of the intended installation.

The radiation which falls on the Earth's surface, very much differs from the radiation coming from the Sun to the Earth atmospheric boundary. The atmosphere sends part of the radiation energy back into space, and absorbs or disperses part of it. Given these conditions, the amount of energy reaching the earth surface depends on the momentary state of the earth atmosphere. Part of the radiation is dispersed or absorbed by molecules of gases forming the atmosphere, or by solid or liquid substances present in atmosphere. All the radiation falling on the earth's surface is called global radiation. It consists of two components - direct and diffuse. The diffuse component is largely contained in the radiation under overcast skies. Under clear skies, slightly less than 1 kW of radiation power falls on a square meter, whereas under overcast skies the radiation power only amounts to tens of watts.

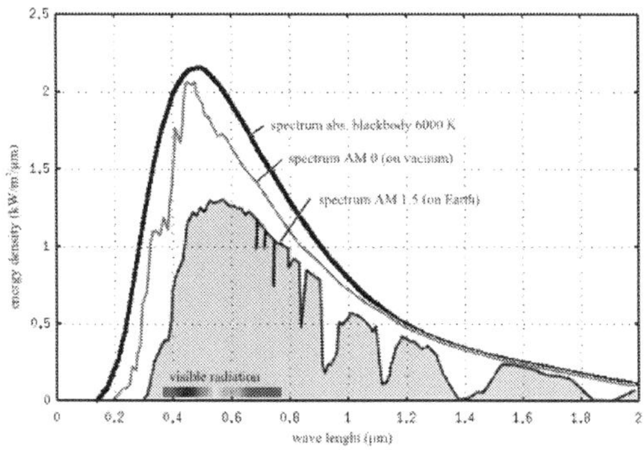

Fig. 1. Solar radiation spectrum after passage of earth atmosphere (Limbra M. and Poulek V. 2006).

Figure 1 shows the spectrum of solar radiation falling on the Earth, with wave length intervals marked that are absorbed by the atmosphere. The wave length representation is changed and the general radiation intensity decreases.

The altitude of the Sun above the horizon affects the radiation intensity. We can tell "the sunshine is stronger" at noon, when the sun reaches zenith, than towards the evening or in the morning, when it is just above horizon. Intensity of solar radiation throughout the day depends on its position in the sky, on the thickness of air layer which the radiation has to pass. A factor known as "Air Mass" is applied, which considers the effective quantity of air lying in the way of sun rays. When the sun reaches the highest point in the sky, an *AM* factor equals one. The amount of solar radiation hitting the ground also depends on altitude, i.e. this again relates to the amount of air sun rays have to pass through. When selecting a locality for photovoltaic panels installation, we have to consider the level of atmosphere pollution which is higher around conurbations than in the country, or the local climate of a relevant location and number of sunny and cloudy days in the year.

To identify the amount of solar radiation falling on the Earth, pyranometers are used, measuring both direct and diffuse radiation. They work on the principle of thermocouple and measure radiation of all wave lengths. If, however, we want to only measure direct radiation, we use a device called pyrheliometer, working on a principle similar to pyranometer. The latter is equipped with a tube which defines an angle under which radiation can enter the sensor. Further data needed to establish the amount of falling radiation can be obtained from meteorological measurements of long term climatic means for the locality in question (Murtinger K., Beranovský J. and Tomeš M., 2007).

1.4 Physical nature of light

Humans perceive light as the most natural of things, helping them recognize objects around them, distinguish between colours. Life without light is unimaginable to most of us. Light forms the visible element of electromagnetic radiation. Its physical nature was proved in the past and electromagnetic radiation has been used in many practical applications, first of all in information transfer.

A time change in the electric field is accompanied by the magnetic field change and vice versa. The vectors of both the fields are perpendicular to each other and oscillation spreads perpendicularly to both vectors (direction of oscillation spreading is obtained through a vector product of the vectors of the electric field intensity and the magnetic field intensity). The wave length spectrum of electromagnetic oscillations is very broad, beginning approx. at 10^{-13} m and reaching across thousands of meters. Of course light (the visible spectrum of the electromagnetic radiation) only forms a narrow interval of the spectrum λ = (380; 760) nm, with each wave length within this interval corresponding with a specific colour of light. The shortest wave lengths correspond with violet, with wave length increasing the colour changes over blue, green, yellow and reaching far as red. We perceive sunlight as white, as it is composed of a continuous spectrum and contains all of its colours. Sunlight can be decomposed using a prism or diffraction grating, all colours of its spectrum being obtained (Limbra M. and Poulek V. 2006).
Overview of wave lengths and their technical specifications:

- Gamma radiation, wave lengths $\lambda < 10^{-11}$ m, use in nuclear physics, defectoscopy.
- X ray, wave lengths $\lambda = (10^{-11}; 10^{-8})$ m, medical diagnostics, defectoscopy, astronomy.
- Ultraviolet radiation, wave lengths $\lambda = (10^{-8}; 10^{-7})$ m, solarium, photochemistry, copying machines.
- Visible radiation spectrum, $\lambda = (3,\ 8{\cdot}10^{-7};\ 7,\ 6{\cdot}10^{-7})$ m, optics, photographic machines, light sources.
- Infrared radiation, $\lambda = (7,\ 6{\cdot}10^{-7};\ 10^{-4})$ m, thermal radiators, grills, remote controls, IR cameras.
- Microwave radiation, $\lambda = (10^{-4}; 10^{-1})$ m, microwave ovens, radar systems, mobile phones.
- Radio radiation, $\lambda > 10^{-1}$ m, radio, television, communication systems.

Electromagnetic waves have a dualistic character. They behave like waves and at the same time corpuscles. The wave character is more pronounced in greater length waves while the corpuscular character in shorter length waves. If a wave's character is corpuscular, energy within it instead of continuously spreads in quanta - parts that we see as ´quasi particles´, i.e. particles/corpuscles with zero rest mass. Because of that, light has a corpuscular character, too, and therefore we can talk about a corpuscular-wave dualism.
Since light has a wave character, we can observe in it properties characteristic of general oscillation such as the laws of reflection and refraction at the boundary of two environments, interference phenomena, diffraction on the diffraction grating, and all the consequences of the Doppler effect such as e.g. a frequency shift toward the red edge of the spectrum known as the "redshift", or frequency changes caused by a star revolving around a centre of gravity alongside a big planet.
Manifestations of corpuscular nature of light belong to the field of quantum mechanics. Radiation is manifested as a flow of particles (quasi corpuscles) called photons. Their energy can be expressed by the relation $E = h \cdot f$, where h is the Planck constant and f is the radiation frequency. According to the relation $\lambda = c/f = c \cdot T$, the wave length is inversely proportional to the radiation frequency, i.e. we may state that photons with a shorter wave length (higher frequency) have more energy than photons with a longer wave length (lower frequency). It is thus possible to calculate that visible radiation photons have energy in eV

order, gamma radiation photons dispose of energy in MeV orders and finally microwave radiation particles possess energies in meV order. An example of corpuscular character manifestation is just the photoelectric effect. At this point Albert Einstein must be remembered, who in 1921 won the Nobel Prize for clarifying this phenomenon (Limbra M. and Poulek V. 2006).

2. Principle of photovoltaics and present-time technology

2.1 Physical nature of semiconductor function

Electromagnetic radiation is converted to electric power in semiconductor photovoltaic cells on the basis of crystalline silicon. This kind of cell belongs to the most widespread. It will suffice for the purpose of explaining the principle of energy transformation. We can divide semiconductors into two groups, i.e. intrinsic and impure. Impure semiconductors may have N type conductivity - prevalence of negative charge carriers (free electrons), or P conductivity - prevalence of positive charge carriers (holes - a place of a missing electron). A silicon atom has a structure of diamond and contains 14 electrons, its last (valence) layer containing four electrons, forming covalent bonds with their neighbours. Under normal circumstances the energy of a free electron may acquire any random values. In a silicon crystal, however, an electron energy may only acquire specific values as a result of motion in a periodic potential. These energy levels are divided into "bands of permitted energies"- i.e. energies an electron may acquire, and " bands of forbidden energies" - energies an electron may not acquire.

The following bands are relevant to the semiconductor function:

- Electron valence band (under very low temperatures it is the last to be occupied by electrons). It contains energy levels of valence electrons. The quantity of the levels is the same as that of valence electrons, they will be occupied under very low temperatures and will create covalent bonds with neighbouring atoms.
- Forbidden band of energies: an electron cannot acquire these energy levels.
- Conductivity band (under very low temperatures it is the first unoccupied). If an electron gets in this band, it may move freely through the crystal lattice space between atoms.

The energy level consistent with the last level of the valence band is marked E_V, the lowest level of conductivity band is marked E_C, the width of the forbidden band $\Delta E_G = E_C - E_V$. Another important energy level is the Fermi energy E_F. In an intrinsic semiconductor (e.g. pure silicon) the Fermi energy lies in the middle of the forbidden band.

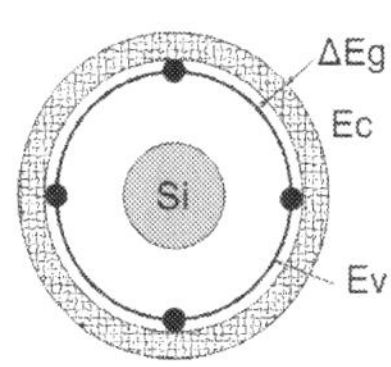

Fig. 2. Atom of silicon with represented energy levels.

As a result of supplying energy, e.g. warming (by phonon), or lighting (by photon), an electron breaks the valence bond and moves to the conductivity bond. This is manifested in the energy diagram by electrons occupying higher energy levels. The valence and conductivity bonds thus become partially occupied. The electrons in the conductivity bond thus become electrons capable of carrying electric current. The free areas left by electrons in valence bands of some atoms, may be occupied by electrons jumping to them from other atoms. This is how the free areas move towards other atoms. In the electric field, free valence electrons move opposite the direction of electric field intensity, as they have negative charges. That is, free areas shift in the direction of field intensity. This place therefore behaves as a positive charge particle. The particle is called a "hole". In an intrinsic semiconductor the number of holes is equal to the number of free electrons. As a result, the crystal appears to be electrically neutral on the outside. If an electron - hole pair is generated as a result of a photon impact, the photon's energy must be greater than or equal to the forbidden band width. Photons with a smaller amount of energy will pass, while those with greater or equal energies are absorbed. the forbidden band width of silicon is approx. 1.1 eV. It is therefore transparent for photons with a wave length greater than 1100 nm.

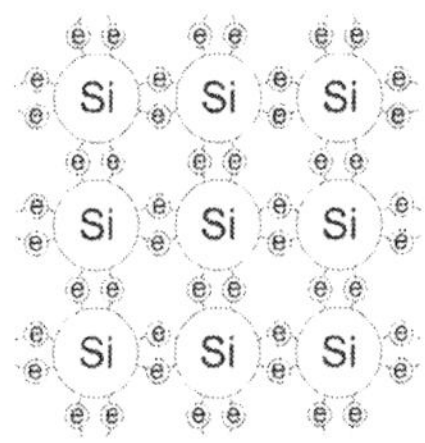

Fig. 3. Atoms of silicon inside intrinsic semiconductor.

Figure 4 shows a function which represents density of states - number of states per unit interval of energy for intrinsic semiconductor.

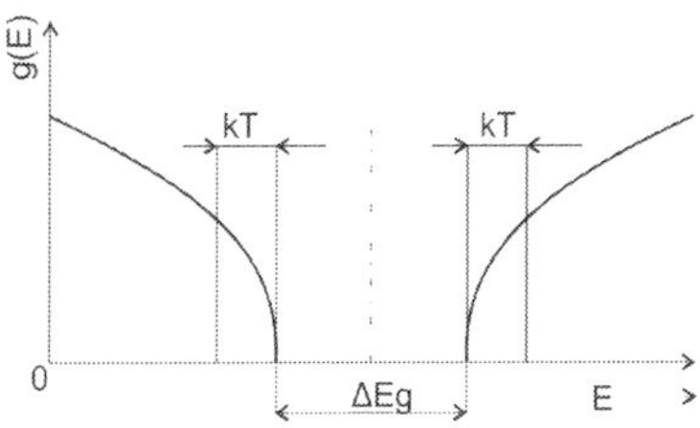

Fig. 4. Density of states for intrinsic semiconductor per unit interval of energy (Limbra M. and Poulek V. 2006).

Figure 5 represents a dividing function f(E) giving probability of energy level occupation by an electron. The probability that all the valence electrons will be occupied is high, as opposed to the probability that this will be in the conductivity band. This is consistent with the above mentioned theory saying that the valence band for the intrinsic one will under very low temperatures be the last occupied and the conductivity band the first to be not occupied. Dependence 1 - f(E) represents the probability of an energy level not being occupied by an electron. Both curves cross at a value of 0.5, which agrees with the value of the Fermi energy which for intrinsic semiconductor is in the middle of the forbidden band. Electrons pertain to the group of particles called fermions and behave in agreement with the Termu – Diracov statistics.

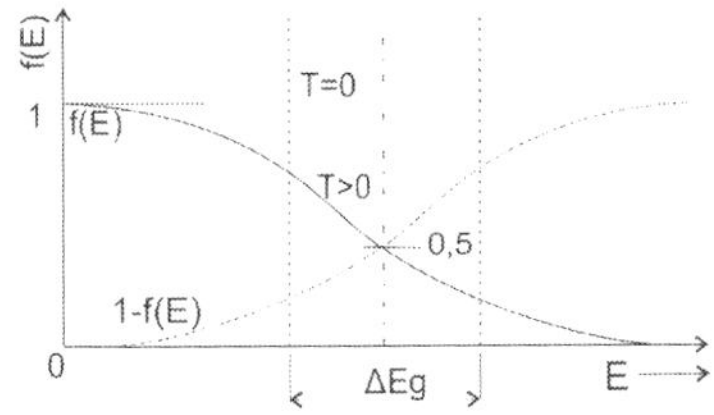

Fig. 5. Probability function f(E) and 1 - f(E) (Limbra M. and Poulek V. 2006).

The function f(E).g(E) = n(E) in Figure 6 gives concentration of electrons in the conductivity band and the function (1 − f(E)).g(E) = p(E) concentration of holes in the valence band at a non-zero temperature. The areas below the curves are the same in intrinsic semiconductor. If we add a pentavalent element atom (a Group V element in the periodic table) to a pure silicon crystal, e.g. phosphorus, arsenic or antimony, a Type N semiconductor is produced. An element of admixture having an extra electron is called donor (i.e. donating one electron). Four valence electrons of arsenic create a covalent bond with the adjacent atoms of silicon, but one of them does not find a partner for bonding. That electron is only weakly attached to its atom and only a small amount of energy suffices to breaking it away. These free electrons effect electron conductivity (Type N non-intrinsic conductivity) through their motion. Presence of such admixture atoms is manifested in the energy diagram by local energy levels arising, lying in the forbidden band near the bottom limit of conductivity band E_C. There are a great many more donor electrons in the semiconductor than intrinsic electrons of silicon and therefore they are majority charge carriers. Semiconductors of Type N also include holes, though these arise as ´intrinsic charge carriers´. Their quantity heavily depends on temperatures. The holes are minority charge carriers in Type N semiconductors.

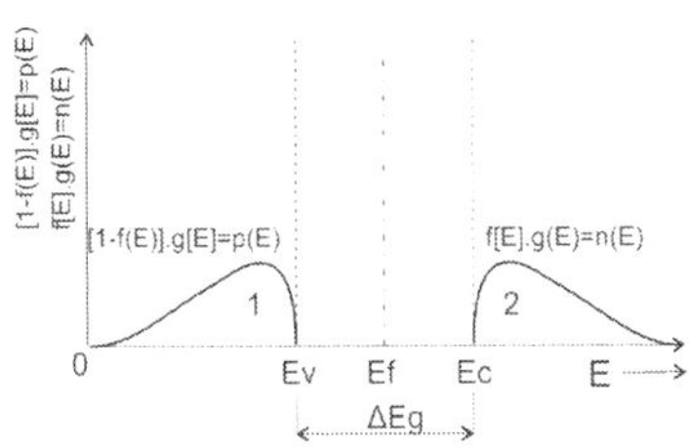

Fig. 6. Concentration of electrons and holes in intrinsic semiconductor (Limbra M. and Poulek V. 2006).

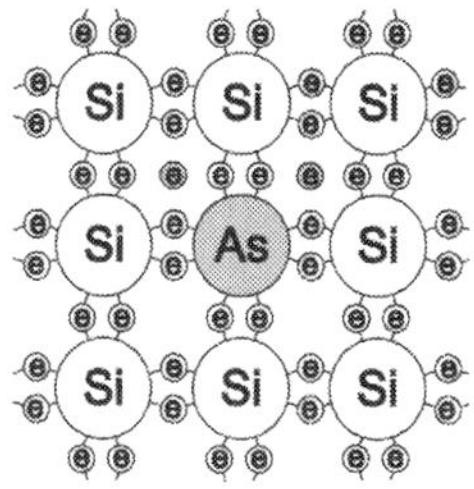

Fig. 7. Atoms of N type semiconductor with As donor atom (Dobiáš, 2008).

Figures 8, 9 and 10 present diagrams similar to those for intrinsic semiconductor, with the following difference for N type semiconductor: the Fermi energy level E_F is shifted towards conductivity band, the donor energy level E_D lies between E_F and conductivity band. ΔE_D energy needed for transfer from this level to conductivity band is very small, of about ΔE_D = 0. 01 eV.

For example, at room temperature an electron can easily go over to conductivity band. Given the small value of this energy, donor atoms are ionized at room temperature and create a positive space charge. The concentration of electrons in conductivity band is much higher than that of holes in the valence band. It can be seen in Figure 10 that area 2 delineating the electrons concentration, is higher than area 1, representing the concentration of holes.

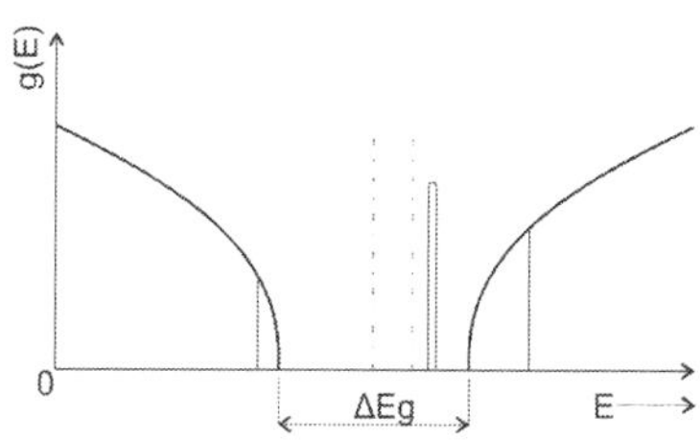

Fig. 8. Density of states in Type N semiconductor N (Dobiáš, 2008).

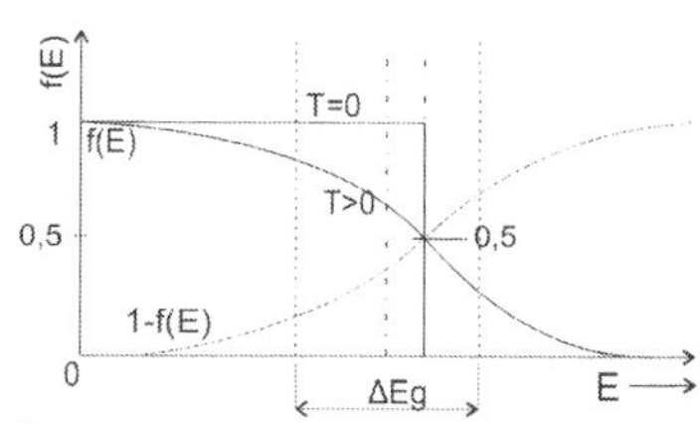

Fig. 9. Probability functions for Type N semiconductor (Dobiáš, 2008).

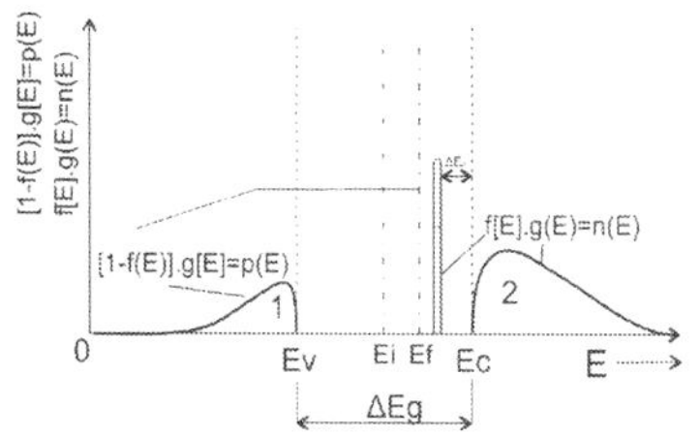

Fig. 10. Amounts of electrons and holes in Type N semiconductor (Dobiáš, 2008).

If in a tetravalent silicon crystal we replace some atoms by trivalent atoms of Group III elements in the periodic table (e.g. B, Al, Ga), Type P semiconductor is produced. Admixture atoms have one valence electron less than silicon, therefore it is called acceptor (it accepts one heat released electron in its valence layer). A trivalent element is not capable of creating a covalent bond of four pairs of electrons. The free place behaves like a hole (defective electron). Due to thermal energy effect a valence electron may jump over to the unfilled bond from the adjacent Si atom and the hole may move through the crystal. These holes of an alien atom create the hole conductivity in semiconductor (non-intrinsic conductivity of Type P).
Holes that are supplied to Type P semiconductor are many more than holes arising in intrinsic semiconductor. Therefore, in Type P semiconductor the hole is a majority charge carrier. There are also free electrons in P semiconductor, arising due to temperature effect, which are intrinsic charge carriers, while in Type P semiconductor they are minority charge carriers. Its amount depends on the semiconductor temperature.

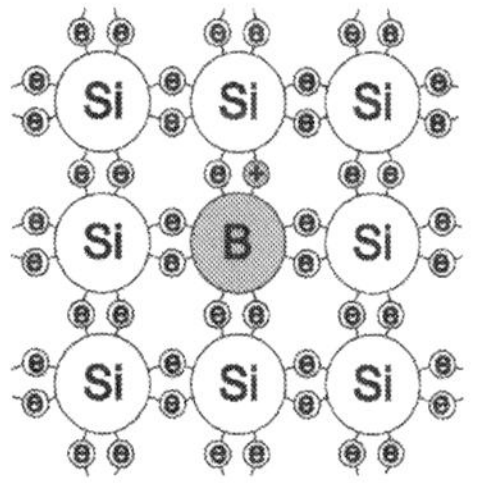

Fig. 11. Atoms in Type P semiconductor with acceptor atom B (Dobiáš, 2008).

Supply of trevalent element (Figure 11) is manifested in the energy diagram in a similar way as in Type N semiconductor. Acceptor level E_A arises in the forbidden band, near the valence band. Only a small amount of energy ΔE_A is sufficient for electron jump-over from the valence band to this level. Such an acceptor atom appears outward as negatively charged. A solid negative space charge is produced. A hole thus generated in the valence band is free to move around. In Type P semiconductor, concentration of loose holes is much higher than concentration of free electrons, the Fermi energy E_F level shifts towards the valence band. If one part of the crystal is doped by one type of admixture and the other one by an opposite type, the free charge carriers (electrons and holes) should based on the laws of diffusion escape from areas of higher concentration to areas of lower concentration. In this way an even spread is created. The escaping free charge carriers leave at the original sites firmly bonded charge carriers of opposite polarity, that will create a space charge. There are relatively strong local fields inside the crystal, which prevent further escape of admixture. The whole system will thus stabilize in a balanced state and appear as electrically neutral on the outside.

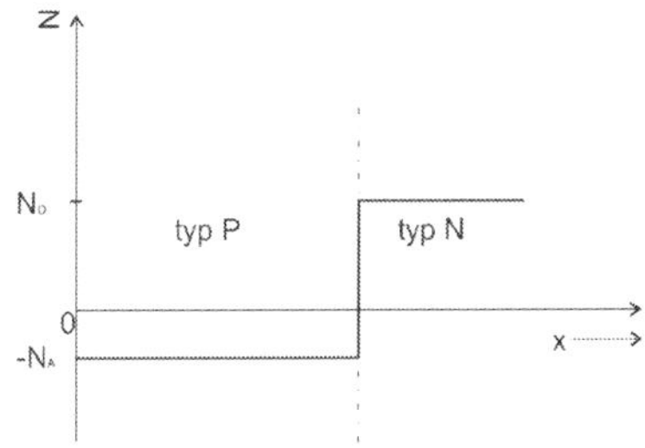

Fig. 12. Sharp division between the areas N and P (Dobiáš, 2008).

An example of such a differently doped crystal may be a semiconductor diode including a sharp divide between the areas P and N. Streams of free charge carriers flow over the PN junction in both directions, Figure 13 shows a situation for free electrons. As for holes, it will be similar. Some free electrons from area N pass to area P where they recombine with free holes (recombination flow of electrons). But then, as mentioned before, electron in area P is the minority charge carrier which is attracted to area N, where is the firmly bound positive charge. This electron flow is called thermal. Without an added external voltage the two currents are balanced and not shown on the outside.

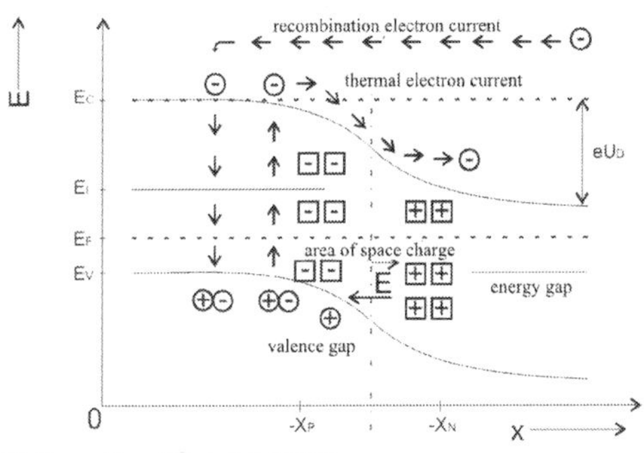

Fig. 13. Unpolarized PN junction (Dobiáš, 2008).

If we apply the external voltage on part P with the positive pole and on part N with the negative pole, the band will curve as a result and the internal diffusion voltage U_D, will be reduced, whereby the recombination electron flow from Type N semiconductor to Type P while the flow of holes from P to N prevail. PN junction has thus been polarized in the transparent direction. When applying external voltage of opposite polarity, potential barrier U_D increases, recombination flow drops and thermal flow of free charge carriers prevails.

2.2 Principle of the photovoltaic cell function

The photovoltaic cell can be imaged as a large-area PN junction positioned perpendicularly to the face of the cell between the front and back walls, while Type N semiconductor faces the front wall and Type P semiconductor the back wall of the cell. The falling photons with an energy greater than corresponding to the forbidden band width ΔE_G, generate an electron-hole pair. This is how photons deliver their energy and are absorbed in PN junction. They give the remainder of their energy if any to the crystalline grating, where it is transformed to heat. Electron-hole pairs are separated from each other by an internal electric field at PN junction. Holes are accelerated in the direction of while electrons in the opposite direction to the internal field $\vec{E}$. Electric voltage appears between the cell poles and after connecting the cell to circuit, electric current starts passing through it.

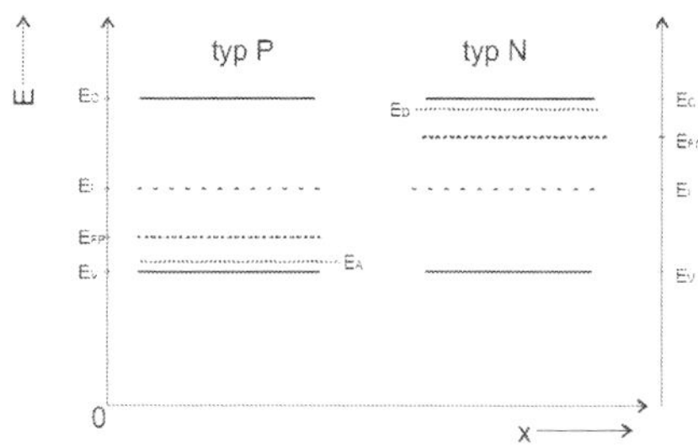

Fig. 14. Diagram of semiconductor cell energy levels (Limbra M. and Poulek V. 2006).

Figure 14 represents a diagram of energy levels in types P and N semiconductors. Figure 15 shows equalization of the Fermi energy and the band curve in an unlit photovoltaic cell. Also the recombination and thermal flows in a balanced state are represented. I.e. semiconductor cell behaves in the dark like a semiconductor diode.

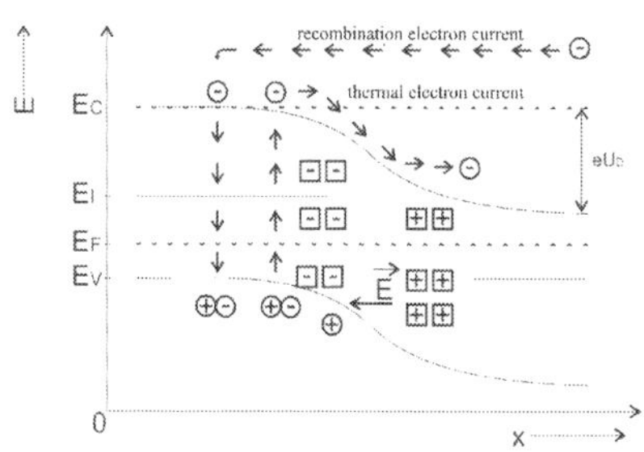

Fig. 15. Unlit photovoltaic cell (Limbra M. and Poulek V. 2006).

Figure 16 presents a diagram of energy levels of the lighted semiconductor cell, not connected to circuit. The falling electrons break the internal balance of charges, or exactly create a different one. The generated holes are accelerated in the direction of the internal field $\vec{E}$ (which corresponds to thermal flows in the reverse direction). Type P semiconductor is charged positively and Type N semiconductor negatively. Potential barrier U_D decreases, the Fermi levels in areas P and N are divided and their difference equals photovoltaic voltage U_P. The voltage can reach a maximum of $U_P \approx 0,6$ V, which equals the original bands equalization. By further increasing intensity of the cell lighting the voltage does not increase. This is because reducing the potential barrier by voltage U_P the recombination current and the system enters the new stable condition.

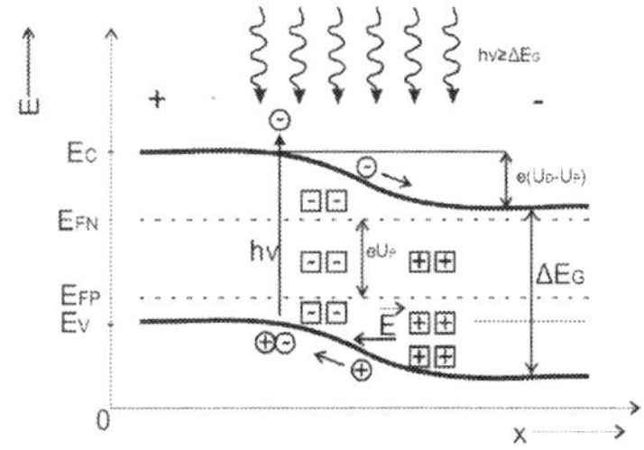

Fig. 16. Lighted photovoltaic cell (Limbra M. and Poulek V. 2006).

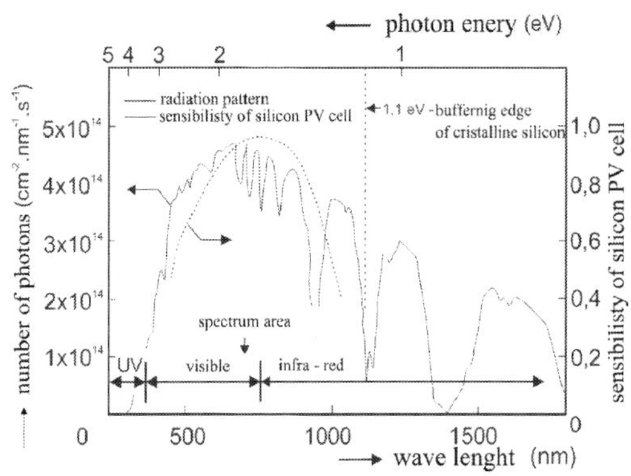

Fig. 17. A solar radiation spectrum after passing through atmosphere, which the photovoltaic cell is able to absorb (Limbra M. a Poulek V. 2006).

After connecting the lit semiconductor cell to the electric circuit the photovoltaic voltage (electromotoric voltage of the source) decreases, the band curve changes and the potential barrier increases again. This effects decrease in recombination flows and thermal flows predominate as a result of separation of the generated free charge carriers by the internal firmly bonded space carrier. The sum of recombination and thermal flows will no longer be zero and the resulting flow will run through the connected electric circuit. As mentioned above, the forbidden band width equals ΔE_G = 1,1 eV, photovoltaic panels are therefore sensitive to radiation of wave lengths shorter than λ = 1100 nm. Figure 17 presents a solar radiation spectrum after passing through atmosphere with a marked area which the silicon photovoltaic cell is able to absorb.

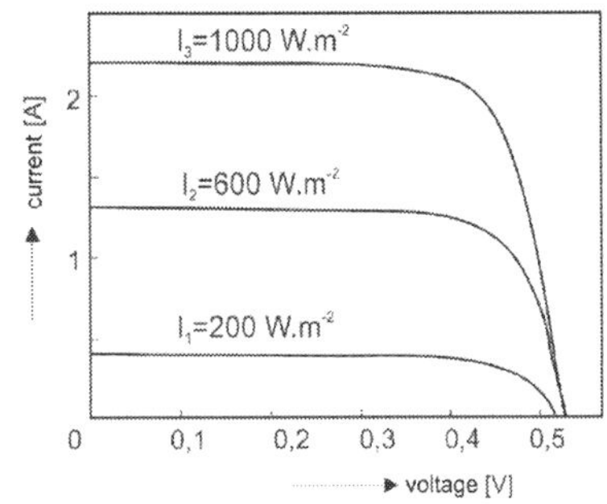

Fig. 18. Loading characteristics of photovoltaic cell for different lighting intensities (Dobiáš, 2008).

Figure 18 represents the loading characteristics of a lit cell at different lighting intensities. Points of intersection of curves and the vertical axis specify swift currents while off-load voltage is specified by horizontal axis points of intersections. An optimum load for semiconductor cell has a value at which the working point lies in such an area of the loading characteristics, where the product of voltage and current reaches the highest value. This being reached, the cell delivers a maximum power to the circuit.

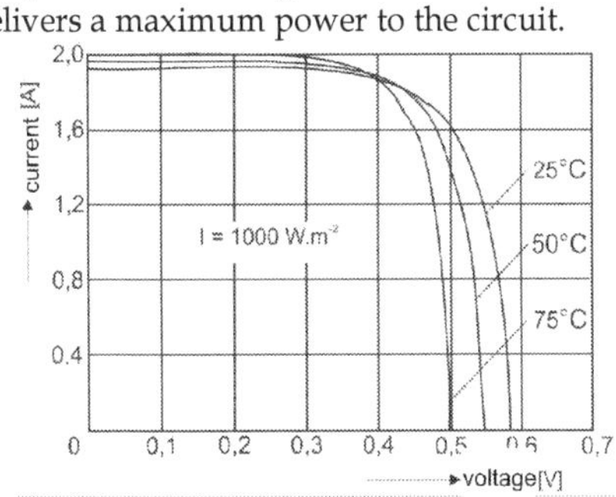

Fig. 19. Change in load curve in relation to temperature (Dobiáš, 2008).

Figure 19 shows how load curves change in relation to heat. At higher temperatures the semiconductor cell is able to deliver higher ***swift*** current, however, its off-load voltage will drop. With temperatures rising and constant intensity of solar radiation the maximum power the cell is able to supply to the circuit is falling. I.e. the efficiency of photovoltaic energy transformation decreases.

2.3 Production of photovoltaic cells

The silicon panels are among the most widespread types and can be divided into 3 basic groups: monocrystalline, polycrystalline panels and panels on the basis of amorphous silicon. Silicon is found most abundantly on the Earth, that is why it is inexpensive, it is not toxic and up to now has been the most widely used and explored semiconductor. Silicon is also suitable due to its forbidden band width of ΔE_G = 1,1 eV. In the natural environment, silicon most frequently occurs in the form of silica or silica sand (SiO2). The raw silicon, suitable for technological purposes, is produced in an arc furnace through reduction with carbon SiO2+C → Si+CO2. Silicon gained in this way is 99% pure, amount of impurities is 1%. Purity needed for use in semiconductor technology is much higher. Today one of the most widely used technologies to increase silicon purity is Siemens with chlorine cycle (Figure 20). The chlorine process, though the most commonly used at present, is technologically extremely demanding and dangerous for operator staff as well as the environment. New technologies of pure silicon production are in development, now in the phase of testing and trial production. If proved effective, they may bring cuts in energy demand and price of pure silicon. The production process typically yields lumps of pure silicon, further processed in creation of monocrystalline or polycrystalline ingots.

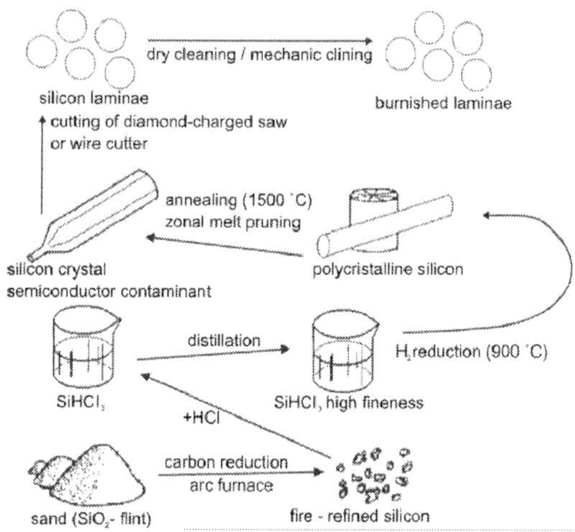

Fig. 20. Silicon production cycle (Limbra M. and Poulek V. 2006).

The production of polycrystalline ingots is substantially simpler than that of monocrystalline. The molten material is simply poured in a mould and left to cool down at a defined rate. Cooling must be as slow as possible, in order for the largest possible monocrystalline grains to be created and a minimum dislocations reached in the material. Monocrystalline ingots are produced with the Czochralski method. A small monocrystal nucleus is dipped in a melting of an approximate temperature of 1415 °C. The nucleus is turned around very slowly and drawn out of the melting. The whole process of drawing takes place in an inert atmosphere, in order to ensure silicon purity.

2.4 Construction of photovoltaic panels

The produced monocrystalline or polycrystalline ingots are cut into wafers that create the basis for semiconductor cells. Wafers are usually cut 100 m thick. Semi-finished products for semiconductor production from different materials are made using technologies similar to those for silicon. As mentioned before, the semiconductor cell is a large-area diode with PN junction in a perpendicular position toward the facing. The admixture elements are added to semiconductor in diffusion furnaces. On the cell's face is applied antireflex coating in order to minimize reflection, i.e. make use of a maximum amount of the falling radiation. Cells of greatest efficiency therefore appear to be black. Sometimes a thin transparent coating is applied on the face for decoration purposes, to enhance the reflected radiation of a specific wave length as a result of wave interference on this coating. Such cells then produce a sort of colour shade in the reflected light. The contact on the front side is usually shaped as a lattice or comb in order to cover the smallest possible surface of the cell face. The back contact on standard panels typically covers the whole surface. Contacts are applied by serigraph onto cells. Only for more demanding applications vacuum technologies are used such as vacuum fomentation. For special cells with high radiation concentrations contacts embedded in semiconductor material are made. This enlarges the contact surface between contact and semiconductor because of great densities of current flowing through such cells (Bařinka R. and Klimek P., 2007).

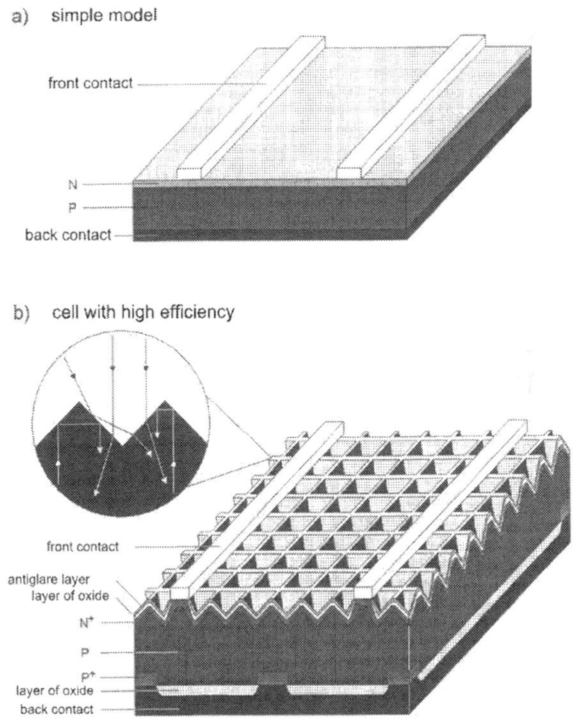

Fig. 21. Simple cell and high efficiency cell (Limbra M. a Poulek V. 2006).

The conventional semiconductor cell is shown in the upper part of Figure 21. This simple cell works based on the above described principle, but its efficiency of photovoltaic transformation is lower mainly because of recombination losses. In order to increase the efficiency, in higher quality cells (Figure 21b) constructional improvements are used to minimize reflection caused losses. The antireflex coating on their surface reduces reflection so as to allow the largest possible amount of photons to penetrate into the area of PN junction. The layers of nonconductible SiO2 chemically deactivate the cell surface while the layers are only etched at some points and it is just through them the electric charge is conducted away. The structure of small pyramids etched in the cell face causes photons easily entering the cell, however if they have passed through PN junction and have not been transformed, they are reflected from the back electrode, but do not pass back out through the cell face because of the total reflexion. Photons are reflected back to PN junction and probability of photovoltaic transformation thus increases. In double-sided panels, the structure of pyramids is etched on both sides, i.e. photons can fall from both sides, e.g. on the back side by reflection from the ground. Thus the amount of energy produced is higher, although photovoltaic efficiency of transformation on the back side is lower, as photons of shorter wave lengths are absorbed in the thicker layer of the semiconductor substrate and will not reach PN junction.

In polycrystals, boundaries of grains weaken the transport qualities of semiconductor, in amorphous semiconductors the conditions are even worse. The efficiency of present-time quality monocrystalline cells is around 20 %. In laboratory conditions, efficiency of up to 30 % can be reached. Cell efficiency on the basis of other semiconductors (GaAs, InP) is even higher. It reaches values around 25% as for cells series production, though their price is several times higher. That is why these cells are mainly used in cosmic applications where the price is not a limiting factor. They are also used in "tandem alignment", where cells on various bases are aligned one after another, each of them using a different part of the spectrum. Besides silicon, CuInSe or CdTe cells or cells with heterojunctions between different kinds of semiconductors are used in thin-layer applications. The cells are relatively cheap, however, they reach smaller efficiency of around 10 %, and have very unstable parameters. A cell with about 15% efficiency can be produced, but its efficiency will drop below 10 % in a short time.

Photovoltaic panels are produced by placing individual cells inside them in a series-parallel combination, so that they deliver the required uniflow voltage and current at defined lighting. The most common photovoltaic panel is made of a front tempered glass which withstands great impacts and is resistant to hail. Onto the glass is applied an EVA (ethyl vinyl acetate) foil onto which individual interconnected cells are laid next to each other. The cells are covered by another EVA foil and the back wall of the cell is usually made of PVF-PET-PVF laminate composition. Next, air is exhausted between the two layers and the panel is warmed up to the temperature of EVA foil melting. EVA foil will melt away as sealing compound and fill the space between the front and back panel sides. Finally, panels are framed and sealed with a silicon sealant in aluminium sections and have a box with outlet contacts fitted. In this way panels are sealed tight to prevent penetration of water and dirt. Panels on the crystalline silicon basis have a service life of 20 to 30 years.

Double-sided panels consist of double-sided semiconductor cells and taking into account efficiency of photovoltaic transformation, they are more advantageous than the one-sided panels now in use. Curiously, their difference in price is not so great. The production technology of double-sided panels only differs in transparent laminate used in its back wall, and the lattice-shaped back contacts not covering the whole surface, as mentioned before. The panel is optimized to take radiation from the front side, still the efficiency of photovoltaic transformation with radiation reaching the back side is not much smaller. Double-sided c-Si based panels are within the infrared area of the spectrum for radiation of a wave length higher than λ = 1100 nm, i.e. with energy smaller than that of the forbidden

band width. This radiation amounts to over 20% of solar radiation energy and proportionally to that, less energy transformed to heat is absorbed in double-sided panels as compared with single-sided panels. Double-sided panels therefore have a lower temperature and the relating higher efficiency of photovoltaic energy transformation.
Innovative is the connection of monocrystalline balls with spherical PN junction. The balls can be rigid with a glass front wall, or flexible, enclosed in plastic with a teflon face. The balls of silicon have an approx. 0.6 mm diameter, with Type P semiconductor inside and Type N semiconductor on the surface. That is, they have spherical PN junction under the surface. Balls are produced by dropping a doped admixture from the capillary tube, which stiffens during fall. After that diffusion takes place and the ball surface is redoped to Type N semiconductor. Balls are then placed onto an aluminium foil with holes smaller than the ball diameter. The foil creates negative electrode, as it touches Type N semiconductor. On the other side of the ball part N is etched away and a contact installed in Type P semiconductor, which creates positive electrode. The whole cell is then enclosed; there are rigid or flexible panels of different sizes made of these cells. The efficiency of such panels has reached around 14 % so far (U.S. Department of Energy, 2006).

2.5 Basic types of photovoltaic cells

The photovoltaic cell has a development history of 50 years during which a large number of construction types using different materials have been developed. The earliest type is photovoltaic cells on the basis of monocrystalline silicon, with a large-area PN junction created inside. The cells are characterized by good efficiency and long-term performance stability. They are still used in large-area installations. Their disadvantage is great consumption of very pure silicon and relatively demanding production.
The great consumption of silicon and high demands on its production stimulated the use of thin-layer cells. Cells of monocrystalline, polycrystalline or amorphous silicon are among the most commonly used. Their main disadvantage consists in their significantly smaller efficiency and smaller stability of parameters (efficiency continues to decrease over time). These cells are beginning to be used in applications where flexibility is required. Owing first of all to the interest of the army there are cells in development that are part of clothes or backpack and enable supply for pack sets.
There are systems that use methods other than PN junction for charge separation, and frequently they are not even semiconductors. These include e.g. photoelectric (photogalvanic) cells or polymer cells. Nanostructures begin to be used, in the form of carbon nanotubes or nanosticks, or application of quantum dots on a suitable pad. These structures are liable to easy influencing their electrical and optical properties. Still, their practical use has been barely noticeable so far, the problem again being their low efficiency and small stability of properties and life. Possibly the closest to commercial use are now the flexible photovoltaic modules based on organic polymers.
The greatest potential for future use is seen in the composite photovoltaic cells which are able to effectively use a broad range of the solar spectrum. Each layer can only use light in a specific wave length range. The radiation that a layer is not able to use will pass to the next layers where it is utilized.

3. Dependence of PC output current and efficiency on its temperature

3.1 Substitute circuit of photovoltaic cell

The output power of a photovoltaic cell and / or a cells module primarily depends on two factors, i.e. the cell heat and the intensity of solar radiation falling onto its surface.

The relation between the cell temperature and the output current and consequently the power has been described in literature using a substitute circuit, see Figure 22. Figure 22 represents a photovoltaic cell substitute circuit consisting of a current generator in parallel connection with a resistor and diode (Luque A, Hegedus S., 2002).

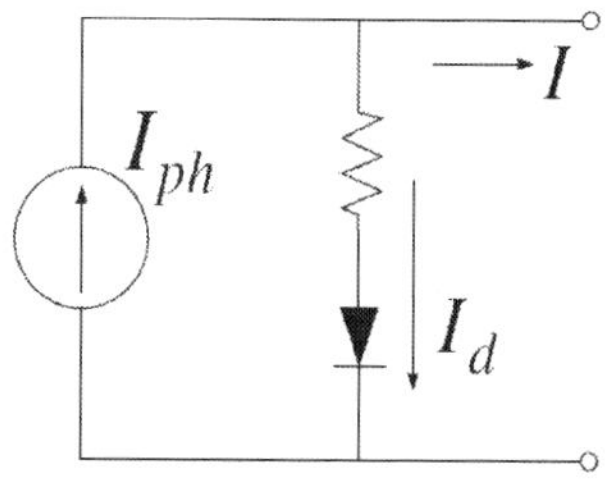

Fig. 22. Substitute circuit of photovoltaic cell.

Photovoltaic current I_{ph} is directly proportional to the intensity of solar radiation S (kW/m^2) and can be expressed as (Markvart T. Solar electricity., 2000):

$$I_{ph} = \left[I_{scr} + C_i \cdot \frac{T - T_r}{1000} \right] \cdot S \tag{1}$$

where T is a cell temperature, T_r is reference temperature, I_{scr} is swift current under reference operation conditions (1 kW/m^2, AM 1.5, 298 K). C_i is a temperature coefficient of swift current. The following relation applies to estimation of a cell temperature from the ambient temperature T_a:

$$T = T_a + \frac{NOCT - 20}{0.8} \cdot S \tag{2}$$

where $NOCT$ is nominal operating temperature. Let us assume a parallel shunt resistance is interminable, therefore the current through this resistance can be omitted, and the current through diode I_d and the reverse saturation current I_{sat} can be expressed as follows:

$$I_d = I_{sat} \cdot \left[\exp\left(\frac{q \cdot V}{KAT} \right) - 1 \right] \tag{3}$$

$$I_{sat} = I_{rr} \cdot \left(\frac{T}{T_r} \right)^3 \cdot \exp\left[\frac{q \cdot E_g}{KA} \cdot \left(\frac{1}{T_r} - \frac{1}{T} \right) \right] \tag{4}$$

where q is an electron charge value, V is output voltage, K is the Boltzmann constant, A is ideality factor, I_{rr} is reverse saturation current at reference temperature, E_g is size of gap between the conductivity and the valence bands. The following relation applies to output current:

$$I = I_{ph} - I_d \tag{5}$$

As follows from the above derivation of output current dependence on temperature, if the temperature of photovoltaic cell increases, swift current will decrease, which results in the decrease of maximum cell output power and efficiency of energy transformation. By contrast, with the growing intensity of falling radiation, swift current as well as maximum output power increase and conversion efficiency rises.

The above conditions, unfortunately, act against each other, and moreover at the least favourable point in time. It is at the point of the greatest sunshine and falling of the greatest intensity of solar radiation on photovoltaic panel surface, i.e. the time of the expected maximum power gain, when due to sunrays effect the warming culminates and the panel electrical specifications change, which results in the above mentioned decrease in conversion efficiency and consequently decrease in the supplied power. The resulting power under these operating conditions is then smaller by percent units than possible maximum power which the panel might provide at the current intensity of falling radiation. The following figures and characteristics represent the described problems graphically. Figure 23 shows VoltAmper and power characteristics of a monocrystalline solar cell under reference operating conditions (Chan DSH, Phillips JR, Phang JC., 1986).

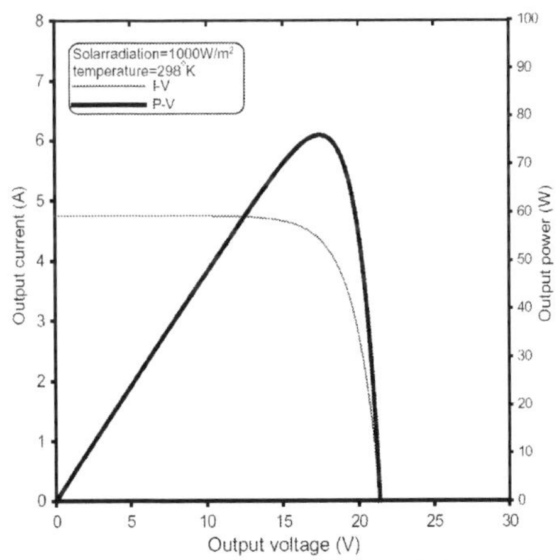

Fig. 23. Specific VoltAmper and power characteristics of photovoltaic cell (Chen YM, Lee CH, Wu HC., 2005).

The level of efficiency change in relation to temperature depends on several factors. It depends e.g. on material composition and cell structure of crystalline silicon of photovoltaic cell. Typical decrease of efficiency in standard cell types ranges between 0.4 to 0.6 % of power per degree Celsius of the growing cell temperature. As to new types of cell that are very thin and have a lower temperature coefficient, the power drops by only approx. 0.2 % per degree Celsius.

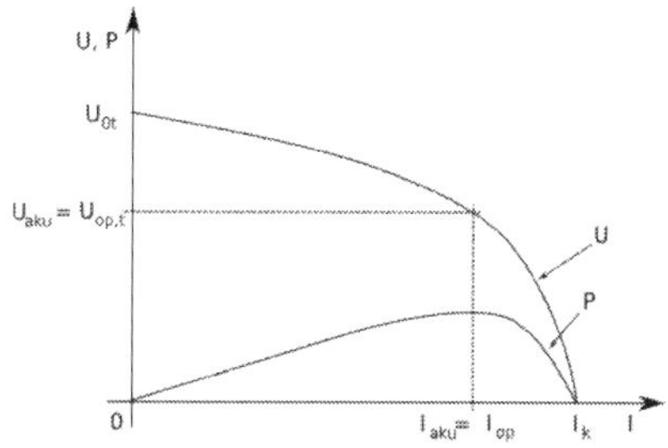

Fig. 24. Load and power characteristics of a typical photovoltaic cell.

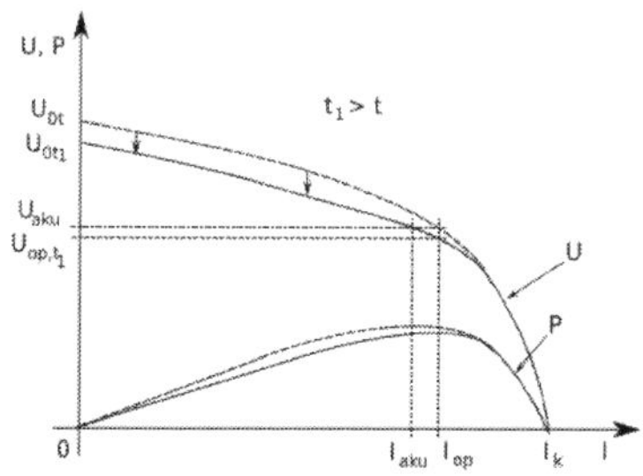

Fig. 25. Effect of temperature change on VoltAmper characteristics of a typical photovoltaic cell.

Figures 24 and 25 demonstrate the impact of drop of falling solar radiation intensity, where alongside the decreasing intensity the potential maximum power which a photovoltaic cell is able to produce decreases too, and the impact of the mentioned temperature on the working point shift on the power curve and the drop of an output voltage optimum value.

The level of operating temperature of a photovoltaic cell or cell array depends on the climatic conditions of a relevant locality, the degree of intensity of falling solar radiation on the one hand, and on the design, construction and assembly of the panel on the other. These factors actually have a substantial impact on the module's ability to cool and transmit heat to the ambient environment, in most cases air, and on determining the definitive panel operating temperature under specific climatic conditions. The measure of efficiency of the construction for panel cooling is known as Nominal Operating Cell Temperature - *NOCT*, which is determined for the defined intensity of falling solar radiation, temperature, weather and atmospheric conditions (Tian Pau Chang, 2008).

3.2 Possibilities of energy use optimization

An optimization device can be used to compensate for this effect, working on the principle of an increasing *DC/DC* of the converter. The solution is based on the electronic power regulation of photovoltaic generator loads with regard to their current operating conditions. The function of an optimization circuit is shown in Figure 26.

Connection (see Figure 26) is the basic form of a non-insulated increasing converter. It can be classed as indirect converter of input-to-output energy transfer, as the energy is accumulated in the coil's magnetic field. During the transistor closing the current I_zap is increasing. Energy is thus gradually accumulated in the coil's magnetic field, the coil therefore behaving like an appliance. Now current is only supplied to the load from Condenser C1. Condenser C1 is thus being discharged and its current drops in the same way as the voltage on it.

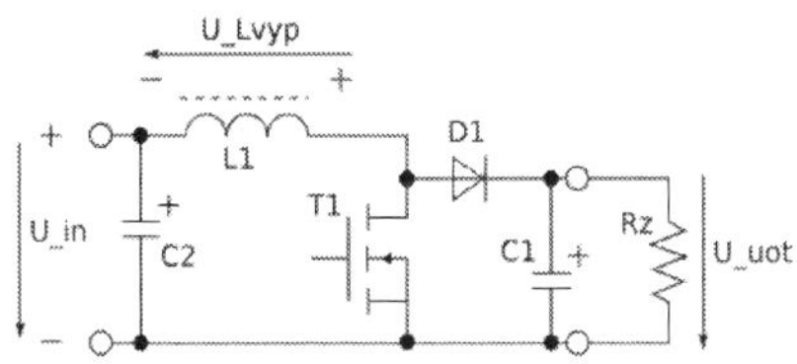

Fig. 26. Increasing converter diagram.

When opening the transistor current flows from the source of input voltage U_in and coil L1 to the load. Coil L1 keeps the flow direction of its current, but it reverses its polarity of voltage (passes from the appliance mode to the source mode), its induced voltage U_Lvyp adds up to voltage of the supply unit U_in and diode D1 passes to conducting state. The coil now behaves like a source series-connected to the source of supply voltage U_in. In this activity phase, current flows to the load and output condenser from these series-connected voltage sources. That is, Condenser C1 is recharged by current I_vyp and voltage increases on it as well as Load Rz. Diode D1 prevents discharging of Condenser C1 through the transistor in the case of the transistor being closed. If the output voltage is determined by the level of accumulator charging, input voltage on Condenser C2 can be reduced through the pulse control of Transistor T1, so that it meets just the optimum operating voltage of the photovoltaic generator. The benefits of this increasing connection are as follows: simplicity, low price, ability to reach voltage increase on the output as against the input voltage without the necessity to use the transformer.
In order to verify the function of optimization, the increasing converter connection has been carried out, whose functional diagram is shown in Figure 26. The complete optimization connection consists of the power and control parts (see Figure 27). The power part is made up of an increasing converter which is connected between the photovoltaic cell and a pair of 12V series-connected accumulators. The control part is supplied directly from PV cell and comprises a processor and two magnifiers. The processor used (PIC16F877) is an 8-bit RISC microcontroller, time controlled from a 20MHz crystal. The processor contains many modules, of which only A/D converter and the module for pulse-width modulation (PWM) are used, the latter being set for 78 KHz frequency at 8bits differentiation (also 39 kHz setting is being tested at 9bits differentiation). A/D converter has a supply of external reference voltage added. The value of charging current supplied to A/D converter, goes over the magnifier and the filter. The process has an array of 8 diodes connected, telling us which pulse-ratio PWM is just using. PWM signal controlling the increasing converters goes over the digital magnifier.

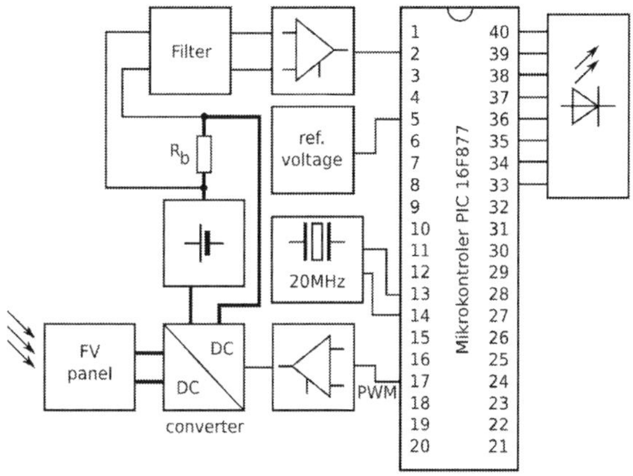

Fig. 27. Optimization circuit.

4. Measurement system for long-term measurement, archiving, processing and visualization of data

In order to verify the above described conditions and for other experimental purposes, a "measurement" system has been set up at the Department of electrical power engineering of BUT, Brno, which records, further processes and displays the measured data. Viewing of data and a great amount of other relevant information is effected through the Internet interface on the Department's site at http://www.ueen.feec.vutbr.cz/laboratory-of-unconventional-energy-conversion.

Unfortunately, the system is currently out of operation because of its planned relocation and mainly its upgrading to a more advanced version; that is why online display of values is not available at present, still viewing of values measured in the past is functional.

Online measurement alone is taken on a PV panel by Solartec (SP17-100, P_{mpp} = 100W_p, U_{nom} = 24/12V, Max.sys.vol = 750W, S = 0.845m^2), installed on a roof terrace of Building U2 within the Dean's office precincts of Brno University of Technology, No. 53 Údolní Street, Brno. The measurement programme reads values from several sensors (see diagrams). In the first place intensity of solar radiation and immediate panel power output are read, the measured data of primary relevance to us, and air temperature and panel temperature are supplementary measured quantities. A temperature-sensitive element for measuring air temperature is installed in a protective "house", in order to eliminate the influence of airflow, rain and direct sunshine. PV panel temperature is another measured value. A temperature-sensitive element is actually placed in the middle of PV panel, on its upper (lighted) side.

For interest's sake we measure the temperature of the lower (unlit) side of the panel too, though these values are not included in the diagrams. Immediate power output of the panel is measured both as actual and optimized values. The actual power is measured in the standard PV panel - accumulator connection. All values are saved in the database and immediately displayed on the Internet.

Data flow through the system is approx. the following: measured quantities are transferred using a separator multiplexer to the measurement system formed by NI-6023E plug-in board. Signals from all the measurement sensors are brought to analogue inputs of the measurement board within a 180-second time loop. The board digitalizes the signals and transmits the raw data to the control programme (script in the Matlab). The script not only communicates with the measurement board thus controlling the whole measurement, furthermore it controls the operation of the optimizing circuit through the card's digital outputs. In order to archive the data in an organized way and further process it, the script sends it to MySQL database on the server and the whole loop is repeated. The data saved in the database can be further processed independently of the ongoing measurements, i.e. carry out all kinds of sorting and additional computations etc. The data arranged and prepared in this way can be presented on-line in practically real time.

In our case, we make several supplementary calculations. In relation to the measured powers, intensity of falling solar radiation and temperatures we further calculate the following:

- panel produced daily energy,
- energy gain from equipment optimization,
- cumulative and immediate efficiency of energy transformation,
- monthly sum of panel produced energy,
- monthly sum of energy gain from equipment optimization,
- monthly sum of cumulative and immediate efficiency of PV panel energy transformation,
- yearly sum of panel produced energy,
- total of panel produced energy.

4.1 Results of long-term measurements

This chapter presents several sample diagrams with measured or calculated values of daily characteristics and further diagrams representing already long-term results within time-periods of up to one year.

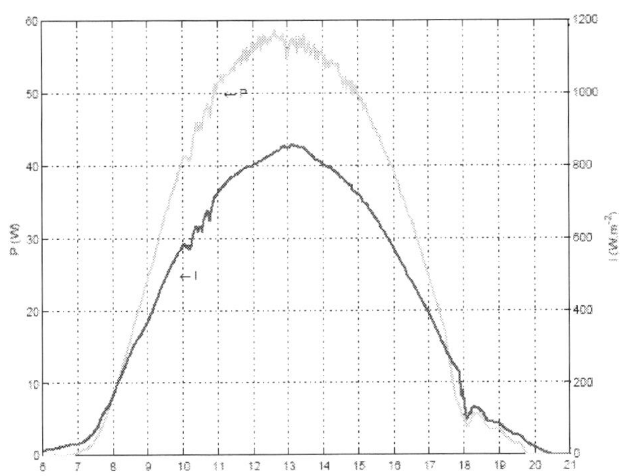

Fig. 28. Power characteristic of PV panel and intensity of falling radiation.

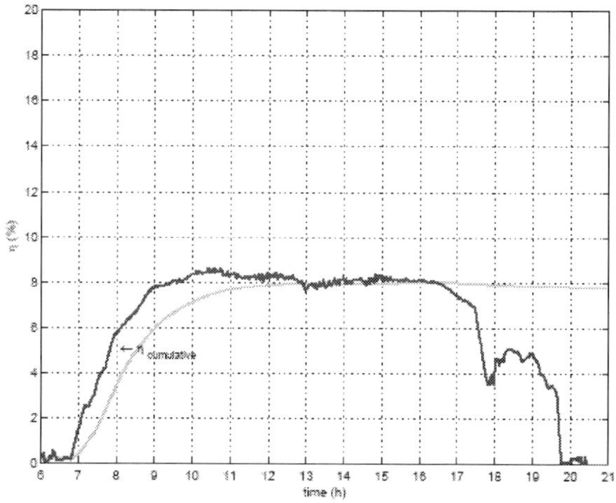

Fig. 29. Average and immediate efficiency of PV panel energy transformation.

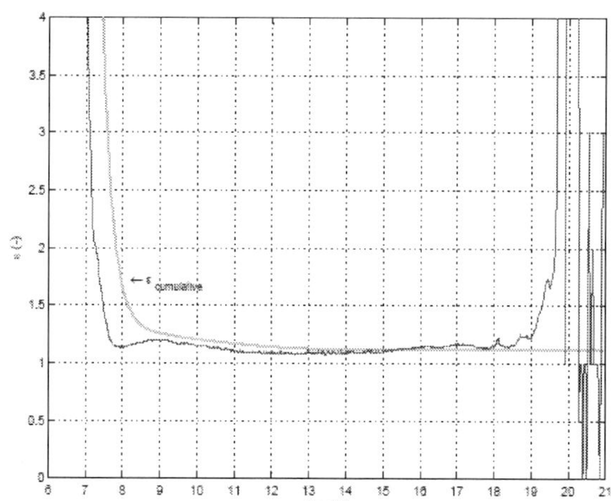

Fig. 30. Average and immediate gain from optimization of PV panel load.

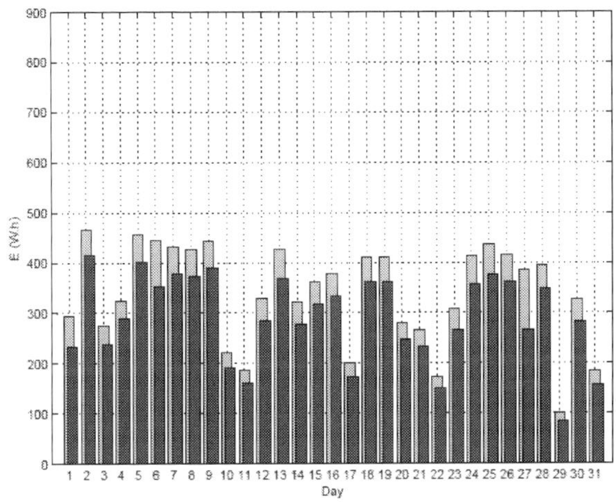

Fig. 31. Daily energy produced on individual days of a specific month.

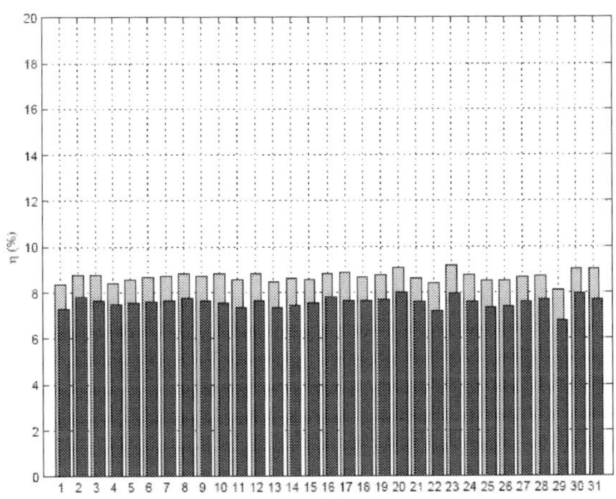

Fig. 32. Efficiency of PV panel energy transformation on individual days within a month.

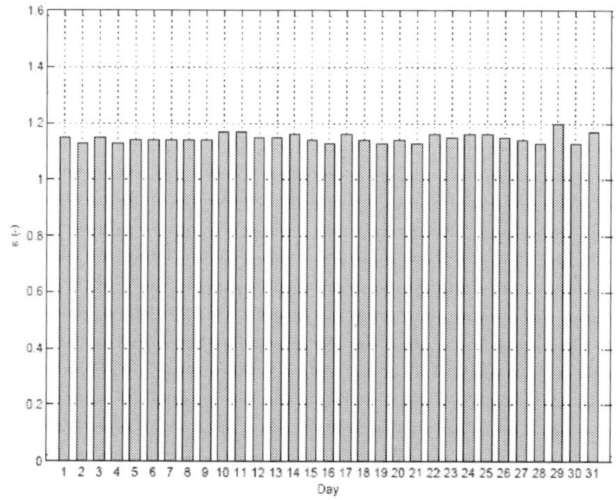

Fig. 33. Gain from optimization of PV panel load in individual days in a month.

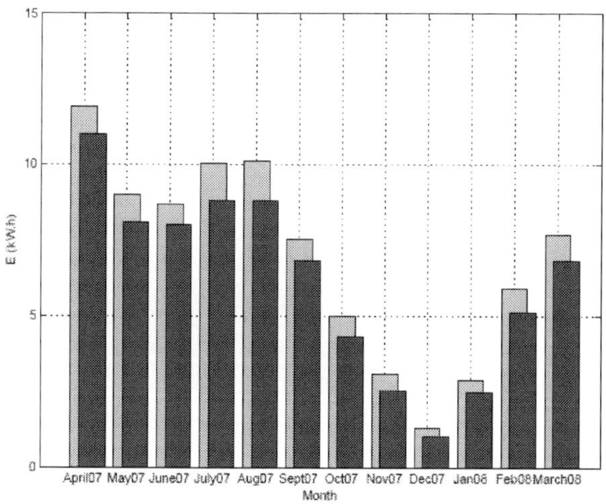

Fig. 34. Energy produced in individual months in a year.

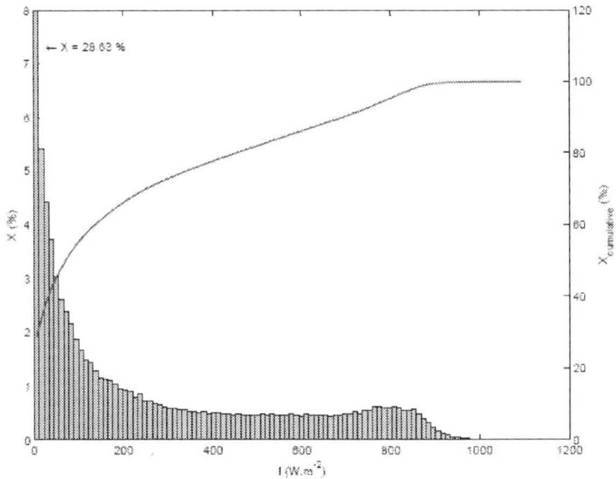

Fig. 35. Histogram of falling radiation intensity from all annual measured values.

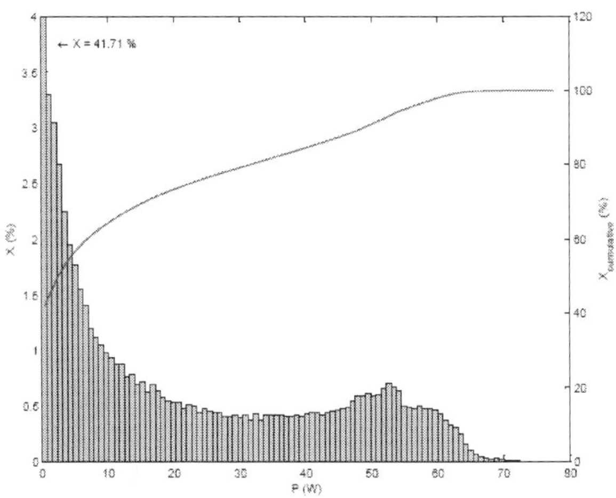

Fig. 36. Histogram of panel power from all annual measured values.

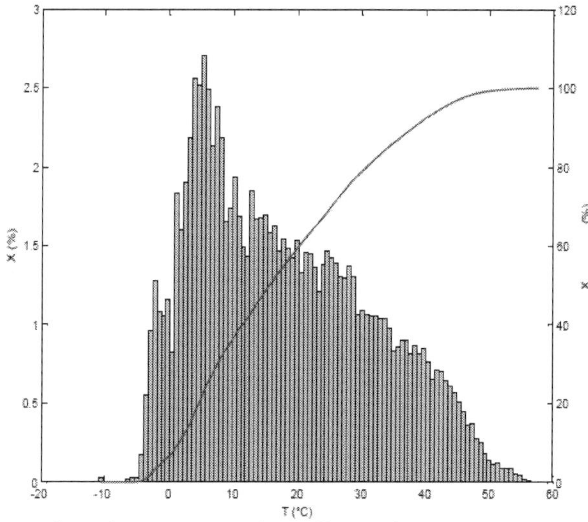

Fig. 37. Histogram of panel temperatures from all annual measured values.

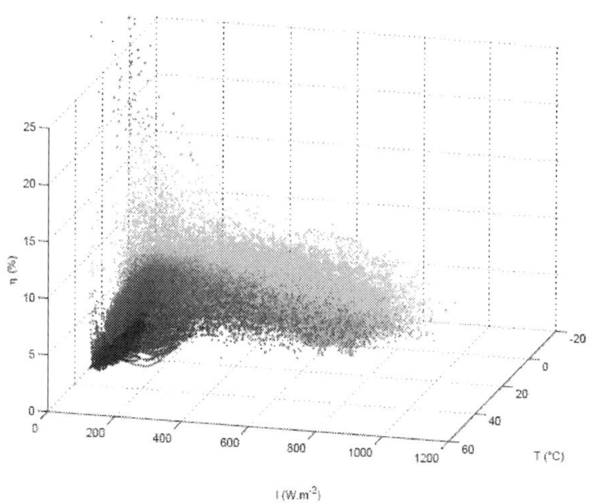

Fig. 38. Function of dependence of energy transformation efficiency on temperature and intensity of falling radiation.

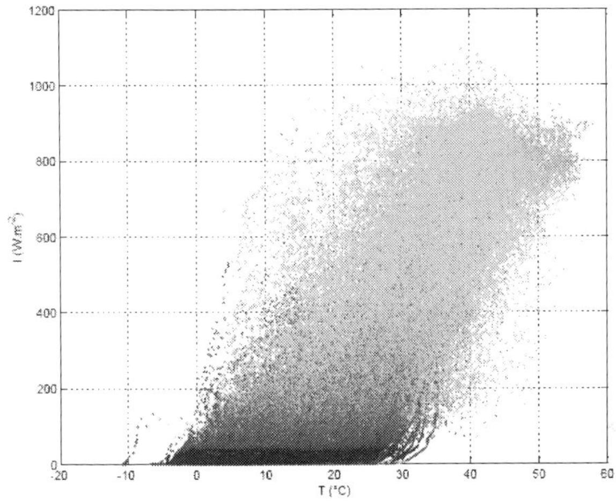

Fig. 39. Projection of previous dependence in 2D x / y axes. Dependence of temperature on intensity of falling radiation.

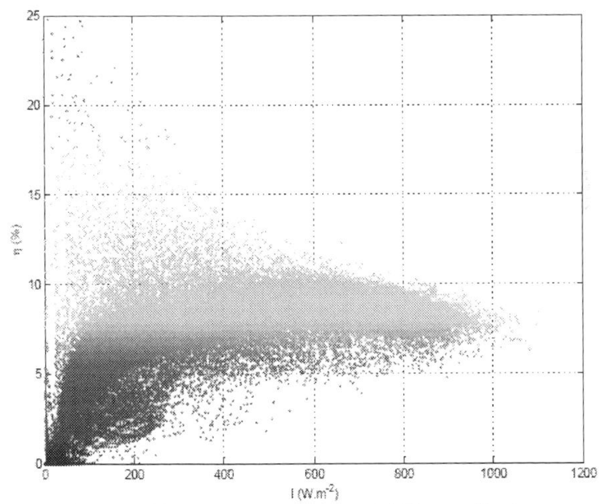

Fig. 40. Dependence of panel efficiency on intensity of falling radiation.

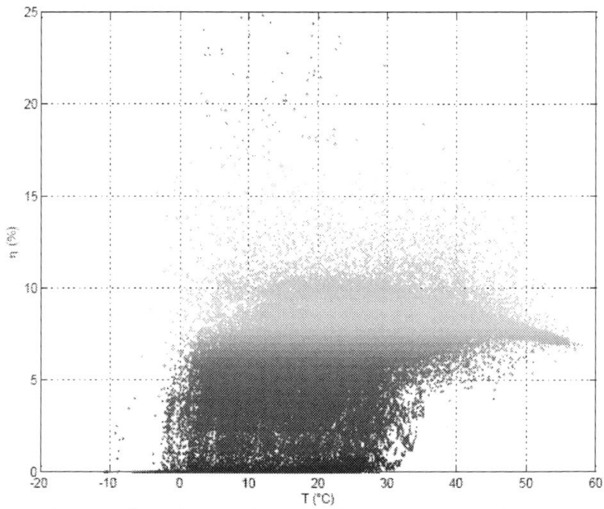

Fig. 41. Last dependence is the relation of panel efficiency and panel temperature.

The included diagrams present an outcome of the statistical analysis of long-term measurement, as could be possibly conducted for such measurements taken over long enough periods of time. Of course the more measurements will be processed within the analysis, the more relevant the results will be. Our system has been in operation, with several smaller breaks, since February 2007, which now represents two years of measurements, i.e. hundreds of thousands of values put into the database.

Besides the results presented there is a possibility of making more additional calculations and drawing further dependences e.g. within random representative time periods, for selected ambient temperatures and panel temperatures etc., regrettably there was no space left for them in this paper. Nonetheless, we are planning to place these and further assessments on the measurement system webpages, so this will comprise comprehensive measurements with not just immediate assessment and display of data, but also with the complete analysis of the whole measurement completed so far, conducted from the long-term point of view.

5. Conclusion

The article is aimed at making the reader aware of quite important problems relating to increasing the efficiency of solar panels, long-term monitoring of their operation and impact of changes in operating conditions on the load characteristics.

Today, disinformation is frequently spread in the community as well as among professionals, concerning particularly the efficiency of PV panels and their actual usability. Measurements like those described in this paper should contribute to making decisions on introducing photovoltaic converters and not only these, but the renewable energy sources in general, where their operation makes real sense.

Our measurement system applied in the analysis of operating conditions impact on the load characteristics of solar energy converters with optimizing their load, has now been in full operation for two and a half years - providing that its announced restructuring and upgrade have been completed by now - and is also presented on the Internet site of our laboratory (http://www.ueen.feec.vutbr.cz/laboratory-of-unconventional-energy-conversion), where the measured values are provided on-line. By doing so we are trying to verify the function of the system alone as well as that of the units constructed by us within long-term measurements. Great amounts of data, approx. thousands of values, allow us to make calculations for which long-term data is needed, such as monthly and yearly summaries.
An advantage of this system of measurement and collection of data is its simplicity and easy adaptation to other measurement tasks, which involves a simple modification of a Matlab control script, creation of a new database for storing measured values and creation of a webpage for presentation of measurement results.

6. Acknowledgements

This paper includes results of the research financed by the Ministry of Education, Youth and Sport of the Czech Republic within Project MSM0021630516.

7. References

Bařinka R. and Klimek P. Obnovitelné zdroje energie pro ČR - část Postupný rozvoj využití sluneční energie fotovoltaickou technologií ("Renewable energy sources for CR - Part Gradual Development of solar energy use by means of photovoltaic technology") [online]. Praha: ČEZ, 2007. 186 pp. Available from URL: <http://www.cez.cz/cs/vzdelavani/pro-profesionaly/odborne-publikace/8.html>.

Chen YM, Lee CH, Wu HC. Calculation of the optimum installation angle for fixed solar-cell panels based on the genetic algorithm and the simulated annealing method. IEEE Trans Energy Converse 2005;20(2):467–73.

Chan DSH, Phillips JR, Phang JC. A comparative study of extraction methods for solar cell model parameters. Solid State Electron 1986;29:329–37.

Dobiáš, P. Fotovoltaika v městských aglomeracích. ("Photovoltaics in conurbations") Brno: Brno University of Technology, Faculty of electrical engineering and communication technologies. Department of electrical power engineering, 2008. 90 pp., Appendices 13 pp. Bachelor's degree thesis supervisor: Ing. Petr Mastný, Ph.D.

Limbra M. and Poulek V. Solární energie (Solar Energy) [book]. Praha: ČZU, 2006. xxx s ISBN: 80-213-1488-5.

Limbra M. and Poulek V. Solární energie (Solar Energy) [book]. Praha: ČZU, 2005. 122 pp. ISBN: 80-213-1335-8.

Luque A, Hegedus S. Handbook of photovoltaic science and engineering. Berlin: John Wiley & Sons Ltd.; 2002. pp. 87–111.

Markvart T. Solar electricity. 2nd ed. Berlin: John Wiley & Sons Ltd.; 2000. p. 5–18.

Murtinger K., Beranovský J. and Tomeš M. Fotovoltaika. Elektřina ze slunce ("Photovoltaics. Electricity from the Sun.") [book]. Brno: ERA Publishers, 2007. 100 pp. ISBN: 978-80-7366-100-7.

Tian Pau Chang, Output energy of a photovoltaic module mounted on a single-axis tracking system, Applied Energy, Volume 86, Issue 10, October 2008, Pages 2071-2078, ISSN 0306-2619,DOI: 10.1016/j.apenergy.2009.02.006. (http://www.sciencedirect.com/science/article/B6V1T-4VT5CRD-3/2/0fa247e9cc03116bc5ef8b87fdd79930)

U.S. Department of Energy - Energy Efficiency and Renewable Energy, Solar Energy Technologies Program PV Connected to the Utility Grid [online]. 2006, last update 5 January 2006 [quot. 24 March 2008]. Available from URL: <http://www1.eere.energy.gov/solar/grid connect.html>.

The use of Switched Reluctance Generator in Wind Energy Applications

Eleonora Darie*, Costin Cepişcă** and Emanuel Darie***
*Technical University of Civil Engineering Bucharest, Electrotechnical Department
** University Polytechnic of Bucharest, Electrotechnical Department
***Police Academy Bucharest, Engineering Department
Romania

1. Introduction

Climate change is a contemporary issue. World are concerned about protecting the environment. Energy savings are a main strategy to get good results in the efforts to preserve the planet. Therefore, is required to develop and use new technologies. Wind power is a recent renewable primary energy source that must be considered in the future global energy. Wind power is one of the renewable energy power sources what help in reducing the carbon dioxide from the atmosphere. Various schemes for generating electricity from the wind have been proposed (Rim & Krishan, 1994). Wind power requires some special electric generator to work under continuously variable speed. In this context, the Switched Reluctance Generator (SRG) is investigated for to maximize wind power efficiency. This machine is robust, easy to construct and to maintain. In many aspects this kind of machine, now available, works better than others like the asynchronous machine. The absence of permanent magnets in the stator and windings in the rotor; low manufacturing costs; low maintenance; robustness; reliability; high efficiency; increased power density are some advantages of SRG side by side with a large range of operational speed. Despite they are somewhat noisy and its control is essentially non-linear the SRG was largely benefited with the recent advances on power electronics and micro processing. Torque, current and voltage ripples are inherent to the SRG, but the modern power electronic minimized these disadvantages. The principle of operation of this machine is known since the beginning of the electrical machines development. Soon the driving of these machines proved to be somewhat complex (Gail & Hansen, 2006). The advances on power electronics and micro processing in the last decade renewed the expectancies. In fact, modern power electronic converters associated with micro processed control hardware have brought competitiveness to SRG driven systems, allowing their efficient and reliable use. The SRG has shown great developing potential and study value in the area of wind power generation (Torrey, 2002). The SRG is a synchronous generator with a doubly salient construction, with salient poles on both the stator and the rotor. Excitation of the magnetic field is provided by the stator current in the same way as it is provided for the induction generator. The SRG is considered inferior to the Permanent Magnet Synchronous Generator

(PMSG) because of its lower power density (Darie & et al., 2007). The SRG requires a full-scale power converter in order to operate as a grid-connected generator. Moreover, the SRG has a lower efficiency than a PMSG and a lower power factor than asynchronous generators (Hansen, 2001). Through these advantages have already been confirming, some aspects of these machines must be mentioning here: there is a strong magnetic discontinuity providing current, voltage and torque ripples that should be properly controlling. The power electronics requirements to control a SRG are sometimes quoting as disadvantages of this kind of machine. Furthermore, they are a little noisy. Due to its advantages, SRG are considering as a special generator for wind power (Pan Zai-Ping et al. 2003).
The switched reluctance machine SRM is an unsuccessful old idea now renewed by recent power electronics and microprocessors developments. It is a doubly salient pole and works as a motor or as a generator depending on the firing angles (Tadashi, 2001).

2. Linear mathematical model of SRG

Figure 1 shows the variation of inductance with rotor position for one phase winding, idealized in that magnetic saturation and the rounding effect of the fringing fields are neglected (Miller et al., 1990). According to the above suppositions, the mathematical model of SRG is ideal and linear (Gail & Hansen, 2006).

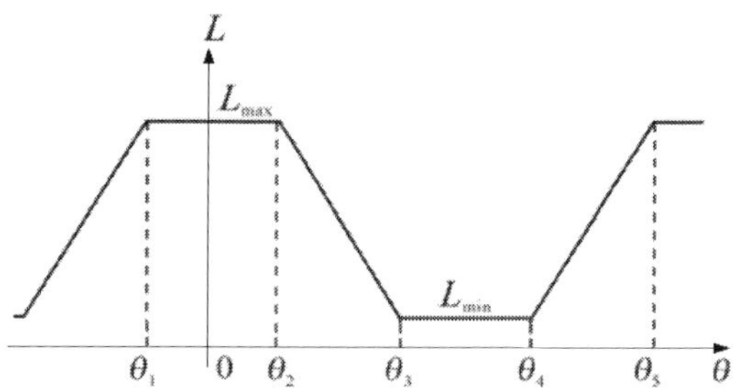

Fig. 1. The relationship between inductance and the position of the rotor in ideal and linear mathematical model.

The iron cores of the stator and the rotor in SRG are all protruding poles. The distribution of magnetic field is different when the relative position between rotor pole and the electrified phase on the stator is different. So the winding inductance L will change along with the change of the relative position between rotor pole and stator pole. When the rotor is turning, the inductance of windings will change from the maximum L_{max} to the minimum L_{min} periodically (Pan Zai-Ping et al. 2003). The inductance reaches maximum when the axes of rotor and stator pole are in the same position and reaches minimum when the axis of the rotor pole and the center of the stator pole are in the same position. The relationship between inductance $L(\theta)$ and the position of the rotor θ can be shown through the following function (Chen et al., 2001):

$$L(\theta) = \begin{cases} L_{\max} - K(\theta_1 - \theta) & \theta_{on} \le \theta < \theta_2 \\ L_{\max} & \theta_1 \le \theta < \theta_2 \\ L_{\max} - K(\theta - \theta_2) & \theta_2 \le \theta < \theta_3 \\ L_{\min} & \theta_3 \le \theta < \theta \\ L_{\max} - K(\theta - \theta_4) & \theta_4 \le \theta < \theta_5 \end{cases}. \tag{1}$$

Among the functions, θ_{on} is the turn-on angle $K = (L_{\max} - L_{\min})/(\theta_3 - \theta_2)$.
Suppose $\theta_{on} \le \theta_1, \theta_2 \le \theta_{off} \le \theta_3, \theta_4 \le (2\theta_{off} - \theta_{on}) \le \theta_5$.

The change of phase current can be divided into six phases: (1) Starting phase: $\theta_{on} \le \theta \le \theta_1$; (2) Rising linearly phase: $\theta_1 \le \theta \le \theta_2$;(3) Rising continuously phase: $\theta_2 \le \theta \le \theta_{off}$; (4) Generating phase: $\theta_{off} \le \theta \le \theta_3$; (5) Falling linearly phase: $\theta_3 \le \theta \le \theta_4$; Falling continuously phase: $\theta_4 \le \theta \le \theta_5$.

The six phases of the phase current's change can be shown by uniform function below (Pan Zai-Ping et al. 2003):

$$i(\theta) = \begin{cases} \dfrac{u}{\omega L(\theta)}(\theta - \theta_{on}) & \theta_{on} \le \theta \le \theta_{off} \\ \dfrac{u}{\omega L(\theta)}(2\theta_{off} - \theta_{on} - \theta) & \theta_{off} \le \theta < (2\theta_{off} - \theta_{on}) \end{cases}. \tag{2}$$

The analytic function of the flux ($\psi(\theta) = L(\theta)i(\theta)$) can be obtained as:

$$\psi(\theta) = \begin{cases} \dfrac{u}{\omega}(\theta - \theta_{on}) & \theta_{on} \le \theta \le \theta_{off} \\ \dfrac{u}{\omega}(2\theta_{off} - \theta_{on} - \theta) & \theta_{off} \le \theta < (2\theta_{off} - \theta_{on}) \end{cases}. \tag{3}$$

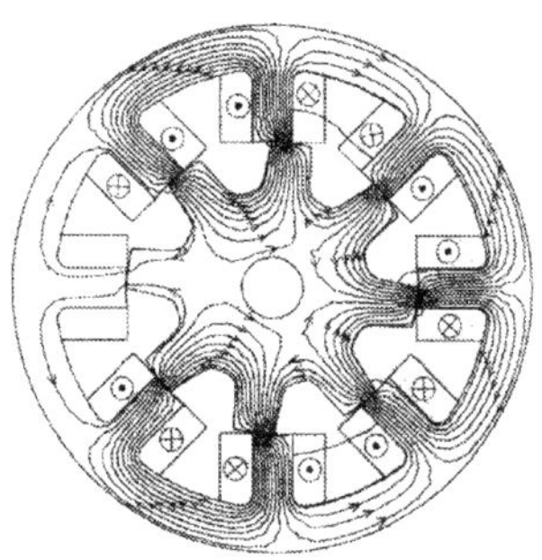

Fig. 2. Flux plots when one of eight coils is unexcited.

Figure 2 shows the flux distribution when one of eight coils is unexcited. Under this condition, unbalanced lateral forces will produce, and this may lead to mechanical failure.

3. Analysis of energy

Neglecting the resistance loss, medium loss and mechanical loss, the following equation can be obtained:

$$\pm ui = Li\frac{di}{dt} + i^2\frac{dL}{d\theta}\omega \tag{4}$$

For generating state, the input electrical energy and the output mechanical energy are all negative, which means mechanical energy is converted into electrical energy. Based on the linear mathematical models of phase current and flux established above, the change of magnetic field energy state can be drown as shown in Figure 3.

The turn-on angle in Figure 3, $\theta_{on} = \theta_1$ and the process of the magnetic field energy's variation in one period can been seen very clearly in the figure. The change of magnetic field energy in one period is zero and magnetic field just act as medium in the whole energy changing process. Thereby, the variation of mechanical energy and electrical energy in one period can be known through reviewing the variation of the magnetic field energy (Pan Zai-Ping et al. 2003).

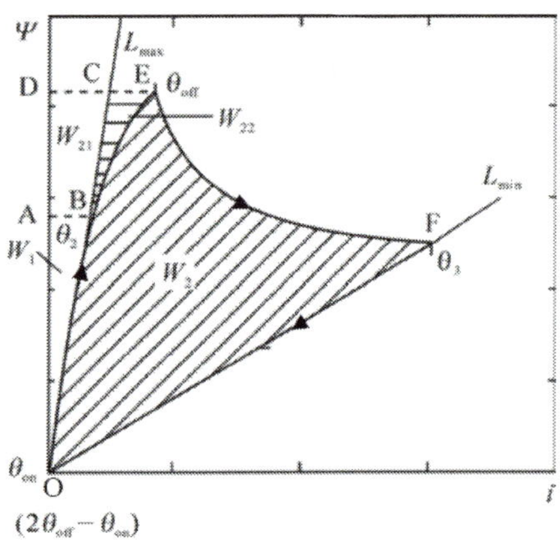

Fig. 3. The change of magnetic field energy state.

As shown in Figure 3, electrical source electrifies from θ_{on} to θ_{off} in one period. From θ_1 to θ_2, the whole input electrical energy from the source is converted into magnetic field energy W_1 and stored in the magnetic field because $dL/d\theta=0$ and its magnitude is equal to the area of OAB. From θ_2 to θ_{off} the input electrical energy from the source W_{21} (the area of ABCD) and the mechanical energy W_{22} (the area of BCE) are converted into magnetic field energy, because $dL/d\theta<0$. The phase current is attenuates to zero at: 2 θ_{off}- θ_{on} (Pan Zai-Ping et al. 2003).

In this process, from θ_{off} to θ_3, $dL/d\theta<0$, the mechanical energy is converted into magnetic field energy constantly. From θ_3 to $2\theta_{off}$-θ_{on} $dL/d\theta=0$, there is no input of mechanical energy and all the magnetic field energy is converted into electrical energy for output, which is shown as W_2 (the area of OBEFO) in Figure 3. The whole magnetic field energy converted

from mechanical energy in one period is: W_2+W_{22}, which is the effective electromagnetic energy (the area of OBCEFO). Enhancing the generating ability of the generator should start with enhancing the effective electromagnetic energy (Pan Zai-Ping et al. 2003).
The way for enhancing the effective electromagnetic energy, should be considered in two parts below: (1) With increasing the maximum inductance or decreasing the minimum inductance, the area of W_2+W_{22}, will be increased; (2) The main way to improve the generation ability is to adjust the turn-on and turn-off angle (Pan Zai-Ping et al. 2003).

4. Mode of operation by SRG

According to the existing research, most of the small wind turbine generators use a permanent magnet machine which has a cogging torque (T_c) due to the existence of a permanent magnet. Cogging torque is the force that created between permanent magnet (PM) and a metal due to the PM characteristics (Lobato & Pires, 2003).

4.1 SRG Characteristics

In electrical drives with variable reluctance (Figure 4), the torque is function of the regular position of the rotor due the double salient poles. The operation of the machine as a generator is obtained by energizing the windings of the stator when the salient poles of the rotor are away from their aligned position due to the rotating motion of the prime mover.

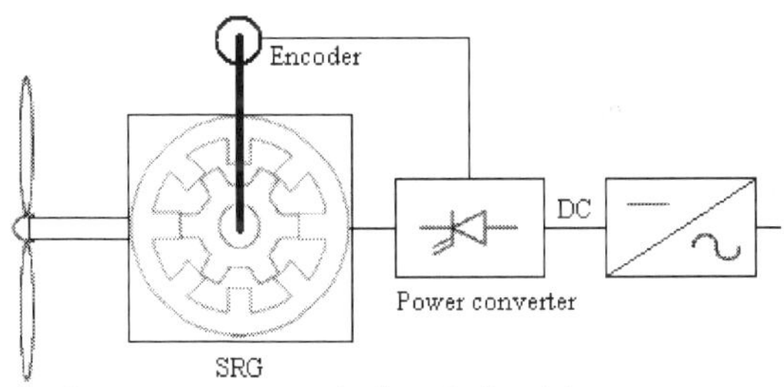

Fig. 4. The Switched reluctance generator in the wind turbine.

The SRG is characterized by the mode of controlling its phase current. For this problem the power electronic converter is used, which functions in a way that the phase currents of the machine are imposed for certain positions of the rotor. In this work is used the standard topology of the converter usually applied in SRM drives, given that it provides a greater flexibility regarding its control and better fault tolerance. The control system of this converter must regulate the magnitude and even the wave shapes of the phase currents to fulfill the requirements of torque and output power available and to ensure safe operation of the generator. This implies that the electronic switches associated with the controller are fully controlled devices. The topology (Figure 5) used power transistors (IGBT or MOSFET) that work as electronic switches. The capacitor shown in this topology prevents fluctuations in the voltage Vs. If losses are neglected the output energy over each stroke exceeds the excitation by the mechanical energy supplied (Hansen, 2001). On considers that there is no magnetic saturation and each phase is magnetically independent from others.

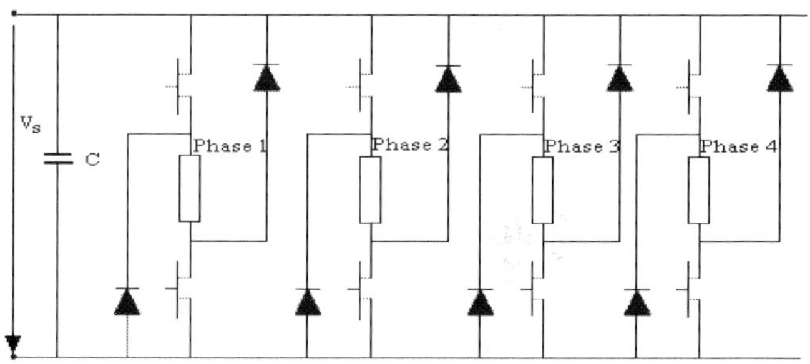

Fig. 5. The Switched reluctance generator in the wind turbine.

The SRM is characterized by the mode of controlling its phase current. For this problem the power electronic converter is used, which functions in a way that the phase currents of the machine are imposed for certain positions of the rotor. In this work is used the standard topology of the converter usually applied in SRM drives, given that it provides a greater flexibility regarding its control and better fault tolerance.

If losses are neglected the output energy over each stroke exceeds the excitation by the mechanical energy supplied (Hansen, 2001). On considers that there is no magnetic saturation and each phase is magnetically independent from others.

In these terms, the expression of the instantaneous power, p, available in the SRG is expressed as follow:

$$p(\theta, i_1, i_2, \cdots, i_n) = \frac{1}{2}\left[\sum_{j=1}^{n} \frac{dL_j(\theta)}{d\theta} i_j^2\right]\omega \, , \tag{5}$$

where: n - the number of phases; j – the phase number; θ – rotor position; ω – rotor speed; i_j – the current phase, $Lj(\theta)$ – the inductance of phase j as the function of θ.

The average of power available P, resulting from the operation of the machine as a generator, is (with excluding the losses) equal to the mechanical power. The values can be obtained from the expression of the average value of the torque T_m using (6) and (7):

$$P = T_m \, \omega \, , \tag{6}$$

$$T_m = \frac{N_r}{2\pi} \int_0^{2\pi / N_r} \left[\sum_{j=1}^{n} \frac{1}{2} \frac{dL_j}{d\theta} i_j^2\right] d\theta \, , \tag{7}$$

where: N_r is the number or rotor poles.

The above equations enable us to infer that the obtained power is approximately constant and it reaches a maximum when the dwell angle is located, in the descending section of the phase inductance profile, which corresponds to the highest average torque (Akhmatov et al., 2001), (Henao & Bassily, 1997).

For this type of machines the torque ripple appears mainly in the commutation zones related with the sequential process of establishing and removing the phase currents.

The imposition of phase current waveform using the current control with an adjusted hysteresis band and a sufficient input voltage, allow the torque ripple reduction.
In this way the ripple can be minimized, thus controlling the phase's currents commutation precisely phased relative to the rotor position. For that effect, the current control is done is done using the trapezoidal phase reference torque model (Grauers, 1997), two adjacent phases can be supplied at the same time to ensure the continuity in the generated torque.
The SRM is capable of operating continuously as a generator by keeping the dwell angle so that the bulk of the winding conduction period comes after the aligned position, when $\frac{dL_j}{d\theta} < 0$. The waveforms of the phases reference current i_j^*, results from the desired torque T^* and is calculated by the following equation (Darie & et al., 2009, 2008):

$$i_j^* = \sqrt{\frac{2T_j^*}{\frac{dL_j(\theta)}{d\theta}}} \quad (8)$$

and are themselves the reference signals to be treated using the feedback pulse with modulation (PWM) with adjusted hysteresis band.

4.2 SRG – The Current Control

The block diagram from the Figure 6, indicates the current control with the torque reference applied to the 8/6 SRG.

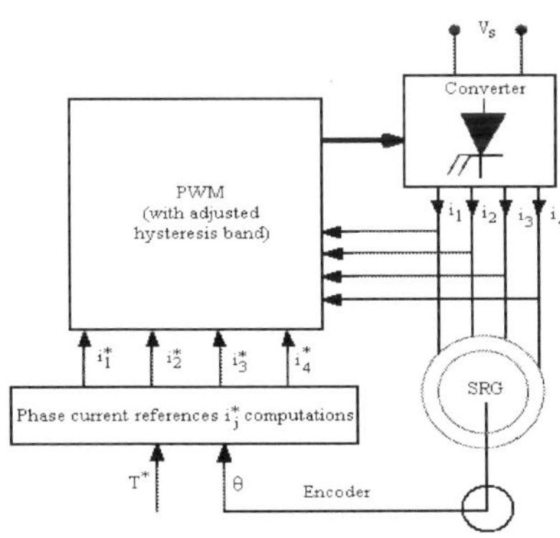

Fig. 6. The current control with the torque reference applied to 8/6 SRG.

The waveforms of the reference currents, i_1^*, i_2^*, i_3^* and i_4^*, on calculated using the trapezoidal model torque associated to each phase, T_1^*, T_2^*, T_3^* and T_4^*.

4.3 SRG – Simulations

On used for simulations an 8/6 SRG, with P_n=2.4 kW, 4 phase. In these simulation examples of the SRG operation, the converter voltage used was V_s=800 V, which allow reduced torque ripple and the rotor speed is 1000 rpm.

Figure 7 shows the phase current resulting from the trapezoidal phase torque.

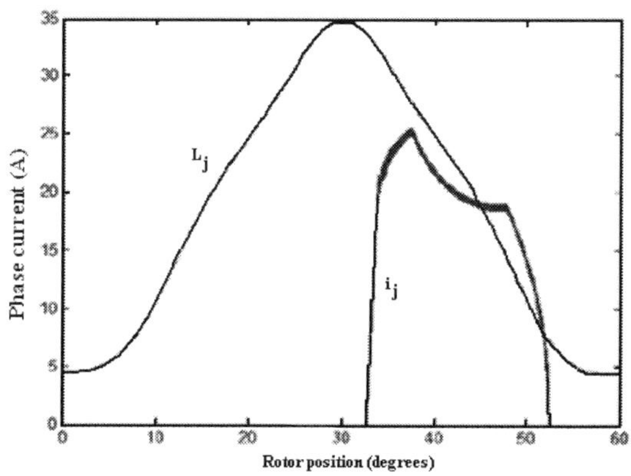

Fig. 7. Phase Current.

In Figure 8, is indicated the total instantaneous torque for 8/6 SRG.

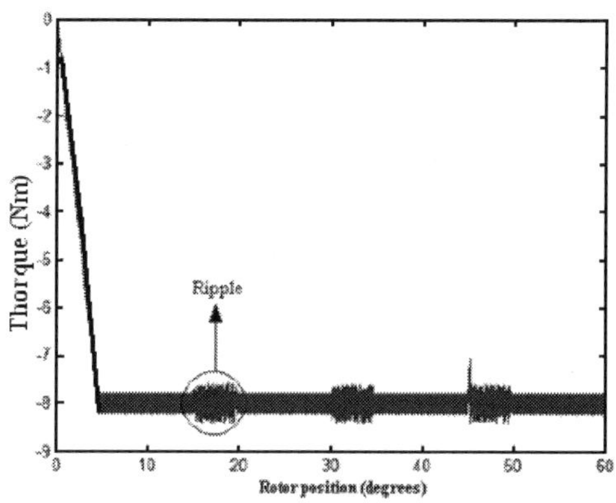

Fig. 8. Total Torque.

In order to achieve higher performance in SRG operation and higher efficiency in the conversion on includes optimal dwell angle control to further reduce the torque ripple.

5. Energy Conversion

In a SRG mechanical power achieved from a prime mover through a shaft is converted into electrical power. When a pole of the rotor is aligned with the excited pole of the stator, there is a state of stable equilibrium. Thus, in the SRG there is a natural tendency to align the rotor and the stator active poles in order to maximize the inductance of the phase. When an external mechanical agent forces the rotor to leave the stable equilibrium position, the electromagnetic torque produced results in a back electromotive force that increases the applied voltage. In this way the machine generates electrical power.
The electrical equation for a phase of the SRG is:

$$U = Ri + L\frac{di}{dt} + E \cdot \tag{9}$$

Where the back electromotive force EMF is given by equation (10):

$$E = i\omega\frac{\partial L}{\partial \theta}, \tag{10}$$

where:

$$\omega = \frac{d\theta}{dt}.$$

The stator is fed in DC. As ω and i are both positive, the sign of E is the same as that of $\partial L/\partial\theta$. From the equation (2), it can be seen that when $(\partial L/\partial\theta)>0$. In this case, electrical power is converted to mechanical power and the machine works as a motor. But when $(\partial L/\partial\theta)<0$ the back electromotive force is negative and it increases the current converting mechanical power into electrical power.
In (9) and (10), U - applied voltage; i - phase current; R - phase resistance, L - phase inductance; E - back electromotive force; ω - rotor angular speed; θ - rotor angular position; t - time.
The dynamic mechanical equation for the SRG is given by (11). It is to be noted that the electromagnetic torque C_{emag} comes as a negative quantity, acting against the rotor mechanical speed.

$$C_m + C_{emag} - J\frac{d\omega}{dt} - D\omega = 0, \tag{11}$$

where: C_m - applied mechanical torque, J - moment of inertia; D - coefficient of friction.
The co-energy of a phase of this machine is given by follow:

$$W^{co} = \int_0^i \lambda \, di \cdot \tag{12}$$

The corresponding electromagnetic torque for a phase SRG is given by:

$$C_{emag} = \frac{\partial W_a^{co}}{\partial \theta} + \frac{\partial W_b^{co}}{\partial \theta} + \frac{\partial W_c^{co}}{\partial \theta}. \tag{13}$$

The mathematical model of 6x4 SRG is shown below:

$$\begin{bmatrix} U_a \\ U_b \\ U_c \\ C_m \\ 0 \end{bmatrix} = \begin{bmatrix} R_a & 0 & 0 & 0 & 0 \\ 0 & R_b & 0 & 0 & 0 \\ 0 & 0 & R_c & 0 & 0 \\ -\frac{r_a}{i_a} & -\frac{r_b}{i_b} & -\frac{r_c}{i_c} & D & 0 \\ 0 & 0 & 0 & -1 & 0 \end{bmatrix} \cdot \begin{bmatrix} i_a \\ i_b \\ i_c \\ \omega \\ \theta \end{bmatrix} + ,$$
$$+ \begin{bmatrix} L_a & 0 & 0 & 0 & i_a \frac{\partial L_a}{\partial \theta} \\ 0 & L_b & 0 & 0 & i_b \frac{\partial L_b}{\partial \theta} \\ 0 & 0 & L_c & 0 & i_c \frac{\partial L_c}{\partial \theta} \\ 0 & 0 & 0 & J & 0 \\ 0 & 0 & 0 & 0 & 1 \end{bmatrix} \cdot \begin{bmatrix} \dot{i}_a \\ \dot{i}_b \\ \dot{i}_c \\ \dot{\omega} \\ \dot{\theta} \end{bmatrix} \tag{14}$$

where:

$$r_a = \frac{\partial W_a^{co}}{\partial \theta}; \quad r_b = \frac{\partial W_b^{co}}{\partial \theta}; \quad r_c = \frac{\partial W_c^{co}}{\partial \theta}. \tag{15}$$

The matrix of states in equation (14) completely describes the dynamic behavior of the SRG.

6. Conversion methods of Wind energy

The wind is an intermittent and variable energy source both in magnitude and in direction. There are several components in wind speed (W_{wind}). The capture of the wind energy, in an efficient way, requires the existence of a constant wind flow sufficiently strong (Henao & Bassily, 1997).

Currently wind turbines are designed to achieve a maximum power at wind speeds above 10 m/s. However, they can be adjusted to the local wind profile.

The maximum theoretical efficiency for the wind to energy conversion is 59.3% (Betz's Limit). The effective efficiency conversion is given by the Power Coefficient (C_p), which is expressed by the following, where P_{mec} is the mechanical power of the turbine and P_w is the available wind power.

$$C_p = \frac{P_{mec}}{P_w}. \tag{16}$$

The power P_w is related with the wind speed V_w calculated by (17),

$$P_w = \frac{1}{2}\rho A V_w^3, \tag{17}$$

where ρ is the air density (ρ = 1.225 kg/m³) and A is the cross-sectional area of the turbine rotor.

When considering the generator efficiency (η), the output power is given by (18).

$$P_{out} = \frac{1}{2}\rho A V_w^3 (\eta C_p)\cdot \tag{18}$$

The power coefficient C_p is the fraction of the wind kinetic power that is captured by the wind turbine blades (Lobato & Pires, 2003). It is efficiency of the rotor. This coefficient changes from turbine to turbine and its value is given by (19).

The power coefficient C_p varies with the Speed Ratio (λ), given in (20).

$$C_p = 0{,}22\,\rho\, V_w^3 (\eta\, C_p), \tag{19}$$

$$\lambda = \frac{r\omega}{V_w}, \tag{20}$$

where: r is the rotor radius, ω is the rotor speed.

The low rotor speeds of the turbine bring about small turbulences in the air flow. With high speeds the turbine behaves as a wall for the wind. Therefore the priority is to adapt the wind speed to the rotor speed with the purpose of obtaining a greater conversion efficiency, which results in a maximum C_p (Gail &Hansen (2006).

The power coefficient C_p versus speed ratio, for a generic turbine is shows in Figure 9.

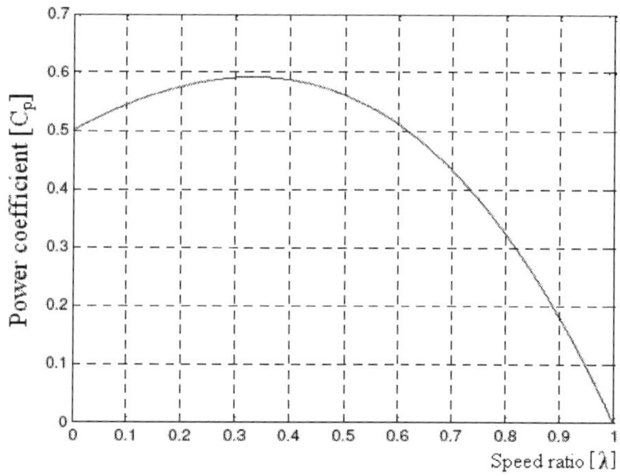

Fig. 9. Power coefficient versus speed ratio for a generic wind turbine.

It can be seen that this coefficient is maxim at around 0.59, then the maximum transfer of energy takes place with almost 60% of the initial value. For turbines of three blades and low speed, the efficiency of the rotor is between 0.2 and 0.4.

It is consensual appreciation that the wind speed in a certain site follows the Weibull probability distribution function like this:

$$p(v_i) = \left(\frac{k}{c}\right)\left(\frac{v_i}{c}\right)^{k-1} e^{-\left(\frac{v_i}{c}\right)^k}, \tag{21}$$

where $p(v_i)$ is the fraction of time where wind speed v_i and $v_i+\Delta v_i$, divided by Δv_i, c is a scale parameter and k is a shape parameter.

Generally $p(v_i)$ is expressed in hours per year per m/s. On most places c varies from 5 to 10 m/s and k varies between 1.5 and 2.5.

Figure 9 shows the curves of Weibull probability distribution for the shape factor k=2 where the scale parameter varies between 5m/s and 13 m/s.

Figure 10 shows that the wind speed is low most of the time. The rotor speed has the same behavior of the wind speed.

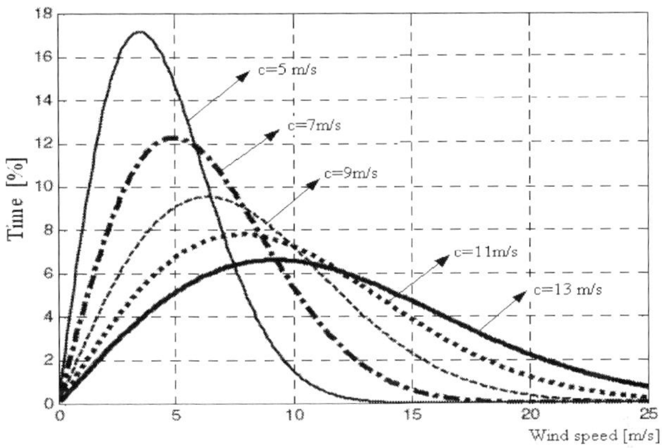

Fig. 10. Wind speed permanency curves.

A generation has to follow this profile should start with low wind speeds and increase of wind speed.

7. Wind system simulation

On presents two modes of mechanical coupling of the turbine to the generator: the direct coupling to the turbine shaft, direct - drive wind turbine (Figure 11) and the SRG coupling to the turbine shaft through a gearbox (Figure 13), (Gail & Hansen, 2006).

7.1 Turbine generator direct coupling

The rotor speed ω of approximately 100 rad/s is too high and not compatible for this type of wind turbines, in normal wind conditions.

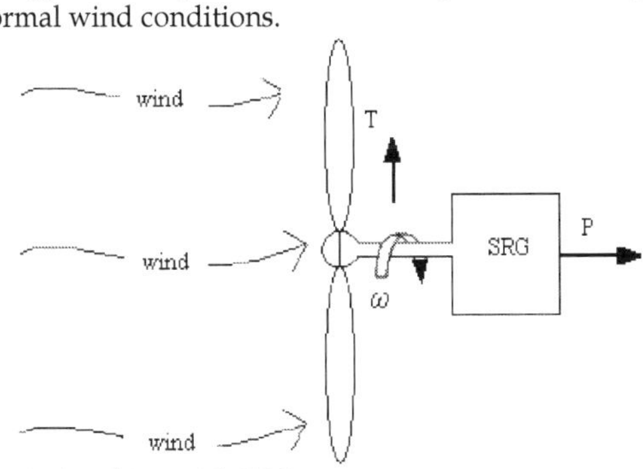

Fig. 11. Direct drive wind turbine with SRG.

Figure 12 shows the electric power generated by the machine coupled with this turbine, where its average power value corresponds to the power of the system excluding losses in the generator.

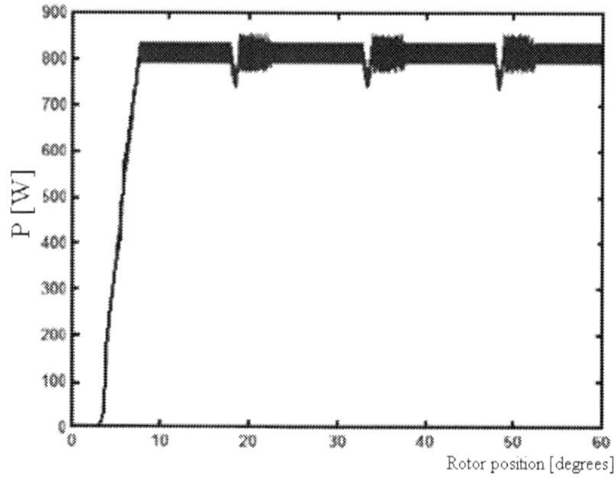

Fig. 12. The 4-phase SRG instantaneous power versus rotor position.

Associated with the required high rotor speed for the good performance of the SRG, the fact that the rotor diameter is small brings about the problem that the wind speed is not sufficient to overcome the combined turbine-generator inertia, namely at the starting stage (Lobato & Pires, 2004).

7.2 Indirect coupling with gearbox

Figure 13 indicates the SRG coupling to the turbine shaft through a gearbox.

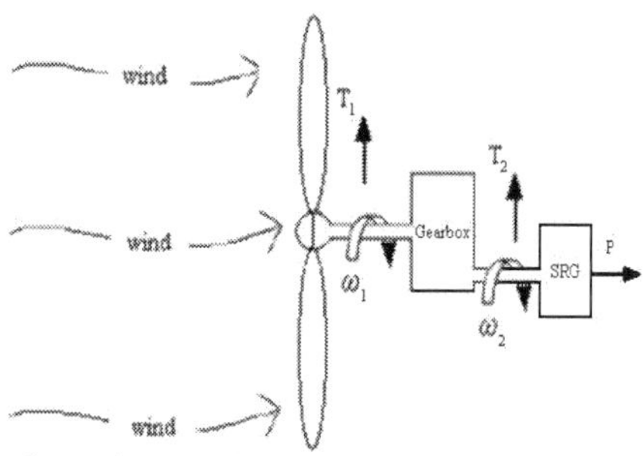

Fig. 13. SRG coupling to the turbine with gearbox.

Assuming that the losses in the gearbox are negligible, and given that the input and output power ($\omega_1 T_1 = \omega_2 T_2$), the transmission ratio rt, varies in the inverse of the torque's ratios:

$$r_t = \frac{\omega_1}{\omega_2} = \frac{T_2}{T_1}. \tag{22}$$

Figure 14 shows the behavior of the electric power generated by the machine, when coupled with a turbine having a rotor diameter of 5m for a constant wind speed of 8m/s.

With the gearbox the rotor speed of the turbine was reduced to less than half of the value obtained in the first case and the power available in the turbine is close to the rated power in the generator (Lobato & Pires, 2004).

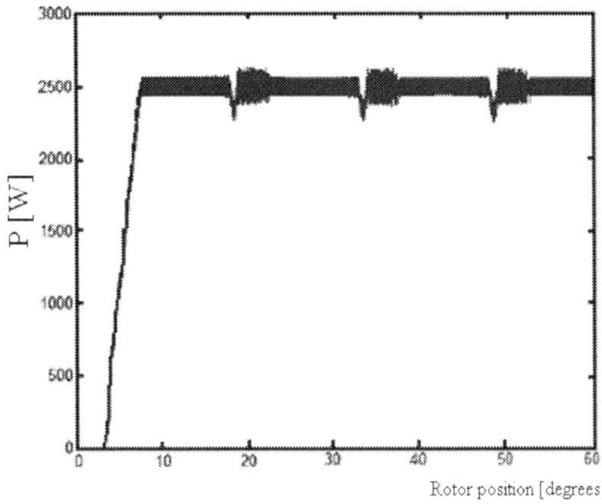

Fig. 14. The 8/6 SRG instantaneous power versus rotor position.

8. Conclusion

Considering this SRG is a low power machine and therefore produces low torques, it was expectable that too high rotor speeds would develop, which are hardly compatible with the wind speeds typical of this type of energy conversions and with the wind turbines available for these applications (Grauers, 1997).

9. References

Darie El., Cepişcă C., Darie Em. (2009). Advantages of Using a Switched Reluctance Generator (SRG) for Wind Energy Applications, *Proceedings of The International Conference on Ecological Vehicles and Renewable Energies, EVER'09,* Monaco, March 26-29, 2009.

Darie El., Cepişcă C., Darie Em. (2008). The use of Switched Reluctance Generator in wind applications, *Proceedings of the 13 th International Power Electronics and Motion Control Conference, EPE-PEMC 2008,* pp. 1986-1989, Poznań, Poland, September 1-3, 2008, ISBN: 978-1-4244-1742-1, IEEE Catalog Number CFP0834A-CDR.

Darie El., Darie Em., Tcacenco V. (2007) About the permanent magnet Synchronous Machine in wind power applications, *Proceedings of CIEM 2007, 3-rd International Conference on Energy and Environment CIEM 2007,* University Politechnica of Bucharest, November, 22-23 2007.

Gail G., Hansen A.D. (2006). Controller design and analysis of a variable speed wind turbine with doubly fed induction generator, *Proceedings of the European Wind Energy Conference,* pp. 500-508.

Lobato P., Pires A. J. (2004). Methodology based on energy-conversion diagrams to optimize switched reluctance generators control, *Proceedings of ICEM 2004*, pp. 700-705.

Chancharoensook P., Rahman M.F. (2003). Control of a Four-Phase Switched Reluctance Generator: Experimental Investigations, *Proceedings of IEMDC 03*, vol. 2, pp. 842-848.

Lobato P., Pires A. J. (2003). A New Control Strategy Based on Optimized Smooth-Torque Current Waveforms for Switched reluctance Motors. *Proceedings of Electromotion'03*, vol. 2, 610-615.

Pan Zai-Ping, Jin Ying, Zhang Hui (2003). Study on switched reluctance generator, *Journal of Zhejiang University*, pp. 594-602, ISSN 1009-3095.

Torrey D.A. (2002). Switched Reluctance Generator and their control, IEEE Trans. on Industrial Electronics, pp. 3-14, Vol. 49, No. 1.

Zhang H., Pan Z. P. (2003). *Nonlinear inductance mathematical model of Switched Reluctance Generator and its applications*. Proceedings of IEEE International Conference on Small and Medium Electric Machines, pp. 6-9, Vol. 30, No. 3.

Akhmatov V., Nielsen A. H. (2001). Variable speed wind turbines with multipole synchronous permanent magnet generators, *Proceedings of Wind Energineering*, vol. 27, no. 6, pp. 531-548.

Chen H., Mang C., Zhao X. (2001). Research on the Switched Reluctance Wind Generator System, Proceedings of IEEE International Conference on Systems, Man and Cybernetics, pp. 1936-1941, Tucson USA.

Hansen A. D. (2001). Wind models for predictions of power fluctuations from wind farms, *Proceedings of APCWE V*, no.89, pp. 9-18, 2001, Kyoto, Japan.

Tadashi S. (2001). The switched reluctance generator, Electronic Control of Switched Reluctance Generator, *Newness Power Engineering Series*, pp. 227-251, Ed. T. J. E. Miller, Oxford.

Henao H., Bassily E. (1997). A new control angle strategy for switched reluctance motor, *Proceedings of EPE '97*, vol.3, pp. 613-618, September 1997, Trondheim, Norway.

Grauers A. (1997). Efficiency of three wind energy generator systems, *IEEE Transactions on Energy Conversion*, vol. 11, no. 3, pp. 650-657.

Rim G., Krishnan R. (1994). Variable Speed Constant Frequency Power Conversion with A Switched Reluctance Machine, *Proceedings of Applied Power Electronics*, pp. 63-71, APEC'94, Orlando, USA.

Miller T.J.E., Stephenson J. M., MacMinn, S.R. (1990). Switched Reluctance Drives, *IEEE Tutorial*, IEEE 25 Th IAS Annual Meeting, Seattle, USA.

Tidal Energy Technologies: Currents, Wave and Offshore Wind Power in the United Kingdom, Europe and North America

T. J. Hammons
UK Sustainable and Renewable Energy Group, University of Glasgow, United Kingdom
T.Hammons@btinternet.com

1. Introduction

This Chapter examines renewable energy potential and sources of the oceans in Europe and North America: tidal, wave, underwater currents and wind. It considers ocean wave and tidal power projects in San Francisco, wave power technologies (oscillating water column, overtopping devices, float systems, and hinged contour devices), and cost. Feasibility assessments of offshore wave and tidal current production are described, and wave project results are given. U.S wave energy resources, feasibility definition study sites, feasibility study for wave energy conversion (WEC) devices, demonstration scale plant design, commercial-scale plant design, learning curves, economics, and recommendations are discussed. Recent progress in offshore energy technology development including wind-power is evaluated. Examined is the generation of electricity by tidal power in the UK with respect to complying with the Kyoto and the Bali Protocols. Approximately 40% of the UK's electricity will have to be generated from renewables (wind, tidal/wave, and plant energy) by 2020 as a result of legally binding EU target under the Bali Protocol. It is likely to mean a six-fold increase in the amount of onshore wind turbines and a 50-fold increase in the number of offshore wind turbines. This is because the 20% target for all renewables by 2020 applies to energy across the board, including transport and heating, where the scope for renewables is less, implying the electric sector must do more. By 2050, the UK is planning to reduce its CO_2 emissions by at least 60% compared with its emissions in 1990.

The Chapter focuses primarily on the proposed Severn Barrage considering potential benefits, conditions for sustainable development, energy policy context and compliance with environment legislation. UK tidal resource is reviewed: stream resource (that is KE contained in fast-flowing tidal currents), and tidal range resource (that refers to gravitation potential energy). The top tidal range and tidal stream sites in the UK with the resource (in TWh/year) are indicated.

2. Tidal Power of the Severn Estuary

Generating electricity from tidal range of the Severn Estuary has the potential to generate some 5% of UK electricity from a renewable indigenous resource. A study is underway to:

- Assess in broad terms the costs, benefits and impact of a project to generate power from the tidal range of the Severn Estuary, including environmental, social, regional, economic, and energy market impacts;
- Identify a single preferred tidal range project (which may be a single technology/location or a combination of these) from the number of options that have been proposed;
- Consider what measures the Government could put in place to bring forward a project that fulfils regulatory requirements, and the steps that are necessary to achieve this; and
- Decide, in the context of the Government's energy and climate change goals and the alternative options for achieving these, and after public consultation, whether the Government could support a tidal power project in the Severn Estuary and on what terms.

The study is expected to last roughly until 2010. Under consideration is tidal range, including barrages, lagoons and other technologies, and includes a Strategic Environmental Assessment of plans for generating electricity from the Severn Estuary tidal range to ensure a detailed understanding of its environmental resource recognising the nature conservation significance of the Estuary. The scheme would use proven technology of a hydroelectric dam but filled by the incoming tide rather than by water flowing downstream. The Severn Estuary has some of the best tidal potential in the world and could more than double the current UK supply of renewable electricity and contribute significantly to targets for renewable energy and CO_2 emissions reduction. The scheme would have a capacity of 8640 MW and produce roughly 17 TWh/year with a load factor of 0.22. Other development of similar potential under consideration in the UK is tidal stream power. Tidal Stream Projects worldwide include the Enermax Project (Italy), the Blue Energy Project (Canada), and the Gorlov Helical Turbine (GHT) (USA).

Discussed is the role of tidal power in the United Kingdom to reduce greenhouse gas emissions. Here, the proposed Severn Barrage is focused on considering potential benefits, conditions for sustainable development, energy policy context and compliance with environmental legislation. UK tidal resource is reviewed: stream resource (that is KE contained in fast-flowing tidal currents) and tidal range resources (that refers to gravitation potential energy). A feasibility study for tidal range in the Mersey Estuary and other schemes in the UK including Loughor Estuary (Wales), Duddon Estuary (Cumbrian Coast), Wyre Barrage (Lancashire, and Thames Barrage. is summarized. Also given is a strategic overview of the Severn Estuary resource, electric output and characteristics, carbon emissions (carbon payback and carbon reduction) and physical implications of a barrage.
The electricity transmission system in the UK in the Severn area is evaluated where system constraints and upgrades and implications of tidal power are considered. The awareness of energy sources (wind, solar, coal, nuclear, gas, tidal/wave and bio-energy) that can generate electricity in the UK is outlined.

Concerns on Environment Impact considering the protected status of the Severn Estuary (Habitats Directive and Nature 2000), the Birds Directive defining biodiversity objectives, habitats and ecology are considered. Potential carbon savings for the two Severn proposals are then reviewed.

A consensus view is given on tidal power in the UK (tidal stream long-term potential {policy improvements, strategic planning and consenting}, tidal lagoons, and tidal barrages). Conditions for a sustainable Severn barrage (energy policy context, ensuring public interest, apportionment of risks and benefits, avoiding short-termism, regional impacts and priorities) complying with environmental legislation (applying environmental limits and providing compensatory habitats) is given. The final decision on whether this project that will contribute to the UK fulfilling its greenhouse gas emission targets will be given the go-ahead is reviewed.

3. Renewable Energy Sources from the Oceans

Renewable energy sources from the oceans include offshore wind, wave energy, tidal energy, Ocean Thermal Energy Conversion (OTEC) and underwater currents. Harvesting ocean energy is not a new concept, yet it has remained a marginal resource. Today there is serious interest in offshore technology in Europe and Asia but funding for projects in power from the oceans at this time in USA is lacking. Wind farm technology has moved offshore where the prevailing winds can be more consistent and out of sight. Offshore wind energy is the fastest growing sector in renewable energy.

Areas of great tidal differences can produce regular and predictable tidal currents of 5 knots or more, creating large energy potentials. However, there are many areas with great tidal differences and only very slow currents. Tidal currents of 5 knots or more typically require a somewhat narrow passageway between a bay or estuary and the ocean and a somewhat shallow depth of the passageway. Tidal range, therefore, is not the only factor in speed of the tidal current. The Physics of Tidal Power is reviewed in Reference [1].

France has had a 240MW tidal power generating facility for 40 years. It is a tidal head plant and the technology is quite different [1].

Projects harnessing tidal currents have shifted toward capturing tidal-driven coastal currents. A study of 106 possible locations in the EU countries for tidal turbines showed that such sites could generate power of the order of 50 TWh per year. The power density of a marine current is approximately four times that for a wind generator, so this marine current resource is potentially large. However, there are fewer places in the world where marine currents are usable [1] compared with access to winds.

Wave energy can be considered as a concentrated form of solar energy. Winds are generated by the differential heating of the earth's surface, and, because they blow over large spans of water, part of their energy is converted into waves. The first commercial-scale wave power facility turning wave energy into compressed air was established in Scotland. Some proposed schemes involve hinged pontoons with hydraulics, while others appear like floating pistons that rise and fall with the wave action. Several prototype demonstrations are planned in the next few years. Growth in this sector is anticipated to reach $100 million per annum by 2010.

The difference in temperature between the surface waters and the deeper ocean waters can produce significant thermal energy.

Ocean-based renewable energy development lag land-based systems because of significant capital requirements and difficulty in obtaining the necessary financing due to risk and market barriers. The technical capabilities, both in engineering and management, exist in the offshore sector appear ready to undertake the size and scope of projects envisioned.

4. Physics of Tidal Power

Tidal energy is derived from the gravitational forces of attraction that operate between a molecule on the earth and moon, and between a molecule on the earth and sun. The force is $f = K\,M\,m / d^2$, where m is the mass of the molecule on the earth, M is the mass of the moon or sun, d is the distance between the bodies, and K is the universal constant of gravitation. The attractive force exerted by the sun is about 2.17 times less than that due to the moon due to the mass and much greater distance that separates the earth and sun. As the earth rotates, the distance between the molecule and the moon will vary. When the molecule is on the dayside of the earth relative to the moon or sun, the distance between the molecule and the attracting body is less than when the molecule is on the horizon, and the molecule will have a tendency to move away from the earth. Conversely, when the molecule is on the night side of the earth, the distance is greater and the molecule will again have a tendency to move away from the earth. The separating force thereby experiences two maxims each day due to the attracting body. It is also necessary to take into the account the beating effect caused firstly by difference in the fundamental periods of the moon- and sun-related gravitational effects, which creates the so-called spring and neap tides, and secondly the different types of oscillatory response affecting different seas. If the sea surface were in static equilibrium with no oscillatory effects, lunar forces, which are stronger than solar forces, would produce tidal range that would be approximately only 5.34 cm high.

4.1 Types of Tide: Tidal phenomena are periodic. The exact nature of periodic response varies according to the interaction between lunar and solar gravitation effects, respective movements of the moon and sun, and other geographical peculiarities. There are three main types of tide phenomena at different locations on the earth.

Semidiurnal Tides with Monthly Variation: This type of tide has a period that matches the fundamental period of the moon (12 hr 25 min) and is dominated by lunar behaviour. The amplitude of the tide varies through the lunar month, with tidal range being greatest at full moon or new moon (spring tides) when the moon, earth, and sun are aligned. At full moon, when moon and sun have diametrically opposite positions, the tides are highest, because the resultant centre of gravity of moon and earth results in the earth being closer to the sun, giving a higher gravity effect due to the sun. At new moon, maximum tidal range is less. Minimum tides (neap tides) occur between the two maxims and correspond to the half-moon when the pull of the moon and sun is in quadrate, i.e., the resultant pull is the vector sum of the pull due to moon and sun, respectively. In this case, the resultant gravitation force is a minimum. A resonance phenomenon in relation to the 12 hr-25-min periods characterizes tidal range.

- *Diurnal Tides with Monthly Variation.* This type of tide is found in the China Sea and at Tahiti. The tidal period corresponds to a full revolution of the moon relative to the earth (24 hr- 50-min). The tides are subject to variations arising from the axis of rotation of the earth being inclined to the planes of orbit of the moon around the earth and the earth around the sun.

- ***Mixed Tides.*** Mixed tides combine the characteristics of semidiurnal and diurnal tides. They may also display monthly and bimonthly variation. Examples are of mixed tides are those observed in the Mediterranean and at Saigon.

4.2 Major Periodic Components. The following periodic components in tidal behaviour can be identified: (i) a 14-day cycle, resulting from the gravitational field of the moon combining with that of the sun to give maxims and minima in the tides (called spring and neap tides, respectively); (ii) a ½ year cycle, due to the inclination of the moon's orbit to that of the earth, giving rise to a period of about 178 days between the highest spring tides, which occur in March and September, (iii) the Saros, a period of 18 2/3 years required for the earth, sun, and moon to return to the same relative positions, and (iv) other cycles, such as those over 1600 years which arise from further complex interactions between the gravitational fields.

Maximum height reached by high water varies in 14-day cycles with seven days between springs (large tide range) and neaps (small tide range). The spring range may be twice that of the neaps. Half-yearly variations are +/-11%, and over 18 2/3 years +/- 4%. In the open ocean, the maximum amplitude of the tides is less than 1 m. Tidal amplitudes are increased substantially particularly in estuaries by local effects such as shelving, funnelling, reflection, and resonance. The driving tide at the mouth of the estuary can resonate with the natural frequency of tidal propagation up the estuary to give a mean tidal range of over 11 m in the Severn Estuary, UK and can vary substantially between different points on the coastline[1]. The physics of tidal range is examined by Baker in more depth in [2].

5. Ocean Wave and Tidal Power Projects in San Francisco

Ocean waves and Bay tides interacting with the Sacramento River flowing from the Sierra Nevada Mountains combine to create two excellent renewable energy resources, Long rolling ocean waves are a condensed form of wind energy. Tidal currents are driven primarily by the gravitational pull of the moon and are independent of local weather conditions [3].

San Francisco is undertaking two renewable energy projects: a pilot demonstration for tidal power that was commissioned in 2005 and a first U.S. commercial installation for wave energy that is proposed to produce up to 750kW. Both projects could be expanded in prudent phases to provide an important part of San Francisco's current 840MW peak demand. San Francisco is modelling these technologies for environmentally safe implementation in coastal and reverie communities round the world.

Wave power can be harvested by a variety of devices, with several unique approaches nearing commercialisation. Most wave energy devices use air or seawater, and one devise produces electricity directly from motion, in addition to taking advantage of hydraulic pumps to generate power [4-10].

---

[1] Tidal range is the tidal height between high-tide and low tide. Typical tidal ranges are Bay of Fundy (Canada) 19.6 m; Granville (France) 16.8 m; La Rance (France) 13.5 m.

The Pelamis device for San Francisco [8] consists of a total of four cylindrical sections, which are connected together by three hydraulic power conversion modules. The total length of the device is 120m and device diameter is 4.6m. The power conversion module comprises four hydraulic rams (two heave, two sway), high-pressure accumulators for power smoothing storage, two variable displacement motors for power conversion, and two 125kW generators, with integrated transformer cabling AC power to shore. The device is secured by a compliant slack moored anchoring system. At a later date, a series of these devices can be installed to comprise a wave farm. By employing the Pelamis device, a generation capacity of 22 MW/km is anticipated. However, additional device optimisation is anticipated.

The HydroVenturi tidal power generation device is comprised of a cube of venturi tubes attached to the marine bottom on a rack sited safely some 18m below the surface and outside the navigation channel in the Golden Gate Passage. The tidal current flowing through the device is accelerated through the Venturi tubes to create a 2.5Kg pressure drop, thus creating suction enough to pull air down to an air storage tank integrated into the cube below the Venturi tubes. The compressed air is then pushed through a pipe to an on-shore air turbine to produce electricity. A 150kW device has been demonstrated in England.

A 1MW Venturi commercial installation is expected for grid intertie at this time (2009), with expansion phased in 5-20MW increments. The air turbines in the commercial installation may be housed in a secure area on land under the Golden Gate Bridge. Several other tidal power generation locations have been identified in order to serve neighbouring communities such as Marin County and Oakland.

San Francisco's interest in Hydro Venturi technology is because it has no moving parts underwater. A technology with no moving parts underwater makes tidal power attractive

6. Wave Power Technologies

Wave power technologies are now reviewed. The oceans contain a vast amount of mechanical energy in form of ocean waves and tides. The high density of oscillating water results in high energy densities, making it a favourable form of renewable energy. The total U.S. available incident wave energy flux is about 2,300 TWh/yr. The US Department of Energy (DOE) Energy Information Authority (EIA) estimated that 2003 hydroelectric generation in USA was about 270 TWh, which is a little more than a tenth of the offshore wave energy flux into the U.S. The fact that good wave and tidal energy resources can be found in close proximity to population centres and technologies being developed to harness the resource have a low visual profile makes this an attractive source of energy. Recent advances in offshore oil exploration technology and remote management of power generation systems have enabled significant progress in advancing technology development by simple technology transfer. However, despite enormous progress over recent years, underwater current and wave power conversion technologies are at an immature stage of development. This is because of a lack of accepted standards, a wide range of technical approaches, and large uncertainties on performance and cost of these systems. Further Research, Development and Demonstration (RD&D) and the creation of early adopter markets through government subsidies is required to move these technologies into a competitive market place.

Feasibility assessments of wave and tidal current power are made in Section 7.

6.1 Wave Power Conversion Devices and Technologies

Wave power conversion devices are installed either **on-shore** and embedded in a cliff or an existing harbour wall, **near-shore** in close proximity to shore standing on the seabed or **off-shore** in deep waters. Similar to offshore wind, a wider applicability and more consistent and concentrated resource of energy can be found offshore and is more suitable for large-scale deployments. Installing such devices away from the coastline solves many issues such as visual impact, permitting and environmental impact.

The device must be able to handle a wide range of incident wave power levels, from near-flat seas to the most extreme storm conditions (which produce waves power levels more than an order of magnitude above the average). Waves typically have a low frequency of the order of 0.1 Hz, while power generation equipment runs at hundreds of rpm. The device must change the slow-acting, multi-directional wave force into a high-speed, unidirectional force capable of powering a generator. Short-term storage becomes an important consideration to maintain consistent power output.

Technologies to convert ocean wave power into electricity are many. The most promising ones are summarized below:

- **Oscillating Water Column** - (OWC) systems consist of a partially submerged structure, which forms an air chamber, with an underwater opening that allows the seawater to flow into the chamber. The volume of air inside the chamber is compressed as the water rises inside the chamber, driving air through a turbine. As the water level in the chamber subsides, the air is drawn back through the turbine. Both directional and self-rectifying air turbines have been developed. The axial-flow Wells turbine is the best-known turbine for this kind of application and has the advantage of not requiring rectifying air valves.
- **Overtopping Devices** - guides incoming waves up a ramp and up into a reservoir raised slightly above sea level. The water trapped in the reservoir flows back to the sea through a conventional low-head hydroelectric generator.
- **Float Systems** - Their common feature is a buoy that sits on the ocean's surface. The motion of this buoy is converted into electricity typically by a hydraulic power take off such as a hydraulic ram. These float systems come in different shapes and forms.
- **Hinged Contour Devices** - contains different floating sections, which are hinged together. As the wave passes, the sections move relative to each other and the hinges produce power. The power conversion uses hydraulic elements.

As part of a nationwide collaborative program to demonstrate offshore wave power technologies, the Electric Power Research Institute (EPRI) reviewed available technology options in 2004. Some of the results are outlined in Table 1. The wide range of different specifications is a clear indicator of the immaturity of this emerging market. Average power output was assessed for a typical Oregon wave climate with an incident wave power level of 21kW/m. This is a typical US west coast wave power level.

The most important criteria assessing these devices are maturity of the development stage. This is indicated in Table 1 as Maturity Rating.

Maturity Rating	Company	Device Width (m)	Device Weight (Tons)	Average Power (kW)	Power Train
1	Ocean Power Delivery	4.6	380	153	Hydraulic
2	Energ-etech	35	450	259	Air Turbine
2	Wave Dragon	260	22,000	1369	Low Head Hydro
2	Wave Swing	9.5	NA	351	Linear Generator
3	Wave Bob	15	440	131	Hydraulic
3	Aqua Energy	6	22	17	Water Pump
3	OreCON	32	1250	532	Air/Hydraulic
3	INRI	5.4	112	16	Water Pump

Source: M. Previsic [11]

Table 1.Technology Comparison

6.2 Electrical Interconnection

Most wave power conversion devices under development incorporate frequency converters and step-up transformers to synchronize onto the grid. As a result, power quality tends to be good and power factors high. Short-term storage is incorporated to account for wave-to-wave variations. Storage options depend on the power take off train, and can incorporate hydraulic accumulators, storage through flywheel effects and capacitor banks. It remains to be seen how well these short-term storage options deal with the large variability of power levels in ocean waves.

Wave farm interconnection voltage levels depend on many variables, but are typically in the range of 12kV to 33kV. Recent offshore wind projects in Europe showed that environmental risks prohibit use of oil insulated cables in the sensitive coastal environment. XLPE insulations have proven to be an excellent alternative, having no such potential hazards associated with devise operation.

6.3 Cost

An Ocean Power Delivery Pelamis wave energy conversion device was used to establish costing models for a commercial scale (300,000 MWh/year) wave farm. Levelized cost components are shown in Figure 1. Cost breakdown shows that impact on cost of electricity of Operation and Maintenance (O&M) is significant and is the one component that has most uncertainty associated to it. The only way such O&M costs can be driven down and confidence established is by building demonstration projects.

An assessment of offshore wave energy conversion devices is made in References [11] and [12]. The methodology, guidelines and assumptions for conceptual design of offshore wave energy power plants is given in Reference [13]. System level design, preliminary performance and cost estimates for Hawaii, Oregon, Main, Massachusetts, and San Francisco Pelamis offshore wave power plants are given in References [14-18], respectively, and system level design, preliminary performance and cost estimate for the San Francisco

Energetech offshore wave power plant is given in References [19]. Further, the state of the art for wave energy conversion is reviewed in Reference [20], and a technical assessment guide for ocean wave power is made in Reference [21]. A wave energy resource assessment for California is given in Reference [22].
Most of the EPRI Wave Power (WP) Reports [12, 14-19] are available on their website (www.epri.com).

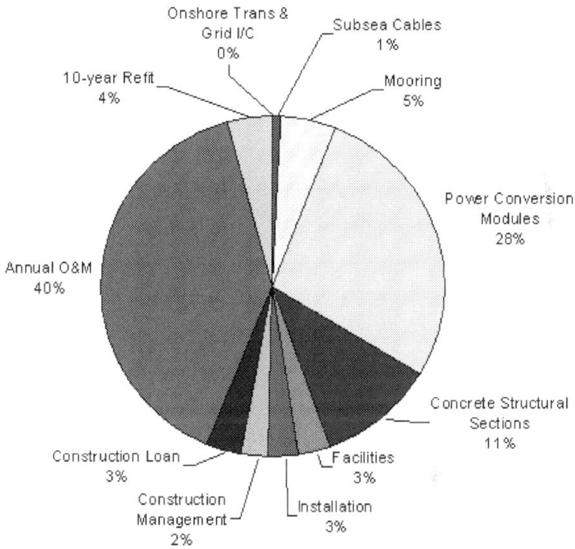

Source: M. Previsic [11]

Fig. 1. Breakdown of Cost: Pelamis Wave Energy Conversion Device

7. Feasibility Assessment of Offshore Wave and Tidal Current Power Production: A Collaborative Public/ Private Partnership

Collaborative power production feasibility definition studies on offshore wave energy and tidal current energy on behalf of a number of public and private entities have been undertaken from 2004. The outcome of the offshore wave study under the Electric Power Research Institute (EPRI) is a compelling techno-economic case for investing in the research, development and demonstration (RD&D) of technology to convert the kinetic energy of ocean waves into electricity
EPRI Wave Power Reports [12, 14-19] and References [23-30] summarize the activities in this area.

7.1 Feasibility of Wave and Tidal Current Energy

The elements of a wave and tidal current energy feasibility study are: a) Identify and characterize potential sites for assembling and deploying a power plant and for connecting the plant to the electric grid; b) Identify and assess wave energy conversion (WEC) devices; c) Conduct a conceptual design of a demonstration- and commercial-scale offshore wave power plant and, based on performance and cost estimates, assess the techno-economic viability of the wave energy source and the energy conversion technology; and d) Identify and assess the environmental and regulatory issues associated with implementing the technology.

Two characteristics of waves and tides important to the generation and dispatch of electricity from wave energy conversion devices are its variability and predictability. While the ocean is never totally calm, wave power is more continuous than the winds that generate it. The average power during the winter may be six times that obtained during the summer; however, power values may vary by a factor of a hundred with the random occurrences of storms. Therefore, the power of waves is highly variable. The predictability of wave energy is of the order of a few days. The waves resulting, for example, from storms that occur off the coast of Japan, will take that long to reach the northwest cost of the United States. The power from tidal currents, on the other hand, typically varies according to a diurnal cycle. The major benefit of tidal power is its high predictability for a given site years in advance, provided there is a thorough knowledge of the site. A drawback of tidal power is its low capacity factor, and that its peak availability misses peak demand times because of the 12.5-hour cycle of the tides.

Ocean waves are generated by the winds that result from uneven heating around the globe. Waves are formed by winds blowing over the water surface, which make the water particles adopt circular motions as depicted in Figure 2. This motion carries kinetic energy, the amount of which is determined by the speed and duration of the wind, the length of sea it blows over, the water depth, sea bed conditions and also interactions with the tides. Waves occur only in the volume of water closest to the water surface, whereas in tides, the entire water body moves, from the surface to the seabed.

The tides are generated by rotation of the earth within the gravitational fields of the moon and sun [1]. The relative motion of these bodies causes the surface on the oceans to be raised and lowered periodically, as illustrated in Figure 3. The physics of tidal power is explained in Reference [1].

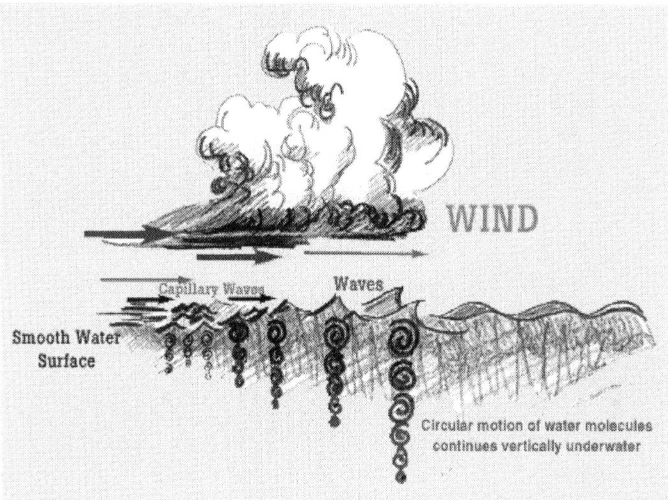

Source: M. Previsic [11]

Fig. 2. Wave Generating Forces based on Wind-Water Interaction

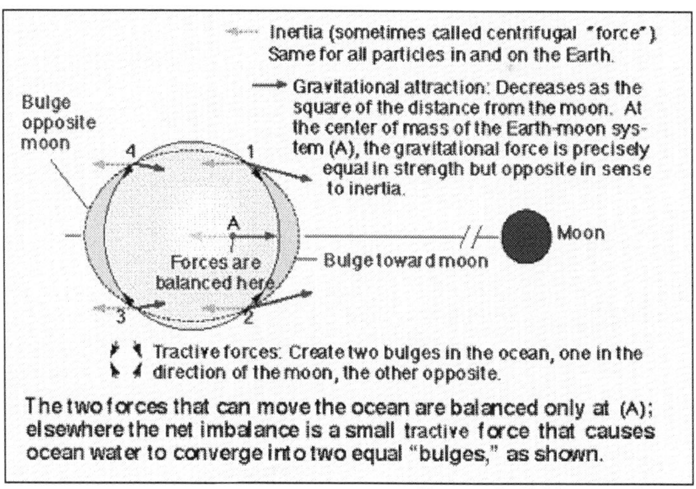

Source: O. Siddiqui and R. Bedard [31]

Fig. 3. Tide-Generating Forces based on Earth-Moon Interactions.

In deep water, the wave power spatial flux (in kW/m of wave front crest) is given by significant wave height (H_s in m) and the peak wave period (T_p in sec). Based on these two parameters, the incident wave power (J in kW/m of wave crest length) associated with each sea state record is estimated by the following equation:

$$J = 0.42 \times (H_s)^2 \times T_p \qquad \text{(kW)}$$

It is significant to note that wave power varies with the square of wave height - that is, a wave whose height is doubled generates four times as much power.
The power of a tidal current is given by the following equation:

$$P_{water} = \tfrac{1}{2}\, r\, A\, V^3 \ \text{(W)}$$

where A is the cross- sectional area of flow intercepted by the turbine device (m^2), r is the water density (kg/m^3) and V is current velocity speed (m/s). The current velocity V varies in a precisely predictable manner as an additive function of period of the different sinusoidal tidal components.

7.2 Wave Project Results

7.2.1 U.S. Wave Energy Resources: An ideal site to deploy, operate and maintain an offshore wave energy power plant must have many attributes. First and foremost is a sufficient native energy and energy spectra potential.[2] The U.S. regional wave regimes and the total annual incident wave energy for each of these regimes are shown in Figure 4 (the total U.S. available incident wave energy flux is about 2,300 TWh/yr). The DOE Energy Information Energy (EIA) estimated that the 2003 hydroelectric generation in USA was about 270 TWh/yr. Therefore; wave energy is a significant resource.

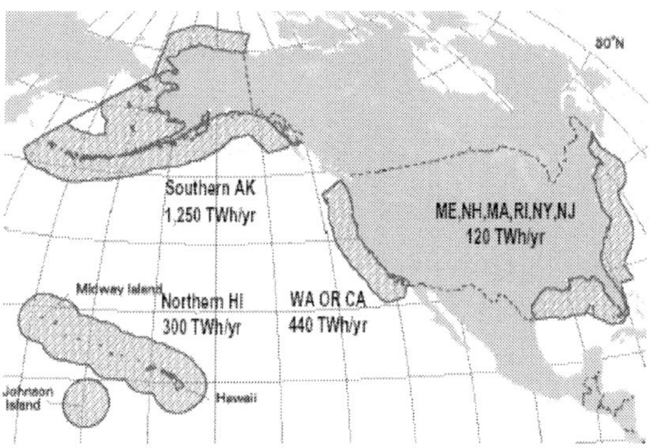

Source: O. Siddiqui and R. Bedard [31]

Fig. 4. U.S. Energy Resources

7.2.2. Feasibility Definition Study Sites: Site attributes characterized by the Project Team included offshore bathymetry[3] and seafloor surface geology, robustness of the coastal utility grid, regional maritime infrastructure for both fabrication and maintenance, conflicts with competing uses of sea space and existence of other unique characteristics that might

---

[2] Energy as function of wave height and wave period or frequency
[3] Bathymetry is the depth of the seafloor below mean water height (i.e., the inverse of a topographic map)

minimize project development costs (e.g., existing ocean outfall easements for routing power cable and shore crossing).
Table 2 identifies the site selected in each of the five states that participated in the study, and also provides a few key characteristics of each selected site.

	HI	OR	CA	Mass	Maine
County	Oahu	Douglas	SF	Cape Cod	Cumberland
Grid I/C	Wai-manalo Beach	Gardner	Wastewater Plant	Well Fleet	Old Orchard Beach S/S
Average Annual J (kW/m)	15.2	21.2	11.2[1]	13.8	4.9
Depth (m)	60	60	30	60	60
Distance from Shore	2	3.5	13	9	9
Cable Landing	Makai Pier	IPP out flow pipe	Water out flow	Dir Drill	Dir Drill

Source: O. Siddiqui and R. Bedard [31]

[1]Sited within the marine sanctuary exclusionary zone

Table 2. Estimated Performances of Pilot Demonstration Plants

7.2.3. Feasibility Study - WEC Devices: Twelve companies responded to EPRI's request for information. An initial screening considered two key issues: 1) technology readiness (i.e. readiness of device for demonstration in the 2006 time period); and 2) survivability in adverse conditions (i.e., sufficiency of technical information provided by the device manufacturer to prove the survivability in storm conditions). The eight devices that passed the initial screening criteria are shown in Table 3.
These eight devices were then assessed with the objective of determining any critical issues and recommending RD&D needed to achieve technological readiness for an at sea demonstration. As a result of this assessment, the eight devices were grouped into one of three levels of development categories:

- Level 1 - Development complete and full-scale testing in the ocean underway
- Level 2 - Development near complete. Only deployment, recovery and mooring issues are yet to be validated. There are funded plans for full-scale at sea testing.
- Level 3 - Most critical RD&D issues are resolved. Additional laboratory and sub-scale testing, simulations and systems integration work is needed prior to finalization of the full-scale design. There are no funded plans for full-scale at sea testing.

	Length (m)	Width (m)	Power (kW)[1]	Type	Rating
Ocean Power Delivery	120	4.6	153	Floating Attenuator	1
Energetech	25	35	259	OWC - Bottom Terminator	2
Wave Dragon	150	260	1369	Floating Overtopping	2
Wave Swing	9.5	9.5	351	Bottom Point Absorber	2
Wave Bob	16	15	131	Floating Point Absorber	3
Aqua-Energy	6	6	17	Floating Point Absorber	3
OreCON	32	32	532	Floating OWC	3
Ind Natural Resources Inc	5.4	5.4	112	Bottom Point Absorber	3

Source: O. Siddiqui and R. Bedard [31]

[1]Based on Oregon average annual wave energy resource

Table 3. Estimated Performance of Pilot Demonstration Plants

At the time of EPRI's analysis (March 2004), only one WEC device manufacturer had attained a Level 1 technology readiness status - Ocean Power Delivery with its Pelamis device.

7.2.4 Demonstration-Scale Plant Design - Oregon Example: Demonstration-scale (as well as commercial-scale) designs were based on the Ocean Power Delivery (OPD) Pemamis WEC device for the five sites listed in Table 2. The Pelamis WEC device consists of four cylindrical steel sections, which are connected by three hydraulic power conversion modules (PCM). Total length of the device is 120m and device diameter is 4.6m. Figure 5 illustrates the device being tested off the Scottish coast.

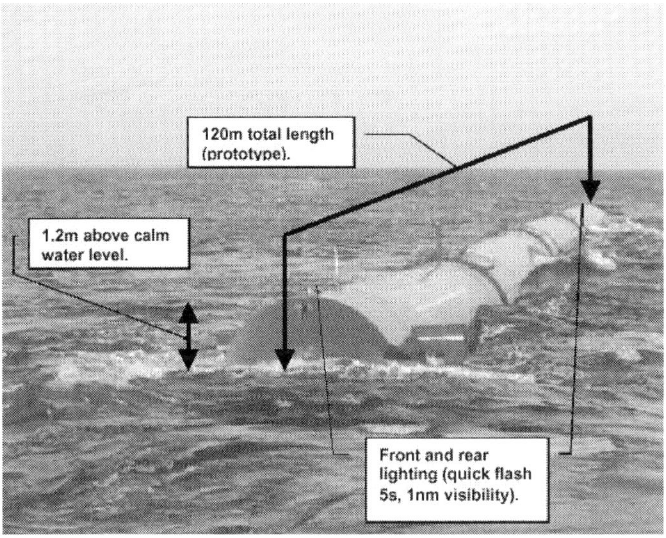

Source: O. Siddiqui and R. Bedard [31]

Fig. 5. OPD Pelamis WEC Device.

A second San Francisco, CA design based on the Energetech OWC WEC device depicted in Figure 6 has also been tested.

Source: O. Siddiqui and R. Bedard [31]

Fig. 6. Energetech WEC Device.

The estimated performance of the single unit demonstration plant at each of the five sites is shown in Table 4.

	HI	OR	CA [1]	Mass	Maine
Device Rated Capacity (kW)	750	750	750	750	750
Annual Energy Absorbed (MWh/yr)	1,989	1,472	1,229	1,268	426
Annual Energy Produced (MWh/yr)	1,663	1,001	835	964	290
Average Electrical Power (kW)	180	114	95	98	33
Number of Homes Powered by Plant	180	114	95	98	33

Source: O. Siddiqui and R. Bedard [31]

[1]Energetech site numbers: 1000 kW, 1643 MWh/yr, 1264 MWh/yr, and 144 kW respectively

Table 4. Estimated Performance of Pelamis Pilot Demonstration Plants

7.2.5 Commercial-Scale Plant Design – Oregon Example: The commercial system uses a total of 4 clusters, each one containing 45 Pelamis units (i.e., 180 total Pelamis WEC devices), connected to sub-sea cables. Each cluster consists of 3 rows with 15 devices per row. The other state designs are organized in a similar manner with 4 clusters. The number of devices per cluster varies such that each plant produces an annual energy output of 300,000 MWh/yr. The electrical interconnection of the devices is accomplished with flexible jumper cables, connecting the units in mid-water. The introduction of 4 independent sub-sea cables and the interconnection on the surface provides some redundancy in the wave farm arrangement.

The estimated performance of the commercial-scale plant at each of the five sites is shown in Table 5.

The device rated capacity has been derated from 750 kW in the demonstration plant to 500 kW for the commercial plant. The performance assessment of the demonstration plants shows that the PCMs are overrated and reducing the rated power to 500 kW per device would yield a significant cost reduction and only a relatively small decrease in annual output (attributed to the fact that the U.S. sites have a lower energy level than UK sites for which the device was originally developed).

7.2.6 Learning Curves and Economics: The costs and cost of electricity shown in the previous section are for the *first* commercial scale wave plant. Learning through production experience reduces costs – a phenomenon that follows a logarithmic relationship such that for every doubling of the cumulative production volume, there is a specific percentage drop in production costs. The specific percentage used in this study was 82%, which is consistent with documented experience in the wind energy, photovoltaic, shipbuilding, and offshore oil and gas industries.

The industry-documented historical wind energy learning curve is shown as the top line in Figure 7 [32]. The cost of electricity is about 4 cents/kWh in 2004 U.S. dollars based on 40,000 MW of worldwide installed capacity and a good wind site. The lower and higher bound cost estimates of wave energy are also shown in Figure 7. The 82% learning curve is applied to the wave power plant installed cost but not to the operation and maintenance part of cost of electricity (hence the reason that the three lines are not parallel).

	HI	OR	CA	Mass	Maine
Device Rated Capacity (kW)	500	500	500	500	500
Annual Energy Absorbed (MWh/yr)	1,989	1,997	1,683	1,738	584
Annual Energy Produced (MWh/yr)	1,663	1,669	1,407	1,453	488
Average Electrical Power at Busbar (kW)	191	191	161	166	56
Number of OPD Pelamis Units Needed for 300,000 MWh/yr	180	180	213	206	615
Number of Homes Powered by Plant	34,000	34,000	34,000	34,000	34,000

Source: O. Siddiqui and R. Bedard [31]

Table 5. Estimated Performance of Pelamis Commercial Plants

Figure 7 shows the cost of wave-generated electricity: low band (bottom curve), upper band (middle curve); and wind generated electricity (top curve) at equal cumulative production volume under all cost estimating assumptions for the wave plant. It shows that the cost of wave-generated electricity is less than wind-generated electricity at any equal cumulative production volume under all cost estimating assumptions for the wave plant. The lower capital cost of a wave machine (compared to a wind machine) more than compensates for the higher O&M cost for the remotely located offshore wave machine. A challenge to the wave energy industry is to drive down O&M costs to offer it even more favourable economically and to delay the crossover point shown at greater than 40,000 MW.

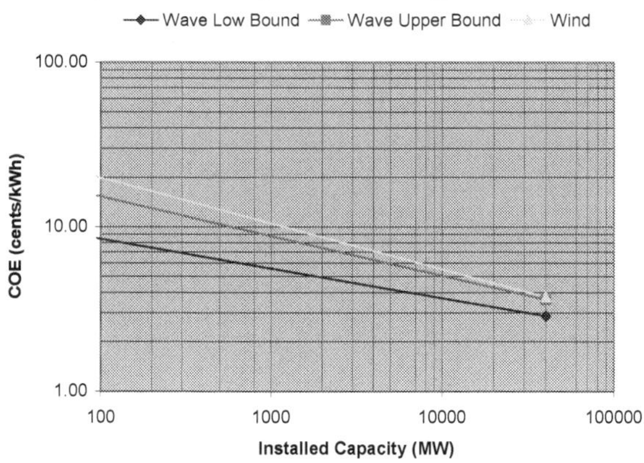

Source: O. Siddiqui and R. Bedard [31]

Fig. 7. Electrical Interconnection of Demo-plant - Oregon Example.

The techno-economic forecast made by the Project Team is that wave energy will first become commercially competitive with the current 40,000 MW installed land-based wind technology at a cumulative production volume of 15,000 MW or less in Hawaii and northern California, about 20,000 MW in Oregon and about 40,000 MW in Massachusetts. This forecast was made on the basis of a 300,000 MWh/yr (nominal 90 MW at 38% capacity factor) Pelamis WEC commercial plant design and application of technology learning curves. Maine was the only state in the study whose wave climate was such that wave energy may never be able to economically compete with a good wind energy site.

In addition to economics, there are other compelling arguments for investing in offshore wave energy technology. First, with proper sitting, converting ocean wave energy to electricity is believed to be one of the most environmentally benign ways to generate electricity. Second, offshore wave energy offers a way to minimize the 'Not In My Backyard' (NIMBY) issues that plague many energy infrastructure projects, from nuclear to coal and to wind generation. Because these devices have a very low profile and are located at a distance from the shore, they are generally not visible. Third, because wave energy is more predictable than solar and wind energy, it offers a better possibility than either solar or wind of being dispatch able and earning a capacity payment.

A characteristic of wave energy that suggests that it may be one of the lowest cost renewable energy sources is its high power density. Processes in the ocean concentrate solar and wind energy into ocean waves making it easier and cheaper to harvest. Solar and wind energy sources are much more diffuse, by comparison.

Since a diversity of energy sources is the bedrock of a robust electricity system, to overlook wave energy is inconsistent with national needs and goals. Wave energy is an energy source that is too important to overlook.

8. Recent Progress in Offshore Renewable Energy Technology Development

Interest in marine renewable energy is at an all-time high, and prospects for ocean-based renewable energy development look brighter all the time. The recent progress in offshore renewable energy technology development is now examined and potential markets for tidal power, wave energy conversion, and offshore wind are considered. The analysis of market potentials for offshore renewable technology is based solely on identified projects. Therefore, the forecasts are relatively conservative, as the prospective markets could expand as technological advances are achieved and as regulatory environments improve.

8.1 Tidal Energy

Historically, tidal projects have been large-scale barrage systems that block estuaries. Within the last few decades, developers have shifted toward technologies that capture the tidally driven coastal currents or tidal stream. The challenge is, *"to develop technology and innovate in a way that will allow this form of low density renewable energy to become practical and economic"* [23].

Two groups of technologies are in operation or planning; these are tidal current turbines and tidal stream generators. Tidal current turbines are basically underwater windmills. The tidal currents are used to rotate an underwater turbine. First proposed during the 1970s' oil crisis, the technology has only recently become a reality. One company, Marine Current Turbines (U.K.) installed the first full-scale prototype turbine (300 kW) off Lynmouth in

Devon, U.K. in 2003. Shortly thereafter, the Norwegian company Hammerfest Støm installed their first prototype device.

There are a great number of sites suitable for tidal current turbines. As tidal currents are predictable and reliable, tidal turbines have advantages over offshore wind counterparts. The ideal sites are generally within 1 km of the shore in water depths of 20-30 m.

Tidal stream generators use the tidal stream to generate power from, for example, the raising and lowering of a hydraulic arm. Several promising devices are at the advanced stage of development. For example, the UK firm, The Engineering Business Ltd. has developed and tested a simple concept of placing hydrofoils in tidal stream to produce an oscillatory motion in the vertical or horizontal plane. The device, known as the Stingray™ Tidal Current Generator, "*transforms the kinetic energy of the moving water into hydraulic power, which turns a generator by means of a hydraulic motor*" [23].

8.1.1 Tidal Forecasts

It is anticipated that multi-megawatt installations will emerge by the end of the decade (Figure 8). By 2008, a forecast of 14.8 MW installed capacity is expected with 65% of the capacity in the United Kingdom. Norway, which already has installed capacity, will be the second dominant player, but lacks defined projects over the next 5 years. Other countries (Canada, France and United States) have a minor role, but could expand prototype devices as the devices progress. Canada and the United States have potential locations, some of which are under negotiation for U.K. tidal generation technology [24].

Almost 70% of forecast capacity by 2009 is anticipated from tidal current turbines with approximately 30% from tidal stream generators. Tidal current turbines represent an extremely important sector for offshore renewables as there are several well-developed devices and such technology, once proven, could be installed in large numbers in the near future. However, a lack of identified projects distorts the forecast near the end of the 5-year period, precisely when significant projects could materialize.

It is conceivable that tidal current turbines such as those of Marine Current Turbines or Hammerfest Støm could eventually be installed in large projects comparable in size to offshore wind farms.

With fewer announced projects, tidal stream generators have a lower forecast. Many of the devices are at earlier stage of design. One system that is generating much interest and has potential is the '*Stingray*' device, designed and built by experienced offshore engineers [26].

Capital expenditures for tidal energy are forecast at $35 million over the next five-year period. The U.K. is the biggest market with $23 million of expenditures to 2008. Forecasts for Norway at a level of $10 million of expenditures over the same period are anticipated. Successful projects could lead to further development later in the period and beyond [24]. Several projects await financial support and could significantly impact the installed capacity as the projects are in excess of 100 MW.

Revised global estimates for capital expenditure in tidal power technology is indicated in Figure 8.

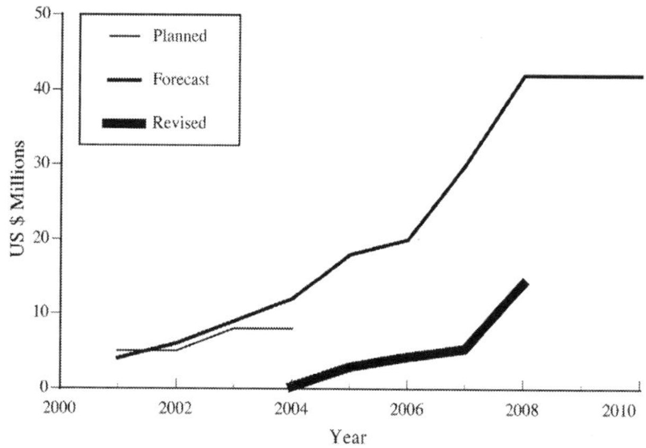

Source: A. T. Jones and A. Westwood [33]

Fig. 8. Revised Global Estimates of Capital Expenditure in Tidal Power Technology (modified from [24]).

Yalu River, China: By creating a tidal lagoon offshore, Tidal Electric has taken a novel approach to resolve environmental and economic concerns of tidal barrage technology [28]. Due to the highly predictive nature of the ocean tides, the company has developed simulation models with performance data from available generators to optimise design for particular locations. The recent announcement of a cooperative agreement with the Chinese government for ambitious 300 MW offshore tidal power generation facilities off Yalu River, Liaoning Province allows for an engineering feasibility study to be undertaken.

Tidal Electric also has plans under consideration for United Kingdom-based projects in Swansea Bay (30 MW), Fifoots Point 930 MW), and North Wales (432 MW).

8.2 Wave Energy

The true potential of wave energy will only be realized in the offshore environment where large developments are conceivable. Nearly 300 concepts for wave energy devices have been proposed.

Modular offshore wave energy devices that can be deployed quickly and cost effectively in a wide range of conditions will accelerate commercial wave energy. In the coming decade, wave energy will become commercially successful through multiple-unit projects

Opportunities for expansion of offshore market are expected to increase. This is because the growth of shoreline wave energy devices will be increasingly limited by the low number of available sites and by high installation costs.

Deployment costs for shoreline wave energy devices are very high because they are individual projects and economies of scale are therefore not applicable. The site-specific demands of shoreline wave energy devices mean a further restriction of growth in this sub-sector. Whereas an offshore 50-MW wave farm is conceivable, and will in time be developed, no shoreline wave energy converter can offer such potential for deployment in this way. As such, individual coastal installations are expected to be few and far between [24].

Shoreline wave energy will, however, continue to be relevant, with approximately 25 percent of the forecast capacity over the next five years. The average unit capacity is generally higher than existing offshore technology. Individual devices can be very effective, especially for remote or island communities where, for example, an individual unit of 4MW could have a big impact [24].

Offshore locations offer greater power potential than shoreline locations. Shoreline technologies have the benefit of easy access for maintenance purposes, whereas offshore devices are in most cases more difficult to access. Improvements in reliability and accessibility will be critical to the commercial success of the many devices currently under development [24].

Most wave energy projects to date have been small, and few are connected to a power grid. Shoreline devices offer the advantage of easier access to a grid. For offshore devices, meeting this need will be challenging and costly, although not prohibitively so.

8.2.1 Wave Energy Forecast: Wave is a most promising sector over the 2004-2008 period and into the long-term future (Figure 9).

The development process for wave energy can be looked at in three phases. First, small-scale prototype devices, typically with low capacity, will be deployed. During the second stage, outside funding from government or private investors is possible for the most promising devices. The final stage is the production of full-scale, grid-connected devices that will in some cases be deployable in farm style configurations.

The United Kingdom is expected to be the dominant player over the next five years. In comparison with other countries, the UK has forecast capacity every year to 2009, whereas installations elsewhere are more intermittent. Australia, Portugal, and Denmark are the next most significant markets and have several projected installations, but they lag far behind the UK. The United Kingdom government has shown reasonable levels of support, which have injected many technologies with valuable grants. The result is a number of advanced wave technologies with good prospects for deployment of prototype devices. Coupled with a world-class natural resource, the United Kingdom could be the undisputed world leader in wave energy by 2009. Prospects after 2009 are even brighter [24].

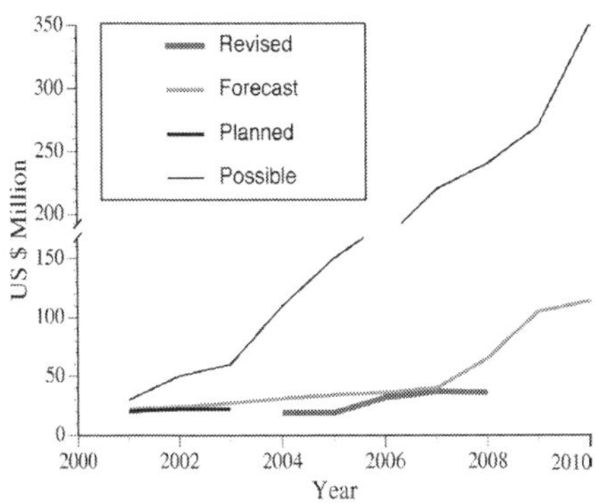

Source: A. T. Jones and A. Westwood [33]

Fig. 9. Revised Estimates for Capital Expenditure in Wave Energy Conversion Technology. (modified from [25]).

The United States market shows encouraging levels of interest in wave technology; however, the market will be affected by the lack of positive government involvement [24].

8.3 Offshore Wind

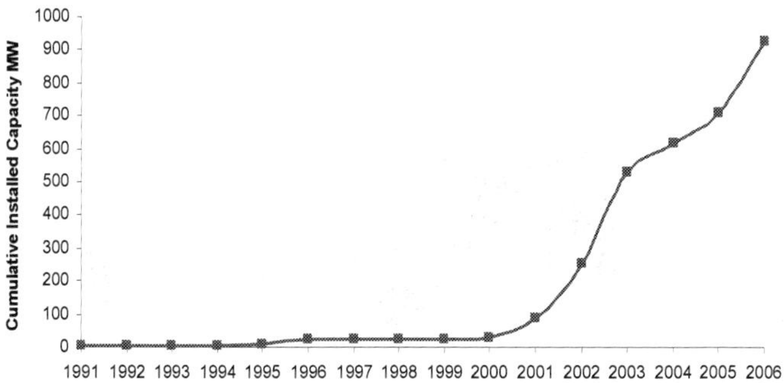

Source: Douglas-Westwood Ltd

Fig. 10. Cumulative Worldwide Offshore Wind Capacity

The total global offshore wind capacity forecast for installation between 2006 and 2010 stands at 7.4 GW (see Figures 10, 11). The UK is the world's largest market for the five-year period 2005-2010. The UK's prospects are expected to be twice those of Germany for this period, although the German market at 1.1 GW is still the second largest in the world. Long-term prospects are excellent off Germany but in the short and mid-term future the industry has much to overcome. Denmark has only two main projects planned for completion by the end of the decade with 200MW each at Horn Rev and Nysted that are now making progress. The Netherlands has just two projects that were commissioned in 2006 and 2007. No firm prospects have emerged from the last licensing round but long-term potential is there.

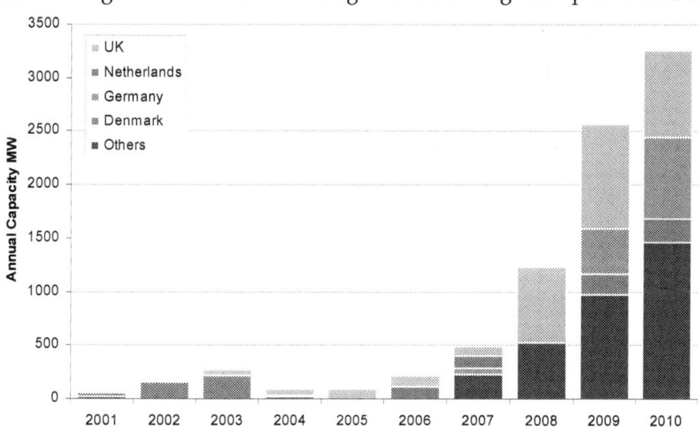

Source: Douglas-Westwood Ltd

Fig. 11. Forecast Global Offshore Wind Capacity

Technological progress is extremely important for the industry, and will drive developments. As better technology is implemented, large strides in capacity will be achieved using proportionally fewer turbines. For example, up to 1,225 turbines will be installed by 2010. Turbine capacity is increasing, from 2000-2003 the average turbine size was 2 MW, current projects are using 3 MW machines and the industry is pushing development of 5 MW turbines for installations from 2009. Prototype installations of these next-generation turbines have already taken place and the first two offshore units were commissioned off the UK at the Beatrice project in 2006.

Long-term signals are good for the UK market, whereas an air of uncertainty hangs over Germany despite its very promising future forecast. The United Kingdom's development is gradual, whereas Germany's depends on large, technologically challenging projects. Denmark's five-year forecast is disappointing, with only two projects scheduled for commissioning in the period, one in 2009 and one in 2010. Although the country showed initial promise for offshore development, a lack of government commitment has been harmful to the industry here. There are no firm plans for future projects after the coming two, so long term prospects are uncertain.

Offshore wind has a potentially large market in North America. Although the United States has considerable offshore wind potential, regulatory uncertainty is a source of concern. The United States has a significant number of projects in the planning stages [30]. These projects, many of which are very speculative, are not expected to arise until the end of the decade.

For the entire marine renewables sector, 7,500 MW of installed capacity is projected between 2006 and 2010. Some 98% of that capacity is in the form of offshore wind farms. Wind farms installed capacity was 213 MW in 2006. By 2010, this will grow to 3,200 MW - over a ten-fold growth within five-years. The value of the market over the next five-years is projected at $16 billion.

Wave and tidal power will only be a small percentage of the total expenditure on offshore renewables, of the order of $150 million in total expenditure between them. However, wave and tidal power currently attract higher expenditures per megawatt. This indicates higher costs of the immature developing industries. These costs will fall as time goes by and the industries progresses. The leading devices should be comparable with, and in some cases more competitive than offshore wind, by the end of the decade.

The dominance of offshore wind does not mean wave and tidal energy are not important, they are just less well developed, and the industry is much younger. If wave and tidal were compared to offshore wind market data from ten years ago, their market share would be much higher. Offshore wind is booming at present. From around 2010, wave and tidal could begin to see this rapid growth.

9. European Energy Potential

The amount of energy available from a tide varies approximately with the square of tidal range. The energy available from a tidal power plant would therefore vary by a factor of four (eight for tidal stream) over a spring-neap tide cycle. Typical variation in output from tidal range and tidal stream power in the Severn Estuary due to the spring-neap cycle is indicated in Figures 12(a) and 12(b), respectively. Approximately 20 suitable regions for development of tidal power worldwide have been identified.

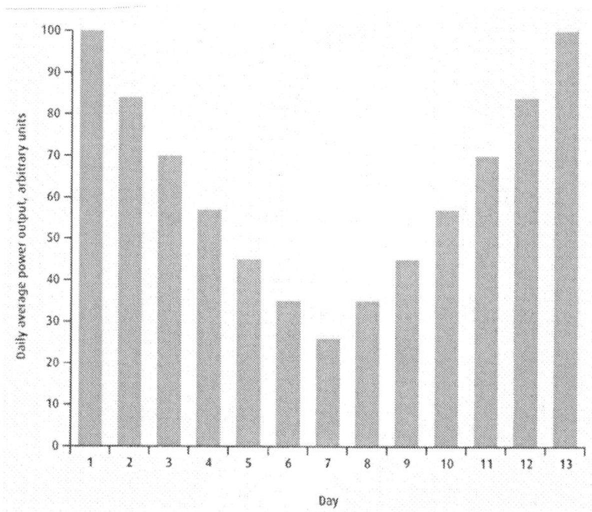

Fig. 12(a). Typical Variation in Output from Tidal Range Power due to Spring-Neap Cycle

A parametric approach [34] has been used to estimate tidal energy potential for appropriate EU countries (Belgium, Denmark, France, Germany, Greece, Ireland, Portugal, Spain, The Netherlands, and UK). An assessment of all reasonably exploitable sites within the EU with a mean range exceeding three meters yielded a total energy potential of about 105 TWh/year. This potential is mainly in the UK (50 TWh/year) and France (44 TWh/year), with smaller contributions in Ireland, The Netherlands, Germany and Spain. Technically available resource for tidal energy estimated using parametric modelling is given in Table 6.

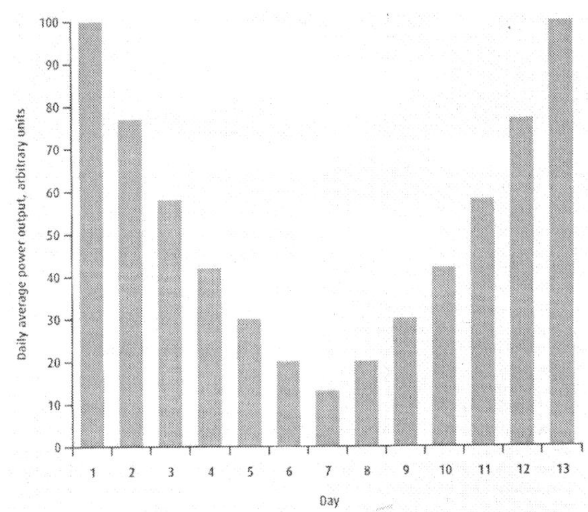

Fig. 12(b). Typical Variation in Output from Tidal Stream Power due to Spring-Neap Cycle.

Country	Technically available tidal energy resource		Percentage of European tidal resource
	GW	TWh/year	
United Kingdom	25.2	50.2	47.7
France	22.8	44.4	42.1
Ireland	4.3	8.0	7.6
Netherlands	1.0	1.8	1.8
Germany	0.4	0.8	0.7
Spain	0.07	0.13	0.1
Other W European	0	0	0
Total W European	63.8	105.4	100.0

Table 6. Technically Available Tidal Energy Resource in Europe Estimated by Parametric Modelling

9.1 Existing Tidal Energy Schemes

Relatively few tidal power plants have been constructed to date. The first and largest is the 240 MW barrage at La Ranch (France) [35], which was built for commercial production in the 1960s. Other tidal power plants include the 17.8 MW plant at Annapolis (Canada), the 400-kW experimental plant at Kislaya Guba (former Soviet Union), and the 3.2 MW Jiangxia station (China).

9.2 Sites Considered for Development Worldwide

Economic feasibility of tidal barrage schemes is dependent on the world market price of fossil fuels, interest rates over scheme expected life, and on level of fossil fuel levies based on the carbon content of fuel and electricity not produced by renewable energy sources, etc. Tidal power sites of capacity above 1GW considered for development with installed capacity and approximate annual output include: (i) Argentina San Jose, 6.8GW, 20.0 TWh; (ii) Canada Cobequid, 5.34 GW, 14.0 TWh; (iii) Canada Cumberland 1.4 GW, 3.4 TWh; (iv) Canada Shepody, 1.8GW, 4.8 TWh; (v) India Gulf of Cambay, 7.0 GW, 15 TWh; (vi) UK Severn, 8.6 GW, 17 TWh; (vii) USA Knit Arm, 2.9 GW, 7.4 TWh; (viii) USA Turnagain Arm, 6.5 GW, 16.6 TWh; (ix) Former Soviet Union Mezen, 15 GW, 50 TWh; (x) Former Soviet Union Tugur, 10 GW, 27 TWh; and (xi) Former Soviet Union Penzhinskaya, 50 GW, 200 TWh.

9.3 Harnessing Tidal Power (flow or basin, existing tidal energy schemes, modes of operation and configuration, adaptation of tide-generated to grid network requirements)

Devises include waterwheels, lift platforms, air compressors, water pressurization, etc. Energy can be extracted either directly by harnessing the kinetic energy of a tide flow, or by using a basin to capture potential energy of a rising and falling mass of water.

9.3.1 Tidal Flow: Tide flows have a poor energy density. Theoretical available power P is given by $P=D\ A\ V^3$, where D is the fluid density, A is the area swept out by the turbine rotor, and V is the undisturbed stream velocity [36]. The energy can be harnessed only with poor maximum efficiency, similar to a windmill, where an efficiency of 59.3% is possible. Directly harnessing power in this way, however, does not require expensive additional structures.

9.3.2 ***Basin:*** This method involves constructing a barrage and forming a basin from a natural bay or estuary. Considerable extra cost is incurred, but this is more than outweighed by the extra energy that is extractable. The energy available from a turbine in an effective barrage is one or two orders of magnitude greater than that from a similar size of turbine in a tide stream of, for example, 2 m/s. The extra cost of constructing the barrage may be only a third of scheme overall cost.

9.4 Modes of Operation and Configuration

The tide is the only factor that affects the generating activity of a tidal power plant that is programmed to produce maximum output. The output at any given time can be accurately calculated as far in advance as is necessary.

9.4.1 ***Single-Action Outflow (Ebb) Generation:*** Barrages can use either one basin or a combination of basins, and can operate by ebb, flood, or two-way generation, with or without pumping. The simplest method is ebb generation using a single basin. The basin is permitted to fill through sluices (gated openings). Generation takes place as the basin is emptied via turbines once the tide level has dropped sufficiently. There are two bursts of generation each day.
Typical day-to-day fluctuations are: (i) there are two bursts of generation activity each day, beginning approximately three hours after high tide and lasting 4-6 hours; (ii) for each cycle production levels rapidly increase with tidal range, the output characteristic therefore displaying a 14-day cycle; (iii) high-water times shift by about 1 hr per day; (iv) in each 14-day period, the generation will not be evenly distributed throughout the 24-hr of the day; (v) output levels will only show slight variation from one fortnightly period to the next; and (vi) annual production levels will show fluctuations of around +/-5% and will follow a cycle of 18 2/3 years.

9.4.2 ***Flood Generation:*** Here, power is provided as the basin fills. The basin empties through sluices as the tide falls. This method is not as efficient as ebb generation since it involves using the basin between existing low tide level and slightly above normal mid-tide level, thus producing less energy. An advantage of this mode is that it facilitates the production of energy out of phase with a neighbouring ebb generation scheme, complementing its output and perhaps providing some firm capacity.

9.4.3 ***Two-way Generation:*** This is a combination of ebb and flood generation, generating as the basin both fills and empties, but with a smaller power output for simple ebb generation (except at the highest tide ranges) due to reduced range within the basin. There is a resultant reduction in efficiency with two-way generation since turbines and water flow cannot be optimised. Two-way generation produces electricity in approximately 6-hr cycles, with smaller power output and a greater plant utilisation factor.

10. Tidal Stream

Tidal current turbines are basically underwater windmills where tidal currents are used to rotate an underwater turbine. First proposed during the 1970s' oil crisis, the technology has only recently become a reality. Horizontal axis turbines are more commonly employed. Marine Current Turbines (MCT) {http//www.marineturbines.com/home.htm} installed the first full-scale prototype turbine (300kW) off Lynmouth in Devon, UK in 2003. Their second project, a 1 MW prototype, is expected soon. It will be followed by an array of similar systems (farm) to be installed in an open sea, where three turbines will be added to provide a total capacity of 5 MW. A similar project is the Hydro Helix project in France.
The Norwegian company Hammerfest Stom installed their first grid-connected 300kW devise that was tested and the concept proven {http//www.e-tidevannsenergi.com/} A tidal stream turbine has been designed for the Pentland Firth between the North of Scotland and the Orkney Islands [37] where the first design was for twin turbines with 20 m rotors and was rated at 1-2 MW depending on current speed. In today's design, the 60 m deep four 20 m rotors cover water flow rather than a pair to keep blade loads within practical limits and the whole power output is 4 MW. The SMD Hydro vision TidEL Project (UK) {http//www.smdhydrovision.com} consists of a pair of contra-rotating 500 kW turbines mounted together on a single crossbeam. The 1 MW units are designed to be mounted in an offshore tidal environment with a peak tidal velocity of 5 knots (2.5 m/s) or more and a water depth of greater than 30 m. The Lunar Energy Project (UK) and the HyroHelix Energies Project (France) {http//www.lunarenergy.co.uk http//www.hyrdohelix.fr/} feature a ducted turbine fixed to the seabed via gravity foundation. A 1/20th model was tested in 2004 and a 1 MW prototype is expected soon. The ideal sites are generally within several kilometres of the shore in water depths of 20-30 m.
There are also vertical axis turbines that are cross flow machines whose axis of rotation meets the flow of the working fluid at right angles. Cross flow turbines allow the use of a vertically oriented rotor that can transmit the torque directly to the water surface without need of complex transmission systems or an underwater nacelle. The vertical axis design permits the harnessing of tidal flow from any direction, facilitating the extraction of energy not only in two directions, the incoming and outgoing tide, but making use of the full tidal eclipse of the flow [38]. In these types of turbines, the rotational speed is very low, of the order of 15 rpm.

10.1 The Enermax Project (Italy) {http//www.pontediarchimede.com}: This uses the Kobold turbine. Its main characteristic is its high starting torque that permits it to start even in loaded conditions. A pilot plant is located in the Straight of Messina, close to the Sicilian shore in Italy, in an average sea tidal current of 2m/sec.

10.2 The Blue Energy Project (Canada) {http//www.bluenergy.com}: Four fixed hydrofoil blades of the Blue Energy tidal turbine are connected to a rotor that drives an integrated gearbox and electrical generator assembly. The turbine is mounted in a durable concrete marine caisson that anchors the unit to the ocean floor, directs flow through the turbine further concentrating the resource supporting the coupler, gearbox, and generator above it. The hydrofoil blades employ a hydrodynamic lift principle that causes the turbine foils to move proportionately faster than the speed of the surrounding water. The rotation of the turbine is unidirectional on both the ebb and flow of the tide. A unit turbine is of the order of 200 kW output power. For large-scale power production, multiple turbines are linked in series to create a tidal fence across an ocean passage or inlet

10.3 The Gorlov Helical Turbine (GHT) (USA) {http//www.gcktechnology.com/GCK/}: The turbine consists of one or more long helical blades that run along a cylindrical surface similar to a screw thread, having a airfoil or airplane wing profile. GHT blades provide a reaction thrust that can rotate the turbine faster than the water flow itself. The GHT is self-starting and can produce power from water current flow as low as 1.5 m/sec with power increasing in proportion to the water velocity cubed. Due to axial symmetry, the GHT always rotates in the same direction, even when tidal currents are reversed. The standard model (1 m in diameter, 2.5 m in length) can be installed either vertically or horizontally to the water current [39]. A single GHT has a rated power of 1.5 kW for 1.5 m/s water speed and 180 kW for 7.72 m/sec. A similar concept to the GHT is the Achard known as the Harvest project (France) {http//www.legi.hmg.inpg.fr/cavit/Deta/Harvest.html}.

11. Adaptation to Grid Network Requirements

The output from a tidal plant displays characteristics that are not compatible with those of conventional generation, transmission, and system load. A pumping system increases average output levels and enhances flexibility of the scheme. This in turn leads to improved economic efficiency as supply times can be varied to match energy cost levels.

Single-action outflow (ebb) generation barrages can use one basin or a combination of basins, and can operate by ebb, flood, or two-way generation, with or without pumping.

12. Proposed Severn Barrage

Few sites worldwide are as close to electricity users and the transmission grid as are the potential sites in the UK.

The Severn is probably the most well known of all potential tidal energy locations, and projects for damming the Severn estuary date back for over a century. The tide range is up to 11 m near the head, being amplified and funnelled by the Bristol Channel. The channel and estuary form a resonator having an effective length equivalent to ¼ of that of the tidal wave. Most attention is focussed on schemes further down the estuary where tide range is reduced and a longer barrage is needed, but where the energy extractable is many times greater. Tidal resonance in the Severn Estuary is illustrated in Figure 13.

A number of different barrage options have been proposed. The Cardiff-Weston scheme is one of the largest and would have a generating capacity of around 8.64GW. The Shoots scheme (which would run near to the two Severn road crossings is 1.05GW with an annual output of around 2.75 TWh. Power output and cost summary for the two options are given in Table 7.

12.1 Potential Benefits

The assumption is that both barrages would be operated on the ebb tide, with the addition of flood pumping to increase *total energy* output. This means that they would be generating electricity for around 7~8 hours on each tide, and output would vary within this period. The annual output of each barrage is less than that implied by their size, around 4.4 % of UK electricity supply, about the same as would be produced by a 2 GW conventional fossil-fuel or nuclear power station.

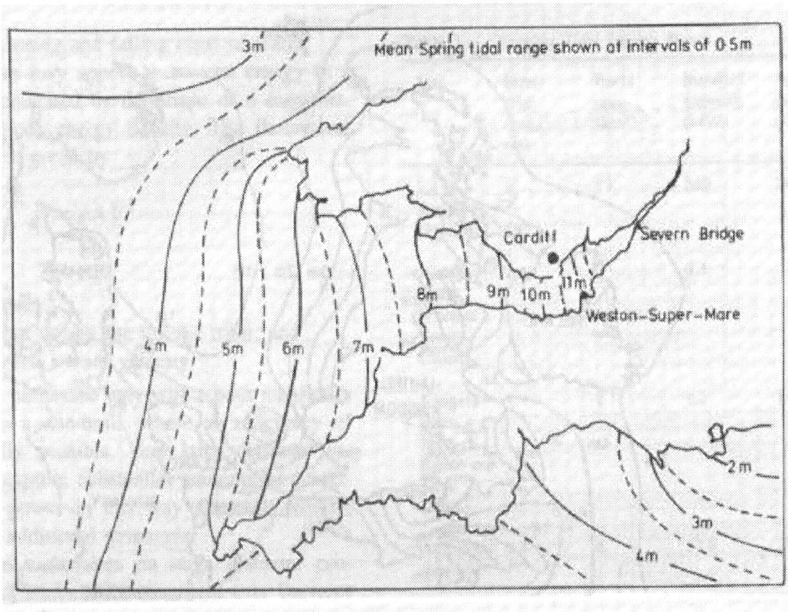

Fig. 13. Tidal Resonance in the Severn Estuary.

The high capital cost of a barrage project leads to a very high sensitivity to the discount rate used (Table 7). At a discount rate of 2 % that could be justified for a climate change mitigation project, cost of electricity from both barrage proposals is highly competitive with other forms of generation. At a commercial discount rate of >8 %, these costs escalate significantly, making private investment unlikely without significant Government market intervention.

There would be substantial flood risk benefits. The timing of output is not optimal, but output is not a major problem for the electricity grid that can be managed at very low cost. The output would displace output from fossil-fuelled plants and would make a genuine and sizable contribution to meeting the UK's targets on renewable energy and on reducing carbon dioxide emissions.

		Cardiff-Weston	Shoots
Length of Embankments		16.1 km	4.1 km
Generating Capacity		8.64 GW	1.05 GW
Annual Average Electricity Output		17 TWh	2.75 TWh
Contribution to UK Electricity Supply (2006 Data)		4.4 %	0.7 %
Estimated Cost of Construction		£15 bn	1.5 bn
Estimated cost of output at various discount rates (high case scenario)	2 %	2.31p/kWh	2.58 p/kWh
	3.5 %	3.68 p/kWh	3.62 p/kWh
	8 %	9.24 p/kWh	7.52 p/kWh
	10 %	12.37 p/kWh	9.54 p/kWh
	15 %	22.31 p/kWh	15.38 p/kWh

Table 7. Power Output and Cost Summary for the two main Severn Barrage Options

12.2 Conditions for Sustainable Development

The issue has been approached from a general position that favours renewable energy under which its development might be sustainable. It has been done within a framework that places a high value on long-term public interest and on maintaining the overall integrity of internationally recognised habitats and species.

12.3 Energy Policy Contexts and Compliance with Environment Legislation

There is risk that development of a barrage might divert Government's attention away from other necessary solutions to the challenge of climate change, including development of a more decentralised energy system and the reduction of energy demand. A Severn barrage has a number of disadvantages that are similar to those of nuclear power, and developing such a large amount of electricity generating capacity in a single location would not itself move the UK any closer to a more decentralised energy system. The Government does not have policies in place at this time to deliver the carbon savings that will be required by 2050, and in particular the delivery of emissions reductions over the next 15 years. A Severn barrage could be pursued as part of a major drive to reduce emissions substantially over both the short- and long-term.

12.4 UK Tidal Resource

Tidal Range Sites		Tidal Stream Sites		
Site Name	*Resource (TWh/year)*	*Site name*	*Area*	*Resource (TWh/year)*
Severn	17	Pentland Skerries	Pentland Firth	3.9
Mersey	1.4	Stroma	Pentland Firth	2.8
Duddon	0.212	Duncans-by Head	Pentland Firth	2.0
Wyre	0.131	Casquets	Alderney	1.7
Conwy	0.06	South Ronaldsay	Pentland Firth	1.5
		Hoy	Pentland Firth	1.4
		Race of Alderney	Alderney	1.4
		South Ronaldsay	Pentland Firth	1.1
		Rathlin island	North Channel	0.9
		Mull of Galloway	North Channel	0.8

Table 8. Top Tidal Energy Sites in the UK with the Resource

Available estimates of the UK's tidal range and tidal stream resource for potential electricity generation are given in Table 8. Estimating potential electricity output requires a number of assumptions on technical constraints of the devises installed, their efficiency, and effect of resource extraction on the remaining resource. This implies that there is a large degree of uncertainty in all resource estimates given in Table 8.

13. Electricity Transmission System

The Electricity Transmission Network in Great Britain is illustrated in Reference [40]. Most generating plants in Great Britain are connected to the transmission system, with some 12 GW of generating capacity connected to the distribution networks. The capacity of the transmission system to connect generation and manage the flows of electricity depends on the capacity of the network. The process of connecting to the network is based around the principle of matching the Connection Entry Capacity (CEC) (the generating capacity of the power station) with the Transmission Entry Capacity (TEC) (the capacity of the network to accept a new generator). Connection offers are made on the basis of an *invest and connect* approach whereby CEC can never exceed TEC, so new lines must be built to connect new generation. At present, there are significant TEC constraints in the north of England and in Scotland, which are preventing the connection of new generation projects. Areas of the transmission network will need to be upgraded to higher voltage levels to increase the TEC.
These issues pose significant challenges for the connection of tidal stream projects. Existing capacity constraints and delays to network upgrades will further delay the date by which tidal stream projects might be connected. If the current approach to transmission connection is not modified, it is unlikely that the UK will see any significant level of tidal stream connection before 2020.
For tidal range, the situation is less significant. Firstly, tidal range resources are generally located in areas where grid constraints capacity transmission lines are less pronounced, and are closer to high capacity transmission lines and to centres of demand. Secondly, tidal

barrages are likely to be larger one-off projects when compared to a tidal stream array, making the incorporation of grid connection costs a smaller part of the overall project cost and therefore more manageable.

The Severn Estuary area has significant network capacity for new generation, with negative transmission network use of system (TNUoS) charges currently in force. This applies for generators in the southwest of England due to a shortage of generation to meet local requirements.

The transmission network around the Severn Estuary is shown in Figure 14. It is quite well developed, with possible connections at both 400kV and 275kV on both sides of the estuary not far from the landing points for a barrage. For the larger Cardiff-Weston scheme, two connections into the 400kV network would be required at both north sides of the barrage (i.e. four connections in total) as the 275kV network on the north side is near to capacity and there is very limited capacity on the 132kV network on the south side.

For the much smaller-rated capacity of the Shoots scheme, all of the connection options seem to have sufficient capacity to accommodate this through just one connection. A connection to the Hinkley Point-Melksham 400kV double-circuit line would be appropriate due to high demand for new capacity. This could further increase with decommissioning of the nuclear power station at Hinkley Point (if it is not replaced by new nuclear generation).

Both schemes would require some new transmission infrastructure to connect into the existing network. This requirement is far higher for the Cardiff-Weston scheme due to its higher rated capacity.

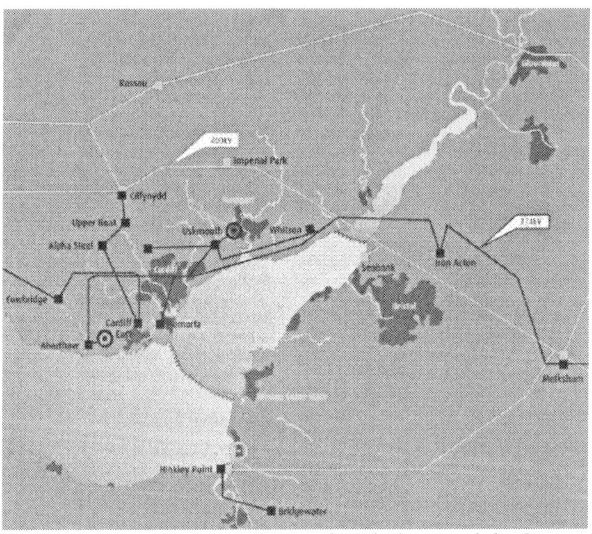

Fig. 14. Electricity Transmission Network (400 and 275kV) around the Severn Estuary.

14. Mersey Barrage

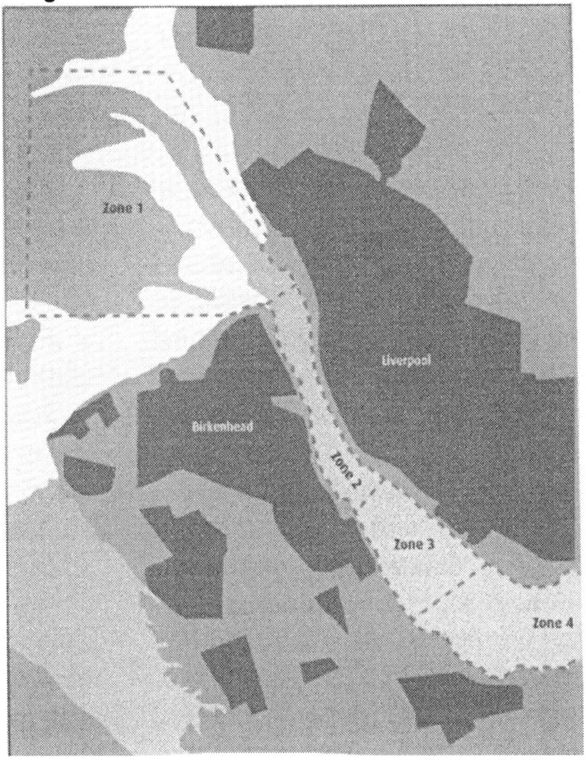

Source: Peel Environmental Ltd.

Fig. 15. Mersey Showing Study Zones

The first stage of the feasibility study was completed in 1988 [41], which included hydraulic and energy modelling together with a preliminary examination of the geo-technical conditions, socio-industrial benefits and likely effects on shipping and the ecology of the estuary. No overriding impediments to the construction of a barrage were identified at that time. The Mersey Estuary has a mean spring tidal range of 8 m and a potential annual resource of about 1.4 TWh. The barrage would have 28 turbine-generators with 8-m turbines rated at 25 MW, giving an installed capacity of 700 MW [42]. The proposed barrage would be approximately 2 km long, with a design life of at least 120 years for the main structure, with two periods of turbine renewal at 40-year intervals.

There is renewed interest as a result of a recent study commissioned by Peel Environmental Ltd in association with the North West Regional Development Agency (NWDA) and the Mersey Basin Campaign [43]. There are a number of potential options for harnessing energy from the Mersey. To assess the options, the estuary was divided into study zones. These are indicated in Figure 15. The only viable option for zone 1 was considered to be a tidal lagoon. This could be operated independently from the other options. For the remaining zones, the two most productive options were two tidal barrage options (one termed as a *tidal gate*), although several tidal stream options were also studied.

The capacity and estimated electricity output from each option is indicated in Table 9. The construction cost of a Mersey Barrage is estimated at £1.5bn (2006 prices). This results in a unit cost of output ranging from 12.27p/kWh to 15.79p/kWh when a commercial discount rate of 8-10% is assumed. This would reduce to about one third if a 2% discount rate were used. The costs using the higher discount rates would result in electricity that is not commercially competitive under current conditions.

Technology Option	Rated Capacity (MW)	Annual Electricity Output (GWh}
Tidal lagoon (Zone 1)	350	650
Tidal Barrage (Zone 2)	700	1200
Central reservation (Zone 2)	20	40
Constrained channel (Zone 2)	50	100
Tidal fence (Zone 2)	35	80
Tidal gate (Zone 3)	380	700
Water wheel (Zone 3)	200	500

Source: Peel Environmental Ltd

Table 9. Comparison of Main Tidal Power Options for the Mersey Estuary.

15. Other UK Barrages

These include the Loughou Estuary in Wales which has an annual spring tide of 3.9 m and could generate 5 MW, the Duddon Estuary located on the Cumbrian coast that has a mean tidal range of 5.8 m and could generate 0.212 TWh/year from ten 10 MW turbines, the Wyre Barrage (Lancashire) with a mean tidal range of 6.6 m and installed capacity of around 60 MW that would generate about 0.131 TWh/year, the Thames Barrier that would form a new flood protection barrier that could generate possibly up to 800 MW, and the Conwy Barrage (North Wales) that would have six 5.5-MW generators giving an installed capacity of 33 MW. Here, the mean tidal range varies from 7.1 m (spring) to 3.5 m (neap) and average energy is 0.0568 TWh/year (0.0602 TWh/year with pumping).

16. Environment Impact

The Severn Estuary is a unique and dynamic environment. It has the second largest tidal range in the world, combined with a high-suspended sediment load, and has a number of special features, including extensive areas of salt marsh, and mobile sandbanks. It is an important site for migratory birds, and for fish movements in and out of the estuary's tributaries, such as the Wye and Usk. For these reasons the Severn Estuary has been designated a protected site under national and international legislation.

The most important pieces of conservation legislation for a prospective Severn barrage are the EU Directives on Birds and Habitats that protect sites designated as Special Protection Areas (SPAs) and Special Areas of Conservation (SACs). Identification of sites is a science-led process that is based on protecting important ecosystem types and threatened bird species. The Severn Estuary is a SPA and a candidate SAC. The aim of designation is to protect against biodiversity loss by conserving a series of important or at-risk habitats and species that make up the Europe-wide Natura 2000 network. The Natura 2000 network is based on the need to conserve biodiversity across Europe, and internationally.

A Severn barrage could lead to a loss of biodiversity, resulting in the need for a compensatory habitats package to maintain overall integrity of the Natura 2000 network. The EU Directives provide a clear and robust legal framework for achieving sustainable development and therefore compliance with the Directive is a central condition for a sustainable Severn barrage. Providing compensatory habitat would be a very significant undertaking on a scale hitherto unprecedented in the UK. It would have to be an integral part of any barrage proposal.

In summary, there is a strong case to be made for a sustainable Severn barrage. Much wider and stronger action on climate change is a pre-requisite for UK Sustainable Development Commission's (SDC) support. There may be an environmental opportunity available by linking a compensatory habitats package to climate change adaptation. A Severn barrage must be publicly led as a project and publicly owned as an asset to ensure long-term sustainability. The Government should consider a range of innovative financing mechanisms that would maintain overall public control and ownership of the project.

17. Carbon Emissions

One of the main arguments for building a Severn barrage is its potential contribution to reducing carbon dioxide emissions and therefore its ability to help the UK meet its national and international obligations on renewables and emissions of greenhouse gases[4].

The reduction in carbon dioxide emissions from a Severn barrage depends heavily on assumptions made on the carbon intensity of the displaced electricity. The output from a tidal barrage is intermittent, is highly predictable, and has very low operational cost. It would be treated as base-load generation, similar to that for nuclear power plants. Therefore, tidal power output is most likely to displace the output from large, centralized, fossil fuel plants.

The long lifecycle of a Severn barrage has a positive impact on the carbon *emissions factor* as the embedded emissions from construction are counter-balanced by 120 years of zero emissions electricity generation. The emissions factor for the Severn Cardiff-Weston barrage is estimated to be 2.42g$C0_2$/kWh and 1.58g$C0_2$/kWh for the Shoots scheme, which translates into a carbon payback of around 5-8 months for the two schemes. It is in the very lowest category for power generation and compares well against other low carbon technologies such as nuclear power (16g$C0_2$/kWh) [44].

[4] Under the recently-agreed EU target for 20% of all energy requirements to come from renewables by 2020 www.defra.gov.uk/news/latest/2007/climate-0309.htm , the UK will need to commit to developing at least this amount. On greenhouse gases, it is assumed that the UK will need to make substantial progress in its goal for a 60% reduction in carbon dioxide emissions by 2050, and that the UK 's commitment will most likely need to rise to a 80-90% cut in line with scientific evidence.

The Severn barrage would displace the need for some other form of new capacity, such as CCGT, as this is currently the preferred choice for new-build base-load generation. New-build gas-fired plant has a carbon intensity of around 90tC/GWh.
Table 10 presents the likely annual carbon savings (as both carbon and carbon dioxide) from the two Severn barrage proposals. Although it is possible to calculate the lifetime carbon savings (over the 120 years expected life of a barrage), the figures are unlikely to be realistic because over this period the generating capacity being displaced will be progressively less carbon intensive.

	Cardiff-Weston		**Shoots**	
	MtC	$MtC0_2$	MtC	$MtC0_2$
Annual carbon savings (based on 90tC/GWh)	1.53	5.60	0.25	0.91
Percentage reduction in UK carbon emissions (1990 baseline)	0.92%		0.15%	
For Comparison				
Annual carbon savings based on *average gas* displacement (100tC/GWh)	1.7	6.22	2.75	1.00
Annual carbon savings based on *grid mix* displacement (131tC.GWh)	2.23	8.15	0.36	1.32

Table 10. Potential Carbon Savings from a Severn Barrage.

18. Physical Implications of a Barrage

The construction, presence and operation of a Severn Barrage would involve major physical changes to water levels, geomorphology, and sedimentary processes. These physical changes underlie and have significant implications for (i) the environment—the estuarine ecosystem, inter-tidal and wetland habitats, birds, fish; and (ii) the economy and society at a local and regional scale—ports and navigation, land drainage and flooding, water quality, infrastructure and transport, employment, industry and recreation.
The changes that a barrage would cause extend well beyond the direct physical footprint of the structure, and involve physical changes to the estuary as a result of reducing the tidal range and changing the water levels within the barrage basin (upstream) and outside the barrage (downstream). The physical barrier across the estuary (Cardiff-Weston barrage is about 16km long), together with the changes to water levels, the tidal currents and the wave regime of the estuary, also mean that the sedimentary and morphological characteristics and processes of the estuary would be significantly changed.
The hyper-tidal nature of the estuary is responsible for creating a series of unique conditions and habitats such as extensive mud flats and mobile sand banks and extracting energy from this dynamic regime in the form of a tidal barrage would fundamentally change the nature of the Severn Estuary. On the whole, a barrage would raise the average water level inside

the basin by raising the low tide levels to around present mean sea level and by reducing high tide levels by up to 1m (up to about 0.5m for the Shoots scheme) The mean sea level in the estuary would be raised by some 2.5m to 3m for the Cardiff-Weston scheme. The overall effect is to reduce the tidal range by about 50%. For the Cardiff-Weston scheme, the range would decline from 11.5m to 4.5m on spring tides, and 5.5m to 2.5m on neaps. For the Shoots scheme, the result would be a similar reduction in tidal range, from 12.5m to 4.5m on spring tides, and 6.5m to 3.5m on neaps. Down stream of a barrage, model predictions for the Cardiff-Weston alignment are that low water level would be raised somewhat and high water levels would be reduced, the effects declining with distance up to 75km seawards.
Morphology refers to the form and development of the landscape. Morphology and the sediment regime have implications for the environment, the engineering of a barrage, and in relation to ports and navigation.

19. Consensus View on Tidal Power in the UK

19.1 Tidal Power

The UK has considerable tidal power resource that could be exploited to produce renewable electricity. Current estimates suggest that the UK total resource is divided roughly equally between tidal range and tidal stream potential, with a combined output equal to around 10% of UK electricity supply. All options for exploiting this resource should be considered as a narrow focus on just one project (a Severn barrage) could be detrimental to the development of a whole class of emerging tidal stream technologies, some of which could be sizeable generators of renewable electricity. There is no conflict between the tidal range and tidal stream technologies that could be deployed in the Severn. Tidal stream devises are unlikely to be viable in the Severn Estuary, but there are more appropriate conditions further out in the Bristol Channel where this might be viable. Small-scale tidal lagoon development could take place alongside a tidal barrage. Large-scale tidal lagoon development in the Severn Estuary would not offer any economic or environmental advantage over a barrage.

19.2 Tidal Stream

The long-term potential of tidal stream technologies [45], subject to constraints that might be imposed due to location-specific impacts upon the environment, natural marine processes, and long-term costs being acceptable, should be exploited. The UK is in a unique position with superior tidal stream resource combined with devices being developed or tested. Tidal stream technologies could make a substantial contribution to the sustainable energies of the UK. Considering the progress that has been made on tidal stream, the objective now should be to *stay the course*. In many ways, the tidal stream industry is the same as wind power was 20 years ago, and the timescale for bringing prototype technologies to large-scale deployment needs to be as fast-tracked as possible.

19.3 Policy Improvements

There are a number of areas where Government policy could be improved. The support and funding structures need to be reviewed and improved in line with circumstances as they develop and change. A flexible approach could be taken on the future of the Department for Business, Enterprise and Regulatory Reform's (BERR) Marine Renewables Deployment Fund (MRDF), which has so far not had any applicants due to delays in getting demonstration projects off the ground. Lessons could be learnt from the success of the Scottish Government's £8 m support package for marine energy technologies, which has had strong interest from both tidal and wave developers. Increased support for marine renewables under a branded Renewables Obligation is also very welcome. This may provide an opportunity to revise the support available under the MRDF so that it focuses on providing grant funding for project development and testing, with the aim of stimulating progress towards initial tidal arrays and pre-commercial schemes.

The European Marine Energy Centre in the Orkney Island is an excellent example of public sector funding being used to stimulate public sector investment and innovation in a strategic and efficient manner. Looking to the future, it is thought that there is potential to exploit the activity entered around the European Marine Energy Centre (EMEC) to develop a regional hub around the Orkney Island and parts of the Caithness coastline, away from the Pentland Firth for commercial testing of devices beyond the prototype stage. In the long-term, lack of transmission capacity would appear to be a serious constraint on development of the UK's tidal stream resource in the north of Scotland. This also impacts on the offshore and onshore wind industry and on wave power devises. There are a number of problems with the current regime for connecting renewable generation and a real absence of long-term thinking on solutions to overcome them. This has serious consequences for the UK's ability to meet its targets for renewable electricity and the more ambitious EU targets that will eventually be implemented.

19.4 Strategic Planning and Consenting

Lack of good baseline information on the marine environment and on effects of large-scale deployment of different devices is a real issue. The gaps have to be filled over time through research of a strategic and generic nature as well as by developers. The Scottish Government is in the process of completing a strategic environmental assessment for marine renewables around the west and north coasts of Scotland, and the Welsh Assembly Government is developing a marine renewable strategy.

19.5 Tidal Lagoons

It is difficult to come to a clear view on the long-term potential of tidal lagoons due mainly to the lack of authoritative evidence and that the concept remains unproven. Government should investigate options to encourage one or more tidal lagoon demonstration projects.

19.6 Tidal Barrages

Extensive information is already available on the Severn Barrage that contains the majority of the UK's tidal range resource, and also for the Mersey. There does not seem to be an extensive overlap between tidal barrages and tidal stream devices, leading to the conclusion that they can, on the whole, be considered separately. The UK's potential for developing different tidal barrage options other than the Severn is extensive, but the reason why this potential has not been developed in the past is that they have not appeared to be economically viable. Further investigation into UK tidal barrage options outside the Severn

Estuary should be considered on a case-by-case basis, as potential benefits will differ considerably.

20. Acknowledgement

The Author is grateful to the UK Sustainable Development Commission, the UK Government independent authority on sustainable development, for information that has been cited in this paper.

21. References

[1] T. J. Hammons. Tidal Power (Invited Paper). Proceedings IEEE (Special Issue on Advanced Power Generation Technoledges), vol. 81, (3), March 1993, pp. 419-433.
[2] A. C. Baker. Tidal Power. United Kingdom: Peter Peregrinus Ltd. 1991.
[3] Barry V. Davis, "A Major Source of Energy from the World's Oceans," IECEC-97 Intersociety Energy Conversion Engineering Conference, 1997,
[4] N.H. Halvorson, "Evaluation of Nova Energy Ltd.'s Hydro Turbine for (Canadian) Ministry of Employment and Investment", N.H. Halvorson Consultants Ltd.
[5] "Renewable Energy: Power for a Sustainable Future; Technology Update: (2001) Tidal Current Power Update & Wave Power Update", Oxford University Press.
[6] Peter Fraenkel, "Renewables; Is the Tide Turning for Marine Current Turbines?", Modern Power System, Marine Current Turbines Ltd, London, UK, June 30, 2001.
[7] R. Bedard, "Final Summary Report: Offshore Wave Power Feasibility Demonstration Project", (E2I EPRI Global WP009 - US), 2005.
[8] Mirko Previsic, "System Level Design, Performance and Costs for San Francisco Pelamis Offshore Wave Power Plant", (E2I EPRI Global - 006A - SF), 2004.
[9] George Hagerman, "Offshore Wave Power in the US: Environmental Issues", (E2I Global EPRI - 007 - US), 2004.
[10] Bonnie Ram, "Wave Power in the US: Permitting and Jurisdictional Issues", (E2I Global EPRI DOE NREL - 008 - US)
[11] M. Previsic, "Wave Power Technologies", Proceedings of the IEEE PES 05 GM, San Francisco, paper 05GM0542, June 2005, pp. 1-6.
[12] E2i/EPRI WP-004-US Rev 1 Assessment of Offshore Wave Energy Conversion Devices
[13] E2i/EPRI WP-005-US Methodology, Guidelines and Assumptions for the Conceptual Design of Offshore Wave Energy Power Plants (Farms)
[14] E2i/EPRI WP-006-HI System Level Design, Preliminary Performance and Cost Estimate - Hawaii -
[15] E2i EPRI WP-006-OR System Level Design, Preliminary Performance and Cost Estimate - Oregon
[16] E2i/EPRI WP-006-ME System Level Design, Preliminary Performance and Cost Estimate - Maine
[17] E2i/EPRI WP-006-MA System Level Design, Preliminary Performance and Cost Estimate - Massachusetts
[18] E2i/EPRI WP-006-SFa System Level Design, Preliminary Performance and Cost Estimate - San Francisco, California Pelamis Offshore Wave Power Plant

[19] E2i/EPRI WP-006-SFb System Level Design, Preliminary Performance and Cost Estimate - San Francisco Energetech Offshore Wave Power Plant

[20] Sea Technology Magazine August 2003, Wave Energy Conversion - The State of the Art

[21] EPRI - RE Technical Assessment Guide Ocean Wave Power Section for the years 2001, 2002 and 2003

[22] California Energy Commission - Wave Energy Resource Assessment for the State of California

[23] A.D. Trapp and M. Watchorn, "EB development of tidal stream energy," in *Proceedings MAREC 2001*, p. 169-173, 2001.

[24] A.T. Jones and A. Westwood, "Economic Forecast for Renewable Ocean Energy Technologies," presented at EnergyOcean 2004, Palm Beach, Florida, 2004.

[25] A.T. Jones and W. Rowley, "Global Perspective: Economic Forecast for Renewable Ocean Energy technologies," MTS Journal, vol. 36. No. 4, pp. 85-90, Winter 2002.

[26] N.J. Baker, M.A. Mueller, M. Watchorn, D. Slee, L. Haydock and N. Brown, "Direct drive power take off for the Stingray tidal current generator," in *Proceedings MAREC 2002*, p. 1-10, 2002.

[27] A. M. Gorlov, "The Helical Turbine and its Applications for Tidal and Wave Power," in *Proc. OCEANS 2003*, p. 1996, 2003.

[28] P.W. Ullman, "Offshore Tidal Power Generation - A new approach to power conversion of the oceans' tides," *MTS Journal*, vol. 36. no. 4, pp. 16-24, Winter 2002.

[29] P. Breeze, "The Future of Global Offshore Wind Power," *Reuter Business Insight* 2004.

[30] L. Coakley, "Long Island Offshore Wind Park - 140 Megawatts of Offshore Wind Energy," presented at EnergyOcean 2004, Palm Beach, Florida, 2004.

[31] O. Siddiqui and R. Bedard, "Feasibility Assessment of Offshore Wave and Tidal Current Power Production", Proceedings of the IEEE PES 05 GM, San Francisco, paper 05GM0538, June 2005, pp. 1-7.

[32] E2i/EPRI WP-08-US- Identification of Permitting Issues

[33] A. T. Jones and A. Westwood, "Recent Progress in Offshore Renewable Energy Development", Proceedings of the IEEE PES 05 GM, San Francisco, paper 05GM0543, June 2005, pp. 1-6.

[34] A. C. Baker, The development of functions relating cost and performance of tidal power schemes and their application to small-scale sites. Tidal Power. London: Thomas Telford, 1986.

[35] M. Banai and A. Bichon. Tidal Power in France: The Ranch tidal power station: Some results after 15 years of operation, Energy Digest, pp. 39-45, October 1982.

[36] A. D.Grant. Power Generation from tidal flows for navigation buoys, 2nd Int. Symp, Wave and Tidal Energy. BHRA, Cambridge, September 1981, pp. 117-128.

[37] I. G. Bryden et.al. Assessing the Potential of a Simple Tidal Channel to Deliver Useful Energy, Applied Ocean Research, Vol. 26, 2003, pp. 198-204.

[38] S. Kiho, et. al. The Power Generation from Tidal Currents by Darrieus Turbine, Renewable Energy, Vol. 9, (1-4), 1996, pp. 1242-1245.

[39] A. M. Gorlov. The Helical Turbine and its Application for Tidal and Wave Power. Proceedings IEEE OCEANCS'03, Vol. 4, San Diego (USA), September 2003, pp. 1996.

[40] NGET (2007). GB 7 year Statement. www.nationalgrid.com/uk/electricity/sys/current/

[41] Tidal Power from the River Mersey: A feasibility Study—Stage 1 Report. Mersey Barrage Co., 1988
[42] N. W. Hammond and P. Wood. Tidal Power from the Mersey: History and Prospects, in Developments in Tidal Energy, Proc. 3rd Conf. Tidal Power, London, UK, November 1989.
[43] Peel Environmental Ltd (2007). Mersey Tidal Power Study. www.merseytidalpower.co.uk
[44] Sustainable Development Commission (SDC). Paper 2, Reducing CO_2 Emissions--Nuclear and the Alternatives. From the SDC Project 'The Role of Nuclear Power in a Low Carbon Economy' {www.sd-commission.org.uk/pages/060306.html}.
[45] T. J. Hammons. Energy Potential of the Oceans in Europe and North America: Tidal, Wave, Currents, OTEC, and Offshore Wind. International Journal of Power and Energy Systems, Paper 203-4142, Vol. 28, (4), 2008, pp. 416-428.

25

Wind Energy Technology

R. Mesquita Brandão[1], J. Beleza Carvalho[1] and F. P. Maciel Barbosa[2]
[1]*ISEP, Porto*
[2]*FEUP and INESC Porto, Porto*
Portugal

1. Introduction

Electricity is regarded as one of the indispensable means to growth of any country's economy. This source of power is the heartbeat of everything from the huge metropolitans, industries, worldwide computer networks and our global communication systems down to our homes. Electrical energy is the lifeline for any economic and societal development of a region or country. It is central to develop countries for maintaining acquired life styles and essential to developing countries for industrialisation and escaping poverty.

Electrical energy is a high quality form of energy which is not found in nature. Instead it has to be obtained from the conversion in power plants of other types of energy supplies.

These energy supplies can broadly be divided into two classes:

- Renewable energy which is obtained from continuous or repetitive currents of energy occurring in the natural environment
- Non-renewable energy which is obtained from static stores of energy that remain boundless released by human interaction

Until the end of last century the industrialised world has mainly satisfied the demand for electrical energy by developing technologies for the generation of electricity which use mainly non-renewable energy supplies such as coal, fuel or nuclear. These environmentally hazardous generation technologies continue to be the backbone of the electricity generation system. The need for harnessing renewable energy supplies is apparent as world population increases and as each individual presses for a higher standard of living in terms of material goods especially in rural and developing regions.

Renewable Energy Systems (RES) and Distributed Generation (DG) have the potential to become the foundation of a future more sustainable energy supply system. Their scale deployment will transform the energy landscape from a system dominated by the centralized combustion of fossil fuels to a new one in which new technologies, environmentally friendly, contribute to a substantial development.

From an investment view point, it is generally easier to find sites for RES and DG than for large central power plants and, in addition, such units can be installed in a short time, near to the end consumer. The widespread integration of RES and DG together with energy efficiency, covering supply and demand, has provided support to achieve the major EU policy objectives:

- Sustainable energy development, combating climate changes and reducing air pollutants is the paradigm of the future power systems. The shift from the large scale combustion of fossil fuels to a more decentralized energy supply based on RES has contributed for meeting the Kyoto commitments, regarding the emission of greenhouse gases, particularly CO2: 8% reduction of emissions from 1990 levels by 2008-2010 and 20% by 2020 compared to 1990;
- Security and diversity of energy supply. Reducing the external energy dependence is crucial for the development of a dynamic and sustainable economy for Europe;
- Increasing the penetration of RES, doubling their share in energy supply quota from 6% to 12% of gross Energy consumption and raising their part in electricity production from 14% in 2001 to 22% is an objective to be attained by 2010.
- Energy marked liberalization, increasing opportunities for smaller scale generators.

The main objective of any generation and distribution utility is to satisfy the demand of customers with a high quality product. This product namely electricity must be supplied continuously round the clock.

Wind generation, photovoltaic's panels, fuel cells and microturbines - just to mention a few- are new forms of electricity under development. They define the so called RES (**R**enewable **E**nergy **S**ystems) and involve the exploitation off distribution sources through the concept of DG (**D**istributed **G**eneration). Today, wind power and CHP (**C**ombined **H**eat and **P**ower) are entering into a competitive level with traditional forms of energy generation. Tomorrow it is expected that one speaks also about microgeneration (microturbines, micro-CHP, photovoltaic systems and fuel cells).

Wind energy is the renewable energy source that had a higher growing in the last decades and can be considered a hope in future based on clean and sustainable energy.
The idea that wind power is going to play a significant role in our energy future has begun to take hold. Clean, emissions free, wind power is now correctly regarded as an increasingly important part of the answer to the twin global crisis of energy security and climate change. We are in the midst of a period of fundamental change as to how we produce and consume energy, and nowhere is this clearer than in the explosive growth in investment in the clean energy sector, with wind power taking by far the largest share of that investment, some 50 billion USdollars in 2007 alone. More wind power was installed in Europe in 2007 than any other technology, some 40% of all new power generation capacity, and it also accounted for 30% of all new generation capacity installed in the United States during that same period. Of equal significance is the fact that for the first time in decades, the majority of the 2007 market was outside Europe, concentrated primarily in the United States and China.
However, the integration of both RES and DG into the overall power systems operation requires that energy generation in both transmission and distribution systems can no longer be considered as passive appendage. Reliability, safety and quality of power are the main issues linked to the large-scale deployment of DER (**D**istributed **E**nergy **R**esources) so that their effect on the European transmission and distribution networks cannot be neglected. Rather, it must be addressed with a comprehensive system approach.
Figure 1 shows global wind power capacity, 1991-2008 (in MW).

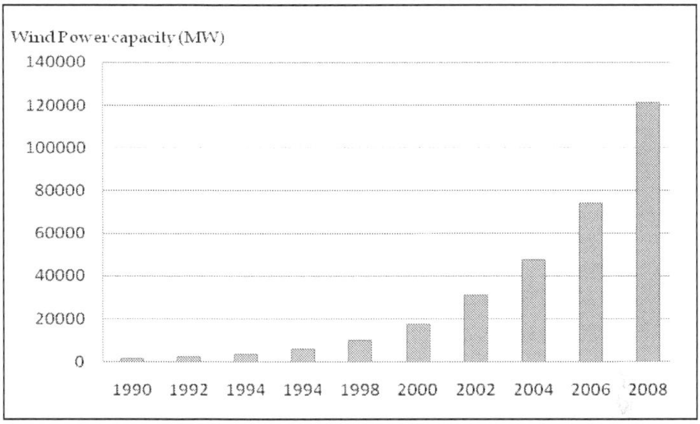

Fig. 1. Wind Power, existing World Capacity, 1995-2008 (in MW) (Source: Global Wind Energy Council, Global Wind 2008 Report)

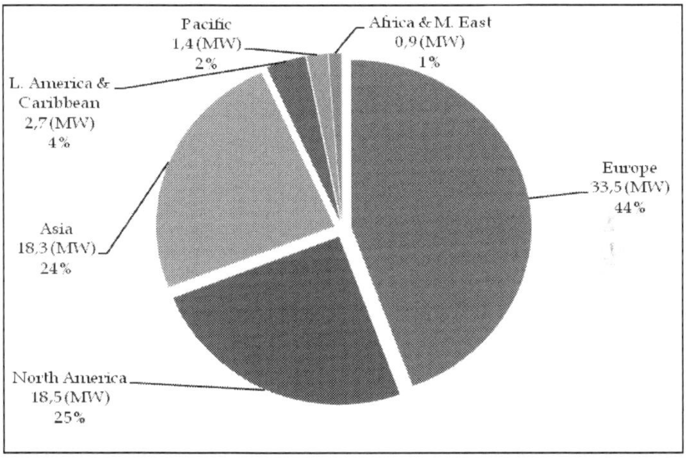

Fig. 2. Targets of wind capacity to be installed by 2010 (Source: Global Wind Energy Council, Global Wind 2008 Report)

2. Europe Wind Power Overview

The two oil crises that occurred in the seventies of the last century led Europe thinking in new ways of energy production based in renewable resources. The wind resource was already very used to produce mechanical work for many years. The presence of windmills is one of the most characteristic features of the Europe landscape, so developments to use this resource to produce electrical energy become to be interesting. National wind power programmes were started in almost all Europe member countries in decade of 1970 and have been continued since then, but it's usual to consider that the modern wind energy

industry had started in the 80's of the last century. Till the early 1980s, Europe possessed only nearly 5% of the worlds installed capacity in wind power. Nowadays Europe is leading in terms of installed capacity and in terms of technology developments.
Taking use of the existence of European programmes and funds, governments of almost European countries have launched programmes aimed to developing the wind energy sector. Using as reference the decade of 1990, because before that wind power installed capacity and the growth per year was almost insignificant, Europe growth from nearly 500 MW in wind power installed in 1990 to 65.933 MW in the end of 2008. In the beginning, countries like Denmark, the Netherlands, Germany, Spain and United Kingdom had the highest degree of development; nowadays Germany and Spain are the countries with the most wind power installed. The power installed in these two countries is more than 60% of the power installed in all Europe.

The wind power potential varies enormously across Europe, as we can see in the figure below. Europe is divides into two wind zones, the north zone where the wind speed is higher, result from the cyclones-anticyclones running in an East-West direction over the North Atlantic Ocean. In the South zone, with more mountainous terrains and where Mediterranean Sea and North Africa creates local wind systems, the wind may have very high speeds.

In addition to the areas of strong potential figured in the map, Figure 3, and in order to make a correct estimate of the potential at a particular site, is necessary to put an anemometer for a minimum period of one year, to collect data of wind speed, frequency and direction, density of air, etc.

Associated to the increase of installed wind power is the development of technologies. In the 90's of last century, wind turbines usually used Europe come in the 100-500 kW range of unit power. Since then, the unit power of the generator, the diameter of rotor blades and the tower height never stop and were always increasing. The evolution of modern turbines is a remarkable story of engineering. In the last 20 years turbines have increase in power by a factor of 100 and coefficient of performance had evolution from 0.4 to 0.5, which is near of the maximum coefficient of performance, determined by Betz limit. During the years seems that there is no limit for the growing of towers height and rotor diameters. In the beginning, small systems had rotor diameters of 12 meters or less, installed in towers of a little more than 20 meters of height and power of 50 kW, nowadays towers can be higher than 120 meters, with rotor diameter of 120 meters and power of more than 5 MW (Figure 4).

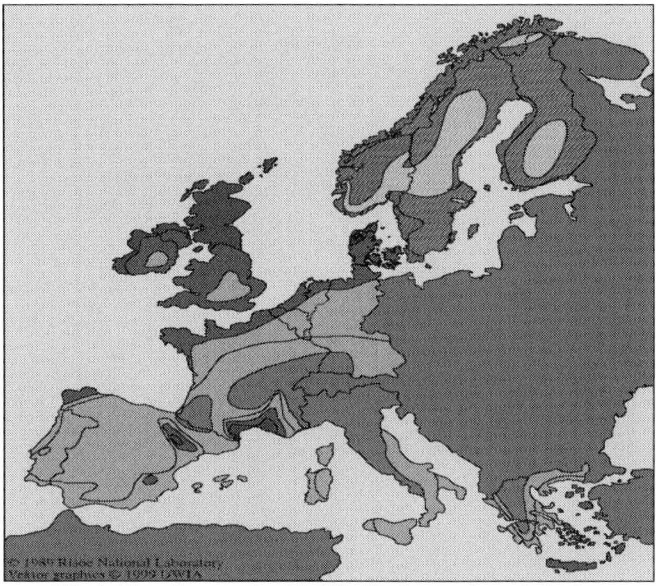

Fig. 3. European Wind Atlas (Source: Danish Wind Industry Association)

	m/s	W/m²	m/s	W/m²	m/s	W/m²	m/s	W/m²	m/s	W/m²
	>6.0	>250	>7.5	>500	>8.5	>700	>9.0	>800	>11.5	>1800
	5.0-6.0	150-250	6.5-7.5	300-500	7.0-8.5	400-700	8.0-9.0	600-800	10.0-11.5	1200-1800
	4.5-5.0	100-150	5.5-6.5	200-300	6.0-7.0	250-400	7.0-8.0	400-600	8.5-10.0	700-1200
	3.5-4.5	50-100	4.5-5.5	100-200	5.0-6.0	150-250	5.5-7.0	200-400	7.0-8.5	400-700
	<3.5	<50	<4.5	<100	<5.0	<150	<5.5	<200	<7.0	<400
			>7.5							
			5.5-7.5							
			<5.5							

Table 1. Wind Speed at 50 m above ground level. (Source: Danish Wind Industry Association)

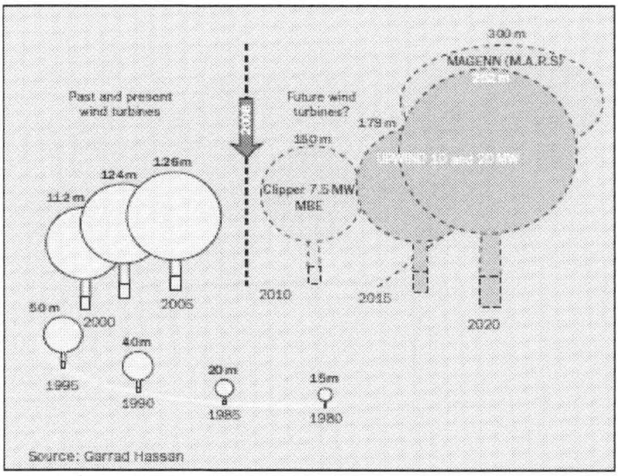

Fig. 4. Growth in size of commercial wind turbine designs (Source: Danish Wind Industry Association)

Figure 4 shows the perspective of growth in size of commercial wind turbine designs from 1980 to 2020.

3. Wind Energy Conversion Systems

The production of electricity from wind energy presents an increased growth and sustained since 1985. Currently, there are wind generators located throughout the world whose power already reaches values exceeding 3000 MW.
The main technologies used in electromechanical conversion of wind energy into electric energy are based primarily on three types of electric machines:

- The direct current (DC) machine
- The synchronous machine
- The induction machine

These machines work on the principles of the electromagnetic actions and reactions. The resulting electromechanical energy conversion is reversible. The same machine can be used as the motor for converting the electrical power into mechanical power, or as the generator converting the mechanical power into electrical power.
Typically, there is an outer stationary member (stator) and an inner rotating member (rotor). The rotor is mounted on bearings fixed to the stator. Both the stator and the rotor carry cylindrical iron cores, which are separated by an air gap. The cores are made of magnetic iron of high permeability, and have conductors embedded in slots distributed on the core surface. Other way, the conductors are wrapped in the coil form around salient magnetic poles. In the Figure 5 is possible to see a cross-sectional view of the rotating electrical machine with the stator with salient poles and the rotor with distributed conductors (Mukund, 1999). The magnetic flux, created by the excitation current in one of the two members, passes from one core to the other in the combined circuit always forming a closed loop. The electromechanical energy conversion is accomplished by interaction of the magnetic flux produced by one member with electric current in the other member. The latter may be externally supplied or electromagnetically induced. The induced current is proportional to the rate of change in the flux linkage due to rotation.

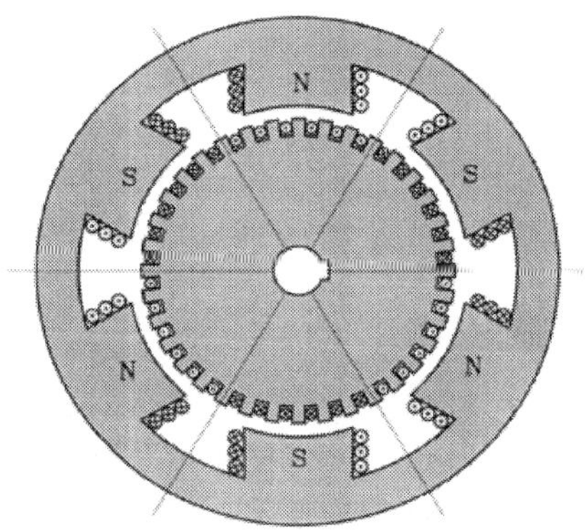

Fig. 5. Cross section of the electrical machine stator and rotor

DC Machine

The conventional DC machine is either self-excited by shunt or series coils carrying DC current to produce a magnetic field. Actually, the DC machine is often designed with permanent magnets to eliminate the field current requirement, hence, the commutator. It is designed in the "inside-out" configuration. The rotor carries the permanent magnet poles and the stator carries the wound armature which produces AC current. This current is then rectified using the solid state rectifiers. Such machines do not need the commutator and the brushes; hence, the reliability is greatly improved. The permanent magnet DC machine is used with small wind turbines, however, due to limitation of the permanent magnet capacity and strength. The brushless DC machine is expected to be limited to ratings below one hundred kW (Mukund, 1999).

Synchronous Machine

Most of the electrical power consumed in the world is generated by the synchronous generator. For this reason, the synchronous machine is a well established machine. The synchronous machine works at a constant speed related to the fixed frequency. Therefore, it is not suitable for variable-speed operation in the wind plants. Moreover, the synchronous machine requires DC current to excite the rotor field, which needs sliding carbon brushes on slip rings on the rotor shaft. This poses a limitation on its use. The need of DC field current and the brushes can be eliminated by the reluctance torque. The reliability is greatly improved while reducing the cost. The machine rating, however, is limited to tens of kW. The reluctance synchronous generator is actually used for small wind generators (Mukund, 1999). In the Figure 6 is possible to see the diagram of connections of wind generators equipped with variable speed synchronous machines.

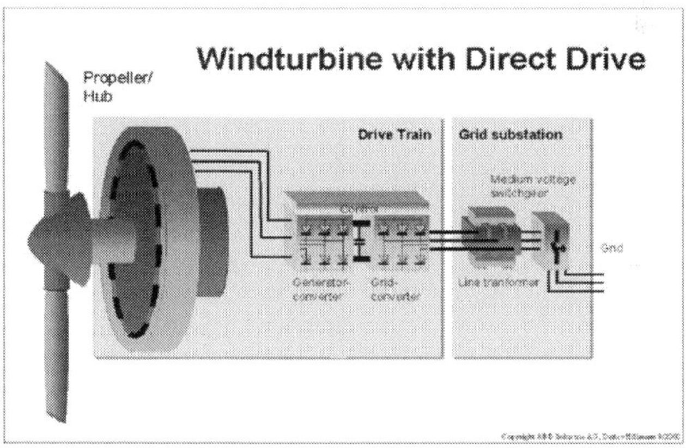

Fig. 6. Connections diagram of a synchronous generator operating the variable speed (Source: ABB Industries).

The systems represented in Figure 6, the synchronous machine is connected through a system of converting ac/dc/ac, as the frequency of stators voltage and currents is different in the frequency of the electrical network.

Such generators typically do not have the gearbox, and the mechanical speed of rotation of the rotor is identical to the speed of rotation of the turbine. Typically the speed of rotation of the turbine (and the rotor of synchronous machine) varies between 17 rpm and 36 rpm, so the machine has a high number of poles.

The stator of the synchronous machine has six phases and is connected to two independent converting systems ac/dc/ac. The parallel between the two conversion systems is made at the outlet of converters dc/ac (grid converters) that is connected to the elevator transformer. Each of the converters ac/dc connected to the generator (the generator converters) consists of one to six pulse bridge converter equipped with thyristors. These thyristors operate with a constant firing angle.

The DC voltage at capacitor terminals, placed in parallel in connection at direct current, must be set to a constant value. However, for low values of the speed of the rotor, the excitation system of synchronous machine is unable to ensure that value, being necessary to use a "chopper" (converter dc/dc) converter installed between the generator and capacitor, which is disconnected when the rotor speed exceeds a certain value.

The grid converter is a six pulse converter bridge equipped with IGBTs, with a control system based in pulse width modulation (PWM). This converter controls the active power injected into the grid and the power factor. The control of active power in the grid converter allows the imposition of electromagnetic torque into generator, thus making it possible to control the rotational speed of the wind turbine-generator group in order to obtain the specific speed of the tip of the blade optimal (λ) for each value of wind speed (Akhmatov, 2002).

Figure 7 illustrates the active and reactive power supplied by the grid converter of such a wind generator according to the rotational speed of the rotor.

Unlike the induction machine, the synchronous machine, when used in the grid-connected system, has some advantages. It does not require the reactive power from the grid. This results in better quality of power at the grid interface. This advantage is more important when the wind farm is connected to a small capacity grid using long low voltage lines. Actually, wind plants generally connect to larger grids using shorter lines, and almost universally use the induction generator (Cigrè, 2001).

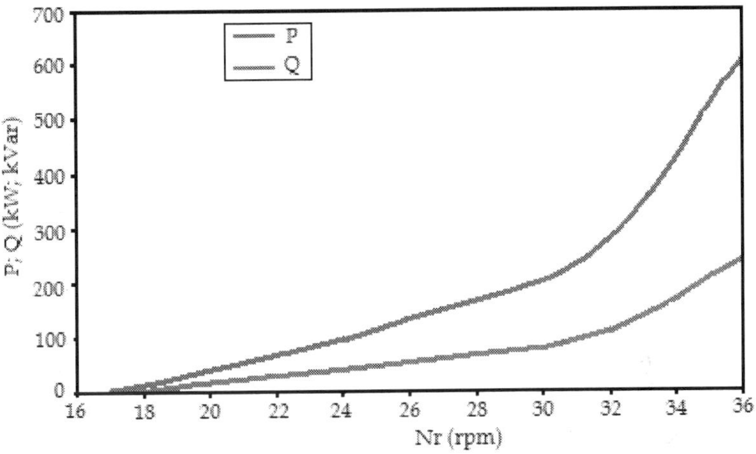

Fig. 7. Active and reactive power supplied by a wind generator equipped with synchronous generator operates the variable speed depending on the speed of the rotor.

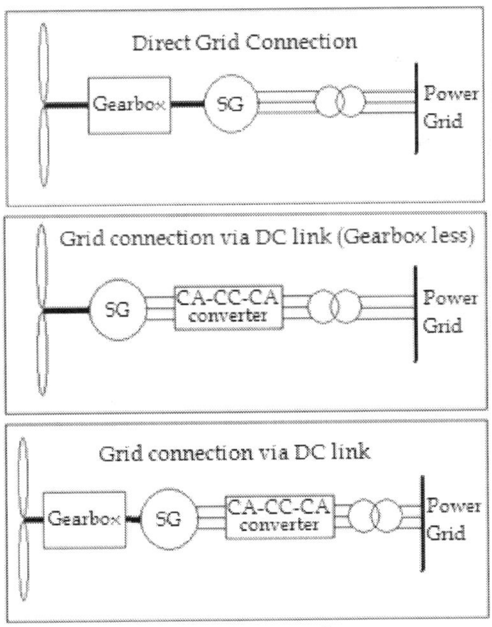

Fig. 8. Settings of synchronous machine used as a wind generator

Induction Machine

The primary advantage of the induction machine is the rugged brushless construction and no need for separate DC filed power. The disadvantages of both the DC machine and the synchronous machine are eliminated in the induction machine, resulting in low capital cost, low maintenance, and better transient performance. For these reasons, the induction generator is extensively used in small and large wind farms and small hydroelectric power plants. The machine is available in numerous power ratings up to several megawatts capacity, and even larger.

The induction machine needs AC excitation current. The machine is either self-excited or externally excited. Since the excitation current is mainly reactive, a stand-alone system is self-excited by shunt capacitors. The induction generator connected to the grid draws the excitation power from the network. The synchronous generators connected to the network must be capable of supplying this reactive power. For economy and reliability, many wind power systems use induction machines as the electrical generator.

Operation of the Induction Generator in Autonomous Mode

The induction machine to function as a generator must be operated at a speed above the synchronous speed and to be provided with a reactive power to produce and keep the machine's magnetic field. This reactive power can be produced by capacitors connected to the machine, as described in Figure 9. Thus, it is possible to achieve self-excitation of the machine in order to feed a load alone.

The capacitors are usually connected in delta because they have the advantage of lower capacity to get the same effect as with capacitors connected in star. Thus, the voltage V1 and the frequency f1 of the generators of induction in empty and laden depend primarily on parameters of the machine, the capacity of condensers and speed n> f1/p, where "p" is the even number of poles.

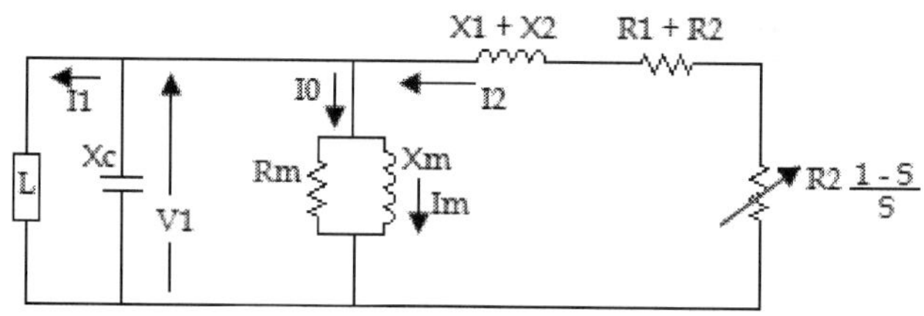

Fig. 9. Approximate equivalent scheme of an induction generator for autonomous load

The existence of residual magnetism in the machine, with the machine to turn, will result in the emergence of power swing in the machine between the stator coils and condensers. Indeed, the coils of inductance L and the capacity C of the capacitors are an oscillating circuit and therefore fluctuations of energy may be damped or amplified.

If the rotor rotates with angular velocity w_r whose frequency is higher than the frequency of own oscillations (given by $1 / \sqrt{LC}$) then the power of the rotor copper losses in the oscillating circuit and the machine turns on. If however there is no residual magnetism or if this is not enough, the oscillations do not occur or cushion quickly and the machine not exciting.

The operating voltage and frequency are determined in terms of the approximate equivalent circuit of Figure 9. On no load, the capacitor current Ic= V1/Xc must be equal to the magnetizing current Im=V1/Xm. The voltage V1 is a function of Im, linearly rising until the saturation point of the magnetic core is reached (Figure 9). The stable operation requires the line ImXc to intersect the V1 versus Im curve. The operating point is fixed where V1/Xc equal V1/Xm, that is when 1/Xc=1/Xm, where Xc=1/wC. This settles the operating frequency in hertz. With the capacitor value C, the output frequency of the self-excited generator is therefore:

$$f = \frac{1}{2\pi X_m} = \frac{1}{2\pi\sqrt{C.L_m}} \tag{1}$$

Under load conditions, the generated power V1I2cosφ_2 provides for the power in the load resistance R and the iron loss in Rm. The reactive currents must sum to zero:

$$\frac{V_1}{X} + \frac{V_1}{X_m} + I_2.\sin\varphi_2 = \frac{V_1}{X_c} \tag{2}$$

This equation determines the output voltage of the machine under load (Mukund, 1999).

As it possible to see in Figure 10, the process of self-excitation requires the presence of a residual magnetism and magnetic saturation of the magnetization curve of the machine to be successful, or to have a clear intersection between the two characteristics (of magnetization and strain in capacitors).

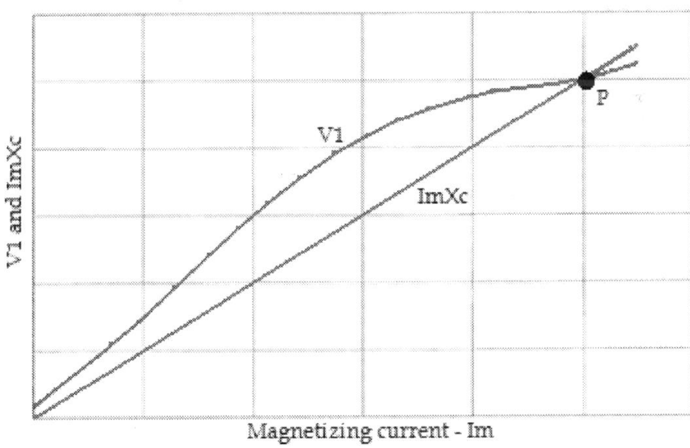

Fig. 10. Operating characteristics of the induction generator with capacitive self-excitation

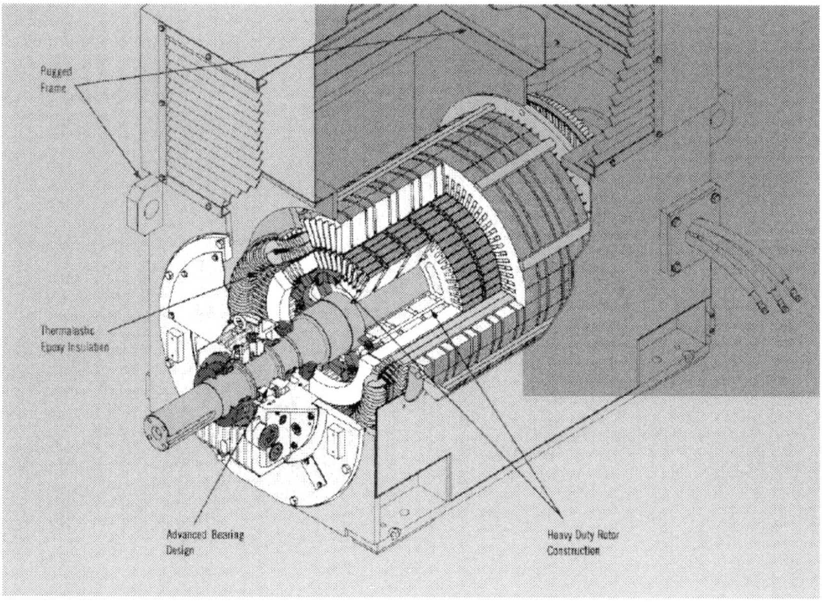

Fig. 11. Two MW induction machine (Source: Teco Westinghouse Motor Company)

Operation of the induction generator connected to the network

The electromagnetic power through the air gap is given by:

$$P_{em} = 3.I_2{}^2.\frac{R_2}{s} \tag{3}$$

that is positive for s> 0 and negative for s <0, where "s" is the slip of the machine. That is, for s <0 the electromagnetic power flow at rotor to the stator. Part of this power is dissipated (by Joule effect) in the copper winding of the stator and the remainder is supplied to the network. This corresponds to the operation of the machine as a generator (Figure 12). In this case, the machine must be operated at a speed n> f1/p and both the power and the electromagnetic torque are negative.

When assessing the performance of induction generator we can use the approximate equivalent diagram of Figure 9 with s <0. The resistance that ((1-s) / s) R2, which reflects the electromagnetic power, depends on the slip, but the reactance X does not depend on the slip, or are always positive. Consequently, the induction machine always absorbs reactive power in whatever condition of operation.

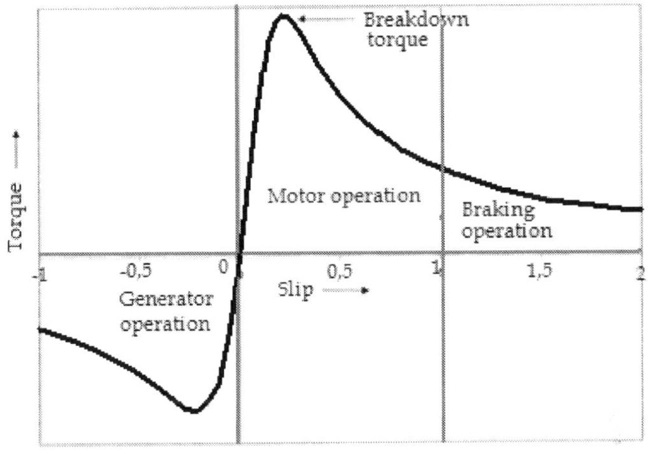

Fig. 12. Torque versus speed characteristic of the induction machine in three operating modes

How is possible to see in Figure 13, if the generator is loaded at constant torque T_L, it has two possible points of operation, P_1 and P_2. Only one of these two points, P_1 is stable. Any perturbation in speed around point P_1 will produce stabilizing torque to bring it back to P_1. The figure also shows the limit to which the generator can be loaded. The maximum torque it can support is called the breakdown torque, which is shown as T_{max}. If the generator is loaded under a constant torque above T_{max}, it will become unstable and stall, draw excessive current, and destroy itself thermally if not properly protected (Manwell et al., 2002).

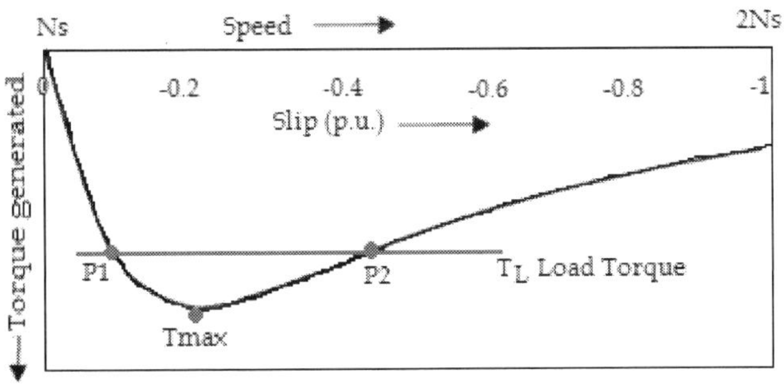

Fig. 13. Torque versus speed characteristic of the induction generator under load

Usual Configuration of the Induction Generator

The induction generators connected to the network or in autonomous mode are mainly used, for constant or variable speeds and a link voltage/constant or variable frequency, in mini-hydro and wind energy systems. Possibilities for the use of double fed induction generators and the squirrel cage rotor are summarized in Table 2.

Induction generator	speed		Network connection	Isolated	Frequency		Voltage	
	Constant	Variable			Constant	Variable	Constant	Variable
Double Fed		X	X		X		X	
Squirrel Cage	X	X	X	X	X	X	X	X

Table 2. Configuration of the Induction Generators.

The principle of operation of the double fed induction machine is based on the ability to control its speed by variation of the resistance of the rotor. Figure 14 illustrates the change curves of torque/slip of the induction machine due to the variation of resistance connected in series with the winding rotor.

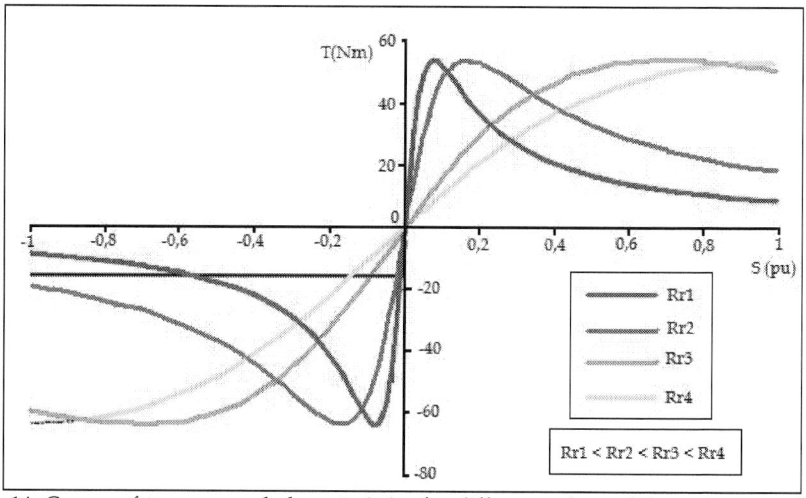

Fig. 14. Curves of torque-speed characteristics for different values of the resistance of the rotor

As shown in the Figure 14, for a given mechanical torque T, can vary the speed of induction machine by varying the rotor resistance. If instead of a variable resistance, if we install a system for converting ac/dc/ac connected to the rotor, it is possible to extract the active power by the rotor of the machine and thus control the speed. This is the principle of energy away from the winding rotor induction machine.

The mode of operation of double fed induction generators based on the principle described above: to negative slips, until it reaches the intensity of the stator rated current of the machine, the power extracted by the rotor of the machine is controlled so as to optimize the speed specified the tip of the blade of the rotor and thereby maximize the value of the coefficient of the power turbine.

For negative slips, higher (in modulus) for which the intensity of the stator current reaches the nominal value, the active power in the stator and the rotor remains constant, like showed for the line in black, on Figure 14 (Manwell et al., 2002).

This principle of speed control by use the slip energy means that this machine can function as a generator for positive slip. To ensure this mode of operation, it is necessary to provide active power to the rotor. In Figure 15 we can see different ways to use the induction machine as wind generator (Cigrè, 2001).

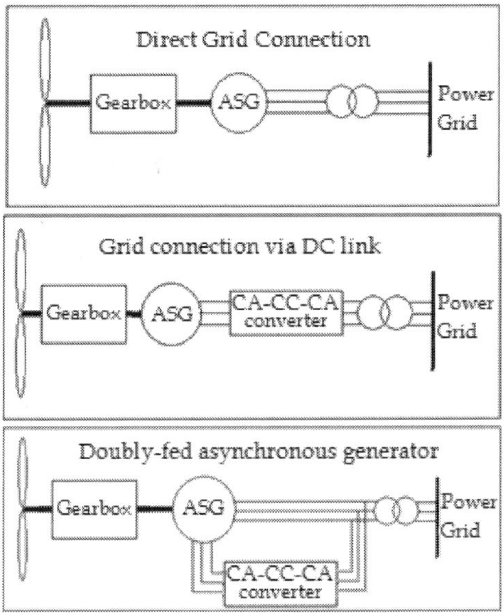

Fig. 15. Settings of induction machine used as a wind generator

The connections of double fed induction machine are shown in Figure 16.

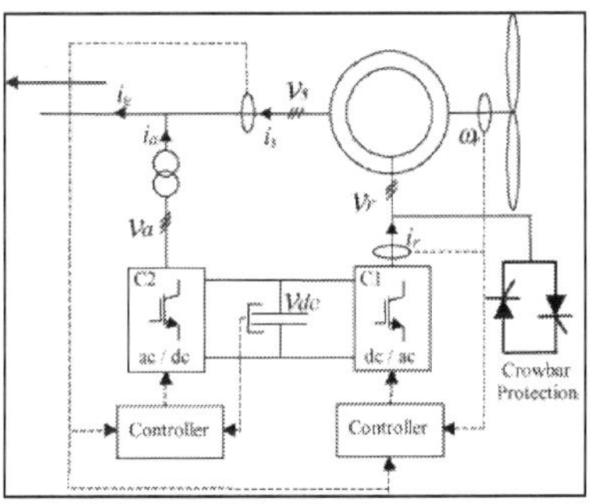

Fig. 16. Scheme of connections of double fed induction machine (Almeida et al., 2004)

The stator of the induction machine is directly connected to electric power. The rotor is connected to the network through a system of converting ac/dc/ac and a transformer.
The converters ac/dc/ac that interconnect the rotor of the machine to the network via the transformer, are bridge-type converters PD3 to six pulses equipped with isolated gate bipolar transistors (IGBTs) controlled by the pulse width modulation. Typically, in double fed induction machine, converter connected to transformer controls the voltage into the terminals of the capacitor in DC current system, and controls the power factor at the point common to the circuits of the rotor and stator. The converter directly connected to the rotor of the induction machine control module and the argument of the intensity of current injected or extracted through the rotor (Ekanayake et al, 2003).

The principle of operation of the control system with pulse width modulation can impose a form of wave approximately sinusoidal with frequency, amplitude and phase adjustable to the AC terminals of the converters.

In Figure 16, the converter ac/cc/ac connected to the rotor of the induction machine, allows the control of the frequency of the wave form applied to the rotor, which is equal to the slip frequency of the machine in a given point. Simultaneously, it also controls the module and argument of the intensity of current in the rotor. Converter ac/cc connected to the terminals of the transformer controls the magnitude of the voltage into the terminals of the capacitor. The frequency of the alternating current frequency is equal to the network frequency with which the converter is interconnected, and the control of the phase impose the power factor of the machine. This feature of the control system of pulse width modulation to adjust the phase of the wave of voltage and intensity of current wave can dispense the use of batteries of capacitors in most cases. Typically, manufacturers provide a control of power factor between 0.9 inductive and 0.9 capacitive to the terminals of the machine (Akhmatov, 2002).

The purpose of the control system of converters ac/cc/ac is to ensure the maximization of the coefficient of the turbine power, especially in the region characteristic of the power depending on the wind and where the power is not controlled. Additionally, the control systems of converters maintain a given value of power factor at the point of interconnection of the doubly fed induction machine with the electric power grid. In region of characteristic where the turbine power is controlled, the control system of converters ac/cc/ac keeps constant the total power, extracted by the stator and rotor of the machine, complemented by the control system of step angle of the rotor blades. It is therefore concluded that the control system of wind generators equipped with double fed induction machine can maximize the electrical power delivered to the network in the range of variation of wind speed (Ekanayake et al, 2003).

4. Grid Integration of Wind Farms

For small penetration levels of wind power in a system, grid operation will not be affected to any significant extent. Presently in European Union (EU), wind energy supplies only about 4% of overall electricity, but with the perspectives of achieving 12-15% of electricity from wind power by the year of 2020, the problem of grid integration must be analysed. For scenery of larger penetration levels, changes in power systems and in the methods of operations to accommodate the further integration of wind energy are needed. In spite of total percentage of wind power in EU being lower, only 4%, in some countries that percentage is higher. Countries as Denmark with 25% or Portugal with 11% are examples of that. Based on data from the WindPower Monthly magazine, March of 2009, the percentage of energy supplied by wind in the year 2020 in each of 27 EU member countries is depicted in Figure 17.

Analysing the data is possible to see that some countries have high targets to reach in terms of wind power. Countries like Ireland, Denmark, Portugal, Germany, Greece, UK and Spain, with more than 20% of power being generated by wind plants, grid integration is an issue that must be ensured. The already established control methods and backup available for dealing with variable demand and supply are more than adequate for dealing with the additional variable supply such as wind power at penetration levels up to around 20% of gross demand (Windpower, 2009).

When power obtained by wind is connected to the distribution network, situation of some countries, Portugal is an example, there is no big problem with the integration of wind in the power system because the network is very meshed and can be very quickly reconfigured.
With the actual penetration of wind power in the Portuguese distribution network, the grid is operated and designed to operate even if there is a partial blackout of wind power. The restrictions to the connection of more wind power to the system are caused by the capacity of power lines.

If the distribution network has weak lines, in terms of capacity of power transmission, making the connection between substations with wind parks and the loads, when there is a lot of wind, is necessary to limit the power produced by the park because of no capacity of energy distribution by the interconnection line. This problem can be solved with more investments in the improvement of power lines.

When energy produced by wind is connected to the transmission level, the problem of network operation is higher. At that level power must be dispatch by the control centre, which means that must be more confidence on the forecast, because power produced must not be off-line instantaneously, causing loss of power. On the other hand, there are few large power stations in a power system, so the shutdown of one of the traditional large power stations, by accident or planned, is a problem to the system operator. In a power system the number of wind generator can be very large, so the system will not notice the shut-down of a 2 MW wind turbine. The problem is when a coal fired plant or a nuclear plant shuts-down instantly. Even in terms of power reserve needs, this issue only is relevant when the magnitude of wind power variations becomes comparable to the load variations. Reserve requirements for wind power can be obtained from the existing conventional power plants, but the allocation and the use of reserves lead to extra system costs.

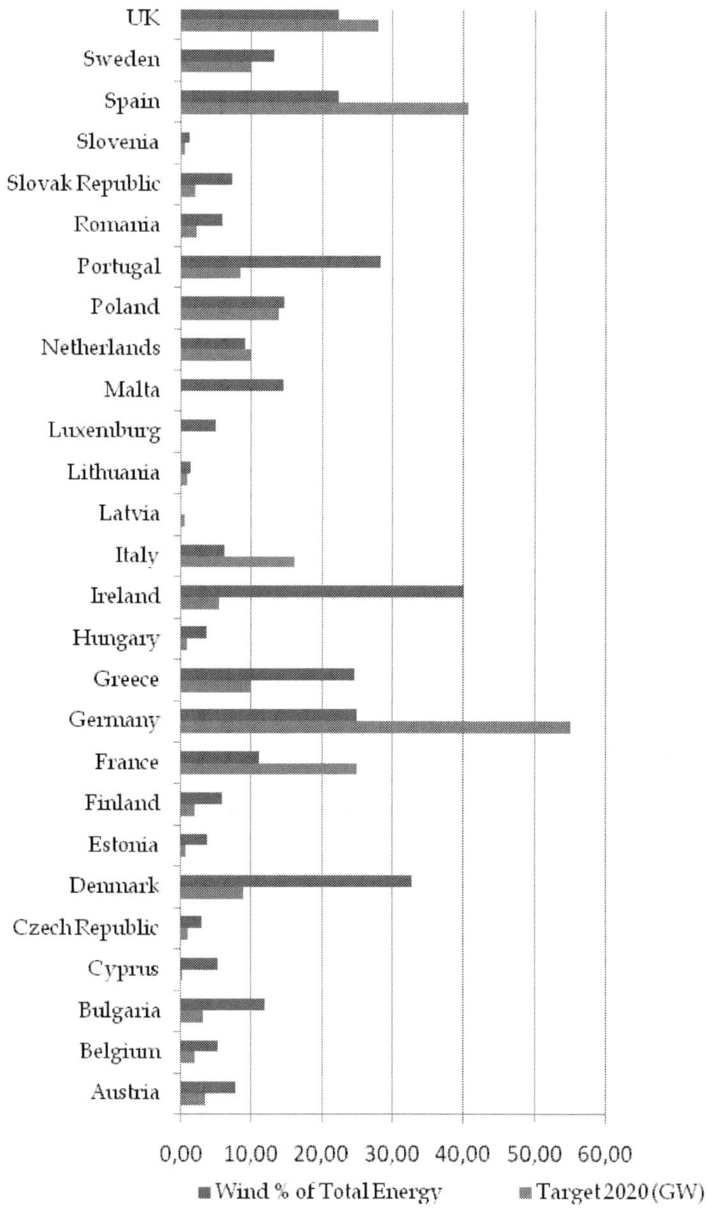

Fig. 17. EU 27 wind development to 2020

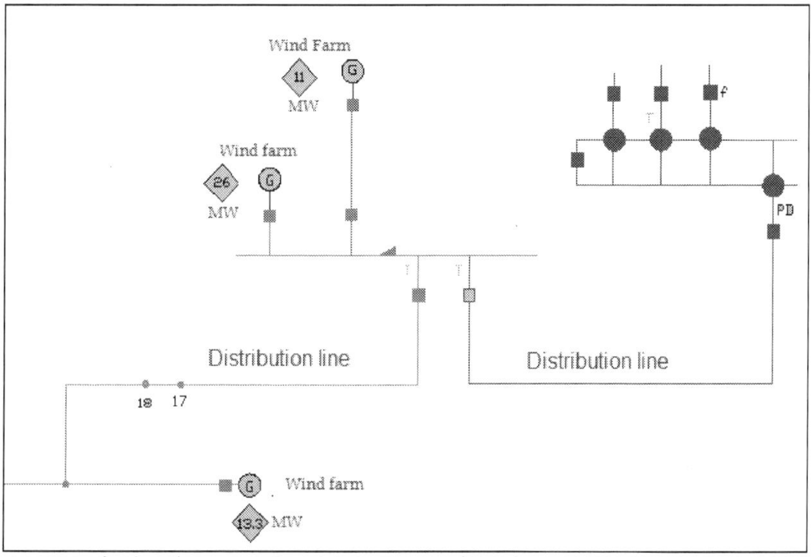

Fig. 18. Part of a distribution grid with integration of wind power generation

The most important principles of operation of any power system is the safety and efficiency. For that reason all the actors, consumers or producers, must comply with technical requirements. Because of that, clear rules are needed to ensure the well, safety and efficient operation of the system. These rules are normally called by "grid codes", and reflect the true technical needs for system operation. The problem is that these technical requirements are not adjusted for all European countries where the level of penetration of wind power is not the same. Normally, grid codes contain high costly requirements sometimes with no technical justification for some actual grids. It is not feasible that grid codes only reflects the technical needs for system operation, based on experience of countries with large penetration of wind energy, in countries with small wind power penetration requirements such as fault-ride trough or primary control are inappropriate. Because of that is important to harmonised grid codes for wind energy at a European level. That seems not to be an easy task, because of the national specificities of each European country in terms of wind penetration level, network robustness, geographical size of the system, etc. Expanding the rules adopted by more developed countries in terms of wind energy, to others where wind energy levels are not so high, is not a good practice.

5. Grid Codes

With the integration of large amounts of energy produced by using the wind resource, in the utility grids, additional requirements in terms of grid codes are needed to integrate wind power with other conventional generation forms. The way as wind farms react to power system disturbances and the way how they contribute to the disturbances, must be a target of regulation. High short-circuit currents, big unbalances on frequency or over- and under-voltages during a fault occurrence can damage wind generators. The relay protection of the

wind farm is designed with the objective of complying with requirements for normal operation and support the grid when a fault occurs and in the system restoration after a fault occurrence. The system relay protection has to secure wind farms against damage from the impacts occurring at faults in the power network. National grid codes, which are from the responsibility of governments and system operators, have the responsibility of imposing rules to maintain power system stability and power quality.
Normally, wind turbines were disconnected from grid when voltage goes below 80 or 90% of nominal value. But a timeless shut-down of wind turbines can have a reverse action, and contribute to system disturbance. To prevent bad actions, the way as wind farms react to voltage dips or frequency deviations, must be regulated.
A voltage dip can be described by a sudden reduction of the voltage, between 10 and 99%, in a point of the network, followed by the restoration of voltage after a short time. Normally time duration is lower than 1 second and amplitude is lower than 60%.

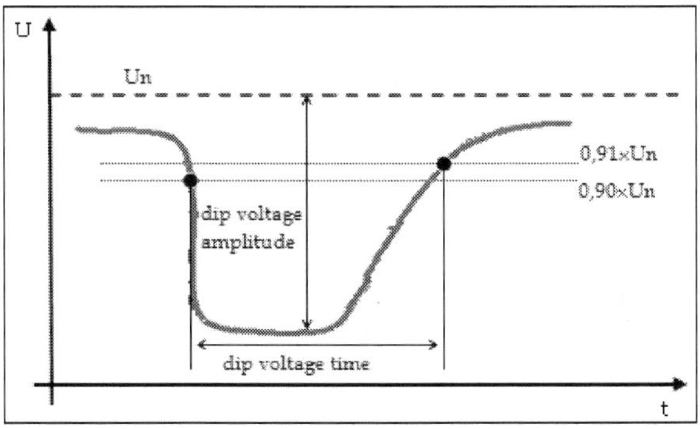

Fig. 19. Dip voltage representation

Most of European countries impose rules to the wind farms grid connection similarly to the existing rules for conventional power plants grid connection. These rules normally contain minimum technical requirements, defined by the system operators to the wind farms owner, as a way to ensure security of supply, reliability and power quality. Next an overview of requirements in the national grid codes for wind turbines is presented. Some countries have different grid codes for transmission and distribution networks (Denmark) while others have focus only on transmission level (Germany and Spain). Some other European countries, even with some considerable penetration of wind generation, doesn't have any specific grid code for wind turbines, using the same rules imposed for conventional power plants connection (Portugal).

For wind farms connected to grids under 100 kV, the Danish system operator impose that they must be maintained connected to the network if a 3-phase short-circuit occurs for 0.1 seconds. In the presence of a 2-phase short-circuit with a duration of 0.1 seconds followed by another one with the same duration occurring 0.3-0.5 seconds after the first one, the wind farm should remain connected. Wind turbines must fulfil these requirements when these situations occur for two times in 2 minutes or six times in 5 minutes (Energinet, 2004a; Iov et al., 2007). For connections to grids with voltages above 100 kV, the same requirements are needed adding the situation for 1-phase short-circuit were wind farm shall not be disconnected in the same conditions described for a 2-phase short-circuit in the distribution network. In the transmission system wind turbine must hold with at least two situations of 1, 2 and 3-phase short-circuit within 2 minutes interval (Energinet, 2004b; Iov et al., 2007).

In Germany the grid code, applied to networks with voltage levels 380, 220 and 110 kV, impose that wind farm shall fed short-circuit current into the grid during the fault period when the fault occurs in the grid, outside the protection range of the generating plant. However wind farms must be disconnected in special situations such as when voltage falls below 85% from the rated voltage. In that situation disconnection should happen with a time delay of 500 msec. If the voltage at the wind turbine terminal is equal, or less, than 80% of minimum permanently voltage permitted, 25% of the generators must be disconnected from the grid after 1.5 sec, 1.8 sec, 2.1 sec and 2.4 sec respectively, if the voltage doesn't increase to the minimum permitted level in the meantime. In a situation of overvoltage, 120% of the maximum permitted voltage value, the affected generator must be disconnected from the grid with a time delay of 0.1 seconds. In a situation of voltage dip wind generators must provide a reactive current to support the grid voltage. This action must take place within 20 msec after fault detection and shall be maintained for more 500 msec after voltage returns to normal (E.ON, 2006; Iov et al., 2007; Tsili et al., 2008).

Spanish requirements related to voltage dips specify that wind farms must support voltage dips without tripping. During the faults and in the recovery period wind farms can't absorb reactive power, must not absorb active power and must provide to the network the maximum reactive current. Consumption of active and reactive power is admitted during a period of 150 msec after the beginning and 150 msec after the clearance of the fault, in specific conditions. This wind farm behaviour is required for balanced three-phase faults and unbalanced two-phase and single-phase faults (REE, 2005; Tsili et al., 2008).

In Portugal wind farms with power installed over 5 MW must remain connected to the grid when voltage in wind farm terminal is over the limit curve illustrated in the Figure 20, for a 3, 2 or single-phase fault. The fault ride through requirements for wind turbines of the three mentioned countries are shown in Figure 20.

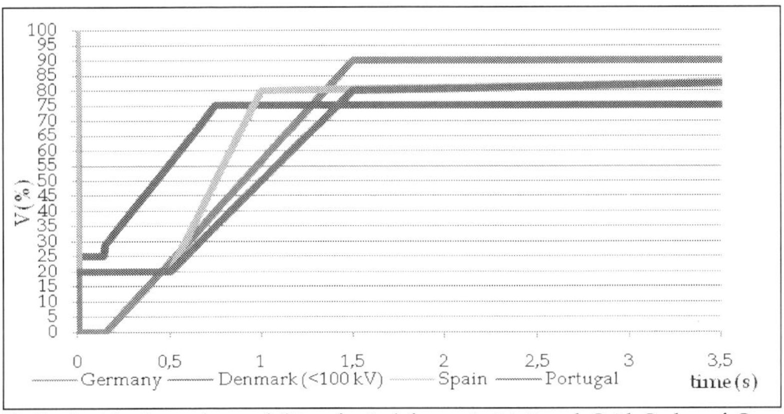

Fig. 20. Fault ride through capability of wind farms in National Grid Codes of Germany, Denmark, Spain and Portugal

During the faults and in the recovery period, wind farms can't absorb active or reactive power. Wind farms must provide reactive power during the voltage dip period as showed in Figure 21, the maximum permitted delay for the beginning of production of reactive power is 40 msec after fault detection. Zone (1) represents the reactive power contribution during the period of fault and restoration, and zone (2) represents the normal operating period.

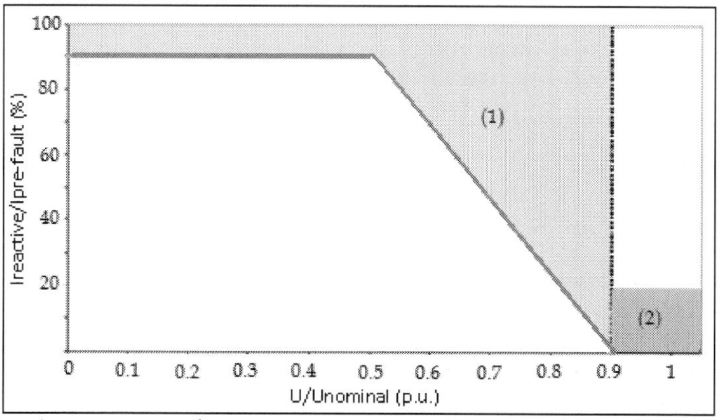

Fig. 21. Reactive power contribution.

Frequency is an indicator of balance between production and consumption. In the European power network frequency is very stable, varies in a range of 50±0.1 Hz.
In terms of frequency, Danish grid code allows a variation between 47Hz and 52Hz, for a short limit of time. Outside these limits disconnection must occur after 0.2 seconds. A continuous variation is permitted between 48,5Hz and 51Hz.
In Germany the frequency range allowed for variation is between 47.5Hz and 51.5Hz, for a short limit of time. Outside the limits the unit must be automatically isolated from the network without delay. Continuous variation is allowed between 49Hz and 50.5Hz (Iov et al., 2007; Tsili et al., 2008).
In Portugal wind farms must support frequency deviations between 47.5 Hz and 51.5Hz without necessity of disconnection from the grid.

6. Typical Faults

Operational costs and an efficient maintenance of wind generators are very important aspects in modern wind farms. Wind generators can't be dispatched efficiently, normally due to the wind characteristics, but worst than don't have wind to generate energy is to have the machine stopped due to a fault. Associated to a fault is a time to repair it and equipment to do that. Because of the wind generators dimension, if is required, for instance, a crane to make the repair, that can be a bigger problem than the original one because of the unavailability of this kind of machines and the high costs associated. Fault detection techniques are becoming indispensable in modern wind parks. Fault detection techniques offer some important benefits, like the prevention of major components failures, the reduction of maintenance costs, detailed information on wind generators performance and vibration characteristics and allow for condition based maintenance schemes to increase maintenance intervals (Caselitz et al., 1996).

There are three typical kinds of faults that can occur in a wind generator, electrical faults, electronic faults and mechanical faults. The electrical faults occur with some frequency but are the most unexpected because all used equipment (electrical machines) is very developed and is well known. Induction generators or transformers are electrical machines used for decades. So, it is not expected to have faults in these equipments but, the reality shows that electrical problems exist. Most common failures are bearing related, due to short circuit currents, and insulation problems. These kinds of problems can be explained by the necessity to reduce the machines dimension and the need to work with new construction materials that aren't well tested. When an electrical fault occurs, the costs involved to solve the problem are very high and normally the problem is solved by the substitution of the fault component.

Electronic faults have a higher occurrence frequency than the electrical faults. These kinds of problems occur frequently in sensors and in electronic cards. The anemometer measurement is one example of a component that has a very high fault rate. Electronic components faults can be provoked by lightning effects or other weather phenomena. Normally electronic components breakdowns occur after storms with lightning hitting the towers. When these problems occur, the solution is to substitute the electronic component by a new one. It isn't necessary any specific and high costly machine to solve the problem, a crane for instance, but is necessary to remember that an electronic component failure can lead to the offline of the turbine and to the costs of not production energy associated to that. There are a lot of sensors installed in a wind generator, for instance, wind generator can give information about errors in yawing, in hydraulics, in ambient surroundings, in rotation, in generators, in pitch system, etc. A wind generator of 2MW can have a list of about 20 errors in generator only. So, a very large number of sensors are present in each wind tower and because of that the probability of bad functioning of the sensors is very high.

The third types of faults, the mechanical, are associated to the gearboxes and to the blades. With the increase of size of wind towers, the rotors velocity must be slowed. So, the gearbox must reduce the velocity applied to the rotor. Because of that, forces applied to the cog-wheel of the gearboxes are very high which can lead to the tooth break. Other problem is when a lightning hits the blades. The blade has a thin metallic layer connected to the mass copper conductor. When a lightning hits the blade, the metallic layer or the copper conductor can be destroyed or the tip of the blade can be damaged. Some of these problems can be solved with no necessity of bring down the blade but others needs that. With the increase of size of the towers and blades, stronger winds are catch. So, the continuous vibrations and centrifugal forces that turbine blades are submitted make this component the weakest mechanical link in the system.

Due to the dimension and to the costs involved, three components of a wind turbine must be monitored. They are the blade system, the gearbox and the generator. Not only are the replacement components expensive, but major expense is also associated with obtaining and mobilizing the large crane needed to repair these components. A fault in any of these systems can cost a lot of money, time and the necessity of specific equipment to make the reparation. Figure 22 depicts the mean time to repair of wind turbines and is clear that the blade system, the gearbox and the generator are the three components that needs more time to repair.

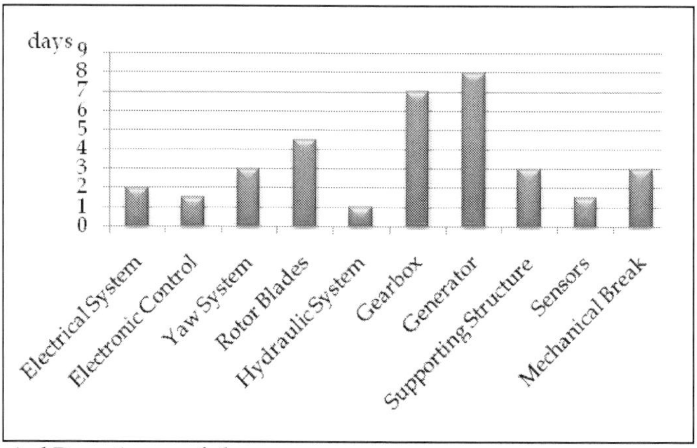

Fig. 22. Typical Downtime per failure

Based on one year data obtained from a real wind farm, equipped with 10 wind generators of 2 MW each, is possible to get some conclusions. With all the data arrived to the control centre was possible to make a graphic with the downtime of each machine and the out of service time of the totality of the wind farm during one year. Figure 23 shows the results.

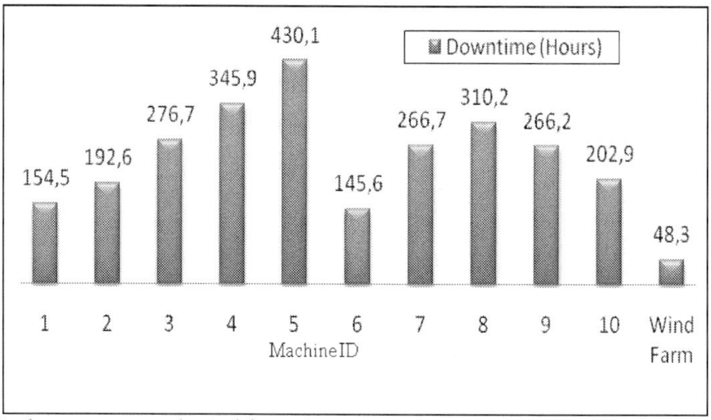

Fig. 23. Wind generators and wind farm downtime, during one year

Making some statistical calculus is possible to see that the average time stopped by each machine was 264 hours/year, approximately. But the totality of the wind farm only was out of service 48.3 hours, in a year. This reinforce what was said before that the probability of blackout of a wind farm is very low and decreases if instead of a wind farm, the analysis is enlarged to a region or to a country.
Looking deeper for the causes of faults that led to the machine stopping and dividing the causes into five groups: not planned (NP), planned (PL), network fault (NF), short time planned (SP) and other causes (OT), some conclusions can be made. The not planned stop

covers all the causes that were not planned, as examples of causes are the substitution of damage equipments; out of communications; stop caused by high temperature of generator, etc. In the planned stop (PL) group are represented the time used for planned maintenance of the wind turbines. In the network fault group (NF) will be grouped all the stops caused by the network, like actuation of protections of maximum or minimum voltage in the substation, high currents in the rotor caused by network problems, etc. In the short time planned stops (SP) is represented the time used for upload software or to make some adjust in parameters. All causes that led to wind turbine stop due to weather conditions, like ice phenomena, strong winds, etc are grouped in the OT group.
Making this analysis is possible to conclude that almost 80% of the stopped time of a wind turbine is due to not planned actions. All the other groups have a small influence in the out of service of the machine. Figure 24 shows the influence of each group in the out of service time of the wind turbines.

As we can see most of the causes that led to a stop of a wind turbine have unexpected causes. An early detection of possible faults in wind turbines assumes a great importance because it allows better maintenance and repair strategies and can prevent major problems in other components. The installation of control centers designed specifically for monitoring, dispatch and control wind farms is a good solution for reduce the downtime of wind turbines or wind farms because the system operator see online what is happening in the installation and can quickly alert the maintenance team when a problem occurs in a machine or in all wind farm.

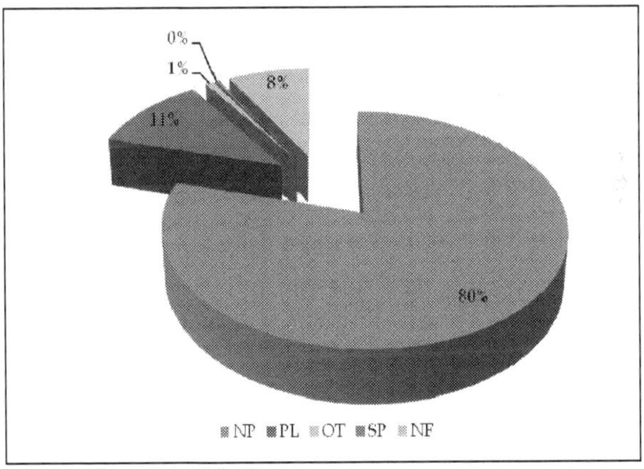

Fig. 24. Percentage of each group of faults

Computer tools that analyzes data with the objective of detect and prevent possible faults are interesting. Techniques of current signature analysis, or vibration analysis, or temperature analysis, are techniques that are taken into account by wind turbine manufacturers. In the control centre it is possible to see information of the temperature of the nacelle, temperature of generator and bearings, transformer temperature, gear oil temperature, and more. It is possible to access to the generators speed, wind speed, power produced, voltage and currents. All this available information can also be used to give some hints of developing failures in a wind turbine.

7. References

Akhmatov (2002). Variable-Speed Wind Turbines with Doubly-Fed Induction Generators - Part I; *Modelling in Dynamic Simulation Tools. Wind Engineering,* (2002) Vol 26, n°2, pp 85-108.

Almeida,R. G.; Peças Lopes, J. A. & Barreiros, J. A. L. (2004). Improving Power System Dynamic Behaviour Through Doubly Fed Induction Machines Controlled by Static Converter Using Fuzzy Control. *IEEE Transactions on Power Systems,* Vol.19, No.4, (November 2004) pp. 1942-1950.

Caselitz, P..; Giebhardt, J.; Krüger, T. & Mevenkamp, M. (1996). Development of a Fault Detection System for Wind Energy Converters, *Proceedings of EUWEC'96,* pp. 14-17, Göteborg, Sweden, May 1996.

Cigrè (2001). Task Force 38.01.10, 2001.

Ekanayake,J. B.; Holdsworth, L.; Wu, X. & Jenkins, N. (2003). Dynamic Modeling of Doubly Fed Induction Generator Wind Turbines . *IEEE Transactions on Power Systems,* Vol.18, No.2, (May 2003) pp. 803-809.

Energinet (2004a), *Grid connection of wind turbines to networks with voltages below 100kV,* Regulation TF 3.2.6 (May 2004), Denmark.

Energinet (2004b), *Grid connection of wind turbines to networks with voltages above 100kV,* Regulation TF 3.2.5 (December 2004), Denmark.

E.ON (2006), Grid Code - High and extra high voltage. E.ON Netz GmbH, (April 2006) Bayreuth, Germany.

Iov, F.; Hansen, A. D.; Sorensen, P. & Cutululis, N. A. (2007). Mapping of grid faults and grid codes, In: *Riso - Report,* ISBN 978-87-550-362-2, Roskilde, Denmark.

Manwell, J.; McGowan, J. G. & Rogers, A. L. (2002). *Wind Energy Explained: Theory, Design and Application,* John Wiley & Sons, ISBN 047 1499722.

Mukund, R. P (1999). *Wind and Solar Power Systems,* CRC Press, ISBN 0-8493-1605-7, United States of America.

REE (2005). Requisitos de respuesta frente a huecos de tension de las instalaciones de produccion de regímen especial, *PO 12.3,* Spain, November 2005.

Tsili, M.; Patsiouras Ch. & Papathanassiou, S. (2008). Grid code requirements for large wind farms: a review of technical regulations and available wind turbine technologies, *Proceedings of ewec 2008,* Brussels, Belgium, March-April 2008.

Windpower (2009). Wind on the Way, *Wind Power Monthly Special Report,* (March 2009) pp. 20-21, ISSN 109-7318.

Wind Generation Modelling for the Management of Electrical Transmission Systems

François Vallée
Electrical Engineering Department, Faculté Polytechnique de Mons
Belgium

1. Introduction

Each investment scenario on a given electrical transmission system must ensure a quality service at the lowest cost (services continuity, system exploitation). In order to answer this major issue of modern networks, it is therefore necessary to compute a faithful representation of the transmission system. In that way, statistical analysis by means of a Monte Carlo simulation (Billinton et al., 1996) - (Papaefthymiou et al., 2006) permits electrical system modeling via the simulation of a large set of representative states. Consequently, Monte Carlo studies permit to obtain coherent exploitation cost and reliability indices for each studied network.

In a near future, stochastic electrical production, and more specially wind generation, is expected to play an important role in power systems. It is therefore imperative to study the impact of this decentralized production source on transmission systems operation constraints. Actually, adequacy studies taking into account wind generation have been extensively developed for the HLI level (load covering with always available transmission system) (Billinton & Bai, 2004) - (Wangdee & Billinton, 2006). From the transmission system point of view, a first reliability study taking into account transmission constraints has been introduced in order to evaluate transmission reinforcement planning associated to large scale wind farms integration (Billinton & Wangdee, 2007). However, that approach was not feasible using normal personal computers and was requiring the use of multiprocessors. Moreover, that proposed study was not considering eventual operation constraints (fatal production, nuclear or high powered thermal units that the producer does not want to stop during the nights…) on classical generation parks.

In the present paper, as reliability and reinforcement analysis are long term studies, stochastic wind generation models are proposed and introduced into an HLII (bulk power system) (Billinton & Wangdee, 2007) - (Allan & Billinton, 2000) non sequential Monte Carlo simulation tool. Thanks to the utilization of that non sequential approach, computing requirements are reduced without worsening the precision of the obtained global indices. Moreover, classical machines and transmission constraints can also be simultaneously considered when facing an increased penetration of wind generation. Therefore, it is believed that the proposed tool will assist system planners and transmission system operators to qualitatively assess the system impact of wind production and to provide adequate input for the managerial decision process in presence of increased wind penetration.

This chapter is organized as follow. In a first part, the methodology used to efficiently introduce wind generation in the HLII simulation tool is explained. In that way, a discussion about the most effective manner to sample wind production in transmission systems analysis is proposed. Then, hypothesis, based on real observations, is made in order to introduce wind generation into an economic dispatch with classical parks and transmission constraints. In a fourth section, wind impact on reliability and reinforcement analysis for transmission systems is computed for an academic test system: the *Ray Billinton Test System (RBTS)* (Billinton et al., 1989). Finally, a conclusion is drawn and points out the major results collected thanks to the introduction of wind generation into HLII analysis taking into account transmission systems constraints.

2. Presentation of an existing HLII Monte Carlo simulation tool: Scanner©

2.1 System states generation

The simulation tool Scanner© is the property of *Tractebel Engineering (Gaz de France – Suez).* Its main objective is to provide technical and economical analysis of development alternatives on a given electrical system. In that way, acceptance (or rejection) criterion is generally based on the following assessment: **"Each investment scenario must ensure a quality service (system exploitation, healthy behavior when facing unexpected outages, continuity of services...) at the lowest cost"**. To answer this issue, a complete analysis of the given transmission system (HLII) is required. Consequently, Monte Carlo simulation tool Scanner© analyses the system evolution as a set of static representative states.

To generate the different system states, the model loops on the 52 weeks of the year (cf. fig.1). During each week, a given number (defined by the user according to the required accuracy on the calculated indices) of system states are generated by mean of the following procedure:

- ✓ *Definition of the system state hour during the considered week:* random generation by use of uniformly distributed numbers on the following fixed interval [0, 168] (168 hours during a week);
- ✓ *For each generated hour:* uniformly distributed random numbers (V) on the interval [0, 1] are sampled for each element (classical generation units, transformers, lines…) in order to decide its operation state, using the following procedure (Billinton et al., 1996):

  If $V \leq$ *Forced Outage Rate (FOR) (Billinton et al., 1996),* the element is considered as unavailable;

  If $V > FOR$, the element is considered as fully available;

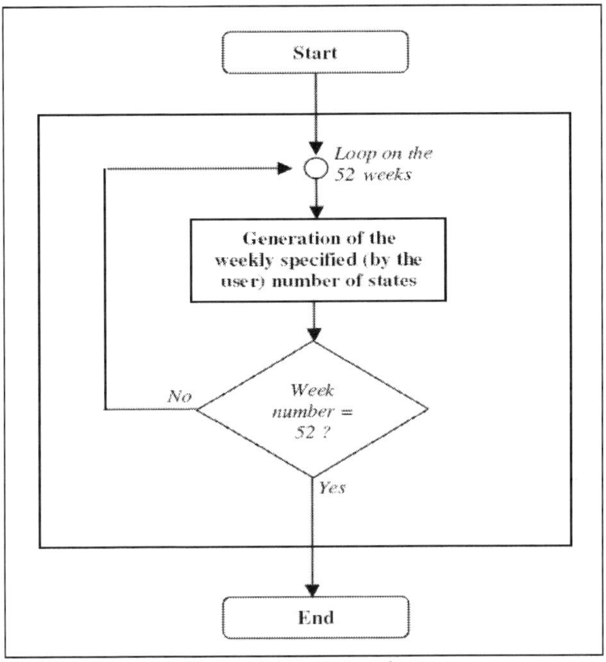

Fig. 1. Algorithm of the system states generation procedure.

Concerning the hourly load at each node of the system, its determination is based on the use of the annual peak load value at the considered node; this last one being combined with two modulation diagrams:

- ✓ *Diagram of weekly modulation of the annual peak load:* this last one permits to calculate the peak load of the actual week on the basis of the annual peak load value for the considered node. This diagram contains thus 52 modulation rates of the annual peak load value;
- ✓ *Diagram of the hourly modulation of the weekly peak load:* it permits to calculate the hourly load for each hour of the week. This diagram contains thus 24 modulation rates of the weekly peak load value.

The introduction of the load in the investigated simulation tool can thus be considered as 'sequential' (whereas the generation of element states is non sequential). Indeed, thanks to the applied methodology, no random sampling is used to generate the hourly load at each node of the system. More simply, the program just considers, in the weekly modulation diagram, the rate corresponding to the actual week during the simulation process (cf. fig.1); then, it associates to the generated weekly peak load the rate of the hourly modulation diagram corresponding to the investigated hour of the day.
Finally, as the consumption during one week can change from one day to the other (days of the week, Saturday or Sunday), several diagrams of hourly modulation can be associated to

each node during one week. Moreover, seasonal aspects can also be taken into account by defining periods during the year and by changing the set of hourly modulation diagrams associated to each node from one period to the other (a period being defined by a given number of weeks during the year).

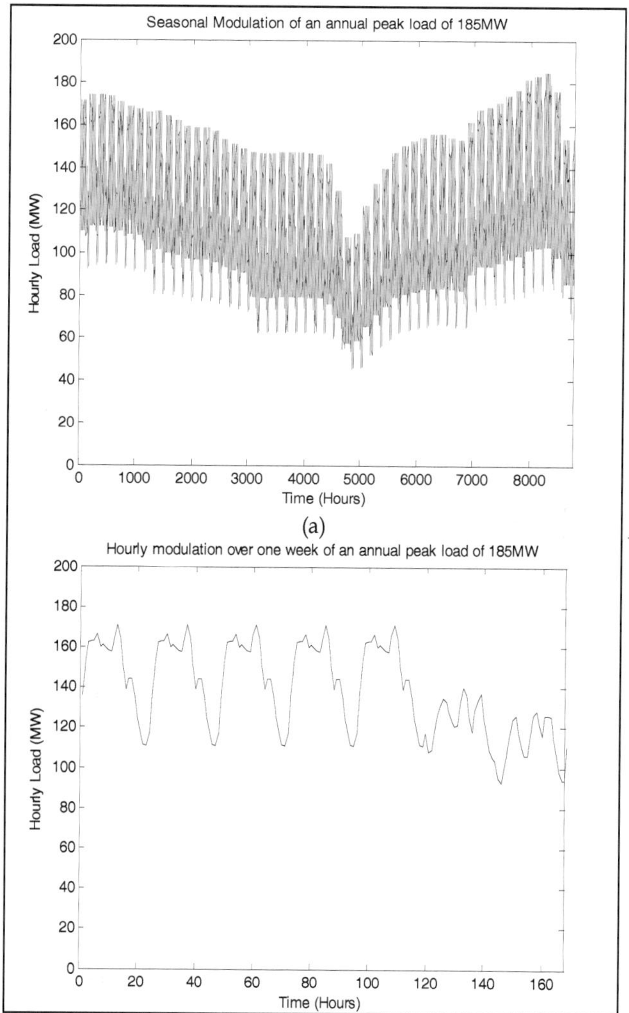

(b)

Fig. 2. Seasonal modulation for 185MW peak load value (a) and hourly modulation over one week of the year for the same peak value (b) based on a Belgian real case (Buyse, 2004)

Figure 2.a illustrates the load behavior during one year (based on a Belgian real case for the year 2000 (Buyse, 2004)) for an annual peak load of 185MW at the considered node. In

Figure 2.b, a zoom is made on one week of the year for the same node and illustrates the consideration of possible change of consumption from one day to the other inside the week.

2.2 System states analysis

Each generated system state must then be analyzed. To proceed to this stage of the process, three steps are consecutively realized for each system state:

1) *Economic dispatch:* this last one is based on the available production units and is done **without considering transmission facilities availability**. The objective is thus to ensure, at the lowest cost, the hourly load with the available production. Note that **the economic dispatch is taking into account possible constraints on classical units operation**. Consequently, several types of production parks are considered in Scanner© among which:

   -Hydraulic production and pumping stations: they are considered as zero cost production in the algorithm and are managed at a weekly time scale;
   -Thermal production: three types of constraints are considered for this kind of production. Firstly, **technical minima** (threshold under which the producer does not want to run, for technical reasons, its unit) can be considered. Secondly, **forced units** can be defined by the user. Those entities represent units (such as cogeneration) that have a threshold over which they must always operate when they are available. Finally, in order to take into account high powered thermal or nuclear units that the producer does not want to stop during the week, those machines are considered as **long term units**. They are managed at the weekly time scale and must always run at their technical minimum value during the actual week when they are needed to cover the reference peak consumptions of the week (in the other case, they are supposed entirely unavailable during the entire week).

   Finally, the algorithm conducted during the economic dispatch proceeds as follows. In a first step, hydraulic production is used to cover the load (following the orders of the weekly management). Then, technical minimum values of forced and long term (if they are required to cover the reference peak loads during the week) thermal units are considered to satisfy the load (minus hydraulic production). Finally, an economic dispatch of the thermal production (minus the technical minimum values of already considered forced and long term units) is realized to cover the remaining load.

2) *DC Load Flow:* this step realizes the computation of active power flows in transmission lines without considering reactive power.
   In order to solve the DC load flow problem, generated active powers calculated during the economic dispatch are introduced at the connection nodes of the concerned machines. Moreover, the generated hourly consumption for the current state is also taken into account at the required nodes. DC load flow then computes active power flows over the transmission system and permits to take into account transmission constraints. In case of line overflow, step 3 is started. On the opposite, if the optimal solution does not involve overloaded lines, this next step is avoided.

3) *Production rescheduling or load shedding:* this step is only started if the optimal solution of the economic dispatch leads to overloaded lines during step 2. In that case, the solution is "disoptimized" by modifying the production plan (production rescheduling). If this first stage is not sufficient to relieve the overflows, a load shedding procedure is then started in order to limit active power flows.

Finally, note that those three presented steps are repeated to analyze each generated system state.

2.3 Calculated indices and productions

Thanks to the Scanner© tool, it is possible to compute several reliability indices for the studied system.

Among them, the most significant ones are certainly the *Loss of Load Expectation* (*LOLE* in hours/year) (Billinton et al., 1996) (HLI index calculated without considering transmission facilities) and the number of hours (per year) of load shedding (due to transmission overflows). Moreover, hours of overflows are also computed for each transmission line in order to point out the weakest points of the system.

Next to those indices, the annual cost of production (with and without considering elements unavailability) is computed. Moreover, mean production and annual energy generated by each classical unit can also be calculated. Finally, histograms of production can be printed out in order to analyze the utilization of each classical unit.

3. Wind generation modelling for HLII non sequential Monte Carlo simulation

Scanner© does not currently take into account wind production in its algorithm (as well as for the system states generation as for the analysis of those states). However, given the major increase of wind penetration in some countries (like Germany) (Ernst, 2005), this variable kind of production can no more be neglected in technical (and economical) transmission system analysis. Therefore, in the present work, wind production has been implemented in the simulation tool Scanner©. In order to achieve that step, modifications related to the introduction of wind have to impact both major stages of the simulation process: **system states generation and the analysis of these states.**

3.1 Introduction of wind power in system states generation

Before taking into account wind power in the system states generation process, the user has to define three entities related to wind production:

-Entity 1 (wind parks): each wind park is practically characterized by its installed capacity, production cost, *FOR* of one turbine, associated wind speed regime and P-W conversion characteristic (Vallée et al., 2008);

-Entity 2 (wind speed regimes): they are characterized by *Cumulative Distribution Functions (CDF)* representing different statistical behaviors for wind speed in the studied territory. Those *CDF* can be classical *Weibull* distributions (Vallée et al., 2008) or arbitrary ones. In the latter case, distributions are linearly interpolated in the program on the joint basis of the wind speed step and probability intervals defined by the user. Finally, a name is associated to each different wind speed regime and the user can freely associate a wind speed regime to a wind park by defining the name of this wind speed regime in the characteristics of the wind park (cf. entity 1);

-Entity 3 (P-W conversion characteristics): they transform wind speed into production. Practically, the conversion characteristics are linearly interpolated on the joint basis of the wind speed step and the power intervals defined by the user. An example of the linear interpolation related to a classical doubly-fed asynchronous generation structure (Al Aimani, 2004) is given in Figure 3. Note that a single conversion characteristic is practically applied to an entire park. Finally, the link between wind parks and associated P-W conversion characteristics is made identically as for the wind speed regime case (cf. entity 2).

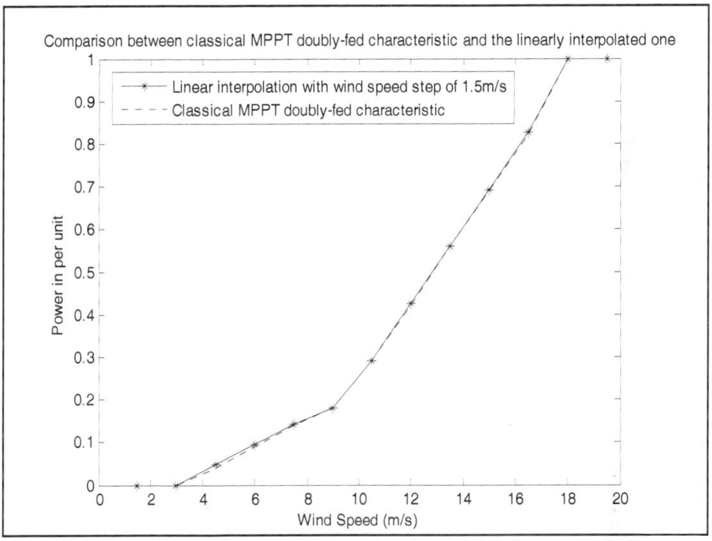

Fig. 3. Comparison between classical *Maximum Power Point Tracking (MPPT)* conversion characteristic (Al Aimani, 2004) and the interpolated one (1.5m/s wind speed step) in p. u.

Once the three basic kinds of entities related to wind production have been defined by the user, the generation process can be started. The algorithm, executed when wind generation is concerned, is presented in Figure 4. The applied methodology is the following one. During each generated system state, a first loop is started over the defined wind speed regimes and a wind speed per defined characteristic is generated by use of the classical inverse transform method (Vallée et al., 2008). By applying this methodology, it is supposed that wind parks subject to the same wind speed regime are entirely correlated. Based on (Vallée et al., 2008) and (Papaefthymiou, 2006), this approach will lead to the most fluctuating wind power and, thus, to the worst case for adequacy studies. Finally, a second loop is made over the defined wind parks. For each wind park, the associated wind speed is decided by taking the one sampled for the wind regime associated to the considered wind park. Then, each wind park production is calculated by introducing the sampled wind speeds in the associated P-W characteristics. Practically, P-W characteristics will be introduced in per unit and the real wind productions will be obtained by multiplying per unit quantities by the *Maximal Available Wind Park Capacity* (*MAWPC*). The *MAWPC* is related to the *Installed Wind Park Capacity* (*IWPC*) by:

$$MAWPC = (1 - FOR) \cdot IWPC \tag{1}$$

By applying (1), possible outages of wind turbines inside a park are considered. Equation (1) supposes that a wind park is made of a sufficiently large number of turbines to consider that the FOR related to one turbine is the same as the one existing for the entire park. This hypothesis is well funded as high powered wind parks connected to transmission (HLII) systems will be practically composed of a large number of turbines.

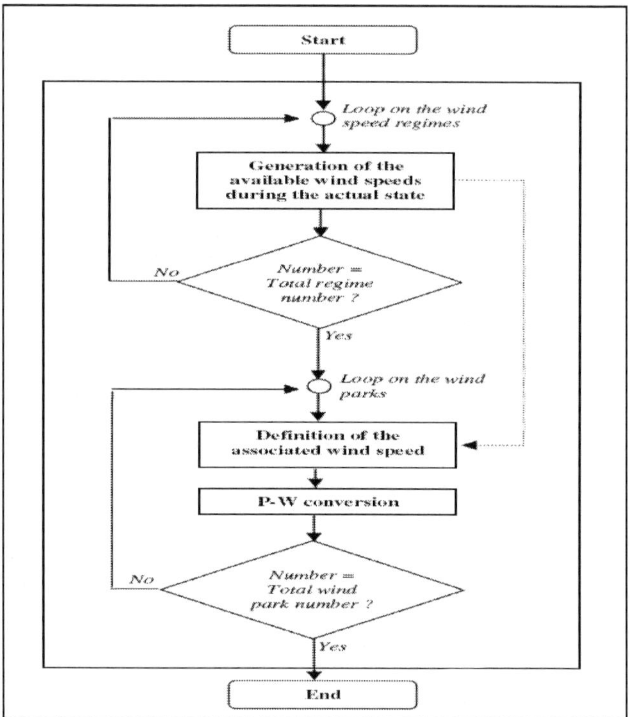

Fig. 4. Algorithm of wind generation during each system state

Finally, note that the wind production generation process is started back for each system state. Consequently, a *Generated Wind Production Distribution* can be plotted for each wind park *'i'*.

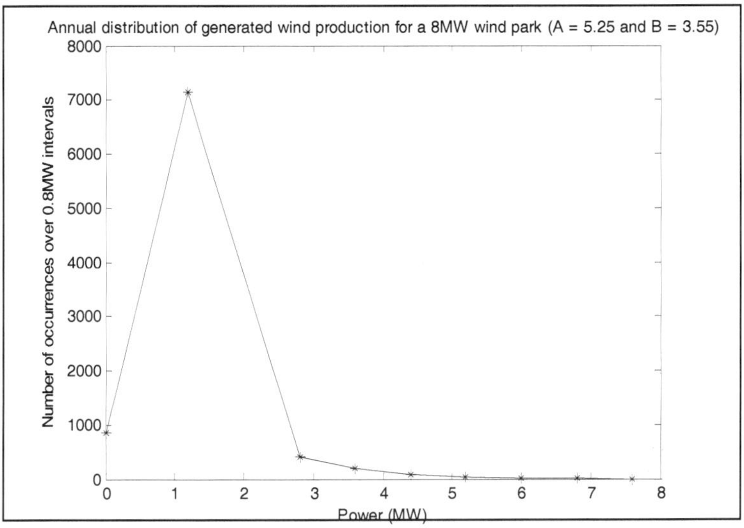

Fig. 5. Simulated annual distribution of generated wind production for a 8 MW wind park (A = 5.25 and B = 3.55)

Figure 5 illustrates the obtained wind production distribution for one 8MW wind park subject to a *Weibull* wind speed regime (with scale parameter A = 5.25 and shape parameter B = 3.55) and using the P-W conversion characteristic of Figure 3. This production distribution is logically concentrated over limited wind production as the considered wind speed parameters were quite low in the present case.

3.2 Introduction of wind power in system states analysis

The *Generated Wind Production* (GWP_i) represents thus, for each defined park *'i'*, the sampled wind power during the simulated system state. This production must then be taken into account in the system state analysis.
The introduction of wind production into the economic dispatch of Scanner© has so been based on several starting hypothesis:

-**Hypothesis 1**: it has been considered that wind power was not accurately predictable at the weekly time scale (Ernst, 2005) and could therefore not impact the management of hydraulic and long term thermal (nuclear) units. Those classical units are thus still processed at the weekly time scale without wind impact;
-**Hypothesis 2**: wind power is considered as a must run production with zero cost. This hypothesis is based on the multiple encouraging policies that generally support wind production (Mackensen et al., 2007) - (Maupas, 2006). Consequently, in the economic dispatch, wind production will be directly considered **after** the technical constraints related to **forced and 'having to run' long term thermal units**;
-**Hypothesis 3**: in case of increased wind penetration, the *Transmission System Operator (TSO)* can be forced (like it has already been the case in some German places (Sacharowitz, 2004))

to cut some wind production when facing classical machines constraints (Sacharowitz, 2004). In the proposed algorithm, when encountering such situations, wind production is decreased, for each wind park, proportionally to its available generated power.

The existence of some transmission system operation constraints can thus lead to a reduction of the real produced wind power. Therefore, in the developed algorithm, two quantities related to wind production have been defined **for each wind park**:

- *Real Wind Production (RWP):* it represents the real produced wind power after having taken into account the economic dispatch. A single RWP_i value is associated to each wind park *'i'*;
- *Lost Wind Production (LWP):* it defines the difference between the generated wind production and the real produced one for each considered wind park. A single LWP_i value is thus calculated for each wind park *'i'*;

The algorithm, implemented in order to take into account wind production in the economic dispatch associated to each generated system state, is described in Figure 6.

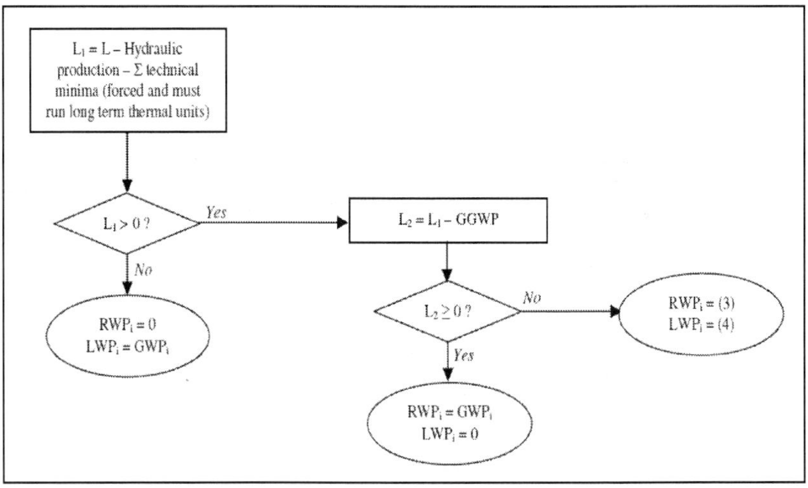

Fig. 6. Algorithm implemented in order to take into account wind production for the economic dispatch of each generated system state

Based on hypotheses 1 and 2, wind production is thus used to cover the remaining load after that hydraulic production and technical minima of forced and 'must run' long term thermal parks have been taken into account. If this remaining load L_1 is equal to zero, all the hourly load has already been covered before considering wind generation. In that case, real transmitted wind production RWP_i is set to zero for each wind park *'i'* and their associated lost wind production LWP_i equals their initially generated wind power GWP_i during the considered system state.
On the other hand, if the remaining load L_1 is greater than zero, *Global Generated Wind Production (GGWP)* is taken into account before the remaining classical thermal production

(without constraints) and is entirely taken off from load L_1. If the obtained load value L_2 (after consideration of hydraulic production, of forced and 'must run' long term thermal parks technical minima and of wind production) is greater or equal to zero, RWP_i equals the generated wind production GWP_i for each park '*i*' and LWP_i is set to zero (and the classical economic dispatch is pursued). In the case of remaining load value L_2 is negative, it is then necessary to reduce real transmitted wind production by following hypothesis 3. In order to apply this reduction of wind power, the remaining load L_1 (before introduction of the generated wind production) is taken back. RWP_i and LWP_i associated to each defined wind park '*i*' are obtained via equations (2), (3) and (4):

$$GGWP = \sum_{i=1}^{N} GWP_i \tag{2}$$

$$RWP_i = \frac{GWP_i \cdot L_1}{GGWP} \tag{3}$$

$$LWP_i = GWP_i - RWP_i \tag{4}$$

At the end of the actual system state economic dispatch, real and lost wind productions (RWP_i and LWP_i) are thus defined for each wind park. In order to take into account wind production impact over transmission constraints, calculated RWP_i are then injected at the adequate nodes and the DC load Flow (section 2.2) is launched. If no line overflows are recorded for the computed system state, RWP_i and LWP_i stay unchanged for each wind park '*i*'. On the opposite, in case of overloaded lines, the algorithm of reschedduling/load shedding (section II.B) is used and RWP_i et LWP_i can have to be modified in order to ensure a safe behavior of the transmission system.

The system state analysis leads thus now to the computation of RWP_i and LWP_i for each wind park '*i*'. This process is set back for each generated system state. Consequently, at the end of the Monte Carlo simulation, histograms of GWP_i, RWP_i and LWP_i can be drawn for each defined wind park '*i*'. Moreover, mean values of generated (GWP_i) and real (RWP_i) exchanged wind powers are calculated, for each wind park, in order to point out the impact of transmission system constraints on wind production. Also note that reliability indices defined in section III are now taking into account wind production.

4. A simple study case

In order to validate the implemented algorithms and to simply point out the impact of transmission system constraints on wind production, a first study case has been simulated (Figure 7).

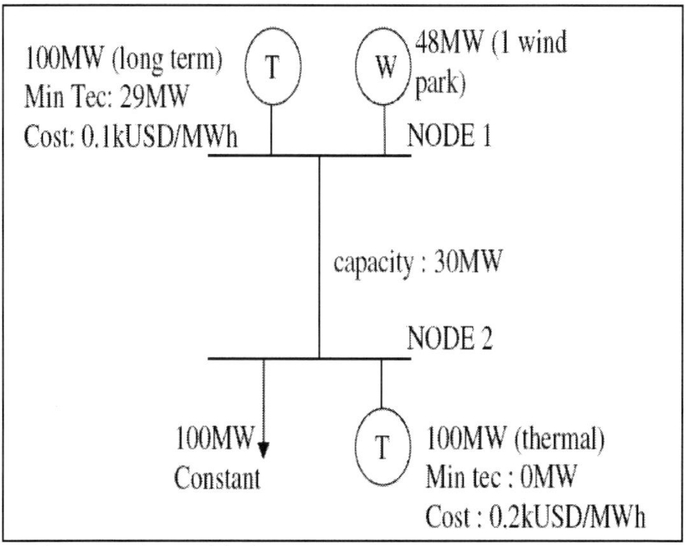

Fig. 7. First study case with simplified transmission system

The investigated system is composed of:

- 2 nodes: node 1 and 2 with base voltage of 132 kV;
- 1 line between nodes 1 and 2 with transmission capacity of 30 MW;
- 1 long term thermal park at node 1 with nominal production equal to 100 MW and technical minimum value of 29 MW. The production cost of this unit is taken as: 0.1 kUSD/MWh;
- 1 wind park at node 1 with installed capacity equal to 48 MW, *Weibull* wind speed regime (A = 8.25, B = 3.55) and P-W conversion characteristic of Figure 3. Following hypothesis 2 (section III.B), wind power is supposed to be a zero cost production;
- 1 classical thermal machine at node 2 with nominal production equal to 100MW and technical minimum value of 0 MW. The production cost of this unit is: 0.2 kUSD/MWh;
- 1 load at node 2 with constant value equal to 100 MW;
-The classical units *FOR* is taken as 0.2 and the one associated to each wind turbine of the considered wind park is set to 0.03.

Two simulation cases have been launched for this simplified transmission system:

- *Healthy case:* possible thermal parks outages are not considered and both parks are thus always available;
- *Outage case:* thermal parks FOR are here taken into account and non-expected outages of those classical production parks can happen.

Practically, in the healthy case, both thermal units are available. Therefore, at the weekly time scale (section 2.2), the cheaper long term thermal machine will be selected to cover the load. During each hour of the week, this machine has thus to operate at its technical minimum value (29 MW). After having taken into account this production, there remains 71 MW load to be covered. In the optimal dispatch, this production ideally has to be produced at the lowest cost and, thus, by the available wind production (hypothesis 2 in section 3.2) and, eventually (if the wind production is not able to cover the remaining 71 MW), by the remaining power of the long term thermal unit (71MW). However, this optimal solution would lead to an active power flow of 100 MW on the transmission line. That is greater than its capacity (30MW). Consequently, the solution has to be 'disoptimized'. In that way, only 30MW can be produced by the parks connected at node 1. It means thus that the long term thermal unit will imperatively produce 29 MW (and, sometimes, more if wind production is insufficient to cover the remaining 1 MW) and that the remaining MW will mainly come from the wind park. Due to the transmission line constraint, 70MW will come from the more expensive classical unit at node 2.

By studying this first healthy case, it can be concluded that the available production will always be greater than the load (no outages on classical units) and, therefore, lead to a LOLE value of 0 h/year. Moreover, no load shedding will be required as the thermal unit at node 2 will always be able to cover the remaining load ('disoptimized' solution). Finally, due to transmission line capacity and long term thermal park constraint, available wind production will practically have to be reduced; leading thus to a mean production lower than the mean generated one. Those results are clearly confirmed in table 1.

	Healthy Case	Outage Case
LOLE (h/y)	0	359.0
Load shedding (h/y)	0	1751.4
Mean production: long term thermal park (MW)	29	29
Annual energy: long term thermal park (GWh/y)	253.7	203.5
Mean production: classical thermal park (MW)	70	74.5
Annual energy: classical thermal park (GWh/y)	611.5	519.0
Annual operation time: long term thermal park (h/y)	8736	7005.6
Annual operation time: classical thermal park (h/y)	8736	6984.6
Mean real production: wind park (MW)	1	2.2
Annual real energy: wind park (GWh/y)	8.4	18.8
Mean generated production: wind park (MW)	7.1	7.1
Annual generated energy: wind park (GWh/y)	60.8	60.8
Annual operation time: wind park (h/y)	8504.9	8504.9

Table 1. Simulation results for the simplified transmission system study case

Moreover, Figure 8 illustrates the obtained annual distribution histograms of *GWP*, *RWP* and *LWP*. It can obviously be observed that, beyond 1 MW, real wind power (*RWP*) can not be transmitted. On the opposite, as *RWP* is reduced compared to the initially available *GWP*, a lost wind power (*LWP*) distribution, different from 0, is consequently computed in that healthy case.

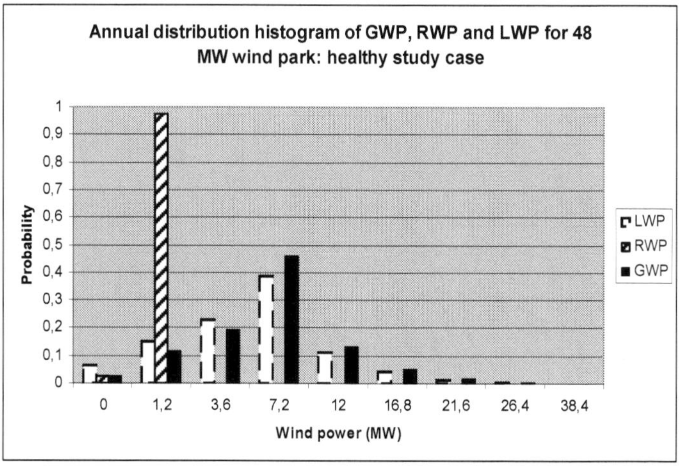

Fig. 8. Obtained annual distribution histograms of *GWP (black)*, *RWP (dashed)* and *LWP (white)* in the healthy study case (abscissa are the power intervals mean values)

Another scenario, considering thermal parks non-expected outages, has also been computed. In this outage case, situations of load non-recovering can appear when both classical parks are simultaneously unavailable. Indeed, wind production (48 MW installed capacity) is then unable to cover the required constant load (100MW). Consequently, a non zero *LOLE* value has been computed for the outage case.

Moreover, when the classical thermal unit connected at node 2 is unavailable (approximately 1747.2 hours of the year as the associated FOR is *0.2*), load shedding is imperative as only 30 MW (line capacity) of the required load can be transmitted on the line. Both LOLE and number of load shedding hours are confirmed in table 1.

Next to those reliability indices, wind power behavior has also been investigated. In the present outage case, several situations can influence wind production:

- When both thermal machines are simultaneously available, wind production behavior is the same as the one computed in the healthy case. Consequently, real transmitted wind production has always to be reduced to 1MW in that case; leading thus to increased lost wind power;
- When thermal unit connected at node 2 is unavailable, load shedding is required and only 30 MW are produced. This generation comes from the technical minimum of the long term thermal machine (29MW) and from wind park (1MW). Real transmitted wind power is thus still reduced in that case;
- Finally, when thermal unit connected at node 1 is unavailable and that the one connected at node 2 is available, all wind production can be transmitted (as there are no operation constraints over thermal machine at node 2) if it stays lower than the line transmission capacity (30 MW).

Consequently, due to situations where long term thermal park at node 1 is unavailable (and thermal park at node 2 is available), real transmitted wind production will be annually less decreased than in the healthy case previously investigated. Those results are confirmed in table 1 and *RWP* annual distribution plotted in Figure 9 points out existing *RWP* beyond 1 MW in the investigated outage case.

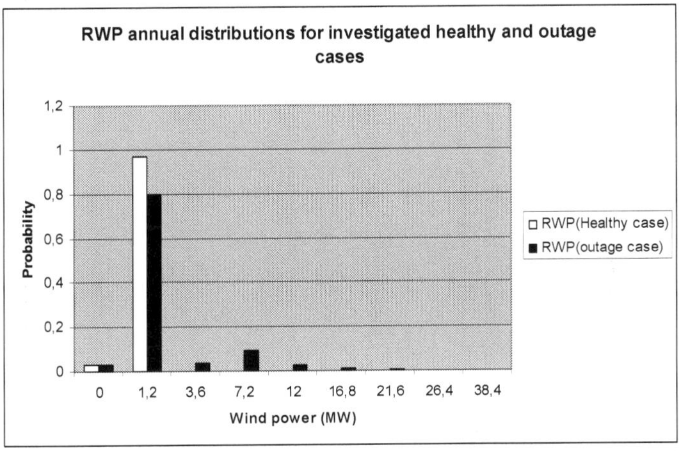

Fig. 9. Annual *RWP* distributions in the investigated outage (black) and healthy (white) cases

This first approach has thus pointed out that, in case of increased wind penetration, the *TSO* may have to reduce wind production in order to ensure a safe operation of the transmission system in presence of machines and lines constraints.

5. Simulations results on a modified RBTS test system

5.1 Wind parks introduced in RBTS generation nodes (Billinton et al., 1989)

In order to point wind generation impact in transmission system management, a slightly modified version of the academic *RBTS* test system (Billinton et al., 1989) has been considered. The implemented version of *RBTS* is described in Figure 10 and differs from the original one for some aspects:

- Peak load is always 185MW but the weekly and hourly modulation diagrams have been based on the Belgian real case (Buyse, 2004) given in Figure 2;
- Wind generation (48MW) is firstly introduced in generation nodes 1 (24MW) and 2 (24MW) by following the data given in tables 2 and 3. Wind speed regimes are supposed to be *Weibull* ones but differ from node 1 to 2. P-W conversion characteristic is the one depicted in Figure 3;
- The *RBTS* number of lines has been reduced from 9 (Billinton et al., 1989) to 7 in the present simulation. Indeed, both parallel lines between nodes 1-3 and nodes 2-4 (Billinton et al., 1989) are suppressed in order to point out the interest of Scanner© tool in showing the weak points of the studied system;

- Compared to the initial version of the RBTS (Billinton et al., 1989), operation constraints can be added, here, for classical parks (section 2.2). Consequently, more realistic system states management can be studied in the present case;
- No *FOR* is considered for lines and transformers as their availability is not under the scope of the present study. On the opposite, transmission constraints are well taken into account;
- *FOR* is considered for classical production parks and for wind generation.

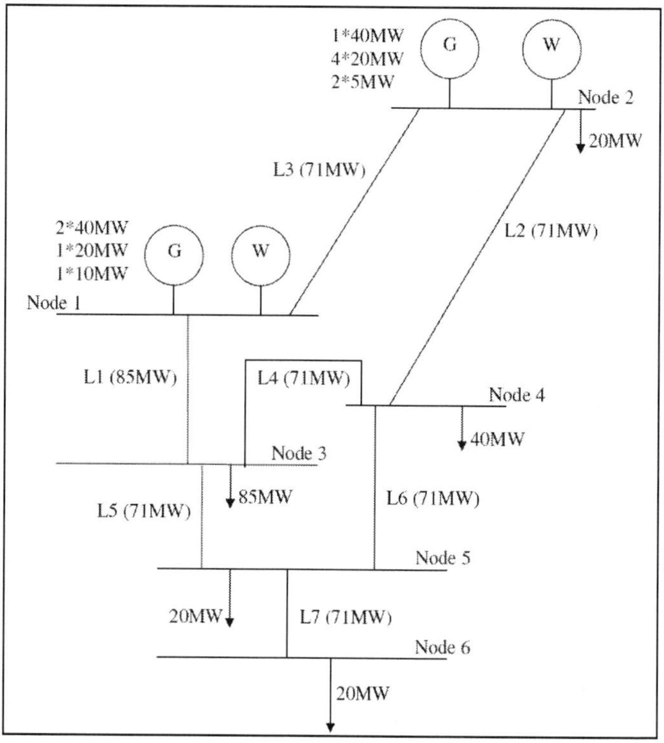

Fig. 10. Implemented version of the RBTS

	Wind regime	A	B
Node 1	*Weibull*	7.10	2.85
Node 2	*Weibull*	9.95	2.75

Table 2. Wind speed regimes considered at nodes 1 and 2 of the *RBTS*.

	Installed capacity (MW)	Connection node
Wind park 1	8	Node 1
Wind park 2	6	Node 1
Wind park 3	12	Node 2
Wind park 4	1	Node 1
Wind park 5	3	Node 2
Wind park 6	4	Node 1
Wind park 7	5	Node 2
Wind park 8	4	Node 2
Wind park 9	5	Node 1

Table 3. Wind generation considered for the modified *RBTS*

In order to face wind generation and transmission system operation constraints, two cases have been simulated:

-*Case 1*: no operation constraints are considered for the classical park: all parks are supposed to be thermal ones without technical minimum value (and are neither forced nor long term ones);
-*Case 2*: operation constraints are associated to the defined classical parks. Here, the sum of technical minima is supposed to be 75MW. This value represents almost 30% of the installed classical capacity (240MW) in reference to the Belgian real case (Buyse, 2004). Those 75MW are divided in nuclear machines (30MW), long term thermal parks (30MW) and forced ones (15MW).

Both cases have been firstly investigated for a 48MW installed wind capacity (table 3). With this first wind penetration, table 4 summarizes real transmitted annual wind energy for each park. It can clearly be observed that, with this reduced penetration, classical park operation constraints have a limited impact on wind as collected energies are similar in both investigated cases.

	Case 1	Case 2
Annual energy wind park 1 (GWh/y)	7.5	7.5
Annual energy wind park 2 (GWh/y)	5.5	5.5
Annual energy wind park 3 (GWh/y)	25.0	25.0
Annual energy wind park 4 (GWh/y)	0.9	0.9
Annual energy wind park 5 (GWh/y)	6.3	6.3
Annual energy wind park 6 (GWh/y)	3.7	3.7
Annual energy wind park 7 (GWh/y)	10.4	10.4
Annual energy wind park 8 (GWh/y)	8.2	8.2
Annual energy wind park 9 (GWh/y)	4.8	4.8

Table 4. Annual wind energies with (case 2) and without (case 1) classical park operation constraints for 48MW installed wind capacity

Moreover, Figure 11 compares *RWP* for wind park 1 in both investigated cases and clearly confirms an identical behavior of this wind park with or without classical operating constraints (this result can be extended to all the other considered wind parks). For information, *GWP* (identical in both cases as wind data are the same from one simulation scheme to the other) is also plotted and points out that all available wind power is transmitted in the system when wind penetration is low. Finally, note that, due to the limited transfer capacity of line *L1*, 340 annual hours of load shedding (node 3) are computed here but have no impact on wind power. In fact, as computed load shedding situations are quite seldom and not severe in the present case, wind generation is not modified by transmission constraints. Consequently, with the version of *RBTS* presented in fig.10, only classical units operation constraints can have an impact on wind generation.

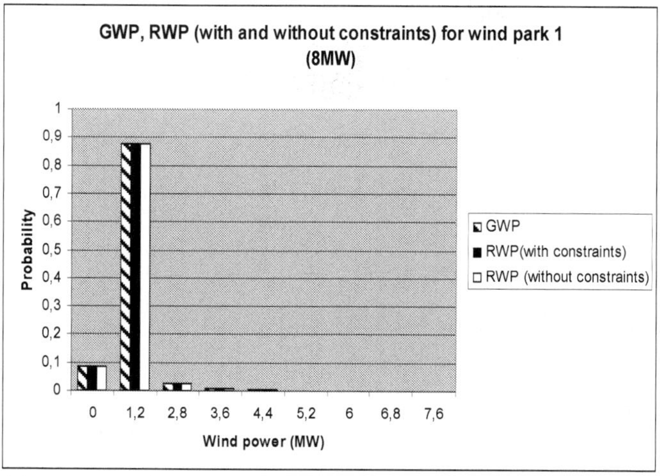

Fig. 11. *GWP* and *RWP* for wind park 1 (8MW) with and without classical operating constraints

A second simulation has then been investigated by doubling wind production at node 2 to 48MW and by setting to zero the one of node 1. Again, simulations have been realized with and without operation constraints on classical parks (cases 1 and 2). In table 5, it can be concluded that, if all wind production is installed in node 2 (greater wind speeds; see. Table 2), some situations of wind power excess can be computed. By comparing the results of tables 4 and 5, it can thus be concluded that, with the same global wind penetration, smoothing effects due to wind generation dispersion can have a positive effect on the electrical transmission system management.

	Case 1	Case 2
Annual energy wind park 1 (GWh/y)	0	0
Annual energy wind park 2 (GWh/y)	0	0
Annual energy wind park 3 (GWh/y)	49.2	48.8
Annual energy wind park 4 (GWh/y)	0	0
Annual energy wind park 5 (GWh/y)	12.4	12.3
Annual energy wind park 6 (GWh/y)	0	0
Annual energy wind park 7 (GWh/y)	20.5	20.3
Annual energy wind park 8 (GWh/y)	16.2	16.1
Annual energy wind park 9 (GWh/y)	0.0	0.0

Table 5. Annual wind energies with (case 2) and without (case 1) classical park operation constraints for 48MW installed wind capacity at node 2 (only)

5.2 Adequate repartition of wind parks in the transmission system

In the previous paragraph, as transmission line capacities were sufficiently high, only classical parks operation constraints had an impact on wind generation. In order to also take into account transmission lines constraints, the *RBTS* test system has been voluntarily weakened by reducing the capacity of line *L1* (Figure 10) to 40 MW.

Figure 12 presents the simulation results collected, in that case, with the wind parks defined in table 3 (48 MW installed in nodes 1 and 2). If compared with table 4 (case 2), it can be observed that the power of all wind parks connected to node 1 must be reduced due to the limited capacity of *L1*.

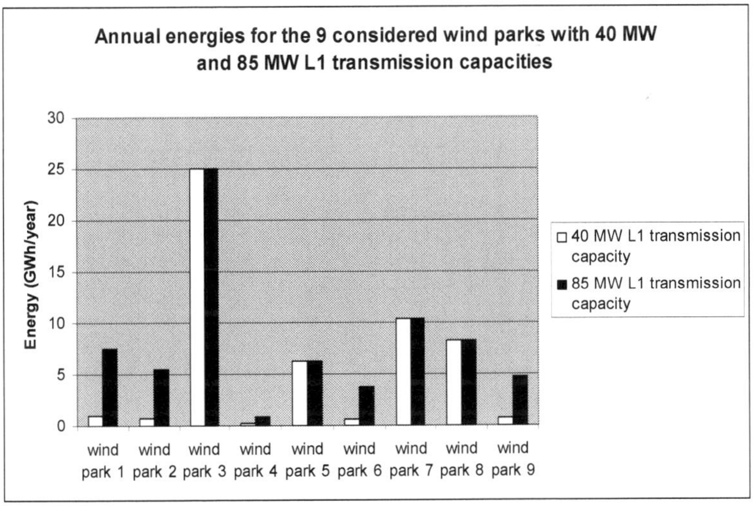

Fig. 12. Annual energy (GWh/year) for the 9 considered wind parks with 40 MW (white) and 85 MW (black) *L1* transmission capacities

In the case of a weakened transmission system, the connection nodes of wind parks take thus a major importance. Indeed, if the considered 48MW of wind generation are now distributed between nodes 2 and 4 (see table 6), the limited transmission capacity of *L1* does no more impact wind power and this last one can be entirely transferred in the network (see table 7). This complete use of wind production was not feasible when some of the defined wind parks (24MW) were directly connected at *L1* (via node 1; see Table 3 and Figure 12).

	Installed capacity (MW)	Connection node
Wind park 1	8	***Node 4***
Wind park 2	6	***Node 4***
Wind park 3	12	Node 2
Wind park 4	1	***Node 4***
Wind park 5	3	Node 2
Wind park 6	4	***Node 4***
Wind park 7	5	Node 2
Wind park 8	4	Node 2
Wind park 9	5	***Node 4***

Table 6. Wind generation considered for the modified *RBTS* test system

Annual energy wind park 1 (GWh/y)	7.5
Annual energy wind park 2 (GWh/y)	5.5
Annual energy wind park 3 (GWh/y)	25.0
Annual energy wind park 4 (GWh/y)	0.9
Annual energy wind park 5 (GWh/y)	6.3
Annual energy wind park 6 (GWh/y)	3.7
Annual energy wind park 7 (GWh/y)	10.4
Annual energy wind park 8 (GWh/y)	8.2
Annual energy wind park 9 (GWh/y)	4.8

Table 7. Annual wind energy for wind parks located in nodes 2 and 4 with limited transmission capacity of *L1* (40MW)

This result points out the utility of the developed tool in order to improve the management of wind generation. Indeed, thanks to the proposed software, the transmission system operator will now be able, not only, to quantify the maximal wind penetration in a given network, but also, to propose an adequate distribution of wind parks connection nodes. However, for this last point, note that environmental concerns for the establishment of wind parks must still be taken into account.

In conclusion, it is believed that the proposed tool will assist system planners and transmission system operators to qualitatively assess the system impact of wind production and to provide adequate input for the managerial decision process in presence of increased wind penetration.

6. Conclusion

In this chapter, wind generation has been introduced into a transmission system analysis tool. This last one was composed of two parts: system states generation (Monte Carlo simulation) and analysis (economic dispatch, DC load flow and eventual load shedding). In order to take into account wind generation in this simulation tool, each part had thus to be modified. Finally, a useful HLII analysis software taking into account wind generation has been developed and has permitted to study the impact of wind generation not only on reliability indices but also on the management of the classical production park. In that way, situations of forced wind stopping were pointed out due to increased wind penetration and transmission system operation constraints. In that way, the interest of the proposed software was demonstrated by adequately determining maximal wind penetration for a given electrical system. Moreover, the importance of wind parks connection nodes was also pointed out in order to limit active power flows over the most restrictive (lowest capacities) transmission lines.

7. References

Al Aimani S. (2004). Modélisation de différentes technologies d'éoliennes intégrées à un réseau de distribution moyenne tension, *Ph.D. Thesis*, Ecole Centrale de Lille, chap.2, pp.24-25, Dec. 2004.

Allan R.N., Billinton R. (2000). Probabilistic assessment of power systems, *Proceedings of the IEEE*, Vol. 22, No.1, Feb. 2000.

Billinton R., Kumar S., Chowdbury N., Chu K., Debnath K., Goel L., Kahn E., Kos P., Nourbakhsh, Oteng-Adjei J. (1989). A reliability test system for educational purposes - Basic data. *IEEE Trans. On Power Systems*, Vol. 4, No. 3, Aug. 1989, pp. 1238-1244.

Billinton R., Chen H., Ghajar R. (1996). A sequential simulation technique for adequacy evaluation of generating systems including wind energy. *IEEE Trans. On Energy Conversion*, Vol. 11, No. 4, Dec. 1996, pp.728-734.

Billinton R., Bai G. (2004). Generating capacity adequacy associated with wind energy. *IEEE Trans. On Energy Conversion*, Vol. 19, No. 3,Sept. 2004, pp. 641-646.

Billinton R., Wangdee W. (2007). Reliability-based transmission reinforcement planning associated with large-scale wind farms. *IEEE Trans. On Power Systems*, Vol. 22, No. 1, Feb. 2007, pp. 34-41.

Buyse H. (2004). Electrical energy production. *Electrabel documentaion*, available web site: www.lei.ucl.ac.be/~matagne/ELEC2753/SEM12/S12TRAN.PPT, 2004.

Ernst B. (2005). Wind power forecast for the German and Danish networks. *Wind Power in Power Systems*, edited by Thomas Ackerman, John Wiley & Sons, chap.17, pp.365-381, 2005.

Mackensen R., Lange B., Schlögl F. (2006). Integrating wind energy into public power supply systems - German state of the art. *International Journal of Distributed Energy Sources*, Vol. 3, No.4, Dec. 2007.

Maupas F. (2006). Analyse des règles de gestion de la production éolienne : inter-comparaison de trois cas d'étude au Danemark, en Espagne et en Allemagne. Working paper, *GRJM Conference*, Feb. 2006.

Papaefthymiou G. (2006). Integration of stochastic generation in power systems. *PhD. Thesis*, Delft University, chap. 5 & 6, June 2006.

Papaefthymiou G., Schavemaker P.H., Van der Sluis L., Kling W.L., Kurowicka D., Cooke R.M. (2006). Integration of stochastic generation in power systems. *International Journal of Electrical Power & Energy Systems*, Vol. 18, N°9, Nov. 2006, pp. 655-667.

Sacharowitz S. (2004). Managing large amounts of wind generated power feed in - Every day challenges for a German TSO and approaches for improvements. *International Association for Energy Economics (IAEE)*, 2004 North American Conference, Washington DC, USA, 2004.

Vallee F., Lobry J., Deblecker O., (2008). System reliability assessment method for wind power integration. *IEEE Trans. On Power Systems*, Vol. 23, No. 3, Aug. 2008, pp. 1288-1297.

Wangdee W., Billinton R. (2006). Considering load-carrying capability and wind speed correlation of WECS in generation adequacy assessment. *IEEE Trans. On Energy Conversion*, Vol. 21, No. 3, Sept. 2006, pp. 734-741.

Variable speed pumped storage hydropower plants for integration of wind power in isolated power systems

Jon Are Suul
Norwegian University of Science and Technology
Norway

1. Introduction

Energy storage is a key issue when integrating large amounts of intermittent and non-dispatchable renewable energy sources into electric power systems. To be able to maintain the instantaneous power balance and to compensate for the influence of power fluctuations that might result from renewable sources, flexible capability for power control is needed. Therefore, sufficient energy storage with suitable interface to the electricity grid is considered to be a necessity for development towards sustainable energy systems based on renewable energy sources (Dell & Rand 2001). Energy storage is also an important issue with respect to power system stability, load balancing and frequency control in power systems with large share of nuclear or thermal power plants that need to run at almost constant power (Shimada & Mukai, 2007; Faias et al., 2007).

Many types of energy storage have been proposed for power system applications, and the different technologies have large variety in storage capacity, power rating and cycling capability. There is continuous development going on to increase the performance, reduce the cost and improve the interface to the electric power system for different technologies like battery storage, hydrogen storage, capacitors, flywheels, superconducting magnetic energy storage (SMES), compressed air energy storage (CAES) and pumped storage hydropower. There also exist large scale test facilities, and a few commercial applications, with most types of relevant energy storage technologies, but until now only pumped storage hydropower is considered to be a mature technology with long history of large scale commercial application (Chen et al., 2009; Dell & Rand, 2001; Hammerschlag & Schaber, 2007; Ibrahim et al., 2008; Kondoh et al., 2000; Ribeiro et al., 2001).

Even if pumped storage hydropower systems have been used for more than hundred years, new improvements are still being introduced and development is going on with respect to rating, application areas and control. One of the most important advances during the last decades has been the development of variable speed systems to allow for controllable power in pumping mode (McClear & Meisel, 1984; Scherer, 2005; Terence & Schäfer, 1993). Such systems have been constructed on commercial basis in Japan since the beginning of the 1990s for load levelling in the power system, with more recent introduction also in Europe, and are usually based on a reversible Francis-turbine with a power electronically controlled machine (Mori et al., 1995; Taguchi et al., 1991; Bocquel & Janning, 2003). With increased introduction of fluctuating renewable energy sources like wind power and photovoltaic in the large scale power systems, pumped storage systems are expected to gain even more importance (Sick & Schwab 2005)

Another important trend in research on pumped storage systems during the last decades has been the focus on application as energy storage in hybrid systems with wind power or other intermittent renewable energy sources as the main energy supply (Bueno & Carta 2006; Kaldellis et al., 2001; Katsaprakakis et al., 2008; Sommerville, 1989; Theodoropoulos et al., 2001). Such hybrid systems have mainly been considered for application in isolated power systems with limited ratings, as an important contribution towards a sustainable electricity supply without dependency on fossil fuels. The consideration of pumped storage schemes for hybrid power systems in isolated electricity grids have until now been mainly focused on simple and robust solutions where the main purpose has been to improve the energy balance of the systems when increasing the share of renewable energy. Although the controllability of variable speed units can be equally important in such hybrid systems as in larger systems using high-capacity pumped storage power plants for load-levelling, little attention has until now been focused on smaller scale variable speed units for isolated grids.

Starting from available scientific literature, this chapter will first briefly review the development history of hydroelectric pumped storage systems. The main focus will be on power electronic solutions for variable speed operation and on application of pumped storage systems to integrate renewable energy sources in weak or isolated power systems. After reviewing the historical development and the state of the art for pumped storage systems from the electrical perspective, the utilization of power electronic control and variable speed operation for grid integration of fluctuating renewable energy sources will be discussed. In specific, the use of a full-scale voltage source converter for control of the pumped storage will be suggested, since this might be a relevant solution in isolated systems with limited ratings. The characteristics of the proposed system will be discussed, and one possible control system for the power electronic converter will be described. This will be presented as a background for illustrating the short-term performance of the proposed system by time-domain simulations and discussing how the variable speed pumped storage system can be used to improve the power system operation and allow for larger shares or fluctuating renewable energy sources.

2. Brief review of pumped storage systems

Energy storage by water reservoirs is a conceptually simple type of energy storage that has been well known and utilized for a long time. The first hydroelectric pumped storage system was constructed in Switzerland already in 1909, and used separate pump and turbine units (EB, 2009). When the Rocky-River Pumped storage hydroelectric station was commissioned in 1929, as the first of its kind in USA, it was well recognized, although not utilized, that the installed pumps could be operated as turbines to generate electricity at reduced efficiency (ASME, 1980, Coleman et al., 1976). In the same time period, development and design improvements of reversible Francis-turbines was going on, and from the 1950s, this has become the standard solution used for almost all new, large scale, pumped storage systems (Coleman et al., 1976; Wikipedia, 2009).

This basic concept of pumped storage systems as sketched in Fig. 1 requires two water reservoirs and a reversible pump-turbine with a grid connected electrical machine. The machine must operate as a motor in pumping mode and as a generator by changing the direction of rotation when the system is operated in turbine mode. Such systems can be constructed in almost any power range with energy storage capacity only limited by the size of the reservoirs, and the round-trip efficiency is usually in the range of 75-85% (ESA, 2009). In this basic form, pumped storage is a mature technology that has been implemented in large scale on commercial basis with more than 90 GW of installed rating worldwide. An extensive, although not necessarily complete, list of the main pumped storage implementations in the world can easily be found from public sources like Wikipedia (Wikipedia, 2009).

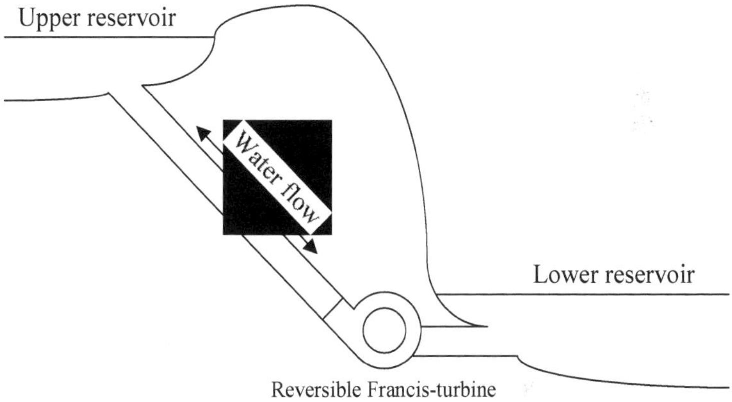

Fig. 1. Basic structure of pumped storage hydro power plant with reversible pump-turbine (Suul et al., 2008a)

2.1 Technological development and introduction of variable speed pumped storage

There has been a continuous development to increase the power rating and the maximum head of reversible pump-turbines, and systems exceeding 400 MW with more than 700 m of pumping head are now in operation (Ikeda, 2000). For the development towards increased heads and higher ratings of pump-turbines, the electrical equipment has not been the main limitation, and this issue will therefore not be further discussed here.

Regarding development of electrical solutions for pumped storage systems, the main issue during the last 3-4 decades has been the introduction of power electronic equipment, and the main achievement in this respect has been the development of power electronically controlled variable speed pumped storage systems. There are many factors that have motivated the drive for variable speed operation of pump-turbines, both regarding the operation of the pump-turbine itself and from the power system point of view.

2.1.1 Motivations for variable speed operation of pumped storage systems

One important motivation for variable speed operation has been the possibility to improve the efficiency of the pump-turbine, since the speed corresponding to maximum efficiency is different for pumping-mode and turbine operation, and is also changing with the water head (Galasso, 1991; Kerkman et al., 1980 a; Lanese et al., 1995; Merino & López, 1996;). Variable

speed operation can also be a necessity in pumped storage systems constructed for large variations in water head.
Even more important is the possibility for power control in pumping mode, since traditional pump-turbines with synchronous machines connected directly to the grid will operate at constant speed and by that constant power in pumping mode. Variable speed operation has therefore been strongly motivated by the possibility to obtain similar power controllability in pumping mode as when operated as a generator (Lung et al., 2007; Schafer & Simond, 1998; Taguchi et al., 1991, Kuwabara et al., 1996). Another benefit by variable speed operation is that the allowable operation range in generator mode can be extended, and that problems with water hammering and other secondary effects in the turbine can be more easily controlled (Kuwabara et al., 1996; Gjengedal, 2001).
From the power system point of view, the possibility for power control in pumping mode is also one of the most important benefits obtained by variable speed operation of pumped storage systems. The power electronic drive system can also be used to increase the response time for power control by utilizing the inertia of the pump-turbine and the electrical machine, both in generating mode and in pumping mode. The fast response can allow for compensation of power fluctuations and damping of power oscillations, and by that improve the stability of the power system (Bocquel & Janning, 2005; Erlich & Bachmann, 2002; Goto et al., 1995; Grotenburg et al., 2001; Schafer & Simond, 1998).

2.1.2 Power electronic solutions to obtain variable speed operation

The possibility for variable speed operation of pump-turbines by use of electrical drives has been considered since power electronic systems like thyristor-controlled HVDC-links were introduced in the power system on a commercial basis. Some of the first investigations of variable speed operation of pump-turbines were therefore considering full-scale thyristor converters based on the configuration indicated in Fig. 2 (Kerkman et al., 1980 a and b). The same configuration has also been considered for variable speed operation of hydropower stations connected directly to an HVDC-link (Naidu & Mathur 1989; Arrilaga et al., 1992). Although this configuration is simple and based on using a traditional synchronous machine, the use of a full-scale converter has been considered a main drawback with respect to cost and losses for pumped storage systems with high total ratings. Therefore only a few pumped storage systems with high demands on operating range have used this topology for continuous variable speed operation until now (Lanese et al., 1995). However, thyristor converters with reduced rating have become a common solution for starting of pumped storage systems running at constant speed (Chiang et al., 1997; Fostiak & Davis 1994). Converters with reduced ratings have also been proposed for improved transition between different operating conditions of pumped storage systems with constant speed (Magsaysay, 1995).

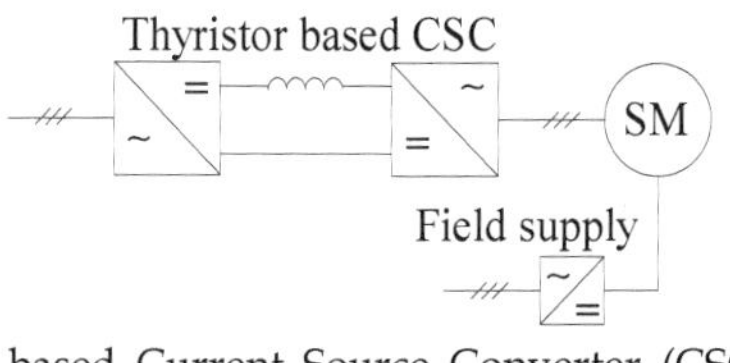

Fig. 2. Full-scale thyristor based Current Source Converter (CSC) driving a synchronous machine

For most applications of pumped storage systems, only a limited controllable speed range is needed during normal operation (Bendl et al., 1999; Boquel & Janning 2003; Mori et al., 1995; Schreirer et al., 2000). This allows for obtaining variable speed operation by utilizing the concept of a doubly fed asynchronous machine (DFAM) and a power electronic converter with reduced converter rating compared to the total machine rating. This topology was considered at an early stage of the investigations on variable speed pumped storage systems and has later been preferred in most large scale implementations to limit the converter ratings. With this concept, the industry has been able to build units with total ratings in the range of several hundreds of MVA (Gish et al., 1981; Hayashi et al., 1988; Lanese et al., 1995; Lung et al., 2007; Sugimoto et al., 1989; Taguschi et al., 1991). In addition to the reduced converter rating compared to the full scale current source converter, this configuration has as another advantage that the reactive power exchange with the grid can be controlled. This can be utilized for voltage control in the grid and contribute to improving the stability and the operating conditions in the rest of the power system (Boquel & Janning 2005; Schafer & Simond, 1998; Taguchi et al., 1991).
When the first commercial, large scale, implementations of variable speed pumped storage systems were investigated, the power electronic converters had to be based on thyristors to achieve sufficient ratings. Since the required frequency for the rotor circuit in the doubly fed machine is given by the deviation from synchronous speed, it is usually limited to a few Hz. Therefore configurations with cycloconverters as shown schematically in Fig. 3 a) has been considered suitable solutions that can be made with rugged designed for high capacity and low losses (Boquel & Jannig, 2005; Furuya et al., 1993; Kuwabara et al., 1996; Taguchi et al., 1991).
As the voltage and current ratings of gate-controlled switches like GTOs, GCTs, IGCTs and IGBTs have increased, topologies based on back-to-back voltage source converters have become relevant for feeding the rotor windings of the doubly fed machine. This configuration is shown in Fig. 3 b), and usually a two-level or three-level neutral-point-clamped voltage source converter is considered as the preferred converter topology. The voltage source converter topology is gaining even more relevance as the development of high power voltage source converters for other drive applications is continuing, and is being used in some of the most recent pumped-storage implementations (Furuya et al., 1995; Hodder et al., 2004; Hämmerli & Ødegård, 2008; Mitsubishi (2009); Sapin et al., 2000; Toshiba, 2008).

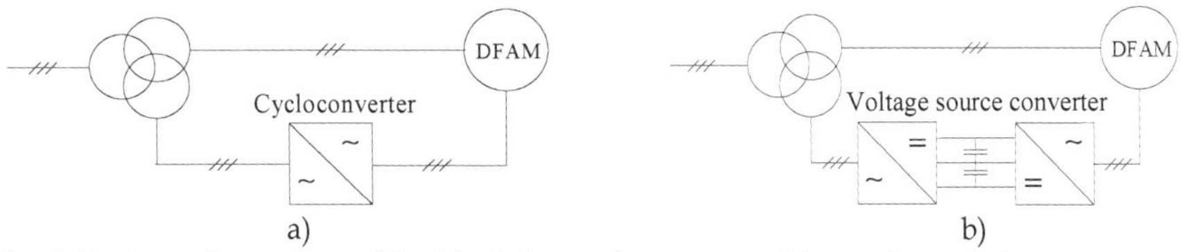

Fig. 3. Basic configurations of doubly fed asynchronous machines where a) shows a system with a cycloconverter feeding the rotor windings while b) shows a system with a back-to-back voltage source converter and capacitive DC-link

The configurations presented in Fig. 2 and Fig. 3 are the basic schemes that are currently used for variable speed pumped storage systems, and such systems can of course also be considered for normal hydropower plants at sights were variable speed operation is considered to be beneficial (Ansel 2006; Arrilaga et al., 1992; Gjengedal 2001, Naidu & Mathur 1989; Sporild et al., 2000).

The alternative solution of using a full scale voltage source converter to drive the machine has not yet been used for commercial applications although it has been suggested as relevant for small scale hydropower stations (Abbey & Joos 2005; European Commission 2000; Fraile-Ardanuy et al., 2006). This topology can be applied both for cage induction machine and for synchronous machined, but except for the lowest power ratings, the synchronous machine might be the most relevant. Configurations with full-scale power electronic converters should however be further investigated for use in pumped storage systems and for applications in weak and isolated grids, and will therefore be the topic of discussion in section 3 (Suul 2006, Suul et al., 2008 a and b).

2.2 Trends in application of pumped storage systems

As explained, hydroelectric pumped storage systems have been constructed for more than 100 years, and until the last two decades the pump-turbine systems were operated at constant speed by synchronous machines directly connected to the power system. For such systems, bulk energy storage has been the main motivation since pumping occurs at constant power. Pumped storage systems can in this way be economically beneficial by utilizing cheap electricity for pumping water and regenerating the electricity at a higher price during peak hours (Allen, 1977; ESA, 2009; Wu et al., 2007). In generation mode, the power of the pumped storage will however be controlled like in a normal hydropower station and this will contribute to improve the controllability and maintain the power balance of the power system.

One of the main limiting factors for implementation of pumped storage systems is the dependency on available sites with possibility for suitable reservoirs. Therefore, several strategies for utilizing the principle of pumped storage systems have been proposed. One of the earliest suggestions has been the use of underground pumped storage systems utilizing suitable geological formations (Allen, 1977). This concept has recently been proposed for development into commercial applications (Riverbankpower, 2009). Also systems utilizing seawater and the sea as a lower reservoir for pumped storage systems have been developed and put in operation (Fujihara, 1998). Recently, pumped storage systems with the sea as upper reservoir, and artificial lakes below sea-level has been suggested as an alternative configuration (KEMA, 2009) Even systems utilizing natural depressions below sea-level as lower reservoirs have been proposed (Murakami 1995). Another possibility that has been proposed is to combine pumped storage systems with utilization of tidal streams (MacKay, 2007).

2.2.1 Large scale power system applications

As described in section 2.1, one important trend in development of pumped storage systems during the last two decades has been the introduction of variable speed systems. The first commercial systems with large doubly fed asynchronous machines were commissioned in Japan around 1990, and the first large scale implementation with a full-scale thyristor converter for operation with wide speed range and wide range of water head was constructed in China around the same time (Lanese et al., 1995; Taguchi et al., 1991;Terens & Schafer 1993). In Japan, the development of pumped storage systems was intended for both energy storage and to improve the controllability of the Japanese power system because the dominating energy source was nuclear power plants running with almost constant production. Therefore, pumped storage power plants with high degree of controllability both in generating mode and in pumping mode were needed to improve the daily balance between production and load and to improve the frequency control of the system. For utilizing the benefits with respect to power system operation, large units were needed, and several units with ratings up to the range of 400 MVA have been developed (Kuwabara et al., 1996; Mori et al., 1995; Mitsubishi, 2008). More projects are still under development and machines up to 475 MVA of total rating are being planned (Toshiba 2008).

In Europe, only smaller scale test facilities were constructed at the time when the first commercial units were commissioned in Japan (Merino & López, 1996; Terens & Schafer, 1993). Recently, one large scale variable speed pumped storage system with machines rated at 300 MVA and rotor circuit cycloconverters rated for 100 MVA has been implemented in Europe and two new units are now under constriction in Slovenia and Switzerland (Alstom, 2009; Bocquel & Janning, 2003; Mitsubishi, 2008). The pumped storage systems in Europe have had the same main motivation as the systems in Japan, and the operation of the pumped storage has been intended to improve the controllability of the power system in presence of a large share of nuclear or coal fired power plants (Erlich & Bachmann, 2002; Schafer & Simond, 1998; Simond et al., 1999). As attention towards utilization of renewable energy has increased, it has also become clear that pumped storage systems have additional value with respect to the ability to compensate for variations in both production and load in the system. Pumped storage systems as a suitable complement to wind power installations and other renewable energy sources have therefore received significant attention (Allen et al., 2006; Bose et al., 2004; Jaramillo et al., 2004; Lilly et al., 1991; Papathanassiou et al., 2003; Sick & Schwab, 2005).

2.2.2 Applications in small and isolated power systems

Hybrid systems, where energy storage is used to increase the utilization of renewable energy sources in isolated power grids has also received significant interest during the last years. Many of the presented projects and studies have been directed towards utilization of wind power in isolated systems and the use of pumped storage systems to increase the annual share of the energy supply that can be covered by the wind turbines (Bakos, 2002; Bueno & Carta, 2006; Bueno & Carta, 2005 a and b; Ceralis & Zervos, 2007; Chen et al., 2007; Kaldellis et al., 2001; Katsaprakakis et al., 2007; Katsaprakakis et al., 2008; Protopapas & Papathanassiou, 2006; Protopapas & Papathanassiou, 2004).

Most small and isolated power systems on islands and in remote areas have been based on diesel generator sets, and many of the proposed hybrid systems are intended for substituting the fossil fuels needed for the existing power supply by renewable energy sources (Bakos, 2002; Bueno & Carta, 2006; Jensen, 2000). Several studies considering pumped storage systems for energy storage and utilization of wind power have therefore been presented regarding small islands with different locations as for example Mediterranean islands, Shetland, and Hawaii (Bollmeier II et al., 1994; Kaldellis &Kavadias, 2001; Kaldellis, 2002; Katsaprakakis et al., 2006; Sommerville, 1989; Taylor, 1988; Theodoropoulos et al., 2001). Some of the presented projects, like the plans for the electricity supply on the Spanish island El Hierro, have even more ambitious goals of obtaining sustainable energy supplies bases entirely on renewable energy sources (INSULA, 2008; Piernavieja et al., 2003). With the goal of increasing the share of the annual electricity consumption provided by renewable energy sources, the energy balance over the year, sizing of the storage capacity and selection of ratings for the different units in the system has been the main focus of many of the presented studies (Anagnostopoulos & Papantonis, 2008; Brown et al., 2008; Bueno & Carta, 2006; Katsaprakakis et al., 2008). Therefore most of the available literature on hybrid power systems with a pumped storage power plant is discussing investigations based on economical considerations and stochastic methods applied for long time periods, to assess the operability and suitable sizing of the components with expected variations in weather conditions over the years.

When the energy balance has been in focus, and because small systems have been considered, simple and robust practical solutions for the pumped storage schemes have been assumed. Utilization of Pelton turbines for the hydropower station appears to be the preferred solution in most small systems, and by that configurations with separate pumps are necessary. The pumping stations is usually assumed to consist of a number of pumps with specific ratings that can be operated in parallel to control the total power in steps, although it has been shown that variable speed operation of at least one unit will be the most flexible solution (Anagnostopoulos & Papantonis, 2007, Bueno & Carta 2006, Bueno & Carta, 2005a).

From the available literature, it appears that dynamic control of the instantaneous power balance and operation of hybrid systems including wind power and pumped storage schemes has not been extensively investigated and documented in the available literature. When introducing controllable power electronic converters to the pumped storage systems, it is therefore relevant to investigate the dynamic control and operation of a hybrid wind and pumped storage system, as will be presented in the following sections.

3. Variable speed pumped storage topology for operation in weak and isolated grids

From the presented discussion, it is clear that pumped storage units with a full-scale voltage source converter controlling the stator windings of the machine can be a possible configuration. With the increased rating of self-commutated semiconductor switches and available high-power motor drives based on the voltage source converter topology, this configuration could be an attractive solution, especially relevant for pumped storage units in weak and isolated systems (Suul, 2006). The schematic layout of such a configuration is shown in Fig. 4, and the machine is considered to be a wound field synchronous machine with static excitation system. A suggested voltage level of 3.3 kV is indicated in the figure although any other standard voltages for medium voltage drive systems could be chosen. The voltage source converters are considered to be standard industrial drives based on the three-level neutral-point clamped topology, although any other voltage source converter topology with ratings suitable for a specific implementation can be used.

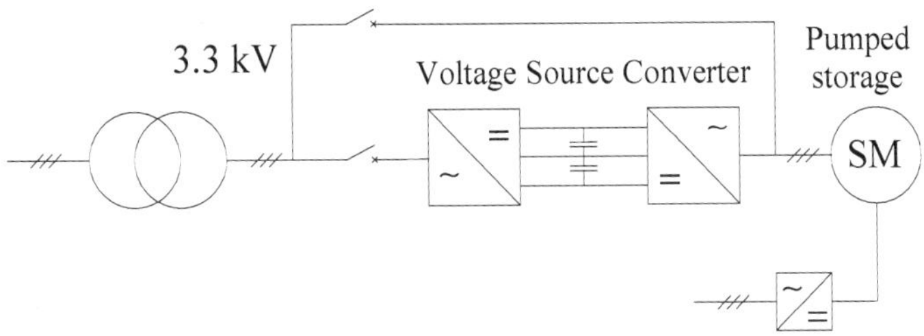

Fig. 4. Suggested topology for variable speed operation of pumped storage

Since the suggested configuration is based on a normal synchronous machine with field windings, it will be possible to bypass the converter and operate the machine directly connected to the grid as indicated in Fig. 4. This can allow for a kind of redundant operation of the system in case of problems with the converter, and can also be utilized to reduce the losses during normal generator operation. The topology of Fig. 4 is also of interest because of the possibility for operating the system as a conventional hydropower plant with traditional solutions available for the operators in case of operational problems. For a small power station in a remote area it can for instance be important to have the possibility to carry out a normal black-start of the power system with the pumped storage unit as a generator, even if the main converter is out of operation (Holm, 2006).

With the presented configuration, it will be possible to use a diode rectifier for the grid side converter and have variable speed operation only in pumping mode (Suul et al., 2008a). This can reduce the losses during pumping, but will not allow for variable speed operation in generator mode. Variable speed operation in generator mode will also make it possible to run at optimal speed for a wide range of water head, or to increase the controllability and the speed of response for the system. A configuration with an active-front-end converter for reversible power flow, will also allow for voltage control in the grid by use of reactive current. Since such a system is capable of operation with any power factor, the grid side converter can be used to control reactive power and grid voltage (Suul et al., 2008b). The grid side converter can be operated for control of grid voltage even in the case when the pump-turbine is not in operation, and will then function as a Static Synchronous Compensator (STATCOM). A controllable grid side converter could also make stand-alone operation and frequency control of the grid possible, even without any other production units based on synchronous generators in operation.

3.1 Control system overview for investigated configuration

The main structure of one possible control system for operating the suggested configuration in pumping mode is included in Fig. 5 (Suul et al., 2008 b). The figure shows how the grid side converter, connected to the main transformer through a LC-filter, can be controlled by a voltage oriented vector current control system in a synchronously rotating dq-reference frame. The estimate of the voltage phase angle used for the park transformation is obtained by a Phase Locked Loop (PLL) that is also tracking the grid frequency and the voltage components in the rotating reference frame (Chung, 2000, Suul, 2006). The d-axis of the rotating reference frame is aligned with the grid voltage vector, and the q-axis is leading the d-axis by 90°. The current controllers are PI-controllers in the rotating reference frame with feed-forward from measured grid voltage and decoupling terms depending on the filter inductance and the grid frequency (Blasko & Kaura, 1997). To avoid oscillations in the LC-filter, an active damping routine can be added to the function of the current controllers (Mo et al., 2003). The output from the current controllers is divided by the DC-link voltage to decouple the current controllers from the dynamics of the DC-link. After transformation into phase coordinates and adding third harmonic injection, the reference voltages are used for PWM modulation of the switches of the converter. Since the d-axis is aligned with the voltage vector, the input reference to the d-axis current controller is generated by an outer loop DC-link voltage controller that is maintaining the power balance of the system. The q-axis current reference can be generated by an outer loop controller for grid voltage or flow of reactive power.

The grid frequency from the PLL is also used for the power control of the pump turbine. Different structures for controlling the power flow of the pumped storage system in pumping mode, and for generating the power reference to the drive system of the synchronous machine will be discussed in the next section. The details of the drive system of the synchronous machine are not of main importance to the characteristics of the pumped storage system as seen from the grid if the response is fast and precise. In this paper a similar vector control structure as for the grid side converter is used, but since the machine has salient poles, a stator flux oriented ml-reference frame is used for the control of torque and flux while a rotor oriented dq-reference frame is used for the current controllers (Alaküla, 1993, Suul, 2006). Basically the same control structure can be used for controlling the synchronous machine drive in generating mode, but an additional speed controller and the hydraulic control system of the turbine will have to be included in the model.

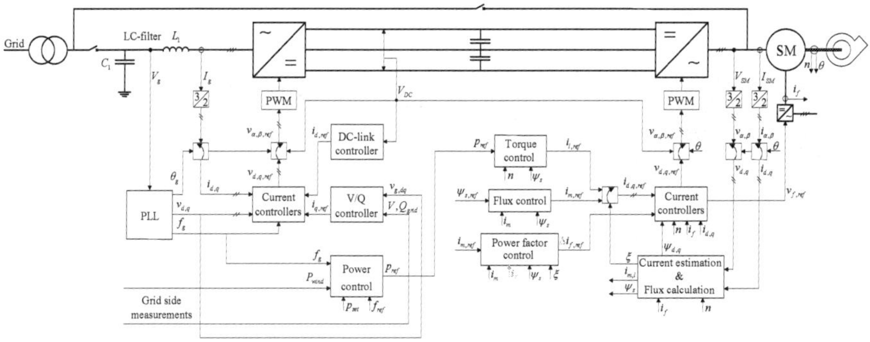

Fig. 5. Overview of possible control system for the investigated converter topology (Suul et al., 2008b)

3.2 Power control strategies for the pumped storage system

For power control of the pumped storage, an external power set-point, p_{set}, as shown in Fig. 5 is assumed as the main input to the control system. This set-point can be provided by the system operator based on scheduled operation and restrictions on operation due to limitation in reservoir capacity and the situation in the power system. The set-point value can have a relatively slow rate of update, and be in the time frame normally used for investigations concerned with optimization of the energy balance of the system. The main interest in this context is however how the local control system can respond to power fluctuations from a wind farm and control the power of the pumped storage system to mitigate negative influences on the rest of the power system.

3.2.1 Balancing power fluctuations by Load Following

If a wind farm or another source of power fluctuations is assumed to be in the end of a radial line and close to the pumped storage unit, the power flow in the grid can be measured and the controllability of the pumped storage unit can be used to compensate these fluctuations directly. This way, an almost constant and adjustable power flow is seen from the rest of the grid as long as the pumped storage system is within its limits of operation (Suul et al., 2008a). Assuming a forecasted power output $p_{w,set}$, and a measured power of p_{wind} from a wind farm, the additional power command Δp_w to the pumped storage unit can be calculated by (1).

$$\Delta p_w = p_{wind} - p_{w,set} \tag{1}$$

It can be noticed that the operation of the pumped storage in pumping mode will result in more production capacity on line during high wind and low-load conditions, and by that increase the control capability and the frequency response of the power system compared to the case without any pumped storage. Still, a power control structure for the pumped storage that is only including load following based on measured output of wind farm will not fully utilize the capability for improvement of the power system operation that is available with the high degree of controllability introduced by the variable speed operation (Suul, 2006).

3.2.2 Frequency Droop Control

Controlling a variable speed pumped storage unit to take directly part in the primary frequency control of the power system, can be of significant importance also in a weak or isolated system. Especially when the load in the system is small and fluctuating power production from a wind farm is dominant, this can be an important way of improving the response of the total power system to disturbances. By controlling the pumped storage power consumption in pumping mode based on a frequency droop, like in a normal hydropower station, the frequency response of the system during low load conditions will be increased. In this case, the frequency response will increase not only by the increased amount of production that will be on line to keep the pumped storage unit running, but also by the frequency response of the pumped storage unit as a frequency controlled load. This way, the introduction of the variable speed pumped storage will help improving the system performance to all kind of disturbances and changes of load or production (Suul, 2006, Suul et al., 2008a). Basically, this will be a similar way of operation as described for the large variable speed pumped storage units that have been installed in Japan and Germany to help balance production and load in systems with large share of nuclear or coal-fired power plants running at almost constant production. With a simple droop, the additional power command Δp_f to the pumped storage control system will be given by (2) as the product of the droop constant K_{Droop}, and the difference between the reference frequency f_{ref} and the grid frequency f_{grid}.

$$\Delta p_f = -K_{Droop}\left(f_{ref} - f_{grid}\right) \tag{2}$$

The introduction of frequency droop control for the pumped storage system can be considered as a contribution to the closed loop control of the grid frequency. Operation of the pumped storage with fixed or slowly varying power to operate the pump-turbine at the maximum efficiency, or operation with load following as the only purpose can be considered as an open loop of feed-forward way of control with respect to influence on the grid frequency. It is therefore clear that the introduction of frequency dependent control of the pumped storage system will give a conceptually different behaviour with more potential for utilization of the controllability to the benefit of the power system operation.

3.2.3 Combination of power control strategies

To utilize the controllability of the pumped storage to the benefit of the power system operation, several strategies for power control can be combined. This is illustrated by Fig. 6, where additional power commands from both a load following controller and from frequency control are summed (Suul et al., 2008a). The possible control routines and algorithms for calculating the long term or stationary power control set-point to the system are also illustrated in the figure, although not of importance for this investigation.

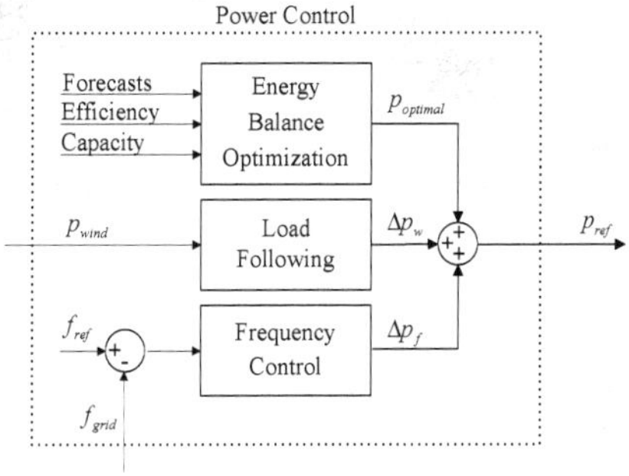

Fig. 6. Possible methods for power control of pumped storage system (Suul et al., 2008a)

The frequency control can also be made more sophisticated than just the simple droop from (2). As an example, the controllability of the pumped storage unit can be further utilized by adding stabilizing signals to the power reference, to damp modes of oscillations in the power system (Goto et al., 1995, Suul 2006). For this investigation, a limited derivative term with time constant of T_d and a low pass filter with time constant T_{filt} can be added to the droop function, so that the power command from the frequency control will be given by (3). Other additional damping structures could be more relevant depending on possible critical modes and the desired influence on the system.

$$\Delta p_f = -\left(K_{Droop} + \frac{sT_d}{\left(1+sT_d\right)\left(1+sT_{filt}\right)}\right)\left(f_{ref} - f_{grid}\right) \tag{3}$$

4. Operation of proposed topology for compensation of wind power fluctuations

To illustrate the operation of the proposed topology and the described control system for a variable speed pumped storage system, time domain simulations of an isolated power system with a large wind farm will be presented. The power control strategies described in section 3.2 will be used as input to the drive system from section 3.1 to investigate how the operation of the pumped storage system can mitigate the influence of wind power fluctuations, and relieve the other generators in the isolated grid in case of changes in production or load. Since a full-scale back-to-back voltage source converter is used, it will also be shown how the voltage or the reactive power flow of the system can be controlled by the drive system of the pumped storage power plant.

4.1 Description of simulated case

A simplified model of an isolated power system described in (Suul, 2006) is used as a starting point for the presented simulations. This case is taken as an example of an isolated system that can significantly benefit from a combination of wind power production and a variable speed pumped storage power plant, but here the energy balance and the possibility for reduction of diesel consumption for electricity production are not further investigated. The minimum load of the system is specified in the range of 14 MW, while the maximum load can reach 70 MW. Introduction of a wind farm rated for 10 MW and a pumped storage power plant in the same power range is considered for the simulations. The most challenging situations for this system will be operation at minimum load when there is a high average power production from wind turbines, and this situation will be the starting point for the simulations.

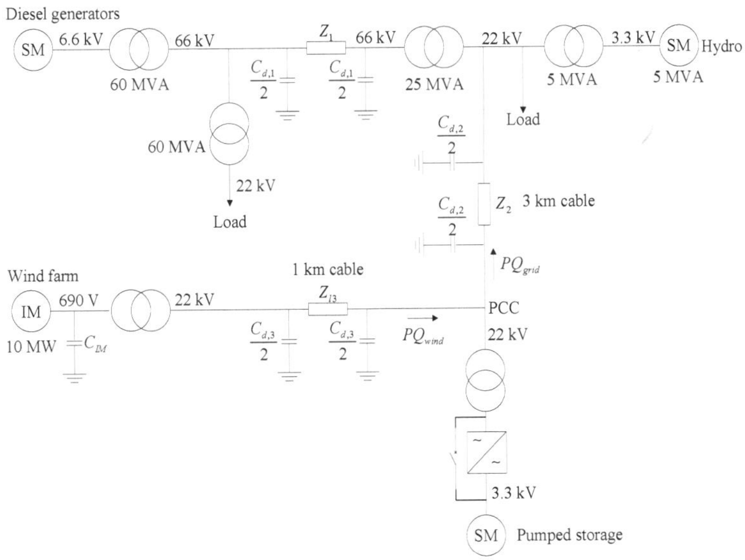

Fig. 7. Simplified grid model of the isolated system used for the presented simulations (Suul, 2006)

The system is simulated by use of a model of the power system including the proposed converter topology and the corresponding control system of the pumped storage unit, developed in PSCAD/EMTDC. All control loops for the drive system of the pumped storage is included in the model, but average models of the converters are used so the instantaneous PWM pattern is not simulated. The basic configuration of the isolated grid, including the ratings of the different units is shown in Fig. 7 while more detailed information about the system is given in Table 1. The system is simulated for 80 seconds and the wind speed input used for simulation of the wind farm is based on the Kaimal power spectra developed for PSCAD simulation from (Sørensen et al., 2002). The power output from the wind turbine model is almost independent of the operation of the pumped storage system, and can be considered equal to the power series given in Fig. 8 for all investigated situations. After 40 seconds of simulation, the hydropower plant is tripped without reconnecting it to the system.

Wind farm	- 10 MW aggregated model - Induction generators directly connected to the grid - Constant capacitors for reactive power compensation
Hydropower plant	- 5 MW synchronous machine with DC-machine exciter system - Power set-point 0,8 pu = 4MW - Static droop; 25 pu = 2.5 MW/Hz
Diesel generators	- 18 MW aggregated model - Power set-point 0.7 pu = 12,6 MW - Static droop; 25 pu = 9 MW/Hz
Pumped storage	- Power control range 4-12 MW - Static droop; 25 pu = 5 MW/Hz

Table 1. Parameters of simulated system (Suul et al., 2008b)

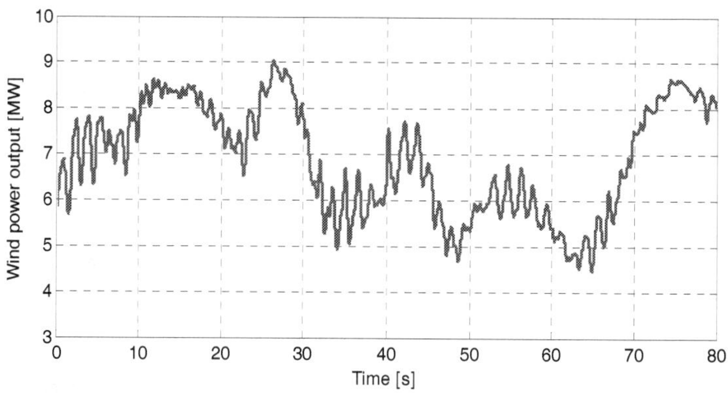

Fig. 8. Power output from wind farm during simulated time series (Suul et al., 2008b)

4.2 Power control

To investigate the power flow and the frequency control of the isolated system from Fig. 7, simulations with different power control strategies are presented. With reference to the power control strategies described in section 3.2, three different cases are compared (Suul et al., 2008 a and b):

1. Constant power input to the pumped storage system as a reference case
2. Load following for direct compensation of wind power fluctuations by the pumped storage power plant
3. Frequency droop with derivative term as given by (3), combined with load following.

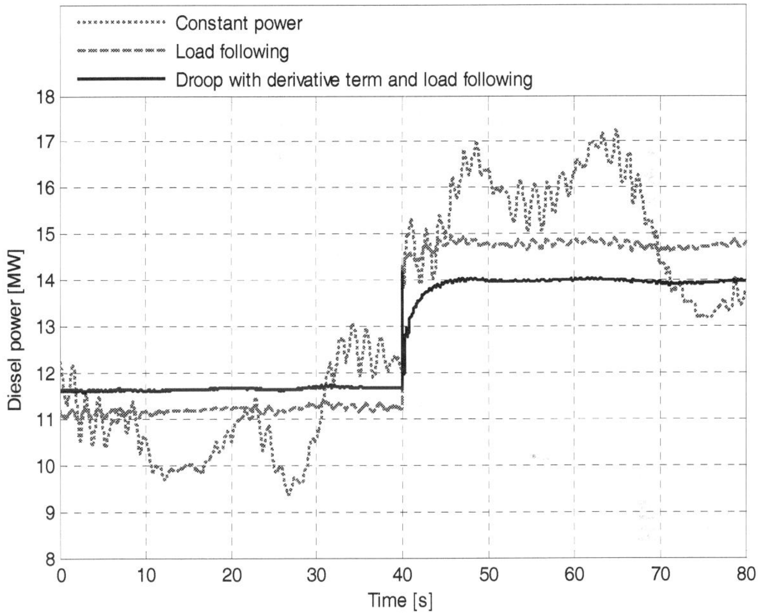

Fig. 9. Power output from diesel generators with different power control strategies for pumped storage power plant (Suul, 2008b)

The response of the diesel generators in the system to the wind power series from Fig. 8 is shown in Fig. 9. It can be seen that with constant power to the pumped storage, the diesel generators have to cover most of the power fluctuations from the wind farm. This result in both large variations in output power that will reduce the efficiency of the diesel generators, and in smaller short term fluctuations that might introduce extra tear and wear in the system. The diesel generators will in this case also have to cover almost all the loss of production when the small hydropower plant in the system is tripped. When controlling the pumped storage power plant to balance the power fluctuations from the wind turbine by load following, it can be seen that the diesel generators are relieved from covering most of the power fluctuation, but that they still have to cover all the loss of production when the hydropower plant is tripped. There are also some small remaining oscillations in the system, mainly because there is some delay in the measurement of the power flow that influences the accuracy of the load following. Adding the frequency droop to the power control of the pumped storage system, the diesel generators are relieved also from some of the steady state frequency control, and the derivative term added to the frequency control is damping the remaining power oscillations in the system.

The response in speed of the pump-turbine in the pumped storage power station is shown in Fig. 10. This figure shows how the short term power fluctuations from the wind farm are filtered by the large inertia of the electrical machine and the pump-turbine, so that mainly the slower power variations are reflected in the speed of the system. It is also seen how the frequency droop control is reducing the power input to the pumped storage system, as seen by a reduction in speed of the pump-turbine, when the hydropower station is tripped.

The grid frequency curves plotted in Fig. 11 show how the rest of the system is relieved from the influence of the variable power output from the wind turbines when the fluctuations are compensated by the pumped storage system. It can also be seen how the frequency response of the power system is improved when the pumped storage is used for frequency control. The results in Fig. 9, Fig. 10 and Fig. 11 indicate how the control of the pumped storage can limit the necessary operating range of the diesel generators, and by that also limit the fluctuations in grid frequency. This can allow for having less diesel generator capacity on line, and since the remaining units in operation can be operated at a higher average load, the efficiency of the diesel generators can be increased, contributing to further reduction in the fuel consumption of the electricity supply system.

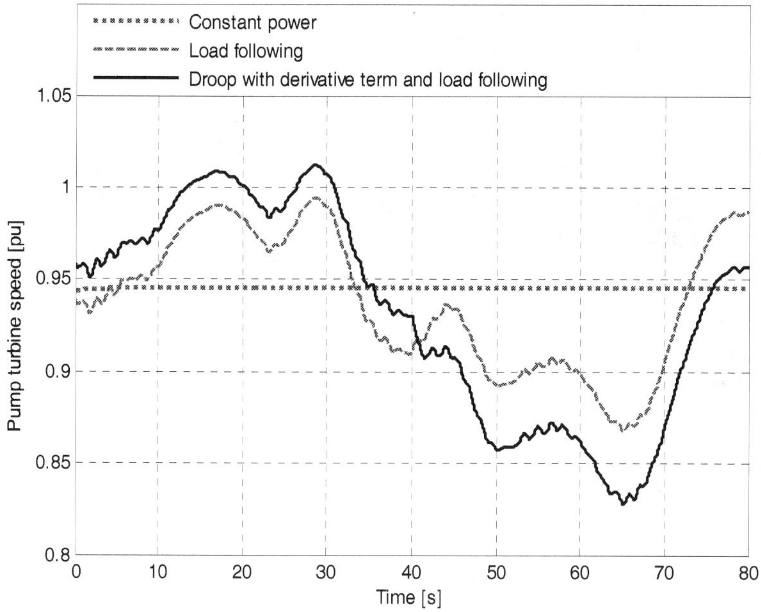

Fig. 10. Speed of pump-turbine with different power control strategies (Suul, 2008b)

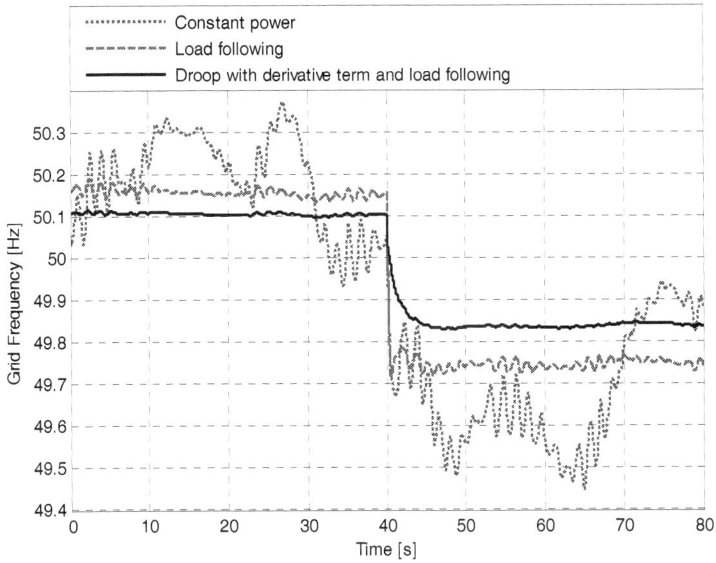

Fig. 11. Grid frequency with different power control strategies for the pumped storage unit (Suul, 2008b)

4.3 Voltage and reactive power control

With the back-to-back voltage source converter, the reactive current on the grid side can be controlled independently of the active power flow. As long as the total current rating is not exceeded, the grid side converter can therefore be used for controlling reactive power flow in the grid or for taking part in the voltage control. In contrast to frequency, that can be considered a global variable in steady state, the grid voltage is a local variable and the design of the control loops for voltage or reactive power will therefore be dependent on the configuration of the local grid. The control objective can also be different depending on what kind of grid the system is located in and what are the most critical challenges of the specific location. The presented topology has the flexibility to easily implement different control structures for voltage or reactive power control.

In the investigated model, the grid is mainly consisting of high voltage cables, and is therefore quite strong with respect to voltage variations and possibilities for voltage collapse. Still voltage flicker can be a problem with large amount of wind power in the system, and control for mitigation of voltage fluctuations is thus relevant.

The influence of different voltage control strategies on the grid voltage, and the flow of reactive power, is illustrated with the simulations shown in Fig. 12 and Fig. 13 (Suul et al., 2008b). The simulations are carried out with the 3rd power control strategy from section 4.2, and results obtained with the following four different control strategies for voltage or reactive power are shown in the figures:

1. Zero reactive current in the converter
2. Grid voltage controlled to 1.0 pu by PI-controller.

3. Reactive power flow to the grid from the point of common coupling (PCC) in Fig 2 is controlled to 0 by PI-controller.
4. Grid voltage control with PI-controller and droop from filtered reactive current on the voltage reference.

With reactive current from the converter controlled to zero, it can be seen that the voltage is fluctuating with the power variations from the wind turbines, and that these fluctuations in reactive power have to be provided by the other generators in the grid. If the grid voltage is controlled to 1.0 pu, the converter for the pumped storage is supplying the reactive power consumed by the wind farm, and also delivering reactive power to the grid to boost the voltage.

The presented figures also show how the reactive power exchange with the grid can be controlled directly, and that the converter in this way can easily compensate all the reactive power fluctuations caused by the wind farm. The voltage in the grid is also stabilized at a level close to the rated value, depending on the characteristics and the response of the rest of the system.

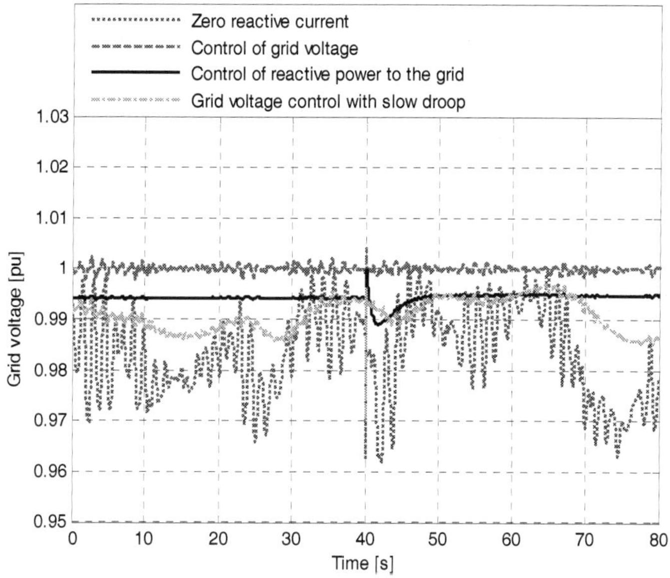

Fig. 12. Grid voltage at point of common coupling with different control strategies (Suul et al., 2008b)

Since pure integral effect in outer loop controllers can lead to unintended interaction between different equipment, steady state droop characteristics should be allowed for the voltage or reactive power controller. This will be especially important if several units with voltage control capability are operating in parallel. In Fig. 12 and Fig. 13, one case is shown where reactive current from the converter is filtered and used to generate a droop function. It is seen from the figures how this makes it possible to mitigate the short term fluctuations in voltage and reactive power flow, while the system has a droop characteristic for the longer term voltage variations. The voltage source converter is providing full flexibility with respect to control of voltage or reactive power, and the functionality and response of the system can easily be designed according to the most relevant control objectives for a specific implementation.

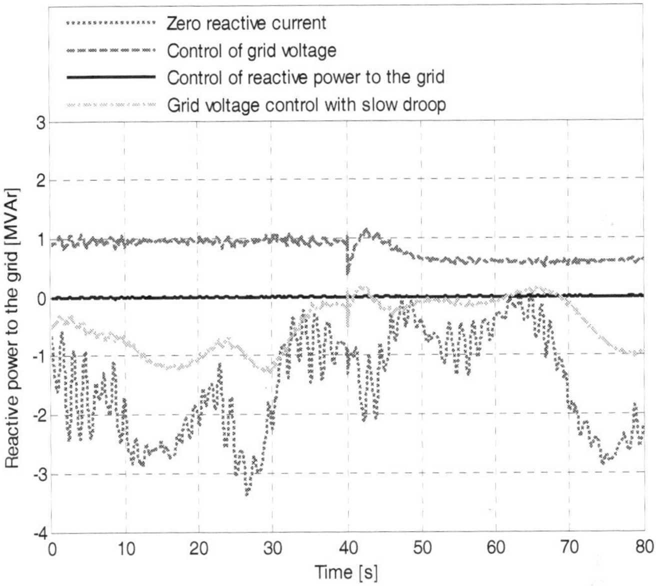

Fig. 13. Reactive power flow from point of common coupling to the rest of the grid (Suul et al., 2008b)

5. Conclusion

This chapter has provided a brief review of the application history of pumped storage systems. Starting from to the development and application of traditional pumped storage systems, the introduction of power electronic drive systems to obtain variable speed operation of pump-turbines has been discussed. Reviewing the different possibilities for power electronic drive systems, it is shown that little attention has until now been directed towards systems with a full-scale voltage source converter and a synchronous machine for driving a variable speed pump-turbine. Although this topology was not considered relevant when the development of variable speed pumped storage systems started, recent advances in semiconductor components and voltage source converter drive systems has made this an interesting topology for small and medium size pumped storage systems.

Introduction of pumped storage systems as energy storage in combination with fluctuating renewable energy sources like wind power has also been reviewed. It has been found that even if the benefits of variable speed operation are well recognized for pumped storage systems operating in large interconnected power systems, most of the smaller hybrid systems with wind power and pumped storage that have been proposed for isolated power systems are based on water pumping at fixed speed. Therefore it is of interest to investigate the dynamic control systems for variable speed pumped storage systems together with the instantaneous power balance of an isolated power system.

From the presented background, the full-scale voltage source converter driving a synchronous machine is suggested as a suitable configuration for a variable speed pumped storage system operating in an isolated grid with high penetration of wind power production. The proposed topology and its main operational characteristics are discussed, and a possible control system is described. With the proposed topology, a pumped storage unit can also be operated with the machine directly connected to the grid, running at constant speed like a conventional power plant, such that the operation of the system will be less dependent on the converter reliability. Use of a back-to-back voltage source converter provides full flexibility in control of active and reactive power. The system can therefore be operated with variable speed and controllable flow of active power both in pumping mode and generating mode, while the grid voltage or the reactive power flow can be simultaneously controlled. Since reactive current in the grid can be controlled independently of the pump-turbine operation, the utilization of the grid side converter for voltage control is only limited by the total available current rating of the converter. If the pump-turbine is stopped, the full current rating can therefore be used to control the grid voltage, and the grid side converter will then operate as a STATCOM.

The operation and control of the suggested topology is illustrated by simulating a model of an isolated grid by using the PSCAD/EMTDC software. The presented simulations illustrate the dynamic operation of the system and show how the variable speed pumped storage can be controlled to limit the influence of wind power fluctuations and at the same time contribute to increased frequency response of the power system. The simulation results therefore indicate how the variable speed pumped storage can allow for more wind power to be introduced to an isolated grid without undermining the frequency control and the instantaneous power balance of the system. The controllability introduced by the suggested configuration for a pumped storage plant can then be used both to compensate for the consequences of fluctuating power production from wind turbines and to improve the general operation of an isolated power system. This can make it possible to allow for higher wind power penetration and significantly reduce the dependency on fossil fuels for electricity production in isolated power systems.

Results showing how the suggested topology can be utilized to control the grid voltage or the flow of reactive power in the system are also presented. Controlling the grid voltage, or compensating for the fluctuating reactive power consumption of a wind farm, can allow for a better distribution of reactive power flow in the system and by that reducing the power losses. Voltage control by the grid side converter can also mitigate possible power quality problems related to voltage flicker and improve the voltage stability of the system.

The model used to generate the presented simulations is primarily intended for representation of the electrical drive systems of the pumped storage power plant. However, more detailed representation of hydraulic limitations and the control of the turbine should be included in the model for investigation of the system when operating in generator mode. Also practical limitations of the diesel generators should be studied in more detail and included in the model to verify the behaviour of the system under faults and other extreme operating conditions. Further on, it would be relevant to study how the presented pumped storage system can interact with a wind farm with variable speed wind turbines in the best possible way. The presented pumped storage system should also be investigated with respect to fulfilment of local grid codes or other requirements, like international standards and norms for a specific case where the ratings and the operating conditions of different units have been optimized with respect to energy balance and power system operation.

6. Acknowledgement

This work has been partly based on the author's Master thesis from the Norwegian University of Science and Technology (NTNU), Trondheim, Norway (Suul, 2006). This thesis was supervised by Professor Tore Undeland (NTNU) and Senior Research Scientist Kjetil Uhlen (SINTEF Energy Research). Results of further work based on the Master thesis has later been published in (Suul et al., 2008 a and b). Investigation of the pumped storage system presented here was initially suggested as a topic for studies by Øyvind Holm in Voith Siemens Hydro Power Generation, Trondheim, Norway. Voith Siemens in Trondheim has also provided some of the parameters needed for the simulation model used to generate the presented results.

7. References

Abbey, C. & Joos, G. (2005). "Energy Management Strategies for Optimization of Energy Storage in Wind Power Hybrid System," *Proceedings of the 36th IEEE Power Electronics Specialists Conference, PESC 2005*, 12-18 June 2005, Recife, Brazil, pp. 2066-2072

Alaküla, M. (1993). "On the Control of Saturated Synchronous Machines," PhD-thesis, Lund Institute of Technology, Lund, Sweeden, 1993

Allen, A. E. (1977). "Potential for Conventional and Underground Pumped-Storage," *IEEE Transactions on Power Apparatus and Systems*, Vol. PAS-96, No. 3, May/June 1977, pp. 993-998

Allen, G.; McKeogh, E. J. & Gallachóir, B. Ó. (2006). "Modelling of a wind-pumped hydro scheme within the Irish liberalized electricity market," Proceedings of the European Wind Energy Conference, EWEC 2006, 27 February - 2 March 2006, Athens, Greece

Alstom (2009). Press Release, "Alstom awarded €125 million contract to supply cutting-edge technology to Switzerland's new Nant de Drance hydropower station", 4 May, 2009, Available from: http://www.ch.alstom.com/pr_corp/2008/ch/57862.EN.php?languageId=EN&dir=/pr_corp/2008/ch/&idRubriqueCourante=15165

Anagnostopoulos, J. S. & Papantonis, D. E. (2007). "Simulation and size optimization of a pumped-storage power plant for the recovery of wind-farms rejected energy," *Renewable Energy*, Vol. 33, No 7, July 2008, pp. 1685-1694

Anagnostopoulos, J. S. & Papantonis, D. E. (2007). "Pumping station design for a pumped-storage wind-hydro power plant," *Energy Conversion and Management*, Vol. 48, No 11, November 2007, pp. 3009-3017

Ansel, A.; Nasser, L. & Robyns, B (2006). "Variable Speed Small Hydro Plant Connected to Power Grid or Isolated Load," *Proceedings of the 12th International Power Electronics and Motion Control Conference, EPE-PEMC 2006*, 30 August - 1 September 2006, Portoroz, Slovenia, pp. 2064-2069

ASME (1980). Rocky River Pumped-Storage Hydroelectric Station, A National Historic Mechanical Engineering Landmark, The American Society of Mechanical Engineers, New Milford, Connecticut, 1980, Accessed May 2009, Available from: http://files.asme.org/asmeorg/Communities/History/Landmarks/3137.pdf

Arrilaga, J.; Sanakr, S.; Arnold, C. P. & Watson, N. R. (1992), "Characteristics of unit-connected HVDC generator-convertors operating at variable speed," *IEE Proceedings C, Generation, Transmission and Distribution*, Vol. 139, No. 3, May 1992, pp. 295-299

Bakos, G. C. (2002). "Feasibility study of a hybrid wind/hydro power system for low-cost electricity production," *Applied Energy*, Vol. 72, No. 3-4, July-August 2002, pp. 599-608

Bendl, J.; Chomát, M. & Schreier, L. (1999). "Adjustable-Speed Operation of Doubly-fed Machines in Pumped Storage Power Plants," *Proceedings of the Ninth International Conference on Electrical Machines and Drives*, 1-3 September 1999, pp. 223-227

Blasko, V. & Kaura, V. (1997). "A New Mathematical Model and Control of a Three Phase AC-DC Voltage Source Converter," *IEEE Transactions on Power Electronics*, Vol. 12, No. 1, January 1997, pp. 116- 123

Bocquel, A. & Janning, J. (2003). "4*300 MW Variable Speed Drive for Pump-Storage Plant Application," *Proceedings of the 10th European Conference on Power Electronics and Applications, EPE 2003*, 2-4 September 2003, Toulouse, France

Bocquel, A. & Janning, J. (2005). "Analysis of a 300 MW Variable Speed Drive for Pump-Storage Plant Applications," *Proceedings of the 11th European Conference on Power Electronics and Applications, EPE 2005*, 11-14 September 2005, Dresden, Germany

Bollmeier II, W. S.; Huang, N. & Trenka, A. R. (1994). "Wind/pumped-hydro integration and test project: preliminary system test results," *Proceedings of Energy-Sources Technology Conference*, 23-26 January 1994, New Orleans, USA

Bose, S; Liu, Y.; Tayla, S.; Vyas, P.; Videhult, S.; Bjerke, M. & Børresen, B. (2004). "A methodology for sizing and cost optimization of wind power with pumped-hydro storage," *Proceedings of the International Conference - RES and RUE for Islands - Sustainable Energy Solutions*, 30 - 31 August 2004, Larnaca, Cyprus

Brown, P. D.; Lopes, J. A. P. & Matos, M. A. (2008). "Optimization of Pumped Storage Capacity in an Isolated Power System with Large Renewable Penetration," IEEE Transactions on Power Systems, Vol. 23, No. 2, May 2008, pp. 523-531

Bueno, C. & Carta, J. A. (2005) a. "Technical-economic analysis of wind-powered pumped hydrostorage systems. Part I: model development," *Solar Energy*, Vol. 78, No. 3, March 2005, pp. 382-395

Bueno, C. & Carta, J. A. (2005) b. "Technical-economic analysis of wind-powered pumped hydrostorage systems. Part II: model application to the island of El Hierro," *Solar Energy*, Vol. 78, No. 3, March 2005, pp. 382-395

Bueno, C. & Carta, J. A. (2006). "Wind powered pumped hydro storage systems, a means of increasing the penetration of renewable energy in the Canary Islands," *Renewable and Sustainable Energy Reviews*, Vol. 10, No. 4, August 2006, pp. 312-340

Ceralis, A. & Zervos, A. (2007). "Analysis of the combined use of wind and pumped storage systems in autonomous Greek islands," *IET Renewable Power Generation*, Vol. 1, No. 1, 2007, pp. 49-60

Chen, H.; Cong, T. C.; Yang, W.; Tan, C.; Li, Y. & Ding, Y. (2009). "Progress in electrical energy storage system: A critical review," *Progress in Natural Science*, Vol. 19, No. 3, March 2009, pp. 291-312

Chen, F.; Duic, N.; Alves, L. M. & Carvalho, M. G. (2007). "Renewislands - Renewable energy solutions for islands," *Renewable and Sustainable Energy Reviews*, Vol. 11, No. 8, October 2007, pp 1888-1902

Chiang, J.-C.; Wu, C.-J. & Yen, S.-S. (1997). "Mitigation of Harmonic Disturbance at Pumped Storage Power Station with Static Frequency Converter," *IEEE Transactions on Energy Conversion*, Vol. 12, No. 3, September 1997, pp.232-240

Chung, S.-K. (2000). "Phase-locked loop for grid-connected three-phase power conversion system," *IEE Proceedings - Electric Power Applications*, Vol. 147, No. 3, May 2000, pp. 213-219

Coleman, R. S.; Brennan, F. L.; Brown, P. G. & Cooper, E. A. (1976). "Survey of Pumped Storage Projects in the United States and Canada to 1975," IEEE *Transaction on Power Apparatus and Systems*, Vol. PAS-95, No. 3, May/June 1976, pp. 851-858

Dell, R. M., & Rand, D. A. J. (2001). "Energy storage - a key technology for global energy sustainability," *Journal of Power Sources*, Vol. 100, No. 1-2, Nov. 2001, pp. 2-17

EB (2009). Encyclopædia Britannica, History of water turbine technology, accessed May 2009, available from: http://www.britannica.com/EBchecked/topic/609552/turbine/45676/History-of-water-turbine-technology

ESA (2009), Electricity Storage Association, accessed May 2009, available from: http://www.electricitystorage.org/site/home/

Erlich, I. & Bachmann, U. (2002). "Dynamic Behavior of Variable Speed Pump Storage Units in the German Electric Power System," *Proceedings of the 15th Triennal World Congress of the International Federation of Automatic Control, 2002 IFAC*, 21-26 July 2002, Barcelona, Spain

European Commission (2000), "New Solutions in Energy; Status report on variable speed operation in small hydropower," 2000, Accessed February 2008, available from: http://ec.europa.eu/energy/res/sectors/doc/small_hydro/statusreport_vspinshp_colour2.pdf,

Faias, S.; Sousa, J. & Castro, R. (2007). "Contribution of Energy Storage Systems for Power Generation and Demand Balancing with Increasing Integration of Renewable Sources: Application to the Portuguese Power System," *Proceedings of the 12th European Conference on Power Electronics and Applications, EPE 2007*, 2-5 September 2007, Aalborg, Denmark

Fostiak, R. J. & Davis, H. R. (1994), "Electrical Features of the Rocky Mountain Pumped Storage Project," *IEEE Transactions on Energy Conversion*, Vol. 9, No. 1, March 1994, pp. 206-213

Fraile-Ardanuy, J.; Wilhelmi, J. R.; Fraile-Mora, J. J & Pérez, J. I. (2006). "Variable-Speed Hydro Generation: Operational Aspects and Control," *IEEE Transactions on Energy Conversion*, Vol. 21, No. 2, June 2006, pp. 569-573

Fujihara, T.; Imano, H. & Oshima, K. (1998). "Development of Pump Turbine for Seawater Pumped-Storage Power Plant," *Hitachi Review*, Vol. 47, No. 5, 1998

Furuya, S.; Taguchi, T.; Kusunoki, K.; Yanagisawa, Y.; Kageyama, T. & Kanai, T. (1993). "Successful Achievement in a Variable Speed Pumped Storage Power System at Yagisawa Power Plant," Conference Record of Power Conversion Conference 1993, PCC '93, 19-21 April 1993, Yokohama, Japan, pp. 603-608

Furuya, S.; Wada, F.; Hachiya, K. & Kudo, K. (1995). "Large Capacity GTO Inverter-Converter for Double-Fed Adjustable Speed System," *CIGRE Symposium on Power Electronics in Electric Power Systems*, Tokyo, May 1995, Paper ID 530-04

Galasso, G. (1991). "Adjustable Speed Operation of Pumped Storage Hydroplants," *Proceedings of the 1991 International Conference on AC and DC Power Transmission*, 17-20 September 1991, pp 424-427

Gish, W. B.; Schurz, J. R.; Milano, B. & Schleif, F. R. (1981). "An Adjustable Speed Synchronous Machine for Hydroelectric Power Applications," *IEEE Transactions on Power Apparatus and Systems*, Vol PAS-100, No. 5, May 1981

Gjengedal, T. (2001). "Application of Adjustable Speed Hydro (ASH) Machines in The Norwegian Power System," *Proceedings of IEEE Porto Power Tech*, 10-13 September 2001, Porto, Portugal, Vol. 2

Goto, M.; Shibuya, A.; Inoue, T.; Ishizaki, M. & Tezuka Y. (1995). "Power System Stabilizing Control by Adjustable Speed Pumped Storage Power Station Using Stabilizing Signals," *CIGRE Symposium on Power Electronics in Electric Power Systems*, Tokyo, May 1995, Paper ID 510-01

Grotenburg, K.; Koch, F.; Erlich, I. & Bachmann, U. (2001). "Modeling and Dynamic Simulation of Variable Speed Pump Storage Units Incorporated into the German Electric Power System," *Proceedings of the 9th European Conference on Power Electronics and Applications, EPE 2001*, 27-29 August 2001, Graz, Austria

Hayashi, S.; Haraguchi, E.; Sanematsu, T.; Takahashi, N.; Yasaka, Y. & Nogura, O. (1988). "Development of Adjustable Speed Generator," *CIGRE, Proceedings of International Conference on Large High Voltage Electric Systems, 1988 Session*, 28 August - 3 September 1988, Paper ID 11-03

Katsaprakakis, D. A.; Christakis, D. G.; Voumvoulakis, E.; Zervos, A.; Papantonis, D. & Voutsinas S. (2007). "The Introduction of Wind Powered Pumped Storage Systems in Isolated Power Systems with high Wind Potential," International Journal of Distributed Energy Resources, Vol. 3, No. 2, 2007, pp. 83-112

Hammerschlag, R. & Schaber, C. P. (2007). Energy Storage Technologies, In: *Energy Conversion*, Goswami, D. Y & Kreith, F. (Ed.) pp. 12.1–12.22, CRC Press, 2007, ISBN 1420044311/9781420044317

Hodder, A; Simond, J.-J. & Schwery, A. (2004). "Double-Fed Asynchronous Motor-Generator Equipped With a 3-Level VSI Cascade," *Conference Record of the 2004 IEEE Industry Applications Conference and 39th IAS Annual Meeting*, 3-7 October 2004, Vol.4, pp. 2762-2769

Holm, Ø. (2006). Departmental Manager - Small Hydro Power, Voith Siemens Hydro Power Generation, Trondheim, Norway: Personal communication, spring 2006

Hämmerli, A & Ødegård, B (2008). "AC excitation with ANPC - ANPC converter technology tailored to the needs of AC excitation equipment for pump storage plants," *ABB Review*, No. 3, 2008, pp. 40-43

Ibrahim, H; Ilinca, A. & Perron, J. (2008). "Energy Storage systems - Characteristics and comparisons," Renewable and Sustainable Energy Reviews, Vol. 12, No. 5, June 2008, pp. 1221-1250

Ikeda, K.; Inagaki, M.; Niikura, K. & Oshima, K. (2000). "700-m 400-MW Class Ultrahigh Head Pump Turbine," *Hitachi Review*, Vol. 49, No. 2, 2000

INSULA (2008), International Scientific Council for Island Development, information page about the El Hierro project for 100 % renewable energy supply, http://www.insula-elhierro.com/english.htm, Accessed July 2008

Jaramillo, O. A; Borja, M. A. & Huacuz, J. M (2004). "Using hydropower to complement wind energy: a hybrid system to provide firm power," *Renewable Energy*, Vol. 29, No. 11, September 2004, pp 1887-1909

Jensen, T. L (2000). *"Renewable energy on small islands,"* 2nd edition, Forum for Energy and Development, ISBN: 87-90502-03-5, Denmark, August 2000

Kaldellis, J. K. (2002). "Parametrical investigation of the wind-hydro electricity production solution for Aegean Archipelago," *Energy Conversion and Management*, Vol. 43, No. 16, November 2002, pp. 2097-2113

Kaldellis, J. K. & Kavadias, K. A. (2001). "Optimal wind-hydro solution for Aegan Sea islands' electricity-demand fulfilment," *Applied Energy*, Vol. 70, No. 4, December 2001, pp. 333-354

Kaldellis, J. K.; Kavadias, K. & Christinakis, E. (2001). "Evaluation of the wind-hydro energy solution for remote islands," *Energy Conversion and Management*, Vol 42, No. 9, June 2001, pp. 1105-1120

Katsaprakakis, D. Al.; Christiakis, D. G.; Zervos, A.; Papantonis, D. & Voutsinas, S. (2008). "Pumped storage systems introduction in isolated power production systems," *Renewable Energy*, Vol. 33, No. 3, March 2008, pp 467-490

Katsaprakakis, D. A.; Dimitris, Pr. & Christakis, G. (2006). "A Wind Parks, Pumped Storage and Diesel Engines Power System for the electric power production in Astypalaia," *Proceedings of European Wind Energy Conference, EWEC 2006*, 27. Feb. - 2 March 2006, Athens, Greece

KEMA (2009). "Energy island is innovative concept for large-scale electricity storage," Information from KEMA, Accessed May 2009, available from: http://www.kema.com/corporate/news/corporate/2007/Q3/energie-eiland.asp

Kerkman, R. J.; Lipo, T. A.; Newman, W. G. & Thirkell, J. E. (1980) a. "An Inquiry into Adjustable Speed Operation of a Pumped Hydro Plant, Part I - Machine Design and Performance," *IEEE Transactions on Power Apparatus and Systems*, Vol. PAS-99, No.5 Sept./Oct. 1980

Kerkman, R. J.; Lipo, T. A.; Newman, W. G. & Thirkell, J. E. (1980) b. "An Inquiry into Adjustable Speed Operation of a Pumped Hydro Plant, Part II - System Analysis," *IEEE Transactions on Power Apparatus and Systems*, Vol. PAS-99, No.5 Sept./Oct. 1980

Kondoh, J; Ishii, I; Yamaguchi, H.; Murata, A.; Otani, K.; Sakuta, K.; Higuchi, N.; Sekine, S. & Kamimoto, M. (2000). "Electrical energy storage systems for energy networks," *Energy Conversion & Management*, Vol. 41, No. 17, November 2000, pp. 1863-1874

Kuwabara, T.; Shibuya, A.; Furuta, H.; Kita, E. & Mitsuhashi, K. (1996). "Design and Dynamic Response Characteristics of 400 MW Adjustable Speed Pumped Storage Unit for Ohkawachi Power Station," *IEEE Transactions on Energy Conversion*, Vol. 11, No. 2, June 1996, pp. 376 - 384

Lanese, J.; Powers, A. & Naeff, H. (1995). "Selection of Large Variable Speed Pumps for the Domenigoni Valley Reservoir Project," *Proceedings. of the 1995 International Conference on Hydropower*, 25-28 July 1995, San Francisco, USA, Vol. 2, pp. 1902-1912

Lilly, P. N.; Radovich, M. & Warshauer, J. (1991). "Improving California Wind Project Dispatchability and Firm Capacity: Coupling Modular Pumped Storage Hydroelectric Technology with Wind Power; Siting, Utility Integration and Regulatory Issues," *Proceedings of Windpower'91*, pp. 28-35

Lung, J.-K.; Lu, Y.; Hung, W.-L. & Kao, W.-S. (2007). "Modeling and Dynamic Simulations of Doubly Fed Adjustable Speed Pumped Storage Units," *IEEE Transactions on Energy Conversion*, Vol. 22, No. 2, June 2007, pp. 250-258

MacKay, D. J. C. (2007), "Enhancing Electrical Supply by Pumped Storage in Tidal Lagoons," Accessed May 2009, Report available from: www.inference.phy.cam.ac.uk/sustainable/book/tex/Lagoons.pdf

Magsaysay, G.; Schuette, T. & Fostiak, R. (1995). "Use of a Static Frequency Converter for Rapid Load Response in Pumped-Storage Plants," *IEEE Transactions on Energy Conversion*, Vol. 10, No. 4, December 1995, pp. 694-699

McCleer, P. J. & Meisel, J. (1984). "Variable speed Operation of Pumped Storage Power Plants," In *Trends in electric utility research*, Bullard, C. W. (Ed.), pp. 113-126, Pergamon Press, New York, ISBN 0080309828

Merino, J. M. & López, Á. (1996). "ABB Varspeed generator boosts efficiency and operating flexibility of hydropower plant," *ABB Review*, Nr. 3, 1996, pp. 33-38

Mitsubishi (2008). "Pumped Storage Power Station with Adjustable Speed Pumped Storage Technology," Presentation available from The Energy and Resources Institute, Accessed May 2009, www.teriin.org/events/docs/present_japan/sess4/yokota_part1-2-malco.pdf

Mo, O.; Hernes, M. & Ljøkelsøy, K. (2003). "Active damping of oscillations in LC-filter for line connected, current controlled, PWM voltage source converters," *Proceedings of the 10th European Conference on Power Electronics and Applications, EPE2003*, 2-4 September 2004, Toulouse, France

Mori, S.; Kita, E.; Kojima, H.; Sanematsu, T.; Shibuya, A. & Bando, A. (1995). "Commissioning of 400 MW Adjustable Speed Pumped Storage System for Ohkawachi Hydro Power Plant," CIGRE Symposium on Power Electronics in Electric Power Systems, Tokyo, May 1995, Paper ID 520-04.

Murakami, M. (1995). "Managing Water for Peace in the Middle East: Alternative Strategies," United Nations University Press, 1995, Accessed May 2009, Available from: http://www.unu.edu/unupress/unupbooks/80858e/80858E00.htm#Contents

Naidu, M. & Mathur, R. M. (1989). "Evaluation of Unit Connected, Variable Speed, Hydropower Station for HVDC Power Transmission," *IEEE Transactions on Power Systems*, Vol. 4, No. 2, May 1989, pp. 668-676

Papathanassiou, S. A.; Tziantzi, M.; Papadopoulos, M. P.; Tentzerakis, S. T. & Vionis, P. S. (2003). "Possible benefits from the combined operation of wind parks and pumped storage stations," *Proceedings of the European Wind Energy Conference, EWEC 2003*, 16-19 June 2003, Madrid, Spain

Piernavieja, G.; Pardilla, J.; Schallenberg, J. & Bueno, C. (2003). "El Hierro: 100% RES, An Innovative Project for Islands' Energy Self-Sufficiency," *Proceedings of the first Island Conference on Innovation and Sustainable Development*, La Palma, Spain, 2003

Protopapas, K. & Papathanassiou, S. (2004). "Operation of Hybrid Wind-Pumped Storage Systems in Isolated Island Grids," *Proceedings of the 4th Mediterranean IEE Conference on Power Generation, Transmission, Distribution and Energy Conversion, MedPower 2004*, 14-17 Nov. 2004, Lemnos, Cyprus

Protopapas, K. & Papathanassiou, S. (2006). "Application of Pumped Storage to Increase Wind Penetration in Isolated Grids," *Proceedings of European Wind Energy Conference*, EWEC 2006, 27. Feb. - 2 March 2006, Athens, Greece

Ribeiro, P. F.; Johnson, B. K.; Crow, M. L.; Arsoy, A. & Liu, Y. (2001). "Energy Storage Systems for Advanced Power Applications," Proceedings of the IEEE, Vol. 89, No. 12, December 2001, pp. 1744-1756

Riverbankpower (2009). Information about proposal and testing of concept for underground pumped storage systems, Accessed May 2009, Available from: http://www.riverbankpower.com/

Sapin, A.; Hodder, A.; Simond, J.-J. & Schafer, D. (2000). "Doubly-fed Asynchronous Machine with 3-level VSI for Variable Speed Pump Storage," *Proceedings of the 14th International Conference on Electrical Machines*, 28-30 August 2000, Espoo, Finland

Schafer, D. & Simond, J.-J. (1998). "Adjustable speed Asynchronous Machine in Hydro Power Plants and its Advantages for the Electric Grid Stability," *CIGRÉ report*, Paris, 1998

Scherer, K. (2005). "Change of Speed," *International Water Power and Dam Construction*, Vol. 57, No. 4, April 2005, pp. 38-41

Schreirer, L.; Chomat, M. & Bendl, J. (2000). "Operation of system double fed machine-turbine in power network," *Proceedings of the 8th IEE Conference on Power Electronics and Variable Speed Drives*, 18-19 September 2000, London, UK, pp. 109-113

Sick, M. & Schwab, A. (2005). "Working with wind," *International Water Power & Dam Construction*, Vol. 57, No. 11, Nov. 2005, pp 38-42

Simond, J.-J.; Sapin, A. & Schafer, A. (1999). "Expected benefits of adjustable speed pumped storage in the European network," *Proceedings of Hydropower in the next century*, 1999,

Shimada, R & Mukai, K (2007). "Load-Leveling and Electric Energy Storage," *IEEJ Transactions on Electrical and Electronic Engineering*, Vol. 2, No. 1, January 2007, pp. 33-38, ISSN: 19314973

Sugimoto, O.; Haraguti, E.; Saikawa, K.; Suzuki, N.; Saito, K. & Yasaka, Y. (1989). "An Adjustable Speed Operation System for Pumped Storage Hydro Power Plant," *Proceedings of the 14th Congress of the Wold Energy Conference*, 17-22 September 1989, Montreal, Canada

Sommerville, W. M (1989). "Wind Turbine and Pumped Storage Hydro Generation on Foula," *Proceedings of the European Wind Energy Conference, EWEC 1989*, pp. 713-717

Sporild, R; Gjerde, J. O. & Gjengedal, T. (2000). "Enhanced Power System Operation by Application of Adjustable Speed Hydro Machines," *Proceedings of International Conference on Electric Utility Deregulation and Restructuring and Power Technologies, DRTP 2000*, 4-7 April 2000, London, UK, pp. 373- 377

Suul, J. A. (2006). "Control of Variable speed Pumped Storage Hydro Power Plant for Increased Utilization of Wind Energy in an Isolated Grid," MSc. Thesis, Norwegian University of Science and Technology, Department of Electrical Power Engineering, 2006

Suul, J. A.; Uhlen, K. & Undeland, T. (2008) a. "Variable speed pumped hydropower for integration of wind energy in isolated grids - case description and control strategies," *Proceedings of Nordic Workshop on Power and Industrial Electronics, NORPIE 2008*, 9-11 June2008, Espoo, Finland

Suul, J. A.; Uhlen, K. & Undeland, T. (2008) b. "Wind power integration in isolated grids enabled by variable speed pumped storage hydropower plant," Proceedings of IEEE International Conference on Sustainable Energy Technologies, ICSET 2008, 24-27 November 2008, Singapore, pp. 399-404

Sørensen, A. D.; Hansen, P.; André, P. & Rosas, C. (2002). "Wind models for simulation of power fluctuations from wind farms," *Journal of Wind Engineering and Industrial Aerodynamics*, Vol. 90, No 12-15, December 2002, pp. 1381-1402

Taylor, J. (1988). "The Foula Electricity Scheme," *Proceedings of IEE Colloquium on Energy for Isolated Communities*, May 1988

Terence, L. & Schäfer, R. (1993). "Variable Speed in Hydro Power Generation Utilizing Static Frequency Converters," *Proceedings of the International Conference on Hydropower, Waterpower '93*, 10-13 August 1993, pp. 1860-1869

Theodoropoulos, P.; Zervos, A. & Betzios G. (2001), "Hybrid Systems Using Pump-Storage Implementation in Ikaria Island," *Proceedings of the International Conference on Renewable Energies for Islands - Towards 100% RES Supply*, 14-16 June 2001, Chania, Greece

Taguchi, T.; Aida, K.; Mukai, K; Yanagisawa, T & Kanai, T. (1991). „Variable Speed Pumped Storage System Fed By Large-Scale Cycloconverter," Proceedings of the 4th European Conference on Power Electronics and Applications, EPE'91, 3-6 September 1991, Firenze, Italy, Vol. 2, pp. 237-242

Toshiba (2008), Adjustable Speed Pumped Storage Experiences, http://www3.toshiba.co.jp/power/english/hydro/products/pump/storage.htm, accessed May 2008

Wikipedia (2009). Pumped-storage hydroelectricity, Accessed May 2008, available from: http://en.wikipedia.org/wiki/Pumped-storage_hydroelectricity

Wu, C.-C.; Lee, W.-J.; Cheng, C.-L. & Lan, H.-W. (2007). "Role and Value of Pumped Storage Units in an Ancillary Service Market for Isolated Power Systems - Simulation in the Taiwan Power System," *Proceedings of IEEE/IAS Industrial & Commercial Power Systems Technical Conference, ICPS 2007*, 6-11 May 2007